Lecture Notes on Data Engineering and Communications Technologies

Volume 122

Series Editor

Fatos Xhafa, Technical University of Catalonia, Barcelona, Spain

The aim of the book series is to present cutting edge engineering approaches to data technologies and communications. It will publish latest advances on the engineering task of building and deploying distributed, scalable and reliable data infrastructures and communication systems.

The series will have a prominent applied focus on data technologies and communications with aim to promote the bridging from fundamental research on data science and networking to data engineering and communications that lead to industry products, business knowledge and standardisation.

Indexed by SCOPUS, INSPEC, EI Compendex.

All books published in the series are submitted for consideration in Web of Science.

More information about this series at https://link.springer.com/bookseries/15362

Ishfaq Ahmad · Jun Ye ·
Weidong Liu
Editors

The 2021 International Conference on Smart Technologies and Systems for Internet of Things

STSIoT2021

Springer

Editors
Ishfaq Ahmad
The University of Texas at Arlington
Arlington, TX, USA

Jun Ye
Hainan University
Haikou, Hainan, China

Weidong Liu
College of Computer Science
Inner Mongolia University
Hohhot, Nei Mongol, China

ISSN 2367-4512 ISSN 2367-4520 (electronic)
Lecture Notes on Data Engineering and Communications Technologies
ISBN 978-981-19-3631-9 ISBN 978-981-19-3632-6 (eBook)
https://doi.org/10.1007/978-981-19-3632-6

This Springer imprint is published by the registered company Springer Nature Singapore Pte Ltd.
The registered company address is: 152 Beach Road, #21-01/04 Gateway East, Singapore 189721, Singapore

Preface

Internet of Things (IoT) is a dynamic network of sensors, cloud storage and multiple embedded electronic devices connected with each other through network connectivity for exchange of data. On the other hand, neural computing and artificial intelligence are bringing paradigm shift in field of intelligent systems as human-like processing can make the system more intelligent. The merging of neural computing algorithms with IoT system can create advance units for various domains.

The real-time health monitoring by sophisticated sensors provides improved life style to patients, and it can be a life savvy in critical situations. With the advent of Internet of Things (IoT) and AI, for example, pioneering work is done in patient health monitoring. IoT is also behind emergence of wearable medical devices. The data from wearable devices or other devices could help in conquering ailments at very early stage. IoT-based architecture improves the reliability of sensors but generates large amount of data for processing. Processing such huge data is always a tedious task for human as well as normal algorithms. Also, security issue can be there due to multiple input points.

We would like to express our thanks to Prof. Ishfaq Ahmad, Prof. Jun Ye, and Prof. Weidong Liu for being the keynote speakers at the conference. We thank the General Chairs, Program Committee Chairs, Organizing Chairs, and Session Chairs for their hard work. The local organizers and the students' help are also highly appreciated.

Our special thanks are due also to editors Dr. Thomas Ditzinger, Prof. Fatos Xhafa, and Jasmine Dou for publishing the proceedings in Lecture Notes on Data Engineering and Communications Technologies of Springer.

Organization

Program Chairs

Ishfaq Ahmad	University of Texas at Arlington, USA
Jun Ye	Hainan University, Hainan, China
Weidong Liu	Inner Mongolia University, China

Publication Chairs

Sulin Pang	Jinan University, China
Deepak Kumar Jain	Oriental University, India

Publicity Chairs

Junyu Xuan	University of Technology Sydney, Australia
Muhammad Alam	University of Aveiro, Portugal

Local Organizing Chairs

Yundong Li	Shanghai DianJi University, Shanghai, China
Pin Lv	Shanghai DianJi University, Shanghai, China

Program Committee Members

Pradeep Singh	ABES Engineering College, Ghaziabad, India
Bhavani Thuraisingham	The University of Texas at Dallas, USA
Manfred Huber	The University of Texas at Arlington, USA
Barbara Carminati	University of Insubria, Italy
Fatemeh Tehranipoor	The Santa Clara University, USA
Jemal Abawajy	Deakin University, Australia
David Andrews	University of Arkansas, USA

Pelin Angin	Middle East Technical University, Turkey
Ilhem Bouchareb	National Polytechnical University of Constantine (ENPC), Algeria
Wenming Cao	Shenzhen University, China
Ling-Jyh Chen	Institute of Information Science, Academia Sinica, Taiwan, China
Juan Corchado	AIR Institute, University of Salamanca, Spain
Elizabeth Diaz	University of Texas at Arlington, USA
Maribel Fernández	King's College, Department of Informatics, UK
Paulo Jorge Sequeira Gonçalves	Polytechnic Institute of Castelo Branco, Portugal
Wei-Chiang Hong	Oriental Institute of Technology, Taiwan, China
Saiful Islam	Institute of Space Technology, Pakistan
Nima Karimian	San Jose State University, USA
Mohammad Khodayar	Southern Methodist University, USA
Hansheng Lei	The University of Texas Rio Grande Valley, USA
Ming Li	The University of Texas at Arlington, USA
Taylor Li	The University of Texas at Arlington, USA
Zhaofeng Ma	Beijing University of Posts and Communications, Beijing, China
Phu Phung	University of Dayton, USA
Thar Baker Shamsa	University of Sharjah, UAE
Vijayan Sugumaran	Oakland University, USA
Marilyn Wolf	University of Nebraska–Lincoln, USA
Junyu Xuan	University of Technology Sydney, Australia
Junchi Yan	Shanghai Jiao Tong University, China
Neil Yen	University of Aizu, Japan
I-Ling Yen	University of Texas at Dallas, USA
Xiaohui Yuan	University of North Texas, USA
Shunxiang Zhang	Anhui University of Science & Technology, China

Keynotes

Some Critical Ideas for Smart Cities

Ishfaq Ahmad

University of Texas at Arlington, USA

Dr. Ishfaq Ahmad is a professor of Computer Science and Engineering, and Director of Advanced Computing Systems Labs at the University of Texas at Arlington (UTA) which he joined in 2002. He earned his Ph.D. in Computer Science and his M.S. in Computer Engineering from Syracuse University, New York, USA, in 1992 and 1987, respectively; and his B.S. in Electrical Engineering from the University of Engineering and Technology, Lahore, Pakistan, in 1985. Prior to joining UTA, he was an associate professor in the Computer Science Department at the Hong Kong University of Science and Technology. He has authored 260-plus publications that include books, papers in peer-reviewed journals and conference proceedings, related to software for supercomputing systems, parallel optimization algorithms, digital video compression and analysis, assistive technologies, and sustainable computing. Professor Ahmad has received numerous international research awards, including five best paper awards at leading conferences and top-tier journals. His research work, according to the Google Scholar, is widely cited with over 20,000 citations of his papers. Aside from being the founding Editor-in-Chief of the Journal, Sustainable Computing: Informatics and Systems, he has served as an editor of six others journals, chaired over 20 computer-related conferences, and delivered more than 150 talks. A senior visiting scientist at the U.S. Air Force Research Laboratory in Rome, New York, and a visiting lecturer at NASA's Johnson Space Center, Houston, TX, he also holds honorary professorships at several universities around the world. Dr. Ahmad is a Fellow of the IEEE.

Discovering the Realistic Paths Towards the Realization of Patent Valuation

Weidong Liu

Inner Mongolia University, China

Dr. Weidong Liu, the associate professor in the college of computer science, Inner Mongolia University. He was born in 1988, China, and received Ph.D. degrees Shanghai University, Shanghai, in 2016. His research interests include intelligent information processing, data mining, machine learning, and financial transaction technology. Currently, he is the principal investigator of several research grants, including 1 National Science Foundation, 1 Provincial Natural Science Foundation, and 1 Postdoctoral Science Foundation. He has published over 20 research papers on refereed journals and conferences.

Securely Outsourcing Services in Cloud Computing: Computation and Retrieval

Jun Ye

Hainan University, China

Dr. Jun Ye received his B.S. degree in Applied Mathematics at Chongqing University; M.S. degree in Cryptography at Guilin University of Electronic Technology; and Ph.D in Xidian University. He is a high-level talent of Hainan Province, and he is working at school of Computer Science and Cyberspace Security of Hainan University. His current research interests include computer science and information security. He has authored or co-authored more than 20 high-level publications, and he is also a reviewer of many well-known journals. He is one of the high-level talents of Hainan Province and got the "First prize" of Science and Technology Progress Prize of Hainan Province in 2019.

Contents

Signal Processing of Ground Penetrating Radar Based on MED Technology

Shuyuan Zhao[1,2], Hu Sheng[3(✉)], and Tianshuang Qiu[1]

[1] Faculty of Electronic Information and Electrical Engineering, Dalian University of Technology, Dalian 116024, Liaoning, China
[2] School of Electrical Engineering, Dalian University of Science and Technology, Dalian 116052, Liaoning, China
[3] School of Electrical and Information Engineering, Dalian Jiaotong University, Dalian 116028, Liaoning, China
vyfang5555@163.com

Abstract. Radar is a modern high-tech technology of radio, communication and control. In our daily life, it can be used as an information transmission medium. With the development of radar, its signal processing technology is continuously improved and perfected, and the most important of which is to improve the radar signal processing capability and speed. This article mainly uses experimental and analytical methods to test the accuracy of various algorithms in GPR signal processing, and uses Internet means and data to understand and study GPR systems. The experimental results show that the estimated signal-to-clutter ratio of the linear prediction method can reach 6.685, and its suppression performance is the best.

Keywords: MED technology · Ground penetrating radar · Signal processing · Linear prediction

1 Introduction

Ground penetrating radar (GPR) uses electromagnetic wave to detect underground targets, and extracts the properties, shapes and other information of targets by analyzing the characteristics of target signals. Among the existing technologies, there are many underground target detection technologies, such as nuclear, gravity field, electromagnetic induction and ultrasonic. However, GPR has obvious advantages in application. Compared with other technologies, GPR has advantages such as low detection cost, fast detection speed, continuous detection process, wide detection target range (non-metal and metal detection), high resolution, convenient and flexible operation, etc. In recent years, ground penetrating radar (GPR) has been one of the hot topics in international academic discussion. It has attracted great attention in many fields such as national defense, highway, urban construction, public security, railway, transportation, mine, airport, archaeology, tunnel, water conservancy and so on. In each field, GPR has shown a very broad application prospect. However, the working environment of GPR is complicated and changeable, which leads to the echo signal received by the receiver contains

I. Ahmad et al. (Eds.): STSIoT 2021, LNDECT 122, pp. 1–10, 2023.
https://doi.org/10.1007/978-981-19-3632-6_1

all kinds of clutter, which makes it very difficult to identify the target echo, so it is necessary to suppress clutter and noise in order to extract the desired target echo signal. Therefore, signal processing is the prerequisite for the normal operation of GPR. GPR system usually consists of electromagnetic signal generation circuit, receiver antenna, signal acquisition circuit and signal processing. As far as the GPR system involved in this paper is concerned, it can be divided into two parts: antenna unit and signal processing unit. The development of radar signal processing technology has made rapid progress in recent years. From the initial simple digitization of fixed waveform receivers, pulse compression and differential can be achieved now, and it also has high sensitivity. Because of the special structure of geomagnetism, there are often some interference signals that affect the receiver through coupling during the detection process. Therefore, in order to eliminate the harm caused by these interference signals to the ground penetrating radar system and improve the performance of the receiver.

2 Related Work

We need to conduct signal processing research. There are many results experiments in the investigation of ground penetrating radar signal processing.

For example, according to Iraklis Giannakis et al., Ground Penetrating Radar (GPR) is a proven tool for detecting and locating rebar (rebar) in concrete structures. However, quantifying the steel bar diameter with GPR is a difficult problem that cannot be solved by current processing methods [1]. According to Lombardi et al., environmental reflections may interfere with the echoes from the mine. One of the main limitations of mine detection using ground penetrating radar (GPR) is clutter. Clutter exhibits similar scattering properties to typical targets, which can significantly improve the detection threshold of the system [2]. Ground-penetrating radar (GPR) appears to be a useful tool for this purpose, according to Abderrahmane Aboudourib et al. The 3D results of the simulations and experiments show the good performance and potential of the proposed treatment [3]. Brian M et al. show that multistatic ground penetrating radar (GPR) signals can be mapped by tomography to produce a 3D distribution of image intensities. In the absence of objects of interest, these intensities can be thought of as clutter estimates [4]. Gianluca Gennarelli et al. proposed a ground penetrating radar survey imaging technique for heterogeneous subsoils characterized by the variation of dielectric constant along depth. The proposed method is based on a linear model of electromagnetic scattering phenomena [5]. S. Wagner conducted a novel study on the effect of receiver timing jitter on the image quality of ground penetrating radar systems. They clearly show the impact of timing jitter throughout the example processing pipeline [6]. A. D. Pambudi et al. provide a robust likelihood ratio test (LRT) for mine and UXO detection with forward looking ground penetrating radar. Robust detectors can significantly reduce false positive rates compared to parametric model-based detectors [7].

GPR technology is very complex. However, because of its wide application field and strong civil and military market, it has always been one of the hot topics in international academic discussion. The research of new theories and technologies in the field of GPR has promoted the rapid development of GPR, and its application fields are constantly expanding. However, there are few researches on the role of MED technology in surface exploration, so this paper puts forward this viewpoint.

3 Ground-Penetrating Radar Signal Processing Based on Med Technology

3.1 Ground Penetrating Radar System

The ground penetrating radar system is a non-destructive testing system that uses ground penetrating radar technology and radar signal processing technology as the core support to detect underground objects or target structures. It requires a certain amount of manpower and material input to conduct reasonable and effective detection and identification of underground regions of interest [8, 9].

(1) System composition of ground penetrating radar

The ground penetrating radar system generally consists of four parts: terminal equipment, such as a control processing unit composed of a microcomputer and a host, is mainly used for data analysis and processing, and the output is displayed on the screen of the terminal equipment. Radar transmitter is a radar transmitter that mainly produces sinusoidal pulses of high-frequency electromagnetic waves. The radar receiver is mainly used to collect and store the echo signal reflected by the interface. Radar antennas, including radar transmitting antennas and receiving antennas, are used for directional transmission and reception of electromagnetic waves. The working principle of the ground penetrating radar system is shown in Fig. 1: the signal received by the ground penetrating radar includes not only the target echo signal reflected underground, but also the wave reflected directly on the ground by the transmitting and receiving antenna and external noise [10, 11].

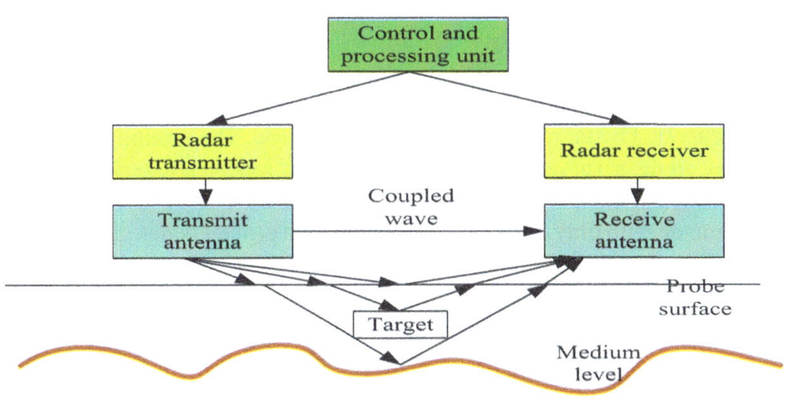

Fig. 1. Ground penetrating radar system workflow

(2) Basic electromagnetic theory of ground penetrating radar

When the ground penetrating radar detects on-site, it will generate high-frequency electromagnetic waves and generate a wave field in the detection area [8, 9]. According to the propagation theory of electromagnetic waves, the propagation of high-frequency electromagnetic waves in underground media follows Maxwell's equations, namely:

$$\nabla \times Q = -\frac{\varepsilon A}{\varepsilon s} \tag{1}$$

$$\nabla \times C = M + \frac{\varepsilon Y}{\varepsilon s} \tag{2}$$

$$\begin{aligned} \nabla \times A &= 0 \\ \nabla \times Y &= D \end{aligned} \tag{3}$$

Among them, Q is the electric field strength, Y is the electric displacement vector, C is the magnetic field strength, and M is the current density. Maxwell's equations describe the laws of dynamics and are important mathematical equations. Ground penetrating radar uses the fluctuating characteristics of electromagnetic fields to identify targets. As the field changes with time, Maxwell's equation describes the relationship between the coupled electric and magnetic fields [12].

During the propagation of electromagnetic waves, electromagnetic waves are composed of electric and magnetic fields, and their motion form is in the form of waves. In the theory of ground penetrating radar, the radar antenna generates a variable line current to form an excitation source to generate electromagnetic waves radiating outward. Regardless of the wave source, any type of wave can propagate.

(3) Data form of ground penetrating radar

The ground penetrating radar system can use a variety of different scanning methods for data collection. According to the theory of electromagnetic wave propagation and reflection, the A-scan signal of the echo signal received by GPR consists of direct waves generated by antenna crosstalk, ground clutter generated by ground reflection, target echo generated by target reflection and white noise. The ground penetrating radar antenna scans along a survey line and continuously receives a series of A-scans to form a two-dimensional data set, called B-scan, which can be described as a two-dimensional image with gray-level B-scan signals. The ground penetrating radar antenna scans along multiple parallel survey lines, and obtains multiple parallel B-scan data to form a three-dimensional data set, called C-scan.

3.2 Ground Penetrating Radar Signal Preprocessing

Ground penetrating radar echoes include direct waves generated by antenna crosstalk, ground echoes generated by ground reflections, echoes generated by uneven underground media, and echo signals generated by false targets. Generally speaking, when ground penetrating radar is used for detection, the position between the transmitting antenna and the receiving antenna is fixed. As for the echo generated by the false target, the general clutter suppression method cannot effectively remove it, and the method of target recognition needs to be used for judgment.

Clutter suppression is a key step in GPR signal data processing. The quality of clutter suppression technology will seriously affect the extraction and recognition of target curves. The clutter suppression methods of ground penetrating radar signal include: mean filter method, median filter method, low-pass filter method, PCA method, linear prediction method, Kalman filter, independent component analysis, etc. The specific process of improving clutter suppression is shown in Fig. 2:

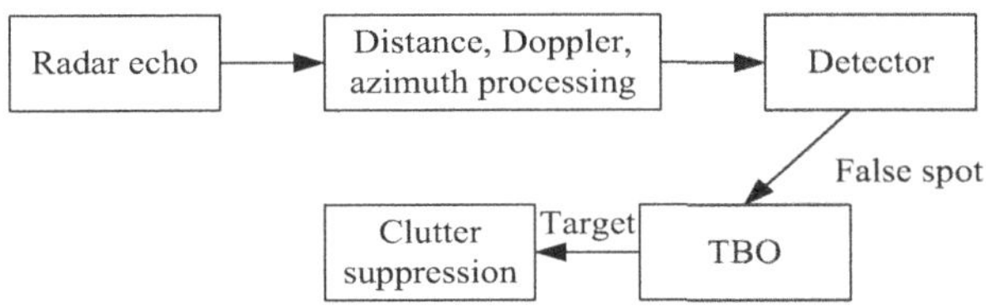

Fig. 2. Clutter suppression process

(1) Mean filtering algorithm

The mean filtering algorithm is the most classic and also the most commonly used clutter suppression algorithm. In the detection environment, some clutter signals such as direct waves and some ground clutter will appear on a series of A-Scan waveforms with the same time and amplitude data, so that a horizontal line will appear in the formed B-Scan echo image. The mean filtering method can simply and effectively remove these clutters. The principle and method of the mean filtering algorithm are very simple, and the processing effect of the direct wave and the level uniform ground clutter is good, but the ground in the actual measurement is often not uniform in the level. At this time, the clutter suppression using the mean filtering algorithm cannot be obtained ideal processing result.

(2) Median filtering method

The median filter method is a technique for smoothing directly in the spatial domain, and it is also a local smoothing technique. Under certain conditions, this method can remove the blurring of image details caused by linear filtering, and remove the interference of noise signals in the detection process. The actual operation is relatively simple, and the mathematical image processing is widely used. Two-dimensional median filtering has better noise reduction performance than one-dimensional filtering. According to the shape of the window, the two-dimensional median filter window has a variety of methods to filter noise. The median filtering method can overcome the blurring of image details caused by linear filtering, attenuate the noise signal generated during the detection process, and better retain the original echo signal.

(3) Principal component analysis

Principal component analysis is an analysis method based on the optimal principle. The PCA method can decompose the signal into independent components with dimensionality reduction on the basis of retaining the main target components, remove the clutter in the signal, highlight the effective target, and facilitate the extraction of target features. This facilitates the extraction of target features, which is essential for expressing target attributes in the radar target recognition system.

(4) Linear prediction method

It is assumed that there is a linear variation characteristic relationship between clutter. In the measurement area, the underground medium characteristics should be relatively close, and the clutter of adjacent sampling channels will not have great mutation. Therefore, the clutter of the next channel is predicted by linear prediction method to achieve the purpose of suppressing clutter. The linear prediction method can well preserve the micro target information, but the target features are not prominent and the resolution is not high. Based on the linear variation of clutter correlation, this paper proposes a new algorithm to suppress clutter effectively: improved linear prediction algorithm. If the address characteristics of the measurement area are relatively close, the clutter of adjacent tracks should not change suddenly. As long as it has linear variation characteristics, the clutter reaching the target is calculated by linear prediction method combined with PCA – this can extract the weak signal in the target signal while extracting the main attributes of the target signal, which can effectively remove the clutter. It also ensures the integrity of the target hyperbolic information.

3.3 Minimum Entropy Deconvolution Technology

The minimum entropy method is a simple mathematical analysis algorithm. It uses the correlation between various elements in the system to construct a "small" word, so that more information can be obtained with less data. Minimum entropy deconvolution is to treat the samples with different scales according to some rules. If there is a local minimum, it can be regarded as that all the information at this point is zero. A global optimal solution can be obtained by using this method. The advantage of this method is that all the original information can be extracted. This method can be used to analyze complex problems. The basic idea of minimum entropy deconvolution is to decompose all elements in the system according to linear law, topology and internal properties. Simplify complex problems through local. There are many methods to solve matrix sparsity. Without considering the relationship between various factors, an algorithm which is most suitable for the amount of calculation is adopted. When the scale of the system is large, the linear recursive minimum entropy method can be used to deal with the whole system. When dealing with a system, first take the known or unknown local minima of each factor as the reference data. If the parameter is uncertain, it can be calculated by the least square method.

There are two methods of minimum entropy deconvolution: iterative method and direct calculation method. Heuristic algorithm is often used in the process of random initialization. It simulates the human brain to deal with natural phenomena, and realizes the operation by using computer technology to replace the training of artificial neural network.

The purpose of minimum entropy deconvolution is to minimize the amount of information in the data. In the process of practical application, the random errors caused by various unpredictable factors often make the calculation results unsatisfactory. Therefore, a probabilistic method is needed to eliminate this randomness.

Because the construction method of minimum entropy deconvolution is to approach the matrix step by step without considering the influence of boundary information, window outside the window and other factors, and obtain a certain membership function by using the dependency relationship between the matrices in the system. The algorithm can be determined by calculating the weight of each element.

The construction of minimum entropy deconvolution is that there are multiple units in a system. Each element can be regarded as an equal matrix, that is, it is composed of each layer after the decomposition of each independent block. The number of layers is related to each sub element. When all independent components are deleted, it can be considered that the whole system belongs to one or more layers, which is regarded as the thin sub part of the lowest level. If all two or more adjacent unrelated components on this level are the minimum, the construction is said to be a local optimization problem.

The minimum entropy deconvolution method uses the convolution effect of FIR filter to design FIR filter based on the principle of minimum entropy. In the implementation process, the objective function of the signal before and after filtering is compared to judge whether the cycle continues, and finally converges to an optimal value to output the filter parameters. Therefore, the main influencing factors of minimum entropy deconvolution algorithm are filter parameter setting, cycle times and convergence error.

4 Ground Penetrating Radar Signal Processing Experiment

4.1 Simulation Experiment

In order to verify the feasibility and accuracy of the improved linear prediction clutter suppression method in the proposed minimum entropy deconvolution technology, this paper uses two sets of simulation data and measured data for data processing and analysis.

In the scattering modeling, the space domain is 0.8 m × 0.6 m, and the whole space is filled with dry sand. The dielectric constant of dry sand is 2, the conductivity is 0.002 S/m, the space step is 2 mm, and the sampling window is 8 ps. Radar transmission waveform is Ricker wavelet, its center frequency is set to 2 GHz, 50 measuring points are set, that is, there are 50 channels of data, the distance between measuring points is 0.01 m, the antenna is arranged at a position of 0.20 m above the ground, and the transmitting and receiving antennas are co-located. Place a metal pipeline in the middle of the space, where the radius of the metal pipeline is $r = 0.04$ mm, and the buried depth $d = 0.2$ mm.

The radar model parameters and media parameters are shown in Table 1:

Table 1. Radar model parameters and media parameters

Transmit signal broadband	8 GHz
Transmission signal start frequency	4 GHz
Transmit signal scan period	22 ms
Transmit signal frequency	500 GHz
Dielectric constant of underground medium layer	1, 2, 3, 4
Depth of underground medium layer	6, 3, 6
Signal-to-noise ratio	12 dB

4.2 Experimental Grouping

In order to fully verify the effectiveness of the algorithm, the simulation data is divided into two groups of simulation data, simulation one and simulation two, among which the simulation data one is the original data of modeling and simulation. The second simulation data is the result of adding Gaussian noise during the simulation process.

5 Simulation Data Analysis

5.1 Signal Clutter Analysis of Simulation Data Processed by Different Clutter Suppression Methods

The estimated values of the signal-to-clutter ratio of the simulation data 1 and simulation data 2 processed by different clutter suppression methods are shown in Table 2. The results show that the improved linear prediction method has the highest signal-to-noise ratio.

Table 2. The estimated value of signal-to-clutter ratio of simulated data processed by different clutter suppression methods

		Mean filtering	Median filter	PCA method	Improved linear prediction method
Simulation 1	SCR1	0.723	0.745	0.759	0.768
	SCR2	6.231	6.452	6.587	6.685
Simulation 2	SCR1	0.241	0.485	0.512	0.531
	SCR2	2.113	4.246	4.367	4.614

As shown in Fig. 3, we can see that the value of SCR1 is within 1 in experiment 1 and experiment 2, while the value of SCR2 is above 2. Among the different algorithms, the improved linear prediction method data is the most obvious. Compared with other clutter suppression methods, the improved linear prediction method has the best clutter suppression performance.

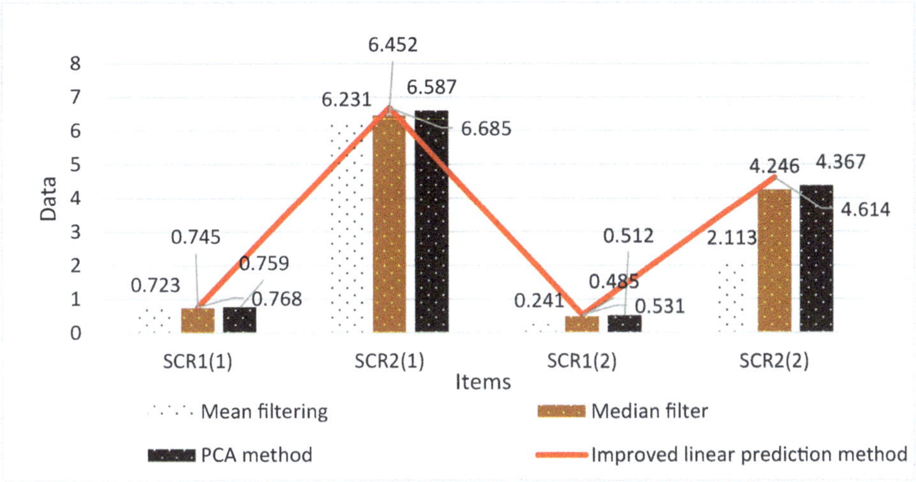

Fig. 3. The estimated value of signal-to-clutter ratio of simulated data processed by different clutter suppression methods

6 Conclusion

Radar signal processing technology is a new subject, and its development and application play a very important role in social economy, military construction and national defense construction. With the continuous progress and maturity of hardware fields such as computers and communication equipment, radar uses more new materials in the process of information collection and transmission. Regarding the signal processing problem of radar collection, this paper proposes the application of MED technology in it. The minimum entropy deconvolution has an incomparable calculation effect in the signal processing of ground penetrating radar.

Acknowledgements. National Natural Science Foundation of China (Grant No. 61671105)

Natural science research project of Liaoning Provincial Department of Education (Grant No. JDL2019014)

References

1. Giannakis, I., Giannopoulos, A., Warren, C.: A machine learning scheme for estimating the diameter of reinforcing bars using ground penetrating radar. IEEE Geosci. Remote Sens. Lett. **18**(3), 461–465 (2021)
2. Lombardi, F., Griffiths, H.D., Lualdi, M., Balleri, A.: Characterization of the internal structure of landmines using ground-penetrating radar. IEEE Geosci. Remote Sens. Lett. **18**(2), 266–270 (2021). https://doi.org/10.1109/LGRS.2020.2970249
3. Aboudourib, A., Serhir, M., Lesselier, D.: A processing framework for tree-root reconstruction using ground-penetrating radar under heterogeneous soil conditions. IEEE Trans. Geosci. Remote. Sens. **59**(1), 208–219 (2021)

4. Worthmann, B.M., et al.: Clutter distributions for tomographic image standardization in ground-penetrating radar. IEEE Trans. Geosci. Remote Sens. **59**(9), 7957–7967 (2021)
5. Gennarelli, G., Catapano, I., Dérobert, X., Soldovieri, F.: A ground penetrating radar imaging approach for a heterogeneous subsoil with a vertical permittivity gradient. IEEE Trans. Geosci. Remote. Sens. **59**(7), 5698–5710 (2021)
6. Wagner, S., Worthmann, B.M., Pham, A.-V.: Minimizing Timing Jitter's impact on ground-penetrating radar array coupling signals. IEEE Trans. Geosci. Remote Sens. **59**(6), 4717–4724 (2021). https://doi.org/10.1109/TGRS.2020.3019976
7. Pambudi, A.D., Fauß, M., Ahmad, F., Zoubir, A.M.: Minimax robust landmine detection using forward-looking ground-penetrating radar. IEEE Trans. Geosci. Remote Sens. **58**(7), 5032–5041 (2020). https://doi.org/10.1109/TGRS.2020.2971956
8. Moalla, M., Frigui, H., Karem, A., Bouzid, A.: Application of convolutional and recurrent neural networks for buried threat detection using ground penetrating radar data. IEEE Trans. Geosci. Remote. Sens. **58**(10), 7022–7034 (2020)
9. Tajdini, M.M., Morgenthaler, A.W., Rappaport, C.M.: Multiview synthetic aperture ground-penetrating radar detection in rough terrain environment: a real-time 3-D forward model. IEEE Trans. Geosci. Remote Sens. **58**(5), 3400–3410 (2020)
10. Srivastav, A., Nguyen, P., McConnell, M., Loparo, K.A., Mandal, S.: A highly digital multiantenna ground-penetrating radar (GPR) system. IEEE Trans. Instrum. Meas. **69**(10), 7422–7436 (2020)
11. Abdelkader, E.M., Marzouk, M,M., Zayed, T.M.: Mapping ground penetrating radar amplitudes using artificial neural network and multiple regression analysis methods. Int. J. Strateg. Decis. Sci. **10**(2), 84–106 (2019)
12. Le Bastard, C., et al.: A linear prediction and support vector regression-based debonding detection method using step-frequency ground penetrating radar. IEEE Geosci. Remote Sens. Lett. **16**(3), 367–371 (2019)

Application Analysis of Information Security Technology in Credit Card System

Yuanyuan Sun[1](✉) and Trini Canals[2]

[1] Pacific Credit Card Center of Bank of Communications, Hefei, Anhui, China
15004869@qq.com
[2] University of Frankfurt, Frankfurt, Germany

Abstract. With the continuous advancement of computer science and technology and internationalization, global competition has become increasingly fierce, especially in the banking industry. To maintain and improve their competitiveness in the market, various banks have expanded the scope and content of the banking industry under the background of continuous improvement of their operating environment, and have continuously adopted new information technologies. Currently, the bank's credit card system has become a place that every bank attaches great importance to. The credit card system will have an increasingly important impact on consumption concepts and customer living in the future development. It can be said that understanding the development of the credit card system will affect the future development of the bank to a certain extent. This article analyzes the application of information security in the credit card system. First, it uses the literature research method to summarize the problems in the credit card system and the information security elements of the development of the credit card system, and uses the questionnaire survey method for the application status of information security in the credit card system. According to the investigation, about 45% of the bank's credit card systems have information leakage problems, and about 26% of them have hacked attacks. For information leakage, banks must not only strengthen external management, but also deal with internal problems. Management also needs to be strengthened. Among the suggestions given, 42% of staff member chose to strengthen computer intrusion detection technology, and 32% chose to improve hardware equipment.

Keywords: Information security · Credit card system · Banking industry · Future development

1 Introduction

With the widespread application of information technology and the rapid development of information technology in the financial field, commercial banks increasingly rely on information technology [1, 2]. Today's world is in a new era of mobile Internet, smart phones, mobile terminals, and the Internet have become the basic elements of modern life [3, 4]. At present, our country has fully entered the era of Internet funds. The Internet has brought new technologies and new opportunities for commercial banks, and it has

I. Ahmad et al. (Eds.): STSIoT 2021, LNDECT 122, pp. 11–19, 2023.
https://doi.org/10.1007/978-981-19-3632-6_2

also brought greater financial risks and challenges to commercial banks [5, 6]. With the rapid growth of Internet financing, threats to the security of personal information and financial assets are increasing, and bank information security has become a major issue. Many factors affect the information security of commercial banks. Security incidents such as information leakage will cause serious consequences to customers and the bank itself. The lighter ones will affect the bank's customer satisfaction, the more serious ones will affect the bank's daily activities. Therefore, in recent years, domestic commercial banks have taken information security management research as a research hotspot and focus [7, 8].

In the research on the application and analysis of information security in the credit card system, many scholars have studied it and achieved good results. For example, Omelyanenko introduced the information security management ideas, management methods, and management content of my country's banking industry. It is believed that the banking industry should pay attention to the importance of information security management, make full use of information security technology, and establish a bank's information security management system [9]. Based on information security, Kachyn elaborated on the concepts, principles and methods of information security risk assessment and domestic information security related policies, and emphasized the importance of information security risk assessment in information security management [10]. Domestic research on information security management theories tends to focus on the summary analysis of foreign information security management theories or simply discuss the construction of information security systems from a certain aspect. There are relatively few systematic and comprehensive theoretical analyses. It is still in its infancy.

This paper analyzes the application of information security in the credit card system. First, it uses the literature research method to summarize the problems in the credit card system and the information security elements of the development of the credit card system, and uses the questionnaire survey method for the application status of information security in the credit card system.

2 Research on Information Security and Credit Card System

2.1 Research Methods

2.1.1 Literature Research

Reading books and articles about the application of information security in the credit card system in literature, the advantage is that you can understand the development process of the research object from the source, and understand the development status of the research object, and provide a clear and structured theoretical basis for in-depth thesis development.

2.1.2 Investigation and Research Method

The questionnaire survey method is that this article conducts a survey through pre-prepared questions and analyzes the answers of the interviewees to draw the necessary conclusions. By designing a questionnaire, it's objective to understand the status quo of the application of information security in the credit card system.

2.1.3 Quantitative Analysis

Qualitative analysis is related to quantitative analysis. Quantitative analysis refers to the analysis of mathematical hypothesis determination, data collection, analysis, and testing [11].

Qualitative analysis refers to the process of conducting research through research and bibliographic analysis based on subjective understanding and qualitative analysis [12].

2.2 Information Security Issues Faced by Credit Card Systems

2.2.1 Threats from the Internet System

The complexity of the Internet system provides a starting point for malicious behavior. To control the information security risks of Internet financing, the challenges they face include malware and phishing sites.

2.2.2 Security Threats to User Privacy

In the era of big data, the widespread popularity of cloud computing, Map-Reduce, No-SQL, and other technologies enables Internet companies to quickly, effectively pass the geographic location, IP address, and access device types left by customers when visiting websites. Browsing behavior, consumption behavior and other information on customers' consumption habits, interest preferences and other information. This residual information is stolen by criminals and poses a major threat to the credit card system.

2.2.3 Threats to the Availability of Network Platforms

Availability is an important guarantee for the development of financial platforms. Once the availability is restricted, it will cause huge losses. For example, in 2013, bank cash registers, ATMs and online banking services in many areas of the country failed and lasted for nearly 1.5 h. According to the information on the bank's homepage, e-banking transactions reached 170 trillion US dollars in the first half of this year. Following this estimate, a one-hour failure could result in at least 30 billion transaction losses. It can be seen that the availability of the platform is the foundation and guarantee for the development of Internet financial services.

2.2.4 Threats from Inside the Company

There are two main threats to the information security of credit card systems, one is from the outside in, and the other is from the inside out. Currently, most multi-service network security methods are aimed at external threats, and greater trust in internal servers often ignores internal management of business information security. Since internal personnel can easily receive various information within the company, it will lead to the leakage of key information and the violation of electronic contracts.

2.3 Information Security Elements of Credit Card System Development

2.3.1 From a Strategic Point of View, Give Full Attention to Information Security

Internet finance companies need to recognize the importance of information security to Internet finance companies, pay attention to strategic information security issues, and plan information security and financial security risk management as a whole. Not only need to strengthen network security protection, take measures at the level of security and communication protocols, and solve the security problems of the Internet itself through a variety of technical means, but also need to strengthen innovation at the business level following business rules and innovation. Risk control and paying attention to the characteristics of innovative business are key factors to develop a smart security strategy.

2.3.2 Seek Multi-party Cooperation to Jointly Build an Internet Financial Ecological Environment

From the perspective of the industry, the information security of Internet financial companies is by no means a problem of a certain company, but requires close collaboration across the industry; from the perspective of the Internet industry, information security issues require the cooperation of relevant parties. Only the government, enterprises, and users can work together to build a safe Internet financial ecological environment.

2.3.3 Carefully Protect the Privacy and Data of Users by Observing the Law and Self-Discipline

A large number of users is the foundation of the long-term effect of Internet financial products. If the security of users' personal information cannot be effectively ensured on the Internet financial platform, the reputation of the platform will be damaged, which will result in the loss of platform users. Therefore, ISPs must strictly abide by the basic information security rules for data storage and analysis, effectively protect customer privacy, and ensure the security of users' private data.

2.3.4 Integrate Multiple Channels to Enhance User Information Security Training

Insufficient user security awareness is an important reason for information security risks. The dissemination and education of user information security awareness cannot be based solely on social resources. Internet finance companies need to be fully aware of the importance of publicizing and disseminating information security to users, integrate publicity and information security dissemination into product design, and accept general evaluation. Provide users with financial risk tips and information security training, and emphasize the importance of information security issues to users in various ways, so that users can carefully develop the habit of paying attention to information security when using credit card system applications.

2.4 Information Security Data Evaluation Model

2.4.1 Establish a Judgment Matrix

Drawing out the target elements at each level and clarifying the relationship between them, we successfully established an interrelated hierarchical structure. Assume that the set consisting of n elements A1, A2, ..., An corresponds to the corresponding criterion B. To obtain the model of the judgment matrix, it is necessary to pass the criterion of the upper layer as the calculation and derivation criterion. Under the guidance of the criterion, it is determined by comparing the n elements with each other.

2.4.2 Calculate the Product Mi of Each Row Element of the Matrix

Calculate M_i the power root of W_i^0

$$W_i^0 = \left(\prod_{j=1}^{n} a_{ij}\right)^{\frac{1}{n}} \quad \ldots i = 1, 2, \ldots\ldots n \tag{1}$$

2.4.3 Normalize the Vector

$$W_i^0 = \frac{W_i^0}{\sum_{i=1}^{n} w_i^0} \quad \ldots i = 1, 2, \ldots\ldots n \tag{2}$$

3 Investigation on the Status Quo of the Application of Information Security in the Credit Card System

3.1 Research Purpose

Investigate the application status of information security in the credit card system through the questionnaire survey method, and give corresponding suggestions based on the problems in the existing credit card system of the respondents, and analyze the measures that should be taken for information security based on the results.

3.2 Questionnaire Survey

3.2.1 Number of Questionnaires

According to the minimum sample size formula in statistics, the author sets the confidence level of the questionnaire to 8%, and the allowable error does not exceed 8%. Calculate the minimum sample size as

$$n_0 = \left(\frac{t_a}{2\Delta p}\right)^2 = \left(\frac{1.645}{2 \times 0.075}\right)^2 = 120 \tag{3}$$

That is, the minimum sample size of this questionnaire is 120 copies.

3.2.2 Data Source

This paper investigates the credit card system and the limited scope in the banking industry. Therefore, three banks in this city are randomly selected to be replaced by Bank A, Bank B, and Bank C. According to the minimum number of questionnaires, the number of questionnaires distributed by the three banks is respectively: 40, 50, 60, the number of questionnaires returned is 39, 49, 60.

4 Data Analysis

4.1 Investigation of Problems in the Credit Card System

This article uses a questionnaire survey method to investigate the problems of the bank's credit card system. The results of the survey are shown in Table 1.

Table 1. There is a problem with the credit card system

	A bank	B bank	C bank
Hacker attack	26%	25%	28%
Information leakage	45%	47%	46%
Data is changed	19%	18%	16%

As can be seen from Fig. 1, among the problems in the bank's credit card system, about 45% of the information leakage problems occurred, and about 26% of the problems occurred with hacker attacks. Regarding information leakage, banks must not only strengthen external management, but also strengthen internal management.

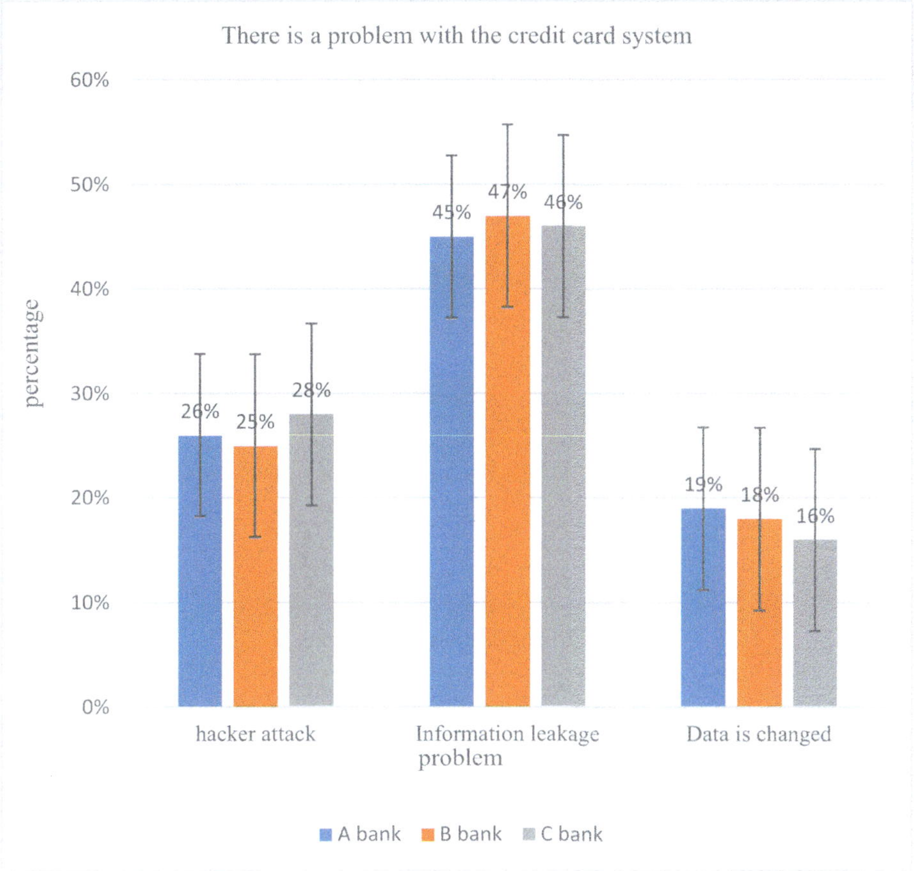

Fig. 1. There is a problem with the credit card system

4.2 Relevant Suggestions for Problems

This article uses a questionnaire survey method to investigate the problems that occur in the bank's credit card system. The results of the survey are shown in Table 2.

Table 2. Suggestions for problems

	A bank	B bank	C bank
Strengthen computer intrusion detection Application of measurement technology	40%	44%	42%
Improve hardware quality	30%	32%	34%
Database management security	30%	24%	24%

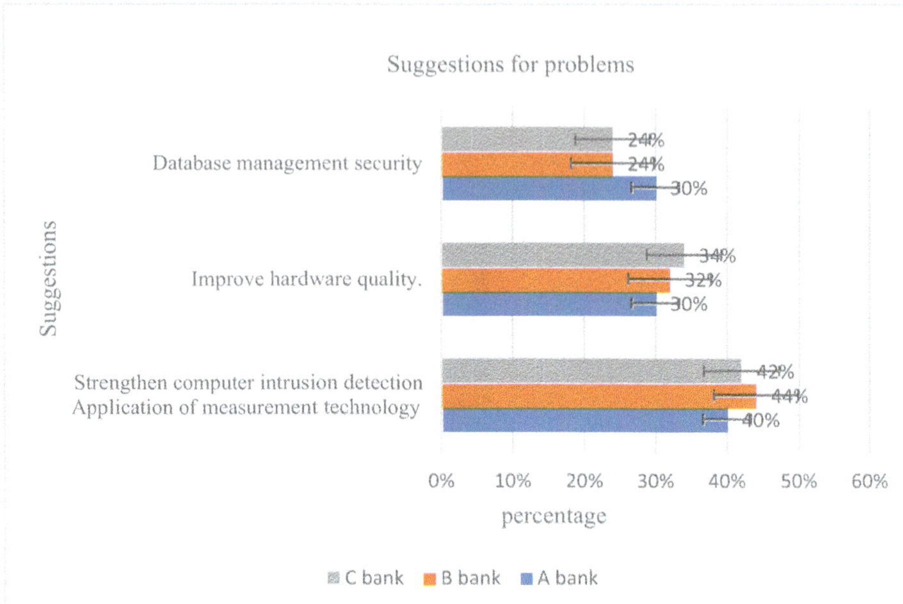

Fig. 2. Suggestions for problems

It can be seen from Fig. 2 that among the suggestions given, 42% of people choose to strengthen computer intrusion detection technology, and 32% of people choose to improve hardware equipment.

5 Conclusion

In recent years, with the rapid development of the mobile phone industry, many traditional industries have gradually changed. Funds were initially used as equivalent means of exchange. Due to economic globalization, the single economy has gradually integrated into the global environment. The bank credit card system is constantly evolving. With changes in consumer perceptions, credit cards have begun to grow in China. With the continuous development and growth of credit card sales, information security is an important guarantee for the credit card system. It can be seen from the survey results of this article that, first, among the problems in the bank credit card system, about 45% of the information leakage problems occurred, and about 26% of the problems occurred with hacker attacks. Second, among the suggestions given, 42% of people choose to strengthen computer intrusion detection technology, and 32% of people choose to improve hardware equipment. What's more, the internal management system of the bank should also be paid attention to. The user's information can be viewed and should be authorized.

References

1. Hao, S., Lü, Y., Jie, L., Yue, L., Xu, D.: Application of classified protection of information security in the information system of air pollution and health impact monitoring. Wei Sheng Yan Jiu J. Hygiene Res. **47**(1), 103–107 (2018)
2. Yong, Q.L.: Application analysis of artificial intelligence in library network security. J. Phys. Conf. Ser. **1744**(3), 032024 (7 pp) (2021)
3. Triana, Y.S., Pangabean, R. Risk analysis in the application of financore information systems using FMEA method. J. Phys. Conf. Ser. **1751**(1), 012032 (10pp) (2021)
4. Kim, S.B., Min, G.K., Park, J.H.: Simple credit card payment protocols based on SSL and passwords. J. Korea Inst. Inf. Sec. Cryptol. **26**(3), 563–572 (2016)
5. Khabarlak, K., Koriashkina, L.: Mobile access control system based on RFID tags and facial information. Bull. Nat. Tech. Univ. KhPI Ser. Syst. Anal. Control Inf. Technol. **2**(4), 69–74 (2021)
6. Demeshko, V.S.: Application of convolutional neural networks in the intelligence security system subsystem. Syst. Anal. Appl. Inf. Sci. **2**, 46–53 (2020)
7. Li, Y., Lu, Y.: Multimodality data analysis in information security ETCC: encrypted two-label classification using CNN. Sec. Commun. Netw. **4**, 1–11 (2021)
8. Borisova, D.E.: Improvement of wage system in health care as a factor in increasing the economic security of the region. Econ. Prof. Bus. **4**, 34–39 (2020)
9. Omelyanenko, V.: Analysis of information asymmetry in innovation system security ensuring. Mark. Manag. Innov. **4**, 199–208 (2017)
10. Kachynskyi, A.B., Styopochkina, I.S.: A systematic approach to the analysis of phenomena of the information and cyberspaces. Reports of the National Academy of Sciences of Ukraine (11), pp.16–23 (2020)
11. Cheng, L.I., Wang, J.J.: Quantitative and qualitative analysis of student tutors as near-peer teachers in the gross anatomy course. Ann. Anat. **210**(2), 147–154 (2017)
12. Maragkaki, A.E., Kotrotsios, T., Samaras, P., Manou, A., Lasaridi, K., Manios, T.: Quantitative and qualitative analysis of biomass from agro-industrial processes in the central macedonia region, Greece. Waste Biomass Valori. **7**(2), 383–395 (2016)

A Job Recommendation System Based on Student and Category Similarity Computation

Yang Tan[1], Jiapeng Zhu[1], Chunxia Leng[1](✉), and Salvatore Gaglio[2]

[1] Department of Computer Science and Engineering, East China University of Science and Technology, Shanghai, China
cxleng@ecust.edu.cn

[2] Free University of Berlin, Berlin, Germany

Abstract. The expansion of enrollment in Chinese universities has further aggravated the pressure of graduates' employment competition. Students have to spend lots of time seeking satisfactory jobs. There is a growing uncertainty in finding a job because education is biased toward basic knowledge, students lack both work experience and social background, and job market changes rapidly. Students tend to look for jobs that match their interests and skills. It's difficult to apply traditional collaborative filtering recommendation algorithm directly. In this case, we adopt the method of recommendation based on students' career preferences. Data collected from IT industry on 51job website are divided into 15 first-level categories and 246 second-level categories through a semi-manual method. After student information and job information are cleaned and captured by keywords, the degree of matching between the student and the second-level category is calculated, and interpretable recommendations are realized.

Keywords: Job recommendation · Similarity calculation · SCSC-JRS · Students employment

1 Introduction

Due to the expansion of enrollment in Chinese colleges and universities, the surge in employment has created tremendous competitive pressure for college graduates in employment [1]. Meanwhile, the impact of COVID-19 and other multiple factors lead to a historical decrease in the number of recruits [2]. In order to cope with the fierce job market, some students join ideal company by internal referrals, others increase work experience by off-campus internships or get jobs from university employment department. Though the methods above are useful, most students still spend lots of time searching and identifying jobs [3]. Therefore, it is important to help students find jobs effectively and identify the degree of association between jobs and students.

In order to solve the problem, a series of solutions are introduced. The collaborative filtering method is effective but heavily relies on users' interact with items. So lacking in job search records makes it difficult to use directly in students' recommendation [4].

I. Ahmad et al. (Eds.): STSIoT 2021, LNDECT 122, pp. 20–27, 2023.
https://doi.org/10.1007/978-981-19-3632-6_3

Some studies are based on the similarity calculation and matching of students' resume and post information [5], but it needs much private and accurate information which is hard to obtain. Some studies combine college students' geographic information and education background to achieve recommendations [6], yet they ignore students' own interests and skills.

After summarizing the above situation, we use the IT industry as a recommended sample, and the main contributions are as follows. Firstly, we calculate the similarity between students' information and 246 secondary IT industry categories, reduce the impact of individual posts on student recommendation, blur the subjectivity of student information. Secondly, we pay more attention to job-hunting intentions. Therefore, recommendations are more independent rather than lay emphasis only on personal private information, or just use information such as social networks as recommendations.

2 Methodology

2.1 Framework Overview

In order to solve the problem of identifying the degree of association between job categories and students, we design a framework called SCSC-JRS (Student and Category Similarity Computation – Job Recommendation System), as shown in Fig. 1.

- Data layer: The data including user's personal information and original post data obtained by the crawler stored in relational database.
- Keyword extraction layer: Extract keywords from the classified and cleaned job data through a variety of algorithms, and compare the results with manual word segmentation to obtain the most suitable algorithm for job keyword extraction.
- Similarity calculation layer: Vectorize the extracted keywords and information submitted by the students in the system, then calculated the similarity by LSI, finally the results are normalized and stored in a Json file as model.
- Recommendation layer: Acquire the preference probability from the Json file mentioned above to recommend. Considering the long tail effect of system [7], we further optimize the probability to recall ratio, and adjust the ranking according to the optimized ratio. Finally, we achieved top-n recommendation.

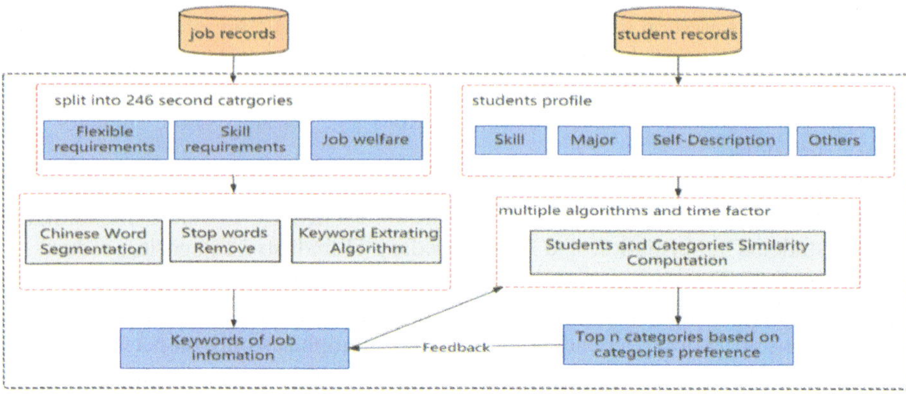

Fig. 1. SCSC-JRS framework

2.2 Data of Jobs and Students

We collected more than 150,000 school recruitment and internship posts in the IT industry from 2020 to 2021 on 51job through distributed crawlers, and divided them into 15 first-level categories (mobile development, etc.) and 246 secondary categories (Hadoop engineer, etc.). After being cut and cleaned, data were saved in database, and mainly retained the following information:

- Flexible requirements: It describes the soft requirements of each post on the job applicant's character, and whether they have job experience. For example, student who achieves a scholarship, and those who love programming are preferred.
- Skill requirements: It describes the detailed requirements for the job applicant's skills or hard conditions in this post. For example, academic qualifications, proficient use of Python.
- Job welfare: It describes the treatment of the job-seekers in this post. For example, work from 9 to 5, provide training or exists a reasonable promotion mechanism.

2.3 Extracting of Keywords

First, get the job information with category tags, then extracting keywords. we use the following three algorithms are adopted. The format of the data after reading is defined as follows:

$$D = [C_1, C_2, \ldots, C_n] \tag{1}$$

C_i means to read all the information of the ith category. D represents the corpus formed after reading the data.

2.3.1 TF-IDF (Term Frequency–Inverse Document Frequency)

TF-IDF is a method widely used in information retrieval and data mining. The main principle is that the more times a word appears in an article and the fewer times it appears in other articles, it means that the word is distinguishable and is the key word of the article [8]. The algorithm is defined as follows:

$$tf \times \mathrm{id}f\,(i,j) = \frac{n_{ij}}{\sum_k n_{kj}} \times \log\left(\frac{|C|}{1+|C_i|}\right) \tag{2}$$

n_{ij} represents the frequency of occurrence of word i in job category j. $|C|$ represents the total number of categories. $|C_i|$ represents the total number of job categories where word i appears, adding 1 to the denominator uses Laplacian smoothing.

2.3.2 TextRank

The TextRank algorithm can extract keywords without the corpus background. It establishes a window after segmentation of the text to realize the connection between words, and then scores each word (node), and selects the highest score as a keyword [9]. The score calculation formula is as follows:

$$WS(V_i) = (1-d) + d \times \sum\nolimits_{j \in In(V_i)} \left(\frac{1}{|Out(V_j)|} \times WS(V_j)\right) \tag{3}$$

d is the damping coefficient, representing the probability of pointing from a node to any node in the graph. $In(V_i)$ represents the word corresponding to the entry edge of i node. $Out(V_j)$ represents the word corresponding to the outgoing edge of j node.

2.3.3 Topic Model

The above two formulas use the relationship between the word and the post text to extract keywords, while the topic model believes that there is no direct relationship between the word and the text, but one or more topics are connected in series, and the text is composed of some topics. The theme is the probability distribution of words [10], and the formula for its core idea is as follows:

$$p(w_i|C_j) = \sum\nolimits_{k=1}^{K} p(w_i|t_k) \times p(t_k|C_j) \tag{4}$$

w_i is the word i. C_j is category document j. t_k is topic k.

Currently, the commonly used methods for calculating distribution information are LSI/LSA and LDA. LSI/LSA is based on SVD decomposition, while LDA uses Bayesian methods to fit distribution information [11].

2.3.4 Students and Categories Similarity Computation

In this part of work, after obtaining the keywords of post information, the LSI model is mainly used to calculate the similarity [12]. The main steps are as follows:

– Use the TF-IDF model to represent each post category as a vector, and concatenate all post category word vectors to form a word-category matrix $A_{(m \times n)}$.

- Use SVD decomposition for matrix A.

$$A_{m\times n} = U_{m\times m} \cdot \Sigma_{m\times n} \cdot V_{n\times n} \tag{5}$$

- The SVD decomposition result is mapped to the approximate SVD result of the lower dimension k, so as to obtain the correlation between the word and the topic, and the correlation between the category and the topic.

$$A_{m\times n} \approx U_{m\times k} \cdot \Sigma_{k\times k} \cdot V_{k\times n} \tag{6}$$

- The text topic matrix is used to calculate the similarity through cosine similarity. The student information and the post information in each category are calculated and summed to obtain the similarity of the post category. After normalizing it, the result is returned to the database.

$$sim(u, C_i) = \frac{\sum_{k=1}^{n} sim(u, c_k)}{n+1} \tag{7}$$

2.3.5 Job Recommend

After obtaining the similarity between users and all categories, we select Top n as the users' most preferred categories by sorting the similarity. Then we select posts published within 7 days with a random probability percentage from each secondary categories as recommendation results.

$$N(u, C_i) = sim(u, C_i) \times N_{C_i} \tag{8}$$

N represents the number of recommended lists. u represents students. C_i represents category i. N_{Ci} represents the number of posts in category i in the database.

Although the recommended method can be used directly, because the employment of college students is different from that of experienced job seekers, the uncertainty of their employment is higher. For example, students who have studied java may also be engaged in python work due to complicated reasons, which means that we need to punish the categories with too high similarity. After testing, the improved similarity probability is adopted, and the similarity calculation is adopted in the n secondary categories, and the posts in the similarity probability percentage are taken out as recommendations to achieve better results.

$$N(u, C_i) = \log(sim(u, C_i)) \times N_{C_i} \tag{9}$$

Algorithm 1. The process of using category-based similarity recommendation algorithm.

Algorithm 1 Category Similarity Recommend

```
Input:U,J
Output:R`
1:  for student in U do
2:      Calculate similarity of categories and student
3:  end for
4:  Sort sim(u,C) in descending order
5:  Obtain the top n categories and its value
6:  for c in categories do
7:      Obtain each job from a job set J where category is c
8:      Calculate similarity of each job to form a recommend list R
9:      Sort R in descending order
10:     Obtain the number N by the value multiply the amount of the job set J where
category is c
11:     Obtain the top N positions from R to form a recommended list R`
12: end for
13: Remove duplicate position from R`
14: Return R`
```

3 Results

3.1 Experiment of Keywords Extraction

In the keyword capture experiment, the method of manually constructing a test set is used for evaluation, and the accuracy and recall rate are used for evaluation. We compared the four algorithms used in the system, each category extracted about 3 post information, a total of 738 pieces as test data. We also manually annotated 23084 keywords as the test set of keyword capture. The experimental results are shown in Table 1.

Table 1. Experiment of keywords extraction

Method	Extract		Right		Precision (%)	Recall (%)
	Total	Average	Total	Average		
TF-IDF	24655	100.22	12472	50.70	50.59	54.03
TextRank	4703	19.12	2078	8.45	44.18	9.00
LSI/LSA	28213	114.69	13976	56.81	49.54	60.54
LDA	28741	116.83	12811	52.08	44.57	55.50

3.2 Experiment of Recommendation

We collected the information of 46 students during the trial operation of the system at East China University of Science and Technology and the interactive records. The results of the experiment are shown in Fig. 2.

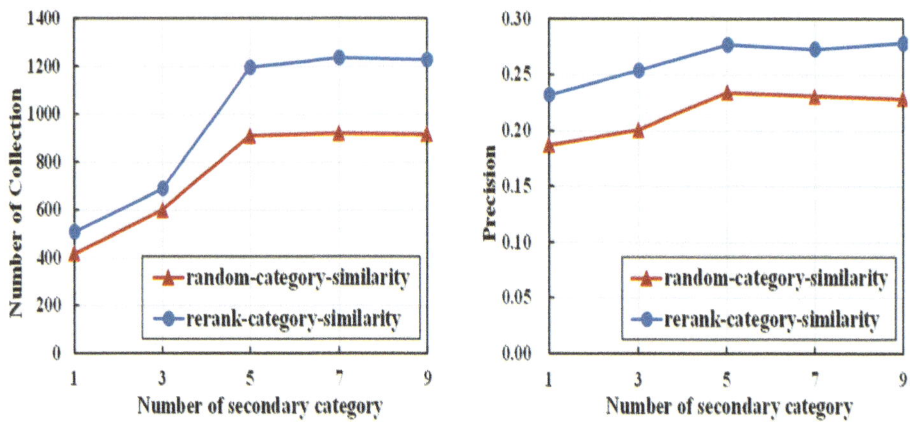

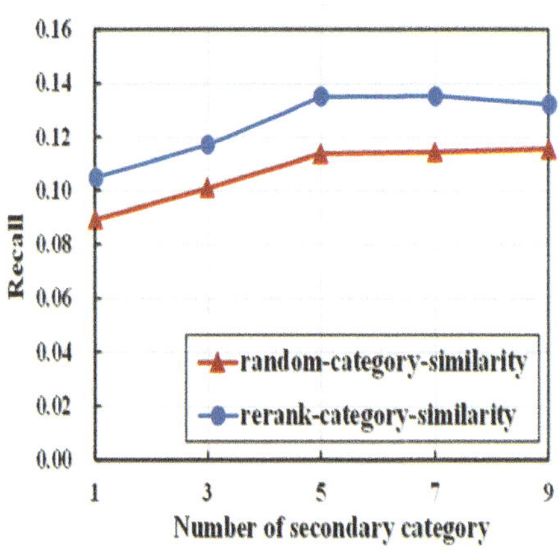

As the number of secondary categories grows, the pros and cons of rerank and random methods are measured from the three perspectives of collection, accuracy, and recall.

Fig. 2. Experiment of keywords extraction.

3.3 Analysis and Discussion

From Table 1 that the accuracy and recall rates obtained by the LSI method are both high, and it can play a better role in job keyword capture.

From Fig. 2 that the improved method contributes more click rates and collections overall, but relatively speaking, it increases the computational time overhead.

In general, SCSC-JRS is more effective than recommendation system based on job similarity and does not treat any specific post as the unit of recommendation. It blurs students' information and pays more attention to category tendencies, which makes it more suitable for inexperienced college students.

4 Conclusion

In summary, we design a position recommendation system based on similarity calculation of student information and job category. Firstly, dividing the IT industry data into multiple categories, and using multiple methods to test and find the most suitable algorithm. Then calculating the similarity between the student information and the post category information to get the post category that best matches the student information, and proposed an improved recommendation method to improve the accuracy of the system. The experimental data comes from real user records when the system is running in ECUST, which reflects the effectiveness and reliability.

Acknowledgements. Supported by the Undergraduate Training Program on Innovation and Entrepreneurship grant 202110251052.

References

1. Chen, J.: Problems and countermeasures of university students' employment under the background of popularization of higher education. Int. J. Intell. Inf. Manag. Sci. **10**(2) (2021)
2. Mok, K.H., Xiong, W., HuYe, Y.: COVID-19 crisis and challenges for graduate employment in Taiwan, Mainland China and East Asia: a critical review of skills preparing students for uncertain futures. J. Educ. Work. **34**(3), 247–261 (2021)
3. Zhou, Q., Liao, F., Chen, C., Ge, L.: Job recommendation algorithm for graduates based on personalized preference. CCF Trans. Pervasive Comput. Interact. **1**(4), 260–274 (2019). https://doi.org/10.1007/s42486-019-00022-1
4. Dhameliya, J., Desai, N.: Job Recommendation system using content and collaborative filtering based techniques. Int. J. Soft Comput. Eng. (IJSCE) **9**(3), 8–13 (2019)
5. Chou, Y.C., Yu, H.Y.: Based on the application of AI technology in resume analysis and job recommendation. In: 2020 IEEE International Conference on Computational Electromagnetics (ICCEM), IEEE (2020)
6. Giabelli, A., et al.: Skills2Job: a recommender system that encodes job offer embeddings on graph databases. Appl. Soft Comput. **101** (prepublish) (2020). https://doi.org/10.1016/J.ASOC.2020.107049
7. Zhao, K., Pi, J.: A stable collaborative filtering algorithm for long tail recommendation. J. Phys. Conf. Ser. **1176**(2), 022033 (2019)
8. Yu, H., et al.: Student sentiment classification model based on GRU neural network and TF-IDF algorithm. J. Intell. Fuzzy Syst. **40**(2), 2301–2311 (2021)
9. Xiong, C., Li, X., Li, Y., Liu, G.: Multi-documents summarization based on TextRank and its application in online argumentation platform. Int. J. Data Warehous. Min. **14**(3), 69–89 (2018)
10. Danny, V., et al.: On mining words: the utility of topic models in health education research and practice. Health Promot. Pract. **22**(3), 309–312 (2021)
11. Sin, B.K.: Infinite latent topic models for document analysis. J. KIISE **45**(7), 701–707 (2018)
12. Anto, S., Ibrahim, S.P.S., Devi, S.S.: Sentimental analysis and LSI similarity measure for efficient page ranking. Int. J. Eng. Adv. Technol. **8**(6s3), 1143–1146 (2019)

Correct Modeling of SH 50ETF Option Implied Volatility Based on Neural Network

Jiawei Gao[1(✉)] and Patimah Sprtuk[2]

[1] School of Economics, Shanghai University, Shanghai, China
18800203236@163.com
[2] University of the Philippines, Quezon City, Philippines

Abstract. We revealed some structural problems with the implied volatility surface (IVS) of Shanghai Stock Exchange (SSE) 50ETF options. First, there is a mismatch between call and put implied volatilities. Secondly, the SSE 50ETF lack short mechanism. Thirdly, the option data is sparse. With these problems, we are not able to apply some of the well-known models derived from the US market or other mature markets to the SSE 50ETF option directly. To solve this problem, we proposed several new modeling methods including a Neural Network approach, replacing the SSE 50ETF spot price with Shanghai 50 Index futures and interpolation methods. And after comparison, we find that, both interpolation methods and replacing ETF data with futures can improve the performance of traditional models and DNN methods performance the best in all scenarios.

Keywords: Implied volatility · Deep neural networks · Volatility surface

1 Introduction

The Black-Scholes formula gave a theoretical model for European option prices, enriching the options pricing. However, there is an inconsistency between the theory assumption on volatility and the real implied volatility in practice. For a given maturity, the implied volatility across different moneyness often exhibit a smile pattern. And there is also a certain pattern across different time to maturity when moneyness is fixed. While when pricing the option with Black-Scholes model, the implied volatility is critical. There are more and more researchers are engaging in the modeling of implied volatility.

Researcher have done a lot of research on the implied volatility modeling, there are three basic models, Sticky Strike Rule and Sticky Delta Rule proposed by Derman[1]. Stationary Square Root of Time Rule proposed by Daglish [2]. All of these models make a assumption that the volatility surface doesn't change over time, however, this is not true according to a lot of research, thus a lot of researcher engaging in extend these models to capture the dynamics of volatility surface models [3–7]. Gonçalves and Guidolin proposed two step methods which combines the sticky delta rule and VAR approach, through VAR they could predict the coefficients tomorrow to improve the forecast results [5].

I. Ahmad et al. (Eds.): STSIoT 2021, LNDECT 122, pp. 28–35, 2023.
https://doi.org/10.1007/978-981-19-3632-6_4

However, there is something we need to point out about the SSE 50ETF options. First, the number of options traded is lower compared to its counterparts, there are only short-dated options which expires on spot, next month and next two calendar quarter months. Second, you can't short the underlying of the options. So there exits lots of shorting opportunities in theory. Third, there is a mismatch between the implied volatility of call and put, the difference increases as the time to maturity increase. To solve the problem, we proposed two methods. First, we use a sampling method to solve the first problem, where we use a quadratic sample bar interpolation. Second, because ETF does not have short mechanisms, so using index future can better reflect the future price of ETFs.

In recent years, a lot of researchers have used the machine learning models to model the implied volatility [8–10]. In this paper, we also use DNN models to examine whether machine learning methods will improve the modeling and prediction accuracy on the SSE 50ETF option implied surfaces.

2 Data and Methods

2.1 Data

The data used in this study is from the JointQuant Dataset, which is a free financial dataset. The data consist of daily trading information of SSE50 option data and the underlying SSE50 ETF and SSE50 index future data from February 9, 2015 to August 1st, 2020. Same as a lot of researchers, we use a lot of filters, First, only options with less than 180 days to maturity and longer than 7 days to maturity are selected. Second, options which have 0 volumes on a certain day are eliminated. Third, options whose price contradicts the pricing interesting vales under 0 are eliminated. And we also delete the options which have dividends.

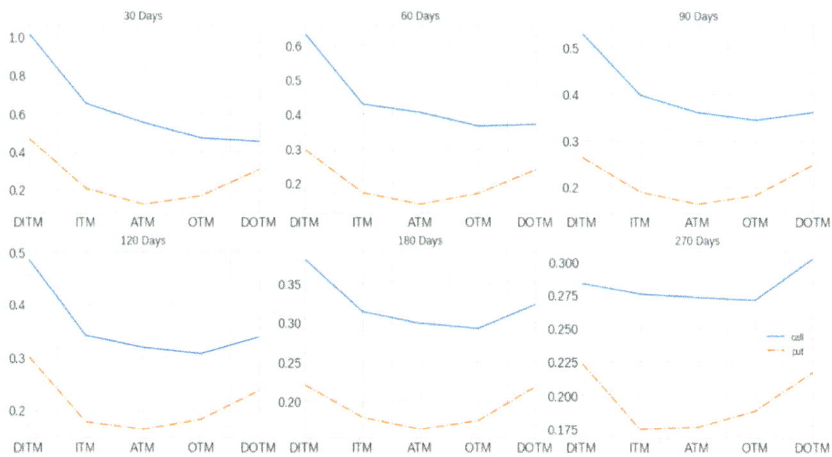

Fig. 1. Implied Volatility gap between SSE 50ETF Call and Put with ETF as underlying

A put contract is classified as deep in the money (DITM) if the moneyness[1] > 0.06; in the money(ITM) if $0.06 \geq m \geq 0.01$, at the money (ATM) if $0.01 \geq m \geq -0.01$, out of the money (OTM) if $-0.01 \geq m \geq -0.06$, and deep out of the money(DOTM) if $-0.06 \geq m$. Equivalent definitions apply to calls, with identical bounds but with m replaced with $-m$ in the inequalities. The classification based on time to expiration follows Bakshi et al. [11].

Table 1. Summary Statistics of SSE 50ETF Options

		Average Price		Average Implied Volatility		Counts	
Moneyness	τ	CALL	PUT	CALL	PUT	CALL	PUT
ATM	<30	0.1260	0.0262	0.5910	0.1109	864	696
	30–60	0.1441	0.0469	0.4248	0.1306	999	898
	60–90	0.1661	0.0676	0.3801	0.1476	418	395
	90–120	0.1748	0.0853	0.3286	0.1577	424	422
	120–180	0.1932	0.1020	0.3061	0.1582	717	708
	>180	0.2090	0.1297	0.2796	0.1725	817	815
DITM	<30	0.4369	0.4434	1.0550	0.5239	2125	234
	30–60	0.4223	0.3227	0.6622	0.2760	2174	283
	60–90	0.4647	0.2677	0.5628	0.1937	1707	256
	90–120	0.4706	0.4159	0.4968	0.3004	1564	684
	120–180	0.4494	0.3376	0.4013	0.2135	2468	1583
	>180	0.3767	0.3604	0.2978	0.2223	1368	1972
DOTM	<30	0.0153	0.0030	0.4639	0.3005	5583	2204
	30–60	0.0328	0.0082	0.3781	0.2320	5391	2235
	60–90	0.0472	0.0157	0.3725	0.2346	2877	1714
	90–120	0.0568	0.0277	0.3438	0.2359	2640	1582
	120–180	0.0848	0.0351	0.3282	0.2119	4155	2517
	>180	0.1172	0.0658	0.3048	0.2154	3067	1372
ITM	<30	0.1969	0.0905	0.6928	0.1818	1851	362
	30–60	0.2062	0.1060	0.4592	0.1576	2006	1006
	60–90	0.2311	0.1274	0.4196	0.1679	981	593
	90–120	0.2334	0.1410	0.3559	0.1705	962	718
	120–180	0.2522	0.1589	0.3233	0.1729	1672	1311

(continued)

[1] $moneyness = ln\left(\frac{K}{Se^{(r-q)(T-t)}}\right)$

Table 1. *(continued)*

		Average Price		Average Implied Volatility		Counts	
Moneyness	τ	CALL	PUT	CALL	PUT	CALL	PUT
	>180	0.2605	0.1776	0.2842	0.1705	1708	1839
OTM	<30	0.0720	0.0106	0.4959	0.1595	2124	1862
	30–60	0.0940	0.0271	0.3820	0.1641	2431	2006
	60–90	0.1206	0.0429	0.3615	0.1717	998	981
	90–120	0.1291	0.0601	0.3137	0.1782	996	963
	120–180	0.1511	0.0730	0.2990	0.1704	1670	1674
	>180	0.1686	0.1020	0.2760	0.1849	1976	1719

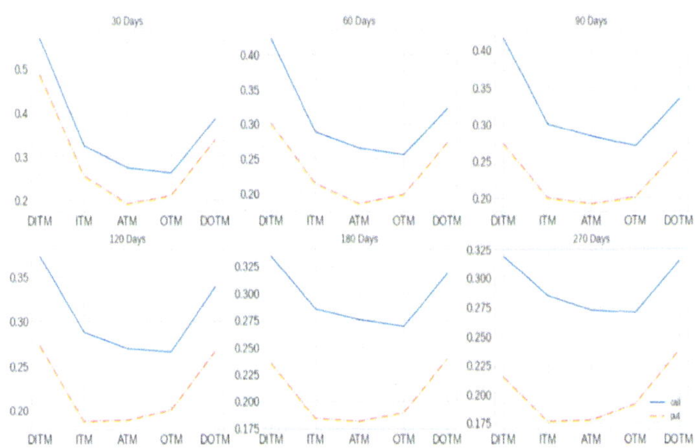

Fig. 2. Implied volatility gap between SSE 50ETF call and put with index future as underlying

From Table 1 we can see that, there is a gap between implied volatility of call and puts. We believe this is partly due the missing of short mechanism of SSE 50ETF options. The SSE 50ETF is not priced appropriately enough because of the lack of a shorting mechanism. And even though there are arbitrage opportunities in the market, we still can't gain from arbitrage because we can't replicate options.

From Fig. 1 and Fig. 2 we can see that, when we use futures price as underlying price instead of ETF close price, the difference between the implied volatility of call and put options becomes smaller. Also, the volatility smile pattern is more pronounced. From Fig. 1 and Fig. 2 we can see that, when we use futures price as underlying price instead of ETF close price, the difference between the implied volatility of call and put options becomes smaller. Also, the volatility smile pattern is more pronounced.

2.2 Methods

We use two methods to model the daily implied volatility surface. First, the traditional linear method are applied to the implied volatility surface each day, then we can estimate a set of coefficients every day.

$$ln\sigma_i = \beta_0 + \beta_1 M_i + \beta_2 M_i^2 + \beta_3 \tau_i + \beta_4 (M_i \times \tau_i) + \epsilon_i$$

where σ_i is the implied volatility and M_i is the moneyness, τ_i[2] is the time to maturity.

$$M_i = ln\left(\frac{K}{Se^{(r-q)\tau}}\right)$$

After having the coefficients, we can use VAR models to model the daily variation. Let $\beta = (\beta_1, \beta_2, \beta_3, \beta_4, \beta_5)$, we consider the following multivariate model for the vector of estimated coefficients

$$\widehat{\beta_t} : \widehat{\beta_t} = \mu + \sum_{j=1}^{p} \Phi_j \hat{\beta}_{t-j} + \mu_t.$$

Second, we can see that the smile of the volatility of SSE 50ETF options is not a perfect shape of parabola, thus we think that DNN networks can model and forecast the shape of implied surface better. We use a simple DNN network which has 5 input variables and 3 hidden layers, The structure of Neural Networks is as Fig. 3.

$$\sigma_i = \sigma\left(m_i, \tau_i, m_i^2, \tau_i^2, m_i \times \tau_i\right)$$

3 Results

The data was categories into two samples, data from February 9, 2015 to February 9, 2016 were used to train the model, data from February 10, 2016 to are used to measure the performance of the model. Every day, we estimate a model when new data comes and then we update our parameters. The following tables reports the in sample and out of sample forecasts of our model.

From Table 2 and Table 3, we can see that, after applying interpolation method, the linear-VAR methods do improve both in call and put scenarios, its MSE smaller

[2] $\tau_i = T - t$

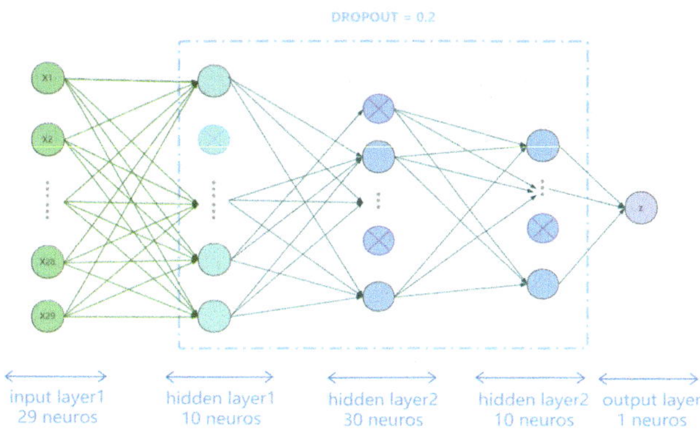

Fig. 3. DNN structure used to model the implied volatility surface

Table 2. Modeling and prediction results of call options

Panel A: Modeling and Prediction result of call options with ETF as underlying assets							
		Raw Data			Interpolation		
		DNN	LINEAR	VAR	DNN	LINEAR	VAR
Modelling	mean	0.0008	0.0251		0.0007	0.018	
	std	0.0019	0.0482		0.0017	0.0288	
	max	0.0742	0.4904		0.0409	0.445	
	proportion	0.0029	0.0901		0.0025	0.0644	
Prediction	mean	0.0143	0.0297	0.0344	0.0119	0.0231	0.0349
	std	0.0665	0.0526	0.2693	0.0504	0.0368	0.3151
	max	4.7613	1.5489	19.7958	5.3698	1.5489	31.7758
	proportion	0.0513	0.1065	0.1234	0.0426	0.0829	0.1252
Panel B: Modeling and Prediction result of Call options with Index future as underlying assets							
		Raw Data			Interpolation		
		DNN	LINEAR	VAR	DNN	LINEAR	VAR
Modelling	mean	0.001	0.0072		0.0006	0.0071	
	std	0.0017	0.011		0.0011	0.0113	
	max	0.0431	0.5077		0.0216	0.5077	
	proportion	0.0041	0.0284		0.0025	0.0281	
Prediction	mean	0.0174	0.0144	5.2609	0.0148	0.0144	0.7787
	std	0.1289	0.0228	503.1064	0.0827	0.0228	61.948
	max	6.1146	0.5732	59304.43	4.6334	0.5662	7735.929
	proportion	0.0689	0.0569	20.8074	0.0585	0.057	3.0798

Table 3. Modeling and Prediction results of Put options

Panel A: Modeling and Prediction result of Put options with ETF as underlying assets							
		Raw data			Interpolation		
		DNN	LINEAR	VAR	DNN	LINEAR	VAR
Modelling	mean	0.0006	0.0051		0.0009	0.0042	
	std	0.0021	0.011		0.0027	0.0056	
	Max	0.0778	0.3875		0.0652	0.2254	
	Proportion	0.0029	0.0261		0.0044	0.0217	
Prediction	mean	0.0111	0.0104	0.0344	0.0084	0.0084	0.0349
	Std	0.0834	0.0234	0.2693	0.0305	0.0108	0.3151
	Max	5.7488	1.3394	19.7958	3.3852	0.1925	31.7758
	Proportion	0.0568	0.0531	0.1762	0.0429	0.0429	0.1788
Panel B: Modeling and Prediction result of Put options with Index future as underlying assets							
		Raw data			Interpolation		
		DNN	LINEAR	VAR	DNN	LINEAR	VAR
Modelling	mean	0.0005	0.0047		0.0006	0.0047	
	Std	0.001	0.0108		0.001	0.011	
	max	0.0283	0.3874		0.0178	0.3874	
	proportion	0.0026	0.0231		0.0031	0.0232	
Prediction	mean	0.0104	0.0094	5.2609	0.0128	0.0094	0.7787
	std	0.0623	0.013	503.1064	0.118	0.0131	61.948
	max	7.7361	0.2339	64498.34	4.3329	0.1925	7735.929
	proportion	0.0514	0.0463	25.964	0.063	0.0465	3.8431

on interpolation data, but for DNN methods, the performance does not improve. The DNN methods performance better than other models both in interpolation and raw data scenario.

We do a rolling prediction to measure the predictive power of the models, we estimate a set of coefficients every day, and then apply these coefficients to next day. The same results hold as in sample modeling. Both interpolation and DNN methods can performance better than modeling directly with raw data and linear models. It is important to note that, the VAR methods performance was bad in both scenarios, and when we propose the extreme values, its performance was still not very good.

Then we estimate the effect of replacing SSE 50ETF with SSE 50 index futures, the results are reported in Panel B of Table 2 and Table 3. Because the mean of volatility is different. Thus, we compare the proportion[3] of different errors, we can see that, use

[3] $proportion = \frac{mse}{mean\ of\ implied\ volatility}$

futures to substitute SSE 50ETF as the underlying do improve the performance of the linear models both in in the sample modeling and out of the sample forecasting.

4 Conclusion

Most of the existing forecasts of SSE 50ETF options are based on mature foreign methods, which ignore the uniqueness of the Chinese market. However, due to its unique characteristics in the Chinese market, models applicable to other markets generally not applicable in China.

To solve this problem, this article adopts three methods, 1. Modeling call and put separately; 2. Sampling; 3. Using futures to replace ETF data. We prove that separate modeling of call and put can better solve the difference between call and put option surfaces, the use of interpolation can make the transition of the volatility surface smoother and avoid overfitting of the model and DNN methods performance better than any other methods.

References

1. Demeterfi, K., Derman, E., Kamal, M., Zou, J.: A guide to volatility and variance swaps. The Journal of Derivatives **6**(4), 9–32 (1999)
2. Daglish, T., Hull, J., Suo, W.: Volatility surfaces: Theory, rules of thumb, and empirical evidence. Quantitative Finance **7**(5), 507–524 (2007)
3. Bates, D.S.: The crash of '87: was it expected? the evidence from options markets. J. Financ. **46**(3), 1009–1044 (1991)
4. Bollen, N.P.B., Whaley, R.E.: Does net buying pressure affect the shape of implied volatility functions? J. Financ. **59**(2), 711–753 (2004)
5. Gonçalves, S., Guidolin, M.: Predictable dynamics in the S&P 500 index options implied volatility surface*. The Journal of Business **79**(3), 1591–1635 (2006)
6. Mixon, S.: Factors explaining movements in the implied volatility surface. J. Futur. Mark. **22**(10), 915–937 (2002)
7. Le, V., Zurbruegg, R.: Forecasting option smile dynamics. Int. Rev. Financ. Anal. **35**, 32–45 (2014)
8. Horvath, B., Muguruza, A., Tomas, M.: Deep learning volatility: a deep neural network perspective on pricing and calibration in (rough) volatility models. Quantitative Finance **21**(1), 11–27 (2021)
9. Liu, S., Oosterlee, C.W., Bohte, S.M.: Pricing options and computing implied volatilities using neural networks. Risks **7**(1), 16 (2019)
10. Zeng, Y., Klabjan, D.: Online adaptive machine learning based algorithm for implied volatility surface modeling. 16 (2019)
11. Bakshi, G., Cao, C., Chen, Z.: Empirical performance of alternative option pricing models. J. Financ. **52**(5), 2003–2049 (1997)

International Trade Strategy of SMEs Based on Blockchain Technology

Yixin Cai[1](✉) and Guido Maria Cortelazzo[2]

[1] School of Economics, Shanghai University, Shanghai, China
yxcai@shu.edu.cn
[2] University of Munich, München, Germany

Abstract. Blockchain has the characteristics of decentralization, openness and transparency, privacy protection, and traceability. It is suitable for international logistics supervision, processing trade supervision, and cross-border payment in the international trade of small and medium-sized enterprises. At present, the application of blockchain technology is in its infancy, facing problems such as lack of universal standards, security risks, and cross-jurisdictional jurisdictions. It is necessary to further promote the true implementation of blockchain technology in international trade through technological innovation, talent cultivation, and improvement of the standard system.

Keywords: Blockchain · International trade · Enterprise strategy · SMEs

1 Introduction

With the continuous development of economic globalization, multi-polarization and liberalization, international trade is playing an increasing significant role in regulating the total factor productivity, adjusting economic structure and improving the relationship between international supply and demand. However, issues such as cross-border payment, logistics traceability, information security, and product quality have become increasingly prominent in the current international trade model, which have greatly hindered the wide participation of enterprises from various countries, especially small and medium-sized enterprises in the global economy.

The blockchain is a new and widely concerned technology with broad application prospects, which may set off a new industrial technological revolution after internet technology. The essence of blockchain is a publicly distributed ledger, which has the characteristics of decentralization, openness and transparency, privacy protection, and traceability, and is suitable for solving problems faced in international trade [1]. This article takes small and medium-sized enterprises as main bodies and constructs an international trade strategy with blockchain technology as the core. While promoting small and medium-sized enterprises to participate in the global economy, it creates new models and new formats for the high-quality development of international trade.

I. Ahmad et al. (Eds.): STSIoT 2021, LNDECT 122, pp. 36–42, 2023.
https://doi.org/10.1007/978-981-19-3632-6_5

2 The Blockchain Technology Overview and Advantages

2.1 Overview

The blockchain was invented by Satoshi Nakamoto in 2008 to serve as the public transaction ledger of the cryptocurrency bitcoin [2]. It is managed by a peer-to-peer electronic cash system, so that payments could be directly transacted without going through a third-party financial institution. As the underlying support technology of Bitcoin data blocks, the essence of blockchain is an unalterable, decentralized distributed digital transaction record, also known as a transaction ledger, guaranteed by cryptography transmission and access security. Therefore, blockchains can be consistent storing data and are resistant to modification of their data once recorded [3]. Different from traditional database managed by a central entity, the blockchain relies on peer-to-peer network and determines that no party can completely control, achieved through encryption and a mathematical consensus protocol. A key feature of this protocol is that it allows collaboration between participants who do not have specific certification in each other, rather than relying on a trusted intermediary. In this way, fictions between participants can be reduced and a door to a higher level of transaction automation can be subsequently opened.

At present, the application modes of existing blockchain networks can be divided into public blockchains, private blockchains, hybrid blockchains and sidechains [4]. A public blockchain is a completely open blockchain system, such as Bitcoin and other cryptocurrencies. Any node can freely join and participate in the reading, writing, verification and consensus of the ledger data. A private blockchain is similar to a database, which is a kind of centralized and permissioned. This blockchain form is suitable for the internal data management and specific institutions auditing. A hybrid blockchain has a combination of centralized and decentralized features. By dividing the roles of participants, different nodes are given different permissions, which is suitable for organizations and alliances composed of multiple entities. A sidechain is a separate blockchain that is attached to its parent blockchain using a two-way peg. It is emerging mechanisms that allows participants from one blockchain to be securely used in a separate blockchain and then be moved back to the original blockchain if needed.

2.2 Advantages

The decentralization of blockchain is its most basic feature that distinguishes it from other technologies. There are no intermediaries such as exchanges and central data platforms in the blockchain. Which means, any participant can use the network to directly read, write and transact. Moreover, another feature of blockchain is transparency. In other word, the documents and transaction information stored in the blockchain can be shared among all relevant parties of the transaction. This greatly helps participants to avoid the frequent transmission of unconfirmed business documents and transaction information between different parties in the traditional international trade model. Therefore, the decentralization and transparency of the blockchain can flatten the international trade process, and achieve the goal of simplification of the trade process and raising processing efficiency.

The consensus mechanism is generally regarded as the key engine of blockchain [5]. Based on this mechanism, the execution of all business links of any international trade requires the consensus confirmation of the relevant transaction parties, which can effectively curb fraudulent activities such as forgery of official seals and signing of false contracts by illegal businessmen. At the same time, the blockchain also has anonymity, that is, only the addresses of the parties involved in trade activities are stored in the blockchain. These addresses are all composed of a string of hash values that cannot be decrypted. Without special authorization, the true identity of any participant cannot be identified, which can promote the neutrality of the transaction among all participants.

The data in blockchain is unalterable and traceable. The record information in the blockchain is stored in blocks and linked according to chronological order, a timestamp. Any modification of any information in any block will result in the change of all subsequent block data. In a blockchain with highly consistent data, the modification of any data can be easily monitored by comparison. At the same time, the data stored in the blockchain cannot be deleted. An effective data modification is only an additional operation after the original timestamp. In this way, any data can be retrieved in the blockchain based on the hash value of the data.

3 Application of Blockchain Technology in the International Trade Strategy of SMEs

3.1 International Logistics Supervision Based on Blockchain Technology

Logistics transportation is the backbone of modern international trade, including transportation, material procurement, warehousing, inventory control and distribution. A complete logistics transportation process is completed by customs, banks, and multiple participants from all over the world. Due to the problems of information asymmetry, poor standard compatibility, and poor data liquidity in trade, effective coordination of various participants often takes a large amount of operating costs of small and medium-sized enterprises.

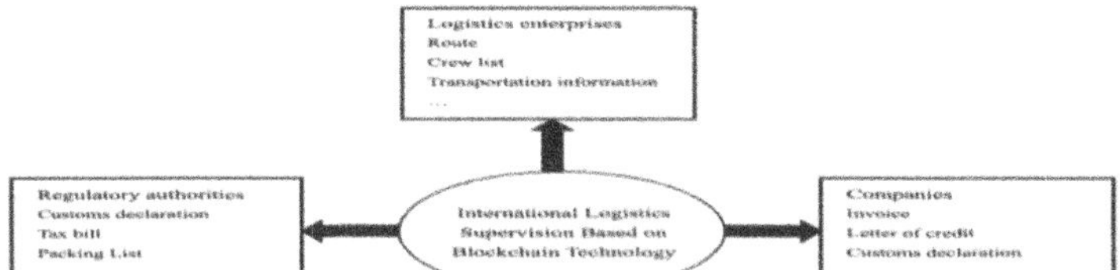

Fig. 1. International logistics supervision based on blockchain technology

As shown in Fig. 1, blockchain technology with its decentralized feature mainly improves the convenience of SMEs in international logistics from the following two aspects: 1. Real-time query of logistics process. Utilizing the characteristics of unalterable and traceable of the blockchain, each block is given specific access rights, and information can be effectively shared on the blockchain. This method can enable real-time tracking of cross-border product logistics status. For example, if you find that the distribution status does not meet the order requirements, you can directly consult with the international logistics company to accomplish the purpose of prior control. Combined with big data and artificial intelligence algorithms, logistics routes can be optimized and distribution efficiency can be raised. 2. Improved the efficiency of customs clearance. Using ledger information sharing on the blockchain, goods information can be submitted for review in the customs system in advance. The system can automatically analyze and calculate according to preset customs clearance standards, and quickly give the review results.

3.2 Processing Trade Supervision Based on Blockchain Technology

The application of blockchain technology can effectively supervise the processing trade chain. Its main application is to ensure the authenticity of traceability information, so that product errors can be detected and corrected in time [6]. Combining the Internet of Things and blockchain technology can obtain a series of validly identified product data, including the source of raw materials, production dates, sales dates, transportation methods, customs declaration time, customs clearance time, etc. The blockchain technology can trace the information of each node in series through the information block to form a complete data chain, which is greatly different from the currently manipulable tag system. The use of blockchain technology in international trade not only provides comprehensive information services, restricts counterfeiting in international trade, but also invisibly strengthens the control of product quality by small and medium-sized enterprises.

The proper use of blockchain technology can provide tremendous help to the innovation and development of small and medium-sized enterprises. Only by continuously using new technologies for transformation and upgrading, improving the technical content, increasing the added value of products, and enhancing business strength can small and medium-sized enterprises provide the basic force for the high-quality development of international trade. Through real-time collection and analysis of processing trade contracts, account books, materials and other data, small and medium-sized enterprises can conduct data analysis and forecasting of the subsequent processing trade, adjust production processes, and upgrade production equipment without delay. In addition, as an advanced technology, blockchain technology can accurately grasp the allocation of production factors among different companies in the same industry. Thus, the use of blockchain technology can promote the rational flow of element resources, raise the efficiency of resource allocation, and promote the transformation and upgrading of small and medium-sized enterprises (Fig. 2).

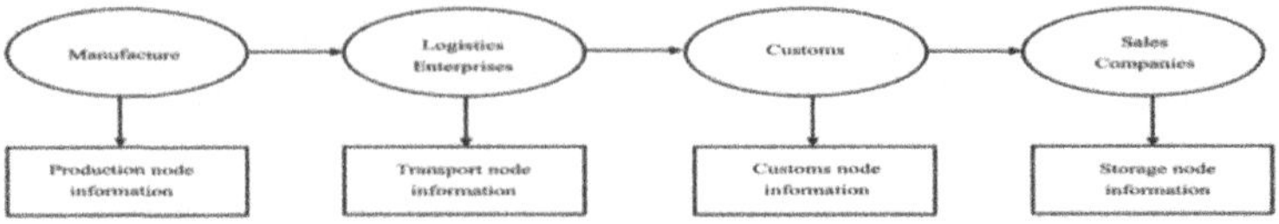

Fig. 2. Processing trade supervision based on blockchain technology

3.3 Cross-Border Payment Network Based on Blockchain Technology

The use of blockchain technology for cross-border payments guarantees real transactions to the greatest extent, reduces the cost of cross-border payments, and ensures that transactions are safe and controllable. Since using blockchain technology in cross-border payments does not require a third-party organization, the buyer and seller are directly peer-to-peer linked, which changes the payment process, solves the dilemma in inefficiency of traditional cross-border payments, and speeds up the settlement. As a result, using blockchain technology for cross-border payments provides a fast, low-cost and secure payment method, which can perfectly replace traditional bank payment methods [7].

Based on the decentralized nature of the blockchain and the consensus mechanism algorithm that buyers and sellers do not need to be authenticated, the use of blockchain payment can store each transaction in a secure distributed ledger in time. The cross-border payments by using digital currency can be more accurate, reliable and lower in cost than traditional payments. At the same time, there is no need for multiple parties to process transactions, which improves the efficiency of one-time exchange between all currencies [8]. In cross-border payments, the introduction of blockchain technology has changed the original credit model and will greatly reduce the cross-border transaction costs of SMEs.

4 Difficulties and Countermeasures of Blockchain Technology in the International Trade of SMEs

4.1 Realistic Difficulties

At present, the lack of universal process standards is the primary problem in the application of blockchain technology in international trade. A valid use of blockchain technology to exchange documents and data in international trade requires clear standards, templates and data formats [9]. If these standards are fragmentary, it is likely to delay the widespread application of blockchain technology in international trade. From the perspective of blockchain, it is necessary to deal with difficulties such as blockchain modules, blockchain application interfaces, utilized encryption algorithms, configuration languages, access process control, and standardization of key blockchain functions and data formats. These are essential for blockchain technology to play an important role in key areas of cross-border trade and supply chain management.

The application of blockchain technology may encounter various issues related to encryption and security mechanisms. Such mechanisms may be attacked or destroyed by hackers due to negligence in implementation or human error in operation [10]. Therefore, in order to apply blockchain-based solutions to international trade, a large amount of security analysis and security testing is required. Specifically, it refers to the security certification platform of blockchain nodes, the penetration testing of nodes and protocols, risk analysis and security certification of nodes, components, and the entire platform.

In addition, the nature of distributed ledger technology has global cross-jurisdictional deployment [11]. In order to maximize the role of blockchain technology in international trade, existing regulations and management processes need to be adjusted. This requires regulators and legislators to cooperate across national borders, coordinate laws and regulatory systems, and manage potential risks, including monopoly, market manipulation and other issues.

4.2 Potential Solutions

In terms of technology, it is necessary to encourage research and development to break through the technical threshold. Compared with mature Internet application, the application of blockchain technology mostly stays at the conceptual stage at present, and there are few products that can be directly put into use. To promote the effective implementation of blockchain technology in international trade, it is necessary to increase investment, continuously improve technology research and development capabilities, and create products that satisfy users.

In terms of talents, it is necessary to perfection the talent cultivation mechanism. For example, the country can encourage universities to set up blockchain-related majors and cultivate blockchain related talents professionally. At the same time, it is also meaningful to cultivate a large number of mixed talents that integrate algorithms, cryptography, and cross-chain protocols to meet the broad needs of blockchain applications in international trade. Small and medium-sized enterprises should carry out skill training for technical personnel to enhance their awareness of technological change. At the same time, technical personnel should also actively learn knowledge about blockchain and strive to build a trading platform based on blockchain technology.

The real landing of blockchain technology in international trade is also closely related to a well-regulated standard system. We must promote unified industry standards, integrate advantageous resources, and lead the healthy development of the industry. It is urgent for us to give full play to the leading role of the International Blockchain Alliance and gradually form a unified international industry standard to help the development and promotion of applications of blockchain technology in international trade.

5 Conclusion

In the future, there is no doubt that blockchain technology will have huge application scenarios and irreplaceable effects in international trade. The characteristics of blockchain, such as decentralization, unalterable, strong plasticity, open sharing, timestamp and smart contract, make blockchain technology always have a lot of application space in international trade.

Taking advantages of blockchain technology, SMEs can greatly improve their competitiveness in international trade. However, the current application of blockchain technology in the international trade of small and medium-sized enterprises is still in its infancy. The widespread application of blockchain technology is still facing issues like standardization, security, and legal regulations. Enterprises will also face new challenges and risks, which need to be further studied. How to maximize the advantages of blockchain technology in the international trade of SMEs and smoothly transform theoretical research into practical applications are the directions of our future efforts.

References

1. Zheng, Z., Xie, S., Dai, H.N., Chen, X., Wang, H.: Blockchain challenges and opportunities: a survey. Int. J. Web Grid Serv. **14**(4), 352–375 (2018)
2. Nakamoto, S.: Bitcoin: A peer-to-peer electronic cash system, Decentralized Business Review 21260 (2008)
3. Belotti, M., Božić, N., Pujolle, G., Secci, S.: A vademecum on blockchain technologies: When, which, and how. IEEE Comm. Surv. Tutori. **21**(4), 3796–3838 (2019)
4. Risius, M., Spohrer, K.: A blockchain research framework. Bus. Inf. Syst. Eng. **59**(6), 385–409 (2017)
5. Efanov, D., Roschin, P.: The all-pervasiveness of the blockchain technology. Procedia Comp. Sci. **123**, 116–121 (2018)
6. Namasudra, S., Deka, G.C., Johri, P., Hosseinpour, M., Gandomi, A.H.: The revolution of blockchain: state-of-the-art and research challenges. Archi. Computat. Methods Eng. **28**(3), 1497–1515 (2021)
7. Treleaven, P., Brown, R.G., Yang, D.: Blockchain technology in finance. Computer **50**(9), 14–17 (2017)
8. Morkunas, V.J., Paschen, J., Boon, E.: How blockchain technologies impact your business model. Bus. Horiz. **62**(3), 295–306 (2019)
9. Bodkhe, U., et al.: Blockchain for industry 4.0: A comprehensive review. IEEE Access **8**, 79764–79800 (2020)
10. Aste, T., Tasca, P., Di Matteo, T.: Blockchain technologies: the foreseeable impact on society and industry. Computer **50**(9), 18–28 (2017)
11. Janssen, M., Weerakkody, V., Ismagilova, E., Sivarajah, U., Irani, Z.: A framework for analysing blockchain technology adoption: integrating institutional, market and technical factors. Int. J. Inf. Manage. **50**, 302–309 (2020)

Information Collection, Analysis and Processing of Digital Substation Based on Artificial Intelligence

Wei Xiong[1(✉)], Kang Gong[1], Wei Shi[1], Yaming Wang[1], and Masashi Akao[2]

[1] State Grid Yichang Power Supply Company, Yichang 443000, Hubei, China
sxq69827288@163.com
[2] University of Leicester, Leicester, UK

Abstract. With the rapid development of information technology and smart grids, in the daily operation of the power system, there are hundreds of pieces of information describing the production, maintenance, and operation of the system every minute. How to proceed from these large-scale, high-dimensional massive data effective information collection provides analysis and processing for power system operators, which has great research value. The research content of this paper is to research the digital substation information collection and analysis processing based on artificial intelligence. This article first summarizes the development and research status of substations, and then outlines the electricity consumption information collection system on this basis. The electricity consumption information system includes system design requirements, business requirements, and functional requirements. Finally, the acquisition system is designed, and the function realization and testing are completed. This article uses field research methods, comparative analysis methods and other research methods to study the theme of this article. Experimental research shows that compared with the traditional information collection and analysis system, the performance of the artificial intelligence-based digital substation information collection and analysis system studied in this paper is more excellent, especially in terms of information collection exceeding 20%, which has higher efficiency and sufficient reflects the feasibility of this study.

Keywords: Artificial intelligence · Digital substation · Information collection · Analysis and processing

1 Introduction

In recent years, digitization, communication and database technologies have been widely used in the power industry. Dispatching automation systems, DMS, SCADA, WAMS, EMS, etc. Collect and record the operation of the power system and generate information in real time at all times [1, 2]. As a result, a large amount of operating data information will be generated, which puts forward new requirements for the power system data storage, especially the ability of information collection and analysis.

I. Ahmad et al. (Eds.): STSIoT 2021, LNDECT 122, pp. 43–50, 2023.
https://doi.org/10.1007/978-981-19-3632-6_6

At present, domestic substations mainly adopt integrated automation systems for substations such as Beijing Sifang Relay, Guodian Nanzi, NARI Relay, etc. These systems have problems such as huge amount of alarm data, unclassified information, and large differences in point tables between different systems [3, 4]. Moreover, the current collected substation operation information lacks the correlation between the alarm signals, which hinders the use of the above-mentioned real-time production operation to make intelligent judgments and assist decision-making. When the substation is under equipment maintenance or equipment failure occurs, a large amount of alarm information will be generated in the monitoring system of the dispatching end. The resulting data flood will cause the operation staff to find it difficult to find the key information that characterizes the failure in time. It is impossible to accurately infer the cause of the failure and delay the analysis and processing of the accident, which is easy to cause major losses [5, 6].

This article aims to improve the operation efficiency of substations, and aims to study the summer and analysis and processing of digital substation information based on artificial intelligence. By combining the traditional substation information acquisition system with the artificial intelligence-based digital substation information acquisition analysis A systematic comparative study is carried out to analyze the feasibility of the content of this article [7–10].

2 Application Research of Digital Substation Information Collection and Analysis Processing Based on Artificial Intelligence

2.1 Design of Intelligent Substation Information Collection and Analysis System

(1) System requirement analysis

As shown in Fig. 1, The intelligent digital substation information collection system includes terminal management, automatic meter reading, orderly power usage, and file management. Terminal management needs to complete the life process management of the terminal, including terminal installation, terminal maintenance, terminal replacement, terminal removal, and smart card replacement business processes; automatic meter reading needs to complete the life cycle of customer meter automatic meter reading business applications, including automatic meter reading putting into operation and cancellation of putting into operation; orderly electricity use needs to complete the life cycle of orderly electricity plan application, including the formulation, modification, execution, and cancellation of orderly electricity plan; file management needs to respond to marketing that can affect terminal collection in a timely manner file changes, including CTPT changes, meter changes, and account cancellation business processes [11, 12].

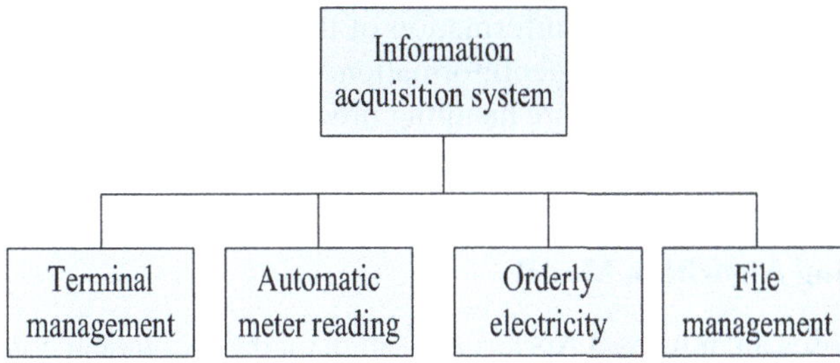

Fig. 1. Intelligent digital substation information acquisition system

(2) System architecture design

The depth of the system model design is clear to the system architecture design, data services, functional applications, implementation ideas, and interface associations with related systems, but does not involve specific implementation methods and data modeling and other design work. The data design requirements only clarify the required static data items (ie user-related archive data, etc.) and collected dynamic data items (ie real-time data collected through the terminal), and do not specifically involve database modeling and data table design.

The system model design receives the file information of the "SG186" marketing business application system, and transmits the meter reading information to the SG186" marketing business application system. The new master station system is a data collection to data from the perspective of a single business requirement for energy information collection. The applied complete system consists of three parts: the main station, channel, and terminal. The main station application is divided into link management, data acquisition (terminal operation control), basic application (load control, meter reading), and comprehensive application (power consumption). Management, anomaly analysis, statistics) and other parts, when built independently, these constitute a complete system.

(3) Functional module design

This time the system mainly includes three functional modules: event query and statistics module, expert analysis module and historical fault library module.

1) Time query and statistics module

The event query and statistics module is mainly used to implement information query and statistics. The dispatcher only needs to select the query range and query conditions to perform the information query function. The query is divided into the basic query and the advanced query of the application rules. The advanced query of the application rules is mainly based on the basic query to perform a secondary query on the information database application of the fixed rules. In addition, the system operator can also use it during use. The query results of interest are saved and used as a data basis for later review records or the establishment of expert rules.

2) Historical failure module

During the long-term use of the regional power grid, a large amount of equipment failure data will be accumulated. In order to effectively use these data, this article specially builds a historical fault library of equipment for a certain regional power grid, which is used to record the historical fault information of the equipment, mainly including the equipment's fault information. Ledger information, equipment failure start and end time, equipment failure description, failure handling process, historical alarm records related to failure, etc.

2.2 Data Mining Algorithm Model

This paper proposes an improved Apriori algorithm on the theoretical concept and algorithm of association rules. This improved algorithm can effectively solve the problem of large amount of data collected in electric power information.

(1) Basic concepts of association mining

Association rules are implicit expressions of the form (A ⇒ B), where A ⊂ I, B ⊂ I, and A ∩ B = Φ. Rule A ⇒ B appears in transaction set D, with support s, where s is the percentage of transactions in D that include A ∪ B (that i bh A and B). It is the probability P(A ∪ B). Rule A ⇒ B has confidence c in the transaction set. If the percentage of D that contains A transaction and also contains B is c, then it is the conditional probability P(B|A). That is, the degree of support is:

$$support(A \Rightarrow B) = P(A \cup B) \tag{1}$$

The confidence level is:

$$confidence(A \Rightarrow B) = P(B|A) = \frac{support\_count(A \cup B)}{support\_count(A)} \tag{2}$$

The conviction of a rule is as follows:

$$conu(X \Rightarrow Y) = \frac{1 - \sup p(Y)}{1 - conf\,(X \Rightarrow Y)} \tag{3}$$

(2) Apriori algorithm

Among all the Frequent Itemsets algorithms of Boolean association rules, the Apriori algorithm proposed by R. Agrawal et al. in 1993 is the most influential. According to the Apriori algorithm theory, the discovery of association rules is mainly divided into the following two steps:

1) According to the minimum support (minsupport) of the frequent item set set by the user, iteratively identify all frequency

   Frequent Itemsets (Frequent Itemsets);
2) Construct a strong association rule greater than the minimum confidence (minconfidence) set by the user in the project set. There are many types of association rule algorithms, but most of them are optimized and improved based on the Apriori algorithm.

(3) Improved Apriori algorithm

Step 1. Apply the DIC algorithm to the substation operation information database and divide the database into I sections according to the number of substation nodes I;

Step 2. According to the sampling algorithm, extract the sample specific subset S_I from each independent area I, and generate the item frequent set C_S. After verification, the individual area frequent set C_{SI} is obtained, and then the repeated algorithm is performed on each area to obtain the global item frequent Set C_I;

Step 3. Perform the second step of the Apriori algorithm, calculate the confidence level, and generate rules from frequent sets.

3 Experimental Research on Digital Substation Information Acquisition and Analysis Processing Based on Artificial Intelligence

3.1 Experimental Protocol

In order to make this experiment more scientific and effective, this experiment carried out research by going deep into a substation in a certain place and using a questionnaire survey. A total of 20 substation workers were surveyed this time. This experiment conducted an investigation on the current status of substations. After that, the traditional substation information collection and analysis system and the artificial intelligence-based digital substation information collection and analysis system studied in this article are run on the simulation software using the same data set, and the results obtained are analyzed and counted by the analytic hierarchy process.

3.2 Research Methods

(1) Questionnaire survey method

This experiment designed a questionnaire about information collection, analysis and processing of substation information, and distributed questionnaires to substation employees. The gender ratio of the employees in this survey was one to three to ensure the validity of the experimental data.

(2) Field research method

This experiment goes deep into a certain substation, and investigates and collects data on the status quo of its information collection and analysis and processing procedures. These data provide a reliable reference for the final research results of this article.

(3) Interview method

This research conducted face-to-face interviews with substation staff on information collection and analysis and processing, recorded the results of the interviews, and sorted out and analyzed them. These data provide data support for the topic selection of this article.

(4) Ahp

This article uses the analytic hierarchy process to analyze the final research results of this article to make the research results more scientific and effective.

4 Experimental Analysis of Digital Substation Information Collection and Analysis Processing Based on Artificial Intelligence

4.1 Analysis of Current Situation of Substation Information Collection and Analysis

In order to make this experiment more scientific and effective, this experiment conducted a survey on the status quo of substation information collection and analysis through a questionnaire survey. The data obtained are shown in Table 1.

Table 1. Analysis of current status of information collection and analysis in substations

	Information collection	Information analysis	Information processing	Others
1	5.63	6.21	5.71	4.95
2	5.21	5.47	5.32	4.85
3	5.12	5.84	5.66	5.12
4	4.69	5.39	5.17	5.33
5	5.30	5.28	5.20	6.07
6	5.41	5.77	5.33	5.20
7	4.48	6.10	5.78	4.98
8	4.96	6.21	6.04	4.67
9	4.72	5.71	6.50	5.36
...				
20	5.00	6.08	6.35	5.01

It can be seen from Fig. 2 that the evaluation of the traditional substation information collection and analysis system is maintained at a point above 5, which shows that the substation staff recognizes the current information collection and analysis, but there are still many problems that need to be solved urgently.

4.2 Comparative Analysis of Information Collection and Analysis Systems

In order to further research and analyze the artificial intelligence-based digital substation studied in this article, this experiment compares and analyzes the traditional substation information acquisition system with the system studied in this article. The data obtained is shown in Table 2.

It can be seen from Fig. 3 that compared to the traditional information collection and analysis system, the performance of the artificial intelligence-based digital substation information collection and analysis system studied in this article is more excellent, especially in terms of information collection exceeding 20%, which has higher efficiency, fully embodies the feasibility of this article.

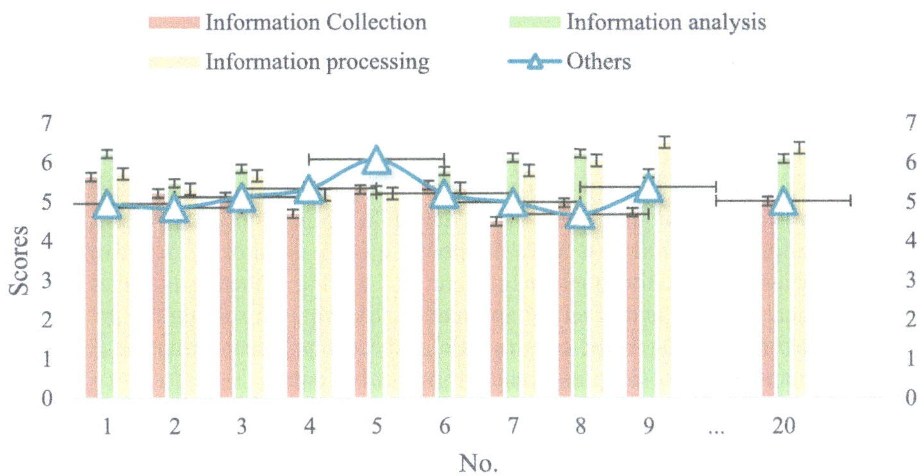

Fig. 2. Analysis of current status of information collection and analysis in substations

Table 2. Comparative analysis of information collection and analysis system

	High efficiency	Convenience	Scalability	Others
Digitizing	71.2%	68.3%	66.7%	62.1%
Traditional	50.8%	53.2%	55.1%	53.7%

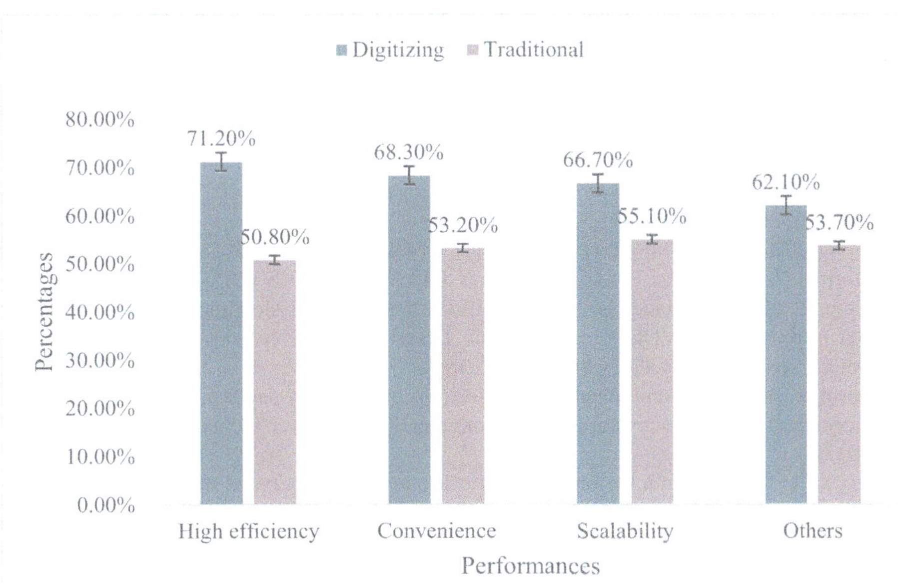

Fig. 3. Comparative analysis of information collection and analysis system

5 Conclusion

With the rapid development of power consumption information collection system in recent years, the technical key points that need to be paid attention to when constructing power consumption information collection system have been put forward. Researched and developed a platform for power consumption information collection system, which has functions such as remote meter reading, data collection, abnormal power consumption information alarm, line loss analysis, power quality monitoring, reactive voltage management and load monitoring management. The writing of the thesis strives to combine theoretical analysis and practical application, focusing on solving the problem of low functional positioning in the construction of electric power information collection system and incapability of adapting to new forms of big data.

References

1. Mcdonald, J.D.: Substation automation. IED integration and availability of information. IEEE Power and Energy Magazine **99**(2):22–31 (2016)
2. Lin, W.M., Tsay, M.T., Wu, S.W.: Application of geographic information system for substation and feeder planning. Int. J. Electr. Power Energy Syst. **18**(3), 175–183 (2016)
3. Paul, D., et al. Transmission system substation refurbishment. Transmission & Distribution World: The Information Leader Serving the Worldwide Power-Delivery Industry **68**(6), 50–52 and 54–56 (2016)
4. Lu, H., et al.: Regular expressions based information analytic method for substation centralized monitoring. Dianli Xitong Zidonghua/automation of Electric Power Systems **41**(5), 78–83 (2017)
5. Jing, L., Wang, L.: Research on data mining technology of intelligent substation information integration platform. IPPTA: Quarterly Journal of Indian Pulp and Paper Technical Association **30**(7), 466–475 (2018)
6. Zhai, H., et al.: Acquisition of state information of electrical equipment in substation based on RFID. Int. J. RF Technol. Res. Appl. **11**(1), 31–44 (2020)
7. Zhou, Y., Li, X., Shi, S.: Analysis of the influencing factors of technical transformation project cost of AC substation based on information entropy-VIKOR method. Electricity and Energy **040**(003), 288–292 and 350 (2019)
8. Weiming, T., et al.: Communication service strategy of smart substation based on message information label. Auto. Electric Power Sys. **042**(003), 124–129 (2018)
9. Wang, Z., et al.: A distance protection method for AC transmission lines based on substation area information in HVDC convertor station. Dianli Xitong Baohu yu Kongzhi/Power System Protection and Control **47**(1), 101–107 (2019)
10. Zhirui, K., et al.: Development and application of remote video monitoring-commanding information system in substation. Inner Mongolia Electric Power Technol. **035**(006), 100–103 (2017)
11. He, X., et al.: Research on online monitoring of service-oriented electricity information collection system. Auto. Technol. Appl. **037**(003), 134–138 (2018)
12. Changbao, X., et al.: Fuzzy comprehensive evaluation of intelligent substation relay protection system state based on information trend prediction and combination weighting. Electric Power Auto. Equip. **038**(001), 162–168 (2018)

Computer Network Monitoring and Analysis Method Based on Petri Net

Tongfei Shang(✉), Tianqi Wang, and Jing Liu

College of Information and Communication, National University of Defense Technology, Xi'an, China
340445698@qq.com

Abstract. Petri nets are an important tool for modeling and analysis of asynchronous concurrent systems. Graphical mathematical tools can be used to complete the description, verification, performance evaluation and testing of the system during the entire life of the information processing system. This article uses Petri nets to describe and analyze the communication protocol of the data acquisition system, describes the protocol by constructing a communication protocol model and guides the realization of the protocol by computer programming, and then describes the local area network client/server system, establishes a GSPN model and Conduct a more detailed analysis.

Keywords: Petri Net · Computer network · Monitoring analysis

1 Introduction

Petri net is one of the powerful tools for the description of information processing system. As a graphical tool, Petri nets can simulate the dynamics and activities of the system through the flow of tokens in addition to the visual description functions of similar flowcharts, block diagrams and net diagrams [1–3]. Therefore, Petri nets are dynamic graphics tools that can establish state equations, algebraic equations and other mathematical models to describe the behavior of the system [4, 5]. This paper analyzes and studies computer network monitoring through Petri nets, which has good application value.

2 Basic Concepts of Petri Nets

Petri nets are graphical and mathematical model tools that can be applied to many systems and fields. During the entire life of the information processing system, Petri nets use graphical mathematical tools to complete the description, verification, performance evaluation and testing of the system [6].

Definition 1. The triple N = (P, T; A) If the following conditions are met, then N is called a directed net, or net for short [6]:

I. Ahmad et al. (Eds.): STSIoT 2021, LNDECT 122, pp. 51–56, 2023.
https://doi.org/10.1007/978-981-19-3632-6_7

a). PTΦ and PT = Φ;
b). A PT TP, only from PT to TP, not from PP, TT;
c). dom(A) cod(A) = PT, where dom(A) = {xl y: (x, y)A}, cod(A) = {xl y: (y, x)A} They are the domain and value domain of A respectively.

Definition 2. Suppose x∈X is any element of the network N = (P, T; A), then' x = {y | (y, x) A} is called the previous set (or input set) of x, x' = {y | (x, y) A} is called the posterior set (or output set) of x.

Definition 3. Let N = (P, T; A) be a directed network:

a). If K is the mapping from P to Z + ω, K: PZ + ω, where Z + = {1, 2, 3,…}, it is said that K is a capacity function on N. K(p) = ω means that the capacity of p is infinite.
b). If K is a capacity function on N, the necessary and sufficient condition for M: PZ0 to be called an identification of N is: ∀p∈P, M(p)K(p), where Z0 = {0} Z +.
c). The mapping W: AZ + is called the weight function of N, and the value of W on the arc (x, y) is represented by W (x, y).

Definition 4. Let = (P, T; A, K, W, M0) be the network system, and M is an identifier on the base network:

a). For t∈T, 't∪t' is called the extension of t.
b). The condition that t has the right to occur under M is: for all p_P, at this time, it is also said that M authorizes t to occur, which is recorded as M [t >.
c). If t has the right to occur under M, then t can occur. The result of this is that M becomes the new identifier M' defined as follows: For all p_P,

$$M'(p) = \begin{cases} M(p) - W(p,t) & p \in' t - t' \\ M(p) + W(p,t) & p \in t' -' t \\ M(p) - W(p,t) + W(t,p) & p \in' t \cap t' \\ M(p) & p \in' t \cup t' \end{cases} \tag{1}$$

If the value of K in the six-tuples = (P, T; A, K, W, M_0) is always infinite and the value of W is both 1, a Petri net is obtained.

Definition 5. Petri nets are composed of four-tuples (P, T; A, M0), denoted as: PN= (P, T; A, M0). in:

$P = \{p_1, p_2, \ldots, p_n\}$ is the position set;
$T = \{t1, t2, \ldots, t_m\}$ is the transition set;
$A = (A_i \cup A_o)$, $A_i \subseteq$ (P´T), $A_o \subseteq$ (TP);
$M_0 = \{m_1, m_2, \ldots, m_n\}$ is the initial marking state.

3 SPN Model of Communication Protocol

In the data acquisition system communication protocol, the sending part can be represented by an action and a state. The action is to send a command frame, and the state is waiting to receive data (i.e. response); the receiving part can also be represented by an action and a state, and the action is to process the command frame. (Including sending response), the state is waiting for command frame; the communication line is the data transmission part, it has no action, under ideal conditions there are only two states, namely command frame on the line and data frame (including sending response) on the line. Using the transition in the Petri net to represent the action and the position to represent the state, the Petri net representation of the three components of the communication protocol in an ideal state can be obtained [7, 8].

The initial state of the SPN model is that there is a Token in each of P1 and P8. From this, the reachable state set is shown in Fig. 1, and the Markov state chain diagram is shown in Fig. 2.

	P_1	P_2	P_3	P_4	P_5	P_6	P_7	P_8
M_0	1	0	0	0	0	0	0	1
M_1	0	1	0	0	0	0	0	1
M_2	0	0	1	0	0	0	1	1
M_3	0	0	0	1	0	0	1	0
M_4	0	0	0	0	1	0	1	1
M_5	0	0	0	0	0	1	1	1

Fig. 1. The reachable state set of the SPN model

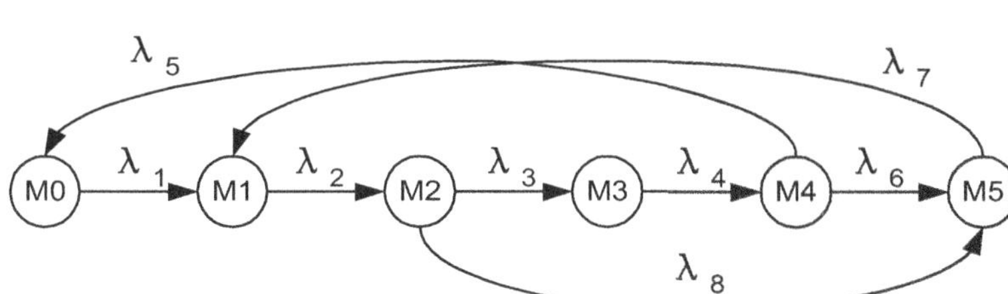

Fig. 2. State chain diagram of Markov

In Fig. 1, suppose that the data transmission rate is 9600bps, the length of the data frame is bytes, the rate of sending data is v times/sec, the rate of generating new frames is times/s, the channel error rate is e, and the timeout rate is t times/s.

$$Q = \begin{bmatrix} -\lambda_1 & \lambda_1 & 0 & 0 & 0 & 0 \\ 0 & -\lambda_2 & \lambda_2 & 0 & 0 & 0 \\ 0 & 0 & -\lambda_3 - \lambda_8 & \lambda_3 & 0 & \lambda_8 \\ 0 & 0 & 0 & -\lambda_4 & \lambda_4 & 0 \\ \lambda_5 & 0 & 0 & 0 & -\lambda_5 - \lambda_6 & \lambda_6 \\ 0 & \lambda_7 & 0 & 0 & 0 & -\lambda_7 \end{bmatrix} \quad (2)$$

$$\Pi = [\pi_0, \pi_1, \cdots, \pi_5]^T \quad (3)$$

Therefore, according to the equilibrium equation of the stochastic Petri net, the steady-state probability of each state and related performance indicators can be calculated. Among them, the state M_0 represents the probability of successfully receiving the data frame, and its steady-state probability is

$$\pi_0 = \frac{1}{1 + \frac{(11-e)\lambda}{8(1-e)^2 v} + \frac{e(2-e)\lambda}{t(1-e)^2}} \quad (4)$$

As shown in Fig. 3, the data transfer between the client and the server relies on the communication between processes, and the server manufacturer provides a high-level language interface for writing client programs [9, 10]. Through these interfaces, the client program can communicate with the corresponding server.

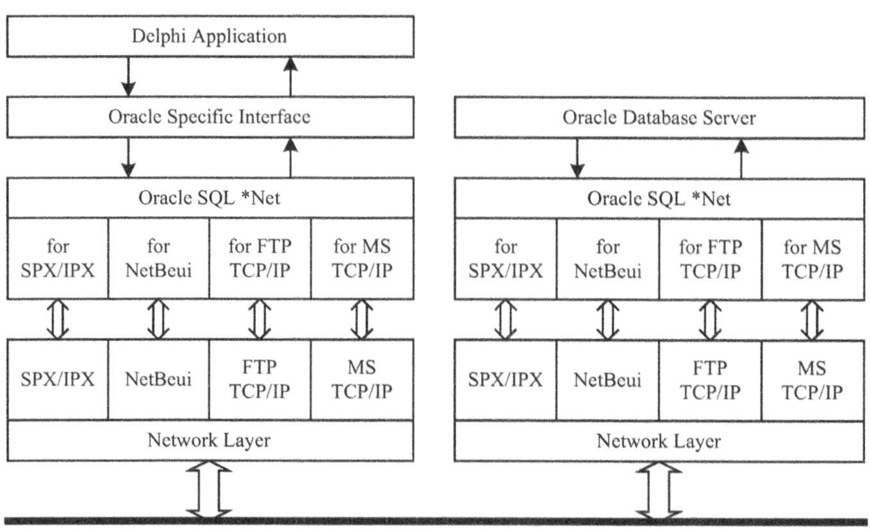

Fig. 3. Client/server communication mechanism

But only through the network layer to connect with the database, and the network has different protocols and manufacturers. In order to ensure that customer applications can run normally in various network environments, Oracle's SQL *Net provides a public and transparent interface for Oracle applications. At the same time, Delphi also needs to provide a dedicated interface to the Oracle database.

4 Simulation Analysis

After completing the GSPN modeling of the client/server system, according to the solution method of the generalized stochastic Petri net, the steady state mark number and the maximum mark number of each position can be easily obtained. Figure 4 shows the relationship curve between the system information throughput rate and the number of buffers. It can be seen from the figure that the number of information buffers N and the information arrival rate have a significant impact on the steady-state throughput S of the system.

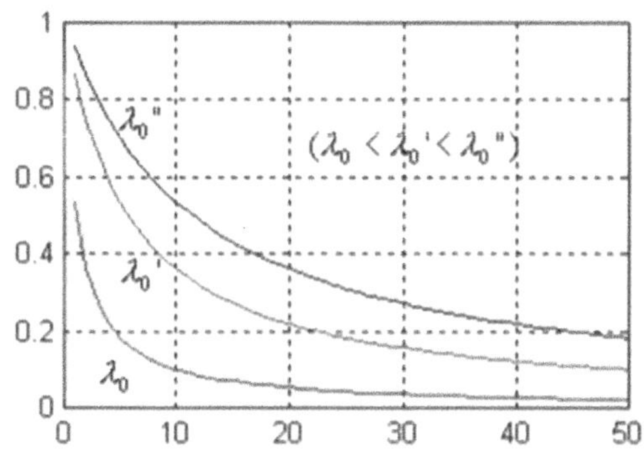

Fig. 4. The relationship curve between average buffer value and throughput rate

Figure 5 shows the relationship curve between the system's rejection rate of the envelope and the number of buffers. An increase in the rejection rate indicates an increase in the risk of normal system operation, a decrease in system overhead, and an increase in the utilization of the buffer. It can be seen that improving the utilization of the buffer zone and reducing the risk of the system are contradictory.

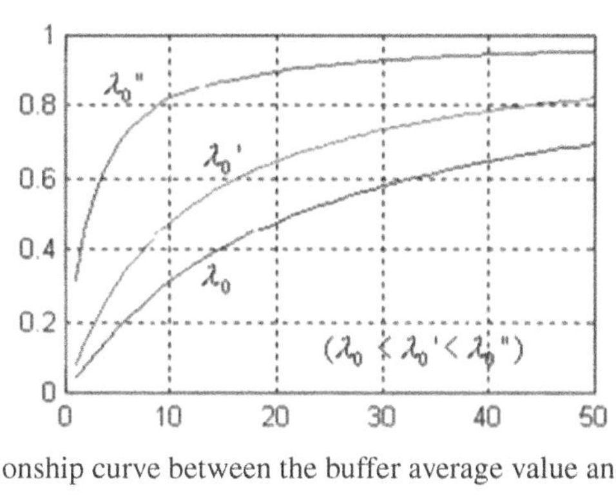

Fig. 5. The relationship curve between the buffer average value and the rejection rate

5 Conclusion

Since the research goal of Petri nets is to model the organizational structure and dynamic behavior of the system, as well as the various state changes that may occur in the system and the relationship between the changes, it is not easy to express the data values or attributes in the system with Petri nets. The specific changes or calculations can easily indicate the conditions under which the changes occur and the state of the system after the changes occur. This paper uses Petri nets to describe and analyze the communication protocol of the data acquisition system, describes the protocol by constructing the SPN model of the communication protocol, and guides the realization of computer programming of the protocol. The simulation proves that the method has certain application value.

References

1. Murata, T.: Petri *Nets*: Properties, Analysis and Application. Proceedings of the IEEE **77**(4), 541–580 (2016)
2. Marson, M.A.: An Introduction to generalized stochastic petri nets. Microelectron. Reliab. **31**(4), 699–725 (2017)
3. Voss, K.: A net modal of a local area network protocol. LNCS, Advanced in Petri Nets **188**(2019), 413–437 (2009)
4. Jefferson, D.R.: Virtual Time. ACM Trans. Program. Lang. Syst. **7**(3), 404–425 (2014)
5. Perumalla, K.S.: μsik — A Micro-Kernel for Parallel/Distributed Simulation Systems. In: Workshop on Principles of Advanced & Distributed Simulation. IEEE (2015)
6. Jha, V., Bagrodia, R.L.: A unified framework for conservative and optimistic distributed simulation. Acm Sigsim Simulation Digest **24**(1), 12–19 (2014)
7. Collier, N., North, M.: Parallel agent-based simulation with Repast for High *Performance* Computing. Simulation **89**(10), 1215–1235 (2013)
8. Mason, S.J., et al.: Quantitative assessment of an agent-based simulation on a time warp executive. In: Winter Simulation Conference. DBLP (2018)
9. Presley, M.T., Reiher, P.L. Bellenot, S.F.: A timewarp implementation of sharks world. In: Proc. of the 22nd Conference on Winter Simulation (WSC' 90), 199–203. IEEE Press (2015)
10. Aaby, B.G., Perumalla, K.S., Seal, S.K.: Efficient simulation of agent-based models *on* multi-GPU and multi-core clusters (2013)

Fuzzy Control Method Based on Dynamic Self-optimization

Yong Li[1], Tianqi Wang[2], and Tongfei Shang[2](✉)

[1] Tenth Telecommunications Institute of Science and Technology, Xi'an, China
[2] College of Information and Communication, National University of Defense Technology, Xi'an, China
340445698@qq.com

Abstract. The most commonly used in industrial control is the digital PID control method. For most control objects, the use of digital PID control can achieve satisfactory control results. But in the ball mill controller, the PID control will produce time lag effect, resulting in decision-making mistakes. This paper designs a fuzzy control method based on dynamic self-optimization, which can use the sensitive output of the controlled process as the input of the fuzzy controller in the fuzzy control loop, and the output of the fuzzy controller is the adjustment value of the controlled system. Input. Another input of the fuzzy controller is the set value input. The simulation results show the effectiveness of this method.

Keywords: Dynamic self-optimization · PID · Fuzzy control

1 Introduction

The fuzzy controller of the ball mill is a controller with a single input-single output structure. The input is the vibration signal on the rear shaft of the ball mill detected by the vibration transmitter, and the output control value is converted into a 0–10 mA direct current after D/A conversion to act on the vibrating coal feeder. The commonly used PID method has poor control performance and poor effect[1–3]. The dynamic self-optimization proposed in this paper can start self-optimization after the system enters the steady state (entering the 5% error band), increase the given R, and wait for the system to approach again after the steady state, check the value of the vibration signal change and repeat the above steps. After several adjustments, the system can run stably near the maximum output point, ensuring that the ball mill always runs near the maximum output point, which really has the effect of energy saving[4, 5].

2 Control Strategy

For objects with pure hysteresis characteristics, there are Smith predictive control algorithm and Darling algorithm to choose from. But Smith predictive control needs to know the mathematical model of the controlled object to construct the compensation function.

I. Ahmad et al. (Eds.): STSIoT 2021, LNDECT 122, pp. 57–61, 2023.
https://doi.org/10.1007/978-981-19-3632-6_8

If the mathematical model of the controlled object is not accurately identified, it will have a great impact on the control effect.

The basic block diagram of Smith predictive control algorithm can be shown in Fig. 1. In the figure, it is the transfer function of the controller, the transfer function of the controlled object, the Smith compensation function, and

$$G_L(s) = G_0(s) \cdot \left(1 - e^{-\tau s}\right) \tag{1}$$

If the transfer function of the controlled object is known, the compensation function can be constructed according to the above formula. The transfer function is the simulation result obtained by applying the Smith predictive control shown in Fig. 1 to the object shown in formula (1).

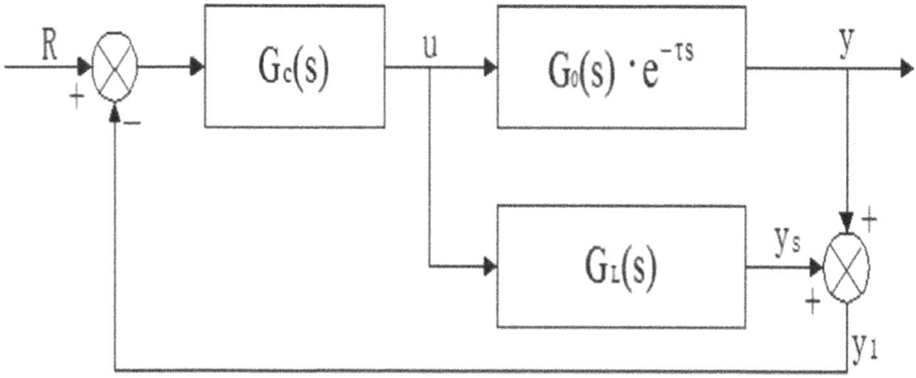

Fig. 1. Smith predictive control block diagram

3 Self-optimization-Fuzzy Cntroller

As the number of inputs of the fuzzy controller increases, the control rules that need to be established will increase exponentially. For example, for a two-dimensional fuzzy controller, assuming that each input variable takes 7 linguistic values, there are only 7 × 7 = 49 control rules, but for a three-dimensional fuzzy controller, the control rules will reach 7 × 7 × 7 = 353. For complex systems, the establishment of control laws (knowledge accumulation) is very difficult. The exponential growth of the control law greatly increases the workload of controller design [6, 7].

$$\mu_{ZR}(x) = \begin{cases} 0 & -80 \le x < -10 \\ \frac{x+10}{10} & -10 \le x < 0 \\ \frac{10-x}{10} & 0 \le x < 10 \\ 0 & 10 \le x < 80 \end{cases} \tag{2}$$

Fuzzy language value is actually a fuzzy subset, which is finally described by a membership function defined in a certain universe. Therefore, it is necessary to determine the scope and membership function of the universe of discourse. According to the graph of each membership function, its mathematical expression can be written. For example, for the deviation, the membership function of the fuzzy subset ZR is[8–10]:

$$\mu_{NB}(x) = \begin{cases} 1.0 & -100 \leq x < -75 \\ -\frac{x+50}{25} & -75 \leq x < -50 \\ 0.0 & -50 \leq x < 100 \end{cases} \tag{3}$$

The continuous realization of fuzzy control means that the input of the controller does not need to be discrete first, but is directly sent to the fuzzy controller realized by the software, and the fuzzy theory is carried out according to the fuzzy theory, fuzzy inference and defuzzification, and the control is calculated online by the program The system has high precision and versatility, as shown in Fig. 2. To illustrate the implementation steps of fuzzy inference in continuous implementation, take two control rules in the fuzzy controller of a ball mill as an example:

if e = PS and e = ZR then u = NS
or if e = PM and e = PS then u = NM

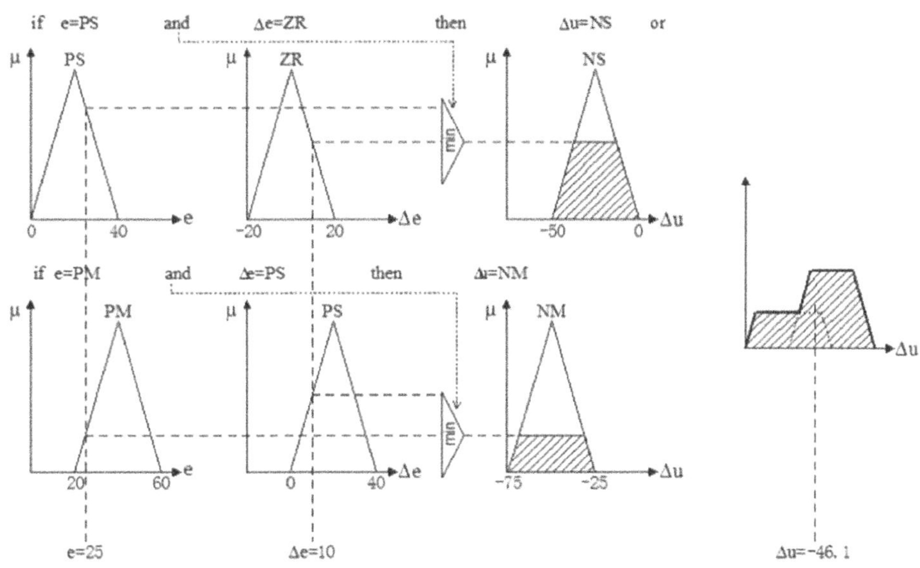

Fig. 2. Continuous realization of fuzzy controller

Finding the best working point of the ball mill depends on the in-depth analysis of the working characteristics of the ball mill. The vibration signal decreases as the amount of coal in the ball mill increases. For convenience, the vibration signal is simply processed in the vibration transmitter, that is, the actual measured vibration signal value is subtracted from a fixed value, so that the decreasing vibration characteristic curve becomes an increasing curve. At this time, A large vibration signal indicates a large amount of coal stored in the ball mill.

4 Simulation Analysis

The ball mill self-optimizing-fuzzy controller designed in this paper has been installed and debugged on site, and has been put into actual production and operation in the power plant. In the measurement of the amount of coal in the ball mill, a large amount of data was measured in the power plant, and it was found that when the acceleration sensor was installed on the front shaft of the ball mill, there were many spikes in the signal from the transmitter. A typical curve is shown in Fig. 3. The abscissa in the figure is the running time, and the ordinate is the output m_A value of the vibration transmitter. It can be seen from Fig. 3 that before t1 and after t4, the output value of the transmitter is relatively small, indicating that the amount of coal stored in the ball mill is small; during this period of time between t2 and t3, the transmitter The output value of the device is relatively large, indicating that the amount of coal stored in the ball mill during this period is relatively large.

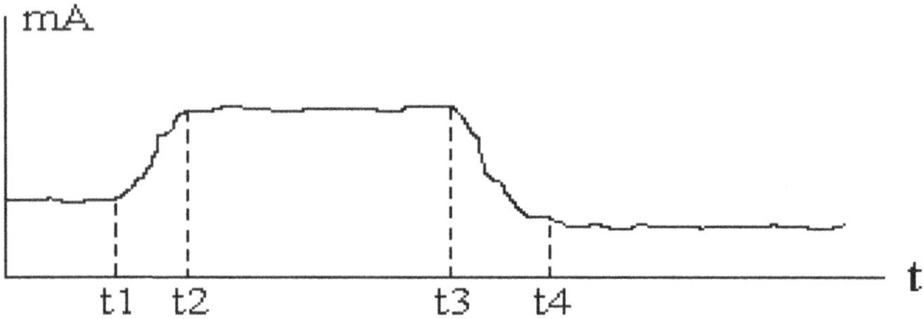

Fig. 3. The output waveform of the vibration transmitter when the acceleration sensor is installed on the front axle

The result has an impact. However, the ball mill in Yan'an Power Plant is relatively small, so this interference signal is clearly reflected in the measurement results. If the acceleration sensor is moved to the rear axle of the ball mill, this phenomenon can be avoided. The measured curve is shown in Fig. 4.

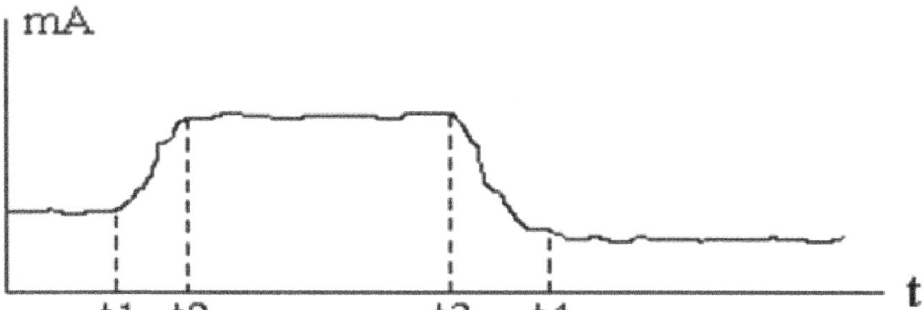

Fig. 4. The output waveform of the vibration transmitter when the acceleration sensor is installed on the rear axle

5 Conclusion

The intelligent optimization and energy-saving controller of the ball mill finally adopts self-optimization-fuzzy control. Because its energy-saving effect takes a long period of operation to be reflected, it is not appropriate to directly analyze the operating curve, and it is better to use statistical data to illustrate the problem. Through practical tests, the controller has obvious energy-saving effects and can produce good economic benefits.

References

1. Kim, J., Zeigler, B.P.: Designing fuzzy logic controllers using a multiresolu-tional search paradigm. IEEE Trans. Fuzzy Sys. **4**(3) (Aug 2016)
2. Constantin Virgil Negoita: Expert Systems and Fuzzy Systems. Menlo Park, California: The Benjamin/Cummings Publishing Company Inc. (2015)
3. Jin, Y.C., et al.: Implementing self-organizing fuzzy controller with hybrid pi-sigma neural networks and its application. In: 3rd Inter. Workshop on Advanced Motion Control. Unive. of California at Berkeley, pp. 893–902 (1994)
4. Sugeno, M. Kang, G.T.: Structure indentification of fuzzy model. Fuzzy Sets and Systems **28**, 15–33 (2016)
5. Takagi, T., Hayashi, I.: NN—driven fuzzy reasoning. Int. J. Approximate Reasoning **5**(3), 191–212 (2017)
6. Takagi, T., Sugeno, M.: Derivation of fuzzy control rules from human operator's control actions. In: Proceedings of the IFAC symposium on fuzzy information, Knowledge representation and decision analysis, pp. 55–60 (July 2016)
7. Liu, C., Kong, L., Zhong, W.: Multi-information fusion based tumor cell of bone marrow involvement, Medical Imaging and Augmented Reality, 2013. In: Proceedings. International Workshop on 10–12, pp. 211—215 (June 2001)
8. Advantech Co. Ltd.: PCM-3718H/3718HGPC/104 12-bit DAS module with programmable gain. User's manual, pp. 2–5 (2014)
9. Javadpour, R., Knappa, G.M.: Fuzzy neural network approach to machine condition monitoring. Comput. Ind. Eng. **45**, 323–330 (2013)
10. Yeo, S.M., et al.: A novel algorithm for fault classification in transmission lines using a combined adaptive network and fuzzy inference system. Electr. Power Energy Syst. **25**, 747–758 (2015)

Application of Data Encryption Technology in Computer Software Testing

Yizhi Wu(✉)

Guangzhou College of Technology and Business, Guangzhou, Guangdong, China
w19124303278@163.com

Abstract. As the society's demand for data sharing has gradually increased, computer software and computer technology have received extensive attention from all walks of life, and data security issues have also attracted the attention of people from all walks of life. Shared data in the use of computers, viruses and hackers pose a threat to the data security in the computer, and there are risks of user information leakage and data loss. Starting from the concept and main characteristics of data encryption technology and computer software testing, this paper studies the impact of using data encryption technology on computer software testing, and uses a questionnaire survey method to study the importance of data security by Internet users. It can be seen from the survey results that there are more college students who use computer software and use it for a long time. There are 142 people who use computer software for learning, which is more than the number of people who use computer for entertainment; the third year students use computers to access the Internet for 5–6 h a day. They face the pressure of postgraduate entrance examinations and public examinations, and use computers to quickly obtain learning data for postgraduate entrance examinations; the proportion of people who approve of data security in computer software is 89%, indicating that users are highly satisfied with network security.

Keywords: Data encryption · Data encryption technology · Computer software · Computer software testing

1 Introduction

With the increasing progress of science and technology, the development of computer software and data sharing have received extensive attention from various industries [1]. Data sharing lays the foundation for the openness of network data information. It can reduce the time cost for people to search for data and information, and improve the efficiency of people's search for data and information [2]. Although the use of computer software is very convenient for people's daily work and study, there are risks of data loss and leakage during use. Data security vulnerabilities not only affect the stable operation of computer software and systems, but also affect the user's personal information and property safety [3]. Computer software is regarded as a bridge connecting computers and users. The purpose of developing computer software is to meet people's production

I. Ahmad et al. (Eds.): STSIoT 2021, LNDECT 122, pp. 62–70, 2023.
https://doi.org/10.1007/978-981-19-3632-6_9

needs. Before computer software is put into use, it is necessary to test its safety and applicability to ensure the data security of computer software [4, 5]. In order to ensure the security of the computer software use process, data encryption technology can be used in computer software testing, targeted analysis of possible data security issues in various aspects, and corresponding data security precautions [6].

The network is a main method of data information transmission. The greater the amount of information it transmits, the greater the hidden data security risk in the transmission process [7]. In order to improve data security, it is necessary to perform data encryption technology processing on the text information, picture information, video information and sound information of the transmitted data during computer software testing [8]. Studies have shown that in order to effectively reduce the processing time of data information and increase security, data encryption and data compression can be organically combined. Data encryption technology is used in Huffman encoding and LZW encoding, and the encoding process can be regarded as an encryption process [9]. When testing computer software, it is necessary to test the structure of the software, the safety of the environment in which it is used, and its performance in order to ensure the safety and practicability of the computer software [10]. The increase in the number of people using computer software technology increases the probability of viruses invading computer software programs. Using data encryption technology to encrypt data information in computer software can effectively prevent hackers from invading computer programs [11]. Common algorithms for data encryption of computer software data information are: DES algorithm, MD5 algorithm, and RSA algorithm developed by IBM [12].

We mainly study the impact of the use of data encryption technology on the security of data and information in the process of computer software testing, and understand people's views on the security of computer software programs. Software designers modify and improve the security of computer software by collecting users' opinions. This article has conducted research from the following points: First, the concepts and characteristics of data encryption technology and computer software testing are described; second, analyze the impact of data encryption technology on computer software testing, and the application of data encryption technology in computer software testing; third, analyze the importance of data security by investigating the importance of data security when using computer software by Internet users.

2 Introduction to Data Encryption Technology and Computer Software Testing

2.1 The Concept of Data Encryption Technology and Computer Software Testing

Data encryption technology refers to a technology that uses a computer system as a carrier to provide security protection for data and information. First, we need to transform the data information we want to encrypt into meaningless ciphertext through encryption technology for transmission, and then transform it into original data information through decryption technology when receiving data information. In the specified data encryption and data decryption process, a key is needed to allow the data receiver to obtain accurate data information, effectively ensuring the security of data transmission and the efficiency of software operation. There are three types of data encryption technology: node technology, link technology, and end-to-end technology. Computer software testing refers to ensuring that the tested software meets the designer's expected requirements for the software through software testing, and that the software can meet the needs of the public when it is put on the market. Computer software testing is mainly to compare the functional differences before and after software design, using data encryption technology to verify the software system reliability of computer software, software environment security testing, and software practicability testing. Testing can be conducted from two perspectives: (1) The software user's perspective, (2) the software designer's perspective.

2.2 Security Vulnerabilities and Preventive Measures of Computer Software Testing

There may be many security vulnerabilities in the process of computer software testing, mainly the following three points: (1) In the process of computer user data sharing, the computer needs to meet the needs of users, and designers will develop new functions and new applications according to user needs. These new functions and applications will have certain risks. Hackers are very likely Invade this new feature, thereby stealing the user's personal information. (2) When using the database on the computer, due to the legality of the set parameters, the user cannot be reasonably verified by the computer when inputting incorrect information, which may easily lead to virus intrusion into the program. (3) When the computer uses the network to complete the communication with the TCP protocol, the virus is likely to use the loopholes in the TCP protocol to attack the network joined by the computer software. When discovering security vulnerabilities in the process of computer software testing, data encryption technology can be used to protect the computer software as follows: First, we need to strengthen virus intrusion detection in the operation of computer software; second, we need to regularly back up important data information to avoid the loss and damage of data information; finally, in computer software, attach importance to the role of data encryption technology, use network firewall technology to isolate viruses, eliminate illegal data intrusion into the computer in time, and improve the security of computer software.

2.3 Application of Data Encryption Technology in Computer Software Testing

In our work and study life, the process of using computers mainly relies on the operation of computer software; so many hackers attempt to attack computer programs to steal user information. Computer software is particularly vulnerable to hacker attacks and implanted viruses, and data encryption technology needs to be used to improve the security of computer software. The application of data encryption technology in computer software testing has the following points: (1) The use of data encryption technology can monitor the encrypted data information. If a virus intrusion is found when testing the data in the computer software, the designer can be reminded to kill the program in time to prevent the virus from invading the computer software and ensure that the computer software is normal run. (2) The use of data encryption technology can make the computer software testing process stable, and can also prevent hackers from inserting computer software programs into computer software programs, and reduce the probability of the system being invaded by viruses. Data encryption technology is considered to be a key step in computer software testing, which greatly guarantees the safety and reliability of computer software operation. The data encryption technology we are familiar with is the anti-virus software and computer housekeeper on the computer. They can comprehensively encrypt the various software on the computer. Starting from the source of the data information transmission, the Huffman coding technology is used to protect the data during the transmission process, reduce the possibility of data information leakage. When using Huffman encoding to encrypt data information, the calculation formulas for the amount of information h and the information entropy H are shown in formulas (1) and (2), In order to calculate the gain value formula (4) of information, the conditional entropy is subtracted from the information entropy. The conditional entropy formula is shown in Formula (3):

$$h(a_i) = \log_2 \frac{i}{w_i} \tag{1}$$

$$H(A) = \sum_{w_i \succ 0} w_i \log_2 \frac{1}{w_i} \tag{2}$$

$$H(Y|X) = \sum_{x} p(x)H(Y|X = x) \tag{3}$$

$$Gain(X) = H(A) - H(Y|X) \tag{4}$$

Among: a_i is a character whose probability is not 0, w_i is the weight value, and A is the input condition, Y, X, and x represent variable values.

3 Questionnaire Survey on the Degree to Which Users Attach Importance to Network Security During the Operation of Computer Software

3.1 Purpose and Target of Investigation

The purpose of using data encryption technology in the computer software testing process is as follows: 1. To improve the reliability of computer software; 2. To enhance the protection of the operation of computer software; 3. To protect the security of user data and information. The main purpose of this survey is to study how much users attach importance to data security when using computer software, and the network security issues they care about. The questionnaire survey selected 400 non-computer majors from a university in Beijing, including 100 freshmen, sophomores, juniors, and seniors. On the premise of ensuring that the number of men and women is balanced, 100 non-computer majors of each grade are randomly selected.

3.2 Investigation Method

In this paper, in the questionnaire survey on the importance of network security by users during the operation of computer software, the survey content is formulated according to the purpose and object of the survey, and 400 copies of the survey are printed out after completion, the questionnaire about network security was distributed to students of different grades who were not majoring in computer science at a certain university in Beijing, and the questionnaire was returned after they had completed it.

3.3 Survey Results

This article is a questionnaire survey on the importance of users to network security during the operation of computer software. After the questionnaire on network security is recovered, the person in charge will sort out the data. Data collation results are analyzed in the fourth section of this article, and this section only briefly elaborates.

4 Analysis of Survey Data on the Degree to Which Users Attach Importance to Network Security During the Operation of Computer Software

4.1 Analysis of the Use of Computer Software by Students in Different Grades

Students in colleges and universities who use computers can learn and entertain. Some of them use computers to download learning software to learn their professional knowledge, and some use computers to download game software for entertainment. The situation of students in different grades using computer software for learning and entertainment is shown in Fig. 1: The number of freshman students who use computer software to study or entertain is not much different, and 41 of them use computers for both learning and entertainment; the number of second-year students who use computers for both study and entertainment is 5 more than that of first-year students, and the difference between the other two lines is small; among the juniors, there are 60 people who use computer software to study, which is the largest number of people who use computer software to study in all grades, and only 16 people use computer software for entertainment, which is the least number of people who use computer software for entertainment in all grades; the number of seniors who use computer software to study is the smallest of all grades, with only 20. The number of students who take into account both learning and entertainment is basically the same as that of freshmen and sophomores. It shows that the academic burden and life pressure of students will affect their use of computer software. Freshmen have just entered the transitional stage, sophomores have adapted to university life, and juniors are facing the pressure of postgraduate entrance examinations and public examinations. Four students are facing employment pressure.

Based on the analysis in Fig. 1, student stress not only affects the computer usage of students in different grades, but also affects the length of time they spend online. The statistics are shown in Table 1: Among all grade students, the longest online time is juniors, with an average of 5–6 h a day. According to Fig. 1, it can be known that they mainly use computers for professional learning; Among all grade students, freshmen spend the least on the Internet, with an average of 2–3 h a day. Freshmen have more professional courses and need to listen to teachers in the classroom and have less spare time; the number of seniors who use the Internet is the largest, with 97. The employment pressure is increasing and they need to inquire about work information through computer software.

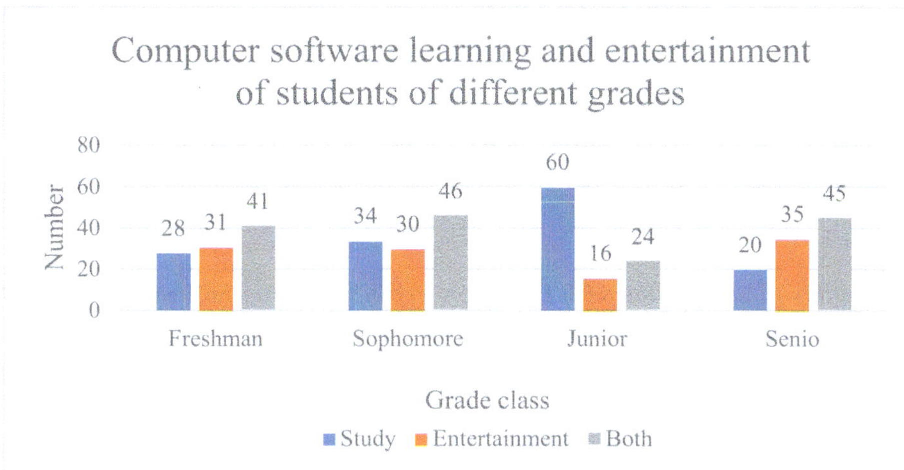

Fig. 1. Students in different grades use computer software to study and entertain

Table 1. Statistics on the number of students in different grades who are online and how long they spend online each day

Grade	Internet users	Online hours per day
A freshman	90	2–3
A sophomore	93	3–4
Junior year	83	5–6
Senior year	97	4–5

In order to intuitively see the average daily online time of students of different grades, it is shown in Fig. 2: It is obvious that juniors spend the most online time every day, with an average of 5.5 h online. The average freshman is online for 2.5 h a day.

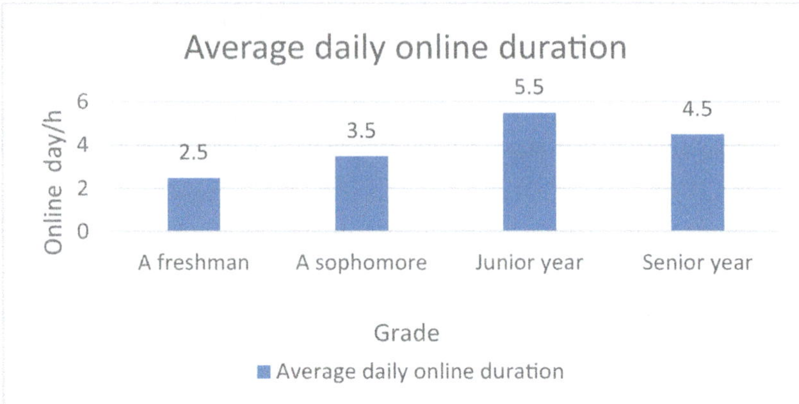

Fig. 2. Average daily online hours of students of different grades

4.2 Survey of Satisfaction with Data Security in Computer Software Among Students of Different Grades

The increase in the number of people surfing the Internet has made network security more and more important. A survey of non-computer majors from a university in Beijing was conducted around the issue of network security. The results of the survey on computer software's satisfaction with network security and the number of people are shown in Table 2 and Fig. 2: It can be seen from the figure that the number of people who are very satisfied with the data security when using computer software accounts for 37% of the total; the number of people who are satisfied with the data security when using computer software accounts for 52% of the total; the number of people who expressed dissatisfaction with data security when using computer software accounted for 7% of the total; the number of people who expressed great dissatisfaction with data security when using computer software only accounted for 4% of the total. The number of people who approve of data security in computer software accounts for 89% of the total, and only 11% are dissatisfied. This shows that our country's network users' satisfaction with network security has increased significantly. The data security of computer software is highly recognized by users, and designers consider more comprehensive security factors

Table 2. Statistics of students' satisfaction with data security in computer software in different grades

Satisfaction	Number of people
Very satisfied	148
Satisfied	208
Not satisfied	28
Very dissatisfied	16

when designing computer software, and have achieved good results in data information protection (Fig. 3).

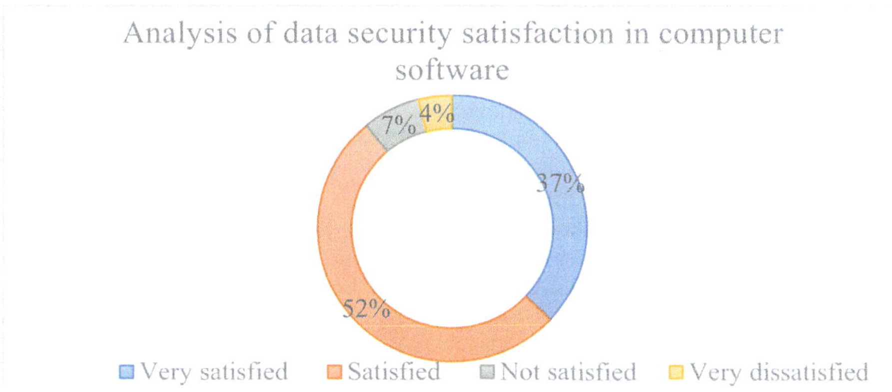

Fig. 3. An analysis of the satisfaction of students in different grades on data security in computer software

5 Conclusions

Data encryption technology is regarded as a key technology for computer software design and webpage security, and it is widely used in all walks of life. Therefore, it is very important to apply data encryption technology to computer software testing. It helps to improve the security and confidentiality of computer software programs, protect users' personal information from leaking, and create a safe network environment. From the analysis of the fourth part of the survey results: the number of college students who use computer software to study is large, and there are 142 non-computer students who only use computer to study, indicating that it is more convenient to use computer software to study; the longest time that junior students use computers to surf the Internet is 5–6 h a day, and senior students have the highest number of Internet users, 97% of the total number of seniors; the number of people who recognized data security in computer software accounted for 89% of the total, and only 11% were dissatisfied, indicating that users are highly satisfied with network security. It is expected that data encryption technology will become more and more mature in the near future, and people will not worry about network security issues when using the Internet.

References

1. Zhang, H.Y.: Application of data encryption technology in computer network communication security. Digi. Technol. Appl. **036**(012), 168–169 (2018)
2. Yucai, Z.: Application of data encryption technology in computer network communication security% computer network communication security data encryption technology application. Comm. Pow. supply Technol. **036**(008), 200–201 (2019)

3. Zhou, X.: Research on the application of data encryption technology in computer security. J. Phys: Conf. Ser. **1648**, 032095 (2020)
4. Sawle, P., Baraskar, T.: Survey on data classification and data encryption techniques used in cloud computing. Int. J. Comp. Appl. **135**(12), 35–40 (2016)
5. Neelima, U., Noorbasha, F.: Data encryption and decryption using reed-muller techniques. Int. J. Eng. Technol. **8**(1), 83–91 (2016)
6. Denisov, E.Y., Voloboy, A.G., Biryukov, E.D., Kopylov, M.S., Kalugina, I.A.: Automated software testing technologies for realistic computer graphics. Program. Comput. Softw. **47**(1), 76–87 (2020). https://doi.org/10.1134/S0361768820080034
7. Hassan, N.A., Hijazi, R.: Data hiding using encryption techniques. Data Hiding Techniques in Windows OS, pp. 133–205 (2017)
8. Namasudra S.: An improved attribute-based encryption technique towards the data security in cloud computing. Concurrency & Comput. Prac. Exper. **31**(3), e4364.1-e4364.15 (2019)
9. Al-Rummana, G.A., Shinde, G., Al-Ahdal, A.: MapReduced based: a new stream cipher technique for data encryption. Int. J. Eng. Adv. Technol. **9**(5), 763–769 (2020)
10. Lee, Y.U.: Secure visible light communication technique based on asymmetric data encryption for 6G communication service. Electronics **9**(11), 1847 (2020)
11. Dhar, S., et al.: An efficient data compression and encryption technique for PPG signal. Measurement 533–542 (2017)
12. Pushpa, B.R.: A new technique for data encryption using DNA sequence. Int. Conf. Intel. Comp. Cont., pp. 1–4 (2017)

New Rural Intelligent Pension Model Based on Big Data Technology

Lina Xiao[1], Jinfen Ye[1](✉), and Joanne Pransky[2]

[1] School of Management, Wuhan Donghu University, Wuhan, China
vivid123jun@126.com
[2] University of Coimbra, Coimbra, Portugal

Abstract. With the aggravation of China's aging, the traditional family pension model has been unable to meet the pension needs of the growing elderly group. With the rapid development of Internet economy, more and more countries apply big data technology and artificial intelligence technology to pension services, resulting in a new intelligent pension model. The development of intelligent pension industry can effectively alleviate the challenges brought by population aging. Starting from the current pension needs of the elderly, this paper investigated the willingness of the elderly to provide for the elderly in rural areas through a questionnaire survey, and proposes to build a new rural intelligent pension model, which guides the urban elderly to flow into rural pension, so as to promote the development of rural pension industry.

Keywords: Big data technology · Intelligent pension · New rural pension

1 Introduction

China has become one of the countries with the most serious degree of population aging in the world. There has been an unprecedented speed and trend of China's aging. It is expected that by 2050, the number of China's elderly population will reach a peak of 487 million, accounting for 34.9% of the total population.

With the aging becoming more and more serious, the pension problem of the elderly is imminent. In order to cope with the livelihood problems brought by the aging population, governments from the national to local have launched support policies, covering industrial development, day care centers, bed construction of pension institutions, employment of elderly care service talents, tax incentives and so on. The decision of the Fourth Plenary Session of the 19th CPC pointed out that we should actively deal with population aging and take aging governance as an important part of promoting the modernization of national governance system and governance capacity.

Compared with cities, rural areas have fresh natural environment, relaxed lifestyle and more suitable pension environment. However, at present, China's rural pension industry is still in the initial stage of development, and the consumption needs of the elderly cannot be met. Therefore, this paper mainly studies the development trend of China's rural pension industry, analyzes the demand for rural pension services and

I. Ahmad et al. (Eds.): STSIoT 2021, LNDECT 122, pp. 71–76, 2023.
https://doi.org/10.1007/978-981-19-3632-6_10

the problems existing in the further development, and finally puts forward a new rural intelligent pension model based on the era of big data.

2 Related Work

Through the retrieval, sorting and analysis of relevant literature on aging, health of the elderly and intelligent pension at home and abroad, many scholars' research mainly focused on intelligent pension products, remote scene design, spiritual education for the elderly, community cooperative pension, pension operation and so on.

Lee-En Kao and Chien-Hsiung Chen designed the user interface of intelligent nursing products from a technical point of view, mainly for the needs and use habits of new elderly people, so as to provide intelligent services more simply and conveniently through the user interface [1]. Xian Wu, Lauren C. Nix et al. payed attention to the privacy and information security of elderly users, they built, developed and evaluated remote telepresence interfaces from the perspective of usability [2]. Lara Gitto understand the ability and frequency of the elderly to use ICT through research, and put forward the strategy of constructing active aging education through the research on the learning ability of the elderly [3]. Dawn Sakaguchi-Tang and other scholars believed that an effective pension model should integrate community pension and promoted the development of elderly communities through communication and cooperation [4]. Daryoush Daniel Vaziri focused on analyzing the needs of the elderly for health technology design, put forward the development model of healthy aging technology, and promoted the integration of intelligent device technology and the daily life of the elderly [5].Valeria D'Amato and other scholars discussed the proposal of personal pension products in combination with the background of the times [6]. Mary Jane C and others suggested using optical character recognition to build a social pension management system for the elderly [7].

Scholars at home and abroad have done more research on national pension and personal pension, mainly focusing on national pension plan, big data security, personal pension products and decision-making. Some scholars also studied intelligent pension product design, community cooperative pension, pension operation and other aspects.They have less research on rural intelligent pension, so it is necessary to study the problem of rural intelligent pension.

3 Development Trend of Rural Pension Industry in China

3.1 China's Aging is Becoming Serious, and the Pension Industry is in a Sunrise Industry

China is accelerating into an aging society. Fewer children and empty nests are the basic reality that China will face. By 2030, China's elderly population will reach 371 million, accounting for 25.3% of the total population. At that time, there will be one elderly in every four people. At the same time, China's elderly population has reached 25 million. Therefore, there is a huge market demand for the domestic elderly care industry.

3.2 The Pension Model is Mainly at Home, and There is a Large Gap in Rural Pension

With the aggravation of population aging, "pension industry" has become a hot word in society. At present, China's pension models mainly include home-based pension, community pension, institutional pension, housing pension and so on [8]. Home-based pension refers to the elderly care with the family as the core and relying on children or relatives. It is the main pension model in China at present. Institutional pension generally refers to the elderly living in public or private nursing homes, welfare homes, nursing homes and nursing homes at their own expense to obtain corresponding medical and care services. Community pension is the intermediate form of the first two, which means that the elderly live at home at night, continue to be taken care of by their families, and enjoy day care, housekeeping, meal delivery and other services provided by community pension institutions during the day.

3.3 The Scale of Rural Pension Increased, and the Growth Rate Showed a Downward Trend

By the end of 2020, there were 38000 registered pension institutions and 8.238 million pension beds in various institutions and communities, including more than 20000 pension institutions in rural areas, which can provide more than 1.94 million beds. More than 90% of the institutions have the ability to provide integrated medical and pension services. With the increase of pension service institutions and facilities, the number of beds tends to rise as a whole, but the growth rate gradually declines. With the strengthening of pension service infrastructure, the supply of pension beds will be improved.

3.4 The Big Data Has not Been Applied in Pension, and Intelligent Pension Continues to Develop

Compared with other consumer groups, the elderly groups have obvious differences in physiology, psychology and experience. At present, the big data platform related to China's rural pension services have not been established, and the existing data sharing mechanism of rural pension services is not perfect.

Meanwhile, in 2019, the scale of intelligent pension industry was close to 3.2 trillion yuan, but the supply of intelligent pension products and services still could not meet the multi-level needs of the elderly. Especially the rural elderly have low consumption cognition and consumption ability. In the long run, the development of intelligent pension products and services is still the focus in the future.

4 Market Research on the Demand for Rural Pension Services

4.1 Design and Distribution of Questionnaire

This questionnaire needs to collect the respondents' cognition of rural pension and their demand for material, spiritual and environmental support for rural pension. The questionnaire includes three parts: the basic personal situation; the cognition and willingness of rural pension; the demand and opinions on rural pension. 558 questionnaires were sent out and 523 were recovered, of which 508 were valid.

4.2 Sorting and Analysis of Questionnaire Data

(1) Personal Profile Analysis

Through the statistics of the collected effective questionnaires, 318 people felt that they could transform rural houses into houses suitable for the elderly, and attached great importance to their environment, medical service level and supporting facilities. According to the analysis, most rural areas have idle houses that can be used for early construction, but they have high requirements for environmental health care and other aspects.

(2) Cognition and Willingness of Rural Pension

From the survey, 89% believed that the natural environment in rural areas was more suitable for the aged than that in cities, and 77% had greater interest in entering rural areas for the aged. In the survey, 67% said they don't know what to communicate with the elderly at home, and said they don't have time to accompany the elderly because of work. Therefore, it can be seen that the elderly need a pension place with beautiful environment and complete medical treatment to enjoy their old age.

(3) Demand and Opinions on Rural Pension

According to the survey feedback data, 62% of people like simple and warm architectural style for living style; In terms of the demand for rural pension, 49% paid attention to the environment, 78% wanted complete supporting implementation, 85% required good medical service level, and 97% wanted rich and colorful entertainment venues.

5 Building a New Rural Intelligent Pension Model

5.1 Introduction to the New Rural Intelligent Pension Model

The new rural intelligent pension model explored in this paper is to establish a pension base in rural areas where the environment and climate are suitable for the elderly, providing them to the urban elderly who are willing to provide for the elderly in rural areas, and provide them with specific services, such as: providing agency purchase service for daily necessities; providing medical services; establishing sound health data management; providing entertainment services.

5.2 Strategies for Constructing the New Rural Intelligent Pension Model

5.2.1 Improving Pension Service Facilities and Building a Personalized Pension Environment

The new rural pension pays attention to the personalized needs of the elderly. On the basis of preliminary market research, according to the local natural environment, living conditions and cultural traditions, it makes full use of tourism and agricultural resources, and promote the coordinated development of "sightseeing tourism", "leisure agriculture", "health home stay" and "health care town". By signing lease agreements with local governments, it can carry out rural transformation to promote local economic development.

During the construction period, the rural idle houses shall be re planned to eliminate potential safety hazards and carry out reasonable construction and development. At the same time, the gymnasium, yoga room and other medical facilities, as well as the rehabilitation facilities, such as the gym, yoga room and other medical facilities, will be provided. At the same time, entertainment places will be added in the later stage, such as gym, calligraphy room, chess and card room, yoga hall, etc.

5.2.2 Encouraging Social Capital to Enter and Cultivating Professional Service Talents

There are very good prospects for the development of rural pension institutions. Local governments in various regions should introduce corresponding preferential measures according to the development degree of pension industry, guide social capital into the rural elderly care industry, formulate access standards, strengthen the professional ability evaluation of pension institutions, establish and improve the service quality standards and evaluation system, so that more elderly people can enjoy high-quality pension services as soon as possible.

High quality rural pension service is inseparable from high-level talent team. At present, the number of rural pension institutions in China is small, and the professional level of employees is uneven, lacking corresponding operation ability and nursing knowledge. In the future development, we should strengthen the training of professional service talents. The trained talents should have basic gerontology, geriatric psychology and communication skills, teamwork, service awareness, information technology, etc. [9]. At the same time, we should carry out reeducation training for the existing employees, encourage them to work with certificates, and form a diversified, unique and personalized nursing service system.

5.2.3 Strengthening the Grass-Roots Health Services and Promoting Medical Care

Compared with cities, the biggest problem of rural pension is that the grass-roots health service system is not perfect. During the 14th Five Year Plan period, we should: first, increasing investment and further improving the service level of grass-roots health institutions; second, strengthening the construction of grass-roots health service team, organizing regular and irregular standardized training to form a large number of qualified general practitioners; third, improving the treatment standards of grass-roots health service personnel, continuously enhancing the attraction of posts, establishing a mobile working mechanism, and stabilizing the team of grass-roots health personnel.

During the 14th Five Year Plan period, we should further improve the service system and management mechanism of the combination of medical care and maintenance, and further promote the service of the combination of medical care and maintenance.

5.2.4 Improving the Pension Information Platform and Promoting the Sharing of Pension Big Data

The application and popularization of big data and artificial intelligence have brought great opportunities to the traditional pension service industry. At present, many developed countries have established relatively perfect basic pension service information platforms to meet the needs of the elderly with different types and different payment capabilities. China should also further strengthen the construction of pension information platform, establish and improve health information database, preference and demand database and pension service enterprise database, open up the isolation between families, communities, medical institutions, enterprises and governments, and realize data sharing [10].

At the same time, in order to prevent the leakage of personal information caused by the application of big data, we should pay close attention to the security of users' private data and establish standardized institutional constraints and system supervision as soon as possible.

6 Conclusions

Since entering the aging society in 2000, China's aging has developed rapidly. On the basis of market research, this paper deeply analyzes the pension needs of the elderly. For the elderly who want to go to rural areas for pension, this paper expounds from the aspects of pension facilities and environmental construction, pension professional services, grass-roots medical services and big data sharing, and puts forward a new rural Intelligent pension model.

Acknowledgements. This research was supported by the grants from Youth Foudation WuHan Donghu University (project number: 2017dhsk007).

References

1. Kao, L.-E., Chen, C.-H.: User interface design of smart IoT care products for the new-age elderly. AHFE (17), 1107–1112 (2021)
2. Xian, W., et al.: The design, development, and evaluation of telepresence interfaces for aging adults: Investigating user perceptions of privacy and usability. Int. J. Hum. Comput. Stud. **156**, 102695 (2021)
3. Gitto, L.: Ability and frequency of ICTs use in an older adults' sample: implications for developing an active aging educational strategy. Int. J. Knowl. Learn. **14**(1), 86–100 (2021)
4. Sakaguchi-Tang, D., et al.: Co-design with older adults: examining and reflecting on collaboration with aging communities. Proc. ACM Hum. Comput. Interact. **5**(CSCW2), 1–28 (2021)
5. Vaziri, D.D.: Facilitating daily life integration of technologies for active and healthy aging - understanding demands of older adults in health technology design, pp. 1–160. University of Siegen, Springer Vieweg, Germany (2018). ISBN 978-3-658-22874-3
6. D'Amato, V. et al.: New challenges in pension industry: proposals of personal pension products. Neural Advances in Processing Nonlinear Dynamic Signals, 253–263 (2020)
7. Mary, J.C. et al.: Senior citizen social pension management system using optical character recognition. ICTC 456–460 (2019)
8. Gao, Y. et al.: The current situation of the development of domestic and international wisdom pension service products and models. ISAIMS 252–256 (2021)
9. Cui, Y., et al.: Design of intelligent home pension service platform based on machine learning and wireless sensor network. J. Intell. Fuzzy Syst **40**(2), 2529–2540 (2021)
10. Mao, J., et al.: Interactive design of products suitable for the elderly under the smart pension. AHFE **5**, 450–457 (2021)

Application of Dual-Loop Control Algorithm Simulation Technology in Power Regulation of New Energy Grid

Chaoshan Xin[1(✉)], Jin Yu[1], Jiayu Bian[1], Guokang Yu[1], and Xiaowei Li[2]

[1] State Grid Xinjiang Electric Power Co., Ltd. Economic and Technical Research Institute, Urumqi, Xinjiang, China
ppso210706@126.com

[2] Beijing Jinghang Tianli Technology Co., Ltd., Beijing, China

Abstract. With the development of economy and society, the utilization rate of energy is getting higher and higher, and the energy problem is becoming more and more serious. Therefore, new energy has received extensive attention from all walks of life. New energy occupies a relatively high proportion of the power grid, which brings new challenges to the operation of the power grid, especially the power generation power system connected to the power grid by the inverter to regulate the power of the power grid. Under these conditions, this paper proposes to apply the dual-loop control algorithm simulation technology to the power regulation of the new energy grid, aiming to realize the rapid regulation of the power of the new energy grid. This article conducted a questionnaire survey on the impact of new energy grid output power on residential electricity consumption. The survey results showed that: Residents in the community have the highest annual electricity consumption in August, at 24880 KW/h, and less electricity consumption in a suitable weather month, at about 9100 KW/h; among the household appliances investigated, the air conditioner has the highest power, with an average use time of 4-8h/day, and the average use time of the hair dryer with the lowest power is 0.6 h/day; among the 300 users, 59% have higher requirements for the stability of the input electric energy, and only 0.5% have no requirements for the stability of the input electric energy.

Keywords: Dual-loop control algorithm · Simulation technology · New energy · Grid power regulation

1 Introduction

With the rapid development of economy and society, people's quality of life is getting higher and higher, and the increasing quality of life is inseparable from electric energy [1]. Because of the scarcity of fossil energy sources such as oil, natural gas, and coal, our country has gradually developed from a traditional fossil energy power generation model to a new energy power generation model. Therefore, all walks of life believe that the new energy grid has great value [2]. In order to maximize the power utilization of the new

I. Ahmad et al. (Eds.): STSIoT 2021, LNDECT 122, pp. 77–85, 2023.
https://doi.org/10.1007/978-981-19-3632-6_11

energy grid, the dual-loop control algorithm simulation technology is used to explore the main operating rules of the new energy microgrid; at the same time, the inverter is used to connect to the power system of the new energy grid to achieve the power of the new energy grid [3, 4]. Based on this, the economic applicability and stability of the new energy grid can be improved, laying the foundation for the in-depth study of the new energy microgrid [5].

Research shows that the traditional PID algorithm is widely used in inverters, but the accuracy of its current and voltage adjustment needs to be further improved [6]. Based on this, the dual-loop control algorithm simulation technology is applied to the inverter, and the dual-loop control parameters of current and voltage can be designed to obtain a dual-loop control structure and construct a dual-loop control inverter for the new energy grid. The dual-loop control algorithm is used to simulate and prove this algorithm can effectively regulate the power of the new energy grid [7, 8]. New energy grids use dual-loop control algorithm simulation technology for power regulation, which can increase the output rate of power and improve the quality of output power [9]. The traditional single current or voltage control structure algorithm has a long cycle and has certain errors. The dual-loop control algorithm simulation technology can synchronize the control and the voltage input cycle. It is a new type of PFC control algorithm based on a digital model [10, 11]. For the utilization of electric energy, it is necessary to consider the coordinated control of the grid storage system while considering the consumption rate of new energy, so as to reduce energy costs while ensuring the dispatch of the new energy grid [12].

This paper studies the application of dual-loop control algorithm simulation technology in new energy grid power regulation, discusses the impact of dual-loop control algorithm simulation technology on the current and voltage in the new energy grid system, and discussed how it can adjust the power of the new energy grid by constructing a digital algorithm simulation model. This article will elaborate on the following points: First, understand the basic principles of new energy and dual-loop control algorithms; second, study the power adjustment strategy of the new energy grid during the operation; third, study the interference suppression during the operation of the dual-loop control algorithm simulation technology; fourth, aiming at the new energy grid's output power for residents in a community in Hubei Province A questionnaire survey was conducted on the electricity situation.

2 Introduction to Dual-Loop Control Algorithm Simulation Technology and New Energy Grid Power Regulation

2.1 Dual-Loop Control Algorithm Simulation Technology and Basic Principles of New Energy

New energy is a renewable energy that is being developed by researchers but has not yet been promoted. Traditional power generation uses fossil fuels, which not only wastes resources but also pollutes the environment; the use of new energy power generation reduces the production cost of electric energy and saves a lot of energy. Therefore, new energy power grids have attracted social attention. The dual-loop control system includes two parts: a current loop and a voltage loop. The current and voltage dual-loop control is mainly used in the inverters of the new energy grid. The new digital algorithm simulation technology utilizes the power state information of the system, which can not only improve the dynamic performance but also Can improve accuracy. The digital system of the dual-loop control algorithm simulation technology is realized by computer programming. Its algorithm is relatively simple, and no error table is needed, which greatly saves memory and cost. Applying the dual-loop control algorithm simulation technology to the power adjustment process of the new energy grid can avoid the error of using a single current or voltage closed-loop algorithm, resist the disturbance of the dynamic process of electrical energy output, and quickly adjust the power of the new energy grid.

2.2 New Energy Grid Operation and Power Regulation Strategy

The operating characteristics of the new energy grid are different from the traditional grid, and its inverter is the key to controlling the stable output of electrical energy. This paper mainly uses dual-loop control algorithm simulation technology to study new energy grid operation and power regulation strategies. According to the loadability and power quality of the new energy grid, it is necessary to expand the inverter system and adopt the form of an external inverter to adjust the output power, voltage value and frequency of the new energy grid in time. Ensure that the new energy grid can automatically adjust the voltage value and frequency stability without the grid reference value; at the same time, the output power of the new energy grid is adjusted according to the capacitance of the new energy grid, and the load is reasonably distributed to ensure uninterrupted power supply for residents and various enterprises. According to the actual operation of the new energy grid, in addition to using the dual-loop control algorithm simulation technology to ensure the stable output of the system, the constant power control technology (PQ) and the constant voltage and constant frequency control technology (V/f) can also be used at the same time.

2.3 Interference Suppression in the Operation of Dual-Loop Control Algorithm Simulation Technology

Under different load conditions, the inverter's voltage output presents periodicity and regularity, accompanied by periodic disturbances. The dual-loop control algorithm simulation technology is used to control the inverter of the new energy grid. It can be divided into two components: the voltage control module of the outer loop and the current control module of the inner loop. The current includes the capacitive current and the load current. In an emergency, the inner loop control module's suppression of the current will affect the current output. If you want to enhance the stability of the inverter dual-loop control in the new energy grid, it is necessary to effectively control the current, voltage and various interference factors in the electric energy output process. Among them, interference factors include natural interference factors and man-made interference factors. The larger interference amplitude is spikes with higher noise frequency. The mathematical model of spikes is shown in formula (1), and the complex number expression of spectral functions is shown in formula (2); PI is used to control position closed-loop system, which is simplified as formula (3); PID formula is shown below:

$$f(t) = \{^{tg\theta \cdot t, 0 \le t \prec T/2}_{tg\theta \cdot (T-t), T/2 \le t \prec T} \tag{1}$$

$$G(j\omega) = \int_{-\infty}^{\infty} f(t) e^{j\omega t} dt \tag{2}$$

$$Pwm = Kp * e(k) + Ki * \sum e(k) \tag{3}$$

$$Pwm = Kp * e(k) + Ki * \sum e(k) + Kd[e(k) - \mathrm{e}(\mathrm{k} - 1) \tag{4}$$

Among them, T represents the duration of the spike pulse; t is the correlation time; tgθ is the hypothetical condition, the hypothesis tgθ $= U_{max}$; $j\omega$ is the frequency spectrum function;E (k) represents the current deviation, e(k − 1) represents the last deviation, and Pwm represents the output.The main way to eliminate the impact of noise on the inversion of the new energy grid is to eliminate noise. An instrument with voltage stabilization and filtering functions can be installed in the system to reduce the impact of noise on electrical energy output.

3 Questionnaire Survey on the Impact of New Energy Grid Output Electric Energy on Residential Electricity Consumption

3.1 Purpose and Target of Investigation

The purpose of this questionnaire is to study the impact of the dual-loop control algorithm on the power regulation of the new energy grid, aiming at the output power of the new energy grid, and discuss the power consumption of the residents by the different output power of the new energy grid when the voltage is unstable. The subject of this questionnaire is for residents aged 18–55 in a community in Hubei Province. Under the condition of ensuring a balanced ratio of men and women, 200 community residents of different occupations were randomly selected to fill in the questionnaire.

3.2 Investigation Method

This questionnaire on the impact of the electric energy output from the new energy grid on the electricity consumption of residents in a community in Hubei Province is based on the unstable performance of the electric energy output from the new energy grid under the premise of clarifying the purpose and objects of the survey. This questionnaire survey is completed by offline questionnaires. The investigators will print out the questionnaires and distribute them to residents aged 18–55 in a community in Hubei Province. The time for the questionnaire is selected at 6–8 pm In between, after they completed the questionnaire survey, the investigators took it back.

3.3 Survey Results

In this article, the questionnaire survey report on the impact of new energy grid output on residential electricity consumption will be retrieved by the investigators. The investigators will conduct detailed data analysis on the results of the questionnaire survey. The analysis results will be systematically analyzed and discussed in the fourth part of this article, this part does not analyze the results.

4 Analysis of the Results of the Questionnaire Survey on the Impact of the Electric Energy Output from the New Energy Grid on Residential Electricity Consumption

4.1 Survey on Electricity Consumption of Residents in a Community in Hubei

According to the questionnaire, the statistics of the electricity consumption of residents in a certain community in Hubei in 2020 are obtained. Among them, the electricity consumption of residents is mainly in winter and summer. In winter, they use air conditioners, heaters and other heating equipment to resist the cold. In summer, they need to use air conditioners and electric fans to cool down. The power of these household appliances is generally large, which will increase the electricity demand of residents in the community. The electricity consumption statistics are shown in the Fig. 1, it can be seen from the figure: The electricity consumption in August is 24880 KW/h, which is the month with the highest electricity consumption of the year; the electricity consumption is basically the same in January, February, July, September and December, all around 18000 KW/h; in April, the electricity consumption in May, October, and November is basically the same, and the annual electricity consumption in these four months is all around 9100KW/h. It shows that people's requirements for the quality of life are getting higher and higher. With the improvement of people's living standards, the more household appliances people buy, the greater the demand for electricity usage, which affects the amount of electricity consumption of residents in the community; in summer, it is a common phenomenon to use air conditioners and electric fans to reduce the indoor temperature to prevent heatstroke. It is also a common phenomenon to use electric heaters, electric heaters and other heating equipment in winter, both of which will increase residents' electricity consumption.

According to the questionnaire, the load parameters and use time of household appliances commonly used by residents in the community are obtained, as shown in Table 1, from the table we can see: The power of the air conditioner in household appliances is the largest, and its nominal power is 2200 W; the household appliance with the longest average use time is also the air conditioner, and its average power consumption time is 4–8 h/day; the household appliance with the smallest power is a hair dryer, and the nominal power is 1500 W, and its average use time is also the shortest, and its average power use time is 0.6 h/day. It shows that the higher the power of large household appliances, the greater the electric energy it needs, so the increase in the frequency of using these household appliances will lead to an increase in the electricity consumption of residents.

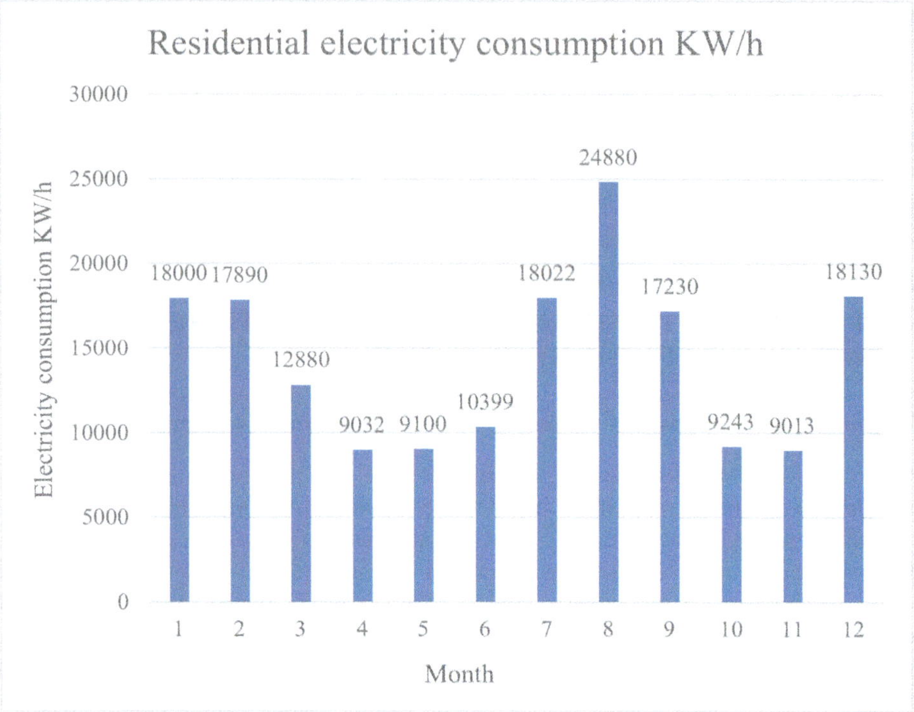

Fig. 1. Statistics of electricity consumption of residents in 2020

Table 1. Household appliance parameters and use time

Household appliances	Specifications/v	Nominal power/W	Average duration h/day
The TV	220	2000	3–5
Air conditioner	220	2200	4–8
Electric grill	220	1700	1.1
A hair dryer	220	1500	0.6
Electric heaters	220	1800	3–5

In order to have a more intuitive view of the service time of common household appliances, we calculate the average service time of each appliance, as shown in Fig. 2. It can be seen intuitively that air conditioning has the longest service time every day, with an average service time of 6 h every day. TV and heater daily use time is the same, other electric appliances use less.

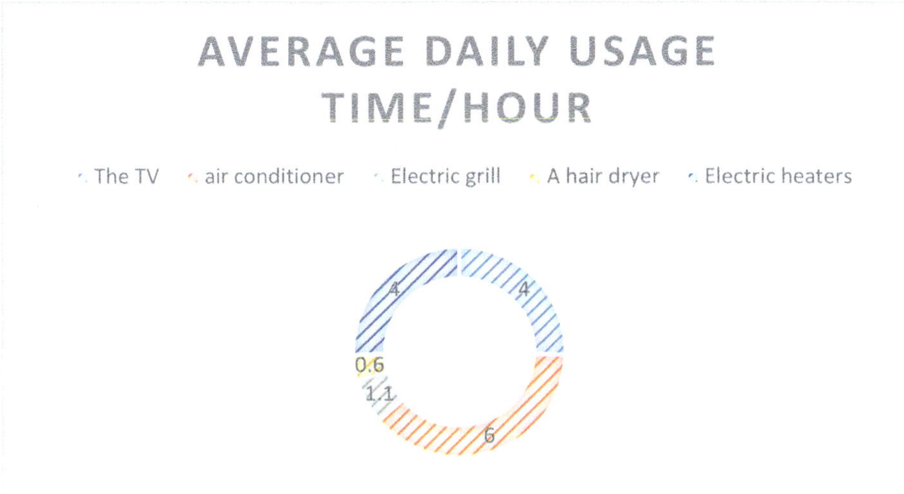

Fig. 2. Household appliance parameters and average daily usage time

4.2 Investigation of User's Requirements for Input Power Stability

The stability of the input power will affect the service life of household appliances, it will also affect the user's sense of use, and it will also affect the working efficiency of household appliances. Therefore, the results of the user's requirements for the stability of the input electric energy are analyzed. Residents in the community can be divided into four levels: higher requirements, general requirements, lower requirements, and no requirements the corresponding survey results are shown in Fig. 2 and Table 2. From Fig. 2 and Table 2, we can get that: 118 users have higher requirements for the stability of the input power, which is the largest number of users, accounting for 59% of the total;68 people have general requirements for the stability of the input power, accounting for 34% of the total; 13 people have low requirements for the stability of the input power, accounting for 6.5% of the total; only one user in the survey has no requirements for the stability of the input power, accounting for 0.5% of the total. It shows that most users have high requirements for the stability of the input electric energy, and few people have no requirements for the stability of the input electric energy (Fig. 3).

Table 2. The proportion of input power stability required by different users

Requirements for input power stability	Proportion of the total population
High requirements	59%
General requirements	34%
Low requirements	6.5%
No request	0.5%

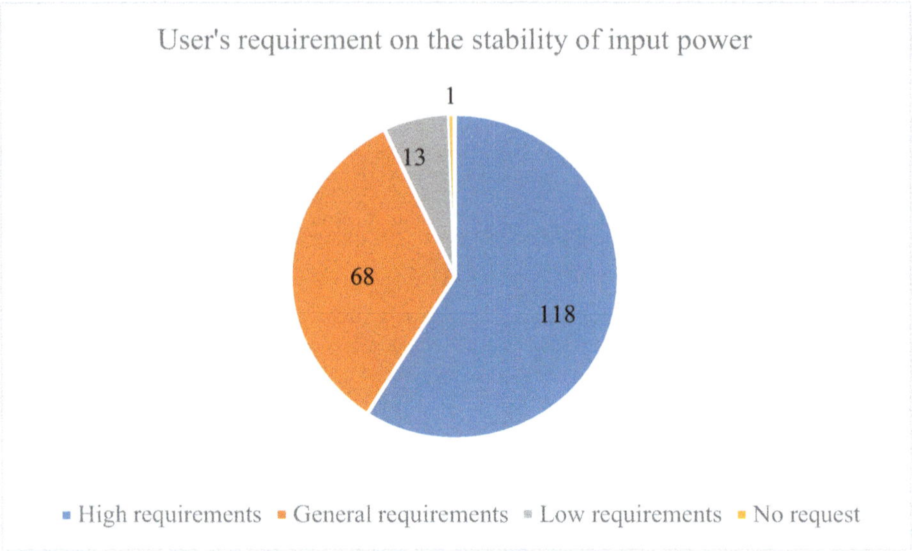

Fig. 3. User's requirement on the stability of input power

5 Conclusions

The rapid development of economy and society has promoted people's living standards. The consumption of fossil energy has made the energy problem more and more serious. The development of new energy sources to alleviate energy pressure has become the focus of social attention. Connect the inverter to the new energy grid, use dual-loop control algorithm simulation technology to adjust the power of the new energy grid, realize the rapid increase/decrease of the power of the new energy grid, and ensure the stable output of electric energy. The use of dual-loop control algorithm simulation technology can reduce errors and reduce production costs, maximize power utilization, and can also increase power output rate and improve the quality of output power. Through the analysis of the results of the questionnaire survey on the impact of the electric energy output from the new energy grid on residential electricity consumption, it can be concluded that the electricity consumption in August in the 2020 electricity consumption survey is 24880 KW/h, which is the month with the highest electricity consumption of the year; the electricity consumption in November is the least in the whole year, which is 9013 KW/h, indicating that climate issues will affect the usage of household appliances. The greater the usage of household appliances, the greater the consumption of electricity by residents. The power of the air conditioner in household appliances is the largest. Its nominal power is 2200 W, and the average power consumption time is 4–8 h/day, which is the electrical appliance with the longest average use time; the home appliance with the smallest power is the hair dryer, which has the shortest average usage time, with an average electricity usage time of 0.6 h/day. In the result analysis of the residents' requirements for the stability of the input electric energy, we can get: There are 118 users who have high requirements for the stability of the input power, accounting for

59% of the total number of users. Only one person has no requirement for the stability of the input power, which only accounts for 0.5% of the total number of users, it shows that most users have higher requirements for the stability of the input electric energy, and very few people have no requirements for the stability of the input electric energy.

References

1. Johnstone, N., Hascic, I., Popping, D.: Erratum to: renewable energy policies and technological innovation: evidence based on patent counts. Environ. Resource Econ. **68**(2), 441–444 (2017)
2. Sasleep, Y., Gupta, S.C., Board, A.K.: Review of hybrid renewable energy systems with comparative analysis of off-grid hybrid system. Renew. Sustain. Energy Rev. **81**(5), 2217–2235 (2018)
3. Wei, X., Liu, J., Wei, T., et al.: High proportion renewable energy supply and demand structure model and grid impaction. Scient. Res. Pub. **4**(2), 1173–1184 (2016)
4. Fan, S., et al.: A novel distributed large-scale demand response scheme in high proportion renewable energy sources integration power systems. Applied Sciences **8**(3), 452–462 (2018)
5. Alpine, R.R., et al.: On the use of probabilistic forecasts in scheduling of renewable energy sources coupled to storages. Appl. Energy **210**(6), 1207–1218 (2018)
6. Ding, L., Chunzhi, Z., Peng, Z.: Research on the optimal dispatch of wind power consumption based on combined heat and power with thermal energy storage. Energy Procedia **15**(2), 978–983 (2018)
7. Ma, H., et al.: A dispatch method of combined heat and power plants with hea017t storage facilities for wind power accommodation. In: 2017 IEEE 6th Internationgal Conference on Renewable Energy Research and Applications(ICRERA). IEEE, pp. 262–272 (2017)
8. Haider, H.T., See, O.H., Elmenreich, W.: Dynamic residential load scheduling based on adaptive consumption level pricing scheme. Electric Power Sys. Res. **133**, 27–35 (2016)
9. Li, H., et al.: Economic dispatch considering the wind power forecast error. Iet Generation Transmission & Distribution **12**(12), 2861–2870 (2018)
10. Ye, L., et al.: Study of assessment on capability of wind power accommodation in regional power grids. Renewable Energy **133**, 647–662 (2019)
11. Qi, Z., et al.: The optimal dispatch with combination of wind power and photovoltaic power systems. Energy Procedia **103**, 94–99 (2016)
12. Li Peng, Y., et al.: Flexible look-ahead dispatch realized by robust optimization considering CVaR of wind power. IEEE Trans. Power Syst. **33**(5), 5330–5340 (2018)

Mine Safety Monitoring and Early Warning System Based on 5G Network Technology

Yigai Xiao[1,2], Zhimou Xie[2,3](✉), and Yunqiu Liu[3]

[1] School of Resources and Safety Engineering, Central South University, Changsha 410083, Hunan, China

[2] Sinosteel Maanshan General Institute of Mining Research Co., Ltd., Maanshan 243000, Anhui, China

xiezhimou123@163.com

[3] Sinosteel Nanjing Huaxin Technology Co. Ltd., Nanjing 211100, Jiangsu, China

Abstract. With the continuous development of industry, the demand for various minerals is also increasing. However, restricted by factors such as environment and equipment, mining safety problems continue to occur. Therefore, research on mine monitoring and early warning has been put on the agenda. The maturity of 5G technology provides technical support for current monitoring and early warning. This article first summarizes the research status of mine safety monitoring system, and combines the advantages of 5G technology to analyze the feasibility of the mine safety detection and EWS based on 5G network technology studied in this article. Afterwards, the functional modules of the mine safety monitoring and EWS were designed in detail. This article systematically expounds the Newton iterative method and other seismic source location methods. Research shows that the mine safety monitoring and EWS based on 5G network technology studied in this article has higher accuracy in monitoring data and positioning personnel.

Keywords: 5G · Mine safety · Monitoring and early warning system · Application research

1 Introduction

With the rapid development of science and technology, the problem of energy shortage has come to the attention of the world people again. The same is true in China. With vast territory and abundant resources, China has a great advantage in mineral resources compared with other countries. Mineral resources are indispensable material basis for improving people's living standard and developing national economy. They are also valuable natural resources in China, which are non-renewable and relatively scarce. Regrettably, in recent years, driven by economic interests and the increase of mining scale and depth, mining safety accidents such as roof caving, deep rock burst, surface collapse, underground water inrush and large area caving emerge in endlessly, making mining safety issues become the focus of attention of the government and enterprises. In the process of mine safety production, as production safety policy of "safety first,

I. Ahmad et al. (Eds.): STSIoT 2021, LNDECT 122, pp. 86–95, 2023.
https://doi.org/10.1007/978-981-19-3632-6_12

prevention first, comprehensive treatment", security is a prerequisite, and is to keep the material basis for the safety in production, but there are still certain contradictions between, when certain input resources, share proportion increase, which means that the relative reducing production resources, on the other hand, Because many enterprises only care about immediate interests and lack long-term strategic goals, they tend to invest most of their resources into production, in exchange for unique profits through the safety of personnel and equipment. In addition, the mining geological environment is complex and the infrastructure is poor, leading to more unsafe mining production.

With the rapid development of my country's economy, the demand for various types of energy is also increasing [1, 2]. The continuous mining of industry has made the traditional mine safety monitoring and EWS unable to meet the increasing demand due to its limited geographical location and difficulty in real-time monitoring of underground operations, which has brought huge losses and casualties to related personnel [3].

This article focuses on improving the efficiency of mine safety construction, and aims to study the mine safety monitoring and EWS based on 5G network technology. The feasibility analysis of the research content of this article is carried out by comparing the data monitoring and personnel positioning with the actual value on the simulation software.

2 Related Work

Mine safety monitoring system is a necessary condition for safe production. The development of foreign coal mining monitoring systems began as early as the 1960s [4]. So far, it has experienced four stages: space division, frequency division, time division system signal transmission and distributed micro-processing technology [5].

Coert introduced a mine safety system that utilizes wireless sensor Networks (WSN). Sensor design and wireless communication in underground mining environment are studied. This information is used to design and implement a powerful hardware-based sensor node with independent microcontrollers that collect data from six different sensors, namely temperature, humidity, airflow speed, noise, dust and gas level sensors, and transmit the processed data to a graphical user interface [6]. Ankit believes the Internet of Things (IoT) can play a huge role in mining operations to improve worker safety and improve productivity. Wireless communication, as a component of Internet of Things, has played an important role in underground mine communication. A mining safety system based on Internet of things is proposed [7]. Ali monitors a person's health data and uses fuzzy logic to calculate the early warning score score. Therefore, a test platform for real-time applications was simulated according to the Health Informatics – Medical/health Device Communication standard [8].

Compared with foreign countries, my country began to study coal mining monitoring systems in the 1970s. At that time, Western countries already had relatively mature coal mine safety monitoring technologies and finished products [9]. Our country has introduced various coal mine safety monitoring systems from the West for many times, so that coal mining monitoring technology has been developed rapidly [10].

To sum up, due to the reasons of natural occurrence conditions and economic entities, there are few studies on mine safety warning abroad. Domestic non-coal mine warning is still in the theoretical research stage, even the developed application system only has the alarm function because of the low degree of system integration, the warning function is relatively single, the accuracy is not high. Therefore, it is of great practical significance to carry out systematic research on the theory and technology of early warning and increase the investment in its application to comprehensively improve the level of mine safety management and the situation of safe production.

3 Application Research of Mine Safety Monitoring and EWS Based on 5G Network Technology

3.1 5G Network Technology Analysis

The safety monitoring and EWS designed in this research is applied to mining operation monitoring. Most of the mining monitoring is in deep mountains or underground, which is greatly affected by the geographical environment [11, 12]. Therefore, the safety monitoring and EWS should include the following characteristics:

(1) Strong anti-interference ability, not affected by weather;
(2) Very long transmission distance;
(3) Fast transmission speed;
(4) High real-time requirements;
(5) Good stability.

3.2 System Requirement Analysis

(1) Feasibility analysis

1) Technical feasibility analysis

The security monitoring and EWS created by the 5G network has high performance requirements for transmission delay, throughput, power consumption, etc. It has the characteristics of ultra-low delay, ultra-low power consumption, high reliability, and large-scale connection, and can completely solve the problem of mines.

2) Economic feasibility analysis

With this system, coal mine safety observers can grasp all kinds of underground data anytime and anywhere, adjust equipment that may cause safety problems in time, avoid production accidents and unnecessary losses, and reduce economic losses, have a very broad market application space.

(2) Role analysis

1) Operation user

The operating user is actually a security monitoring staff. Safety monitoring personnel monitor the production status of the mine in real time, mainly through the safety production, staffing, video monitoring and other modules of the mine safety monitoring and EWS client to solve some unexpected phenomena or factors in the first time. At the same time, safety observers can view these data information through daily production reports and key engineering modules to control mine output or work group tasks.

2) Server system

a. Application server system

The application server system connects the various subsystems of the mine safety monitoring and EWS, obtains the data information of the relevant monitoring objects in time, organizes the data information, and provides data services for the monitoring system. The monitoring objects processed by the application server system include daily production data, key project data, safety production data, and staffing information data.

b. Video server

Video streaming is more professional than regular data information, so a dedicated video server is needed for transmission. Nowadays, most coal mines use more mature video servers to provide different data interfaces for different mobile devices. The video server in this study sends it to the mobile device through a proprietary compression process.

3.3 System Function Design

(1) Monitoring module design

The monitoring system mainly includes information concentrator and wireless sensor nodes.

1) Information concentrator

The information concentrator is mainly composed of supporting modules such as a single chip microcomputer, a wireless radio frequency module, a serial port transceiver module, and an alarm module. Among them, which completes the scheduling and processing of various tasks; the serial port transceiver module completes the data transmission interaction between the information concentrators; the alarm module can complete the sound and light threshold alarm, cross-border personnel are equipped with alarms, warning signs and other functions.

2) Wireless sensor node

The wireless sensor node is mainly composed of supporting modules such as a single chip microcomputer, a wireless radio frequency module, and environmental sensors. The single-chip microcomputer, the wireless radio frequency module and the information concentrator have similar functions, while the wireless sensor node has only one radio frequency module chip, which simultaneously completes the data sending/receiving function. Environmental sensors are responsible for collecting data.

(2) Design of personnel positioning module

The personnel positioning function is mainly composed of the reader (that is, the information concentrator in the monitoring module) and the personnel positioning node. The personnel positioning node is used for the positioning information interaction with the reader. The personnel positioning node (identification card) in the mine can interact with the reader in full duplex. In addition to passively uploading positioning information to the ground command center, you can also actively upload an emergency distress signal to the ground command center through the emergency call button. The ground command can also issue various instructions (buzzers, light warnings of different colors) to underground mine operators, such as evacuation alarms, requiring operators in specific areas to contact the ground. When an emergency occurs underground, the emergency communication system is only installed in the second room and key location of the main motor room, so it cannot cover the work area of all personnel. However, the personnel positioning module covers almost all underground working areas and has a certain range of communication.

3.4 Analysis of Common Source Location Methods-Newton Iteration Method

(1) Principle

Iterative methods are often used to solve approximate roots of nonlinear equations. It is important to be able to determine the iteration function. The classic iterative method is to directly solve the unknown number through the function itself, thereby obtaining the iterative function, but this method requires multiple iterations and is relatively slow. Newton iteration is usually used to calculate the location of the microseismic source, because they choose another iteration format with faster convergence speed, and other iteration forms can be obtained through corresponding transformations.

Let x_n be the approximate root of the nonlinear equation f(x) = 0, and expand f(x) at x_0 through Taylor polynomial expansion:

$$f(x) = f(x_0) + f'(x_0)(x - x_0) + \frac{f''(x_0)}{2!}(x - x_0)^2 + \dots \tag{1}$$

To simplify the problem, linearize the equation and use only the first two terms to approximate f(x), then it can be expressed as:

$$f(x) \approx f(x_0) + f'(x_0)(x - x_0) = 0 \tag{2}$$

Suppose $f'(x_n) \neq 0$, let the solution of the equation be x, then we can get

$$x = x_0 - \frac{f(x_0)}{f'(x_0)} \tag{3}$$

Let $x_1 = x_0 - \frac{f(x_0)}{f'(x_0)}$, similarly, expand f(x) = 0 to a Taylor polynomial of x_1 and take only the linear part:

$$x_2 = x_1 - \frac{f(x_1)}{f'(x_1)} \tag{4}$$

By analogy, Newton's iterative form can be obtained:

$$x_{n+1} = x_n - \frac{f(x_n)}{f'(x_n)}, n = 1, 2... \tag{5}$$

(2) Geometric meaning and advantages and disadvantages

Draw the tangent of f(x) at point x_0, the slope is represented by $f'(x_0)$, and the tangent equation is as follows:

$$y - f(x_0) = f'(x_0)(x - x_0) \tag{6}$$

Set y = 0, the intersection point of the tangent equation and the x-axis can be obtained as x_1, namely:

$$x_1 = x_0 - \frac{f(x_0)}{f'(x_0)} \tag{7}$$

In the same way, draw a point $(x_1, f(x_1))$, that is, the tangent of f(x), and get the intersection point with the x-axis as x_2, and iterate all the time, and finally approach the root of the equation. As shown in Fig. 1.

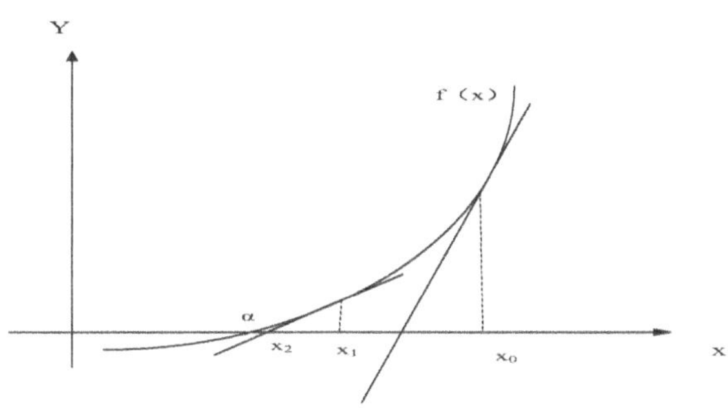

Fig. 1. Schematic diagram of Newton's iteration method

Newton's iteration method converges by a square factor, so it usually only takes a few iterations to obtain a more accurate solution. The advantage of this form of Newton' iteration method: it can solve the roots of nonlinear equations accurately and effectively. However, Newton's iteration method is very sensitive to initial values, and incorrect initial values may not converge. At the same time, it contains the derivative of the function, and the amount of calculation is relatively large.

3.5 Design of the Signal Detection Algorithm in the Mine Based on 5G Technology

The mine signal detection algorithm based on 5G technology mainly eliminates interference, linearly weights the received vector to meet specific detection standards, separates the transmitted signals of different users, and then detects the transmitted signals of different users. This time, a large antenna array is used to form a TDD cell, which is composed of M receiving terminals and K transmitting ports.

The M × 1 dimensional column vector at the receiving end is expressed as:

$$y = \sqrt{p_n} G_x + n \tag{8}$$

$$r = A^n y = \sqrt{p_n} A^n G_x + A^n n \tag{9}$$

r_k, x_k is the k-th element of the K-dimensional column vectors r and x, respectively.

$$r_k = \sqrt{p_u} a_k^H Gx + a_k^H n = \sqrt{p_u} a_k^H g_k x_k + \sqrt{p_u} \sum\nolimits_{i=1, i \neq k}^{k} a_k^H g_i x_i + a_k^H n \tag{10}$$

a_k, g_k is the k-th column vector of the matrices A and G, respectively.

4 Experimental Research on Mine Safety Monitoring and EWS Based on 5g Network Technology

4.1 Experimental Protocol

This experiment uses simulation software for the data monitoring module and personnel positioning module of the mine safety monitoring and EWS based on 5G network technology studied in this paper. In this experiment, 4 groups were set up to ensure the scientific validity of the experimental data. In the data monitoring experiment, this experiment is tested by monitoring the gas concentration. In the personnel positioning experiment, this experiment uses the error between the monitoring and EWS studied in this article and the traditional EWS for experimental analysis, and finally uses mathematical statistics for statistics and analysis of the data obtained.

4.2 Research Methods

(1) Comparative analysis method

In this experiment, the mine safety monitoring and EWS based on network technology 5g studied in this document is compared to the traditional monitoring system. The information received shall provide a reliable reference to the final results of the investigation of this document.

(2) Observation method

In this study, the gas concentration was observed and recorded and the recorded data were classified and analysed. This evidence provides strong support for the final results of the investigation of this document.

(3) Mathematical Statistics

The relevant software is used to compile statistics and analyses of the research results of this document.

(4) Simulation experiment method

In this experiment, the safety monitoring and EWS studied in this article is tested on simulation software, which provides data support for the feasibility analysis of the subject of this article.

5 Experimental Analysis of Mine Safety Monitoring and EWS Based on 5G Network Technology

5.1 Monitoring and Analysis of Gas Concentration

In order to make the experiment more scientific and effective, the mine safety monitoring and EWS based on network technology 5 shall be simulated on simulation software. The results are presented in Table 1.

Table 1. Gas concentration monitoring and analysis

	Test 1.	Test 2.	Test 3.	Test 4.
Detection value	0.46%	0.52%	0.44%	0.48%
Actual value	0.45%	0.54%	0.41%	0.50%

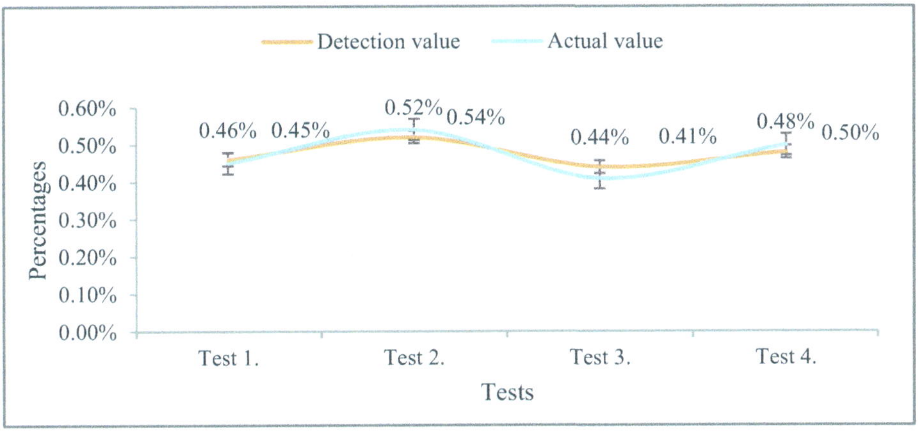

Fig. 2. Gas concentration monitoring and analysis

As shown in Fig. 2, the error of the first test is 0.1%, the error of the second test is 0.02%, the error of the third test is 0.03%, and the error of the fourth test is 0.02%. The average error is 0.02%, less than 0.05%. This shows that the mine safety monitoring and EWS based on network technology 5g studied in this document has the following characteristics: the high gas monitoring rate, which fully reflects the appropriateness of this research content.

5.2 Personnel Location Monitoring Experiment

In order to study and further analyse this experiment, the experiment compares the traditional monitoring system with the monitoring system based on network technology 5g, which has practical value. The results are given in Table 2.

Table 2. Personnel location monitoring experiment

	Test 1.	Test 2.	Test 3.	Test 4.
5G	100.53	151.04	198.79	250.44
Traditional	103.45	147.41	203.70	244.48
Actual value	100.00	150.00	200.00	250.00

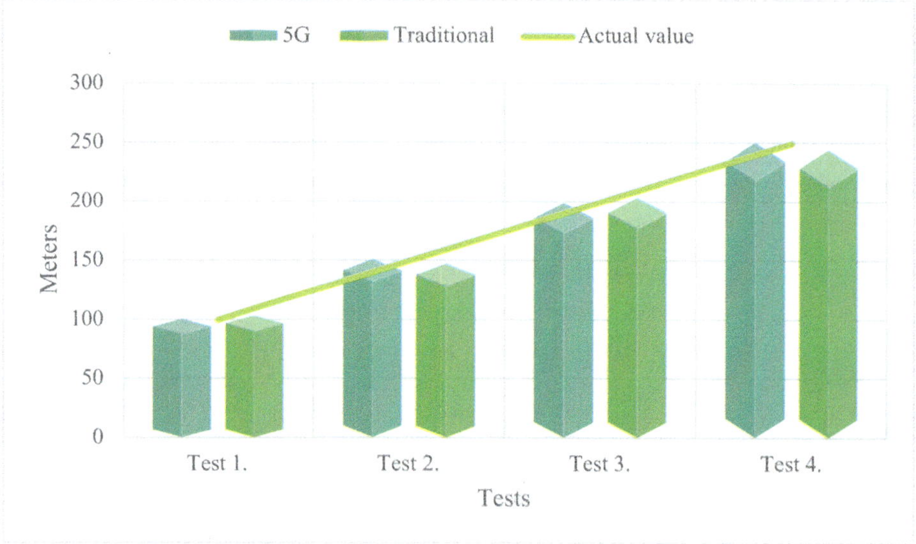

Fig. 3. Personnel location monitoring experiment

As shown in Fig. 3, on the contrary, the traditional safety monitoring system, safety monitoring and EWS, Based on network technology 5 studies the position of personnel at a shorter distance than the actual distance, which is of high precision, which fully reflects the designers.

6 Conclusion

This article aims to study the mine safety monitoring and EWS based on 5G network technology, through the analysis of common seismic source location methods-a detailed overview of the Newton iteration method, and designing the safety monitoring and EWS based on the 5G technology-based signal detection algorithm in the mine. The monitoring module and personnel positioning module are designed and researched. Finally, the simulation operation on the simulation software proves that the system can effectively improve the safety level of mining.

References

1. Yang, Y.: Study on mine safety monitoring system with emergency voice communication based on wireless sensor network. Hans J. Wirel. Commun. **08**(5), 209–215 (2018)
2. Jo, B.-W., Khan, R.M.A., Lee, Y.S., Jo, J.H., Saleem, N.: A fiber bragg grating-based condition monitoring and early damage detection system for the structural safety of underground coal mines using the internet of things. J. Sensors **2018**, 9301873:1–9301873:16 (2018)
3. Henriques, V., Malekian, R., Bogatinoska, D.C.: Mine safety system using wireless sensor networks. In: MIPRO, pp. 515–520 (2017)
4. Wibowo, A., Pratama, C., Sahara, D.P., Heliani, L.S., Rasyid, S., Akbar, Z., Muttaqy, F., Sudrajat, A.: Earthquake EWS using ncheck and hard-shared orthogonal multitarget regression on deep learning. IEEE Geosci. Remote. Sens. Lett. **19**, 1–5 (2021)
5. Cremen, G., Bozzoni, F., Pistorio, S., Galasso, C.: Developing a risk-informed decision-support system for earthquake early warning at a critical seaport. Reliab. Eng. Syst. Saf. **218**(Part), 108035 (2022)
6. Jordaan, C., Malekian, R.: Design of a monitoring and safety system for underground mines using wireless sensor networks. Int. J. Ad Hoc Ubiquitous Comput. **32**(1), 14–28 (2019)
7. Singh, A., Kumar, D., Hötzel, J.: IoT Based information and communication system for enhancing underground mines safety and productivity: genesis, taxonomy and open issues. Ad Hoc Netw. **78**, 115–129 (2018)
8. Çalhan, A., Cicioglu, M., Ceylan, A.: EHealth monitoring testbed with fuzzy based early warning score system. Comput. Meth. Prog. Biomed. **202**, 106008 (2021)
9. Heyns, A.M., Du Plessis, W.P., Curtin, K.M., Kosch, M., Hough, G.: Decision support for the selection of optimal tower site locations for early-warning wildfire detection systems in South Africa. Int. Trans. Oper. Res. **28**(5), 2299–2333 (2021)
10. Debauche, O., et al.: Towards landslides EWS with fog – edge computing and artificial intelligence. J. Ubiquitous Syst. Pervasive Netw. **15**(2), 11–17 (2021)
11. Tubis, A., Werbinska-Wojciechowska, S., Sliwinski, P., Zimroz, R.: Fuzzy risk-based maintenance strategy with safety considerations for the mining industry. Sensors **22**(2), 441 (2022)
12. Pillay, M., Tuck, M.: Permit-to-work systems as a health and safety risk control strategy in mining: a prospective study in resilience engineering. In: AHFE, vol. 5, pp. 145–154 (2017)

The Influence of Fintech on the Performance of Commercial Bank Based on Big Data Analysis

Juan Wang[1(✉)], Na Zhang[1], and Robert Rodes[2]

[1] College of Economics, Bohai University, Jinzhou, Liaoning, China
wj_mulucky@126.com
[2] Moscow University, Moscow, Russia

Abstract. This article chooses 31 domestic commercial banks' panel data between 2011 to 2020 to analyze the impact of Finance-tech on the commercial banks' profitability by using fixed effect model. After investigating the commercial banks through heterogeneity research, this article finds that compared with stated-owned banks and joint-stock banks that have greater market influence, the Finance-tech has greater impact on the city commercial and rural commercial banks whose market influence is relatively weak.

Keywords: Finance-tech · Banks' profitability · Panel data · Heterogeneity

1 Introduction

The Finance-tech has been applied widely in finance sector since 21th century, and its daily changes also accelerate the development of finance industry. The *Financial technology development plan (2019–2021)* published by People's Bank of China planes to establish a well-function financial technology system that will be the new motivation for the high-quality development of finance, raising the application level of Finance-tech and promoting China's Finance-tech to the first level. According to the *pulse of financial technology* (first half of 2020) released by KPMG, China's 2018 Finance-tech investment (including Venture capital, private equity and M & A) reached 30.69 billion dollar, which is 4.6 times of that of 2017, making China become the biggest finance-tech investment market. *50 leading financial technology enterprises in China in 2020*, another report released by KPMG, founds that Finance-tech enterprises are playing an increasing important role in many sectors, including comprehensive financial technology, technological innovation financial institutions, payment technology, credit technology and so on. Apart from that, it also found that Ant Financial Services, Jingdong Digital Technology Co., Ltd., Du Xiaoman Finance and Lufax have already been listed on the global fintech 100 list, which means the Finance-tech market of China has become an oligopoly market. Pressed by the fast development of Finance-tech enterprises, the traditional commercial banks begin to find new profit points by developing Finance-tech.

As the development of Finance-tech imposes great external effect on commercial banks, scholars begin to focus on the external influence of Finance-tech on the profitability of commercial banks in terms of negative effect, such as Crowding-out Effect,

I. Ahmad et al. (Eds.): STSIoT 2021, LNDECT 122, pp. 96–106, 2023.
https://doi.org/10.1007/978-981-19-3632-6_13

and positive effect, such as spillover effect. Besides, some scholars are also interested in whether the influence on the banks with different natures would be different. Therefore, on the basis of reading a large number of relevant literature, this article proposes research assumptions and build models to analyze the influence of Finance-tech on the profitability of different kind commercial banks.

2 Related Work

The new conception of Finance-tech has raised great attention of both domestic and foreign scholars. The early researches about Finance-tech mainly focus on the qualitative researches, including its conception, development history and regulatory issues, for example, Arner and Barberis (2015) pointed out that the Finance-tech was the combination of finance and technology and the practical solutions supported by science technology to the financial problems [1]; Narinder Kumar Bhasin (2021) the Indian banking system is passing through the journey of revolutionary reforms from traditional banking to e-collaboration of digital banking products and financial technologies (Fintech) companies [2] AI and data science are driving new-generation financial technology (FinTech) [3], which profoundly disrupts existing theories of money, investment, credit, market, and regulation, and furthermore empowers new-generation financial innovations, products, services, operations, processes, and ecosystems. We illustrate how an analysis of the shortlisted set can be used to gain understanding of what FinTech innovations are, where and when they emerge, and provide the basis for further work on what their impact is on the companies investing in them, and ultimately on society [4]. The revolutionary changes brought by fintech entrants ("Fintechs") in the financial services sector has caused severe turbulence in the operational and service activities of the incumbent 'traditional' banking organizations ("Banks") [5] (M. Siek and A. Sutanto 2019). Several variables were selected to analyze the impact of fintech on traditional Indonesian banks [6]. Jerene, W., & Sharma, D. (2020) revealed that the quality of service positively influences bank customers' satisfaction and their e-loyalty [7]. Disruptive innovations that shake up traditional financial markets. As financial technology can reduce intermediary costs, expand financing channels, and increase financial inclusion and inclusiveness, a series of traditional banking products, from new technology development to financial service commercialization activities, from payment to investment consulting, are being challenged by innovative fintech products (Lee and Shin, 2018) [8]. Latin America uses Fintech to serve smartphone users, bringing a lot of convenience during the COVID-19 (D. Rojas-Torres et al., 2021) [9]. The banking industry must make strategic transformation to adapt to the disruptive changes in the external environment caused by fintech (Van Loo, 2018) [10], and turn challenges into opportunities for further growth of commercial banks.

This shows that the development of financial technology will certainly change the business function of traditional banks. This paper analyzes the impact of financial technology on bank performance.

3 Research Hypothesis

From the income source perspective of traditional banks, traditional banks mainly make profit through deposit and loan business, however, as Finance-tech decreases the deposit scale, which in consequence decreases the loanable funds scale, and limited the development space of intermediary business, the profitability of traditional banks is influenced to some extent. From the risk prospective of traditional commercial banks, the deterioration of the deposit structure on the liability side and the rise of interest payment cost will heavy the commercial banks' risk burden, which will influence the profitability of the commercial banks. From the prospective of commercial banks' consumer groups that obey the "2:8 law", commercial banks tend to exclude long-tail customers, such as small and micro enterprises, while the Finance-tech could overcome information asymmetry problems and reduce trading cost through big data and cloud computing. which provides a more complete customer profile and improves the matching efficiency of funds, and hence attracts the original long-tail customer of commercial banks, squeezing the profit space of commercial banks. Based on above analysis, this paper proposes assumption H1.

H1: The development of Finance-tech imposes negative impact on the profitability of commercial banks.

As there are many differences in term of capital scale, market influence, customer group and risk defense between banks with different natures, the self-protection abilities and related solutions of banks would be different when facing the hit of the development of Finance-tech. Therefore, the impact of Finance-tech on the profitability of banks with different natures would be different. Compared with city commercial banks and rural commercial banks whose customer groups are relatively small, the state-owned banks and joint-stock banks would suffer the Finance-tech hit less as they have more powerful market influence and diversified customer groups. Based on the analysis above, this paper proposes assumption H2.

H2: The negative impact of Finance-tech on state-owned banks and joint-stock banks would be less than city commercial banks and rural commercial banks.

4 Research Design

4.1 Sample Selection and Data Sources

The Finance-tech grew fast in China with the appearance of P2P in 2011. This paper selected 31 domestic banks' data between 2011 and 2020 from Tonghuashun Ifind database, National Bureau of statistics and annual reports of commercial banks.

The principles of choosing the sample are listed below: Firstly, foreign banks and policy banks are excluded from the sample. Secondly, the banks that founded after 2011 are excluded from the sample. Thirdly, the banks whose variable data are incomplete seriously are excluded from the sample.

4.2 Variable Selection and Descriptive Statistics

(1) Dependent variable

This paper treats the profitability of commercial banks as dependent variable, the return and asset and return of equity as proxy variable.

(2) Independent variable

This paper treats the degree of Finance-tech development as independent variable, because there are no specified ratio to measure the degree of Finance-tech development, this paper chooses the text mining method used by Guo pin and Shen Yue to construct the index, the specified steps are listed below:

Firstly: Establishing the original database. This paper selects 4 dimensions and 14 key wards in the term of the function of modern finance and the underlying technology support of Finance-tech, the details are shown in Table 1.

Table 1. Finance-tech index original database

Dimension	Key words			
Payment settlement	Third party payment	Network payment	Cross border payment	
Deposit, loan and capital raising	Online loan	P2P	Crowd-funding	Internet finance
Wealth management	Intelligent investment advisor	Security brokerage APP	Mobile banking	
Underlying technology support	Big data	Cloud computing	Artificial intelligence	Block-chain

Secondly: Statistical calculation of keyword frequency in Chinese Crucial Newspapers Full-text Database (CCND). Askitas and Zimmermann (2009) found that there are strong correlation between the amount of news report and some economic behaviors. Apart from that, CCND covers 6 albums and 43 special topics that are public published in China, which could better reflect the news report situation of different industries in China. This paper defined the ratio of the frequency of key word in CCND to the same year total words published by CCND as the word frequency of the key word.

Thirdly: Synthetic Finance-tech development index by factor analysis. The key words are analyzed by factor analysis, KMO and Bartlett ball test. Because the KMO is 0.6297, which is bigger than 0.6, and the outcome of the Bartlett ball test is satisfying, it is suitable to use factor analysis to analyze key words. Then, this paper selects 3 common factor whose eigenvalue is greater than 1, and the total variance contribution is 86.65%, which means the three common factors could reflect the information included in the key words

Table 2. The definition of variables

Type of variable	Variable name	Variable symbol	Variable design
Dependent variable	Return on total assets	ROA	Return on total assets
	Return on net assets	ROE	Return on net assets
Independent variable	Finance-tech index	FI	Synthesis by factor analysis
Control variable	Capital adequacy ratio	CAR	Capital adequacy ratio
	Loan ratio of non-performing assets	NPL	Loan ratio of non-performing assets
	Net interest margin	NIS	Net interest margin
	Proportion of non-interest income	NIP	Non-interest income/business income
	Liquidity ratio	CR	current assets/current liabilities
	Inflation rate	IR	Inflation rate
	Degree of financial deepening	GS	Total market value of stock/GDP

well. After that, this paper calculates the score of common factors and standards them by MAX-MIN method to get the Finance-tech index.

(3) Control variables

This paper selects capital adequacy ratio(CAR), Loan ratio of non-performing assets (NPL), Net interest margin(NIS), Proportion of non-interest income (NIP), Liquidity ratio (CR), Inflation rate(IR) and Degree of financial deepening (GS) as control variables. The details are shown in Table 2.

4.3 Model Setting

This paper builds the basic models based on above analysis:

$$\mathrm{ROA}_{\mathrm{it}} = \beta 0 + \beta 1\mathrm{FIt} + \sum \mathrm{j} = 1^{7}\alpha \mathrm{jCONTROL}_{\mathrm{jit}} + \mathrm{ui} + \varepsilon \mathrm{it} \tag{1}$$

$$\mathrm{ROE}_{\mathrm{it}} = \beta 0 + \beta 1\mathrm{FIt} + \sum \mathrm{j} = 1^{7}\alpha \mathrm{jCONTROL}_{\mathrm{jit}} + \mathrm{ui} + \varepsilon \mathrm{it} \tag{2}$$

5 Empirical Analysis

5.1 Descriptive Statistics of Data

The Table 3 shows the descriptive statistics of the model variables, the range of commercial banks' ROA is between 0.25 and 1.63, and the standard deviation of the ROE is 4.457, which means there are relatively large gap between different banks' profitability. Apart from that, the standard deviation of NIP is 11.939, which means the main income resource of different banks different greatly, and this difference may influence the relationship between the Finance-tech and the profitability of commercial banks.

Table 3. Sample descriptive statistics

Variable	Number of samples	Mean	Standard deviation	Minimum value	Maximum value
ROA	309	0.961	0.238	0.250	1.630
ROE	309	15.303	4.457	3.770	26.650
FI	310	0.569	0.307	0.000	1.000
CAR	308	12.902	1.751	9.000	24.860
NPL	310	1.270	0.493	0.030	4.310
NIS	310	2.132	0.441	0.810	3.350
NIP	310	23.205	11.939	3.243	84.560
CR	310	52.515	13.623	27.600	127.220
IR	310	2.510	1.063	1.400	5.400
GS	310	58.423	13.615	40.319	78.500

Note: Except for Finance-tech index (FI) which is standardized to 0–1, other variables are expressed in %

5.2 Regression Analysis

(1) The regression between the development of Finance-tech and the profitability of commercial banks.

The Table 4 shows the regression of model (1) and model (2) using fixed effect model after F test and Hausman test. Column (1) and Column (2) show the regression of Finance-tech on commercial banks' ROA when the control variables are not added and added respectively. Column (3) and Column (4) show the regression of Finance-tech on commercial banks' ROE when the control variables are not added and added respectively. It can be concluded that the Finance-tech has a negative impact on the profitability of commercial banks at the significance level of 1%, which validates the hypothesis H1.

Table 4. The empirical result of the development of Finance-tech's influence on the profitability of commercial banks

Variable	Model			
	(1)	(2)	(3)	(4)
	ROA		ROE	
FI	−0.501***	−0.416***	−10.33***	−7.575***
	(−25.08)	(−10.46)	(−25.30)	(−9.64)
NIS		0.103***		0.701*
		(4.97)		(1.72)
NIP		0.00129*		0.0138
		(1.81)		(0.98)
CAR		0.0210***		−0.366***
		(4.89)		(−4.30)
NPL		−0.111***		−2.919***
		(−6.18)		(−8.22)
CR		−0.00110**		−0.0182*
		(−1.99)		(−1.67)
GS		−0.000715		−0.0133
		(−1.26)		(−1.18)
IR		−0.0430***		−0.762***
		(−6.38)		(−5.73)
Constant term	1.247***	1.027***	21.19***	29.89***
	(96.54)	(12.83)	(80.22)	(18.87)
N	309	307	309	307
Adjusted R^2	0.660	0.757	0.664	0.776

Note: the values in brackets indicate t-test statistics, ***, * *, and * indicate significant at the level of 1%, 5%, and 10%, respectively, the same below

5.3 Heterogeneity Analysis Based on Different Natures of Banks

The Table 5 shows the group regression on different natures banks by fixed effect model. Column (4), (5) and (6) respectively report the impact of Finance-tech development on the ROA of commercial banks under the subgroups of state-owned banks, joint-stock banks, urban commercial banks and rural commercial banks. Column (7), (8) and (9) respectively report the impact of Finance-tech development on the ROE of commercial banks under the subgroups of state-owned banks, joint-stock banks, urban commercial banks and rural commercial banks. It can be concluded that at the significant level of 1%, the negative impact of Finance-tech on joint-stock banks, state-owned banks, city commercial banks and rural commercial banks is gradually increasing, which validates the hypothesis H2.

Table 5. The influence of Finance-tech on the profitability of commercial banks: Heterogeneity analysis

Model	(5)	(6)	(7)	(8)	(9)	(10)
Variable	ROA			ROE		
FI	−0.327***	−0.306***	−0.384***	−7.125***	−6.456***	−8.097***
	(−4.58)	(−3.62)	(−6.86)	(−3.77)	(−5.31)	(−5.07)
NIS	0.165***	0.160***	0.0885***	1.225	0.402	1.517**
	(2.75)	(4.00)	(3.34)	(0.77)	(0.70)	(2.00)
NIP	−0.000429	0.000123	0.000854	−0.00491	0.00610	−0.0137
	(−0.19)	(0.07)	(1.00)	(−0.08)	(0.33)	(−0.41)
CAR	−0.0159**	0.00495	0.0228***	−0.594***	−0.320***	−0.457**
	(−2.16)	(0.41)	(4.20)	(−3.05)	(−2.71)	(−2.00)
NPL	−0.103***	0.00960	−0.178***	−3.626***	−3.813***	−0.598
	(−3.06)	(0.25)	(−7.83)	(−4.08)	(−7.71)	(−0.82)
CR	0.00151	−0.00273**	−0.000862	0.0221	−0.0109	−0.0568***
	(1.23)	(−2.40)	(−1.31)	(0.69)	(−0.76)	(−2.64)
GS	0.00128*	−0.00283***	−0.000770	0.0245	−0.0205	−0.0366*
	(1.75)	(−2.74)	(−0.98)	(1.27)	(−1.20)	(−1.87)
IR	−0.0415***	−0.0447***	−0.0352***	−0.510**	−0.663***	−0.982***
	(−5.34)	(−3.32)	(−3.66)	(−2.48)	(−3.17)	(−3.85)
Constant term	1.282***	1.113***	1.042***	29.73***	29.66***	31.43***
	(5.70)	(6.78)	(10.23)	(5.00)	(13.41)	(10.12)
N	50	110	147	50	147	110
Adjusted R^2	0.940	0.700	0.811	0.924	0.734	0.805

6 Robustness Test

6.1 Endogenous Control

Finance-tech would influence the profitability of commercial banks, at the same time, the well-operated commercial banks may have more willingness to use Finance-tech to impose anti-influence on Finance-tech. To alleviate the endogeneity problem, this paper treats Internet penetration raten (IPR) and Internet penetration rise rate (IPRR) as instrumental variables, and uses the two-stage least square estimation (2SLS) method learned from Xie Xuanli to test the outcomes.

The Table 6 shows the regression on basic model using 2SLS, the result shows that after the endogeneity of instrumental variables is controlled, Finance-tech still has a negative impact on the profitability of commercial banks, proving the results are robust.

6.2 Replace Explanatory Variables

Because there are no specific ratio to measure the development of Finance-tech, this paper chooses text mining method learned from Shen Yue and Guo Pin, this method focuses on the national level development of Fiance-tech. Besides, this paper chooses to

Table 6. Regression result after endogeneity control using instrumental variables

Model	(11)	(12)
Variable	ROA	ROE
FI	−0.427***	−8.132***
	(−10.26)	(−9.88)
NIS	0.101***	0.601
	(4.90)	(1.48)
NIP	0.00135*	0.0168
	(1.91)	(1.20)
CAR	0.0214***	−0.349***
	(5.02)	(−4.14)
NPL	−0.109***	−2.815***
	(−6.09)	(−7.95)
CR	−0.00105*	−0.0161
	(−1.93)	(−1.49)
GS	−0.000638	−0.00949
	(−1.12)	(−0.84)
IR	−0.0435***	−0.787***
	(−6.52)	(−5.97)
N	307	307
Adjusted R^2	0.756	0.775

replace the Finance-tech index with the search index of the word Finance-tech on Baidu search, learned from Liu Mengfei and Wang Qi [12], to make robust test.

Table 7 chooses fixed panel model to check the robustness of the basic model by replacing the independent variables and shortening the sample time period.

Table 7. Robustness test: Replacement of independent variables and shortening of sample time period

Model	(13)	(14)	(15)	(16)
	Substitute explanatory variable		Shorten sample period	
Variable	ROA	ROE	ROA	ROE
FI	−0.000313***	−0.00528***	−0.450***	−7.841***
	(−10.95)	(−9.08)	(−9.39)	(−8.94)
NIS	0.121***	1.118***	0.0645***	0.495
	(6.13)	(2.78)	(2.88)	(1.20)
NIP	0.000683	0.000618	0.00199***	0.0238*

(continued)

Table 7. (*continued*)

Model	(13)	(14)	(15)	(16)
	Substitute explanatory variable		Shorten sample period	
Variable	ROA	ROE	ROA	ROE
	(0.99)	(0.04)	(3.03)	(1.97)
CAR	0.0224***	−0.359***	0.00653	−0.315***
	(5.26)	(−4.14)	(1.09)	(−2.87)
NPL	−0.137***	−3.468***	−0.166***	−2.880***
	(−8.17)	(−10.14)	(−9.46)	(−8.98)
CR	−0.00111**	−0.0206*	−0.000998*	−0.0224**
	(−2.04)	(−1.86)	(−1.96)	(−2.40)
GS	−0.00191***	−0.0372***	0.000436	0.000818
	(−3.70)	(−3.55)	(0.73)	(0.07)
IR	−0.00641	−0.120	−0.00185	−0.521*
	(−0.97)	(−0.90)	(−0.13)	(−1.97)
Constant term	0.847***	27.01***	1.213***	28.28***
	(10.24)	(16.04)	(13.23)	(16.84)
N	307	307	216	216
Adjusted R^2	0.764	0.769	0.711	0.758

6.3 Shorten the Sample Age

As the Fiance-tech became popular in China since 2014 [11], the sample period was shortened to 2014–2020 to test the stability of the model. The column (15) and column (16) show the regression outcome of the shortened period sample. The outcome shows that the Finance-tech could exert negative impact on the profitability of commercial banks at the 1% significant level, which proves the robustness of the model. Apart from that, compared to the regression outcomes between 2011 and 2020, the coefficient of the FI of shortened period sample decreased from −0.416 and −7.575 to −0.450 and −7.841, implying the negative impact of Finance-tech increased since Finance-tech became popular.

7 Conclusions and Suggestions

This paper selects China's 31 commercial banks' panel data between 2011 and 2020 as sample, uses text mining method to construct Finance-tech index, builds models to analyze the impact of Finance-tech on the profitability of commercial banks. It can be concluded two conclusions: Firstly, as the spillover effect may be more significant than crowding-out effect, the Finance-tech exerts negative impact on the profitability of commercial banks; Secondly, the commercial banks whose market influence is more powerful may mitigate the negative effect carried by Finance-tech more effectively, which is proved by the mitigation ability of joint-stock banks, stated-owned banks, city commercial banks and rural commercial banks decreases gradually.

Based on the above conclusions, this paper proposes two suggestions: Firstly, as the Finance-tech has become a new motivation for the development and reformation of traditional finance industry, commercial banks should increase investment in Finance-tech and corporate with Finance-tech companies actively to alleviate the negative impact carried by the Finance-tech. Secondly, commercial banks should apply cutting-edge technology in the business to serve high-quality customers and offer new type financial products and services to the long-tail customers to break the "2:8law".Thirdly, the state-owned banks and joint-stock banks that have powerful market influence could establish Finance-tech sector to develop Finance-tech, while for the city commercial banks and rural commercial banks whose strength is relatively weak, as their customers are mainly limited in a specific area, they could combine local development to offer specialize financial products and services to attract customers and corporate with Finance-tech companies within a manageable risk range. Fourthly, the regulators should establish a complete Finance-tech regulatory system to avoid enormous risks carried by the disorder development of Finance-tech.

References

1. Arner, D.W., Barberis, J., Buckley, R.P.: The evolution of Fintech: a new post-crisis paradigm. Geo. J. Intl. L. **47**, 1271 (2015)
2. Bhasin, N.K., Rajesh, A.: Impact of E-collaboration between Indian banks and Fintech companies for digital banking and new emerging technologies. Int. J. e Collab. 17(1), 15–35 (2021)
3. Cao, L., Yuan, G., Leung, T., Zhang, W.: Special issue on AI and FinTech: the challenge ahead. IEEE Intell. Syst. **35**(3), 3–6 (2020)
4. Caragea, D., Chen, M., Cojoianu, T., Dobri, M., Glandt, K., Mihaila, G.: Identifying FinTech innovations using BERT. In: 2020 IEEE International Conference on Big Data (Big Data), pp. 1117–1126 (2020)
5. Anand, D., Mantrala, M.: Responding to disruptive business model innovations: the case of traditional banks facing Fintech entrants. J. Bank. Financ. Technol. **3**(1), 19–31 (2018). https://doi.org/10.1007/s42786-018-00004-4
6. Siek, M., Sutanto, A.: Impact analysis of Fintech on banking industry. In: 2019 International Conference on Information Management and Technology (ICIMTech), pp. 356–361 (2019). https://doi.org/10.1109/ICIMTech.2019.8843778
7. Jerene, W., Sharma, D.: The effect of e-finance service quality on bank customers' Fintech e-loyalty: evidence from Ethiopia. Int. J. E-Bus. Res. **16**(2), 69–83 (2020)
8. Lee, I., Shin, Y.J.: Fintech: ecosystem, business models, investment decisions, and challenges. Bus. Horiz. **61**(1), 35–46 (2018)
9. Rojas-Torres, D., Kshetri, N., Hanafi, M.M., Kouki, S.: Financial technology in Latin America. IT Professional **23**(1), 95–98 (2021)
10. Van Loo, R.: Making innovation more competitive: the case of Fintech. 65 UCLA Law Review 232. Available at SSRN (2018)
11. Li, G.: The integration of finance and technology: meaning, motivation and risk. Int. Econ. Rev. (03), 91–106+6 (2020)
12. Liu, M., Wang, Q.: The impact of fintech on commercial bank performance: a theoretical and empirical study. Financ. Forum **26**(03), 60–70 (2021)

Research and Design of Soft Switch Technology in New Energy Vehicle Wireless Charging System

Jun Wang[1,2], Qiang Liu[1,2,3](✉), Jinming Huo[1,2], and Rosa Cantero[4]

[1] School of Automotive Engineering of Jilin Engineering, Normal University, Changchun, Jilin, China
liuyushuojldx@126.com

[2] Innovative Research Team of Jilin Engineering, Normal University (IRTJLENU), Changchun, Jilin, China

[3] State Key Laboratory of Automotive Simulation and Control, Jilin University, Changchun, Jilin, China

[4] University of Frankfurt, Frankfurt, Germany

Abstract. Wireless power supply technology as an emerging new energy vehicle power real-time supply technology to the development of new energy vehicles has brought a lot of convenience. The wireless power supply technology of new energy vehicles can reduce the amount of on-board batteries and even do not need to carry batteries. By laying the guide rail under the road, the electric energy is transmitted to the car in a non-electric contact way using electromagnetic technology for real-time use. However, the transmission efficiency of the wireless power supply system for new energy vehicles is not high, which greatly increases the loss of the system and reduces the effective energy supply of the system. The soft switching technology proposed in this paper can control the switching appliance under the zero voltage and zero current environment, so as to reduce the circuit switching loss. Therefore, soft switch technology is widely used in wireless power supply device can effectively improve the efficiency of the whole system.

Keywords: Soft Switching · Wireless power · Phase-shifting control

1 Introduction

Switching power supply is one of the important links in the charging system of new energy vehicles. To realize the lightweight and miniaturization development of power electronic devices, it is necessary to realize the high-frequency circuit. However, in the hard switching and high frequency environment, the electromagnetic interference and switching loss of traditional switching devices are also large, and the emergence of soft switching can effectively solve this problem. This technology can keep the primary inverter in the stable soft switching control state, reduce the loss and prolong the working life of the switch tube. This technology can keep the primary side of the inverter in the stable soft switching control state, reduce the loss and prolong the working life of the

I. Ahmad et al. (Eds.): STSIoT 2021, LNDECT 122, pp. 107–112, 2023.
https://doi.org/10.1007/978-981-19-3632-6_14

switch tube. This scheme design is of great significance for the promotion and large-scale application of wireless power supply technology for new energy vehicles in the future [1].

2 Application of Soft Switch Technology in New Energy Vehicle Wireless Charging System

2.1 Soft Switch Technology Introduction

Various new devices, new topological circuits and better control principles are applied to power supply equipment, which makes the rapid development of power supply technology and accelerates the wide application of high-frequency switching power supply in the field of new energy vehicles. The concept of soft switching is the counterpart of hard switching. There is a clear difference between hard and soft switching. In the process of hard switching controlling the circuit, the current and voltage will change very dramatically. Such drastic changes will not only cause great loss to the switch, but also produce great noise, which will not only reduce the efficiency of the circuit to a large extent, but also cause interference to the work of the peripheral electronic equipment. In addition, with the increase of switching frequency, switching loss will also increase, compared with hard switching, soft switching has a great improvement in this respect. Soft switch is based on hard switch, which is the inheritance and improvement of traditional hard switch. Different from hard switching, soft switching adds some resonant devices, including small inductors, capacitors and so on. The new resonant devices constitute the auxiliary conversion current network, and the condition of the switch has been greatly improved [2]. This problem has also been properly solved. With the support of soft switching technology, switching loss and switching noise are greatly reduced, and the efficiency of the circuit is greatly improved. Soft switch mainly includes two aspects, one is soft open, the other is soft off. Soft on can also be called zero voltage switch, and soft off is zero current switch. During operation, the voltage is usually dropped to zero, and then the current is raised to the on-state value, which is the ideal soft switching process. The ideal soft switching process will not produce switching loss and switching noise, in line with low carbon, energy saving, environmental protection requirements.

2.2 Wireless Charging System for New Energy Vehicles

A typical ev wireless charging system is divided into two parts: ground transmitter (ground terminal for short) and vehicle receiver (vehicle terminal for short), mainly including primary and secondary power converters, compensation network, controller and communication module. The ground terminal is connected with the mains as the power supply part of wireless charging; Vehicle-mounted terminal is connected with vehicle battery to convert the energy received from the ground terminal into the electricity required by the battery. Ground terminal and the vehicle-mounted terminal exchange information through wireless communication. In the process of energy transmission, the energy transmitting end and the energy picking end adopt non-contact mode, and there is no electrical connection completed by using cables, so the safety problems such as leakage caused by bad contact such as open circuit short circuit and so on are prevented. With the rapid development of wireless power supply technology for new energy vehicles, many research institutions and companies have joined in the research and application of this technology [3]. However, in the new energy vehicle wireless power supply system, the inverter switch tube is not in the soft switching state, which not only greatly shorts the working life of the switch tube, but also directly increases the loss of the switch tube, resulting in the low transmission efficiency of the system, large energy loss, and difficult to be widely popularized in the application of commercial.

Based on the background of the wireless power supply system of new energy vehicles, this paper researches the difficult technology of inverter soft switch control in the wireless power supply system of new energy vehicles, aiming at achieving the stable soft switch work of the wireless power supply system of new energy vehicles. The primary transmitting mechanism works in resonant state by soft switching method. And with certain technical means to make the new energy vehicle wireless power supply system to maintain a stable soft switch control state, reduce losses, improve the working life of the system, improve the energy transmission efficiency of the system. The main solutions are:

On the basis of the research status and development trend of switching power supply at home and abroad, the theoretical analysis, simulation calculation, system design and test of high power phase shift full bridge soft switching power supply are carried out. Using the technology of phase shift control and soft switching, the power semiconductor switching tube in the inverter circuit of high-power phase shift full-bridge soft switching power supply can realize the zero voltage switching (ZVS), reduce the switching loss of the power semiconductor switch tube, reduce the voltage and current stress of the power semiconductor switch tube, improve the efficiency of the whole machine. On the basis of the traditional phase-shift full bridge inverter, this paper adopts the scheme of adding "LC" auxiliary network to both bridge arms of the inverter, which can realize the ZVS of all bridge arms power semiconductor switch tubes in the full load range. The "LC" auxiliary network adopts asymmetric design, and appropriate auxiliary inductance is selected for each "LC" auxiliary network [4]. This design can help to reduce the conduction loss, reduce the resonant inductance and leakage inductance and their negative effects, and ensure that the bridge arm can realize the zero voltage switches. The selection basis of each component model and parameter in the circuit is given.

Design control and protection circuit, including phase shift control circuit, input current limiting current, output overvoltage and overcurrent detection and protection circuit, fault output and reset circuit, closed loop; The CHIP UC3875 is used to output PWM signal, and the phase shift Angle of the four-way PWM signal is controlled by closed-loop feedback loop. In view of the possible problems of the power supply, it is protected by the protection circuit.

3 Control Strategy and Implementation of the System

3.1 Basic Structure of Phase Shift Control Converter

Phase shift PWM control mode is a kind of soft switch control mode which is widely used in the whole bridge transformation in recent years, its working principle is actually the resonance technology and the conventional PWM technology combination, It is composed of input DC power supply, two bridge arms (four switch tubes), transformer, transformer output end rectifier diode, filter inductor, capacitor and load. The so-called phase shift control mode is the circuit of the switch tube M1 and M4, M2 and M3 turn on, conduction Angle 180° each; However, M1 and M3, M2 and M4 cannot be conducted at the same time. They need to be separated by an Angle, namely the phase shift Angle. The output voltage can be adjusted by adjusting the phase shift Angle. Since the switch tubes M1 and M3 are shut off before the switch tubes M4 and M2, respectively,M1 and M3 are called advanced bridge arms, and M2 and M4 are called hysteretic bridge arms [5].

In order to realize the soft switching of switching tube, a phase shift control zero voltage switching DC/DC full bridge converter is presented, It is main principle is to use the leakage inductance of transformer and junction capacitance of power switch tube to realize the ZVS switch. The circuit consists of one capacitor in parallel on the switching tube of the original full bridge circuit, When the switch tube is turned off, the primary side current of the transformer charges the shunt capacitor Cl or C4 of the switch tube and discharges C2 or C3 at the same time, in this way, the rise rate of M1 or M4 is limited when the M1 or M4 is turned off, and MI or M4 soft switch is realized.; As the voltage of C1 or C4 rises to Uin, the voltage of C2 or C3 drops to zero, and its inverse parallel diode conduction provides a zero-voltage conduction condition for M2 or M3 [6]. Full bridge converter with phase shift control has the advantages of simple circuit, constant frequency control and so on, which is widely used in medium and high power occasions; But because the power MOSFET tube conduction is equivalent to a resistance, its on-state loss is relatively large, and the frequency is smaller than IGBT, duty cycle loss is serious, based on this, a new phase – shift – controlled zero – voltage – zero – current full – bridge converter is presented.

3.2 Design of Control Circuit

The control circuit mainly includes: the high performance UC3875 as the core of the main controller, output voltage/current sampling module, various protection modules, isolated driver module.

The core technology of high-frequency switching power supply is pulse Width modulation (PWM) phase shift control module, it has a profound effect on the performance of switching power supply, it is main function is to make the full bridge inverter circuit two bridge arm MOSFET control signal conduction Angle stagger, further, multiple duty cycles are obtained, thus adjusting the output voltage, to make it stable. Its core is phase – shifted pulse – width modulation controller. The controller used in this design is UC3875, It is an integrated chip specially used for phase shift control, four phase – shift pulse width modulation control signals can be generated at one time, and its own sampling function, a voltage, current, and over voltage, over current and other protection circuit [7]. The main function of the converter control circuit is to control the phase shift Angle between the two bridge arms and provide some basic protection circuit. The control circuit of the system is mainly composed of the signal detection part, the control signal generation part, the power drive part, the isolated output part and the system protection part. This converter control system uses voltage control mode, control circuit to control the advance bridge arm and lag bridge arm phase shift Angle, thus control the output voltage value; In addition to provide some basic protection circuit, such as: short circuit protection, over voltage protection, current limiting protection circuit, to ensure the safe and reliable work of the converter.

UC3875 chip is the core of the control circuit, which is composed of reference power supply, oscillator, sawtooth generator, error amplifier, soft start, PWM comparator and trigger, output stage, over current protection, dead zone time setting, frequency setting and so on. The reference power supply provides a precision reference voltage source, as the voltage of the given signal is compared to the output voltage, and a resistor and capacitor at the frequency setting end (FREFSET) and the signal ground can set the switching frequency of the output stage. The oscillator oscillation frequency is thus set. The slope of the sawtooth wave is determined by the indirect resistance between the sawtooth wave slope setting end and the power supply to provide a constant current source for the sawtooth wave, and the indirect capacitance between the sawtooth wave pin and the signal ground. The output of 0UTA, OUTB, OUTC and OUTD is used to drive the four switching tubes of the full bridge converter. The dead time of output signals OUTA, OUTB and OUTC and OUTD can be determined by connecting resistance capacitors between the "DLY A/B" and "DLY C/D" terminals and the signal ground respectively [8]. The dead time provides the delay between the closing of one switch tube and the conduction of the other switch tube in the same branch. The setting of the two dead time can provide the respective delay for the two half Bridges to adapt to the difference in the charging current of the resonant capacitor. The SOFT START time is determined by the capacitance between the "SOFT – START" end and the signal ground. Therefore, the resonant switching time of each pair of output stages can be controlled separately. The advantages of phase shift control are fully reflected in the full bridge topology mode. The UC3875 operates in both voltage and current modes and has an over current shutdown for rapid protection from failure [9].

4 Software Design of Control System

In order to improve the response speed of the control system and the output characteristics of the power supply, the overall design structure of the combination of cycle and interrupt is adopted, the program modules of data sampling and processing subroutine, pulse signal generation subroutine, liquid crystal display and parameter setting subroutine, control policy subroutine, fault alarm and delay stop subroutine are designed. In the "AD" conversion subroutine to read the sample value, using power as the outer loop, the current as the inner loop double closed loop control strategy, with the peripheral frequency and voltage conversion and "PWM" pulse generator circuit, to achieve the accurate power supply output regulation [10].

5 Conclusions

Based on the analysis of the working principle of phase-shift control full bridge converter, this paper presents the realization method of new soft switching technology, and designs a practical wireless charging system for new energy vehicles. The integrated controller

UC3875 is adopted in the control circuit, which improves the reliability and efficiency of the wireless charging system. The research content of this paper has a strong application background, well solved the influence of the factors that will appear in the operation of the wireless power supply system of new energy vehicles on the system, and well realized the soft switching control.

Acknowledgements. This paper is used for the conclusion of the university-level scientific research project "Jilin Engineering Normal University" of <<Development and research of buried wireless charging pile for new energy vehicles>>. The project number is "XYBK202015".

References

1. Bhartia, P., Bahl, I.J.: Millimeter Waves Engineering and Applications, pp. 123–134. Wiley, New York (1984)
2. Symons, R.S.: Modern microwave power sources. IEEE AESS Syst. Mag. **17**(1), 19–26 (2002)
3. Tamyurek, B., Torrey, D.A.: A three-phase unity power factor single-stage AC–DC converter based on an interleaved flyback topology. IEEE Trans. Power Electron. **26**(1), 308–318 (2011)
4. Mark Smith, K., Lai, Z.: A new PWM controller with one cycle response. IEEE Tram Power Elect. (14), 142–150 (1999)
5. La, Z., Smedley, K., Ma, Y.: Time quality one cycle control for power factor correction. IEEE Tram Power Elect. **12**(2), 369–375 (1997)
6. Key, T.S., Lai, J.S.: Comparison of standard and power supply design options for limiting harmonic distortion in power systems. IEEE Trans. Ind. Appl. **29**(4), 688–695 (1993)
7. Lee, C.F., Mok, P.K.T.: A monolithic current-mode CMOS DC-DC converter with on-chip current-sensing technique. IEEE J. Solid-State Circuits **39**, 3–14 (2004)
8. Stratakos, A.J., Sanders, S.R., Brodersen, R.W.: A low-voltage CMOS DC-DC converter for a portable battery-operated system. Proc. IEEE Power Electron. Specialists Conf., 619–626 (1994)
9. Hurtuk, P., Radvan, R., Frivaldsky, M.: Full bridge converter with synchronous rectifiers for low output voltage application. In: 2011 International Conference Applied Electronics (AE), IEEE, pp. 1–4 (2011)
10. Rodriguez, J.R., et al.: Large current rectifiers: state of the art and future trend. IEEE Trans. Ind. Electron. **3**(52), 738–746 (2005)

High Dimensional Data Visualization Analysis Based on Unsupervised Laplacian Score

Hao Peng[1,2](✉), Jian Zhou[1], and Shenglan Liu[1]

[1] School of Computer Science and Technology, Dalian University of Technology, Dalian 116024, Liaoning, China

[2] School of Mathematics and Physics, Qingdao University of Science and Technology, Qingdao 266061, Shandong, China

Abstract. With the rapid development of big data technology and information visualization technology, the concept of data visualization is constantly evolving and developing. As one of the classic high-dimensional data visualization methods, the parallel coordinate axis has excellent plane geometric characteristics. However, as the amount of data increases and the dimension of the data feature increases, the number of polylines on the finite plane of the parallel coordinate graph also increases. The crossing and occlusion of lines lead to serious visual redundancy and clutter. This project uses the feature distribution and feature axis arrangement on the parallel axis as the research entry point, and uses two unsupervised feature selection methods (Laplacian Score and SVD-Entropy) to re-arrange the features on the PCP axis to improve parallelism. Phenomena such as data disorder and clutter on the coordinate axis. Furthermore, we proposed a plane geometry optimization CLS algorithm by combining two unsupervised feature selection algorithms and the PCP axis radius coverage calculation method. The proposed algorithm conforms to people's perception characteristics of information and plane space representation, and can help people more quickly analyze and understand data.

Keywords: High dimensional data visualization · Parallel coordinate axis · Laplacian score · SVD-entropy

1 Introduction

With the development of digital multimedia, computer networks, and information network media, the Internet has become an indispensable way for human life to obtain information [1, 2]. Among them, digital information visualization is the main expression vehicle for Internet information dissemination. High-dimensional data mining is a kind of data mining based on high dimensionality [3–5]. The main difference between it and traditional data mining is its high dimensionality. High-dimensional data mining has become the focus and difficulty of data mining [6]. With the advancement of technology, data collection has become easier and easier, leading to larger and larger databases and higher complexity, such as various types of trade transaction data, Web documents, gene

I. Ahmad et al. (Eds.): STSIoT 2021, LNDECT 122, pp. 113–118, 2023.
https://doi.org/10.1007/978-981-19-3632-6_15

expression data, document word frequency data, The dimensions (attributes) of user rating data, WEB usage data, and multimedia data can usually reach hundreds to thousands of dimensions, or even higher. It is worth noting that the dimensional characteristics of high-dimensional data have many influences on data visualization analysis [7]. First of all, due to the continuous increase of sample dimensions, a large number of traditional statistical methods have lost their effects. This is the "curse of dimensionality" that people often refer to. At the same time, people have noticed that when the number of samples is limited, the rapid increase of feature dimensions will lead to a sparse pattern of data distribution in space, and the correlation between multiple random variables that affect the sample will also increase [8]. Although technological progress has made it possible to obtain a large number of characteristics of observed individuals, including discrete, continuous and even dynamic data [9]. However, people have discovered in research that not all recorded data are related to research interests or provide effective information. The purpose of our data analysis is to extract useful information, so extracting the most effective data from the complex data set has become the focus of research [10]. The characteristic variables of the data are screened, and information useful to the research problem is selected from a large number of characteristic changes, and redundant variables and even noise variables are deleted, so that the processed data set tends to be low-dimensional and concise. Choose feasible methods to effectively reduce the dimension, so that the streamlined data characteristics better reflect the process of data generation, have better interpretability for actual problems, and can greatly reduce costs [11].

The existing parallel axis and RadViz dimensional sorting algorithms are mostly a black box process. The final dimensional order is directly given, and users are rarely involved in the algorithm process. This makes it difficult for users to understand the algorithm process and makes it difficult for users to recommend later. The order of dimensions is effectively adjusted to find a better order of dimensions. For example, Zhang et al. used a hierarchical clustering algorithm with variable parameters in consideration of RadViz characteristics to recommend the initial order of dimensions, and provided a dendrogram showing the results of the algorithm to guide users to interactively adjust, select, and delete dimensions, and perform features. Subset selection. Experiments show that the method in this paper has good interactivity, pays attention to user experience, and reduces the overlap problem of projection points in RadViz. Zhou et al. used parallel coordinates to present geospatial multi-dimensional attribute information, introduced mutual information to measure the correlation between geographic spatial clustering and attribute categories, dynamically determined the order of parallel coordinate axes, and further calculated the binding of data lines between the attribute axes and the map. Determine the location, optimize the layout of the data line, and reduce the disorder of the data line distribution between the map and the parallel coordinate system.

The data that needs to be processed in the fields of computer vision, multimedia analysis, etc. often have a very high dimensionality. The processing of high-dimensional data increases the time and space complexity of the operation, and also leads to the over-fitting phenomenon of the learning model. The view of manifold learning believes that due to the limitation of the internal characteristics of the data, some high-dimensional data will have dimensional redundancy. In fact, these data can be uniquely represented as long as they use a relatively low dimensionality. In addition, not all features are related to the learning task. The above two points show that it is necessary and possible to reduce the dimensionality of these high-dimensional data. Feature selection is a commonly used method of dimensionality reduction. It selects a group of features related to classification from a feature set through a certain algorithm, and uses the selected features for model learning. This method does not change the original representation of the data, and when the selected feature is determined, it only needs to simply extract the feature directly from the original feature set. Feature selection methods can be divided into supervised feature selection methods and unsupervised feature selection methods according to whether there is classification information in the training data. There is a large amount of unlabeled data in the real world, and the labeling of the data requires a high price, so the research on unsupervised feature selection methods has great practical significance. This project mainly uses unsupervised feature selection methods Laplacian Score and Singular Value Decomposition (SVD-Entropy) for experiments.

2 Algorithm Formulation

2.1 Calculation of Laplacian Score

The calculation of Laplacian Score is mainly based on Laplacian Eigenmaps and Locality Preserving Projection. To some extent, the Laplacian score of a feature can be regarded as the Rayleigh quotient of the related feature with respect to the Laplacian graph G. The Laplacian score of each feature is calculated by its local retention ability, which can be represented as follows:

Laplacian Score Algorithm	
Step 1:	Constructing similarity matrix: S: $S_{ij} = \begin{cases} \exp(-\frac{\|x_i - x_j\|^2}{t}) & \text{if } x_i \in KNN(x_j) \text{ or } x_j \in KNN(x_i) \\ 0 & \text{otherwise} \end{cases}$
Step 2:	Let $\tilde{f_r} = f_r - \frac{f_r^T D1}{1^T D1} 1$ with $D = diag(S1)$, $1 = [1, 1, \ldots, 1]^T$, $L = D - S$
Step 3:	Computing each LS of the corresponding features: $L_r = \frac{f_r^T L f_r}{f_r^T D f_r}$

2.2 Formulation of SVD-Entropy

Assuming that there is a matrix $M \times N$, an orthogonal basis $\{v_1, v_2, \ldots, v_n\}$, the orthogonal basis is mapped to: $\{Av_1, Av_2, \ldots Av_n\}$ using the mapping matrix A. If you want to make them orthogonal to each other:

$$Av_i \cdot Av_j = (Av_i)^T Av_j = v_i^T A^T Av_j = 0 \tag{1}$$

For singular values, it is similar to the eigenvalues in our eigendecomposition. The singular value matrix is also arranged from largest to smallest, and the singular value is reduced in special blocks. In many cases, the top 10% or even 1% of singularities the sum of values accounts for more than 99% of the sum of all singular values. In other words, we can approximate the matrix with the largest k singular values and the corresponding left and right singular value vectors:

$$A_{m\times n} = U_{m\times m} \sum_{m\times n} V_{m\times n}^T \approx U_{m\times k} \sum_{k\times k} V_{k\times n}^T \tag{2}$$

In summary, the PCP axis obtains feature reconstruction weights as follows:

$$\arg\min_{w_i} RE(w_i) = \left\| x_i - \sum_{j\in N(x_i)} w_{ij}x_j \right\|_2^2 + \xi \left\| \sum_{j\in N(x_i)} w_{ij}d_{ij} \right\|_2^2 \tag{3}$$

where w_{ij} represents the reconstruction weight of reconstruction x_j, and ξ is the regularization parameter.

3 Experimental Results and Analysis

Parallel coordinates are an important technology for information visualization. A significant advantage of parallel coordinates is that it has a good mathematical foundation, and its projective geometric interpretation and duality characteristics make it very suitable for visual data analysis. The parallel coordinate method uses coordinate axes that are parallel to each other. Each coordinate axis represents an attribute of the data. Each high-dimensional data is represented as a polyline connecting its data value points on each coordinate axis (dimension) to form a polyline. Each line segment reflects the value of high-dimensional data on two adjacent coordinate axes. For ease of expression, it is called a data line. Usually between a pair of adjacent coordinate axes, the data lines of all data are overlapped together, and the data correlation in the adjacent dimensions of the data set can be reflected from the overall distribution characteristics of the data lines. In order to overcome the problem that the traditional Cartesian rectangular coordinate system is easy to run out of space and difficult to express data above three dimensions, Parallel Coordinates uses a series of parallel coordinate axes to represent each variable of high-dimensional data, and the value of the variable corresponds to the position on the axis. In order to reflect the trend of change and the relationship between various variables, the points describing different variables are often connected into a broken line. The PCP axis based on LS and SVD can be re-arranged according to the feature score, as shown in the following figure (Fig. 1):

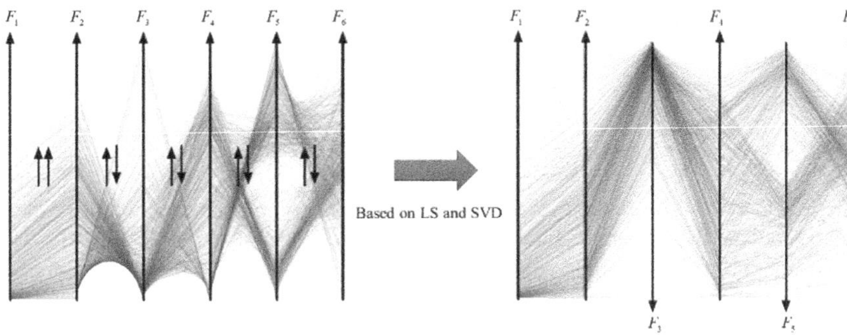

Fig. 1. PCP axis arrangement based on Laplacian score and SVD-entropy

In order to further verify the visualization performance of original PCP with different feature selection methods, we employed four evaluation metrics (Accuracy; Time; Satisfaction; Average R2), which can be seen in Table 1:

Table 1. Visualization comparison between PCP and its variants

Methods	Accuracy (%)	Time (s)	Satisfaction	Average R^2
PCP	84.22	63	4.0	0.327
PCP + LS	88.73	49	5.0	0.376
PCP + SVD	86.45	52	5.0	0.394
PCP + LS + SVD	91.33	37	6.0	0.409

As can be observed in Table 1, PCP + LS + SCD achieved the best performance (Accuracy: 91.33%; Time: 37 s; Satisfaction score: 6.0; Average R: 0.409). Furthermore, we note that the visualization performance with feature selection is more advanced at addressing high dimensional data in PCP, which is nicely demonstrated by the corresponding experimental results (PCP + LS; PCP + SVD; original PCP).

4 Conclusion

In today's data explosion era, the emergence of a large number of high-dimensional unlabeled data makes data processing face great challenges, and so unsupervised feature selection is very necessary.

In summary, this project uses the unsupervised feature selection method to carry out the PCP axis feature rearrangement, which has significant theoretical and practical significance and is worthy of further investigation. Relevant research not only helps to solve the specific problems of high-dimensional data visualization and reveals the basic representation rules, but also can be used as a reference for the further development of data visualization and data mining technology, and promote the cross integration between different disciplines. The planned research content of this project will closely focus on the Laplacian feature selection algorithm (Laplacian Score) and the support vector machine algorithm (SVD-Entropy).

References

1. Chen, X.J., et al.: Local adaptive projection framework for feature selection of labeled and unlabeled data. IEEE Trans. Neural Netw. Learn. Syst. **29**(12), 6362–6373 (2018)
2. Krishnapuram, B., Harternink, A.J., Carin, L., Figueiredo, M.A.T.: A Bayesian approach to joint feature selection and classifier design. IEEE Trans. Pattern Anal. Mach. Intell. **26**(9), 1105–1111 (2004)
3. Liu, H., Yu, L.: Toward integrating feature selection algorithms for classification and clustering. IEEE Trans. Knowl. Data Eng. **17**(4), 491–502 (2005)
4. Wolf, L., Shashua, A.: Feature selection for unsupervised and supervised inference: the emergence of sparsity in a weight based approach. J. Mach. Learn. Res. **6**, 1855–1887 (2005)
5. Liu, S.L., Feng, L., Qiao, H.: Scatter balance: An angle-based supervised dimensionality reduction. IEEE Trans. Neural Netw. Learn. Syst. **26**(2), 277–289 (2015)
6. Yang, D.-H., Li, X., Sun, X., Wan, J.: Detecting impact factor manipulation with data mining techniques. Scientometrics **109**(3), 1989–2005 (2016). https://doi.org/10.1007/s11192-016-2144-6
7. Zhang, J., Luo, Z.M., Li, C.D., Zhou, C.G., Li, S.Z.: Manifold regularized discriminative feature selection for multi-label learning. Pattern Recogn. **95**, 136–150 (2019)
8. Huang, R., Jiang, W.D., Sun, G.L.: Manifold-based constraint Laplacian score for multi-label feature selection. Pattern Recogn. Lett. **112**, 346–352 (2018)
9. Wang, D., Nie, F.P., Huang, H.: Feature selection via global redundancy minimization. IEEE Trans. Knowl. Data Eng. **27**(10), 2743–2755 (2015)
10. He, X.F., Niyogi, P.: Locality preserving projections. Adv. Neural Inform. Process. Syst. **16** (2003)
11. Hoffman, M., Steinley, D., Brusco, M.J.: A note on using the adjusted rand index for link prediction in networks. Soc. Netw. **42**, 72–79 (2015)

Cloud-Edge Cooperation Data Acquisition and Processing Method of Multi-energy Systems

Honggang Wang, Xin Ji(✉), Tongxin Wu, Jianfang Li, Yude He, Chengyue Yang, and Haifeng Zhang

Big Data Center of State Grid Corporation of China, Xicheng District, Beijing, China
xin-ji@sgcc.com.cn

Abstract. There are tens of thousands of power equipment and network equipment in the power grid system, including a large-scale heterogeneous network composed of intelligent terminals, sensors, databases and so on. These heterogeneous networks often belong to different business systems with a different logic and will produce a considerable amount of data at the edge of the network all the time. By studying the big data monitoring and analysis technology for enterprise operation decision-making, we can explore various values in heterogeneous data of power grid. For different types of data, the value in historical data is analyzed through algorithms such as data analysis, data mining and machine learning, and the law and value in real-time data are found through flow calculation.

Keywords: Cloud-edge cooperation · Data acquisition · Energy management

1 Introduction

To further promote green, low-carbon and sustainable development of energy, comprehensive energy services have developed rapidly in the world in recent years, which has triggered profound changes in the energy system and become the focus of new strategic competition and cooperation among countries and enterprises. At present, the power industry has three main characteristics and needs to speed up the service transformation: first, more and more new energy needs unified access to the network such as photovoltaic, wind power and hydropower; second, more and more new businesses adapt to the emergence of various business forms such as energy sales services, information services, market transactions and intelligent power grid value-added services, The third is the increasingly complex power grid dispatching and maintenance, which requires the use of ICT technologies, such as the Internet of things and big data to build an information management and control platform, realize the deep integration of power flow, information flow and business flow, ensure the supply and demand balance of multi-energy and support the transformation of comprehensive energy services. At present, with the development of integrated energy service business, there are unprecedented challenges to energy perception from breadth to depth, as illustrated in Fig. 1.

I. Ahmad et al. (Eds.): STSIoT 2021, LNDECT 122, pp. 119–127, 2023.
https://doi.org/10.1007/978-981-19-3632-6_16

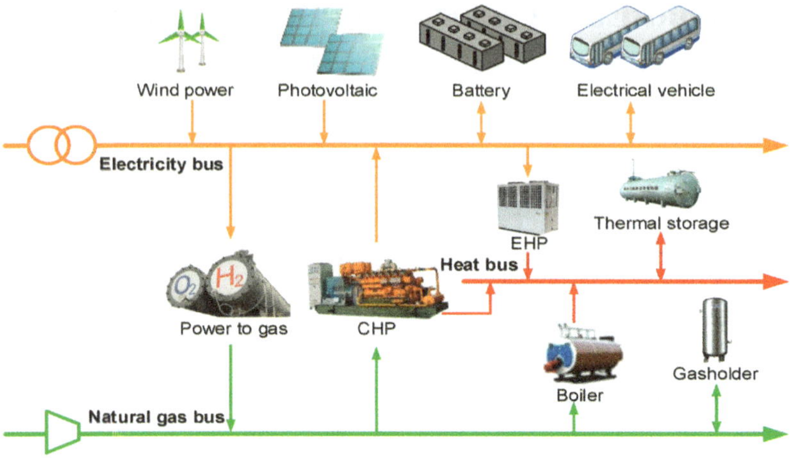

Fig. 1. A typical multi-energy system

At present, energy perception equipment still faces the following problems:

- There are barriers to energy collection. Today's energy collection is characterized by single energy collection, island storage, and chimney data upload. There are barriers to energy collection.
- The existing terminal technology is backward. The existing terminal is bulky, difficult to install, expensive, and single communication mode, which cannot meet the needs of large-scale distribution.
- The networking operation and maintenance costs are high, the communication technology is single, the communication distribution is dense, the equipment and networking cost is high, and the business upgrading needs to replace the hardware or upgrade to the site for debugging, which costs a lot of operation and maintenance.
- The operation management level is low, and the energy management is only supervised without control or manual control, which cannot realize the automatic diagnosis and optimal regulation of energy system utilization strategy, resulting in great waste.
- Low intelligence, unable to make independent decisions, need to upload a large amount of data, but great pressure on communication bandwidth and cloud platform computing, and cannot generate real-time control strategy according to real-time working conditions.
- The security is low. Due to the risk of insufficient security protection of existing equipment, external personnel can tamper with data and control equipment to threaten the security of the power grid.

Therefore, it is urgent to carry out technical research on energy consumption data acquisition and monitoring in the context of comprehensive energy, build data acquisition and monitoring system with comprehensive state perception and processing.

2 Related Work

2.1 Data Acquisition

The work in [1] introduced the structure, acquisition principle, and networking mode of the data acquisition system as a whole, and designs and implements a data acquisition terminal, including detailed hardware design, underlying software design, and application program design of the data acquisition system, The feasibility of the software and hardware design of the data acquisition system is verified by the indoor environment data acquisition system. The study in [2] described the design and implementation of the pulse acquisition circuit of acquisition terminal, as well as the design of each software module of the acquisition terminal. This acquisition terminal can not only collect pulse watt-hour meter but also collect watt-hour meter with RS485 interface. The work in [3] provided a new design method of terminal unified management interface with good flexibility and expansibility, which provides bottom communication support for heterogeneous terminal unified management systems in next-generation networks. For the applicability of the wireless communication module, the authors in [4] designed and developed a LoRa wireless communication module based on the relevant specifications of the power consumption information acquisition system, and uses the actual watt-hour meter and watt-hour meter protocol test software to carry out the actual meter reading test. The experimental results show that the technical indexes and meter reading of the module meets the design requirements, it can be applied in the future power wireless meter reading. The work in [5] analyzes the access mode and security problems of mobile terminals of power grid enterprises, puts forward the security protection countermeasures of mobile terminals of power grid enterprises, and puts forward corresponding feasible schemes for power grid enterprises to widely carry out mobile terminal business and application and improve the security of information intranet of power grid enterprises.

2.2 Multi-energy Data Acquisition and Processing

The studies in [6] have realized enterprise energy consumption data collection through Internet of things technology, global positioning system and wireless sensor network technology. However, there are few types of monitored energy, so the collected energy information can only be subject to simple statistical analysis, rather than deeper data mining. The work in [7] applied the Internet of things technology to the real-time monitoring of energy conservation and emission reduction in industrial parks and describes the real-time monitoring system of energy conservation and emission reduction in detail. However, the monitoring of the system can only be aimed at the location where sensors can be fixed, and the processing and analysis of the collected energy data are not sufficient, so it does not have the ability of energy conservation management. In [8], the Internet of things technology combines the upper computer monitoring platform and network and uses the fault diagnosis system and various monitoring equipment running on-site to obtain the operation data and operation status of high-frequency power supply in real-time from a long distance. However, the system designed in this document adopts C/S architecture.

2.3 Internet of Things Framework

The work in [9] divided the Internet of things into three parts from top to bottom: perception layer, network transmission layer and application layer. The sensing layer of the Internet of things is at the bottom of the Internet of things architecture, which is composed of various sensors, control modules, networking communication modules and intelligent gateways for accessing sensors, aggregating data, and connecting the sensing layer and the network layer. The network layer is composed of the Internet, personal local area network, mobile communication network, network transmission system and data resource integration and open business platform, which provides open network and data resource integration and opening services for the Internet of things. The application layer is the user-oriented "interface" of the Internet of things. Combined with the specific industry needs, it constructs the Internet of things application environment monitoring, natural disaster early warning, smart home and intelligent transportation for all walks of life-based on the perception layer and network layer. In [10], the work described the development process of Lora, introduces three-terminal devices of Lora modulation and LoRa-WAN, finally analyzes the key technologies of Lora, describes in detail and lists some specific examples of Lora application. The study in [6] analyzed and summarized the current RS485 communication interface design scheme of intelligent electric energy meter, and puts forward suggestions and solutions for the RS485 interface design from the aspects of RS485 chip itself, data receiving sensitivity, carrying capacity, communication reliability, and so on. The study in [11] designed a wireless gateway based on ZigBee and WiFi to realize ZigBee data interaction of intelligent terminal in home environment and remote. The authors in [12] introduced the LoRa technology and compared different wireless communication technologies. The work in [13] discussed the NB-IoT enhancement technology under the Internet of things coverage, analyzes the application of Nb-IoT Internet of things coverage enhancement technology, puts forward a method to evaluate NB-IoT technology based on the coverage enhancement technology, and focuses on the coverage enhancement of repeated transmission. The enhancement technologies in the current 3GPP proposal are simulated and compared.

3 Cloud-Edge Cooperation Data Acquisition and Processing

3.1 System Architecture Design

The network system of multi-function acquisition on the user side is shown in Fig. 2.

The network architecture is a hierarchical network structure, including sensor node, gateway node (intelligent acquisition terminal), the local base station (energy controller), and transmission network and finally connected to the cloud service platform. To obtain data more accurately, the distribution of sensor nodes is usually very dense, which may be distributed in different monitoring areas, which constitutes multiple sensor networks. In the practical application of this sensor network system structure, the data is measured by the sensor node, and then transmitted to the network management node (intelligent acquisition terminal), and then the acquisition terminal transmits the collected data to the edge controller.

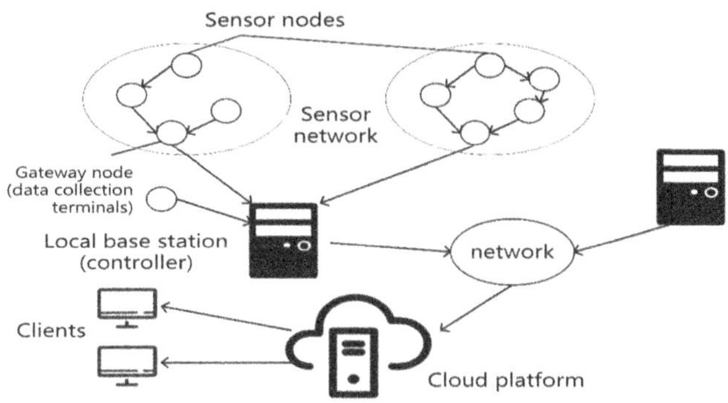

Fig. 2. Cloud-edge cooperative data acquisition and processing framework

Sensor nodes can carry out data calculation and data communication. They can realize data fusion in sensor networks, reduce data traffic and alleviate the forwarding burden of sensor nodes. The gateway node is mainly responsible for collecting the data transmitted from the sensor node. All gateway nodes will connect with the transmission network and transmit the collected data to the superior node. Sensor network includes sensor nodes and gateway nodes. The whole sensor network plays the role of data preprocessing and data uploading.

The base station node is a controller connected to the cloud node with certain computing, storage, analysis and decision-making capabilities. It is mainly responsible for collecting the data uploaded from the gateway node and sending the data to the cloud data processing center through the transmission network. At the same time, it is also a local database, which can cache sensor data in the local database. The base station node can process and analyze local data and realize local decision-making.

3.2 System Functionalities and Specifications

(1) Energy consumption acquisition

Using the existing intelligent electric energy meter, gas meter, water meter and other collection technologies and the collection terminal in the energy efficiency monitoring subsystem of industrial, commercial and residential users, collect the user's energy consumption data and the data of multiple accounting quantities of the intelligent collection terminal by installing the collection module in the user or replacing the intelligent collection terminals.

(2) Multi-energy data analysis and processing

On the one hand, the collected energy consumption data should be intuitively displayed to users to let users understand their energy consumption level; on the other hand, it should be used as the data basis for evaluating users' energy-saving levels. The data collected by the intelligent acquisition terminal is transmitted to the energy controller after multi-energy data aggregation for centralized storage and local processing. When the intelligent acquisition terminal and the energy controller are in the local LAN, it is considered to transmit the sensing layer data to the energy controller through RS-485, WiFi and LoRa. Data analysis is the core of energy data acquisition and monitoring systems. To process, store and display energy efficiency data, analyze whether users' energy consumption behavior is reasonable and feasible through data evaluation methods. Further, the user energy consumption mode can be regulated according to the user energy consumption data, and the optimal regulation strategy can be automatically obtained according to the user energy consumption mode. The manager can evaluate the information of the energy efficiency management scheme provided by the user. Among them, it is necessary to formulate a comprehensive and reasonable evaluation system for user energy consumption. Machine learning techniques can be adopted in the data process, e.g., Random Forest based algorithms, as illustrated in Fig. 3.

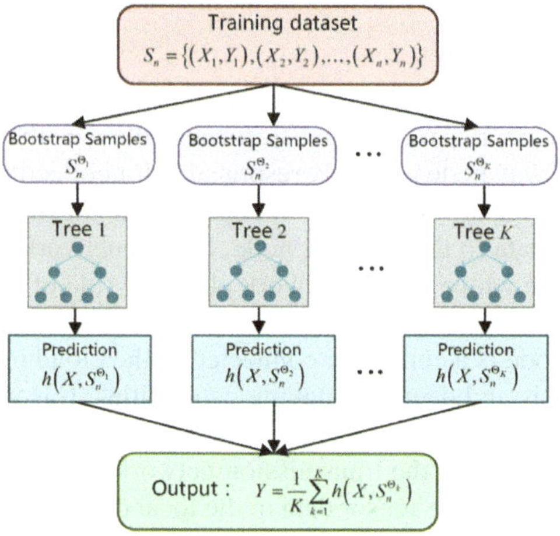

Fig. 3. Basic structure of Random Forest method

(3) Equipment monitoring

At the system terminal, the energy consumption status of each device inside the user is visually presented to the user through the software management system. The user can select to add, change and delete internal energy consumption devices in the user interface of the system according to their own needs. It can not only realize the operation through the web but also control the energy consumption devices through control devices such as terminal software, such as setting timing startup Stop and control the equipment remotely through the terminal. You can view the working conditions of the energy controller, as well as the energy consumption and data analysis results of various energy-consuming equipment in real-time, to obtain the local optimal control strategy. Also, the edge computing devices can be adopted, as shown in Fig. 4 and Fig. 5.

(4) System function realization process

According to the simple description of the system hierarchy and main functions, the functions to be realized by the system mainly include completing the real-time

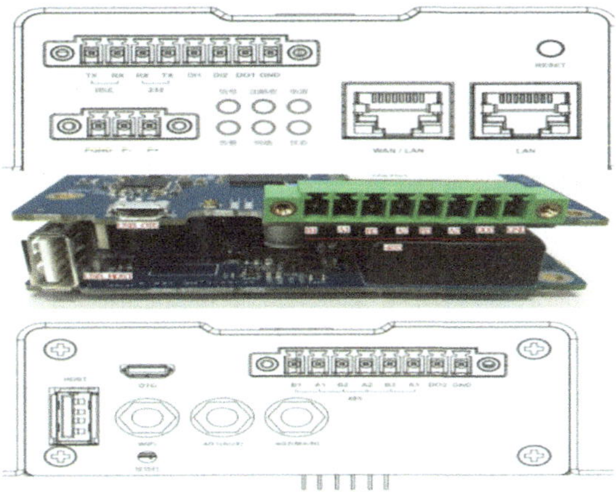

Fig. 4. A basic design of the edge device

Fig. 5. Installation of the monitoring devices

acquisition, processing and analysis of user energy consumption data and completing the local energy efficiency analysis and optimization regulation. For example, the energy consumption of a typical data center can be described as follows:

The components of the server include CPU, memory, hard disk, network, etc. the required components are configured reasonably according to the needs of the service. The existing nonlinear relationship model is given in (1).

$$P_{used} = P_{idle} + \left(P_{busy} - P_{idle}\right) \times u \tag{1}$$

where P_{used}, P_{idle}, P_{busy} and u represent the actual power, idle power, full operation power and utilization rate.

The above can be further simplified as follows:

$$P_{used} = \beta \cdot P_{busy} + (1 - \beta) \cdot P_{busy} \cdot u \tag{2}$$

Further the energy consumption of data center can be described as

$$E = \int_{t_1}^{t_2} P_{used}(t)dt \tag{3}$$

The idle operational energy consumption is

$$DE_j = \int_0^{T_j^{dormancy}} P_j^{dormancy} dt \tag{4}$$

where $P_j^{dormancy}$ and $T_j^{dormancy}$ are the idle energy and idle time.

4 Conclusions and Remarks

In this paper, through the research on the key technologies of data acquisition and monitoring of energy consumption control system, master the technical architecture of energy consumption data acquisition and monitoring in multiple scenarios such as commercial buildings, industrial enterprises and park services, the dynamic perception and online analysis technology of energy consumption information based on Internet of things and edge computing, and the application method of edge computing gateway in power Internet of things, to realize business sinking Reduce the underlying data processing from the cloud to the ground, solve the problem of low resource utilization efficiency and time processing efficiency under a single cloud computing model, and provide support for building a secure, intelligent, professional and integrated Internet of things management platform.

Acknowledgments. This work is supported by the technical project of the state grid corporation of China "Research on power grid big data monitoring and analysis technology coordinated by data center and edge computing" (5700-202055183A-0-0-00).

References

1. Arnaudov, R.I., Dochev, I.N.: Functional generator and data acquisition system controlled by internet. In: 2005 IEEE Intelligent Data Acquisition and Advanced Computing Systems: Technology and Applications, pp. 276–278 (2005)
2. Weng, Y., Wang, X., Qelger, Ma, G.: Data acquisition terminal application design and development based on the Android platform. In: 2012 Fifth International Conference on Intelligent Networks and Intelligent Systems, pp. 257–259 (2012)
3. Zirui, G.: Design and implementation of financial information management system for mobile terminal. In: 2020 5th International Conference on Smart Grid and Electrical Automation (ICSGEA), pp. 354–357 (2020)
4. Ke, K., Liang, Q., Zeng, G., Lin, J., Lee, H.: Demo abstract: a LoRa wireless mesh networking module for campus-scale monitoring. In: 2017 16th ACM/IEEE International Conference on Information Processing in Sensor Networks (IPSN), pp. 259–260 (2017)

5. Hu, H., et al.: The development and application of dynamometer card measurement and analysis system based on Android platform. In: 2016 8th International Conference on Computational Intelligence and Communication Networks (CICN), pp. 150–154 (2016)
6. Li, G., et al.: Energy efficient data collection in large-scale internet of things via computation offloading. IEEE Internet Things J. **6**(3), 4176–4187 (2019)
7. Che Soh, Z.H., Hamzah, I.H., Che Abdullah, S.A., Shafie, M.A., Sulaiman, S.N., Daud, K.: Energy consumption monitoring and alert system via IoT. In: 2019 7th International Conference on Future Internet of Things and Cloud (FiCloud), pp. 265–269 (2019)
8. Ortega, M.G.S., Rodriguez, L., Gutierrez-Garcia, J.O.: Energy-aware data collection from the internet of things for building emotional profiles. In: 2018 Third International Conference on Fog and Mobile Edge Computing (FMEC), pp. 234–239 (2018)
9. Lin, J., et al.: A survey on internet of things: architecture enabling technologies security and privacy and applications. IEEE Internet Things J. **4**(5), 1125–1142 (2017)
10. Wu, F., Miao, Z., He, C.: Remote monitoring system for intelligent slaughter production line based on internet of things and cloud platform. In: 2020 11th International Conference on Prognostics and System Health Management (PHM-2020 Jinan), pp. 538–542 (2020)
11. Pan, G., He, J., Wu, Q., Fang, R., Cao, J., Liao, D.: Automatic stabilization of Zigbee network. In: 2018 International Conference on Artificial Intelligence and Big Data (ICAIBD), pp. 224–227 (2018)
12. Edward, P., El-Aasser, M., Ashour, M., Elshabrawy, T.: Interleaved chirp spreading LoRa as a parallel network to enhance LoRa capacity. IEEE Internet of Things J. **8**(5), 3864–3874 (2021)
13. Chung, H., Lee, S., Jeong, J.: NB-IoT optimization on paging MCS and coverage level. In: 2018 15th International Symposium on Wireless Communication Systems (ISWCS), pp. 1–5 (2018)

Dynamic Equivalence of Power System Based on Artificial Immune Algorithm

Ziliang Qiu(✉)

Shenzhen Power Supply Limited Company, Shenzhen 51800, China
szpsb886@126.com

Abstract. The electrical industry is an important part of the energy industry and an important foundation of the national economy. Therefore, its safe power operation plays a very important role in social production and life. However, the current power system faces challenges in the scale and complexity of simulation calculations, so it is very important to study the dynamic equivalence of power systems. Based on the artificial immune algorithm, this paper studies and designs a power system dynamic equivalent system. First, this article explains the concept of artificial immunity, and discusses the application of artificial immune algorithms in depth; then, a system framework for power system dynamic equivalence is designed, and its dynamic equivalence changes are verified. The final results show that the model based on the equivalent of artificial immune algorithm can reflect the dynamic power characteristics of the external system well, and it can also be suitable for the study of static characteristics.

Keywords: Artificial immunity · Immune algorithm · Power system · Dynamic equivalent

1 Introduction

With the development of social economy, the power industry is playing an increasingly important role in the national economy, and society's demand for total power is also increasing. However, the complexity and instability of the power system make the development of the power industry face many challenges. In-depth study of the dynamic equivalence of the power system has become an inevitable need to improve the reliability and quality of power supply in the power grid [1, 2].

In recent years, many experts have conducted extensive research on artificial immune algorithms, and there are many research results based on dynamic equivalence of power systems. Today's experts divide power dynamic equivalence methods into two categories: one is the equivalence method based on the same concept. By adding the disturbance of the simplified linearized system model and calculating the rotor oscillation curve, it can identify the same group of machines and save a lot of time, thereby realizing dynamic equivalence on a small computer. In fact, the equivalence at this time is recognized in the case of the linearized system model. The other is an equivalent method based on online measurement and parameter artificial immune algorithm. Many scholars believe that the

I. Ahmad et al. (Eds.): STSIoT 2021, LNDECT 122, pp. 128–136, 2023.
https://doi.org/10.1007/978-981-19-3632-6_17

internal interference of the power dynamic system has little influence on the outside, so the linearization of the external system can be realized. This is not only conducive to the large-scale design of the controller of the electrical dynamic system, but also conducive to the simulation and analysis of the system [3, 4].

With the rapid economic development, the power system is also rapidly developing into large-scale multi-machine power grids, AC/DC hybrid power transmission and large-scale regional power grids. Therefore, the dynamic equivalence of the power grid still has defects such as the dimension and complexity of simulation calculation and accuracy. Therefore, it is very necessary and urgent to optimize the system [5, 6].

2 Overview of Power System Dynamic Equivalence Based on Artificial Immune Algorithm

2.1 The Basic Principles and Characteristics of Artificial Immunity

The basic principle of the artificial immune system is derived from the immune system of the higher spine organisms on the earth. The components of the immune system are composed of three major parts, namely immune organs, immune tissues and lymphocytes. They can prevent organisms from invading pathogens and harmful other factors. Among them, lymphocytes play a very important role in the immune system. The artificial immune system has three characteristics: immune stability, immune safety, and immune pertinence. This means that under normal circumstances, immune cells can protect the body from virus infection, and can maintain aging and dead cells in the body, keeping the body in a relatively stable environment. The artificial immune system has a very complex physiological structure and function of the immune system, which can recognize foreign components and mutated endogenous cells entering the human body, respond accordingly and effectively eliminate them [7, 8].

2.2 Development of Artificial Immunity

The biological immune system is the origin of the artificial immune system. Influenced by the biological immune system, human beings have developed and designed learning systems implemented by computers, including artificial neural networks and genetic algorithms from the brain's nervous system. The biological immune system is a large and complex biological system. It is its job to distinguish harmful antigens from the outside world and its own tissue area and eliminate antigens to keep the body stable. The biological immune system has very strong learning and memory and feature extraction capabilities, so the artificial immune system draws on the advantages of the biological immune system, and at the same time has the functions of highly parallel, distributed, self-adaptive and self-organizing. The artificial immune system is still a brand-new branch in the field of computer intelligence in human research. The application of artificial immune system at this stage is slowly expanding to information security, data mining, etc. Artificial immune system is a research object with this huge potential, so it has attracted the attention of domestic and foreign universities, research institutes and their industries, and is a research hotspot in many international journals [9, 10].

2.3 The Purpose of Power System Dynamic Equivalence

For a large interconnected network with hundreds of generators and thousands of lines, it will be very difficult to analyze and calculate using a complete system model. It not only consumes a lot of manpower and material resources, but also consumes a lot of time, especially uncontrollable for online analysis and control. On the other hand, the continuous deepening of market reforms makes it necessary to understand the parameters of each component of the system and the operating state of electricity. With the gradual improvement of operational independence in different regions, trade secret information will inevitably show a certain degree of opacity between them, and data acquisition between different subsystems will become more and more difficult. In addition, the scope of today's subsystems very extensive. There are many variables, and it is very complicated to analyze hundreds or thousands of curves, and it is difficult to obtain effective results. Dynamic equivalence is a process in which no processing is performed in the search system, but only the equivalent simplification of the external system. The premise of equivalence is that the external system has the same or similar impact on the research system before and after the equivalence. The dynamic equivalence research of power system is a very practical research topic [11, 12].

2.4 Function of Artificial Immune Algorithm

From the perspective of enlightenment, artificial immune algorithms can be roughly divided into three categories: based on immune network mechanism, based on clonal selection mechanism and based on negative selection mechanism. Including general immune algorithm, Clone Selection Algorithm, Negative Selection Algorithm, artificial immune network algorithm. The general process of the immune algorithm to solve the problem is as follows:

(1) Determine the field of application (metaphor)
(2) Determine the immune entity

    1) Notation
    2) Affinity measurement
    3) Application of immune mechanism
    4) Algorithm loop

(3) Obtain a problem solution (one or more antibodies)

The abstract morphological space model of artificial immune theory is considered from a mathematical point of view. The morphology of any antibody or antigen m in the morphological space can be expressed as an attribute string of length L. Therefore, in the L-dimensional morphological space, the attribute string m = (m1, m2, ..., mL) can be represented by a point, where SL $\in$ m. The attribute string can be composed of any type of attributes, such as real numbers, integers, characters, etc. In a responsive system, if the cross-reaction threshold ε reaches the lowest value of the response, it can be identified. Where ε is equal to the recognition threshold, affinity threshold or activation threshold.

In the morphological space, the affinity between the antigen and the antibody is related to the distance D between the two, and can be calculated with any distance between the two strings. The smaller the distance, the greater the affinity, and at the same time, the better the match between the two.

The calculation formulas of Euclidean, Manhattan distance and Hamming distance are as follows:

$$D = \sqrt{\sum\nolimits_{i=2} (xy_i - ab_i)^2} \tag{1}$$

$$D = \sqrt{\sum\nolimits_{i=2} |\mathrm{xy}_i - ab_i|} \tag{2}$$

$$D = \sum\nolimits_{i=2} \partial, \quad \partial = \begin{Bmatrix} \partial = 1, \ \mathrm{xy}_i \neq ab_i \\ \partial = 0, \ or \end{Bmatrix} \tag{3}$$

Among them (<xy1, xy2, …, xyi> represents antibody coordinates, <ab1, ab2, …, abi> represents antigen coordinates).

3 Power System Dynamic Equivalence Experiment Based on Artificial Immune Algorithm

3.1 Dynamic Equivalent Modeling of Power System

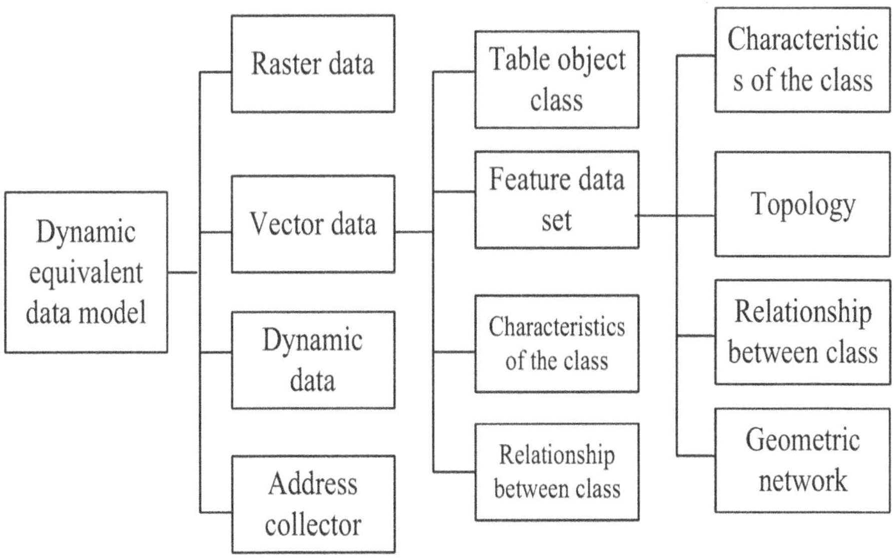

Fig. 1. Dynamic equivalent modeling of the electric power system

It can be seen from Fig. 1 that the dynamic equivalent modeling of the power system is composed of four major parts, namely raster data, vector data, dynamic data and address collectors. These can collect the power value of the power system in each circuit in real time to ensure the safety of system operation and the safety of component operation.

3.2 The Flow of the Dynamic Equivalence Calculation Example

The magnetic winding voltage of the generator is controlled by the generator's dynamic equivalence. The dynamic equivalence is also an important dynamic element of the power system. Therefore, a generator dynamic equivalence model should be established. Dynamic equivalence provides dynamic equivalence power to generators, plays the role of regulating voltage, keeping terminal voltage or pivot voltage constant, and controlling the dynamic equivalence distribution of generators.

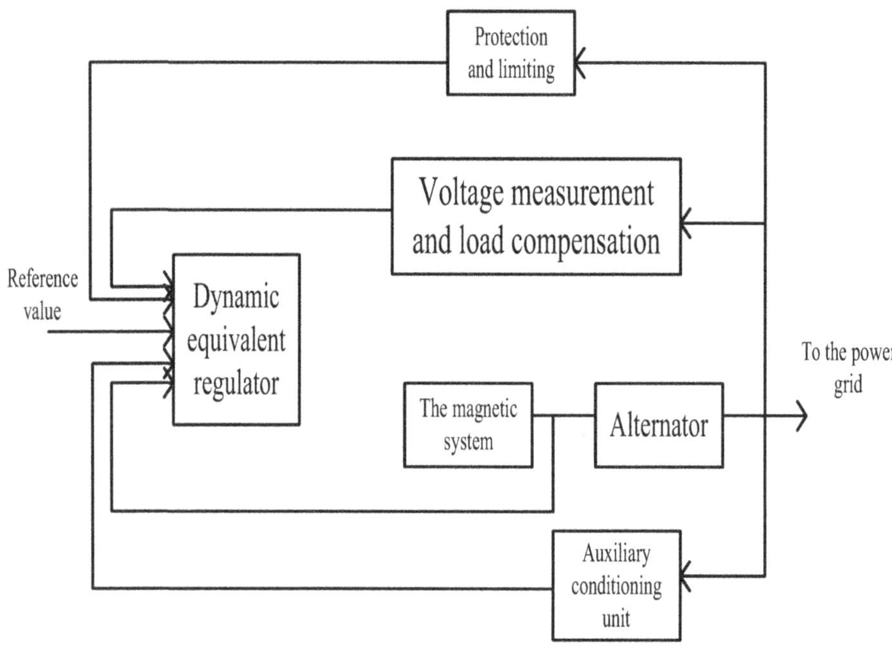

Fig. 2. Dynamic equivalent frame diagram

Under normal circumstances, the general composition of dynamic equivalence is shown in Fig. 2. The dynamic substitute value is the dynamic substitute group of the generator to provide the dynamic substitute current; the dynamic backup controller is used to set or control the dynamic backup power; the generator terminal voltage measurement and load balance connection measure the generator terminal voltage to compensate the generator load current. The auxiliary controller sends an auxiliary control signal to the dynamic standby controller. The most commonly used auxiliary controller is the power system stabilizer. Protection and limit connections ensure that the various operating parameters of the equipment do not exceed their limits.

4 Experimental Analysis of Power System Dynamic Equivalence Based on Artificial Immune Algorithm

Validation of the model obtained by dynamic equivalence is an important step in the practical application of the equivalence model. Only after verification can the model have practical significance. The current load model verification is mainly divided into two categories: one is identification and adaptive verification, that is, the measured or simulated dynamics of the external system and the dynamics calculated by the alternative model; the other is the system area, and the verification includes the verification of the external system. The actual measured or simulated dynamic behavior is compared with the dynamic behavior of the system after being replaced by an equivalent model. Generally, dynamic equivalence uses identification and adaptation testing, because it is relatively simple and easy to compare the dynamic response of the external system with the response data of an equivalent model based on measured or simulated data, and this kind of system-wide review is relatively complicated. Many factors as the dynamic response of the system include many factors related to the dynamic component model of the system. In the equivalence method based on the artificial immune algorithm equivalence method, the identification and adaptive data are first verified, and then the whole system verification is performed, including the system static characteristic verification and the dynamic characteristic verification.

4.1 Static Characteristics

Table 1. Inspection of the static properties

Hybrid	Traditional algorithm results		Results of the artificial immune algorithm		Deviation of the voltage amplitude and the phase angle (%)	
	VRMs(kV)	θ(deg)	VRMs(kV)	θ (deg)	AV(%) 0 (deg)	
1	224.1	−13.6	218.8	−14.48	2.365	−0.884
2	222.1	−15.1	215.56	−16.06	2.9446	−0.961
3	223.7	−14.7	216.46	−15.51	3.2365	−0.809
4	224.8	−12.5	220.07	−13.16	2.1041	−0.655
5	227.3	−10.9	222.96	−11.7	1.9094	−0.801
6	227.8	−10.4	223.57	−11.08	1.8569	−0.675
7	227.7	−10.3	225.46	−10.72	0.9838	−0.419
8	228.2	−10.1	225.55	−10.49	1.1613	−0.388
9	229.3	−8.8	225.38	−9.142	1.7096	−0.342
10	227.9	−9.5	223.31	−10.09	2.014	−0.594

Checking the static characteristics of the system mainly includes checking the power flow of the system. After the dynamic replacement model replaces the external network, the dynamic value of the grid node is simulated and compared with the calculation result of the traditional algorithm. For example, Table 1 lists the calculation results of the artificial immune algorithm for the main node voltage of the entire network under the equivalent model, and compares the calculation results with the calculation results of the algorithm. The table lists the artificial immune algorithm calculation results of the effective power transmission of the main branches of the entire network under the replacement model, and compares the calculation results with the traditional algorithm before the replacement model. When using the dynamic equivalent model, the maximum voltage deviation is 3.2484%, and the maximum phase angle deviation is 0.961°. It can be seen that the main node voltage amplitude and phase angle calculated by the artificial immune algorithm under the equivalent model are close to the calculation results of the traditional algorithm before the equivalent value. The transmission power of 500 kV line and 500 kV–220 kV line calculated by the artificial immune algorithm based on the equivalent model is basically the same as the calculation result of the traditional algorithm before the equivalent value. Therefore, the equivalent model based on the artificial immune algorithm proposed in this paper can well reflect the dynamic performance characteristics of the external system, and may also be suitable for the study of static characteristics.

4.2 Dynamic Characteristics

When the power system is dynamically equivalent, the dynamic characteristics of the system must not be distorted. Therefore, the dynamic characteristics of the system are verified, and the dynamic characteristics of the power angle of the system will be verified. When a three-phase short-circuit fault occurs at the fifth node in the system, it is a large disturbance of the system, which has a good impact on the dynamic characteristics of the entire power grid. In particular, the node is relatively close to the node where the equivalent model is located, so it can be calibrated well. The dynamic characteristics of the system after verification and equivalence are shown in Fig. 3.

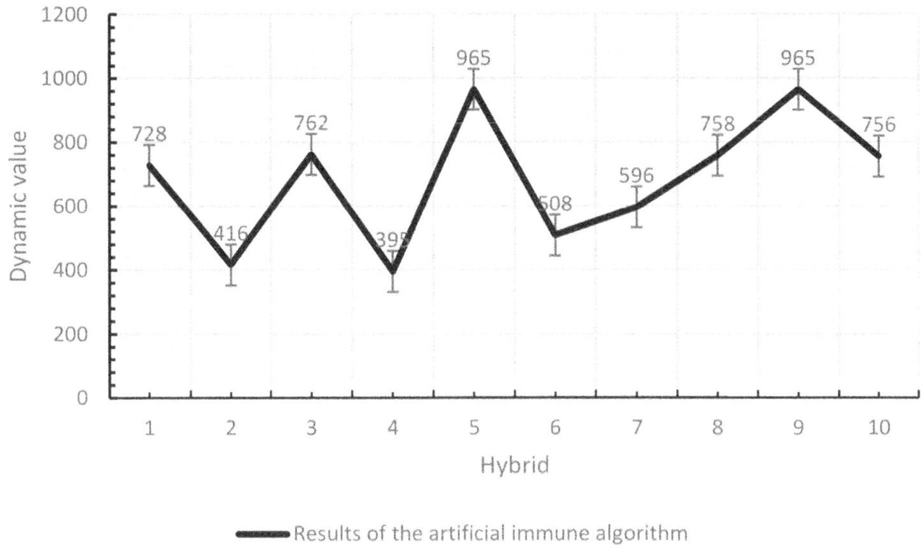

Fig. 3. Dynamic equivalence of the artificial immune algorithm

5 Conclusion

In power systems, there are many parameters and variables which contain many metrics. More accurate analysis of these factors has an important effect on the whole control process. This paper uses artificial immunity algorithm to study the problems and optimization effect under the dynamic equivalent evaluation system of electric power system. At the same time, simulation experiments also further proved that this intelligent method has certain practical significance and value for improving the operation economic efficiency and improving the comprehensive competitiveness.

References

1. Liao, J.: Research on PAGV path planning based on artificial immune ant colony fusion algorithm. J. Intell. Fuzzy Syst. **35**(16), 1–6 (2018)
2. Zhao, N., Tsai, S.B.: Research on prediction model of hotels' development scale based on bp artificial neural network algorithm. Math. Probl. Eng. **2021**(1), 1–12 (2021)
3. Chen, Y., Lin, M., Yu, R., et al.: Research on simulation and state prediction of nuclear power system based on LSTM neural network. Sci. Technol. Nuclear Installations **2021**, 1–11 (2021)
4. Ying, W., Gang, M., Yi, T.: Research on dynamic thevenin equivalence method based on deviation correction. Am. J. Electr. Electron. Eng. **5**(6), 195–201 (2017)
5. Hassen, H.B., Tounsi, J., Bachouch, R.B.: An artificial immune algorithm for HHC planning based on multi-agent system. Procedia Comput. Sci. **2019**(164), 251–256 (2019)
6. Tian, S.: Dynamic equivalence of electric power system based on RTDS. Smart Grid **07**(6), 498–506 (2017)
7. Chen, J., Chen, S., Ma, C., et al.: Fault detection of aircraft control system based on negative selection algorithm. Int. J. Aerosp. Eng. **2020**(2), 1–10 (2020)

8. Wu, H., He, Y., Zhao, B., et al.: Research on dynamic equivalent of wind farm based on improved k-means clustering algorithm. Taiyangneng Xuebao/Acta Energiae Solaris Sinica **39**(11), 3232–3238 (2018)
9. Chen, H.: Research on innovation and entrepreneurship based on artificial intelligence system and neural network algorithm. J. Intell. Fuzzy Syst. **40**(2), 2517–2528 (2021)
10. Gao, Y.-F., Wang, J.-Q., Xiao, T.-N., Jiang, D.-Z.: Fast emergency control strategy calculation based on dynamic equivalence and integral sensitivity. Front. Inf. Technol. Electron. Eng. **20**(8), 1119–1132 (2019). https://doi.org/10.1631/FITEE.1700389
11. Yan, T., Liu, G., Xiao, B., et al.: Voltage stability analysis of power system with DGs based on dynamic equivalent circuit. Dianli Xitong Baohu Yu Kongzhi/Power Syst. Protect. Control **45**(6), 74–79 (2017)
12. Feng, W., Wei, L.: Dynamic equivalence analysis of distribution network integrated with high penetration of distributed photovoltaic generation system. Power Syst. Autom. **041**(009), 65–70, 181 (2017)

Analysis of Value Mining of Engineering Digital Information Based on BIM-DATA

Weijun Wang and Xiaoliang Li(✉)

Department of Economic Management, North China Electric Power University, Baoding, Hebei, China
hdlxl0901@126.com

Abstract. Digital construction and information management have become the development trend of the construction industry. The integration of BIM technology and big data concepts has broad application prospects in engineering digital information management. The application of BIM technology has not only promoted the flow of construction information in the entire life cycle of the project, improved the level of project management informationization, but also brought higher practical value and social benefits. Based on the large amount of historical engineering data accumulated by the BIM digital platform, the value of engineering data information management is analyzed. This paper proposes a value mining method for engineering digital information management based on BIM massive data, which provides a reference for mining the potential value of engineering information, and provides a path basis for the realization of engineering data twinning and digital management. It is expected to provide theoretical and practical guidance for future engineering project decision-making through value mining.

Keywords: Building information modeling · Value mining · Realization path

1 Introduction

As a project information management model for the entire life cycle, BIM is the most cutting-edge technology used to collect, integrate, analyze and store engineering project data. Combining BIM with modern information technology in a construction project can realize the information management of all participants, all goals, all elements, and the entire life cycle of the construction project.

In the process of project construction management, the collection and storage of project data are integrated into the engineering data center, and information management is implemented. However, in practice, information processing and analysis cannot meet the needs of in-depth digital applications and structural requirements, cannot highlight the value of information, and cannot extract valuable information and experience from it. Zhang huixiang et al. proposed a BIM theoretical system suitable for the entire life cycle of engineering based on big data thinking [1]. Qu Yao combines the concepts of BIM, big data technology and engineering quality management to build a big data application

I. Ahmad et al. (Eds.): STSIoT 2021, LNDECT 122, pp. 137–142, 2023.
https://doi.org/10.1007/978-981-19-3632-6_18

framework based on engineering quality [2]. Based on the idea of fusion of BIM and big data, Liu Qian made up for the problems of big data in the field of engineering bidding and procurement, and looked forward to the application of "BIM + big data" in the field of engineering bidding and procurement [3]. Jialin Cao et al. pointed out that the introduction of BIM technology into project cost management can effectively improve cost data analysis and sharing capabilities [4]. In order to achieve better project management goals, Pan Yue and Zhang Limao built a digital twin framework based on a new generation of information technology and data analysis [5]. Zhao Bin et al. used the BIM model as the carrier of big data on subway construction quality, and linked the two to guide the quality control of the construction site [6]. Xie Linlin et al. have established an intelligent management platform for prefabricated building scheduling based on BIM and digital twin technology, so as to process system information more effectively [7].

Based on the concept of big data, this paper analyzes the application value of BIM technology in engineering project information management, studies the mining value of BIM-DATA, and proposes the realization path of value mining of engineering digital information management based on BIM-DATA. This provides a reference for the application of engineering digital information.

2 Engineering Data Management Based on BIM

Due to the large scale and wide range of construction projects, the amount of data generated is extremely large, the sources are scattered, the attributes are diverse, and they are always in dynamic changes. Traditional engineering project information management communication methods are backward, and there are problems such as data fragmentation, inefficient sharing, storage distortion and transmission delay, and data cannot be effectively integrated and utilized. The popularization of information technology and the application of BIM technology have brought new ideas for engineering project information management [8].

Engineering project information management based on BIM technology is based on the integrated application of information technology such as artificial intelligence, Internet of Things, cloud computing, etc., to collect and transmit information throughout the project life cycle. Meanwhile, multi-dimensional attribute information such as all participants, all elements, and all goals of the construction project is edited, stored, transmitted, and integrated, and finally a massive data storage BIM-DATA of the construction project is formed. BIM-DATA is an engineering project data center based on BIM technology, which can be divided into project data center and engineering data center according to the level. The project data center refers to the project-level database, which contains the building entity data and element data in the whole process of the management of a specific engineering project. Architectural entity data is data such as geometric information and component attributes generated after building entities are digitized through BIM technology. The element data is generated during the construction process, including unstructured data such as image data and video collected by monitoring equipment such as sensors and cameras. The engineering data center is an enterprise-level database that contains various data of various historical projects. Based on data search and query, data mining can be carried out according to the needs of different departments to explore the value of information. The data standards in the BIM data center are unified and interrelated. Based on the information of the data center in the different construction phases of the project, it can maintain a high degree of transferability and operability among the participants. At the same time, the information between different stages, different professions, and different elements of the construction project can be organically integrated to realize the collaborative management of all participants.

Based on the BIM information management organically combines the information management of all participants, the entire period, all objectives, and all elements of the construction project, making the construction project information more complete and the information more coordinated. Ultimately realize the best value of the whole life cycle, the collaborative sharing of all tasks and stages of all participants, the whole life cycle assets, and the informationization and intelligence of the whole process. The BIM based engineering project information management framework is shown in Fig. 1.

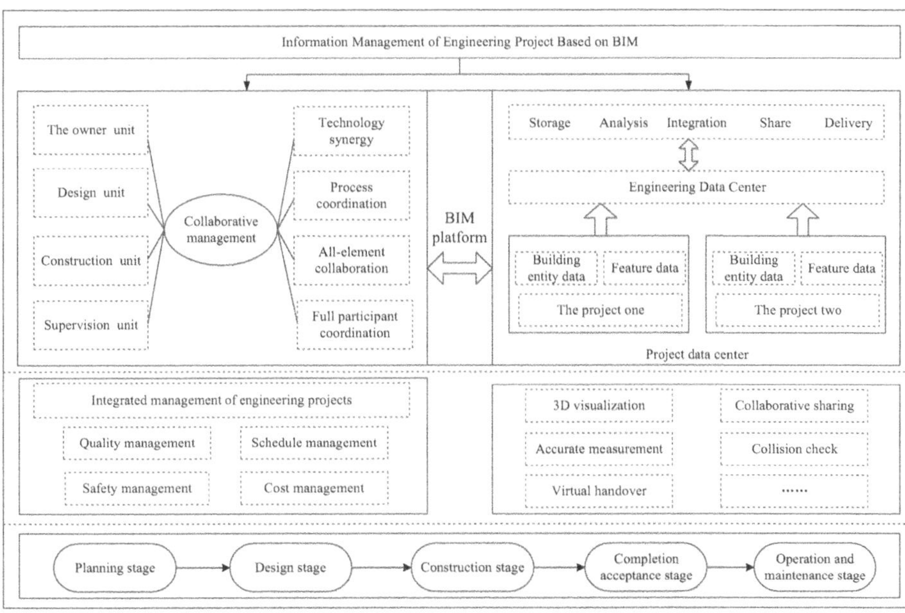

Fig. 1. Based on the BIM engineering project information management framework

3 Digital Information Value Mining Path and Value Realization

In the whole life cycle of engineering projects, digital value mining based on BIM-DATA can effectively improve the independent learning ability and deep thinking ability of the data center [9]. By enhancing the real-time interaction between the three-dimensional model and the physical environment, making full use of the perception monitoring system to build a digital twin in the virtual space, complete accurate mapping and real-time information update, and realize the collaborative sharing of information between professional design and participants. Realize dynamic simulation, analysis and control of physical objects through data drive, such as collision detection, energy consumption analysis, cost prediction and other functional optimization design schemes. Using the digital twin model, virtual construction is realized on the computer, and potential problems such as the function and constructability of the project are predicted, including construction method experiment, construction process simulation, and construction plan optimization. By sorting out multi-source heterogeneous data and digging out the objective connections between data, it is possible to more systematically grasp the multi-agent and multi-requirement collaboration and optimization management in engineering project management. Establish a series of business decision-making models through the data visualization function to realize the assessment of the current state and the prediction of future trends, so as to provide a comprehensive and accurate decision-making basis for business decisions. The BIM-DATA value mining route framework is shown in Fig. 2.

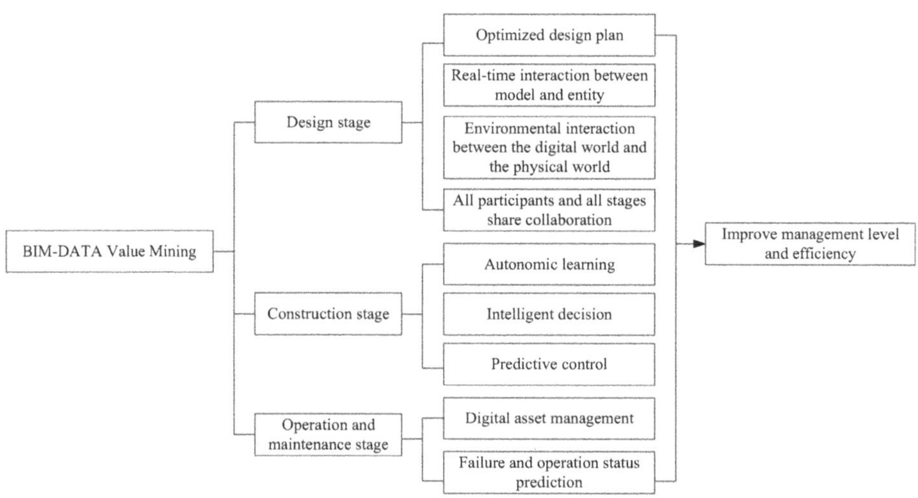

Fig. 2. BIM-DATA value mining route framework

The full mining and utilization of data can better improve the project management level and management efficiency. The integration of BIM and big data thinking, through artificial intelligence, visualization, deep learning, cloud computing and other means to achieve project cluster data mining and application, will promote the management mode to data-driven transformation, thus making BIM-DATA intelligent [10].

The value mining of engineering digital information management based on BIM-DATA is a process of processing data, extracting value and using it, and its core elements are data and algorithms. Data is the foundation of everything, and powerful algorithms can use data to learn, analyze and mine data, and transfer the results to the physical world, which can help managers formulate control and decision-making plans at various stages. In the design stage, the use of project-related data to construct the surrounding environment of the restoration project can not only be interactively designed in a visual environment, but also can expose defects in advance during the virtual design process, so that designers can optimize the model in time. In the construction phase, through the analysis and training of historical data of similar project groups, feasible solutions can be given to specific problems, and the construction process can be controlled. In the operation and maintenance phase, the design and construction data will be fully retained and imported into the synchronized digital platform. Through the construction of a spatiotemporal database, the details of the built object can be displayed in real time, and based on the virtual control reality, remote control and remote maintenance can be realized [11]. Combining on-site product performance data and engineering simulation results, predict the future performance of the product under working conditions, optimize the maintenance schedule, reduce unplanned equipment shutdowns, and improve operational performance. Figure 3 shows the value verification of engineering digital information based on BIM-DATA.

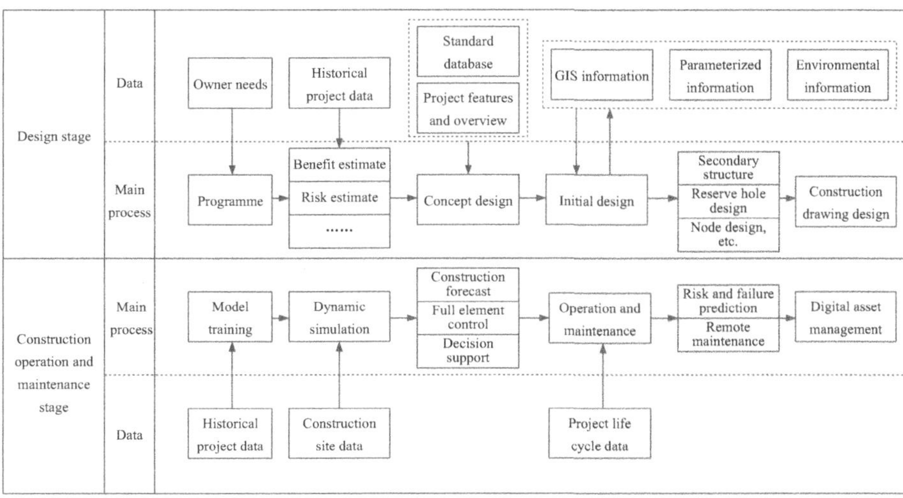

Fig. 3. Value verification of engineering digital information based on BIM-DATA

The integration of the concepts and methods of BIM and big data brings a new experience to engineering project management in the construction field. The full exploration of the value of engineering digital information will also play an active role in promoting the intelligent development of BIM and improving the management level and efficiency.

4 Conclusion

The overall information level and data value density of engineering project management is low, and a large amount of data is in a fragmented state, causing the problem of difficult application of big data in the industry. Under the technical support of digital twins, Internet of Things, machine learning algorithms, etc., fully excavate BIM-DATA, and then extract valuable information from massive engineering data. It can provide important strategic resources for the construction industry, help the construction industry realize the digital transformation of the whole process, all elements and all participants, and promote the high-level development of the construction industry.

References

1. Zhang, H., Gao, Y., Ding, C., Li, J.: Research on BIM theoretical system based on big data thinking. Project Manag. Technol. **17**(05), 47–50 (2019). (in Chinese)
2. Qu, Y.: Application of big data of engineering quality based on BIM. Eng. Technol. Res. **4**(06), 217–218 (2019). (in Chinese)
3. Liu, Q.: The application of BIM + Big Data in engineering bidding and procurement. Constr. Econ. **40**(08), 65–67 (2019). (in Chinese)
4. Liu, Q., Cao, J.: Application research on engineering cost management based on BIM. Procedia Comput. Sci. **183**, 720–723 (2021)
5. Pan, Y., Zhang, L.: A BIM-data mining integrated digital twin framework for advanced project management. Autom. Constr. **124**(1), 103564 (2021)
6. Zhao, B., Tian, X., Li, G., Yan, J., Zhang, H.: BIM construction guidance technology based on big data of metro engineering. China Standardization **2019**(18), 25–26+29 (2019). (in Chinese)
7. Xie, L., Chen, Y.: Research on intelligent management platform of prefabricated building project scheduling based on BIM + Digital Twin technology. Constr. Econ. **41**(09), 44–48 (2020). (in Chinese)
8. Zhou, X., Zhao, J., Wang, J., Huang, X., Xie, P.: Parallel computing-based online geometry triangulation for building information modeling utilizing big data. Autom. Constr. **107**, 102942 (2019)
9. Ansah, M.K., Chen, X., Yang, H., Lu, L., Lam, P.: Developing an automated BIM-based life cycle assessment approach for modularly designed high-rise buildings. Environ. Impact Assess. Rev. **90**, 106618 (2021)
10. Su, T., Li, H., An, Y.: A BIM and machine learning integration framework for automated property valuation. J. Build. Eng. **44**, 102636 (2021)
11. Lu, Y., et al.: Review of BIM and GIS integration: methods, applications, challenges. Build. Sci. **37**(04), 126–134 (2021). (in Chinese)

Dynamic Response Time Measurement Method of Linear Differential Transformer Displacement Sensor

Yongqing Li(✉), Cunwen Peng, Cunxiao Zhang, Anan Shen, Jian Zhang, Yuedong Lu, Lihong Yi, Yue Gao, and Hui Li

Shenyang Academy of Instrumentation Science Co., Ltd., Shenyang, Liaoning, China
info@hb-sais.com

Abstract. There is no authoritative and effective solution for the dynamic response time test of linear differential transformer displacement sensor at home and abroad. This project has carried out special research on the dynamic response time of displacement sensor. Through theoretical analysis and calculation, effective test methods such as free fall test method, spring acceleration test method and excitation coil instantaneous signal excitation method are proposed, and various test methods are effectively verified through relevant experiments.

Keywords: Displacement sensor · Differential transformer · Response time · Eexcitation coil

1 Introduction

The linear differential transformer displacement sensor adopts the principle of electromagnetic induction and belongs to a non-contact and loss free measurement method. It has the outstanding characteristics of high measurement accuracy, good stability and strong environmental adaptability [1]. It can be widely used in equipment engineering, automobile manufacturing, aviation, aerospace, power system, industrial process control and other fields. With the increasing development of the application requirements of displacement sensors, the requirements for dynamic characteristics such as sensor output response time are becoming higher and higher [2, 3]. For linear differential transformer displacement sensor, the measurement accuracy and environmental adaptability of the sensor are directly related to the main working performance of the sensor. Therefore, they are the key technical indexes of the sensor. As an important dynamic characteristic index of the sensor, the response time of the sensor is an important parameter to reflect the synchronization between the sensor and the measured displacement. In most applications of displacement sensors, the sensor is required not only to have good static performance such as measurement accuracy, but also to have ideal dynamic response characteristics, that is, the faster the change speed of the measured displacement is, the better, and the shorter the response time is, so as to realize the real-time and accurate measurement and control of the system and process [4, 5]. Therefore, the response time of displacement sensor is a very key technical index, which needs to be measured accurately and effectively to identify the dynamic response ability of displacement sensor.

I. Ahmad et al. (Eds.): STSIoT 2021, LNDECT 122, pp. 143–148, 2023.
https://doi.org/10.1007/978-981-19-3632-6_19

2 Several Test Methods of Response Time

The response time of the linear displacement sensor is the time required for the signal output of the sensor to detect the change of the displacement of the pull rod. In the actual test process, in order to truly test the response time of the sensor, the movement speed of the detection pull rod should be fast enough that its movement time is negligible relative to the response time of the sensor, so as to avoid that the movement time of the detection pull rod is superimposed on the response time of the displacement output and cannot be effectively eliminated. At present, there has been no effective test method for the dynamic response time of linear displacement sensor at home and abroad. Therefore, this project has carried out the research on several effective test methods for the response time of linear displacement sensor.

(1) **Free fall test method**
The response time free fall test method is to place the detection pull rod of the displacement sensor in its reciprocating displacement chamber and at the electrical zero point, with the movement trend towards the ground, so that the detection pull rod falls in a free fall mode. At the same time, the whole output process of the sensor from zero point to full scale is recorded with a storable oscilloscope. According to the free fall motion formula: $S = \frac{1}{2}g \times t^2$, the time required for the movement of the pull rod can be calculated. Where, s is the displacement of the pull rod, g is the gravitational acceleration, $g = 9.8\ \mathrm{m/s^2}$, and t is the time required to detect the drop of the pull rod from the zero position to the full scale position. For example, the range of a certain type of displacement sensor is 20 mm. According to the free fall formula $S = \frac{1}{2}g \times t^2$, S = 20 mm, t = 63.9 ms is obtained. It can be seen that the movement time of the pull rod in the free falling state is up to 63.9 ms, which is enough to mask the real response time of the sensor. The initial speed of the rod falling freely is 0 m/s, and the speed when moving to the full-scale position is only 0.63 m/s ($V_t = g \times t$). It can be seen that the moving speed of the detection rod under the condition of free falling is still very limited, and the accurate measurement of the dynamic response time of the sensor can not be fully realized. Therefore, the free fall test method is not suitable for the test object with high requirements for sensor response time (within 10 ms).
(2) **Spring acceleration test method**
In order to improve the moving speed of the sensor detection pull rod during the test, a light rigid spring is installed on the pull rod and properly installed and fixed, as shown in Fig. 1.

According to the spring force formula: $F = K \times X$ (K is the elastic coefficient of the spring and X is the shape variable of the spring) and Newton's second law formula $F = m \times a$. The movement time of the displacement detection pull rod from zero position to full scale position under the action of spring can be calculated [6]. In this experiment, the elastic force of the spring at the zero position is 5 kg (49n), and the total mass of a certain type of displacement pull rod and nut is 10 g. The calculated average motion acceleration of the pull rod is $a = 1/2F/m = 2450\ \mathrm{m/s^2}$, which is far greater than

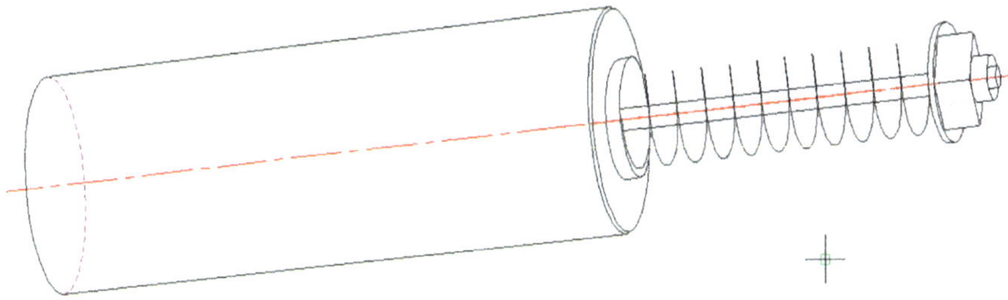

Fig. 1. Add spring to the detection pull rod

the free fall acceleration of 9.8 m/s^2. It is further calculated that under the action of spring, the movement time required to detect the pull rod from zero position to full scale position is about 4 ms. As shown in Fig. 2, the output response time of a certain type of displacement sensor from zero position to full scale position is about 7 ms, including the time occupied by mechanical displacement movement.

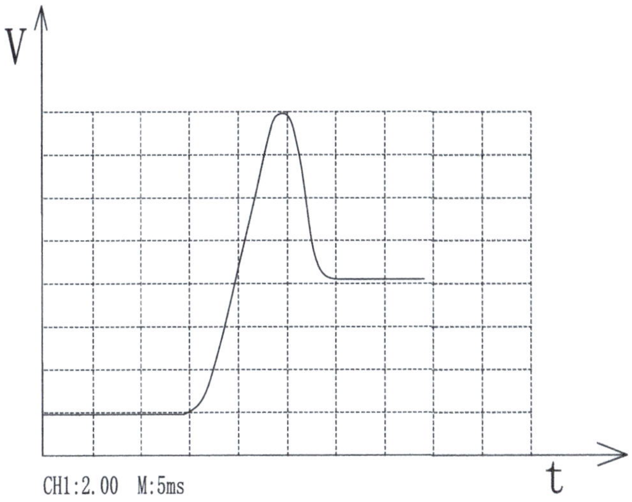

Fig. 2. Sensor output response curve of spring acceleration test method

It can be seen that although the movement speed of the displacement pull rod is greatly improved under the action of the spring, there is still a movement time of 4 ms. When actually testing the response time of the displacement sensor, the time occupied by the movement of the pull rod should be excluded from the output response time of the sensor. Therefore, the actual dynamic response time of the displacement sensor is about 3 MS. Because the actual movement time of the pull rod is difficult to make a very accurate calculation, the measurement accuracy of the spring acceleration test method can only be controlled within 1.5 ms, which can be used to test the dynamic performance of the displacement sensor with a response time of more than 5 ms. In order to test the dynamic response time of displacement sensor more accurately, the experimental test scheme needs to be further optimized.

3 Excitation Test Method of Instantaneous Signal of Excitation Coil

(1) **Experimental principle of excitation signal test method**
When the magnetic flux of the transformer is applied to the differential magnetic displacement sensor, the alternating magnetic displacement sensor is made of the magnetic flux of the transformer φ express. In primary and secondary coils φ It is the same. According to Faraday's law of electromagnetic induction, the induced electromotive forces in the primary and secondary coils are $e_1 = -N_1 \times d\varphi/dt$, $e_2 = -N_2 \times d\varphi/dt$. Where N1 and N2 are the turns of primary and secondary coils. $U_1 = -e_1$, $U_2 = e_2$ (U1 is the effective value of primary coil voltage, U2 is the effective value of secondary coil voltage), let $K = N_1/N_2$, K is called the transformation ratio of transformer. From the above formula, $U_1/U_2 = -N_1/N_2 = -K$, that is, the ratio of the effective values of the voltage of the primary and secondary coils of the transformer is equal to its turns ratio, and the phase difference of the voltage of the primary and secondary coils is π [5] (Fig. 3).

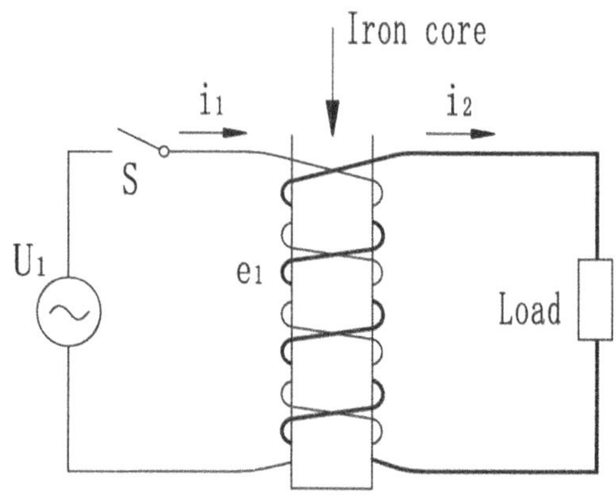

Fig. 3. Working principle of differential transformer

According to the above theoretical analysis and practical test, the phase difference between the primary coil excitation and the secondary coil output of the displacement sensor is π, and the corresponding response time is only tens of subtleties (taking the sensor with excitation frequency of 10 kHz as an example, the secondary lags behind the primary is only 50 μs), compared with the response time of the sensor with the response time of milliseconds, it is basically negligible. Therefore, the factors affecting the output response time of the sensor mainly come from the rectification, filtering, amplification and other links of the circuit [7, 8]. Therefore, the detection pull rod of the sensor can be placed at the full-scale position in advance, which is equivalent to the detection pull rod of the sensor moving from the zero position to the full-scale position at a very fast speed, which completely saves the moving time of the pull rod. Then, the instantaneous excitation signal is applied to the primary coil, and the output signal is monitored synchronously at the signal output end of the sensor. When the output signal reaches the full-scale signal amplitude, the time lag behind the starting time of the high level of the excitation step signal is the response time of the displacement sensor, so that the real response time of the displacement sensor can be measured accurately.

(2) **Excitation test method of instantaneous signal of excitation coil**
In order to accurately test the response time of the displacement sensor, it is necessary to make the sensor operate under reasonable state parameters. The excitation signal selected in this experiment is a sine wave signal with a frequency of 10 kHz and a signal amplitude of 3 V. In the experiment, a fast switch is used to control the output of the excitation signal, and an instantaneous excitation signal is applied at both ends of the primary coil. A storable oscilloscope is used to track and record the complete change process of the output signal waveform of the displacement sensor after the switch is turned on. Track and record the synchronous closing signal of the fast switch with channel 1 of the oscilloscope, and track and record the output signal of the displacement sensor with channel 2 of the oscilloscope. As shown in Fig. 4, the horizontal scanning speed of the oscilloscope is 1 ms/Div. it can be measured from the operation curves of channel 1 and channel 2 of the oscilloscope that the time when the output signal of the sensor lags behind the excitation signal is 2.8 ms, that is, the response time of the displacement sensor is 2.8 ms.

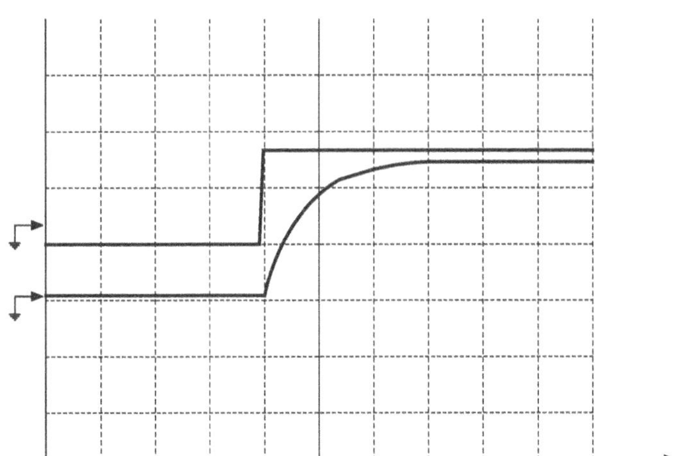

Fig. 4. Response time test curve of primary excitation and output signal

4 Conclusion

In this project, various test methods of response time of displacement sensor are comprehensively, systematically and deeply studied. Through theoretical analysis and practical test, a new response time measurement method - excitation coil instantaneous signal excitation test method is creatively proposed, which effectively improves the measurement accuracy of response time. The invention, popularization and application of the measurement method will provide an accurate and effective measurement scheme for the technical identification of the response time of the displacement sensor, and will play a very positive role in promoting the development of the displacement sensor industry and the progress of technology.

References

1. Lin, X.: Research on Static Characteristics of High Precision Differential Transformer Displacement Sensor. Tianjin University of Technology (2019)
2. Yu, M.: Research on LVDT Displacement Sensor for Aeroengine. Nanjing University of Aeronautics and Astronautics (2020)
3. Yu, M., Li, P., Liu, X.: Study on output voltage linearity of dual redundancy LVDT displacement sensor. Electron. Measur. Technol. **43**(02), 26–32 (2020)
4. Gu, M.: Structural Design and Performance Research of Differential Transformer Displacement Sensor. China University of Mining and Technology (2020)
5. Yan, F.: Measurement Accuracy Analysis and Compensation Method of LVDT for Triaxial Tester. Tianjin University of Technology (2020)
6. Meng, X., Li, W., Zhang, X., Liu, Y.: Design of a low pressure spring safety valve for radome. Eng. Constr. Des. **14**, 136–137 (2020)
7. Mayunga, S.D., Bakaone, M.: Dynamic deformation monitoring of Lotsane bridge using global positioning systems (GPS) and linear variable differential transducers (LVDT). J. Data Anal. Inf. Process. **09**(01) (2021)
8. Saurav, S., Muthuganesh, M., Chaurasia, P.K., Murugan, S.: Analysis, design, and development of a compact LVDT for in-reactor experiments. IETE J. Edu. **60**(2) (2019)

Online Education APP Information Supervision Based on Machine Learning Algorithms

Boyang Yu[1(✉)] and Trumone Sims[2]

[1] School of Public Finance and Administration, Harbin University of Commerce, Harbin, Heilongjiang, China
595234885@qq.com
[2] Ecole Normale Superieure de Lyon, Lyon, France

Abstract. The mobile Internet era has brought a new way of online learning on the mobile terminal, and various online education applications are emerging one after another. The benefits of online education are obvious, such as learning anytime and anywhere, massive network resources, rich and excellent teaching resources, etc., but today's online education apps still have uneven quality, scarce research literature, and narrow educational sections. Students' learning willpower is weak and other issues. Therefore, it is necessary to study online education APP information supervision based on machine learning algorithms. This article first discusses the concept of online education APP and expounds the application of machine learning algorithms, and then designs and develops a system for online education APP information supervision, and tests the performance of the system. The final test result shows that the system response time is basically maintained at about 23/ms, indicating that the system response speed is relatively fast; the system delay time is basically maintained at about 12/ms, it can be said that the delay time is very low, and it also shows that the system response speed is fast. At the same time, the running time of the system is about 45/m, which can save the memory occupied by the system itself, and supervise the user learning situation of online education while saving loss.

Keywords: Machine learning · Online education · Information supervision · Education app

1 Introduction

In recent years, the research of machine learning methods has developed rapidly, and it is the current focus of academic and commercial research in terms of algorithms and applications. In the early days of artificial intelligence, machine learning technology was almost entirely a written theory of character. By the 1990s, statistical machine learning had replaced symbolic learning. From the initial theoretical model research, development to solve practical problems in real life. This is a huge advancement in scientific research in the application of problem-oriented research [1, 2].

I. Ahmad et al. (Eds.): STSIoT 2021, LNDECT 122, pp. 149–157, 2023.
https://doi.org/10.1007/978-981-19-3632-6_20

Machine learning was born with the development of artificial intelligence and is a part of artificial intelligence. The concept of artificial intelligence appeared in the 1950s and was proposed by a group of well-known scientists at that time. Since the 1970s, it has been known as the world's three major advanced technologies along with space technology and energy technology; from 1997, deep Blue defeated a professional chess player in the human-machine challenge, to 2016 Alphago defeated the world champion of Go, which can be described as artificial intelligence. A stage of rapid development has also proved that the future IQ of artificial intelligence is immeasurable. The research of artificial intelligence is developing in the direction of "rationality", "knowledge" and "learning", and the method of machine learning is the embodiment of "intelligence" in the field of artificial intelligence [3, 4].

Although online education has many advantages that offline education does not have, today's online education apps still have problems such as uneven quality, scarce research literature, narrow sections of education involved, and weak learning willpower of students. Therefore, based on the machine learning algorithm, the online education APP information supervision education research is helpful to make up for the existing online education APP shortcomings.

2 Overview of Online Education APP Information Supervision Based on Machine Learning Algorithms

2.1 The Concept of Online Education APP

The full name of APP in English is "Application", which means "application". Recently, it refers to applications downloaded on online devices such as tablets or mobile phones. Educational APP refers to application software with educational nature, mainly involving learning and teaching. The use of educational applications for learning has become a current trend. With the popularity of online devices such as mobile phones and tablets, educational applications can meet the needs of different learners and learning levels [5, 6].

2.2 The Impact of Online Education Apps

Online learning mainly refers to the use of smart phones and tablet computers on the Internet by learners to realize learning without time and space constraints. With the rapid development of smart phones, their comprehensive functions, low cost and portability have created the possibility for online learning, making online learning more common. In modern education and teaching, the advantages of online learning are mainly reflected in the following aspects: First, the education and teaching place is no longer fixed, and learners can rely on the Internet to learn independently and spontaneously anytime and anywhere. Secondly, with the help of smart phones for online learning, users can conduct various learning cooperation and interactions among different learners. Third, using online devices such as smartphones for online learning can provide learners with rich information resources needed for knowledge learning, allowing learners to learn flexibly and meet their own learning needs. Of course, when using online smartphones in school classrooms, please consider the following factors: attitudes of teachers, students, parents and other relevant personnel; teachers' organizational leadership ability; what is teacher's professional development; relevant curriculum activities: whether there are comprehensive technical integration and support measures, etc. Teachers should also consider how to guide students to use mobile phones, which part of the classroom and how to use mobile phones. Therefore, the popularization of online learning in school education is a process in which learning attitudes and behaviors are gradually accepted [7, 8].

2.3 Basic Theory of Machine Learning

Machine learning is a cross-functional and cross-domain topic. The areas covered include probability theory, statistics, approximation theory, convex analysis and computational complexity theory, etc. The content of machine learning is to improve the performance of the system through the intelligent calculation and use experience of the computer, and to generate the experiment of the corresponding algorithm model. The generation process of the algorithm model is actually the process of machine learning. The generation of learning algorithms includes the process of simulating human thinking learning, the process of thinking about incomplete information, the process of constructing and discovering new things, and the process of processing big data under current trends. At present, machine learning algorithms are mainly divided into the following categories: supervised learning algorithms, unsupervised learning algorithms and semi-supervised learning algorithms. Among them, supervised learning is generally divided into regression algorithms and classification algorithms. Regression is an attempt to combine input and output variables with continuous functions; this class maps input variables into discrete categories. Unsupervised learning means that we don't know in advance what will happen to the output. Unsupervised learning has no labels or only the same labels. Semi-supervised learning is a learning method that combines supervised learning and unsupervised learning. In machine learning, there are both labeled data and unlabeled data. Partially supervised learning can improve the efficiency and accuracy of learning, and it will attract more and more attention from machine learning researchers [9, 10].

2.4 Principles of Machine Learning

Machine learning tasks can be divided into supervised learning and unsupervised learning, depending on whether the data in the training sample is labeled. The labeled training data samples are usually called supervised learning. In supervised learning, each sample consists of an input object and an expected output value. For example, "linear regression" in regression algorithms involves trying to learn a linear model to predict the true output value as accurately as possible given the attributes of the labeled data. Linear regression and logistic regression algorithms are commonly used in regression algorithms for supervised learning. Classification in supervised learning involves obtaining a classification function or model to assign data objects of unknown category to a given category. The difference between it and regression is that the predicted output of classification is a discrete category, while regression is a continuous category. Since discrete data and continuous data can be converted to each other, regression algorithms and classification algorithms can be considered the same algorithm. At present, the most commonly used classification algorithms are: linear regression, logistic regression, decision trees, support vector machines, Bayesian networks, etc. [11, 12].

Assuming that n samples of unknown categories X1, X2, ..., Xn are obtained from the sample space A, the grouping process can be described as follows: For each sample Xi(i = 1, 2, ..., n), it belongs to One of the m regions of A, and Xi belongs to only one of them. The distribution of the sample satisfies the following formula:

$$A_1 \cup A_2 \cup A_3 \ldots\ldots \cup \mathrm{A}_n = \mathrm{A} \tag{1}$$

$$A_i \cap A_j = \varphi(\forall i \neq j) \tag{2}$$

The classification results obtained by clustering can clearly distinguish the samples, but the final confirmation of each category must be re-evaluated according to certain standards. There are many clustering methods. According to whether the criterion function is used in the clustering process, it is divided into direct method and indirect method; according to the sequence of the clustering process, it can be divided into two types: aggregation and splitting. When the samples are grouped, determining the similarity between the samples is the key technology of the grouping algorithm. The method of measuring similarity is mainly based on Euclidean distance, Mahalanobis distance, angle and Tanimoto coefficient.

3 Establishment of Online Education APP Information Supervision Based on Machine Learning Algorithms

3.1 The Overall Structure of the System

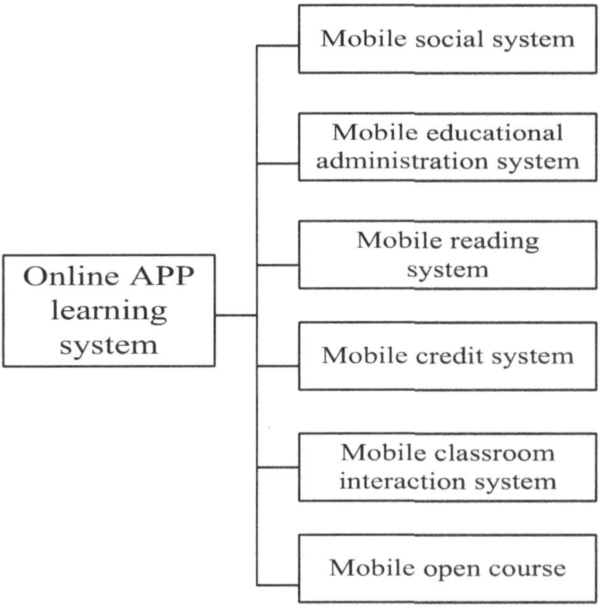

Fig. 1. Online education APP system structure

As shown in Fig. 1, these four systems mainly solve the six main problems faced by schools: promoting the formative evaluation of teaching, improving the teaching effect of teachers, and increasing classroom activity; developing online credit learning, integrating network resources; expanding the dimension of classroom learning; cultivate mobile reading habits. It will provide an open learning space to overcome the bottleneck of school resources; complete education management surveys with one click, open educational management information channels, improve communication, and achieve interaction between students and teachers.

3.2 Online Education APP Information Supervision Process

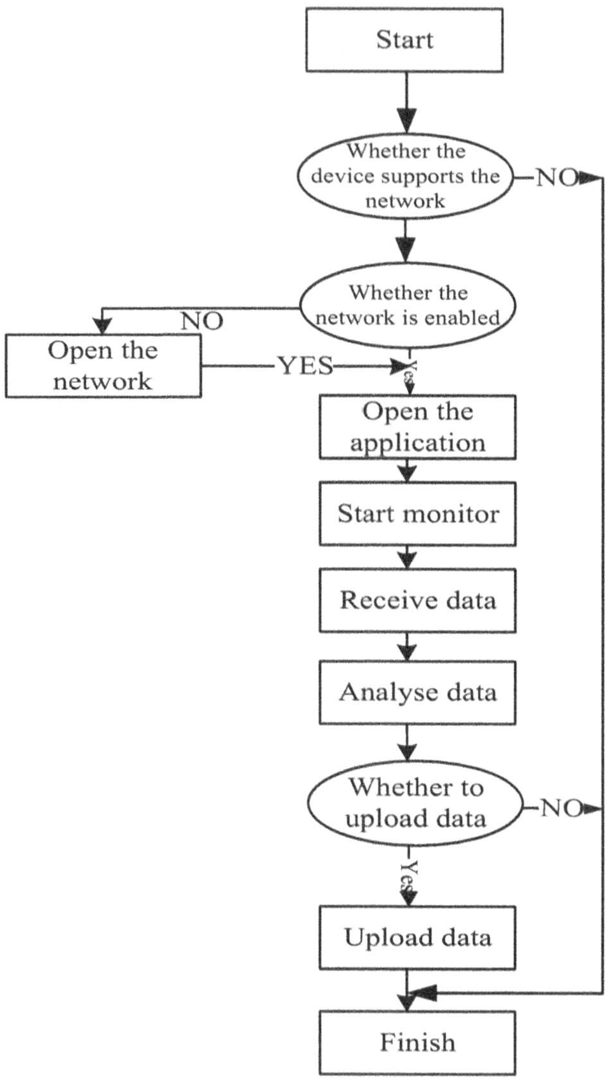

Fig. 2. Online education APP information regulatory flow chart

Online education APP monitoring is the main function of the online education monitoring terminal. It receives the learning data from the user of the learning information collection terminal by connecting to the network, and sends it to the remote server after local analysis. The main implementation process is shown in Fig. 2. The Android platform supports the Internet protocol stack, and the application framework layer provides

APIs for developing the Internet. These APIs can be used to establish an end-to-end wireless interconnection network connection to realize wireless data transmission between two connected network devices.

There are five main steps for communication between connected network devices: connecting network settings, searching for devices, pairing devices, connecting devices, and transferring data. After the online education monitoring terminal starts the networking function, the learning monitoring terminal establishes a connection network connection with the collection terminal, the learning monitoring terminal sends pairing information, and the collection terminal confirms that the pairing is successful. After the online learning monitoring terminal and the education learning information collection terminal are successfully paired, data transmission can be carried out, with the mobile learning monitoring as the server and the education learning information collection terminal as the client. After the training monitoring terminal receives the data sent by the collective terminal, it must perform local analysis according to the data log format to filter the user's training data. So far, the learning monitoring terminal has successfully received the user's learning data and sent the data to the server. Downloading data to the server is indeed an invisible process. In order to improve the user experience and verify the efficiency of data download, the learning monitoring terminal will automatically display the "Download" dialog box when it receives the user's training data. After the user clicks the OK button in the download dialog box, the monitoring terminal downloads the learning data through the machine learning algorithm.

3.3 The Demand for Online Education Apps

Table 1. Online education APP needs

Study	Read	Educational administration	Social intercourse
Study system	Large number of special topics	Credit	Class group construction
Mobile exam system	Literature mining and learning	Achievement query	Share your learning experiences
Operation, discussion	Large number of journals	Active query	Students make real-name friends
Pure network mobile learning	Expand your learning	School timetable	Classroom-based, learning-based social networking
Classroom real-time interactive teaching	Mass electronic books	Notification notification	Create a learning behavior dynamics

The online learning APP is designed to meet the four major needs of teachers and students. It can be seen from Table 1 that pure online mobile learning realizes real-time interaction among classes, assignments, discussions, online exams, online credits in learning. In terms of reading, many e-books and magazines have achieved success. By tapping online learning apps, teachers can discover academic boundaries in time. In terms of education, the school management department will notify information, and students can view credits, activity notifications, lesson plans, etc. In terms of social interaction, students can make friends with their real names, form a learning circle based on classroom learning, and share learning dynamic behaviors.

4 Online Education APP Information Supervision Test Based on Machine Learning Algorithm

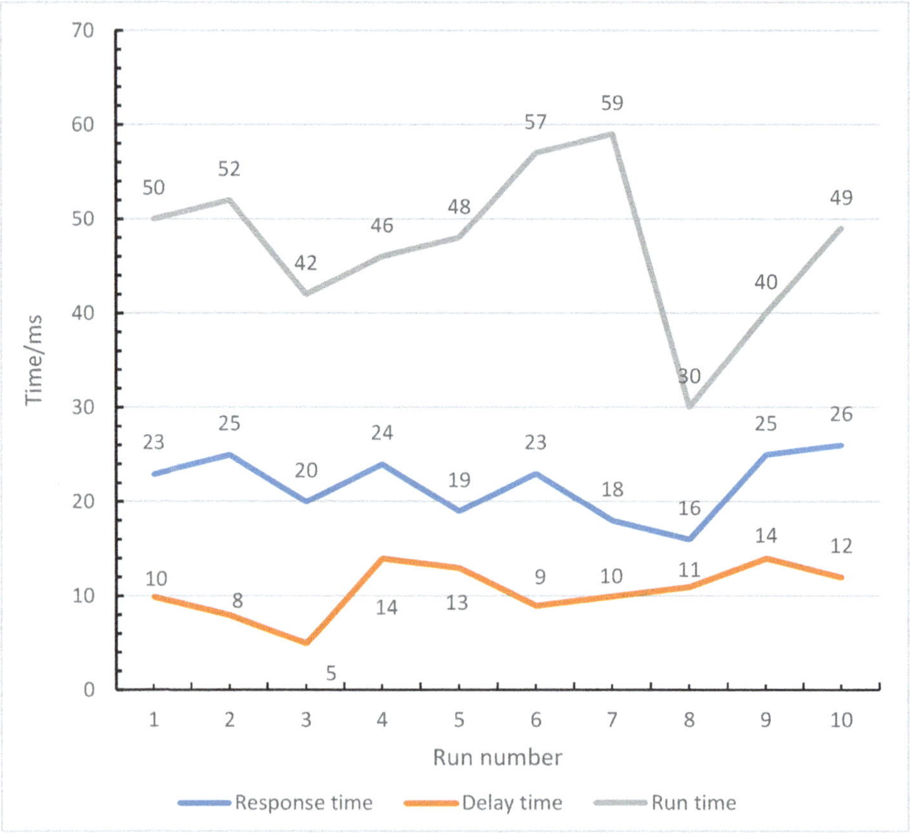

Fig. 3. System performance test

It can be seen from Fig. 3 that the performance test of the system is generally stable, and the system response time is basically maintained at about 23/ms, indicating that the system response speed is relatively fast; the system delay time is basically maintained at about 12/ms, it can be said the delay time is very low, and it also shows that the system has a fast response speed. At the same time, the running time of the system is about 45/m, which can save the memory occupied by the system itself, and also supervise users who supervise online education while saving wastage.

5 Conclusion

With the rapid development of computer technology, people use online terminal devices such as mobile phones and tablet computers to make information exchange more convenient. Accompanied by the increasing popularity of the Internet and the increasing popularity of online smart terminals. As an emerging product market, education apps have grown rapidly and are widely used in the lives of the general public. They have become a media force that cannot be ignored. Through this platform, efficient and valuable and useful information can not only improve the efficiency of users' use of curriculum resources by providing learners with high-quality, but also help teachers effectively understand the effects of students' acceptance of new knowledge and mastering of learning progress.

References

1. Cao, Y.: Internet financial supervision based on machine learning and improved neural network. J. Intell. Fuzzy Syst. **47**, 1–12 (2020)
2. Solmaz, M.E., Mutlu, A.Y., Alankus, G., et al.: Quantifying colorimetric tests using a smartphone app based on machine learning classifiers. Sens. Actuators **255**(2), 1967–1973 (2018)
3. Rückert, P., Papenberg, B., Tracht, K.: Classification of assembly operations using machine learning algorithms based on visual sensor data. Procedia CIRP **97**, 110–116 (2021)
4. Soula, M., Karanika, A., Kolomvatsos, K., et al.: Intelligent tasks allocation at the edge based on machine learning and bio-inspired algorithms. Evolving Syst. 1–22 (2021)
5. Nguyen, T., Raich, R., Fern, X.Z., et al.: IEEE 27th International Workshop on Machine Learning for Signal Processing (MLSP), IEEE 2017, Tokyo, 25–28 September 2017, MIML-AI: Mixed-supervision Multi-instance, pp. 1–6 (2017)
6. Yang, Y., Hu, X., Rong, M., et al.: Characteristic index digging of combat SoS capability based on machine learning. J. Syst. Simul. **031**(006), 1048–1054 (2019)
7. Jiang, S., Yan, G., Li, J., et al.: A quantitative analysis on the "approximatability" of machine learning algorithms. Comput. Res. Dev. **054**(006), 1337–1347 (2017)
8. Nam, J.M.: A study between online entrepreneurship education and entrepreneurship: based on PBL (problem-based learning) and flipped learning. Asia-Pac. J. Bus. Venturing Entrepreneurship **12**(2), 31–40 (2017)
9. Liang, J., Nie, Y.: A hybrid teaching mode based on machine learning algorithm. Open Artif. Intell. J. **06**(1), 22–28 (2020)
10. Khanal, S.S., Prasad, P.W.C., Alsadoon, A., Maag, A.: A systematic review: machine learning based recommendation systems for e-learning. Educ. Inf. Technol. **25**(4), 2635–2664 (2019). https://doi.org/10.1007/s10639-019-10063-9
11. Mourdi, Y., Sadgal, M., Fathi, W.B., et al.: A machine learning based approach to enhance MOOC users' classification. Turk. Online J. Distance Educ. **21**(2), 47–68 (2020)
12. Zhan, Y., Tadikamalla, P.R., Craft, J.A., et al.: Human reliability study on the door operation from the view of deep machine learning. Futur. Gener. Comput. Syst. **99**, 143–153 (2019)

Intelligent System of Scientific and Technological Talent Inquiry Based on Deep Learning Algorithm

Hua Zheng[1(✉)], Linzhi Nan[1], Qi Yang[1], Mengmeng Yang[1], Ting Yang[1], and Turiman Bin Suandi[2]

[1] Haojing College of Shaanxi University of Science and Technology, Xianyang, Shaanxi, China
Cathyzheng0123@163.com

[2] Universiti Putra Malaysia, Seri Kembangan, Malaysia

Abstract. With the rapid development of the socialist market economy, scientific and technological talents play a vital role in all aspects of our country's socialist economic development and social progress. In recent years, with the widespread application of e-government projects in various administrative agencies, business service efficiency and other aspects have been significantly improved. Despite the relatively rapid changes in corporate growth, many employers may still be unable to recruit ideal scientific and technological talents that can help companies survive and develop. In order to better absorb and introduce scientific and technological talents, creating an intelligent system for scientific and technological talents has become an inevitable trend in the development of modern scientific and technological talent management technology. This paper aims to study the intelligent system of scientific and technological talents query based on deep learning algorithms. Based on the analysis of system requirements, non-functional requirements and deep learning algorithms, the functional modules of the intelligent system of scientific and technological talents query are designed and implemented, and finally tested. The performance of the system and the test results show that the intelligent system designed in this paper is available.

Keywords: Deep learning algorithms · Scientific and technological talents · Query systems · Intelligent systems

1 Introduction

With the rapid development of science and technology and the accelerating pace of economy and globalization, human society is gradually transforming from a traditional industrial economy to the development of a modern knowledge economy. As the main body of independent R&D and innovation activities, scientific and technological talents have begun to develop in this development process [1, 2]. In order to have its own advantages in human resources and win the fierce initiative in market competition, many countries and regions in the world are actively investing in the training and management

I. Ahmad et al. (Eds.): STSIoT 2021, LNDECT 122, pp. 158–165, 2023.
https://doi.org/10.1007/978-981-19-3632-6_21

of talents. China also attaches great importance to the training and management of talents [3, 4].

In recent years, research theories related to scientific and technological talents have emerged one after another. Some scholars believe that the cultivation of innovative scientific and technological professionals is the core resource for the construction of national defense science and technology and key laboratories, and the fundamental driving force for the construction and development of high-level laboratories. Under our country's current system and system conditions, how to better give full play to the advantages of the platform, overcome weak links and difficulties, and accelerate the training of more high-level scientific and technological professionals is to better serve the construction of armed equipment and the party's peace the requirements of various major tasks of the country [5]; some scholars say that cultivating innovative scientific and technological talents is also an important means of my country's strategy of strengthening the country by talents. Research on the optimization of the training model of innovative scientific and technological talents will help improve our country's talent training system and cultivate more innovative scientific and technological talents that meet the needs of market economy countries and society [6]; in 2021, a researcher will encourage scientific and technological talents. The existing problems in the mechanism integrate the ideas and methods of human capital theory, behavioral science and incentive theory to construct a new type of incentive mechanism for scientific and technological talents. Combining efforts will fundamentally stimulate the enthusiasm, initiative and creativity of scientific and technological talents [7]. It is necessary and urgent to study the intelligent system of scientific and technological talents inquiry. This system is necessary and important for the expansion of the scientific and technological talent team and maintaining development advantages.

Based on the analysis of system requirements, non-functional requirements and deep learning algorithms, this paper designs and implements the functional modules and database modules of the intelligent system for scientific and technological talents, and finally tests the performance of the system.

2 Research on the Intelligent System of Scientific and Technological Talent Inquiry Based on Deep Learning Algorithm

2.1 System Requirements Analysis

(1) User registration

As an individual user, scientific and technological talents can directly log in to this website and fill in their own relevant information to register; the administrator of the location of scientific and technological talents can directly log in to this registration system by registering a user name, account number, and password, and fill in this registration. Personal information application form, upload the relevant copy information of the code certificate, business license of the relevant scientific and technological organization, or the legal person license of other institutions to register on the website, and attach the paper information of the above copy in time to this time. The official seal of the registered unit is mailed to the New Advanced

Technology Talent Service Center of the Ministry of Science and Technology of the State Council [8, 9].

(2) Fill in and update scientific and technological talent information

Register scientific and technological talents to log in to the system and fill in relevant information online. There are already scientific and technological talents in the library, and the original information must be updated [10].

(3) Organization recommendation department recommendation

Departments and local organizations recommend that each department is responsible for managing the user registration and login system (registered accounts and passwords are uniformly allocated), and according to the requirements of recruitment collection, the relevant scientific and technological professionals submitted by their units are recommended online [11, 12].

2.2 Analysis of System Non-functional Requirements

(1) The principle of standardization. The system design follows national standards and complies with relevant industry regulations.
(2) The principle of reliability. The reliability of the system is directly related to the size of the software. The larger the scale of the software, the more difficult it is to guarantee the reliability. At the beginning of the design, the difficulties and problems encountered should be considered, and the occurrence of failures can be avoided during the operation of the system, and at the same time, it has the ability to solve the failures.
(3) The principle of robustness. When the user enters content outside the system's specifications, the system can give corresponding prompts or make reasonable handling methods.
(4) The principle of scalability. The compatibility of the system and the needs of future development should be considered, as simple and easy to operate as possible, leaving room for upgrades and expansion.
(5) The principle of safety. The system must be able to ensure the security of the user's basic information and ensure that the user's information will not be leaked. At the same time, the system itself can also deal with various problems correspondingly to improve the overall safety performance of the system.

2.3 Deep Learning Algorithm

(1) Neurons

The neuron model mainly refers to a basic structure of a biological neural network and its constituent units. Its basic principle and structure are roughly the same. It is an abstraction based on the specific working principles and structural design of various biological neurons.

First, input x_1, x_2, x_3 and intercept $+1$ into the neuron, and the neuron calculates the output value as shown in formula (1).

$$h_{w,b}(x) = f\left(W^T x\right) = f\left(\sum_{i=1}^{3} W_i x_i + b\right) \tag{1}$$

Among them, the input of the neuron is x_i, the weight of the input layer is W_i, and the input bias of the neuron is b.

(2) Convolution

The essential feature of convolution is to extract feature information of different frequency bands in the image. The principle of convolution operation is to find the inner product of the convolution kernel in a filter matrix and each data window in the image. The basic formula of convolution operation is shown in formula (2).

If the variables of the convolution are the sequence x(n) and h(n), the result of the convolution is:

$$y(n) = \sum\nolimits_{i=-\infty}^{\infty} x(i)h(n-i) = x(n) * h(n) \tag{2}$$

where * represents convolution. When n = 0, the sequence h(−i) is the result of the inversion of the time sequence i of h(i). This process makes h(i) inverted by 180° with the vertical axis as the center, and then the sum is multiplied. The algorithm is called convolution. When the convolution variables become functions x(t) and h(t), the convolution formula becomes as shown in formula (3).

$$y(t) = \int_{-\infty}^{\infty} x(p)h(t-p)dp = x(t) * h(t) \tag{3}$$

Among them, p is the integral variable, the integral summation, t is the amount of displacement of the function h(−p), and * represents the convolution.

3 Experiment

3.1 System Overall Design

The scientific and technological talent inquiry intelligent system is divided into user layer, application layer and data layer in total. Their definitions and functions are shown in Fig. 1:

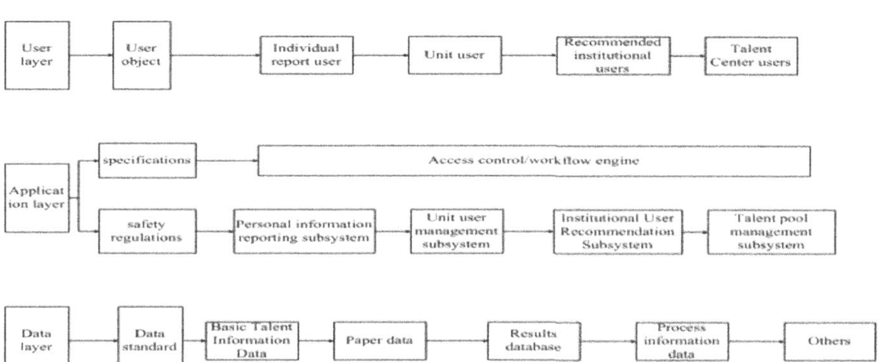

Fig. 1. System overall structure

User layer

The object of user service is enterprise users, including manager users of talent service centers, individual talent users, users of units, and manager users of talent recommendation service organizations. All users can directly conFig and manage their own authority information through the authority management system, and achieve a variety of functional operations such as clear authority division and various duties.

Application layer

They respectively refer to various application functions for managing talents of companies and institutions, including electronic reporting systems and personal information management of institutions, electronic systems recommended by users of management institutions, and management of talent pools of companies and institutions. The functions of these business process applications are mainly constructed and developed based on unified technical specifications and safety performance standards. By controlling various workflow devices and their licenses, the integrated operation of various business processes is realized.

Data layer

It is composed of database tables including basic talent data, paper data, achievement data, process information data, etc. It is composed of a database list including basic talent information, paper information, achievement information, process information, etc. It refers to an information system composed of all the information materials and data table items used in the management of the talent warehouse. The various structured designs within the scope required by the data table items must be in accordance with It is defined by standards related to national technical data and information resource norms. When a standard changes, it can be reconFigd, defined and changed. It realizes the interrelationship between the tables of the relational database, and achieves a tree structure, a through-type structure, and a basic framework for large data volume analysis, query and statistics.

3.2 System Function Module Design

(1) Talent pool management subsystem

The talent pool information management subsystem is mainly responsible for the comprehensive management, query and data statistics of all technological talent pool information by the staff of the science and technology information center.

(2) Personal information reporting subsystem

The personal information filling subsystem is to provide a data collection and management portal for scientific and technological professionals to fill in personal information. After registering through the website, science and technology talents can fill in and submit their personal information after logging in to the system, and the information cannot be modified repeatedly after submission.

(3) Unit user management subsystem

Unit users refer to employers engaged in the cultivation of scientific and technological talents. Unit users can register through the system login method, and their relevant information can be stored in the database after being checked by the talent center. The unit user is responsible for reviewing and supervising the application materials of the unit's scientific and technological talents.

(4) Institutional user recommendation subsystem

The recommended departments are usually national, provincial science and technology departments, and industry academia associations. Their accounts are managed by the talent center. The function of recommending organizations is to supervise and recommend subordinate units and scientific and technological workers.

4 Discussion

4.1 System Test Method

In terms of system testing, our goal is to verify the normal operation of all system modules and meet the requirements of demand analysis by conducting use case testing on the intelligent system for scientific and technological talents. At the same time, we need to know where the system does not meet the requirements through testing. When a problem is found, it is necessary to find out the cause and location of the error, and then propose a solution and fix the problem. In this paper, through use case testing, collect data sets and import them into the system, and then check the execution conditions and data output results to verify whether the system has errors.

4.2 System Performance Test

System performance testing is to test the performance of the system and whether it maintains a stable state through multiple people using the system at the same time. The test tool is LoadRunner. The specific test conditions are shown in Table 1:

Table 1. The system performance test cases

Input action	Desired result	Real result
100 people query talent information at the same time	Display time < 0.7 s	0.6 s
500 people query talent information at the same time	Display time < 1 s	0.8 s
1000 people query talent information at the same time	Display time < 1.5 s	1.4 s
100 registered users at the same time	Registration time < 3 s	3 s
500 registered users at the same time	Registration time < 5 s	4.7 s
1000 registered users at the same time	Registration time < 7 s	6.4 s

It can be seen from Fig. 2 that when 100 people query talent information at the same time, the expected result is less than 0.7 s, and the actual result is 0.6 s; when 500 people query talent information at the same time, the expected result is less than 1 s, and the actual result is 0.8 s; when 1000 people query the talent information at the same time, the expected result is less than 1.5 s, and the actual result is 1.4 s. When 100 people register at the same time, the expected result is less than 3 s, and the actual result is 3 s;

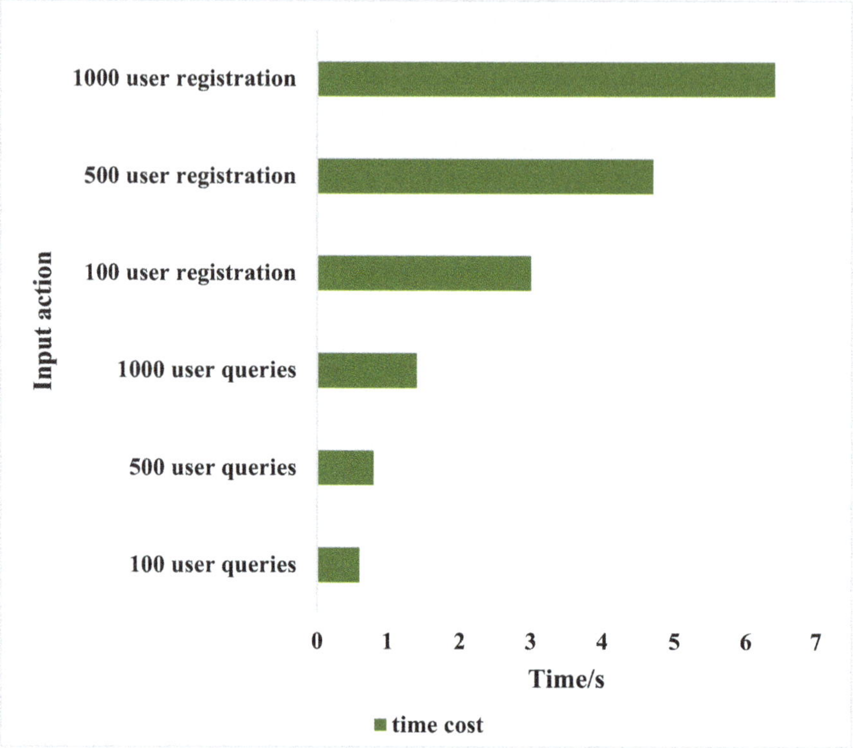

Fig. 2. The system performance test cases

when 500 people register at the same time, the expected result is less than 5 s, and the actual result is 4.7 s; when 1000 people register at the same time, the expected result is less than 7 s. The actual result is 6.4 s. Therefore, the intelligent system for scientific and technological talent inquiry designed in this paper is available.

5 Conclusions

Today's society is a society of talent mobility, and the rational allocation of talents is of the same significance for the second revolution in the industrialization of production equipment and the third technological revolution in high and new technology. Just "make the best use of everything, and make the best use of their talents." Everyone's skills and know-how are different. If we can really put "good use of talents" into practice, it will be social wealth that cannot be underestimated. Establish an intelligent system for scientific and technological talent inquiry, allowing enterprises to find suitable scientific and technological talents according to their own development needs, and realize the reasonable allocation of scientific and technological talents.

Acknowledgements. This work was supported by The Scientific Project of Education Department of Shaanxi Provincial Government, Research of ecological construction of scientific and technological talents in Xixian New Area (No. 21JK0071).

References

1. Wang, F., Liu, P., Wang, P.: An evaluation study of rural scientific and technological talents based on TODIM method with hybrid indicator. J. Intell. Fuzzy Syst. **40**(9), 1–14 (2021)
2. Allal-Chérif, O., Aránega, A.Y., Sánchez, R.C.: Intelligent recruitment: how to identify, select, and retain talents from around the world using artificial intelligence. Technol. Forecast. Soc. Chang. **169**(5), 120822 (2021)
3. Han, S.S., Kim, M.S., Lim, W., et al.: Classification of the clinical images for benign and malignant cutaneous tumors using a deep learning algorithm. J. Investig. Dermatol. **138**(7), 1529–1538 (2018)
4. Hu, X., Li, D.: Research on a single-tree point cloud segmentation method based on UAV tilt photography and deep learning algorithm. IEEE J. Sel. Top. Appl. Earth Observations Remote Sens. **13**, 4111–4120 (2020)
5. Dittberner, A., Sickert, S., Denzler, J., et al.: Intraoperative onlinebildgesteuerte Biopsien auf Basis eines Deep-Learning-Algorithmus zur automatischen Detektion von Kopf-Hals-Karzinomen mittels Real-Time Nah-Infrarot ICG-Fluoreszenz-Endoskopie. Laryngo Rhino Otologie **98**(S 02) (2019)
6. Mallio, C.A., Napolitano, A., Castiello, G., et al.: Deep learning algorithm trained with COVID-19 pneumonia also identifies immune checkpoint inhibitor therapy-related pneumonitis. Cancers **13**(4), 652 (2021)
7. Blaivas, M., Blaivas, L.N., Abbasi, A., et al.: 296 Development of an artificial intelligence deep learning algorithm that utilizes IVC collapse to predict fluid responsiveness. Ann. Emerg. Med. **76**(4), S114 (2020)
8. Fan, Y.: Criminal psychology trend prediction based on deep learning algorithm and three-dimensional convolutional neural network. J. Psychol. Afr. **31**(3), 292–297 (2021)
9. Nikolov, S., Blackwell, S., Zverovitch, A., et al.: Clinically applicable segmentation of head and neck anatomy for radiotherapy: deep learning algorithm development and validation study. J. Med. Internet Res. **23**(7), e26151 (2021)
10. Udristoiu, A.L., Stefanescu, D., Gruionu, G., et al.: Deep learning algorithm for the confirmation of mucosal healing in Crohn's disease, based on confocal laser endomicroscopy images. J. Gastrointest. Liver Dis.: JGLD **30**(1), 59–65 (2021)
11. Lin, C.S., Lee, Y.T., Fang, W.H., et al.: Deep Learning algorithm for management of diabetes mellitus via electrocardiogram-based glycated hemoglobin (ECG-HbA1c): a retrospective cohort study. J. Pers. Med. **11**(8), 725 (2021)
12. Xin, C., Bian, G.B., Zhang, H., et al.: Optical coherence tomography-based deep learning algorithm for quantification of the location of the intraocular lens. Ann. Transl. Med. **8**(14), 872 (2020)

Construction Safety Intelligent Management System Based on BIM and Genetic Algorithm

Dapeng Zhang(✉)

Sinohydro Bureau 1 Co., Ltd., Changchun 130000, Jilin, China
dapeng_zhang6789@163.com

Abstract. With the continuous expansion of the scale of construction projects and the improvement of the level of construction technology, more new scientific and more efficient management methods are needed to meet people's needs for safe production and life. Construction safety management is an important part of engineering projects, and it involves all aspects of construction enterprises. Therefore, the design of the construction safety management system is based on the consideration of reducing accidents. For this reason, this article has carried on corresponding research to this system based on BIM technology and genetic algorithm. This article first applies the data analysis method to research BIM technology and related theories, and then explores the application of genetic algorithms in safety management. Subsequently, experimental and investigation methods were used to conduct in-depth research on the series of problems and designs of the construction intelligent management system, and finally the investigation results and conclusions were drawn. The survey results show that more than 50% of safety accidents are caused by falling from a high altitude. Therefore, in the design of the safety intelligent management system, special attention should be paid to the monitoring and prevention of guardrails.

Keywords: BIM · Genetic algorithm · Construction safety · Intelligent management

1 Introduction

Construction industry is one of the industries prone to safety accidents, and the safety problems in production attract attention. Due to the intensive factors of production and the large number of personnel in the construction site, frequent safety accidents and high casualty rate are caused. According to the statistics of the Labor department of Hong Kong, the accident rate of the construction industry in Hong Kong is 1000:68.1, which shows the risk of the construction industry. With the continuous enhancement of safety awareness, although it has been through strengthening safety management, improving construction safety laws and regulations, in-depth refinement of safety supervision and management work to reduce the occurrence of accidents in the process of construction production, but the current safety problem is still very serious. Because the production

I. Ahmad et al. (Eds.): STSIoT 2021, LNDECT 122, pp. 166–175, 2023.
https://doi.org/10.1007/978-981-19-3632-6_22

process of the construction industry has the characteristics of mobility, singleness, complexity and long-term, it needs to occupy a lot of resources for a long time, and will produce a series of problems such as pollution, energy consumption and safety. Therefore, the transformation and upgrading of China's traditional construction industry is imminent. Compared with the traditional construction industry, the construction industry is safe and reliable, which is in line with the development trend of traditional construction industry. However, the production and construction process of prefabricated components need accurate and complex construction technology, the safety control point has changed fundamentally, and there are higher requirements for the level of safety management. Therefore, the construction safety management of prefabricated concrete buildings cannot be ignored. The traditional safety management method cannot effectively monitor and deal with potential safety hazards, and cannot meet the construction safety management needs of prefabricated concrete buildings. The emergence of BIM Technology is expected to solve this problem. However, due to technical and management reasons, the practical application effect of BIM Technology is not ideal. Therefore, how to promote the practical application of BIM Technology in construction safety management is an important topic that should be studied at present.

With the development of the construction industry, the development of civil engineering is also intensified. From time to time, people will ring a bell for the construction safety of the project. The construction industry is currently a profession with a large number of employees, and its incidence of safety accidents also has a certain number. Therefore, people need to pay attention to the safety issues in the construction process. The application of BIM technology in the construction industry is becoming more and more extensive, and its role in intelligent security management is becoming more and more prominent. Therefore, the research topics in this article are very valuable.

2 Related Work

There are countless studies on construction safety management systems. Each scholar has different research results on the construction safety intelligent management system based on BIM and genetic algorithm.

Phong develops an optimal construction site layout plan by achieving the three main objectives of minimum facility cost, minimum safety facility risk and minimum noise pollution. In addition, a fuzzy swarm optimization (FBCO) algorithm is proposed to adjust the ρ and τ parameters and obtain the feasible optimal solution [1]. Oluwatumininu's research shows that a safe work environment boosts productivity. This was after the results showed that conditions in the work environment significantly affected productivity. The study concluded that ergonomic practices on construction sites are important for safe working conditions and increased productivity [2]. Justine Flore Tchouanguem Djuedja proposes that Building Information Modeling (BIM) is a process for managing construction project information that provides the basis for improved decision-making and collaboration in the construction supply chain. One of the barriers to BIM adoption is the limited interoperability of different BIM systems [3]. Maryam Barzegar et al. argue that IFC has a different process in 3D urban land management, where spatial analysis plays an auxiliary role. In the 3D data model, IFC classes (Industry Foundation Classes) are used to model the physical dimensions of city properties [4]. According to Lea Sattler et al., despite the widespread adoption of BIM over the past decade, cross-domain collaboration through BIM still faces various obstacles [5]. Ahmed Elaksher et al. proposed, 3D building models are in great demand in various applications such as surveying and planning applications in urban areas. This has led to the development of automated algorithms to extract these patterns as they reduce time and cost compared to manual screen scans [6]. Risk management has always been at the heart of most construction projects, according to F. Henry Abanda et al. Building Information Modeling (BIM) provides a way to manage risk in construction projects [7].

At present our country construction personnel safety management, there are still many problems in comprehensive domestic and foreign research results can be seen to take measures for the administration of science technology to BIM and genetic algorithm is introduced into the construction personnel to eliminate the safety hidden trouble in safety management, construction personnel life and property security, reduce the occurrence of safety accidents have positive effects. Therefore, this paper applies BIM and genetic algorithm technology to the safety management of construction personnel, bringing a new development direction.

3 Construction Safety Intelligent Management System Based on BIM and Genetic Algorithm

3.1 BIM Technology and Related Theories

BIM technology is based on 3D modeling technology, which is an important change from the traditional construction industry to the field of 3D modeling technology over the years [8].

(1) Technical advantages

1) Risk resolution. With the help of BIM technology, the accuracy of architectural engineering design has been greatly improved, which not only reduces the possibility of subsequent risks, but also allows designers to make more optimal designs.
2) Visualization. Visualization is the most basic ability that BIM technology can provide, and it is also an important advantage that distinguishes it from traditional methods [9].
3) Information sharing. Using BIM technology, various information can be retained from the initial stage of construction project design, and it can be integrated and integrated with the information in the subsequent construction and operation stages to realize the penetration and sharing of information before and after.
4) Parametric modeling. Parametric modeling is the fundamental basis of BIM technology, and it is also the condition for its series of advantages to appear. In BIM technology, parametric modeling can be realized by means of graphic element components and family libraries [10].

(2) Core principles

BIM is a technology that integrates building information, drawing information, and thinking information and presents them. First, the construction engineering model constructed by BIM technology is based on the data integration of construction engineering information, and the establishment of the model is realized through the penetration of these data, which can finally provide each participant with an accurate and clear three-dimensional architectural model. Second, BIM technology follows the basic principle of continuous model building in all stages of construction engineering.

(3) Necessity in safety management

1) Efficiency necessity

There are several links in the whole project construction process, such as project construction and design, construction, operation, maintenance, etc. This requires not only close cooperation between different disciplines, but also frequent interaction of a large amount of information [11].

2) Scientific necessity

In the process of project construction, scientificity has always been the goal pursued. The scientificity of construction and management can not only help construction companies complete the construction, but also help improve the efficiency of the project [12].

3.2 Application of Genetic Algorithm in Management System

(1) Process

Genetic algorithm is suitable for dealing with optimization problems of complex systems, and it also has good applications for complex problems such as safety management. The basic process is shown in Fig. 1.

(2) Genetic algorithm improvement

It can be seen from the algorithm flow that the first step of genetic algorithm execution is to set various control parameters for algorithm execution, design fitness functions, design selection operators, crossover operators, mutation operators, and retention operators.

The goal of safety management is to make the entire construction process have reliable safety, including labor, materials, machinery, etc. We can choose to treat the objective function as the fitness function, that is:

$$Fit\left(\int(a)\right) = \frac{1}{\int(a)} \tag{1}$$

Among them, $\int(a)$ is the objective function and $Fit(\int(a))$ is the fitness function.

When the population tends to the local optimum, the crossover probability needs to be increased. The specific expression for the crossover probability is as formula (2):

$$P_x = \begin{cases} P_{x1}, \int{}' \prec \int_{avg} \\ P_{x1} - \frac{P_{x1} - P_{x2}}{1+\exp\left(-s\left(\frac{\int{}' - \int_{avg}}{\int_{max} - \int_{avg}}\right)\right)}, \int{}' \geq \int_{avg} \end{cases} \tag{2}$$

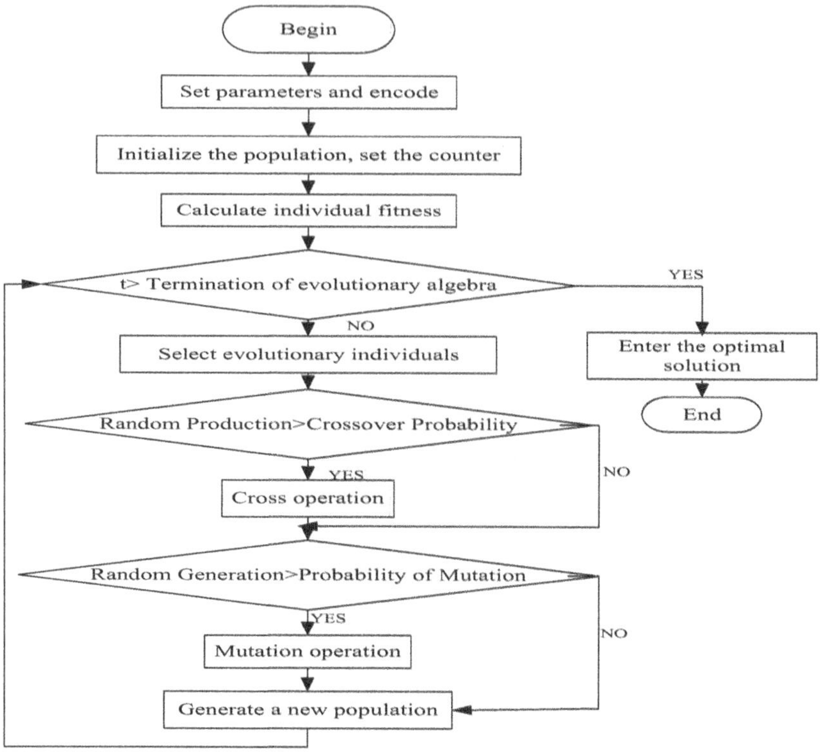

Fig. 1. Genetic algorithm process

The size of the mutation probability P directly determines how many new individuals are produced in a generation of evolution. Like the improved method of crossover probability, the adaptive mutation probability expression is as formula (3):

$$P_r = \begin{cases} P_{r1}, \int ' \prec \int_{avg} \\ P_{r1} - \frac{P_{r1} - P_{r2}}{1+\exp\left(-s\left(\frac{\int ' - \int_{avg}}{\int_{max} - \int_{avg}}\right)\right)}, \int ' \geq \int_{avg} \end{cases} \tag{3}$$

Among them, $\int_{avg}$ represents the average of the fitness values of all individuals in the entire population, and $\int_{max}$ is the fitness value with the largest fitness among all individuals in the current population.

Referring to the adaptive crossover operator method, the expression of the reserved operator P is (4):

$$P_k = P_{k1} - \frac{P_{k1} - P_{k2}}{1 + exp(-s(\int avg_{max}))} \tag{4}$$

Among them, P_{k1}, P_{k2} are constants.

3.3 Construction of Construction Safety Management System

Based on the visual characteristics of BIM technology, the building, structure, and high formwork construction are combined to form a three-dimensional three-dimensional model through Revit software. During the high-formwork construction process, on-site construction is inconvenient, and it is easy to cause safety hazards. BIM technology can plan the spatial layout of the erection of high-supported formwork rods. In the construction drawings of the high formwork, there are many scaffolding members, which can easily make the erector misunderstand the data. BIM can display the construction process in three dimensions and animations. It can also work out corresponding safety management measures based on the actual situation of the project. Combining BIM technology with information monitoring technology can visually manage the entire construction process.

3.4 Subsystems of Intelligent Management System

(1) On-site visual monitoring system

The real-time status of the construction site is one of the important guarantees for auxiliary construction safety.

(2) Dynamic video monitoring system

Install remote control cameras at the required locations on the construction site, while the system continuously saves these video images, and then uses image recognition and tracking technology to analyze relevant changes in a timely manner. The main modules of the system are shown in Fig. 2:

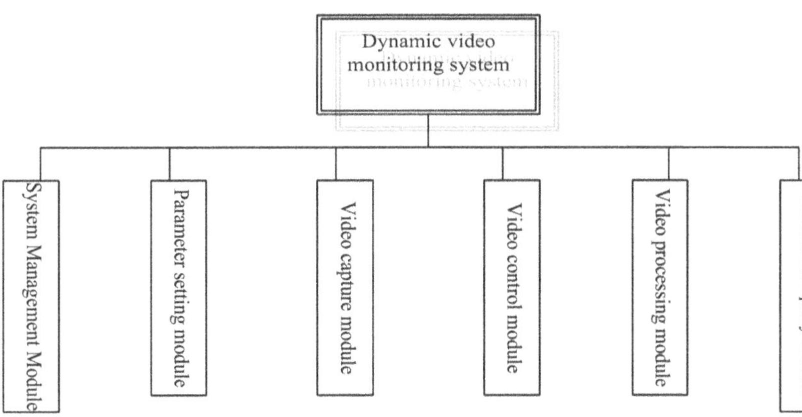

Fig. 2. Dynamic video detection system module

(3) Voice two-way intercom system

The two-way voice intercom can guarantee the real-time contact between the monitoring personnel and the site. Under normal construction conditions, the construction manager and remote monitoring personnel can issue notices or arrange the construction process through it.

(4) Construction quality monitoring system

During the construction process, it is necessary to strictly record every construction process, and unconditionally accept the supervision of the supervision unit and the supervision unit in real time. The system includes two major subsystems: raw material quality inspection system and process quality process control system.

3.5 BIM-Based Construction Safety Management System

(1) BIM-based construction safety management model

1) The data layer includes the building model established by software such as Revit, the construction schedule plan for the key schedule nodes before the construction, and the hazard source information of the project.
2) The model layer refers to the improvement of the construction process, which is divided into construction process improvement and construction safety improvement.
3) The application layer refers to the improvement of the construction process and construction safety management through BIM technology, and finally the optimal plan can be obtained.

(2) Construction safety verification framework model based on BIM-CSKB

One is to use database software for storage and query. The second is to use BIM technology to check the construction plan in the early stage of construction.

(3) Construction safety verification system based on BIM-CSKB

The construction stage is the key stage of production that turns architectural design drawings into engineering objects. The quality of building products largely depends on this stage, and the BIM technology platform integrates and shares information related to project construction safety management.

The logical structure system of the construction safety inspection system includes the data layer, model layer, application layer and client layer, as shown in Fig. 3.

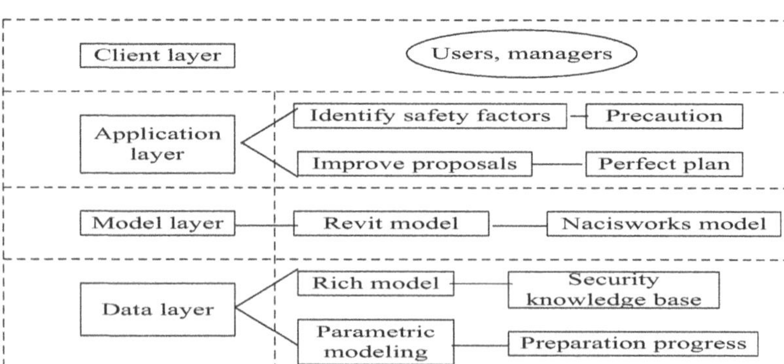

Fig. 3. Framework system for building construction safety verification

4 Realization of Construction Safety Intelligent Management System

4.1 BIM Modeling Software

The project software includes:

Revit: create BIM model, deepen design, and export BIM information
Navisworks Manage: carry out collision check and construction progress simulation
Fuzor: three-dimensional roaming display
CAD: architectural plan design and drawing
Project: preparation of project construction schedule
Photoshop: image processing and editing.

4.2 Construction Simulation

Install 8 tower cranes and arrange the tower cranes with traditional CAD plans. First, use Revit to model, use Navisworks to import the schedule file into the Timeliner tool, and combine the prepared construction organization plan and schedule to simulate the construction process to determine the height. When a problem is detected, immediately modify the original construction plan and re-test until the problem is solved.

4.3 Improvements in Construction Safety Management

(1) Construction preparation

Equipped with safety inspectors, increased safety protection supplies, established a safety production leading group, and conducted construction safety education and training.

(2) Protective measures

During construction operations, stack materials stably without obstructing disassembly and passage. Dispatch full-time safety personnel to carefully check the protective measures for accident-prone parts such as the side of the cave entrance and high-altitude operations. Personnel engaged in high-altitude operations, lifting machinery operations and other special operations shall obtain the corresponding qualification certificates before they can take up their posts.

5 Necessity Analysis of Construction Safety Management System

5.1 Causes of Safety Incidents

According to the data collected by the housing and construction departments of our province in the safety accident notification, the common causes of accidents in our province are as follows: falling from high altitude, collapse, heavy object hitting, lifting injury, etc. The specific situation is shown in Table 1:

Table 1. Causes of construction safety accidents in our province from 2016 to 2020

	2016	2017	2018	2019	2020
Fall from height	43.12	51.29	52.33	54.76	51.37
Collapse accident	13.24	14.21	15.31	13.29	11.63
Object hit	16.32	12.35	11.97	14.64	15.60
Lifting injury	5.39	6.74	8.35	9.27	7.56
Mechanical injury	4.31	5.12	4.36	3.57	3.69
Electric shock	3.69	2.46	2.10	4.12	2.17
Poisoning and suffocation	2.73	1.35	1.54	1.56	2.12

As shown in Fig. 4, we can conclude that the incidence of high-altitude fall accidents is generally as high as 50%, indicating that high-altitude falls often occur during the construction process, so attention must be paid to this.

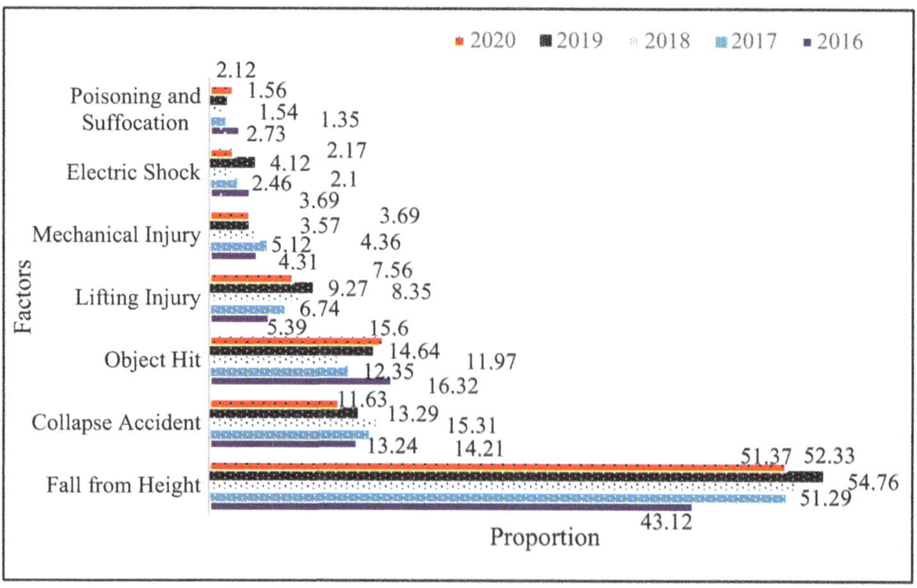

Fig. 4. Causes of construction safety accidents in our province from 2016 to 2020

6 Conclusion

The role of BIM technology in construction is to simulate the construction process based on information and data, find dangerous factors before construction, and provide solutions. For the design of the security intelligent management system, this article involves three-dimensional model, visualization, virtual roaming, etc. for sensory experience. In addition to analyzing the existing problems of the safety management system, this article also describes the directions that need to be considered in the design of the system, such as health monitoring systems, real-time positioning systems, wireless sensing systems, and model parameterization systems. Early warning and disposal of dangers are also needed. Through these designs, continuously improve the safety and intelligent management system and increase its value.

References

1. Nguyen, P.T.: Construction site layout planning and safety management using fuzzy-based bee colony optimization model. Neural Comput. Appl. **33**(11), 5821–5842 (2020). https://doi.org/10.1007/s00521-020-05361-0
2. Ademolu, O., Aigbavboa, C., Thwala, W.: An effective strategy for safety management and enhanced productivity in construction industry. In: Kantola, J.I., Nazir, S., Salminen, V. (eds.) AHFE 2021. LNNS, vol. 267, pp. 86–91. Springer, Cham (2021). https://doi.org/10.1007/978-3-030-80876-1_13
3. Tchouanguem Djuedja, J.F., Karray, M.-H., Kamsu-Foguem, B., Magniont, C., Henry Abanda, F., Smith, B.: BFO-based ontology enhancement to promote interoperability in BIM. Appl. Ontol. **16**(4), 453–479 (2021)
4. Barzegar, M., Rajabifard, A., Kalantari, M., Atazadeh, B.: An IFC-based database schema for mapping BIM data into a 3D spatially enabled land administration database. Int. J. Digit. Earth **14**(6), 736–765 (2021)
5. Sattler, L., Lamouri, S., Pellerin, R., Fortineau, V., Larabi, M., Maigne, T.: A query-based framework to improve BIM multi-domain collaboration. Enterp. Inf. Syst. **15**(10), 1395–1417 (2021)
6. Ahmed, E., Tarig, A., James, B.: High-quality building information models (BIMs) using geospatial datasets. Earth Sci. Inf. **14**(2), 847–860 (2021). https://doi.org/10.1007/s12145-021-00591-9
7. Henry Abanda, F., Musa, A.M., Clermont, P., Tah, J.H.M., Oti, A.H.: A BIM-based framework for construction project scheduling risk management. Int. J. Comput. Aided Eng. Technol. **12**(2), 182–218 (2020)
8. Bimarta, R., Kim, K.-H.: A robust frequency-adaptive current control of a grid-connected inverter based on LMI-LQR under polytopic uncertainties. IEEE Access **8**, 28756–28773 (2020)
9. Mohammed, B.H., Safie, N., Sallehuddin, H., Hussain, A.H.B.: Building information modelling (BIM) and the internet-of-things (IoT): a systematic mapping study. IEEE Access **8**, 155171–155183 (2020). https://doi.org/10.1109/ACCESS.2020.3016919
10. Badran, D., AlZubaidi, R., Venkatachalam, S.: BIM based risk management for design bid build (DBB) design process in the United Arab Emirates: a conceptual framework. Int. J. Syst. Assur. Eng. Manag. **11**(6), 1339–1361 (2020). https://doi.org/10.1007/s13198-020-00993-w
11. Prieto, S.A., Adán, A., Quintana, B.: Preparation and enhancement of 3D laser scanner data for realistic coloured BIM models. Vis. Comput. **36**(1), 113–126 (2020)
12. Rebelo, F., Santos, D., Noriega, P., Figueiredo, C., Oliveira, T., Vilar, E.: Expected architects acceptance of a BIM tool to optimize the building energetic performance. In: Rebelo, F., Soares, M. (eds.) AHFE 2020. AISC, vol. 1203, pp. 249–255. Springer, Cham (2020). https://doi.org/10.1007/978-3-030-51038-1_35

Design of Weather Monitoring and Forecasting System Based on Computer Distributed Network

Jianye Cui(✉), Jian Huang, Youchun Li, and Yingwei Zhu

State Grid Zhejiang Electric Power Co., Ltd., Company Jinhua Power Supply Company, Jinhua 321000, Zhejiang, China
wenrrq911@163.com

Abstract. Meteorological monitoring and forecasting system is a new type of work equipment with real-time, high effectiveness, high degree of automation, convenient equipment maintenance, low economic cost, wide range of use, and long-term operation in harsh environments. Meteorological disaster monitoring and early warning systems have become an important part of modern meteorological services. Therefore, for the design of the weather monitoring and forecasting system, this article discusses from the distributed network. This article mainly uses experimental methods, data analysis methods and other methods to conduct an in-depth discussion on the research of system construction. Experimental results show that the system designed in this paper has a certain effect on weather monitoring and forecasting, and its accuracy can reach 98%.

Keywords: Computer · Distributed network · Weather monitoring · Forecasting system

1 Introduction

With the rapid development of China's economy, the government and the general public have put forward higher requirements for the accuracy and refinement of weather forecast, and also put forward new requirements for the pertinence, popularity and guidance of weather forecast. And meteorological disaster prevention and mitigation has put forward higher and higher requirements for the improvement of the detection and early warning capabilities of severe weather and meteorological disasters. Compared with modern hardware construction, software construction obviously lags behind. Although aiming at the actual business needs, some business systems have been developed and introduced by various meteorological administrations, but different business systems are independent from each other, and the sharing rate is poor, so there are many simple repetitive labor and transactional work in daily work. Meteorological service is also a basic public welfare undertaking for social development and people's life as well as economic and national defense construction. Its development is of great significance to the national economy and people's livelihood. Therefore, it is helpful to improve the quality of meteorological service to make full use of modern technology such as computer, network

I. Ahmad et al. (Eds.): STSIoT 2021, LNDECT 122, pp. 176–186, 2023.
https://doi.org/10.1007/978-981-19-3632-6_23

and satellite to carry out research on meteorological information and forecast information technology. As atmospheric science theory, numerical calculation method and the continuous development of high-performance computer technology, modern weather forecast method have been established from the traditional qualitative theory, mathematical statistics and forecasters in atmospheric semi-empirical method on the basis of experience, development on the basis of the theory of atmospheric science, through the high performance computing platform simulation prediction results of the numerical prediction method. Numerical weather forecast has effectively solved the problems in the past, such as insufficient forecast products, too short possible forecast time, and weak comprehensive processing ability of all kinds of meteorological observation data. It has become an important basis and fundamental way for meteorological departments to make weather forecast, and has irreplaceable status and function of other forecast methods. In management, with the rapid development of science and technology, the position of the computer has been gradually important.

Every year due to meteorological disasters, the modernization of meteorology has been greatly improved with the continuous development of technology. The weather detection business has also shifted from manual to automated. The development of automation can reduce unnecessary manual errors and achieve preventable and understandable effects. Therefore, based on the characteristics of the distributed network and the algorithm rules, this article is necessary for the design and research of the weather monitoring and forecasting system.

2 Related Work

For example, M. Doostmohammadian et al. consider a distributed support vector machine (SVM) binary classification problem. The idea is to train a network of agents with a limited amount of data to collectively learn an SVM classifier from a global database [1]. According to F. Rahimi and H. Rezaei, the error estimation controller design problem studies a class of nonlinear network systems with respect to communication topologies. In real systems, data may be delayed or lost due to unreliable communication channels in exchanges between agents and their neighbors [2]. In recent years, surveillance systems have played an important role in our lives. Therefore, Jamal Mabrouki et al. proposed an automatic weather monitoring system that can realize real-time dynamic climate data in a specific area [3]. Agnes Semwanga Rwashana argues that developing countries can benefit from ICT only if the framework conditions for the use of technology are adapted to their specific needs. The lack of timely, accurate and reliable meteorological data has caused serious loss of life and property in Uganda [4]. M. T. Falconi and F. S. Marzano proposed that backscattering properties allow retrieval of useful geophysical particle distribution parameters when electromagnetic radiation interacts with particle distributions, resulting in absorption and scattering. We describe ground-based weather radar as well as other platforms, such as airborne and space-based configurations [5]. Short-term load forecasting (STLF) plays a vital role in regulated power systems and electricity markets and is often used to predict the outcome of outages. Khalil et al.

Studied and evaluated the potential of combining satellite precipitation data set (MSPD) with rain gauge (RGS) and satellite precipitation data set (SPD) in monitoring meteorological drought in Pakistan from 2000 to 2015 [6]. Mikhail designed an automatic hydrometeorological station (ahms) to monitor high anthropogenic pressure areas along the coast of Lake Baikal [7].

There are so many people researching weather monitoring systems from different aspects, but the research on the monitoring system of the computer distributed network is still a research hotspot, so this article is also based on this for the forecast system design.

3 Weather Monitoring and Forecasting System Based on Computer Distributed Network

3.1 The Necessity of System Construction

Most weather early warning equipment does not have remote monitoring and automatic alarm functions. A connection failure during operation can only be found through telephone consultation, manual troubleshooting, and recording. Key data and technical indicators of the equipment in operation cannot be obtained. If the device has hidden vulnerabilities, serious errors can easily occur. Therefore, to maintain the normal operation of the equipment, in addition to doing routine maintenance, it is necessary to have a substantial and complete system that can monitor the operating status of the equipment in real time to further improve the operation of the equipment. It can grasp the operating status of system equipment in time, and realize remote diagnosis and maintenance of system faults [8].

At present, the monitoring and management capabilities of the weather detection network are obviously insufficient. Traditional maintenance support methods show the following defects: monitoring information is single, monitoring information management is difficult, monitoring status data is not uploaded in real time, and monitoring methods are conservative.

3.2 System Construction Goals, Principles and Requirements

(1) Goal

The overall design goal of the platform is to create a unified sharing system of weather forecast and early warning information centered on the data center, including conventional weather observation data, unconventional weather observation data, monitoring information, forecast analysis data, weather forecast warning information, etc.

(2) Principle

1) Normative. The established system is mainly provided to the meteorological department. Therefore, it must comply with the meteorological business specifications and meet the needs of the meteorological business system. The data collection time, data organization and data storage can all meet the business system specifications.

2) Scalability. Meteorological undertakings are also undergoing continuous development, and new requirements will also be put forward for ground observation systems, such as increasing observation density, adjusting observation time, and new communication methods. We must fully consider the future development direction of ground meteorological observations in system design. It is to give full consideration [9].
3) Practicality. The forecast and early warning information release system needs to have an in-depth understanding of user needs and has a scientific design concept and technical structure. Therefore, it can have sufficient predictability when solving business and technical problems to meet the long-term business development needs in the future.
4) Manageability. The management of the entire system can be realized by the sub-station administrator.

(3) Demand

With the automation of the ground weather observation system, the number of observation sites has increased significantly, and the observation time density has also increased significantly. The remote monitoring and alarm system of the comprehensive meteorological observation system is an industry-specific monitoring business application software designed around the business of monitoring the operation of the ground observation system for the needs of ground observation automation [10].

3.3 The Key Technology Used by the System

(1) MESIS system

MESIS is a new generation of weather service product analysis and production system. It is produced under the background of information technology such as geographic information, database, visualization, multimedia and Web.

(2) UML modeling technology.

UML eliminates unnecessary differences between modeling languages.

(3) Web Service technology.

Web service technology is a technical framework that calls each other in the Internet environment, which can add other functions to WEB applications.

(4) XML document object model.

XML is a universal and adaptable format, it can be used anywhere.

(5) Visual InterDev

Visual InterDev is a Web development tool provided by Microsoft for programmers and developers. It has a powerful integrated database tool. It provides an integrated and visual development environment and is also a complete and detailed website development system. It can be used to achieve Fast and visual development. It has the characteristics of dynamics, powerful database tools, integration with existing business systems, and reusability of components.

(6) Neural network prediction method

The neuron model of the artificial neural network is designed by simulating the information input and output characteristics of biological neurons.

Σ represents the accumulation of all input signals. G(I) represents the response function of the neuron. The expression of the neuron model is:

$$I = \sum\nolimits_{s=1}^{m} A_s W_s - \chi \quad (1)$$

$$R = G(I) \quad (2)$$

For the response function F(S), different functions can be selected according to actual needs, which mainly include three types of response functions, namely linear function, Sigmoid function, and step function. The expressions are as follows:

Linear function

$$G(I) = S * I \quad (3)$$

Sigmoid function

$$G(I) = \frac{1}{1 + f^{-SI}} \quad (4)$$

Step function

$$G(I) = \begin{cases} 1, & I \geq 0 \\ 0, & I \prec 0 \end{cases} \quad (5)$$

3.4 Design of the Monitoring Center of the Early Warning System

(1) Design of data acquisition and preprocessing module

1) Data collection. In the early warning system, data collection must be carried out first, and the performance of the module that intercepts network data is the basis for realizing the high efficiency of the early warning system.
2) Data preprocessing. After the network data packet is captured by the data acquisition device, the data is stored in a table. For the captured data packet, extract the connection record in it. The function of the data preprocessing module is to organize the data in the data packet into a format that can be directly processed by the data mining algorithm for analysis [11, 12].

(2) Improvement of association analysis algorithm and establishment of association rules

The data mining algorithm based on association rules is divided into two steps: find out the set of all elements, and generate strong association rules from the set of elements.

For credibility, the following formula can be used.

$$confidence(Q \Rightarrow S) = P(Q|S) = \text{sup\_}count(Q \cup S)/\text{sup\_}count(Q) \quad (6)$$

Among them, $sup\_count(Q \cup S)$ is the number of transactions that include $Q \cup S$ itemset, and $sup\_count(Q)$ is the number of transactions that include itemsets Q. According to this formula, the association rules can be generated as follows:

For each frequent itemset s, all non-empty subsets of s are generated;

For each non-empty subset k of s, if

$$\text{sup\_}count(Q \cup S)/\text{sup\_}count(Q) \geq \text{min\_}conf \quad (7)$$

Then output rule "$k \Rightarrow (s - k)$". Among them, min_$conf$ is the minimum confidence threshold.

(3) Establishment of classification rule set

In a period of time, there are many connections in the network, and there will be confusion. Therefore, it is necessary to collect and restore all the data packets of the same connection.

1) The selection method of characteristic attributes. It is arranged according to the chronological order of the appearance of the data, and statistics are made for multiple connection records.
2) Mining design of classification rules. The classification attribute of the record is extracted, and the classification model obtained is used to map it to a specific class.

(4) Use mining rules for detection

First, take the detection rules in the detection rule set as the standard, judge the records to be detected, and compare the records to be detected with the rules in the rule set one by one to see if the rules are met:

1) If the detected network connection record conforms to a rule in the rule set, its record conforms to the type of a certain rule, and the classification result can be output.
2) If there is no rule in the detection rule set that matches the detected record, it proves that the record has never appeared before, and it is a new connection.

3.5 Design of Meteorological Data Analysis System

(1) The weather history data analysis system is a comprehensive system that integrates data collection, data filtering, data analysis and structure display. The structure can be divided into 4 modules, as shown in Fig. 1:

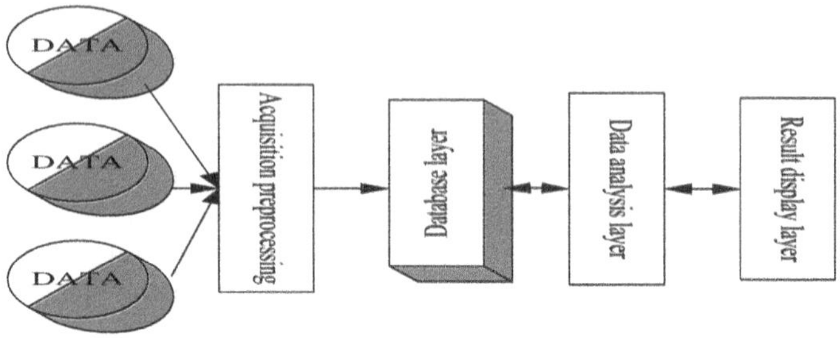

Fig. 1. The overall structure of the meteorological data analysis system

(2) System function

From the user's point of view, the system functions include 6 modules including single-site live display, regional live display, single-site data statistics, regional data statistics, time-dependent data statistics and report statistics.

1) Single station live display: display the forecast data value of each mode to a single meteorological station based on the time distribution curve graph comparison, histogram comparison and data table as needed.
2) Regional live display: According to the needs, display the forecast data value of each mode to the individual weather station based on the curve graph comparison, histogram comparison and data table of the regional distribution.
3) Single-site data statistics: Taking timeliness as the main element, statistically analyze the weather forecast quality distribution of each site under different elements, and its display methods include graphs, histograms and data tables.
4) Regional data statistics: Statistics of the weather forecast quality distribution of all stations in each area under different elements by region. In addition to the three comparison methods of graphs, histograms and data tables, a map comparison mode is also introduced.
5) Time-effectiveness data statistics: With time-effectiveness as the main element, statistically analyze the distribution of weather forecast quality under different elements in each area as a whole, and its display methods include graphs, histograms and data tables.
6) Report statistics: According to the needs of users, generate data reports needed by them to assist them in completing their work.

3.6 Overall System Design

(1) Hardware design

In order to design a suitable monitoring unit, it is necessary to select a suitable microprocessor and design corresponding sensors and communication interfaces. Figure 2 shows the overall structure of the data acquisition sensor hardware system.

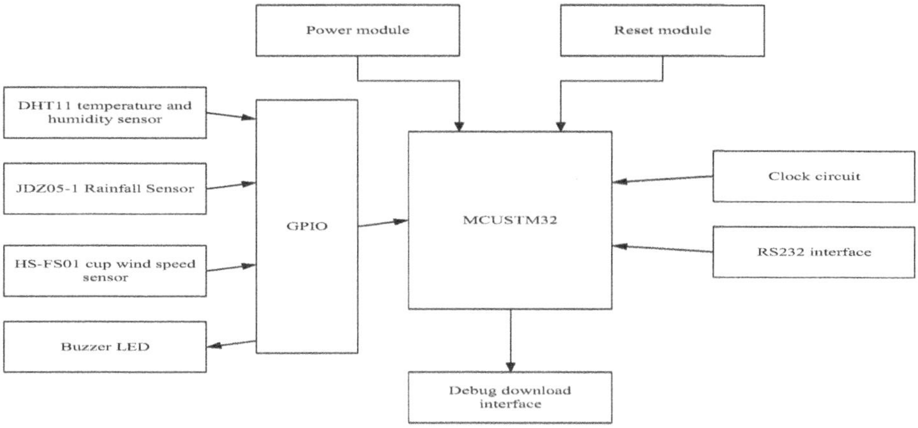

Fig. 2. System hardware overall design

Among them, the MCU uses an interconnected series of microcontrollers. The power module ensures the normal operation of the entire system. Through the RS232 interface, the wireless module can be used to communicate with the host computer.

(2) Program design

The normal operation of the data acquisition sensor terminal requires the support of the program, and a good program design can make the hardware work stably and ensure the normal operation of the entire acquisition system. The entire program is designed around the STM32f103RET6 microcontroller. After the system is powered on, first initialize each module. Then upload the data according to the design requirements of each module, which can be active transmission. According to the design requirements, the data is transmitted every certain time.

(3) Host computer design

Taking into account the requirements of the weather data monitoring system for data storage, a SQL Server database is designed here to build user information tables, temperature data tables, rainfall data tables, wind speed data tables, and humidity data tables.

4 System Function Test

After the development of the meteorological disaster monitoring and early warning and emergency service auxiliary decision-making system is completed, in order to ensure the safe and stable operation of the system, system testing is required.

4.1 Hardware Environment

(1) Server-side configuration
    CPU: Intel(R) Core(TM) i5 2.6 GHz
    Memory: 8G
    Hard Disk: (SCSI) 50G*3
    Network card: 100/1000M
(2) Data collection and processing terminal configuration
    CPU: Intel(R) Core(TM) i5 2.6 GHz
    Memory: 4G
    Hard Disk: (SATA) 100G
    Network card: 10/1000M

4.2 Operating System and Development Environment

The server operating system is Windows Server 2018+SQL Server 2018, the data processing operating system is Windows XP, and the development and debugging tools are Microsoft Visual Basic.

4.3 System Test

After the development of the system is completed, the meteorological information is measured. The main collected data include temperature, rainfall, wind speed, humidity, etc.

5 Analysis of Test Results

Use the host computer serial port for data collection, and get the following data. The host computer has well realized the monitoring of temperature, rainfall, wind speed and humidity data. The details are shown in Table 1.

According to Table 1, we can know that the specific monitoring conditions of temperature, rainfall, wind speed and humidity collected in this article are more in line with the actual weather conditions of the monitored area. From the perspective of temperature changes, the temperature in this area remains cool. From the point of view of rainfall, there is a certain amount of precipitation, so its humidity is maintained at a relatively high level.

Table 1. Test data

	Temperature value (°C)	Rainfall value (mm)	Wind speed value (m/s)	Humidity value (%)
1	23	12	10	65
2	26	17	15	70
3	21	11	17	68
4	20	13	12	73
5	25	16	15	80
6	22	17	16	81
7	20	13	11	69
8	24	15	13	74

6 Conclusion

From ancient times to the present, weather monitoring has been an important indicator task in agriculture and daily life. In ancient times, there were night observation celestial phenomena, seismographs, and weather forecasts based on common sense of life, but today's weather forecasts rely on science and technology, satellites and the movement of the earth. The weather forecast can give people hints, such as whether it is suitable to go out, whether it is suitable to grow crops, and whether it is suitable for production. Therefore, the monitoring and early warning of meteorology is a very basic and important research activity. Especially for the study of meteorological disasters, it is necessary to predict in advance to reduce losses. For the real-time monitoring design of the weather system, the idea proposed in this paper is to highlight the following aspects of the overall design of the system: one is data processing and analysis, and the other is early warning monitoring. The system of this article needs to have excellent data statistics and analysis capabilities, and its early warning capabilities cannot be ignored. In addition, the overall design of the system needs to be supported by sensitive sensors and controller devices. Although the system designed in this paper can basically achieve the functions of monitoring and forecasting, it still has certain shortcomings. The monitoring of extreme weather can also be displayed in a dynamic way to make it more specific and vivid.

Acknowledgements. Fund project: Science and technology project of State Grid Zhejiang Electric Power Co., LTD. "Research on networked historical meteorological simulation and high-precision meteorological forecast technology for improving power dispatching Safety Production" (Project No.: 5211JH180082).

References

1. Doostmohammadian, M., Aghasi, A., Charalambous, T., Khan, U.A.: Distributed support vector machines over dynamic balanced directed networks. IEEE Control Syst. Lett. **6**, 758–763 (2022). https://doi.org/10.1109/LCSYS.2021.3086388

2. Rahimi, F., Rezaei, H.: A distributed fault estimation approach for a class of continuous-time nonlinear networked systems subject to communication delays. IEEE Control Syst. Lett. **6**, 295–300 (2022). https://doi.org/10.1109/LCSYS.2021.3071478
3. Mabrouki, J., Azrour, M., Dhiba, D., Farhaoui, Y., El Hajjaji, S.: IoT-based data logger for weather monitoring using Arduino-based wireless sensor networks with remote graphical application and alerts. Big Data Min. Anal. **4**(1), 25–32 (2021)
4. Rwashana, A.S., Atwine, A.M.: Examining the adoption of ICTs for weather monitoring and climate change adaptation in Uganda. Int. J. ICT Res. Afr. Middle East **9**(1), 67–81 (2020)
5. Falconi, M.T., Marzano, F.S.: Weather radar data processing and atmospheric applications: an overview of tools for monitoring clouds and detecting wind shear. IEEE Signal Process. Mag. **36**(4), 85–97 (2019). https://doi.org/10.1109/MSP.2019.2890934
6. Rahman, K.U., Shang, S., Zohaib, M.: Assessment of merged satellite precipitation datasets in monitoring meteorological drought over Pakistan. Remote Sens. **13**(9), 1662 (2021). https://doi.org/10.3390/rs13091662
7. Makarov, M., Aslamov, I., Gnatovsky, R.: Environmental monitoring of the littoral zone of lake baikal using a network of automatic hydro-meteorological stations: development and trial run. Sensors **21**(22), 7659 (2021)
8. Salheddine, S., Ouissal, S., Benslama, M., Abderraouf, M., Beylot, A.-L.: Development of an intelligent electronic sentinel for the monitoring and detection of meteorological phenomena due to global climate change. In: ICAASE, pp. 1–7 (2020)
9. Bhardwaj, A., Misra, V.: Monitoring the Indian summer monsoon evolution at the granularity of the Indian meteorological sub-divisions using remotely sensed rainfall products. Remote Sens. **11**(9), 1080 (2019)
10. Bebarta, D.K., Das, T.K., Chowdhary, C.L., Gao, X.-Z.: An intelligent hybrid system for forecasting stock and forex trading signals using optimized recurrent FLANN and case-based reasoning. Int. J. Comput. Intell. Syst. **14**(1), 1763–1772 (2021)
11. Fenu, G., Malloci, F.M.: Lands DSS: a decision support system for forecasting crop disease in Southern Sardinia. Int. J. Decis. Support Syst. Technol. **13**(1), 21–33 (2021). https://doi.org/10.4018/IJDSST.2021010102
12. Kumar, G., Singh, U.P., Jain, S.: Hybrid evolutionary intelligent system and hybrid time series econometric model for stock price forecasting. Int. J. Int. Syst. **36**(9), 4902–4935 (2021). https://doi.org/10.1002/int.22495

Path Planning of Indoor Mobile Educational Robot Based on Improved Deep Reinforcement Learning

Weiping Zhu[1(✉)] and Wonchana Katsri[2]

[1] School of Humanities and Education, Nanchang Institute of Technology, Nanchang, Jiangxi, China
zhuweiping1991410@163.com
[2] Suan Dusit University, Bangkok, Thailand

Abstract. With the maturity of artificial intelligence and Internet of Things technology, the research on robots has also become one of the hotspots of artificial intelligence. Indoor mobile educational robots are an important part of machine intelligence. Research on the path of indoor mobile educational robots has become a key point in machine research. The purpose of this paper is to study the path planning of indoor mobile educational robots to improve deep reinforcement learning. This article first summarizes the research status of mobile educational robots at home and abroad. On this basis, the kinematics model of the indoor mobile educational robot is researched and analyzed. This article systematically elaborates the path planning based on the Actor-Critic algorithm and the deep reinforcement learning training model based on the minimum depth of field information. And use comparative analysis method, observation method and other research methods to carry out experimental research on the theme of this article. Research shows that the Actor-Critic algorithm proposed in this paper is shorter in path planning time and path distance than traditional algorithms.

Keywords: Deep learning · Mobile robots · Path planning · Applied research

1 Introduction

With the rapid development of intelligent vision technology, deep learning and other technologies, robots have also begun to be used in the education field [1, 2]. Mobile educational robots have good teaching practicability, compatibility and other characteristics, and are one of the important means for education and scientific research, training students' practical ability [3, 4].

At present, the research on mobile robots is relatively in-depth at home and abroad, but the overall development and research of educational robots is still in its infancy [5, 6]. The research field of foreign educational robots mainly supports language education and skill education through simple body movements [7, 8]. Compared with foreign countries, many universities and research institutes in my country have also carried out research on robot applications. Including the Robot Research Institute of Huazhong University

I. Ahmad et al. (Eds.): STSIoT 2021, LNDECT 122, pp. 187–195, 2023.
https://doi.org/10.1007/978-981-19-3632-6_24

of Science and Technology, Baidu's ABC Robot platform, etc., it shows the charm of robots in China [9, 10]. However, most domestic robots cannot provide detailed services such as human-computer interaction with service objects [11, 12].

The purpose of this paper is to improve the efficiency of indoor robot planning, and to study the path planning of indoor mobile educational robots that improve deep reinforcement learning. By running and comparing the algorithm studied in this paper with the traditional path planning algorithm on simulation software, the research topic of this paper is analyzed.

2 Application Research on Path Planning of Indoor Mobile Educational Robots to Improve Deep Reinforcement Learning

2.1 Kinematics Model Design

The indoor mobile educational robot has two wheels on different axes on the same straight line. By controlling the speed of the two wheels, the robot can rotate on the spot, move in a straight line, or rotate or move according to a specified trajectory. xc, yc, θc() is the pose of the robot's center of gravity in the world coordinate system. Figure 1 shows the pose model of an indoor mobile educational robot.

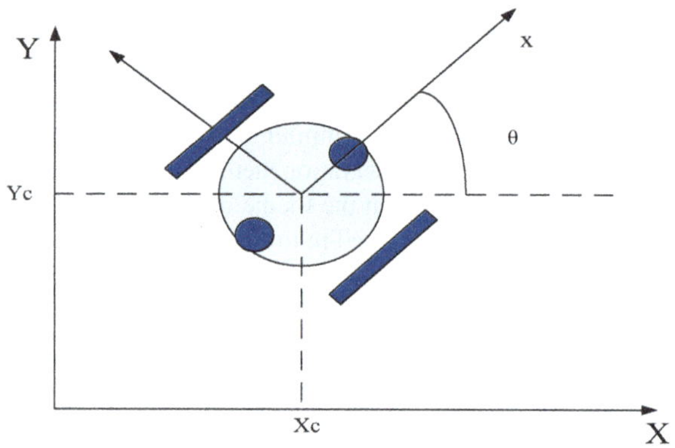

Fig. 1. Pose model of two-wheel differential mobile robot

(1) Forward kinematics equation

Figure 2 shows the pose of the robot at the interval Δt, and the forward kinematics equation is established.

1) Left and right wheel speed

$$V_l = \omega(R - L/2) \tag{1}$$

$$V_r = \omega\left(R + \frac{L}{2}\right) \tag{2}$$

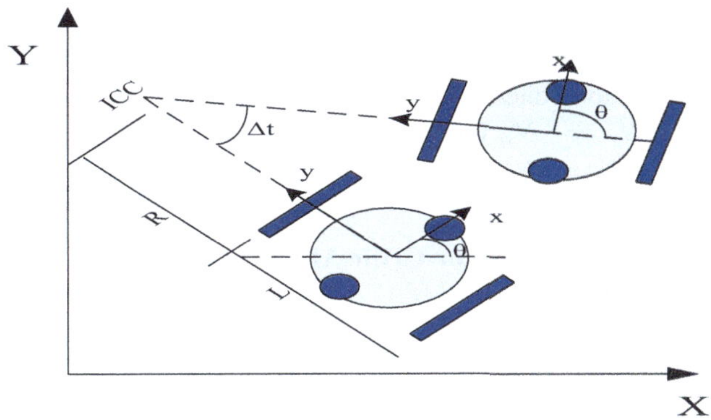

Fig. 2. Mobile robots in different poses

2) Angular velocity

$$\omega = \frac{V_l - V_r}{L} \quad (3)$$

3) Instantaneous radius of curvature

$$R = \frac{V_l}{\omega} + \frac{L}{2} \quad (4)$$

4) ICC coordinates

$$ICC = [ICC_x, ICC_y] = [x - R\sin\theta, y + R\sin\theta] \quad (5)$$

5) Direction angle after moving Δt time

$$\theta' = \omega \Delta t + \theta \quad (6)$$

In practical applications and difficult to accurately measure, they are calculated from the odometer data:

$$v = n \times step/\Delta t \quad (7)$$

So you will get:

$$R = \frac{L(n_l + n_r)}{2(n_l - n_r)} \quad (8)$$

$$w\Delta t = \frac{(n_l + n_r) \times step}{L} \quad (9)$$

In the formula n_l, n_r are the calculated values of the left and right encoders; step≈0.1 mm, from which the posture after Δt time will be obtained:

$$\begin{pmatrix} x' \\ y' \\ \theta' \end{pmatrix} = \begin{pmatrix} cos(w\Delta t) & -sin(w\Delta t) & 0 \\ sin(w\Delta t) & cos(w\Delta t) & 0 \\ 0 & 0 & 1 \end{pmatrix} \begin{pmatrix} x - ICC_x \\ y - ICC_y \\ \theta \end{pmatrix} + \begin{pmatrix} ICC_x \\ ICC_y \\ w\Delta t \end{pmatrix} \quad (10)$$

(2) Inverse kinematics

Substituting formula (10) into formula (6) can determine the special situation under certain motion states. If $V_r = V_l$, it means that the robot is moving in a straight line. If $V_r = -V_l$, the robot is moving in a rotation, and θ can take any value.

2.2 Path Planning Based on Actor-Critic Algorithm

(1) Reinforcement learning method based on value function

Robot navigation is actually the task of interaction between the robot and the environment. In each step of the cycle, the robot starts its movement with a set of set motion actions, such as forward, backward, left, and right. Select action $A = \{1, \ldots, K\}$, and the action command will be passed to the robot to interact with the environment. In general, the environment may change at any time, so navigation based on reinforcement learning can be used to avoid obstacles in a dynamic environment. The controller does not need to know environmental information. On the contrary, viewing the image from the environment, it is a vector representing the original pixels of the current camera. Moreover, it will receive a reward that represents the change in the scoring score. In addition, the rating score may depend on previous actions and overall observations. You can only receive feedback about the operation after thousands of loop steps.

Since the robot only observes the current camera image, it can only observe part of the robot's state and tasks. In other words, it is impossible to get a complete picture of the current state from the current camera x_t. So, learn a series of actions and observations, $s_t = x_1, a_1, x_2, \ldots, a_{t-1}, x_t$ and evaluation strategies that depend on these sequences. Assume that all these sequences in the controller end in finite loop steps. The sequence format is the Markov Decision Process (MDP).

MDP is composed of five key elements $\{S, A, P, R, \gamma\}$, S stands for finite state set, A stands for finite action set, P is the state transition probability matrix, R is the reward function, and γ stands for time lapse conversion factor $\gamma \in [0,1]$. Among them:

$$P_{ss'}^{a} = P[S_{t+1} = s' | S_t = s, A_t = a] \tag{11}$$

The goal of the robot is to maximize the future reward value by choosing actions during the interaction with the environment. We make a standard assumption that the future reward will be affected by the factors of each time step. γ is converted, and the future discounted income is defined in time T. In other words, T is the time step at which the game ends. An important problem in MDP is to find the strategy $\pi(s)$. Select an action in the state s_t to form the function of the action. The purpose of this article is to choose the action function π that maximizes the cumulative return R(T).

$$R(T) = \sum\nolimits_{t=1}^{T} \gamma' R_{a_t}(S_t, S_{t+1}) \tag{12}$$

The most typical algorithm based on value functions is the Q-learning algorithm. In theory, the Q-learning algorithm can be used to train robot path planning.

However, the depth image as the state of the robot greatly increases the dimensionality of the state space. With the current performance of computer hardware, it is almost impossible to process such a large amount of data in real time, and it is even more impossible to complete real-time obstacle avoidance tasks. Therefore, in this article, we will use deep learning to process deep images and optimize the loss function.

(2) Gradient algorithm of improved value function

The strategy gradient method has several advantages: expertise can be directly incorporated into strategy parameterization, and usually requires much fewer parameters than the corresponding value function to represent the optimal strategy; the optimization result can at least converge to a local optimization strategy, which can handle continuous status and actions, as well as incomplete status information. Table 1 shows the comparison between Q-learning and Policy Gradient Algorithms.

Table 1. Comparison of Q-learning and policy gradient algorithms

Q-learning	Policy gradient
Value function parameterization, poor convergence	Simpler strategy parameterization and better convergence
Unable to solve the case where the action is a continuous set	Able to handle continuous states and actions
Single step update, fast convergence	Discrete problems converge slowly
No local optima	Easy to fall into the local minimum problem

It can be seen from Table 1 that the variance based on Policy Gradient is relatively high, and instead of using v^t to estimate the action value function Q, use Critic to estimate Q, namely $Q\pi_\theta(s, a) \approx Q^w(s, a)$, which is called the Actor-Critic algorithm based on the Q value.

2.3 Deep Reinforcement Learning Training Based on Minimum Depth of Field Information

(1) Optimize the construction of state space

At the beginning of learning, the robot was very "unfamiliar" with the unknown environment and had not yet established a stable state space. Random actions need to be performed to obtain rewards for different actions in each state and slowly build up the environment. When determining the movement of the robot, the robot collects the depth image information in the environment in real time as the real-time state at the beginning of the training. Randomly give linear velocity and angular velocity motion, calculate the reward value of the exercise, collect the depth image information after the exercise, and repeat this process until the probability of the reward value of the sufficient state and corresponding action is collected. When the reward value converges to a certain value, the training is finally completed.

In this process of trial and error, when the robot moves away from the obstacle, the state at this time can be called the non-obstacle avoidance request state. Therefore, it is hoped that the robot can establish a state space for obstacle avoidance during the training process. It is used to construct the initial state space and improve the efficiency of subsequent training.

(2) Training based on minimum depth of field information

In order to guide the robot to better construct the state space in the early stage, this paper proposes a selective training mode based on the minimum depth image to increase the training speed and accelerate the network convergence. The minimum value is obtained by traversing the depth image matrix. If the minimum value is greater than the preset threshold, the robot's motion is controlled in a point-stable manner to make the robot move to the target point smoothly. In the movement process, when the minimum depth image is lower than the threshold, the depth image is input to the network for training, and the next action is determined randomly or according to the network. If the minimum value is again greater than the threshold, the robot will move back and forth because it is subject to the kinematics constraints of the next action. In the movement process, when the minimum depth image is lower than the threshold, the depth image is input to the network for training, and the next action is determined randomly or according to the network. If the minimum value is greater than the threshold again, the next movement of the robot is constrained by kinematics and other factors.

3 Experimental Research on Path Planning of Indoor Mobile Educational Robots with Improved Deep Reinforcement Learning

3.1 Experimental Protocol

In order to make this experiment more scientific and effective, this experiment carried out a simulation experiment on the Actor-Critic algorithm researched and designed in this paper in the Matlab simulation software. In this experiment, the default (0, 0) is the starting point on the GUI interface, and the upper right corner (12000, 12000) is selected as the end point. The environment map was constructed by placing regular obstacles. In this experiment, static path rules were performed on two different maps. And use mathematical statistics for statistical analysis of the obtained results.

3.2 Research Methods

(1) Comparative analysis method

This experiment compares and analyzes the Actor-Critic algorithm studied in this article with the traditional path planning algorithm to analyze the feasibility of the subject of this article.

(2) Simulation experiment method

In this experiment, the path planning of the mobile educational robot is simulated on Matlab simulation software and the data is recorded. These data provide a reliable reference for the final research results of this article.

(3) Observation method

In this experiment, the path of the mobile educational robot is observed and data are collected, and the collected data are sorted. These data provide theoretical support for the topic selection of this article.

(4) Mathematical Statistics

Use relevant software to make statistics on the final research results of this article.

4 Experimental Analysis of Path Planning for Indoor Mobile Educational Robots with Improved Deep Reinforcement Learning

4.1 Map 1 Experimental Data Comparison

In order to make this experiment more scientific and effective, this experiment will find an optimal path after iterating from the bottom left to the top right corner, and compare and analyze it with the algorithm studied in this paper. The data obtained are shown in Table 2.

Table 2. Map 1 experimental data comparison

	Actor-critic planning time (s)	Actor-critic actual line length (m)	Traditional planning time (s)	Traditional actual line length (m)
Group 1	2.11	16.03	1.93	15.43
Group 2	1.97	15.62	2.01	25.48
Group 3	2.24	16.24	1.77	15.01
Group 4	2.00	16.95	1.85	15.98

It can be seen from Fig. 3 that compared with the traditional path planning algorithm, the Actor-Critic algorithm proposed in this paper takes less time to find the path planning in the simulation experiment. The walking path is also shorter than the path planned by the traditional algorithm, which fully reflects the excellent performance of the Actor-Critic algorithm studied in this paper.

4.2 Map 1 and 2 Experimental Data Comparison

In order to compare and analyze the algorithms proposed in this paper, this experiment is based on map 1, and another set of simulation map 2 is set up for experiment. Finally, the data obtained from the two experiments are compared. The data obtained as shown in Table 3.

It can be seen from Fig. 4 that in Map 1 and Map 2, the Actor-Critic algorithm studied in this article takes less time to find the shortest path and the distance of the shortest path than the traditional path planning algorithm, which fully reflects the Actor-Critic algorithm studied in this article.

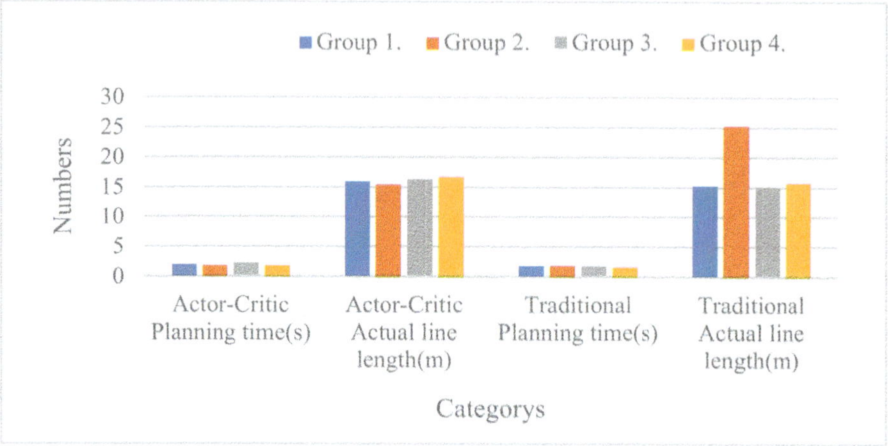

Fig. 3. Map 1 experimental data comparison

Table 3. Map 1, 2 experimental data comparison

	Actor-critic planning time (s)	Actor-critic actual line length (m)	Traditional planning time (s)	Traditional actual line length (m)
Map 1	2.07	16.34	1.87	15.54
Map 2	2.16	17.28	2.00	15.72

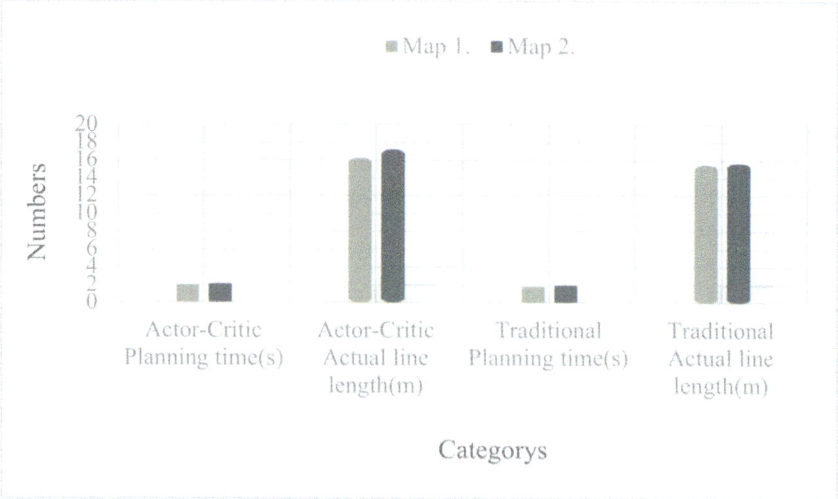

Fig. 4. Map 1, 2 experimental data comparison

5 Conclusion

The purpose of this paper is to study the path planning of indoor mobile educational robots to improve deep reinforcement learning. Through the establishment of the movement model of the mobile educational robot, the path planning algorithm is discussed, and an improved Actor-Critic algorithm is proposed on this basis. Finally, the Actor-Critic algorithm and the traditional path planning algorithm are run and compared on the simulation software. The analysis proves the feasibility of the research in this paper.

Acknowledgements. This work was supported by Jiangxi Provincial Department of Education Science and Technology Research Project-Indoor Positioning Research for Educational Robots (GJJ202119).

References

1. Zhao, M., Lu, H., Yang, S., et al.: A fast robot path planning algorithm based on bidirectional associative learning. Comput. Ind. Eng. **155**(7), 103 (2021)
2. Park, J.H., Huh, U.Y.: Path planning for autonomous mobile robot based on safe space. J. Electr. Eng. Technol. **11**(5), 1441–1448 (2016)
3. Panda, M.R., Dutta, S., Pradhan, S.: Hybridizing InvasiveWeed optimization with firefly algorithm for multi-robot motion planning. Arabian J. Sci. Eng. Section A Sci. **43**(8), 4029–4039 (2018)
4. Lin, H.Y., Huang, Y.C.: Collaborative complete coverage and path planning for multi-robot exploration. Sensors **21**(11), 3709 (2021)
5. Albani, D., Hnig, W., Nardi, D., et al.: Hierarchical task assignment and path finding with limited communication for robot swarms. Appl. Sci. **11**(7), 3115 (2021)
6. Sadiq, A.T., Raheem, F.A., Abbas, N.: Ant colony algorithm improvement for robot arm path planning optimization based on D* strategy. Int. J. Mech. Mechatron. Eng. **21**(1), 96–111 (2021)
7. Wang, W.K., Wu, X.B., Chen, W.B.: Research on path planning of intelligent plant inspection robot. J. Comput. (Taiwan) **29**(2), 174–185 (2018)
8. Guan, Y., Song, C., Dong, H.: Path planning of the free-floating manipulator for capturing a moving target. Jiqiren/Robot **39**(6), 803–811 (2017)
9. Zhu, H.: Motion trajectory planning and simulation of 6-DOF manipulator arm robot. Acad. J. Manuf. Eng. **15**(3), 66–71 (2017)
10. Li, I.-H., Chien, Y.-H., Wang, W.-Y., Kao, Y.-F.: Hybrid intelligent algorithm for indoor path planning and trajectory-tracking control of wheeled mobile robot. Int. J. Fuzzy Syst. **18**(4), 595–608 (2016). https://doi.org/10.1007/s40815-016-0166-0
11. Maeda, G., Ewerton, M., Koert, D., et al.: Acquiring and generalizing the embodiment mapping from human observations to robot skills. IEEE Robot. Autom. Lett. **1**(2), 784–791 (2016)
12. Janutenaite-Bogdaniene, J., Macerauskas, E., Drukteiniene, A., et al.: Cylindrical piezorobot's trajectory planning and control. J. Vibroengineering **19**(4), 2670–2679 (2017)

Optimization Design of Network Information System Based on Big Data Technology

Zejian Dong(✉)

Shandong Vocational College of Light Industry, Zibo 255300, Shandong, China
hanxuebuji_2@126.com

Abstract. With the rapid development of Internet technology, the network information data presents an explosive growth trend, and the society has also entered an era of big data. The era of big data facilitates people's life and work, but it also brings many technical challenges. In order to solve the problems of various formats, form complexity and huge challenges brought to traditional computing technology, this paper optimizes the network information system under the background of big data, analyzes the optimization needs of the system, optimizes the functional module according to the demand content, and finally compares the optimization function. This paper finds that the start and implementation of the information management system is a very important and critical link in the construction of the enterprise information system, and the intelligent network information management system realizes the effect of information security.

Keywords: Big data · Network information · System optimization · Optimization design

1 Introduction

The research and application of network information system optimization design based on big data technology has become an important driving force of global scientific and technological innovation and economic development [1]. In fact, the focus of the word "big data" is far beyond the definition of data scale, which represents the development of information technology into a new era, the new technologies and methods needed for a large number of network information processing, and the new services and value brought by big data applications [2, 3].

Network information system optimization design research based on big data technology has attracted the interest of many experts, and has been studied by many teams. For example, some teams found that many domestic enterprises have begun to learn the development path of foreign enterprises, but many enterprises still use a simple single module to realize part of the business information, lack of road guidance, random business process, irregular process, this aspect of centralized processing of information is not outstanding enough, there is no complete supply chain information system, to conduct comprehensive management and decision support [4]. Most of the companies that implement the supply chain information management system are still branches or

I. Ahmad et al. (Eds.): STSIoT 2021, LNDECT 122, pp. 196–204, 2023.
https://doi.org/10.1007/978-981-19-3632-6_25

joint ventures of foreign multinational corporations, indicating that the supply chain comprehensive information management system is quite extensive and mature in developed countries [5]. Some teams believe that with the development of society, network information products have been strengthened, media resource management system as a resource management platform, must continue to develop, in order to better serve users, combined with widely used solutions, revision for the industry popular media asset management system design part, and propose a more optimized media asset management system design scheme [6]. Although their research results are all very rich, there are still some shortcomings.

According to the characteristics of high data synthesis and strong crossover, this paper conducts systematic research based on big data technology, including distributed storage management, computing system optimization, and machine learning and data analysis algorithm design on big data.

2 Method

2.1 Gathering Model Construction

In the stable operation of the information system, any function can be operated by multiple users, and each user needs work tasks, they sometimes use many functions, therefore, our research perspective is to use "user" and "function" as two nodes in the network, to build a "double-user function" network model [7]. By studying the relationship between users and functions, it reflects the usage behavior and activity rules of the fine-grained system users, but the fine-grained system can obtain individual behavior rules and does not represent a specific case of a person or a module. Therefore, in order to understand the overall situation of the enterprise information system, this paper will take role and function modules as the research perspective, and then explore the general or specificity of user role access rules. Role and function modules corresponding to the upper layer of the user and the upper layer of functional activity, can be obtained through the projection and collection of the "user-function" dual function network. As shown in formula (1–2), R represents the gathered role with m roles where each role is a user reunion at its next level and the number of users in each role is not completely equal to [8].

$$R = \{r_1, r_2, \cdots r_m\} \tag{1}$$

$$r_j = \left\{u_{j1}, u_{j2}, \cdots u_{jm}\right\}(j = 1, 2, \cdots, m) \tag{2}$$

2.2 Relational Equation

Let the initial security probability of the network information system be P_{s_0}, the initial cost is C_0, $\phi(\Delta C)$ for consumption functions, safety probability p_s and has a consumption function $\phi(\Delta C)$ there are relationships as shown in formula (3), in $\Delta C = C - C_0$, C increases the cost of security measures, ΔC for incremental increment, k_p as a proportional factor [9]. Formula (3) is called a relational equation, when the problem of the security probability and initial cost of the initial security are known k_p and consumption function $\phi(\Delta C)$'s questions. At $k_p = 1 - P_{s0}$ the time, it is often ideal, when the relational equation has the following form, as shown in formula (4). From a perspective of safety, consumption function $\phi(\Delta C)$ shall meet the conditions of formula (5), conditions (5) $\phi(\Delta C)$ has such a form as formula (6) [10].

$$P_s = P_{s_o} + k_p\phi(\Delta C) \tag{3}$$

$$P_s = P_{s0} + (1 - P_{s0})\phi(\Delta C) \tag{4}$$

$$\begin{aligned} &P_s = P_{s0}, \Delta C = 0, \phi(0) = 0 \\ &P_s = 1, \Delta C \to \infty, \lim_{\Delta C \to \infty} \phi(\Delta C) = 1 \end{aligned} \tag{5}$$

$$\phi(\Delta C) = 1 - e^{-\frac{1}{a}\Delta C^{\beta}} \tag{6}$$

2.3 Information Management System

In today's economic globalization environment, the market demand is complex and changeable. How to be in the rapidly changing market environment, in a large number of market information and data, we can seize opportunities while avoiding risks, which puts forward higher requirements for the efficiency of the enterprise supply chain. However, many enterprises in the supply chain have difficult to extract useful information, make correct business decisions and lose development opportunities. Therefore, it is very important to use information management system to help enterprises carry out supply chain information reform, help full information sharing among nodes in the supply chain through information management system, and eliminate the obstacles and uncertainty of information transmission in the supply chain [11]. Through the appropriate information strategy, one is to help the enterprise managers to obtain the necessary information in real time, the second is to reduce the information island in the supply chain, promote the cooperation and communication between enterprises, use the concept of supply chain management, eliminate the information bottleneck of enterprises, and improve the operation efficiency of the enterprise information strategy [12].

2.4 Network Information System Security

Network information system security refers to the use of security system, security measures, security strategies, security management, to ensure the normal network information system and its safe operation, which are the results of the comprehensive role of all aspects. It is widely believed internationally that network information systems must take various security protection measures to ensure the integrity, confidentiality and reliability of network information, and combined with effective security management modes, to provide the necessary security support in the transmission, sharing, encryption, access, processing and analysis of network information to make the information service proceed normally. To ensure its integrity, the confidentiality and reliability of the information must be improved through the construction of the relevant network infrastructure and equipment. Network information system security has several main characteristics, especially: integrity, controllability, confidentiality, and availability.

3 Experiment

3.1 Experimental Data Source

The main research content of this paper is to analyze the current situation and practical problems of the network information system, combine the current supply chain management theory and the latest information technology, and optimize and design the network information system.

3.2 Experimental Design

This paper first reviews the relevant theory, information management system and big data theory and application, investigates the supply chain related business process of an enterprise, analyzes the current supply chain information system, finds the optimization direction and solution, analyzes the overall supply chain process, clarifies the optimization direction and design ideas, and determines the overall framework of the network information system.

4 Result

4.1 Analysis of the Viability of the Network Information System

The viability of the network information system refers to the ability of the system to complete the key tasks in time when the virus invasion fails and during the accident. The core goal of the survival of the network information system is to provide the key services in time even if it is damaged or even destroyed. Key services in the viability concept are the functions that the system must be maintained when affected by a harsh environment. In order to have the ability to provide critical services, the network information system must provide four key properties as described in Table 1: resistance, identifiability, recoverability, and adaptability.

Table 1. Key properties of the network information systems

Keyed attribute	Bewrite	Policy instance
Resistance to the attacks	Strategies to resist an attack	Certification Password technology Access control Functional isolation Message filtering Surviability packaging
Identiability of attacks and destruction	Strategies to detect attacks and assess damage	Intrusion detection Integrity detection
Attacked after the critical services and recovery of all services	Limit damage, restore damaged information or functionality Maintain or recover within task-time constraints Critical services, policies to restore all services	Redundant components Data replication System backup and Recovery Accident planning
Adaptive and reduce future attacks The evolution of the effectiveness	Give the knowledge from the invasion to improve the department Strategies for unified viability	New intrusion recognition mode

The impact of four sampling data on the survival of the network information system is described in Table 1 above. It can be seen from the table that with the improved performance of various indicators, the survival performance of the network information system is also significantly improved, jumping from the initial poor state to the good state. When the index changes hourly, the survival performance of the network information system is also slow, and when the index changes greatly, the survival change of the network information system is also more drastic.

4.2 Analysis of the Information Collection System

In order to understand whether the designed information collection system can effectively improve the data collection efficiency, we conducted the experimental data comparison, established the experimental group and the control group, selected the designed network information collection system, and the control group selected the traditional collection method to compare the data collection rate. The results are shown in Fig. 1.

According to the experimental data in Fig. 1, it is clearly seen that under the fixed amount of data, the information collection time of the experimental group is significantly less than the control group, so it can be determined that the optimized network information collection system can effectively improve the efficiency of data collection.

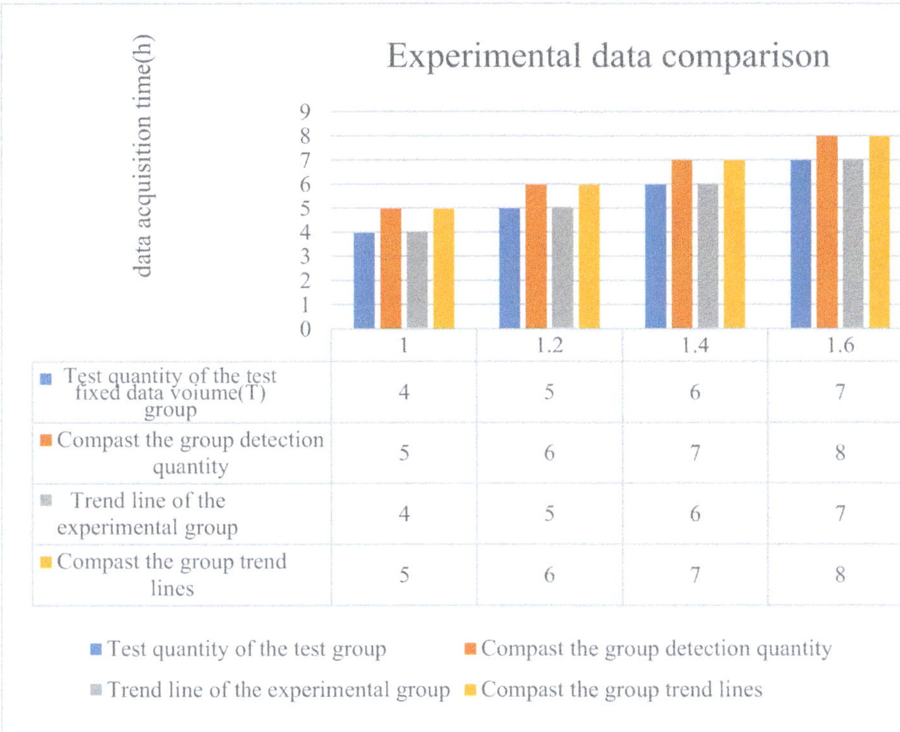

Fig. 1. Comparison of the experimental data

4.3 Analysis the Relationship Between Redundant Design and Service Viability

When the component failure rate is low, the backup component using the cold backup method is low and slightly more productive than the hot backup method; when the system is attacked, because the hot backup has rapid take over service ability, it has higher attack resistance, as shown in Fig. 2.

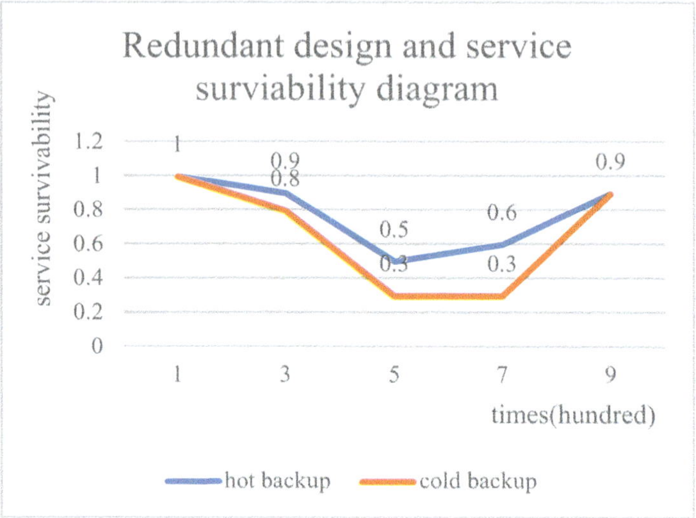

Fig. 2. Redundancy design and service viability relationship diagram

According to the curve in Fig. 2, for information systems with homogeneous backup, network attacks will fail all homogeneous components, service failure, penetration effect, reduced instantaneous availability; for information systems with heterogeneous backup, the information system can effectively prevent the attack, and instantaneous availability is low at the beginning of the attack until other components work.

4.4 Analyze the Object-Oriented Development Methods

Object-oriented development approach emphasizes the data or information in the system as the main line. As shown in Table 2 below, the system information model is comprehensively, systematic and introduced in detail to guide the design of the system. However, the approach is not very mature and requires further improvements to be truly applied in the reality of information system engineering.

Table 2. Object-oriented development features

Starting point	Object-oriented
Realization principle	From bottom to top
Programming method	Object-oriented
Logical tool	Class and object diagram, service state conversion diagram, and data dictionary
The various phases of the life cycle	The stages overlap, grow, cycle, smooth transition
Maintainability	Good
Extendibility	Just modify and add operations, while the basic object structure remains unchanged
Reusability	Good
Theoretical basis	Relatively immature
Flow control	Time-driven

It can be seen from Table 2 above that each development method has its own characteristics, style and limited conditions. In the actual network information system, two or more methods are often combined, so as to get better results.

5 Conclusion

The development of network information system is inseparable from the development of computer and big data. This paper studies the optimal design of network information system based on big data technology and discusses the overall design, heterogeneous database access and system security design. This paper discusses some key problems in its design and solves the system to solve the same or similar problems in other network information systems. Of course, as the design of information system, its design has never changed, but developed with the development of various technologies. For example, the database used in this design is a relational database, using a complete object-oriented database will be the future development direction, therefore, the use of object-oriented database to achieve information system needs further research. Similarly, with the development of network technologies, a variety of multimedia data will inevitably need to appear in big data technology information systems, which also requires further research on how to improve the diversity and selectivity of the software.

References

1. Tang, W., Liu, Y.: University mobile employment network information system in the internet age. In: Journal of Physics: Conference Series, vol. 1881, no. 2, p. 022095, (6pp) (2021)
2. Tyutyunnik, V.M., Gromov, Y.Y., Aleksandrov, E.Y.: Analytical models of parrying negative external influences on a network information system. Autom. Documentation Math. Linguist. **54**(5), 250–254 (2020). https://doi.org/10.3103/S0005105520050040

3. Chai, H., Sun, Q., Zhou, Y., et al.: Design of a digital currency information system based on the UnionPay network. Front. Eng. Manag. **7**(4), 471–484 (2020)
4. Sulaiman, M.H., Aizuddin, A.N., Hod, R., et al.: Acceptance towards social network information system for earlier detection of influenza outbreak. Med. J. Malaysia **76**(2), 145–150 (2021)
5. Yang, B.: Research on network database laboratory instrument management information system based on content query. In: IOP Conference Series Materials Science and Engineering, vol. 740, no. 5, pp. 012126–012133 (2020)
6. Al-Mamany, D.A., Hameed, A.H., Zehawi, R.: Development of pavement management system on road network using geographic information system method-Kirkuk to Erbil highway. Des. Eng. (Toronto) **2021**(6), 37–47 (2021)
7. Rajagede, R.A.: Improving automatic essay scoring for indonesian language using simpler model and richer feature. Kinetik Game Technol. Inf. Syst. Comput. Netw. Comput. Electron. Control **6**(1), 11–18 (2021)
8. Kassaw, M., Asefa, B.: Road network analysis for ambulance transportation service using geographical information system (GIS): a case of Arada Sub-City, Addis Ababa, Ethiopia. Am. J. Traffic Transp. Eng. **5**(6), 65–77 (2020)
9. Noviandi, N., Ilham, A.: Optimization fuzzy inference system based particle swarm optimization for onset prediction. Kinetik Game Technol. Inf. Syst. Comput. Netw. Comput. Electron. Control **5**(1), 5–11 (2020)
10. Wei, K.L., Chiu, C.T.: Probabilistic second-chance broadcasting with/without global positioning system information in wireless ad hoc networks. IEEE Access **8**(3), 212608–212622 (2020)
11. Fudholi, D.H., Fikri, K.: Towards an effective tuberculosis surveillance in Indonesia through Google trends. Kinetik Game Technol. Inf. Syst. Comput. Netw. Comput. Electron. Control **5**(4), 1–3 (2020)
12. Luo, J., Zhao, C., Chen, Q., Li, G.: Using deep belief network to construct the agricultural information system based on Internet of Things. J. Supercomput. **78**(1), 379–405 (2021). https://doi.org/10.1007/s11227-021-03898-y

Hilbert R-tree Space Indexing Based on RHCA Clustering

Keyan Cao[1,3(✉)], Yefan Liu[1], Yukuan Dong[2], and Qiushi Wang[2]

[1] College of Information and Control Engineering, Shenyang Jianzhu University, Shenyang, China

[2] College of Architecture and Urban Planning, Shenyang Jianzhu University, Shenyang, China

[3] Liaoning Province Big Data Management and Analysis Laboratory of Urban Construction, Shenyang, China

Abstract. Spatial indexing is an important research in the field of spatial databases, and plays a key role in how to efficiently perform spatial data retrieval and query. In this paper, a new hierarchical clustering algorithm RHCA is proposed, and accordingly, a Hilbert R-tree index based on RHCA clustering algorithm is proposed. This clustering algorithm is improved in the split stage and merge stage of hierarchical clustering. First, the sample distribution is counted in the split stage to find the appropriate split position, and then the merge strategy with label detection is used in the merge stage, which reduces the amount of calculation and overcome the shortcomings of the traditional hierarchical clustering algorithm that the intermediate results cannot be traced back. The experimental results show that the Hilbert R-tree index based on the RHCA clustering algorithm reduces the execution and query time by about 25%, the coverage and overlap area is reduced by 27%, the performance of the index is greatly improved compared with Hilbert R-tree.

Keywords: Spatial index · Hilbert R tree · Hierarchical clustering

1 Introduction

As the key technology of GIS and spatial database, spatial data indexing technology has been widely used in the field of GIS and computer-aided design. However, the research in spatial indexing in recent years has not yet focused on the indexing structure of GIS. In this paper, a new Hilbert R-tree indexing method is proposed based on CLUBS+ clustering and Hilbert filling curve. Compared with Hilbert R-tree, this algorithm takes less time to generate R-tree, has high space utilization, covers less and overlaps less area, and improves the performance of indexing to a larger extent.

2 Related Work

There have been many new advances in the direction of spatial indexing, but there are various reasons why it is not suitable for GIS systems. Li et al. [1] proposed IR tree, a

I. Ahmad et al. (Eds.): STSIoT 2021, LNDECT 122, pp. 205–211, 2023.
https://doi.org/10.1007/978-981-19-3632-6_26

hybrid data structure that combines inverted files with R-tree to provide spatial filtering, text filtering, but it is less capable of handling large-scale data. Park et al. [2] proposed a distribution-based Z-order lightweight spatial index, but this can make the time required for retrieval. Among the hierarchy-based clustering methods, Guha proposed the Cure [3] clustering method, which uses random sampling and partitioning for local clustering, but this method has a large error when the amount of data is large. Later, he proposed Rock [4] algorithm, which considers the influence of surrounding objects when calculating the similarity of two objects, but does not solve the error problem in the case of large data. The Birch [5] algorithm proposed by Zhang et al. uses a tree structure to process the dataset, and the algorithm is efficient, suitable for convex or spherical clustering types, and insensitive to noise and input data. Other representative algorithms of hierarchical clustering include CLUBS+ [6] and CHAMELEON [7].

3 Related Concepts and Definitions

3.1 Sample Similarity Measure and Related Definitions

In this paper, the similarity definition formula [9] is used as the similarity metric between samples: assume that P is a sample of d-dimensional attributes, where there are n numerical attributes, denoted as P^n, and c sub-types of attributes, denoted as P^c. Then the similarity S(P,Q) between samples P, Q is defined as:

$$S(P, Q) = \frac{n}{d} D_1(P^n, Q^n) + \frac{c}{d} D_2(P^c, Q^c) \tag{1}$$

3.2 Hilbert Curve, Hilbert Code and Hilbert R-tree

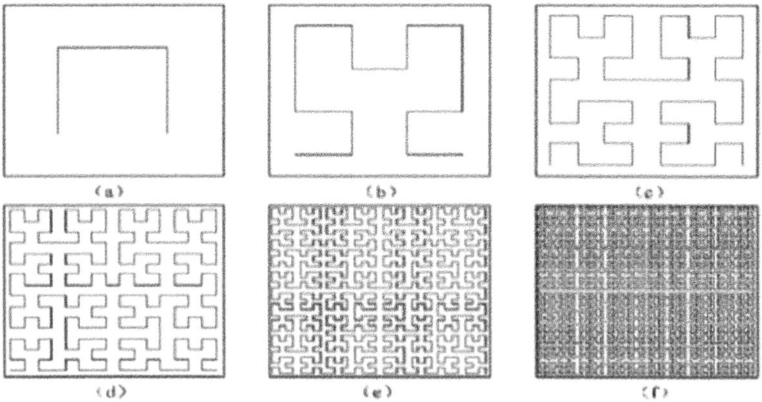

Fig. 1. Hilbert curve

As shown in Fig. 1, The Hilbert curve can linearly traverse every discrete cell in two dimensions or higher, and only once. Each discrete cell is linearly ordered and encoded.

The encoding is a unique identifier for the cell. Kamel proposed Hilbert R-tree [8], whose main idea is to map the high-dimensional spatial data to one dimension and to preserve most of the valid spatial information.

4 Hilbert R-tree Indexing Algorithm Based on RHCA Clustering

4.1 RHCA Clustering Algorithm

4.1.1 Splitting Strategy Based on Statistical Sample Distribution

The overall dataset is first considered as a class, and the class is divided into two according to the classification strategy based on sample distribution to find a suitable position for splitting the dataset. The algorithm uses WCSS as a metric to measure the similarity of samples within a class, counts the distribution of samples within a class in each dimension during the splitting process to find the suitable splitting position.

4.1.2 Merging Strategy with Additional Marker Detection

Each node in Fig. 2 represents a class, and the sub-nodes represent two sub-classes of the node. The number in the node represents the label of the class. Figure b shows the two subclasses split from the initial set, and the subclasses are marked and added with level = 2, i.e. "12", and so on. When merging, the highest level = 4 is detected, at this time the number of bits r = 1, detect whether the last r (r = 1) bits of the subclass marker is equal to 4, that is, the red node in Figure d. If the subclass meets the merging condition, it will be merged and the class marker will be updated and the level value will be reduced by one, that is, level = 3, as shown in Figure c. If the merging condition is not met, the two child nodes will be replaced by the If the merge condition is not met, the two child nodes are replaced by the parent node and the child class marker is updated and the level is reduced by one. The above merging process is iterative until the level is reduced to 0. In this paper, the Calinski-Harabazs index [10] is introduced as an intra and interclass similarity measure. It is defined as:

$$I(C) = (B(C)/W(C)) * ((n - k)/(k - 1)) \tag{2}$$

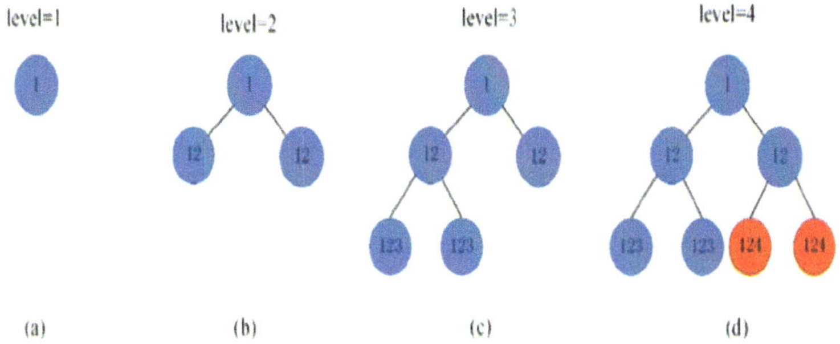

Fig. 2. Diagram of the consolidation process

4.2 Hilbert R-tree Based on RHCA Clustering

4.2.1 Generation of Hilbert R-tree Based on RHCA Clustering

Perform RHCA clustering on the spatial database to divide the data by regions; Calculate the Hilbert codes of spatial objects within each cluster and arrange each object in ascending order based on code value; Calculate the Hilbert codes of each cluster, and arrange each cluster in ascending order based on its code value; build intermediate nodes and root nodes in the bottom-up order to generate Hilbert R-tree.

4.2.2 Query of Hilbert R-tree Based on RHCA Clustering

For point query, the Hilbert code at the center of the MBR of the query object is calculated first, and from the root node, it is compared with the maximum Hilbert code stored in the intermediate nodes of each layer from top to bottom until it locates the corresponding leaf node and returns the leaf node.

4.2.3 Insertion of Hilbert R-tree Based on RHCA Clustering

First, calculate the Hilbert code of the MBR center of the object to be inserted; then, start from the root node and recursively compare the Hilbert code values of each intermediate node downward to find the smallest intermediate node larger than the code value of the object to be inserted up to the leaf node level; finally, insert the object to be inserted into the corresponding leaf node.

4.2.4 Deletion of Hilbert R-tree Based on RHCA Clustering

The leaf node of the object to be deleted is found according to the query algorithm, and then the target object whose geometric information is stored in the leaf node is found and its data items are deleted. Recursively adjust the data rectangle of its parent node up to the root node. If the Hilbert code value of the deleted object is the largest within this leaf node, the corresponding Hilbert code of its parent node also needs to be updated.

5 Simulation Experiments and Result Analysis

5.1 Performance Analysis of RHCA Clustering Algorithm

Through the analysis, the results of the simulation experiments lead to the conclusion that:the algorithm is scalable and robust in handling massive and high-dimensional data, the time complexity of this algorithm is smaller; the algorithm can handle data of various attribute types and has good clustering effect; the processing results do not depend on the input order of data; this algorithm does not require a priori knowledge such as the number of clusters and the radius of clusters to be input in advance. A comparison with common clustering algorithms is shown in Table 1.

Table 1. Comparison of RHCA and classical clustering algorithm

Cluster	Features				
	Parameters	Scalability	Data shape	Data properties	Time complexity
RHCA	——	Higher	Arbitrary	Arbitrary	O(nlogn)
CLUBS+	——	General	Arbitrary	Numerical	O(n*k)
K-means	Number of classes	General	Convex	Numerical	O(n)
DBSCAN	Neighborhood radius	General	Arbitrary	Numerical	O(nlogn)
BIRCH	Threshold of the largest sample	Higher	Convex	Numerical	O(n*k)

5.2 Performance Analysis of Hilbert R-tree Based on RHCA Clustering

5.2.1 Dataset and Performance Indexes

This experiment uses quantitative data extracted from the dataset, the amount of data extracted from the three datasets are 5000, 30000 and 45000 data respectively. The data source is the national basic geographic database of the National Geographic Information Resources Catalogue Service System. To judge the performance of spatial indexing, the execution and query time is an important index. In addition, the coverage and overlap of R-trees are equally critical. This experiment is measured by the total area occupied by nodes in the index and the area occupied by leaf nodes. The algorithms are analyzed and compared through experiments.

5.2.2 Comparison of Experimental Results

Table 2. RHCA compared to traditional Hilbert R index in execution time

Dataset	Number of data points	Execution time of the algorithm/s	
		RHCA	Hilbert R
Dataset1	5000	38.77	45
Dataset2	30000	114.5	155
Dataset3	45000	175	230.4

Table 2 compares this algorithm with the Hilbert R index in terms of execution time, etc. Ten queries are performed on three datasets for this index and Hilbert R index, and the comparison results after taking the average are shown in Table 3. Figure 3 compares the index building time. The results of leaf node area and total area de comparison are shown in Table 4.

Table 3. RHCA compared to traditional Hilbert R index in query time

Dataset	Number of data points	Query time of the algorithm/s	
		RHCA	HilbertR
Dataset1	5000	756	965
Dataset2	30000	1405	1950
Dataset3	45000	1916	2650

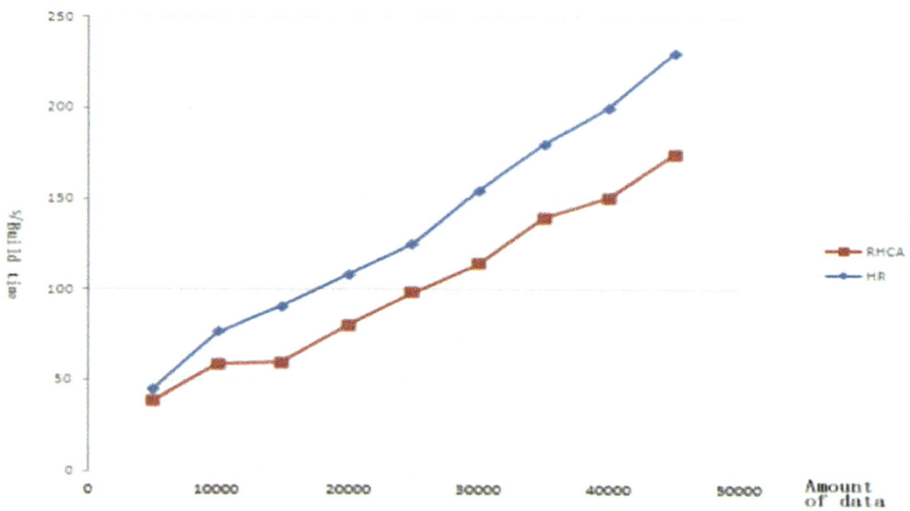

Fig. 3. Building time

Table 4. Compared in Leaf area and total area

		RHCA	HR	RHCA/HR
Dataset1	Leaf node area	1.01	1.32	76.52%
	Total area	1.46	1.94	75.26%
Dataset2	Leaf node area	1.18	1.59	74.21%
	Total area	1.65	2.23	73.99%
Dataset3	Leaf node area	1.29	1.75	73.71%
	Total area	1.94	2.96	65.54%

5.3 Performance Analysis

Compared with the traditional HR tree, the present index not only has higher space utilization, but also has some improvement in generation time and other performance.

6 Conclusion

The main contribution points are as follows: (1) an improved RHCA clustering algorithm is proposed based on the traditional CLUBS+ clustering algorithm, which preserves the internal relationships of geographic datasets; (2) the RHCA clustering algorithm is integrated with the Hilbert R-tree algorithm, and proposes a Hilbert R-tree indexing algorithm based on RHCA clustering; (3) overcomes the problem that the intermediate results of hierarchical clustering algorithm cannot be retraced.

Acknowledgements. This work was supported by National Natural Science Foundation (61602323).

References

1. Li, G., Li, L.: A hybrid structure of spatial index based on multi-grid and QR-tree. In: Proceedings of the Third International Symposium on Computer Science and Computational Technology, pp. 447–450 (2010)
2. Park, K., Joly, A., Valduriez, P.: DZI: an air index for spatial queries in one-dimensional channels. Data Knowl. Eng. (2019). https://doi.org/10.1016/j.datak.2019.101748
3. Guha, S., Rastogi, R., Shim, K.: CURE: an efficient clustering algorithm for large databases. ACM Conf. Manag. Data (SIGMOD). **27**(2), 73–84 (1998). https://doi.org/10.1016/S0306-4379(01)00008-4
4. Guha, S., Rastogi, R., Shim, K.: ROCK: a robust clustering algorithm for categorical attributes. Inf. Syst. **25**(5), 345–366 (2000). https://doi.org/10.1016/S0306-4379(00)00022-3
5. Zhang, T., Ramakrishnan, R., Livny, M.: BIRCH: an efficient data clustering method for very large databases. ACM Conf. Manage. Data (SIGMOD) **25**(2), 103–114 (1996)
6. Mazzeo, G.M., Masciari, E., Zaniolo, C.: A fast and accurate algorithm for unsupervised clustering around centroids. Inf. Sci. **400**, 63–90 (2017). https://doi.org/10.1016/j.ins.2017.03.002
7. Karypis, G., Han, E.H., Kumar, V.: CHAMELEON: A hierarchical clustering algorithm using dynamic modeling. IEEE Trans. Comput. **32**(8), 68–75 (1999). https://doi.org/10.1109/2.781637
8. Kamel, I., Faloutsos, C.: Hilbert R-tree: an improved R-tree using fractals. In: 20th Intemational Conference on Very Large DataPmses, Santiago, Chile, pp. 500–509 (1994)
9. Tong, W., Wang, Y., Zhong, J., et al.: A new weight based density peaks clustering algorithm for numerical and categorical data. In: 13th International Conference on Computational Intelligence and Security (CIS). IEEE, pp. 169–172 (2017)
10. Calinski, T., Harabasz, J.: A dendrite method for cluster analysis. Commun. Stat. **3**(1), 1–27 (1974). https://doi.org/10.1080/03610927408827101

Network Communication Signal Tracking Technology Under Cloud Computing Data

Xiaojun Liu(✉)

Shandong Vocational College of Light Industry, Zibo 255300, Shandong, China
hanxuebuji3@126.com

Abstract. With the development of the times, computer technology is booming, and the degree of association between modern society and computer is higher and higher. Therefore, the rapid development of computer technology has led to earth shaking changes in the whole society. However, due to the rapid development of network information technology, there are unstable factors, hiding their identity in the network world, calling the wind and rain, doing whatever they want, which has caused a certain turbulence to the society. Therefore, in order to solve these potential dangers, we hope to use cloud computing technology with various algorithms to develop a set of network communication signal tracking technology to track everyone's IP address in real time, so as to ensure the security of the network, which is the purpose of this paper. After we reported the project to the school and obtained approval, we borrowed the school's laboratory, used the school's internal network data and the concealment of the school's website, consulted the literature on cloud computing technology and network communication signal tracking technology, and used the improved particle swarm optimization algorithm and K-means algorithm to model and analyze it to determine the data effectiveness of the experiment. The experimental results show that there is a certain correlation between cloud computing technology and network communication signal tracking technology. Because of the complex network, the network communication signal tracking technology needs huge data and computing resources to build an effective network communication signal tracking system. Compared with before, it has increased the speed by about 189% and is more accurate. This experiment is relatively successful.

Keywords: Cloud computing · Network communication · Signal tracking · Particle swarm optimization

1 Introduction

The development of the current era is becoming more and more rapidly changing, and technology has been updated all the time, so we have to keep up with the times and use them to change our lives [1]. Due to the rapid development of times, the current network is becoming increasingly complex, full of all kinds of data. Some highly skilled hackers use sophisticated means and new technologies to conceal their true identity and do whatever they want in the online world, which poses a certain threat to the society [2]. Therefore,

I. Ahmad et al. (Eds.): STSIoT 2021, LNDECT 122, pp. 212–219, 2023.
https://doi.org/10.1007/978-981-19-3632-6_27

our purpose is to conduct a comprehensive research on network communication signal tracking technology by using cloud computing technology to update it, so as to better adapt to the current society and maintain network security [3]. Since the origin and development of target location tracking technology in the 1950s and 1960s in the United States, after decades of in-depth research by universities and research institutions in hundreds of different countries, Now there is a very mature positioning and tracing should be widely promoted by technology. It includes positioning techniques such as global satellite positioning, wireless communication base station positioning, infrared positioning, noise wave positioning and wireless sensor network positioning. These location-based techniques cover a variety of complex target location-based and tracer needs, including single target location-based, multi-target location-based and variable target location-based. In addition, the robustness of the location-tracking system becomes stronger and the accuracy of location-tracking becomes more and more accurate with the in-depth study and optimization of the algorithm. For the original wireless sensor network passive target localization method is mainly based on fingerprint algorithm localization, and its method is similar to the fingerprint algorithm in the application of source localization. All of them are established through the off-line collection point training, and then through the measured value matching into the line target position. The most important difference between the two is the measurement value used in the training and matching of the source positioning situation, which is actually the source without line end of the target carried and the other without line node (such as routing, etc.) communication In fact, the measurement values used in training and matching in passive location configuration are simply the values of the communication chain between nodes without lines.

2 Related Work

Cloud computing data comes from different sensors, controllers and computing terminals. There is no fixed system planning between these devices.

Duschia et al. Proposed an extensive data set containing 260 frames of information bearing UWB signals transmitted and received in 120–140 GHz, 210–240 GHz and 1–1.05 THz bands. This data set will enable the wireless community to experimentally explore solutions related to UWB time, frequency and phase synchronization, channel estimation and equalization, modulation and so on [4]. Pedro et al. Proposed a detector based on neural network to recover the channel occupancy information from the compressed signal, and compared it with the optimal maximum likelihood detector, assuming that the channel state information is perfect and imperfect [5]. This method can minimize the system cost caused by load reduction and solve the problem of long time-consuming resource scheduling process [6]. Mr. Satish briefly summarized the current state of the art of the latest technologies and solutions for location and Tracking (L&T) in wireless sensor Networks (WSN), focusing on RSS-BASED solutions [7]. Therefore, we combined the above problems and two research schemes to obtain a better research scheme, that is, set two schemes for real-time regulation, so as to obtain the experimental data we need [8]. Some scholars have proposed a new passive target localization and tracking method based on the network of wireless sensor. In his research, the sensor nodes without wires were deployed in grid array on the antenna plate in the chamber and connected to the grid. The value of chain measurement of radio frequency signal reflected through ground is used as important information to determine the existence of target [9].

For better protecting data security, we mainly redesign the network communication signal tracking technology [10]. That is, change the original basic technology of network communication signal tracking from the original computer technology to the current cloud computing and big data technology, and change the data source from the automatic collection of some data by the system to the automatic collection of all data by cloud computing, then, by calling huge computing resources to systematically analyze the data, we can find a really useful data from a pile of miscellaneous data, which represents the user's IP address and indicates his real identity [11]. In this way, when we can track all the data and analyze the real effective data, we can determine everyone's identity, and then the network communication signal tracking system will be constructed [12].

These tracking algorithms have not been modified in the original, because the detection rate and tracking precision of the target have to be improved. The simple method of detection and location does not realize the tracking process of the target, because it requires further perfection. In this paper, network communication signal tracking technology and method based on cloud computing data are studied in depth.

3 Improved Particle Group Algorithm

Taking as an example the search for food for flocks of birds, assuming that there are particles in the n-dimensional space environment, the optimal feeding position of the particles can be expressed as, and the historical optimal feeding position found by the particle cluster can be expressed as: then the type of change of speed and position of the d-dimensional component of the ene particle in the target search area are:

$$v_{id}(t+1) = wv_{id}(t) + c_1 r_1 (p_{id}(t) - x_{id}(t)) + c_2 r_2 \left(p_{gd}(t) - x_{id}(t)\right) \tag{1}$$

$$x_{id}(t+1) = v_{id}(t+1) + x_{id}(t) \tag{2}$$

$$w = w_s - (w_s - w_e) \cdot \frac{g}{G_{max}} \tag{3}$$

The update formula of particle velocity is:

$$v_{id}(t+1) = wv_{id}(t) + c_1 r_1 (p_{id}(t) - x_{id}(t)) + c_2 r_2 \left(p_{gd}(t) - x_{id}(t)\right) + c_3 r_3 (s_{id}(t) - x_{id}(t)) \tag{4}$$

4 Experiment Build

4.1 Selection and Treatment of Experiment

We applied for the project to the school before the experiment. After getting the support of the school, used the central server to build a cloud computing center to continuously collect data in the campus network, and then extracted some key data to mark, and then carried out the experiment. We use various algorithms to build different systems to find the key and effective data from a pile of data, count the accuracy and analysis speed, and then process the data.

4.2 Analysis of Experimental Data

Through many experiments, we get a pile of data processed by different algorithms, and then process it systematically. 200 random data are selected and stored in 300 million data as tags, and the experimental results are obtained by data analysis.

5 Evaluation Results

5.1 Analysis of Experimental Results

Table 1. Evaluation results of three algorithms on various data of the system

	Accuracy/%	Inspection time/S	Inspection speed/(million pieces/s)
Improved particle group algorithm	99.6	17.0	17.6
K-means algorithm	98.4	21.9	13.7
Original satellite network algorithm	91.5	49.2	6.1

According to the information given in Table 1 and Fig. 1, the three algorithms have obvious differences in accuracy, inspection time and inspection speed. Compared with the original satellite network algorithm, the improved particle swarm optimization algorithm and K-means algorithm have greatly improved in all aspects. In particular, the inspection speed is 189% higher than the original, with extremely fast inspection speed and extremely high accuracy. Therefore, we choose to use the improved particle swarm optimization algorithm to cooperate with cloud computing technology to establish a network communication signal tracking system, and then establish a cloud computing data center to improve the stability of data transmission and the effectiveness of analysis.

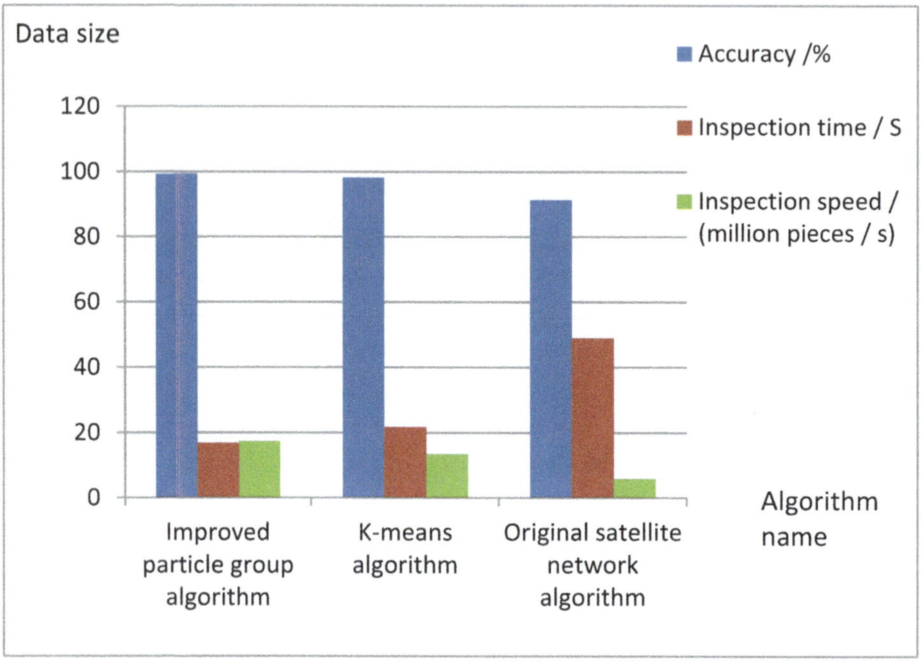

Fig. 1. Comparison results of three algorithms on various data of the system

Table 2. Comprehensive analysis of three methods for data center resource scheduling

	Load degree/%	Scheduling time/S	Power consumption/J
Energy saving resource allocation	62.1	23	820
Joint energy dispatch of geographically distributed mixed use buildings	82.1	17	1200
Mixed scheduling resource allocation according to the actual situation	65.1	19	950

Table 2 shows the system feedback results of processing 300 million data according to the three methods. Due to the low server level, the processing time is long, but the gap can be clearly seen. In order to more intuitively show the comparison of experimental results, we have made Fig. 2 to express the differences of the three methods.

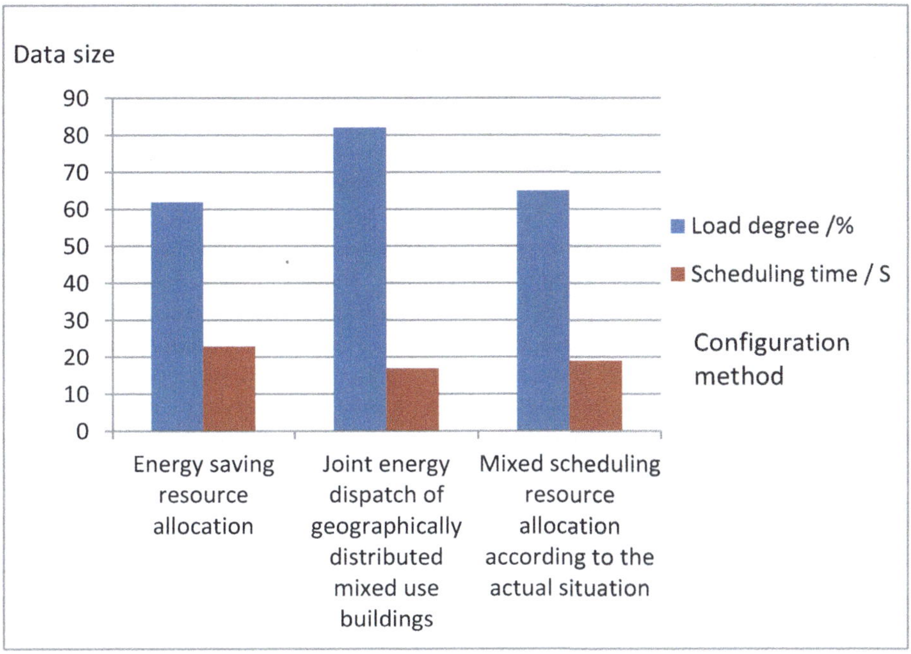

Fig. 2. Comprehensive analysis of three methods for data center resource scheduling

5.2 Development and Security of Cloud Computing

The rapid growth of information data leads to the increasing demand for large-scale data processing. As one of the new computing models and service modes, cloud computing technology came into being. Its data center is composed of massive computing nodes. By assigning computing tasks to the data center, it can realize the demand for computing performance, information services and storage space. With the rapid development of cloud computing, the data scale expands rapidly, which brings great challenges to processing strategies. Among them, the most prominent problem is resource allocation. The key reason why cloud computing can be widely used is the efficiency of resource allocation technology in cloud data center. The important infrastructure of cloud computing is the data center network, which plays an irreplaceable infrastructure role. It is placed in the cloud computing hardware layer and functions to deliver power for Internet applications and services.

Due to the reliability of data, when users upload data to the cloud service provider's server, the storage and use of data are out of control. It is difficult to ensure the credibility, integrity, data ownership and data privacy in the process of data transmission and use, which has become a key problem for effective data sharing and rational use. At present, data owners, users, cloud service providers and third-party platforms mostly carry out data security protection separately, lacking unified data protection means with consistent trust, which is not conducive to the management of the whole life cycle of data and multi-party integration and sharing. In terms of data acquisition, the data obtained and collected by cloud service providers far exceeds the data uploaded by users. However, because cloud data has a huge amount of data and computing resources, we can still make more powerful network communication signal tracking technology based on this.

5.3 Signal Tracking

Network communication technology will produce a certain amount of operation difficult data in the operation process. Therefore, while studying, we should pay attention to the regulation and center search of network signals, and constantly enhance the internal management performance of communication signals. Many scholars study the regulation information function of the collected network information data set, constantly integrate the network communication signal form, obtain more favorable operation data, improve the tracking information function and improve the signal operation performance. The traditional network communication signal tracking technology based on data transmission continuously expands the data operation space, ensures the transmission effectiveness of communication signals, avoids the intrusion of irrelevant factors, increases the strength of data integration, manages and studies the data with strong operational relevance, and can better grasp the state of data information and obtain favorable data information. However, in the actual operation process, the traditional research has little operation force for information, cannot realize the key data transformation, the degree of exchange between information is low, the degree of data correlation is low, and the operability of the obtained data is not strong. Therefore, aiming at the above problems, a network communication signal tracking technology based on cloud computing is proposed.

6 Conclusion

In order to protect data security, we often create multiple nodes on the network to confuse others and cover up the actual data. However, for highly skilled people, the online world is a place to do whatever they want, just like the panda burning incense a few years ago, which has had a great and bad impact on the society. Therefore, we need to increase control over them and do not allow unstable factors in the society. Just as such, network communication signal tracking technology has been developing to keep the whole society in a relatively stable state. With the development of the experiment, we had difficulties and failures in the process, but we also made some gains in the end, which has made a little contribution to the development of network communication signal tracking technology in the direction of cloud computing. I hope it can be helpful to you.

References

1. Li, M.: Research on the mechanism and influence factors of urban style building based on cloud computing logistics information. Clust. Comput. **22**(6), 13873–13880 (2018). https://doi.org/10.1007/s10586-018-2120-4
2. Fabio, A., Vittorio, C., et al.: Analytical modeling of performance indices under epistemic uncertainty applied to cloud computing systems - ScienceDirect. Future Gener. Comput. Syst. **102**(C), 746–761 (2020)

3. Namasudra, S.: An improved attribute-based encryption technique towards the data security in cloud computing. Concurr. Comput.: Pract. Exp. **31**(3), 4364 (2019)
4. Bodet, D., et al.: Data signals for Terahertz communications research. Comput. Netw. **203**, 108628 (2022)
5. De Souza, P.H.C., Mendes, L.L., Chafii, M.: Compressive learning in communication systems: a neural network receiver for detecting compressed signals in OFDM systems. IEEE Access **9**, 122397–122411 (2021)
6. Alasti, H.: Communication-efficient tracking of unknown, spatially correlated signals in ad-hoc wireless sensor networks: two machine learning approaches. Sensors **21**(15), 5175 (2021)
7. Jondhale, S.R., Maheswar, R., Lloret, J.: Received Signal Strength Based Target Localization and Tracking Using Wireless Sensor Networks, pp. 1–202. Springer, Heidelberg (2022). https://doi.org/10.1007/978-3-030-74061-0. ISBN 978-3-030-74060-3
8. Pandita, A., Upadhyay, P.K.: Fault tolerance aware scheduling for brokers in cloud computing datacenters. Recent Patents Comput. Sci. **10**(4), 299–307 (2017)
9. Booranawong, A., et al.: Real-time tracking of a moving target in an indoor corridor of the hospital building using RSSI signals received from two reference nodes. Med. Biol. Eng. Comput. **60**(2), 439–458 (2022). https://doi.org/10.1007/s11517-021-02489-6
10. Wang, Y., Huang, Z.: MEDLL on-strobe correlator: a combined anti-multipath technique for gnss signal tracking. J. Navig. **73**(3), 658–677 (2020)
11. Abbaszadeh, M., Atthanayake, I.U., Thomas, P.J., et al.: Molecular signal tracking and detection methods in fluid dynamic channels. IEEE Trans. Mol. Biol. Multi-Scale Commun. **6**(2), 151–159 (2020)
12. Chacon-Murguia, M.I., Rivero-Olivas, A., Ramirez-Quintana, J.A.: Adaptive fuzzy weighted color histogram and HOG appearance model for object tracking with a dynamic trained neural network prediction. Signal Image Video Process **15**(7), 1585–1592 (2021). https://doi.org/10.1007/s11760-021-01891-9

5G Enhancement Based on User Plane Service Architecture

Shuhua Mao(✉), Xiaoli Xie, Shenghui Dai, Bolu Lei, and Liya Li

Department of Information Engineering, East China University of Technology, Fuzhou, Jiangxi, China
shhmao@ecit.cn

Abstract. This paper studies the enhancement of the architecture based on the optimization of the user plane service architecture, and proposes to extend the service concept from the 5G control plane to the user plane, so that the 5G system provides higher flexibility and better modularity, so as to make it easier and easier. Good support for the automation and high reliability of network functions and services.

Keywords: User plane · Flexibility · Reliability

1 Introduction

The best modularity of the system should improve the flexibility of the system in terms of customizing its functions and features, and increase the flexibility of dynamically adding and deleting services. Through appropriate service modeling, the services can be deployed by themselves without being forced to rely on certain NF. Research to extend the service concept from the 5GC control plane to the user plane function will focus on how to extend the service concept to only the N4 interface, how to integrate the specific resources of UPF, the status of the PDU session and the user plane channel, etc. into the service-based architecture [1, 2]. And ensure that these aspects of UPF are included in the existing principles of SBA.

2 NF Service Interaction Model Framework

2.1 Program Description and Problem Analysis

In the direct interaction model, the NF service itself has many responsibilities that are not part of the business logic of the NF service. This brings some redundant burdens to the implementation of NF services, which may limit the agility and interoperability of development and deployment [3]. But more importantly, entrusting these responsibilities to NF services will result in the limitation of automation flexibility and customer service availability, especially in the case of NF service instances failing, which will have a negative impact on overall system availability and reliability [4].

I. Ahmad et al. (Eds.): STSIoT 2021, LNDECT 122, pp. 220–224, 2023.
https://doi.org/10.1007/978-981-19-3632-6_28

In order to reduce the complexity of NF services, the functions shared by all NF services can be identified and extracted and placed in components other than the actual NF services [5]. The NF service itself is responsible for discovering and selecting peers for inter-service communication, and maintaining the state of these communication relationships for subsequent transactions. This behavior is based on the assumption and previous assumption that the two communication peers have high availability and reliability [6].

Given the conditions, this is no longer valid in a cloud-based deployment environment. In the event of a failure in an NF service instance, it must be notified to each corresponding communication peer and a failover strategy must be executed to find and connect to the replacement peer NF service instance and restore and synchronize the communication and application process state [7]. The solution is to solve how NF service instances communicate with each other, and it is no longer necessary to implement such a failover mechanism as a part of each NF service.

2.2 Improve Program Conditions and Structure

Assuming that there is no long-term binding between the NF service instance and the application process environment, the NF service instance is allowed to send messages to the peer NF service type, the NF service instance is allowed to receive messages from another service instance, and the message can be sent from the peer to peer [8]. The body is delivered to the selected receiving peer, and the message content supports any payload.

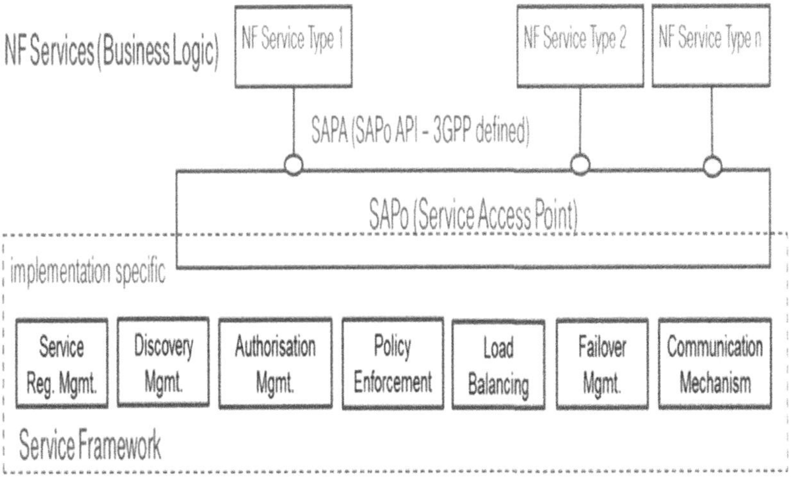

Fig. 1. Architecture diagram of the improvement plan

The improved architecture is mainly composed of the following three building blocks. As shown in Fig. 1. The service access point (SAPo) function element provides access to the service framework functions. These function elements provide the northbound API, namely the service access point API (SAPA), which is used for NF services. There may be multiple SAPo instances, and each SAPo instance allows registration and deregistration of NF service instances and NF service instances to send and receive messages. SAPos can also monitor the existence of registered NF service instances, which defines the API of the common service framework function set. Service Access Point is a functional entity that acts as an adapter between an implementation-specific service framework and NF services that utilize the functions of the service framework. SAPA provides methods for registration/deregistration of NF service instances, as well as methods for sending and receiving messages. It is universal in accessing the public service framework function set, and is agnostic to the content of the messages exchanged between NF services. SAPA must include methods for performing registration, cancellation, and authorization. The routing mechanism takes over the responsibility for the discovery and selection of communication peers, as well as the process of actually transmitting messages between peers. The communication mechanism is an internal protocol used for clear text messaging.

This solution proposes to implicitly manage the failover situation in the service framework and eliminate this burden from the implementation of NF services. It is achieved through the decoupling and non-sticky communication relationship between the NF service instance and the stateless NF service design. The request message is not sent to a specific instance of the NF service, but to the NF service type [9]. This allows NF service instances to be replaced in general without the need to specify a specific recovery process. If the communication mechanism detects that some NF service instance does not respond, the corresponding message will be routed to another NF service instance that can handle it without affecting customer service.

3 NF Service Instance Registration Process

The NF service instance sends a 3GPP-compliant registration message to the service framework to register itself. The integrated SAPo performs all necessary operations to adapt the registration message to the format implemented by the vendor-specific service framework and forwards it to the framework [10]. The vendor-specific service framework implements any steps required to process this registration. The service framework returns the 3GPP-compliant registration response to the NF service instance. As shown in Fig. 2.

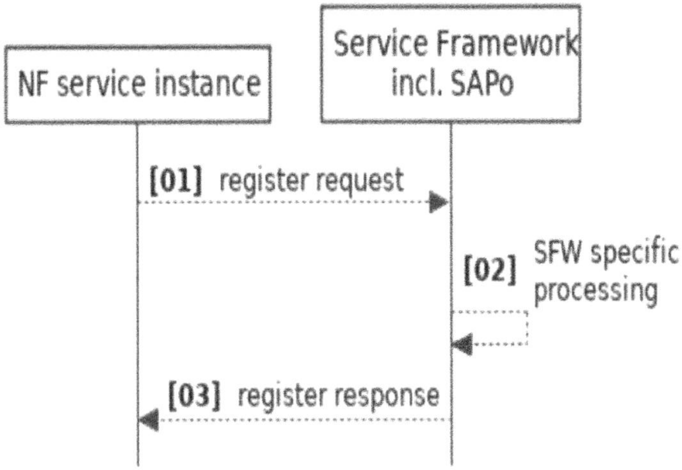

Fig. 2. The NF service instance registration process

When unregistering, the NF service instance sends a 3GPP unregistered message to the service framework to unregister itself. The integrated SAPo performs all necessary operations to adapt the deregistration message to the format implemented by the vendor-specific service framework and forwards it to the framework [11]. The vendor-specific service framework implements any required steps to handle this logout. The service framework returns a deregistration response conforming to the 3GPP standard to the NF service instance.

4 Conclusions

The service implementation of this solution does not need to deal with the discovery of communication peers, the maintenance and potential restoration of communication relationships, and the implementation of communication-related strategies. Service implementations must register and unregister, and send and receive messages through SAPA.

Acknowledgements. This work was supported by Science and Technology Research Project of Jiangxi Provincial Department of Education in 2019: Research on User Plane Redundant Transmission Technology for Highly Reliable 5G URLLC.

References

1. Renzo, M.D., Zappone, A., Debbah, M., et al.: Smart radio environments empowered by reconfigurable intelligent surfaces: how it works, state of research, and road ahead. IEEE J. Sel. Areas Commun. **38**, 2450–2525 (2020). https://doi.org/10.1109/JSAC.2020.3007211

2. Tang, W., Chen, M.Z., Dai, J.Y., et al.: Wireless communications with programmable metasurface: new paradigms, opportunities, and challenges on transceiver design. IEEE Wirel. Commun. **27**(2), 182–184 (2020)
3. Konstantinos, L., Alexander, G., Ray, S., et al.: Use cases and scenarios of 5G integrated satellite - terrestrial networks for enhanced mobile broadband: the Sa T5G approach. Int. J. Satell. Commun. Netw. **37**(2), 96–99 (2019)
4. Albonda, H.D.R., PérezRomero, J.: Reinforcement learning-based radio access network slicing for a 5G system with support for cellular V2X. In: Kliks, A., et al. (eds.) CrownCom 2019. Lecture Notes of the Institute for Computer Sciences, Social Informatics and Telecommunications Engineering, vol. 291, pp. 262–276. Springer, Cham (2019). https://doi.org/10.1007/978-3-030-25748-4_20
5. 3GPP PCG, Draft Summary Minutes, Decisions and Actions from 3GPP PCG Meeting#46-e, 26–27 April 2021
6. Chochliouros, I.P., et al.: Enhanced mobile broadband as enabler for 5G: actions from the framework of the 5G-DRIVE project. In: MacIntyre, J., Maglogiannis, I., Iliadis, L., Pimenidis, E. (eds.) AIAI 2019. IAICT, vol. 560, pp. 31–45. Springer, Cham (2019). https://doi.org/10.1007/978-3-030-19909-8_3
7. Zhang, Z., Dai, L.: Capacity improvement in wideband reconfigurable intelligent surface-aided cell-free network. In: IEEE 21st International Workshop on Signal Processing Advances in Wireless Communications (SPAWC), pp. 1–4 (2020)
8. Rakuten Mobile, RWS-210247 Reconfigurable Intelligent Surfaces, 3GPP TSG RAN Rel-18 workshop Electronic Meeting, 28 June–2 July 2 2021
9. 3GPP. TS 22.261: Service Requirements for the 5G System; Stage 1 V1.1.0 (Release 15) (2021)
10. 3GPP. TR 23.737: Study on Architecture Aspects for Using Satellite Access in 5G V0.9.0 (Release 16) (2020)
11. Van-Dinh, N., Duong, T.Q., Vien, Q.-T.: Correction to: editorial: emerging techniques and applications for 5G networks and beyond. Mob. Netw. Appl. **26**, 1987 (2020)

A Security Model of IoT Device Identity Authentication Based on Digital Identity Certificate and Public Key Encryption

Ping Xia(✉)

Guangzhou College of Technology and Business, Guangzhou, Guangdong, China
asher0308@163.com

Abstract. With the continuous integration and application of the Internet of Things and cloud computing, more and more devices need to connect to the Internet of Things cloud platform for remote access and control. At the same time, the access of illegal devices and abuse of access rights have brought many data security problems. IoT devices have application characteristics such as low power consumption, low cost, small storage, and heterogeneity, which are different from the structure of terminal devices in computer networks. Therefore, applying traditional identity authentication technology to IoT devices will no longer be applicable. This paper adopts the digital identity certificate and public key encryption technical scheme based on cryptographic technology to construct a credible security architecture for identity authentication of Internet of Things devices. Design the initial identity registration and activation process and technical route of IoT devices centered on the trusted identity registration agency, and then propose the network access verification process and security management of IoT devices based on the challenge-response mechanism. This solution uses cryptographic technology to ensure the confidentiality and non-repudiation of information during the identity verification process, which can better meet the actual needs of Internet of Things devices to access the network, and ensure secure device authentication connections, authorized access and identity management.

Keywords: Digital identity certificate · Public key encryption · Internet of Things equipment · Identity authentication · Security model

1 Introduction

With the advent of the 5G era, the Internet of Everything is about to become a reality, and the Internet of Things has entered the development field represented by basic industries and large-scale consumption. The Internet of Things is the fusion application of intelligent perception, radio frequency identification technology, ubiquitous computing, and ubiquitous networks. It is known as the third wave of the development of the world's information industry after computers and the Internet. The most notable feature of the Internet of Things is that hundreds of millions of sensor nodes are connected to the network and distributed in the working environment to sense and collect massive

I. Ahmad et al. (Eds.): STSIoT 2021, LNDECT 122, pp. 225–233, 2023.
https://doi.org/10.1007/978-981-19-3632-6_29

amounts of data, and upload them to the data analysis system of the cloud platform through the smart gateway. Through the processing and analysis of large amounts of data, decision-making applications that will affect system operation and user safety are produced. So as to realize the functions of intelligent and remote identification, positioning, tracking and supervision, so the data plays a key role in the application of the system. Trusted data is the basis for reflecting system application value and security function services, which requires the source of data to rely on trusted IoT devices. The trustworthiness of IoT devices mainly depends on the identity authentication link of the IoT system, which ensures the authenticity and legitimacy of the identity of the access device. This is the first line of defense to authorize legal access to the resources of the Internet of Things system, and it is also an important security component to ensure the security and stability of the system. IoT devices have application characteristics such as low power consumption, low cost, small storage, and heterogeneity, which are different from the network structure of terminal devices in computer networks. Therefore, applying traditional identity authentication technology to IoT devices will no longer be applicable. Aiming at the characteristics of the IoT network and devices, this paper uses digital certificates and public key encryption in cryptographic technology to design a new identity authentication security model for Internet of Things devices to achieve trusted authentication and identity management of access devices.

2 Related Work

With the continuous development of IoT applications, IoT devices play a pivotal role in IoT networks. The identity authentication protocol and scheme for IoT devices has become a hot research topic, and there are many research results at home and abroad and the effect is remarkable [1].

Seungyong Yoonero proposed that the use of PUF-based device authentication technology in complex IoT environments can effectively defend against information leakage and various network attack threats [2]. Some scholars proposed to use PUF-based lightweight protocols to provide mutual authentication and secure communication for IoT devices, and passed the security tests for modeling attacks [3]. S. Gladson Oliver proposed a secure mutual authentication scheme for IoT devices using the CoAP protocol, which improves the efficiency and security of device authentication by reducing the processing overhead of data transmission [4]. Hamza Sajjad Ahmad proposed to use the improved SAFER and S-Box protocols to achieve identity authentication for IoT communication devices, and at the same time design a secure encryption algorithm based on ground computing, low memory and cost to achieve data communication [5]. Some scholars have proposed a blockchain-based IoT identity authentication scheme, which can effectively improve the security and efficiency of identity authentication, but there are still certain difficulties in practical applications [6, 7].

Although there are many protocols and schemes for IoT device authentication, it is still necessary to strengthen the research on this topic to further optimize the authentication scheme and improve the security and practicability of IoT device authentication.

3 IoT Device Authentication Security Architecture

As more and more devices are connected to the Internet of Things system for remote access and control, it is necessary to adopt a safe and reliable device identity authentication mechanism when accessing the system to prevent the abuse and destruction of illegal devices [8]. In the designed security model of IoT device authentication, it mainly includes three main networks of IoT device network, identity registration authority and IoT cloud platform, and an intermediate component, as shown in Fig. 1.

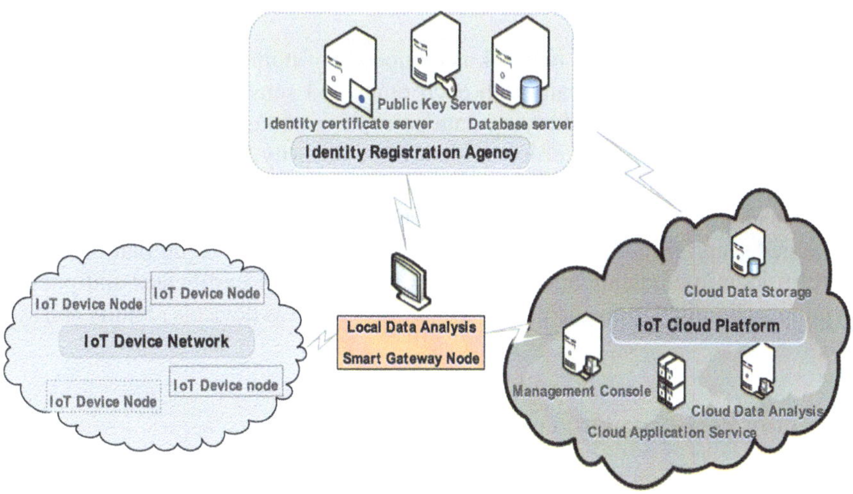

Fig. 1. Security architecture of IoT device authentication

The functions of each module are shown in Table 1. As an intermediate component, the intelligent gateway acts as a wireless routing and intelligent control hub, and is the intermediate link between various networks. Smart gateway is an important sensor node management and control device in the Internet of Things system, and it is also a central component that realizes the interconnection and intercommunication between heterogeneous networks, data protocol and format conversion, and remote data transmission [9]. In the device identity authentication, the smart gateway is mainly responsible for the protocol conversion and data communication between each main network, and realizes the basic analysis and application of local data [10].

Table 1. The functions of each module of the IoT device authentication security model

Module	Core functions
IoT device network	Low-level data perception and collection, preliminary data processing and upload
Identity registration agency	IoT device identity registration and management, key distribution and management, digital identity certificate issuance and verification
IoT cloud platform	Device access authentication and management, secure connection and data collection to the cloud. Remote control and application of the server
Smart gateway	Wireless routing and intelligent control hub

3.1 Internet of Things Device Network

The Internet of Things device network is composed of numerous sensor device nodes. Various nodes can communicate with each other and sense in coordination, and are mainly responsible for the underlying data perception and collection, preliminary data processing and uploading. In order to ensure the security of the underlying network interface and data, a safe and effective identity authentication technology is required to identify the identity of the device and restrict the access and access of illegal devices. In the designed security model, before the device connects to the cloud platform, it needs to perform identity authentication based on digital identity certificates and public key encryption technology to ensure the legitimacy of the access device's identity and system security.

3.2 Identity Registration Agency

The identity registration agency is an important application module for the initial registration and management of the identity of the Internet of Things equipment, which mainly includes an identity certificate server, a public key server and a database server. The functions of each server are shown in Table 2. The identity registration agency is independent of the IoT system platform, and can serve as a fair and trusted authoritative third party to provide secure IoT device identity registration, key distribution and management, digital identity certificate issuance and verification and other services. Identity registration and authentication are the basis of data security and trustworthiness in the Internet of Things system, which requires registration agencies to adopt standardized identity authentication formats and protocols. And follow the device and application login and binding mechanism based on security services to realize the security and accountability traceability of device identity and legal use.

Table 2. Function description of each server module of identity registration agency

Server module	Role and functions
Identity certificate server	Lead the device registration and activation process, verify device identity, generate device device identity code and digital identity certificate
Public key server	Distribute and manage public key pairs (pki, ski) for devices
Database server	Encrypted storage device private information, device code, public key and digital identity certificate and other important data

3.3 IoT Cloud Platform

The IoT cloud platform is the core functional application and service providing module of the IoT system, which mainly includes management console, cloud data storage, cloud data analysis, cloud application service interface and other parts. Connecting to the massive IoT network, the management console is mainly responsible for the identity verification of the device access to the cloud system, unified management and maintenance of the device identity, and supports safe and reliable access connections and data collection to the cloud. The cloud platform provides the cloud application interface upwards to realize the remote control and application of the server.

4 Identity Registration and Authentication Process Design for IoT Devices

4.1 Identity Registration and Activation Process for IoT Devices

After the Internet of Things devices (Di) are powered on for the first time, they need to complete the identity registration and activation process through a trusted third-party identity registration agency (IRA), the specific process is shown in Fig. 2. The registration agency issues a globally unique device identity code Tid, a device public key pair (pki, ski) and a digital identity certificate CAi for the device, the expression is as follows:

$$IRA \rightarrow Di : Tid \| (pki, ski) \| CAi \quad (1)$$

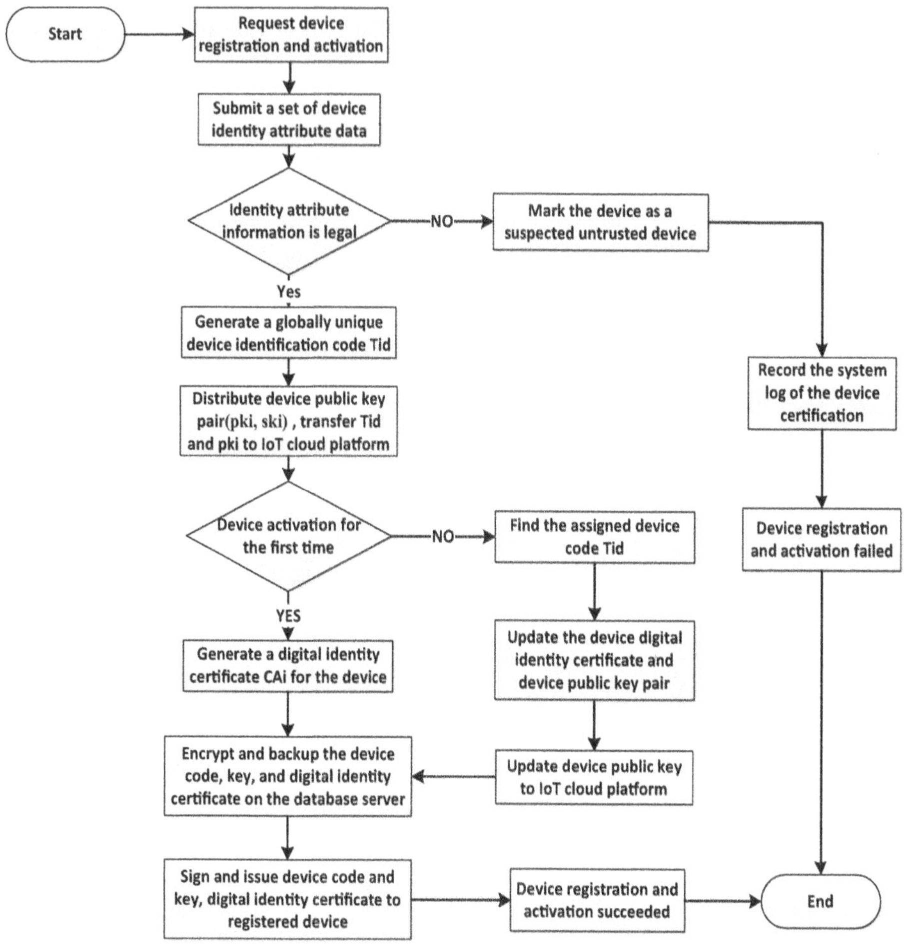

Fig. 2. Identity registration and activation process of IoT devices

First, the device node sends a registration and activation request to the identity registration agency, and selects a set of verifiable device identity attribute data as application data. It can include equipment hardware manufacturer and model, production batch and production password, processor and chipset code, hardware MAC address and IP address, equipment environmental characteristics (such as GPS positioning) and other information. The registration agency performs authenticity verification and registration identification on the device request information, and if the device identity attribute data verification fails, the device is marked as a suspected untrusted device. Record system logs about the hardware attribute identification, registration event information and action identification of the device, feedback registration failure information, and end the verification of this registration request. If the hardware identity information is verified to be true and valid, a globally unique device identity code Tid is generated according to a set of attribute information of the device. At the same time, the public key server interface is called to allocate a public key pair (pki, ski) for the device, and the device code Tid and the device public key pki are transmitted to the IoT cloud platform for the later device identity authentication process.

After the device registration process is completed, the identity certificate server performs the device activation operation. If it is the first activation, the device's key attributes, identity identification sequence, one-time random number, and system timestamp parameters are used to generate the device's digital identity certificate CAi. The registration authority encrypts and backs up the device code Tid, (pki, ski) and CAi link in the back-end database, and issues the signed data item to the device. For security reasons, the activated device identity is only valid during the activation security period and should be reactivated after the expiration date. If the device is repeatedly activated, the identity certificate server matches and searches the assigned device identity code Tid according to the key attribute information presented by the device. Update the storage device digital identity certificate CAi and the device public key pair (pki, ski), and sign and distribute the data to the device to complete the issuance and activation process of the device digital identity certificate.

4.2 Identity Verification Process for Device Access to the IoT Cloud Platform

When IoT devices access the cloud platform, the management console uses a challenge-response mechanism to authenticate the identity of the device, and the interaction process is shown in Fig. 3. After each IoT device is registered and activated by the identity certification authority, the identity code Tid and public key pki will be stored in the cloud platform server, which is mainly used for device identity security authentication services. Use digital identity certificates, public key encryption and signatures to ensure the security and non-repudiation of transmitted data.

(1) The device sends a network access session request containing its own identity Tid.
(2) The management console queries the device code and public key records stored in the cloud server to verify whether the device has been registered. If the device is not registered, the verification ends, the device is prompted to register and activate, and the device authentication event is recorded. When the device code verification has passed, the management console uses a random number generator to generate a random number sequence Ni, and generate a one-time session key Ks. And link the time stamp Ti and the session key Ks, use the device public key pki to encrypt the data item and send it to the authentication device as the challenge value C, the expression is as follows:

$$C = Epki(Ni, Ti, Ks) \tag{2}$$

(3) After receiving the data, the device uses the private key ski to decrypt the challenge value, extracts the random number sequence Ni, the timestamp Ti, and the session key Ks, and stores the session key Ks locally for encryption of data transmitted with the cloud platform. The device must respond within the valid timestamp Ti, link its digital identity certificate CAi, device code Tid, and random number sequence Ni, use the session key Ks to encrypt the link data, and use the device private key ski

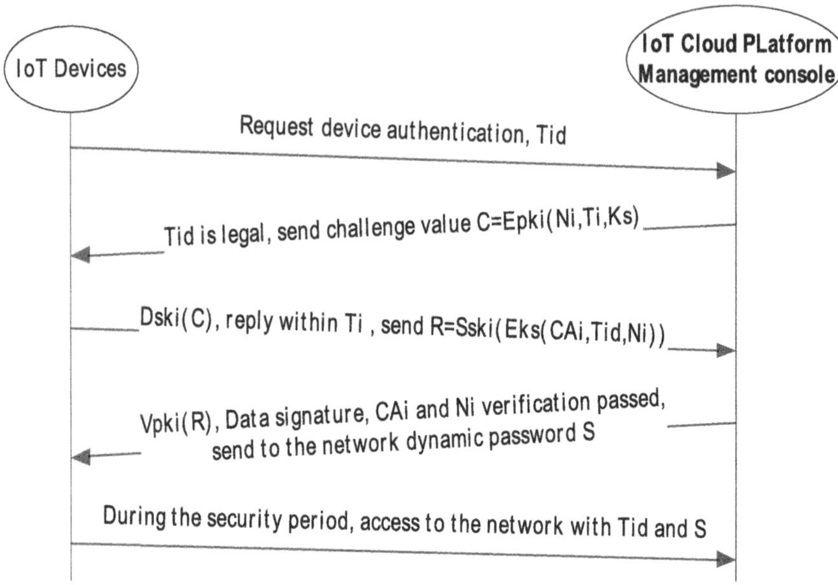

Fig. 3. Identity authentication interaction process of IoT devices based on challenge-response mechanism

to sign and send it to the management console as response data R, the expression is as follows:

$$R = Sski(Eks(CAi, Tid, Ni)) \tag{3}$$

(4) The management console verifies the authenticity of the signature of the received data to achieve the effect of non-repudiation of the data source. If the verification is passed, the shared session key Ks is used to decrypt the data, access the identity registration agency, query the registration information of the device corresponding to the device code, and confirm the authenticity and validity of the digital identity certificate. The digital identity certificate is verified, and the dynamic password S for the device's network access is issued. At the same time, the management console safely evaluates the authentication device, sets the device's network access license security cycle, stores the device's identity authentication information, and regularly performs device identity maintenance and security management.

(5) During the security period of the network access permit, the device uses the identity Tid and the dynamic password S to access the cloud platform and uses the system service, and uses the session key Ks to encrypt private communication data.

5 Summary

This article uses cryptographic technology to achieve credible identity authentication when the device is connected to the Internet of Things cloud platform. Utilize the device's digital identity certificate constructed in the registration and activation process of the identity registration agency to prevent illegal device identity forged access and privilege abuse. High-strength digital identity credentials are used to ensure the unforgeability of authentication device identities and the security of network resource access. The use of a public key encryption algorithm with high security strength, and the addition of random numbers and timestamps, realize the encryption and signature of the transmitted data in the authentication process, which can effectively guarantee the confidentiality and non-repudiation of the session data. The security model proposed in this question can be optimized and perfected. It will be further studied later in order to provide a reference for the identity authentication of the Internet of Things equipment with high security value and unified standardization.

References

1. Oliver, S.G., Purusothaman, T.: Lightweight and secure mutual authentication scheme for IoT devices using CoAP protocol. Comput. Syst. Sci. Eng. **41**(2), 767–780 (2022)
2. Yoon, S., et al.: PUF-based authentication scheme for IoT devices. In: ICTC 2020, pp. 1792–1794 (2020)
3. Idriss, T., et al.: A lightweight PUF-based authentication protocol using secret pattern recognition for constrained IoT devices. IEEE Access **9**, 80546–80558 (2021)
4. Akhundov, H., et al.: Public-key based authentication architecture for IoT devices using PUF. CoRR abs/2002.01277 (2020)
5. Ahmad, H.S., et al.: Device authentication and data encryption for IoT network by using improved lightweight SAFER encryption with S-Boxes. Int. J. Embed. Real Time Commun. Syst. **12**(3), 1–13 (2021)
6. Panda, S.S., et al.: A blockchain based decentralized authentication framework for resource constrained IOT devices. In: ICCCNT 2019, pp. 1–6 (2019)
7. Vivekanandan, M., et al.: BIDAPSCA5G: Blockchain based internet of things (IoT) device to device authentication protocol for smart city applications using 5G technology. Peer-to-Peer Netw. Appl. **14**(1), 403–419 (2021)
8. Alam, A., Molyneaux, H., Stobert, E.: Authentication management of home IoT devices. In: Moallem, A. (ed.) HCII 2021. LNCS, vol. 12788, pp. 3–21. Springer, Cham (2021). https://doi.org/10.1007/978-3-030-77392-2_1
9. Kobayashi, S., Miyazaki, T.: Authentication and trustful communication protocol for IoT devices. In: IMCOM 2021, pp. 1–4 (2021)
10. Sikarwar, H., Das, D., Kalra, S.: Efficient authentication scheme using blockchain in IoT devices. In: Barolli, L., Amato, F., Moscato, F., Enokido, T., Takizawa, M. (eds.) AINA 2020. Advances in Intelligent Systems and Computing, vol. 1151, pp. 630–641. Springer, Cham (2020). https://doi.org/10.1007/978-3-030-44041-1_56

Data Visualization and Practice Platform Based on Data Mining Technology

Jie Kong(✉)

China University of Labor Relations, Beijing, China
kongjiepku001@163.com

Abstract. Modern society is an information society. With the continuous popularization of informatization, people have realized the importance of information literacy. As the talent reserve force of our country, college students have good information literacy has become the requirement of the times. Cloud computing (C C G) provides technical support for information literacy teaching due to its high cost performance, good scalability, and on-demand services. Based on this, the purpose of this article is to design and research a university information literacy teaching platform based on C C G technology. This article first summarizes the current situation of information literacy research, and then extends the current problems and shortcomings of information literacy teaching in my country's universities. On the basis of it, design and analyze the information literacy teaching platform of colleges and universities (C A U), and analyze the strategy of information literacy teaching. This paper systematically expounds the design of the scheduling algorithm of the teaching platform based on C C G, and uses the field research method, questionnaire survey method and other research forms to carry out experimental research on the theme of this article. Research shows that the university information literacy teaching platform based on C C G technology studied in this paper has higher feasibility.

Keywords: Cloud computing · Information literacy · Teaching platform · Design research

1 Introduction

With the maturity of information technology, information has flooded all aspects of society, so the cultivation of information literacy needs to be solved urgently [1, 2]. Information literacy, as one of the necessary qualities of talents in the information society, is an important way for every college student to improve their learning ability and ideological awareness [3, 4]. Therefore, information literacy is being paid attention to by experts and scholars in the field of education [5, 6].

In terms of information literacy research, foreign research methods mainly focus on online courseware teaching methods. Among all online education courseware, the most famous is TILT (Texas Information Literacy Tutorial) [7, 8]. Since then, many universities have begun to develop their own online courseware [9, 10]. Compared with

I. Ahmad et al. (Eds.): STSIoT 2021, LNDECT 122, pp. 234–242, 2023.
https://doi.org/10.1007/978-981-19-3632-6_30

foreign countries, our country mainly starts with the conditions that affect the formation of college students' information literacy, and focuses on the construction of university libraries and the role of libraries. Judging from the research results, my country has made great achievements in information literacy, but there are still many shortcomings in the research in this area [11, 12].

This article aims to improve the information literacy of our country's college students, and aims to design and research the information literacy teaching platform of C A U based on C C G technology. By comparing the high-tech information literacy teaching platform based on C C G technology studied in this article with traditional information literacy teaching methods, the feasibility analysis of the content of this article is carried out.

2 The Design of University Information Literacy Teaching Platform Based on Cloud Computing Technology

2.1 Analysis of Existing Problems in Information Literacy Teaching in C A U

(1) Not getting the attention of relevant parts

At present, the relevant departments of the school are not paying enough attention to information literacy. This is mainly expressed as follows: information literacy teaching methods are relatively scattered, single, lacking exploratory and practical; the content of information literacy education is mainly limited to the training of college students' computer skills and the guidance of the content of the document retrieval course, which is too single; the educational concept is retrogressive, partly C A U do not offer literature retrieval courses or information literacy-related courses, and there is a lack of information literacy training environment.

(2) Library informatization and multimedia education have not received enough attention

In foreign countries, many internationally renowned universities have dedicated information literacy columns on their library websites. This is enough to show that university libraries attach importance to student information literacy. However, many universities in Japan focus on luxurious gates and administrative office buildings, and lack sufficient attention and funding for the library as the university's largest knowledge base. As a result, many school libraries have small area, idle space, and insufficient electronic resource procurement.

(3) The setting of information literacy training courses is unreasonable

As the most important course of information literacy, the literature retrieval course has not received enough attention. Some universities do not even offer literature retrieval courses. Some schools do not offer literature retrieval courses or information literacy related courses. Even if it is opened, only elective courses and lectures are provided. This not only wastes educational resources, but also makes it difficult to achieve good training results.

(4) Insufficient information literacy of college teachers

As a preacher, teachers have a great influence on students' words and deeds. Therefore, their information literacy has a direct and significant impact on students' information literacy level. However, the information literacy of some college teachers has not yet reached the required level. The information literacy of some college faculty and staff is not up to the standard, and it is difficult to achieve the goal of cultivating college students' information literacy. Some teachers have limited educational ability to cultivate information literacy for students.

2.2 Design of Information Literacy Teaching Platform

(1) Demand analysis

1) Functional requirements

a. The system needs the ability to recommend resources and courses for each user.
b. The system must provide an open social platform for all users of the platform to publish and share information literacy perceptions, learning trends and high-quality resources.
c. The system not only provides a community activity platform for schools, classes and clubs, but also combines with traditional education models to support open communities that are self-organized based on common interests.

2) Non-functional requirements

Holistic demand

a. The interface is beautiful, and the menu structure is clear and concise.
b. Easy to use, users of all ages can quickly master it.
c. Support multiple network access methods such as Wi-Fi, 4G, and 5G networks.
d. Minimize the flow consumption during data exchange and reduce the power consumption of terminal equipment when using software.

Scalability requirements

The scalability of the system design is very important. As far as function expansion is concerned, not only horizontal expansion is required, but also system upgrades must be backward compatible. Considering the realization of specific web pages, browser support is also very important. As far as the actual interface is concerned, the difficulty of changing the interface directly affects the stability of the system and the importance of the code.

Data security requirements

In the actual system implementation process, when privacy and sensitive data need to be protected, the key data sent through the network can be encrypted. If the data is persistent and serialized, you need to pay more attention to the encapsulation of sensitive information to prevent data leakage.

(2) System function design

1) Analysis of authentication module

The authentication module contains 2 sub-modules: real-name registration and login authentication. The user needs to fill in real information such as real name and ID card. If the information entered in the information meets the requirements, the system will return a successful registration prompt. Otherwise, registration will fail. Another thing to note is that the system needs real-name authentication and real ID information needs to be entered. Login authentication means that the user enters the correct user name and password, and the system returns a prompt message indicating that the login is successful. Otherwise, the login will fail.

2) Announcement module analysis

Announcements include photo news and important news announcements. Picture news displays information literacy pictures or information literacy advertisements related to announcement news. Important news announcements match the news content on the portal and display the latest announcements or important news.

3) Analysis of recommended modules

The recommendation module displays information literacy education resources that match the student information in the form of a list. Click a course to open the course details page. This page shows the different attributes of the course and the different knowledge units of the course. Click a knowledge unit to view a list of all resources under the knowledge unit.

4) Spatial module analysis

Users can see the status information of themselves and the people they follow in their own space, and understand the dynamics and moods of the followers in real time. You can also click on any circle to enter the circle dynamic page, where you can see various information and information literacy dynamics shared by circle members.

2.3 Analysis of Information Literacy Education Strategies in Universities

(1) Establish specialized courses for cultivating information skills and conduct system information literacy education

Advances in information technology enable people to quickly use computers without programming. In other words, the society's demand for software application talents has gradually exceeded the demand for programming talents. Therefore, when universities offer information technology courses, they should aim at improving the ability of information retrieval and application, and complete the learning process with the help of computer information systems.

(2) Increase investment in information infrastructure

Universities need to establish higher education resource sharing mechanisms, promote higher education informatization construction, and develop and integrate different types of education resources at the same time. The construction of informatization should include the construction of high-quality courses, the sharing of literature resources, and the construction of educational platforms. Transform the scientific research results of university teachers and students into educational resources, adhere to the interaction between education and scientific research, and realize the connection between education and scientific research.

(3) Strengthen the construction of library informatization

The library is a collection place for scientific research results, sharing educational functions with the faculty and staff of C A U, and is an important part of the educational resources of C A U. Library construction is also an important symbol of the comprehensive strength of a university. University library is an important place for cultivating information literacy and an important resource collection place. Its construction has a profound and important influence on the information literacy of college students. Therefore, strengthening library information construction is an inevitable trend for the development of universities.

2.4 Design of Scheduling Algorithm for Teaching Platform Based on C C G

C C G is an on-demand service, which inevitably faces various user service requirements. This article aims at the different needs of different users, so that the teaching platform can meet the expectations of users in the C C G environment.

Based on the characteristics of the C C G business model, this paper selects four goals of job completion time, bandwidth, cost, and reliability to quantify the satisfaction of different users. The following are the weight preferences of the four goals:

$$\omega = \{\omega_1, \omega_2, \omega_3, \omega_4\}, (\sum_{i=1}^{4} \omega_i = 1) \tag{1}$$

Assuming that the amount of resources actually consumed by task T_i is A_i, and the amount of resource consumption expected by the user is E_i, the user satisfaction function of task T_i is:

$$W_i = \theta\, ln(A_i/E_i), (0 \leq \theta \leq 1) \tag{2}$$

The closer the actual resource consumption A_i 1 is to the resource consumption expected by the user E_i, the higher the user satisfaction and the closer the function value is to zero. If $W_i > 0$, it means that the actual resource consumption is higher than the user's expected resource consumption. Conversely, $W_i < 0$ means that the actual resource consumption is lower than the user's expected resource consumption.

Set up the N * P dimension ETC matrix. t_{ETC} (I, j) represents the expected execution time of the i-th task in the j-th resource list. Then the execution time of job J_m can be expressed as:

$$t(J_m) = max_{j=1}^{p} \sum_{i=T_{Total}(m-1)}^{T_{Total}(m)} t_{ETC}(i, j) \tag{3}$$

In this case, the total time required to complete all M operations is:

$$t_{Total} = \sum_{m=1}^{M} t(J_m) \tag{4}$$

According to formula (2), let t_{expt} be the expected completion time of J_m, and then the completion time user satisfaction function of J_m is:

$$W_{Time}(J_m) = \theta\ ln[t(J_m)/t_{expt}] \tag{5}$$

3 Experimental Research on the Information Literacy Teaching Platform of Colleges and Universities Based on Cloud Computing Technology

3.1 Experimental Protocol

In order to make this experiment more scientific and effective, this experiment carried out a questionnaire survey on its information awareness by going to a university in a certain place. This experiment conducted a targeted questionnaire survey on students' use of library electronic resources and databases. On this basis, through in-depth interviews with relevant information technology teachers on the efficiency of the C C G technology-based college information literacy teaching platform and traditional teaching methods studied in this article, and the results obtained are calculated using mathematical statistics.

3.2 Research Methods

(1) Questionnaire survey method

In this experiment, a targeted questionnaire was set up by asking relevant experts, and a semi-closed method was used to conduct an information literacy questionnaire survey on students, the purpose of which is to enable the surveyed students to fill in correctly.

(2) Field research method

This research conducted field investigations and collected data by going to C A U in a certain place, and collecting data for its information literacy teaching. These data provide a reliable reference for the topic selection of this article.

(3) Comparative analysis method

This experiment compares and analyzes the university information literacy teaching platform based on C C G technology studied in this article with traditional information literacy teaching methods to analyze the feasibility of the research content of this article.

(4) Mathematical Statistics

Use related software to make statistics and analysis on the research results of this article.

4 Experimental Analysis of Information Literacy Teaching Platform in Colleges and Universities Based on Cloud Computing Technology

4.1 Analysis of the Level of Information Awareness

In order to make this experiment more scientific and effective, this experiment conducted an experimental analysis on the degree of understanding of college students' information awareness through a questionnaire survey. The data obtained are shown in Table 1.

Table 1. Analysis of the level of information awareness

	Know and understand	Know but don't know much	Know but don't understand	Don't know
Electronic resources	11%	51%	20%	18%
Library data training	6%	27%	27%	40

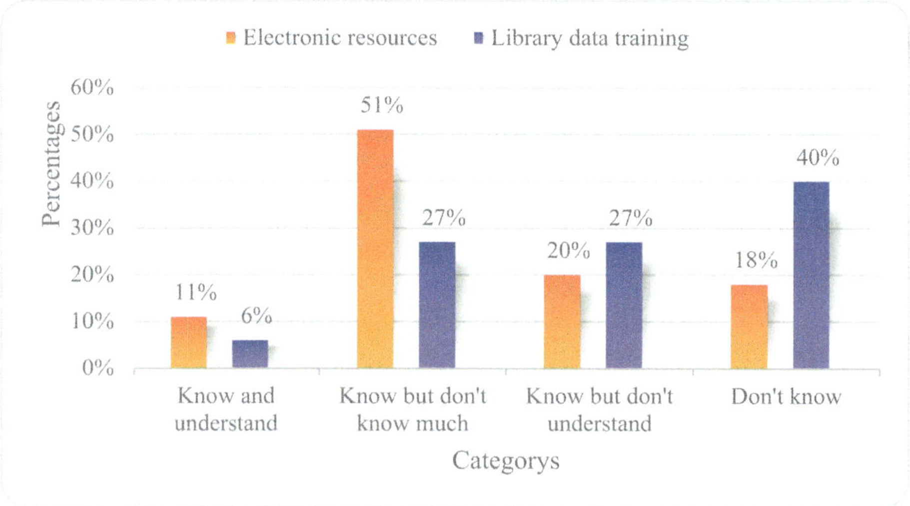

Fig. 1. Analysis of the level of information awareness

It can be seen from Fig. 1 that 62% of the electronic resources on the homepage of the school library know and have used it, but 51% of them only know but rarely use it; in the library training on the use of databases, 67% of people have not attended the lecture, and 40% of them don't even know about it. In summary, C A U still do not pay enough attention to the education of information literacy, and it is difficult to achieve the goal of cultivating college students' information literacy.

4.2 Comparative Analysis of Information Literacy Teaching Methods in C A U

In order to further research and analyze this experiment, this experiment compares and analyzes the university information literacy teaching platform based on C C G technology designed in this article and the traditional information literacy teaching method. The data obtained is shown in Table 2.

Table 2. Comparative analysis of teaching methods of information literacy in C A U

	Teaching methods	Learning efficiency	Learning Resources	Others
C C G	76.2%	77.8%	74.1%	66.7%
Traditional	57.8%	55.2%	56.0%	57.9%

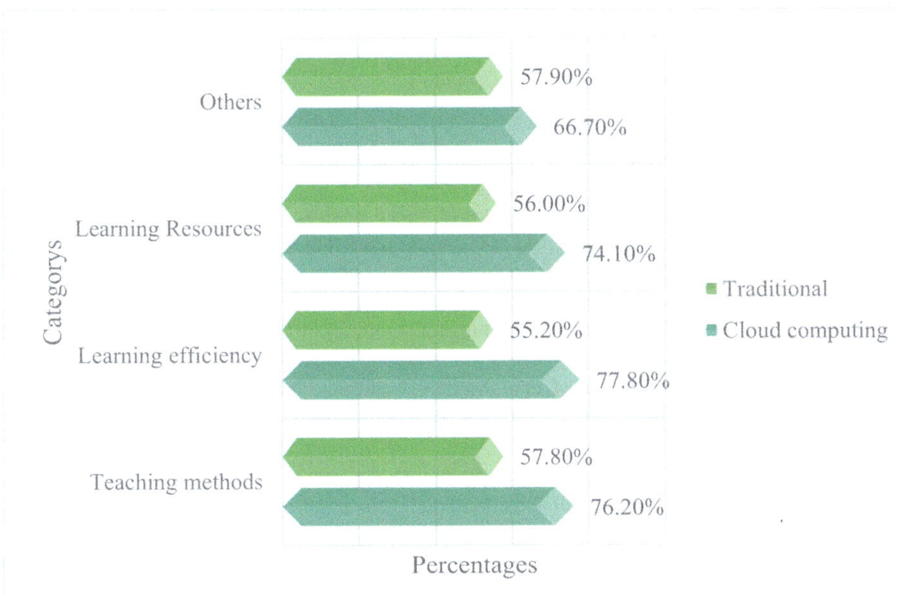

Fig. 2. Comparative analysis of teaching methods of information literacy in C A U

It can be seen from Fig. 2 that compared to the traditional information literacy teaching method, the C C G technology-based university information literacy teaching platform studied in this article is more excellent in many aspects, especially in terms of learning efficiency exceeding 20%, which fully reflects this article studies the feasibility of the content.

5 Conclusion

This paper aims at designing a C C G technology-based university information literacy teaching platform. By analyzing the current status of university information literacy teaching, based on its existing problems and deficiencies, constructing a C C G technology-based university information Literacy teaching platform. It also analyzes the strategy of information literacy training in C A U. This paper designs the scheduling algorithm of a teaching platform based on C C G, and conducts experimental research in the form of a questionnaire survey in C A U. The research shows that C A U still do not pay enough attention to information literacy education, and it is difficult to achieve the training goals of college students' information literacy.

Acknowledgements. Teacher training project about cloud computing and big data management, Project was supported by "University-Industry Collaborative Education Program (Project Number: 202002248013)".

References

1. Schloman, B.F., Gedeon, J.A.: TRAILS: tool for real-time assessment of information literacy skills. Charleston Advisor **17**(3), 43–48 (2016)
2. Weiner, S.: Information literacy and the workforce: a review. Educ. Libr. **34**(2), 7 (2017)
3. Rosman, T., Mayer, A.K., Krampen, G.: A longitudinal study on information-seeking knowledge in psychology undergraduates: exploring the role of information literacy instruction and working memory capacity. Comput. Educ. **96**(5), 94–108 (2016)
4. Gretter, S., Yadav, A.: Computational thinking and media & information literacy: an integrated approach to teaching twenty-first century skills. TechTrends **60**(5), 510–516 (2016)
5. Korsnakova P, Throndsen I, Loi M, et al. Students' computer and information literacy from a European perspective. Findings from ICILS 2013. Part 2. Eng. Struct. J. Earthq. Wind Ocean Eng. **27**(8):1197–1208 (2017)
6. Pettersson, J., Bjorkander, E., Bark, S., et al.: Using scenario-based training to promote information literacy among on-call consultant pediatricians. J. Med. Libr. Assoc. Jmla **105**(3), 262–267 (2017)
7. Chen, D., Liu, H.: Information literacy education of university based on the concept of "Internet+" %. J. Agric. Libr. Inf. Sci. **028**(011), 110–113 (2016)
8. Townsend, L., Hofer, A., et al.: Identifying threshold concepts for information literacy: a delphi study. Commun. Inf. Lit. **10**(1), 1 (2016)
9. Frank, E.P., Pharo, N.: Academic librarians in data information literacy instruction: a case study in meteorology. Coll. Res. Libr. **77**(4), 536–552 (2016)
10. Markless, S., Streatfield, D.: How can you tell if it's working? Recent developments in impact evaluation and their implications for information literacy practice. J. Inf. Lit. **11**(1), 106 (2017)
11. Raish, V., Rimland, E.: Employer perceptions of critical information literacy skills and digital badges. Coll. Res. Libr. **77**(1), 87–113 (2016)
12. Squibb, S.D., Mikkelsen, S.: Assessing the value of course-embedded information literacy on student learning and achievement. Coll. Res. Libr. **77**(2), 164–183 (2016)

Track and Field Image Target Detection Based on Feature Learning

Wei Li(✉)

Baotou Medical College Sports Department, Baotou 014010, Inner Mongolia, China
liwei3117@126.com

Abstract. With the advancement of video target positioning technology, the amount of image data captured by people is increasing rapidly, which puts forward more efficient and automated requirements for data processing methods. As the most popular artificial intelligence technology, deep intelligence has become a research hotspot and has shown potential in the field of mobile image target search. This paper focuses on the research of track and field image target detection based on feature learning. Aiming at some of the problems in video target detection, we use feature learning to solve related problems, so as to improve the accuracy of target detection, and carry out experimental verification. As a result, the target detection method based on feature learning proposed in this paper is 5% more accurate than the traditional target detection method.

Keywords: Feature learning · Target detection · Track and field video · Sports video

1 Introductions

As one of the five human senses, vision is the main means by which people in modern society receive external data and information [1, 2]. With the widespread operation and dissemination of various related technologies such as computer electronics, computers, and communications, as well as the emergence and rapid development of various new-generation information technologies such as computer networks, computer Internet of things, and cloud computing, various digital optical images and videos the data is increasing exponentially every day [3, 4]. How to allow the machine to automatically receive, process, analyze and master this massive amount of visual data is a major subject in the current scientific research of robotic vision in our country, and it is also a major subject in the field of computer science research in our country [5, 6]. Generally speaking, people who are interested in video data usually notice something. Target recognition technology and tracking technology are the most important basis for extracting the required information from unstructured pixel data to video data. Moving target detection and tracking technology mainly uses multiple unstructured video sequences to automatically detect the moving scene we want [7, 8].

I. Ahmad et al. (Eds.): STSIoT 2021, LNDECT 122, pp. 243–251, 2023.
https://doi.org/10.1007/978-981-19-3632-6_31

Aiming at the research of image target detection, a new small target detection technology based on frequency domain saliency extraction and image sparse representation is proposed. First, the features of the Fourier spectrum image and the amplitude spectrum of the Fourier transform are used to roughly extract the salient regions, and use the threshold segmentation system to classify the salient areas in the background to obtain a binary image [9]. Other researchers have proposed that detecting and separating small moving targets in complex and noisy scenes is still a challenging task for monitoring and recognition systems based on moving targets, and proposed a new model to solve the problem in noisy environments. The moving target is difficult to detect in [10]. Some researchers also proposed an adaptive terahertz image target detection algorithm based on a logarithmic model, and verified the effectiveness of the algorithm through experiments. The experimental results show that the background noise of the image can be filtered out and the shape of the object can be extracted more clearly and the edge [11]. Some researchers have proposed a new target detection algorithm based on NS and DMs, and experiments using multiple target SSS images (with or without shadow areas) have shown that no errors or missing detection [12]. Through the above summary, it can be seen that there are still relatively few image targets for feature learning.

This paper studies track and field image target detection based on feature learning, summarizes the problems existing in track and field image target detection on the basis of relevant literature, then analyzes the application of feature learning in track and field image target detection, and then carries out track and field image target detection experiment based on feature learning to verify the effectiveness of the method.

2 Research on Target Detection in Track and Field Images

2.1 Problems in Target Detection of Track and Field Images

(1) The appearance of the target will change with changes in the target's internal factors or external environmental factors. Among them, the internal factors include the change of the target position, the change of the target scale and the distortion of the target itself; the external factors mainly include the change of light, the change of the field of view and obstacles. These factors will affect the target display model's ability to distinguish between targets and backgrounds and different targets. Factors such as illumination changes, clutter and target scale changes make it very difficult to accurately monitor video targets in complex scenes.

(2) The problem of multi-target video tracking includes not only difficult factors such as illumination changes, clutter, and target scale changes, but also inaccurate placement and missed inspections caused by false observations caused by background interference. In addition, the number of targets in a real scene usually changes over time, and the lack of tracking and erroneous observation makes the estimated target number uncertain. Moreover, the target display features in complex scenes also contain some uncertainty information. The existence of the above-mentioned multiple uncertain factors makes the correlation between the target and the observation data difficult, and further increases the difficulty of multi-target online video tracking.

(3) Real-time online video tracking. Real-time is an important reason for online video surveillance applications, such as real-time video surveillance and unmanned driving. However, reliable target display models often need to calculate complex display features or learn from a large number of training samples, while tracking algorithms based on simple target display models usually have low tracking accuracy. The development of real-time tracking algorithms is also one of the research methods for online video tracking targets important difficulty.

(4) The similarity measurement of fuzzy image detection and target recognition. For the last stage of video motion detection, target recognition is based on the output target image and the result of target detection and tracking. Due to the movement of the target, these exported images often contain a certain degree of blur. In the target recognition process, if these blurred images are not eliminated, the target recognition rate will decrease. At present, the focus of many recognition algorithms is on the recognition algorithm itself, while ignoring the research of blur image removal. In addition, to determine the target category in the test video from the database, any recognition algorithm cannot do without similarity measurement. Different methods of measuring similarity will directly result in different recognition percentages. Therefore, choosing a suitable similarity measurement method is very important for identifying the target video.

2.2 Application of Feature Learning in Track and Field Image Target Detection

Artificial intelligence technology is applied to image processing to realize intelligent image interpretation, such as processing visible light image data sources. Moving target detection is an important part of animation interpretation. This link can provide information about the category and destination in the image scene. The targets in the picture include, but are not limited to, athletes, stadiums and other categories. The combination of deep learning theory in artificial intelligence technology and image target detection can automatically identify and extract target features in images, and perform rapid and intelligent processing of image data, so it has attracted more and more attention.

2.3 Target Detection Algorithm

rcnn is mainly divided into three components. The first step is that we use statistical learning methods to find several selected areas in the map according to different characteristics such as shape, color, and texture. Then in the second step, the data output of these selected areas is sent to the convolutional neural network to extract these features. In the third step, the data output of the selected area is sent to the svm471 classifier for classification. The location and size of the selected area are fine-tuned accordingly to facilitate more precise positioning.The related flowchart is shown in Fig. 1:

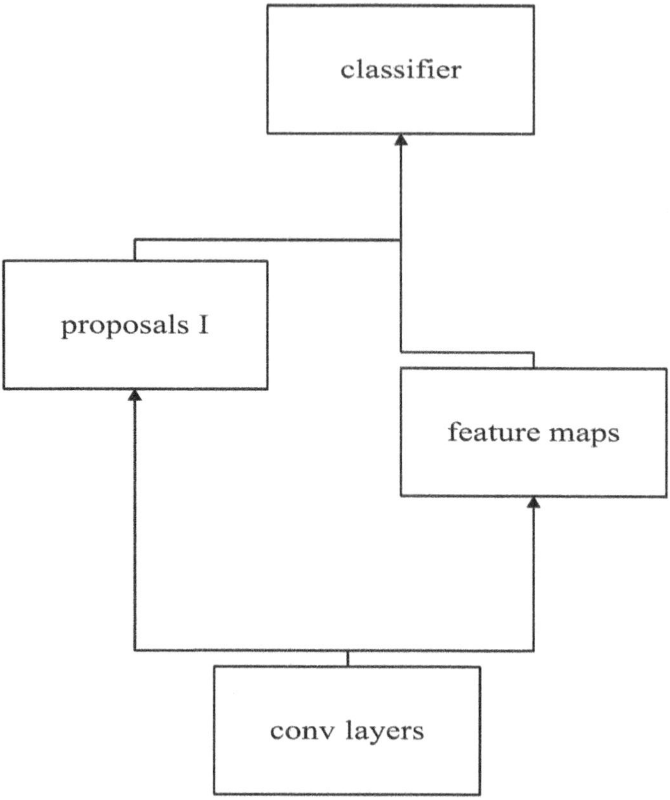

Fig. 1. Target detection algorithm flow chart

(1) The average accuracy of target detection for a single category of objects is expressed as:

$$AP = \int_0^1 P(R)dR \tag{1}$$

R is the call rate.

(2) The average accuracy of target detection and evaluation for multiple categories of objects is expressed as:

$$mAP = \frac{1}{N_{cls}\sum_i AP_i} \tag{2}$$

I is the target number.

In our proposed algorithm, we use GMM as a background modeling algorithm to obtain the background probability of each frame. The background modeling algorithm here can be replaced by other algorithms and does not affect the framework of our algorithm. The definition of the GMM model is as follows:

$$G(x) = \sum \pi_k N(x; \nu_k) \tag{3}$$

Among $N(x; \nu_k)$ is the Gaussian probability density function, Z is the variance, π is the weight of each single Gaussian model, and x is the data sample.

3 Track and Field Image Target Detection Experiment Based on Feature Learning

3.1 Experimental Data

The experiment conducted a comparative experiment on 3 challenging videos to evaluate the results. The three videos in the experiment have a total of more than 40,000 frames, which are public public data sets. The scene involves many dynamic factors such as: athletes on the track, including different sizes, speeds and types of moving objects, so the video is sufficient to evaluate the overall performance of the proposed algorithm and the comparison algorithm.

3.2 Experimental Parameter Design

The training platform used in this article is TTANxPascal, 4 GPUs are used for testing and training, and the framework used is cafe2. This article is based on Detecton, and all experimental settings in this article are based on this. The main networks trained in this article are Iesnet5t and Iesnet101. During training, the minimum edge of the image is set to 600sense data set or S00coco data set, and the maximum edge is set to 1100 or 1433. Momentum is set to 0 and 9, the initial learning rate is set to 0.02, and the penalty rate is 0.0002. This article uses training shards, a total of 370,000 repetitions, and the learning rate drops to 0.0002.

3.3 Image Processing

Because the video is affected by factors such as light changes, camera shake, or bad weather, the image quality deteriorates and the detection object of the target video changes, making it difficult to monitor and target recognition. Among them, lighting changes are the most common factor that affects the quality of video images. The method to solve lighting changes is generally to improve the image. The use of image enhancement technology can deliberately highlight all or part of the image to improve the loss of target information caused by lighting changes, and make the original distorted clear image and improve the effect of tracking and identifying the target.

3.4 Object Detection Learning Module

When faced with complex background issues, it is important to choose important shapes. Therefore, it is particularly important to improve the ability to use shapes in many ways. The actual receiving field of CNN is much smaller than the theoretical receiving field, and the context information is not yet fully available. The typical structure of CNN is the local relationship between learning paths, so it is difficult to capture the relationship in the spatial context. For a unified similarity graph, sufficient environmental information is not enough. Therefore, the learning module is further introduced into the remote environment information receiving framework. Especially in the detection of image targets, it is very important to establish confidence in the environment. In order to obtain the value information of the image and improve the expression ability of the image from a global perspective, the learning module-environment-based GC is introduced.

4 Analysis of Experimental Results

4.1 Detection Accuracy

In this paper, the target detection of track and field images based on feature learning is carried out on the 3 videos of the experiment, and the track and field athletes in the video are detected. 5 repeated experiments are carried out to record the accuracy of the detection. The relevant data results are shown in Table 1:

Table 1. Detection accuracy result

	Video 1	Video 2	Video 3
1	98.42%	99.88%	98.25%
2	96.05%	95.73%	94.15%
3	99.88%	94.46%	95.21%
4	94.73%	95.46%	92.67%
5	94.46%	95.51%	94.29%

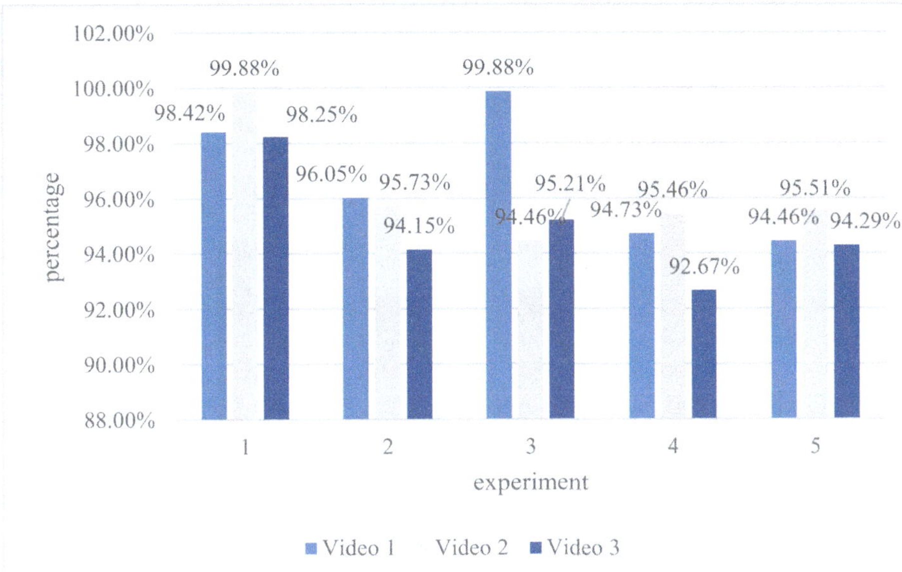

Fig. 2. Detection accuracy result

It can be seen from Fig. 2 that the target detection accuracy of track and field images based on feature learning in this paper is over 90%, and the highest is 99.88%. From this we can see the feasibility of the track and field image target detection based on feature learning in this paper.

4.2 Comparison of Detection Accuracy

This paper compares the accuracy of track and field image target detection based on feature learning with traditional image target detection, repeats 5 times of detection experiments on track and field athletes, and records the accuracy of each time. The relevant data is shown in Table 2:

Table 2. Comparison of detection accuracy

	Target detection in this article	Traditional target detection
1	100.00%	99.70%
2	95.31%	90.79%
3	97.33%	90.60%
4	98.43%	92.90%
5	98.40%	90.28%

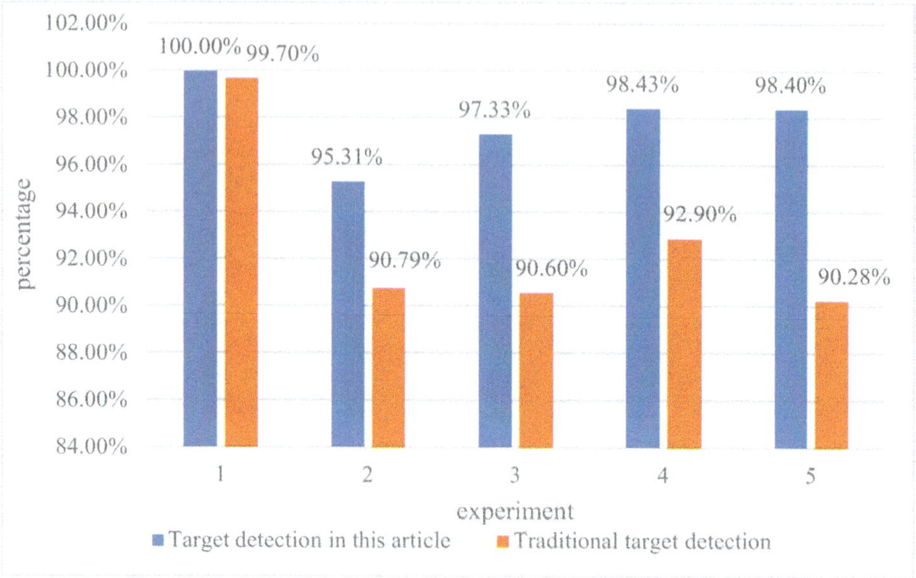

Fig. 3. Comparison of detection accuracy

It can be seen from Fig. 3 that the accuracy of target detection based on feature learning in this paper is higher than that of traditional target detection, and it is 5% higher than that of traditional target detection methods. It can be seen from this that feature learning is introduced in target detection is effectiveness.

5 Conclusions

This paper studies the target detection of track and field images based on feature learning. Aiming at the problem of target detection in track and field images at this stage, feature learning is introduced into video target detection to improve the accuracy of target detection in sports videos. The results are verified by experiments. It is concluded that the detection accuracy of the video target detection method based on feature learning in this paper is higher than that of the traditional detection method, which confirms the effectiveness of the method in this paper, but there are still some shortcomings in the research process of this paper. Only the accuracy of target detection has been verified, and no other experimental verifications have been carried out, which will be supplemented in future studies.

References

1. Wang, Y., Yue, J., Dong, Y., et al.: Review on kernel based target tracking for autonomous driving. J. Inf. Process. **24**(1), 49–63 (2016)
2. Edelson, G.S.: Two-stage active sonar network track-before-detect processing in a high clutter harbor environment. J. Acoust. Soc. Am. **140**(4), 3349 (2016)

3. Hu, Y., Dong, M., Ota, K., et al.: Mobile Target detection in wireless sensor networks with adjustable sensing frequency. IEEE Syst. J. **10**(3), 1160–1171 (2017)
4. Li, Y., Pan, J., Long, J., et al.: Multimodal BCIs: target detection, multidimensional control, and awareness evaluation in patients with disorder of consciousness. Proc. IEEE **104**(2), 332–352 (2016)
5. Zhou, P., Cheng, G., Liu, Z., Bu, S., Hu, X.: Weakly supervised target detection in remote sensing images based on transferred deep features and negative bootstrapping. Multidimension. Syst. Signal Process. **27**(4), 925–944 (2015). https://doi.org/10.1007/s11045-015-0370-3
6. Oliveira-Rodríguez, M., Serrano-Pertierra, E., García, A.C., et al.: Point-of-care detection of extracellular vesicles: sensitivity optimization and multiple-target detection. Biosens. Bioelectron. **87**(5), 38–45 (2017)
7. Hu, Y., Liu, A.: Improvement the quality of mobile target detection through portion of node with fully duty cycle in WSNs. Comput. Syst. Sci. Eng. **31**(1), 5–17 (2016)
8. Shi, C., Salous, S., Fei, W., et al.: Power allocation for target detection in radar networks based on low probability of intercept: a cooperative game theoretical strategy. Radio Sci. **52**(8), 1030–1045 (2017)
9. Borisova, I.V., Legkiy, V.N., Kravets, S.A.: Application of the gradient orientation for systems of automatic target detection. Comput. Opt. **41**(6), 931–937 (2017)
10. Dong, Y., Du, B., Zhang, L.: Target detection based on random forest metric learning. IEEE J. Sel. Top. Appl. Earth Observ. Remote Sens. **8**(4), 1830–1838 (2017)
11. Yang, S., Shi, Z., et al.: Hyperspectral image target detection improvement based on total variation. IEEE Trans. Image Process. **25**(5), 2249–2258 (2016)
12. Xu, J., Liao, G., Lei, H., et al.: Robust adaptive beamforming for fast-moving target detection with FDA-STAP radar. IEEE Trans. Signal Process. **65**(4), 973–984 (2016)

Application of CNC Technology in Mechanical Manufacturing Under Integrated Intelligent Algorithm Analysis

Song Luo[1](✉) and Tetsuzo Tanino[2]

[1] Jiangxi Jianglian Energy and Environmental Protection Co., Ltd., Nanchang, Jiangxi, China
Luoong2021@126.com

[2] University of Pennsylvania, Philadelphia, USA

Abstract. On the basis of traditional mechanical manufacturing, through the effective combination with computer technology, intelligent algorithm, CAM technology and other technologies, NC machining technology has been formed. Compared with traditional machining, numerical control technology has great advantages and has become an indispensable process technology in contemporary machinery manufacturing industry. The purpose of this paper is to study the application of NC technology in mechanical manufacturing under the analysis of integrated intelligent algorithm. This paper takes the CNC machine tool control system as the research object, combined with the integrated intelligent algorithm, designs the system. This paper analyzes the functional requirements of the application of NC machine tool control system in mechanical manufacturing, then designs the overall hardware circuit of NC system, and explains in detail the module circuit design and its working principle, mainly including power circuit design and input-output interface circuit design. Finally, the NC system is tested. The control effect of NC machine tool on workpiece contour error is as follows: the minimum value of contour error CEA is 0.0124 mm and the maximum error is 0.141 mm; the minimum linear error is 0.0113 mm and the maximum error is 0.0146 mm. It can be seen that under the NC system, the contour error of the machine tool is small and the precision is high.

Keywords: CNC lathe · Reliability analysis · Adaptive fuzzy control · Machine manufacturing

1 Introduction

With the rapid development of social science and technology, the high-tech industry is undergoing rapid changes, and the manufacturing industry with CNC machining as the core gradually occupies an important position in the country's comprehensive strength competition [1, 2]. On the basis of traditional processing, through effective combination with intelligent algorithms, numerical control technology, CAD and other technologies, a numerical control processing technology has been formed [3, 4]. Compared with

I. Ahmad et al. (Eds.): STSIoT 2021, LNDECT 122, pp. 252–261, 2023.
https://doi.org/10.1007/978-981-19-3632-6_32

traditional processing technology, it has greater advantages and has become an important processing technology in modern manufacturing [5, 6].

Regarding the research of numerical control technology, many scholars at home and abroad have conducted multi-angle and in-depth discussion on it. For example, Szczesiak studied the parameterized numerical model used to predict the mechanical properties of polymer materials FDM technology [7]. Zheng K An experimental study on the feasibility of a new technology for manufacturing hot stamping dies for prefabricated pipe networks [8]. Peng S, based on previous studies, conducted numerical simulation experiments on surfacing welding technology to create new hot stamping dies [9]. Wang X conducts research on the application of automation technology in automobile machinery manufacturing [10]. It can be seen that since the development of CNC technology, its application has received extensive attention. This research combines integrated intelligent algorithms to study the control system of CNC machine tools, and explores the solutions to some key problems in CNC process design. Therefore, this research has important practical significance.

The purpose of this article is to study the application of numerical control technology in mechanical manufacturing under integrated intelligent algorithm analysis. Starting from the point of view of CNC system control, this article first conducts in-depth research on the real-time monitoring of CNC machine tool status, centralized control of CNC system data, and efficient diagnosis of CNC machine tool faults according to the functional requirements of the CNC basic control system, and then designs an integrated intelligent algorithm CNC machine tool system, and finally verify the system.

2 Application of CNC Technology in Mechanical Manufacturing Under Integrated Intelligent Algorithm Analysis

2.1 Requirement Analysis of CNC Machine Tool Control System Under Integrated Intelligent Algorithm Analysis

(1) Centralized control of data

The data of the numerical control system has the characteristics of multiple dimensions and multiple parameters, including processing procedures, system parameters, PLC parameters, tool compensation parameters, etc. [11, 12]. Among them, PLC parameters also include ladder diagram programs, timer parameters, counter parameters, and so on. Therefore, the system should have the function of centralized management and control of the data of the CNC machine tool. The machine tool manager can remotely view and modify the data of the machine tool CNC system through the system host, and backup and process key data, thereby improving the management and operation of the machine tool.

(2) Machine tool real-time monitoring and fault diagnosis

The CNC machine tool has a complex structure and integrates multiple systems such as mechanical, electrical, and hydraulic. After a machine tool failure occurs, it is urgent to carry out maintenance and treatment quickly, otherwise the processing program will be interrupted, the machine's downtime will be increased, and the utilization rate of the CNC machine tool will decrease. However, the machine tool operator cannot judge and deal with the damage of the machine tool in time, and professional maintenance personnel are required to repair the machine tool. It is inevitable that the diagnosis and maintenance of machine tools will be delayed due to factors such as communication, capital, and transportation. Therefore, the system should specifically have the function of quickly diagnosing faults of CNC machine tools. Among them, the main self-diagnosis and operation diagnosis when the system is connected to the power supply, including: checking whether the communication between the main controller of the upper and lower motors and each device is normal; whether the electrode wire status is normal and whether the drive is operating normally, etc. At the same time, the system software must also update and display the monitored status information and detected abnormal information in real time.

(3) Energy efficiency analysis of machine tool equipment

The working states of CNC machine tools are shutdown, idle, running and alarm. In the same period of time, lengthening the running time of the machine tool can increase the utilization rate of the equipment. The longer the idle time, the lower the utilization of the equipment. The alarm of the CNC machine tool refers to the abnormal operation of the equipment, which requires key maintenance and treatment. Therefore, the system should have the ability to count the duration of different states of CNC machine tools, and calculate the time proportion of each state. In this way, the production task setter can adjust the allocation of production tasks according to the time proportion of the equipment status, thereby increasing the utilization rate of the machine tool and increasing economic benefits.

(4) Online product quality management

In the production process of machinery manufacturing enterprises, the quality data after product processing is very important. Due to raw material problems, process formulation errors, severe tool wear and other reasons, the quality of the machine tool's product processing will be unqualified. The feedback of product quality inspection data is delayed, leading to untimely error processing time, reducing the qualification rate of the product, and ultimately causing significant losses to the enterprise. Therefore, the control system should have the function of online management of product quality, by reading the detection data of machine tool products in real time, and feedback whether the product is qualified according to the data, so as to adjust the parameter settings of CNC machine tools and solve process problems or tool wear problems in time.

2.2 Hardware Design of CNC System Under Integrated Intelligent Algorithm Analysis

(1) The overall hardware circuit design of the numerical control system

The design of the numerical control system adopts the structure of "PC+ programmable motion control card". The upper computer realizes the communication with CPLD through the serial port. The control signal sent by the upper computer is processed by the CPLD and then output to each control port to realize the coordination of the various modules of the system. Mainly include: host computer (industrial control computer), serial communication interface circuit, CPLD motion control card, power supply circuit, CPLD configuration circuit (including JTAG circuit, clock circuit), input and output interface circuit design, threshold voltage control circuit and power failure protection circuit.

The hardware circuit design of the CNC system mainly includes motion control board circuit design and PCB board production. The motion control board circuit design also includes power supply circuit design, CPLD configuration circuit design, threshold voltage control circuit design, input and output interface circuit design, serial communication Interface circuit design and power-down protection circuit design.

(2) Power circuit design

The voltage conversion of the power circuit design of this system selects REG1117, and converts the voltage of 5 V to 3.3 V for use by the board. The NPN output stage driver can help the load current achieve the highest efficiency, so REG1117-3.3 is selected for the design of the power supply circuit.

(3) Design of input and output interface circuit

The interface signal between the CNC machine tool and the control system realizes the communication between the two, which is mainly composed of the control system channel signal, PLC information alarm signal, MCP signal and axis-related signal. The interface signal is composed of two parts: data interface and function interface. The data interface contains the working status signal, control signal and auxiliary function signal of the machine tool. The function interface is used to realize the cyclic transmission of PLC and NCK signals. The input and output switches in this numerical control system mainly include: high frequency, cable transmission, liquid flushing, engine lock, machine tool lighting, etc. Two three-state eight-way manipulation devices SN74LVC245APWLE are selected for the input and output switch signal transmission in the system.

(4) Service-oriented CNC system design and implementation technology

The implementation technologies in the service-oriented open CNC system design framework correspond to the implementation activities at different stages in the design process, as shown in Fig. 1.

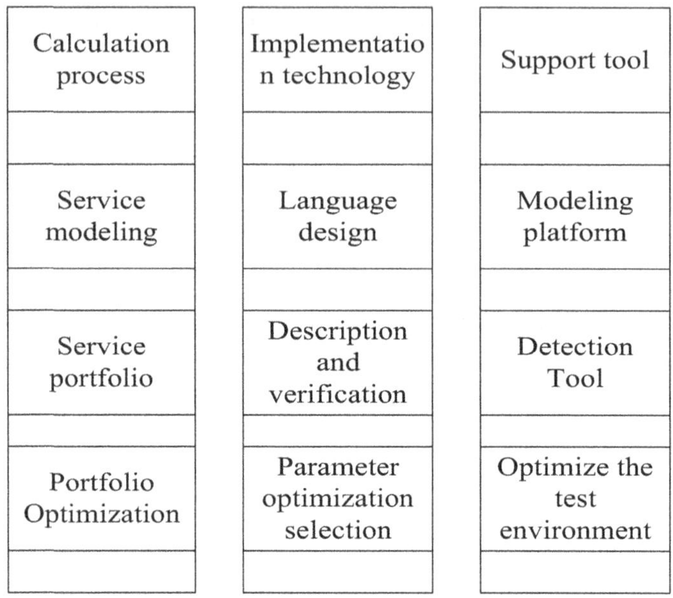

Fig. 1. The technical process of CNC system design and implementation

Among them, the service-oriented domain modeling language design is based on the idea of domain modeling, using meta-building technology to construct a service-oriented meta-model of the CNC domain with multiple information fusion characteristics. The design strategy of the meta-model reflects the granular unit and Hierarchical division mechanism is used to encapsulate control functions as a service, and then through the mapping and integration of services in different levels, a service-oriented application system is realized.

(5) Porosity of repair layer

When performing surface thermal spraying, due to the stacking of deformed particles, there will be pores and slag trapped in it. The coating density is directly measured, and the porosity can be obtained by comparing it with the actual density. According to the formula (1):

$$K = (1 - \frac{\rho_c}{\rho_m}) \times 100\% \tag{1}$$

Among them, K is the porosity; ρ_c is the measured density of the repaired layer; ρ_m is the actual density of the repaired layer.

3 Survey Design

3.1 Test Environment

The experiment selects FANUC0i-TD numerical control panel, Mitsubishi M70 numerical control panel, Siemens 840Dsl numerical control panel to verify the function of the numerical control machine tool control system proposed in this study.

During the system test, the main controller needs to be connected to the equipment, including the main controller power supply, pulse power supply, detection power supply module, various voltage input and output interfaces, display, keyboard and mouse, etc. After the connection is complete, then turn on the built-in Windows XP Embedded system. After the system is started, the control software runs automatically, and at the same time, the connection status displayed on the software homepage is normal.

3.2 Accurate Calculation of Contour Error

In the X-Y coordinate system of the worktable, the contour error ε of the arc motion trajectory of the two-axis motion control system is accurately calculated as shown in formula (2):

$$\varepsilon = \|\overline{OP}\| - r \quad (2)$$

Substitute the coordinates of each point to get:

$$\varepsilon = \sqrt{(X_P - X_O)^2 + (Y_P - Y_O)^2 - r} \quad (3)$$

where: P point is the current position point of the execution end of the CNC machine tool, O point is the center of the arc motion trajectory, r is the arc motion trajectory radius, the shortest distance ε from point P to the linear command trajectory is the contour error, P_x is the actual position point X coordinate, P_y represents the Y coordinate of the actual position point.

4 Data Analysis

4.1 Contour Error Data Analysis

The contour error control effect of the CNC machine tool on the work piece is shown in Table 1: the minimum contour error CEA is 0.0124 mm, the maximum error is 0.141 mm; the minimum linear error is 0.0113 mm, and the maximum error is 0.0146 mm.

Table 1. Contour error data

Work piece number	Contour error CEA (mm)	Straight line error (mm)
1	0.0124	0.0146
2	0.0136	0.0124
3	0.0141	0.0137
4	0.0124	0.0113
5	0.0137	0.0129
6	0.0126	0.0146
7	0.0125	0.0134
8	0.0127	0.0127
9	0.0137	0.0140
10	0.0134	0.0131

Fig. 2. Contour error data

It can be seen from Fig. 2 that under this numerical control system, the contour error of the machine tool is small and the accuracy is high. At the same time, in the processing of large curvature parts and inflection points, the size and direction of the contour error often fluctuate greatly, and the numerical control system proposed in this paper can effectively reduce the fluctuation of the contour error value.

4.2 Verification of Production Management Functions

Table 2. Work piece time statistics

Machine number	Processing time (s)	Current number of work pieces	Work piece average time
1	220	129	1.71
2	247	128	1.93
3	244	129	1.89
4	243	130	1.87
5	242	127	1.91
6	241	128	1.88
7	240	129	1.86
8	243	127	1.91
9	247	126	1.96
10	245	128	1.91

This experiment is set in a gear processing factory, which produces a variety of gear products. In order to realize the efficient management of production data, the manager needs to give a uniform number to the machining programs of the machine tool, that is, the same product is assigned an independent program number, and the same program number is used in different processes and different machine tools. The production management module can inquire about the man-hours consumed for processing the work piece within the selected starting time and the number of pieces produced by the machine tool. Among them, the work-hour statistics of the work piece are shown in Table 2.

It can be seen from Fig. 3 that the shortest average time for each work piece of the machine tool is 1.71 s, and the longest time is 1.96 s. And the CNC system interface can directly display the time spent processing the work piece and the number of pieces produced by the machine tool. It can be seen that the CNC system the production management function can be carried out effectively.

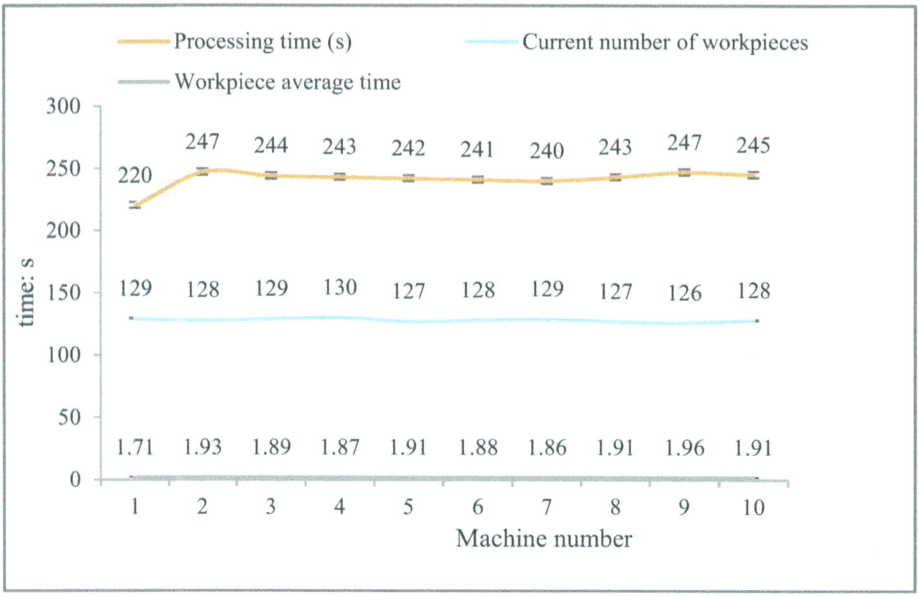

Fig. 3. Work piece time statistics

5 Conclusion

The advanced level of CNC machine tools is an important reference to measure the advanced level of a country's manufacturing industry, and the CNC machine tool control system is the key to the performance of CNC machine tools. In order to maintain the real-time processing capability of the numerical control system, the operating system of the numerical control device is upgraded to a system with a visual operation interface, which effectively improves the sensory experience of the user system. Through research, this article mainly completed the following tasks: analyzed the functional requirements of the system, including requirements for real-time monitoring of the state of CNC machine tools; centralized control of CNC system data, and efficient diagnosis of CNC machine tool faults, and designed a CNC machine tool system with integrated intelligent algorithms. The reliability of the system is verified.

References

1. Serafimovich, P.G., et al.: On a silicon-based photonic-crystal cavity for the near-IR region: numerical simulation and formation technology. Semiconductors **50**(8), 1112–1116 (2016). https://doi.org/10.1134/S1063782616080212
2. Rashwan, S.S., Nemitallah, M.A., Habib, M.A.: Review on premixed combustion technology: stability, emission control, applications, and numerical case study. Energy Fuels **30**(12), 9981–10014 (2016)
3. Viscusi, A., Bruno, M., Esposito, L., Testa, G.: An experimental/numerical study of bonding mechanism in cold spray technology for metals. Int. J. Adv. Manuf. Technol. **110**(9–10), 2787–2800 (2020). https://doi.org/10.1007/s00170-020-06060-9

4. Wang, L., Tan, X., Sun, H., et al.: Numerical study of ceramic catalytic turbine technology for emission reduction during vehicle warm-up. Int. J. Engine Res. **1**, 146808742110360 (2021)
5. Xu, S., He, J., Pei, F., et al.: Numerical study on the denitrification efficiency of selective noncatalytic reduction technology in decomposing furnace. J. Chem. **2020**(7), 1–11 (2020)
6. Xu, S., Jiao, C., Ning, M., Dong, S.: Analysis of Buoyancy Module Auxiliary Installation Technology Based on Numerical Simulation. J. Ocean Univ. China **17**(2), 267–280 (2018). https://doi.org/10.1007/s11802-018-3305-4
7. Szczesiak, R., Kowalik, M., et al.: Parametric numerical model for predicting mechanical properties of structures made with FDM technology from polymeric materials. Polimery **63**(9), 626–632 (2018)
8. Zheng, K., Tong, C., Li, Y., et al.: An experimental and numerical study of feasibility of a novel technology to manufacture hot stamping dies with pre-constructed tube network. Int. J. Adv. Manuf. Technol. **111**(9–10), 2919–2937 (2020). https://doi.org/10.1007/s00170-020-06280-z
9. Peng, S., Zhou, J., Zhang, M., et al.: Fundamental research and numerical simulation of new hot stamping tool manufactured by surfacing technology. Int. J. Adv. Manuf. Technol. **107**(7–8), 3527–3541 (2020). https://doi.org/10.1007/s00170-020-05274-1
10. Wang, X., Wang, Y.: Application of automation technology in automobile machinery manufacturing. J. Phys.: Conf. Ser. **1885**(4), 042056 (2021). (6pp)
11. Liu, H.: Research on computer vision detection technology and applications on machinery manufacturing and automation. Int. J. Technol. Manage. (005), 16–18 (2016)
12. Li, X., Wu, T.: Project-oriented development and practice of the machinery manufacturing technology course. Int. J. Emerg. Technol. Learn. (iJET) **13**(5), 130 (2018)

Data Collection of Electronic Bills Based on Network Printing Simulation

Chunyong Sun(✉) and Bin He

Digital Grid Research Institute, China Southern Power Grid, Guangzhou 510700, Guangdong, China
xiaochongtougao3@163.com

Abstract. With the advent of the Internet age and the rapid development of global information technology, people's life has entered the electronic age. Among them, electronic account not only improves service efficiency, but also has important practical significance for protecting the environment, saving resources and reducing carbon emissions. The electronic bill data collection system supports fast data retrieval and automatically uploads it to the platform during the storage service to realize experimental data collection and data exchange. This paper aims to study the electronic bill data information collection based on network printing simulation. Based on the analysis of system functional requirements, system non-functional requirements, network printing simulation system and data information collection algorithm, the functional modules in the electronic bill data information collection system are designed, and finally the performance of the system is tested. The test results show that the electronic bill data information collection system designed in this paper has achieved good performance and meets the needs of this paper.

Keywords: Network printing simulation · Electronic billing · Data collection · Information collection

1 Introduction

With the rapid development of the national economy, people have more and more bills in their lives. In addition to utility bills, there are different types of bills such as radio and television bills, communication fees, and insurance premiums [1, 2]. Since these bills were originally printed on paper, a large amount of paper was required. It will not only consume a lot of funds and resources, but also destroy the ecosystem and have a negative impact on economic growth [3, 4].

In recent years, domestic and foreign scholars have conducted in-depth research on electronic bills and have achieved great results. In order to solve the prominent problem that the traditional data scalable storage modular solution cannot well meet the need for rapid location access to massive electronic bill data, some scholars have proposed a new type of massive electronic bill scalability that supports rapid data location. Storage architecture, designed a horizontally scalable storage architecture based on the hash

I. Ahmad et al. (Eds.): STSIoT 2021, LNDECT 122, pp. 262–270, 2023.
https://doi.org/10.1007/978-981-19-3632-6_33

modular algorithm to reduce the amount of data that needs to be migrated when adding or deleting data nodes [5]; in terms of the use effect of electronic bills, there are also scholars have made relevant surveys. The survey shows that paperless electronic bills can effectively improve operating efficiency, reduce operating expenses and costs, and can also improve service capabilities and customer satisfaction. The service process is tracked and monitored, and data related to it is calculated to help customers make decisions [6]; in addition, related researchers have designed an electronic bill presentation and payment system, including scalable a customized, data source-independent data layer is used to access the business data repository to process billing information. The data layer module accesses the data in the business data repository and implements a data source-specific method to complete the request [7]. Modern enterprise information technology has become a powerful means of promoting the transformation and upgrading of domestic and foreign enterprises, improving quality and efficiency, and it can continuously improve and optimize through close cooperation with the development of interaction with the market, and obtain a series of valuable research results.

Combined with the needs of electronic bill data information collection system, this paper consulted the literature related to electronic bill and advanced data information collection technology at home and abroad, fully considered the problems of electronic bill and data information collection, analyzed and designed the functional modules of electronic bill data information collection system in detail, and finally tested the performance of the system.

2 Research on the Data Collection of Electronic Bills Based on Network Printing Simulation

2.1 System Function Requirement Analysis

(1) Data collection requirements

The collection staff can connect the account to the platform as an online account promotion center. At the same time, it also supports performance monitoring of system data-driven equipment, such as hardware resource usage, CPU, memory usage, etc. [8, 9].

(2) Management of electronic bills

After uploading the user and bill information, the electronic bill is generated and stored locally. If necessary, customize the electronic bill sent to the user. It can easily and quickly obtain and retrieve relevant information.

(3) Sending of electronic bills

This function is performed by sending custom electronic invoices to different types of users. In the process of designing and implementing this function, when using different communication interfaces, it should pay attention to timely and regular delivery, and support automatic transmission and manual transmission at the same time [10, 11].

2.2 Analysis of System Non-functional Requirements

(1) No single point of failure: consider the reliability and availability of the computing system from the aspects of virtual server configuration, network equipment, and network protocols. Provide overall system redundancy, create and improve advanced storage and backup systems to ensure the integrity, stability and availability of system data. Details include:

Virtual server: CPU, memory, hard disk.

Network line: load balance among multiple lines, network protocol, routing protocol, automatic switching to meet failure.

Storage system: adopts distributed cloud storage to store data, data management control software.

(2) The ease of use and maintenance of the system. If you do not participate in the development of the system, you will not understand the use of the electronic billing data collection system. Therefore, a successful application must have a complete human-machine interface so that the user knows exactly what the software should be. Which button performs a function, so that the user learns to use it faster, and shortens the process for the user to become familiar with the system.

(3) The openness of the system and the scalability of the system. Before developing an electronic billing data collection system, it is important to carefully consider the scalability of the software. This can be achieved through the openness of the system, and certain user specifications must be met. In the subsequent application process, software modules and functions can be added and modified as needed, and the system hardware can be configured [12].

(4) Interface requirements: the electronic billing data information collection system needs to take into account functionality, aesthetics and personalization to achieve the harmony and unity of the system interface and content. This is embodied in the requirements of the electronic bill data collection system for unified interface style, unified content display, unified function layout, and unified function buttons. The display interface of the electronic bill data information collection system strives for functionality and convenience, so that the application of the electronic bill data information collection system fully meets the functional needs of users for query and browsing. On the premise of ensuring easy use and simple operation, the interface should be beautiful and generous. At the same time, the style of the system interface must highlight the focus of business functions, with clear primary and secondary functions and clear parts. The detailed enlargement of the electronic account accounting information collection system should reflect the sense of hierarchy and humanity, which is suitable for users to search and browse for a long time.

2.3 Network Printing Simulation System

The key of the virtual printing system is to intercept the printing task of the application program, which can be realized by modifying the printing subsystem module of the operating system. There are mainly three methods as follows:

(1) Publisher is a DLL user responsible for converting Spooled Data into print jobs that can be sent to Monitor Print. Publisher reads the spool file containing the line data of the print job, converts the data stream and sends the modified data to the spooler, and then the spooler sends the print stream data to the appropriate print supervisor.
(2) Additional functions of Microsoft Provider. Microsoft Render provides a calling device to check the data stream of the type data before sending it to the Spooler.
(3) Monitoring port. The data available in the port monitor includes print service data and a series of controls sent to the publisher.

2.4 Data Collection Algorithm

In the data information collection model, each item or attribute is a variable of a node, and the edges between nodes reflect the relationship between nodes to indicate whether they are together. The network model is as in formula (1).

$$P(X = x) = \frac{1}{Z} e^{\sum_{n=1}^{n} \alpha_i * x_i + \sum_{j,k} \omega_{j,k}^{+} * f^{+}(x_j, x_k)} \tag{1}$$

Among them, the range of i, j and k is from 1 to n, the number of items or attributes in the entire database of n type, and the explanation equation of Z is in formula (2) (3).

$$Z = \sum_{X \in \chi} e^{\sum_{n=1}^{n}} \alpha_i * x_i \tag{2}$$

$$Z = \sum_{j,k} \omega_{j,k}^{+} * f^{+}(x_j, x_k) \tag{3}$$

3 Experiment

3.1 User Login and Registration Module

(1) The only thing that needs attention in the user login system is the user login interface, which is the first step to understand human-computer interaction. During the login process, the user must enter the account information, account name and program password. If you have not filled it out completely, please click the login button. The system will ask this entry not to be left blank. If it is incorrect, please re-enter the words. Whether this function can be successfully implemented has a lot to do with the database. When the information is completely matched, the system prompts the user to log in successfully. If the information fails to match, the system displays that the information entered by the user is incorrect. Please enter it again until it is correct. The user login flow chart is shown in Fig. 1.

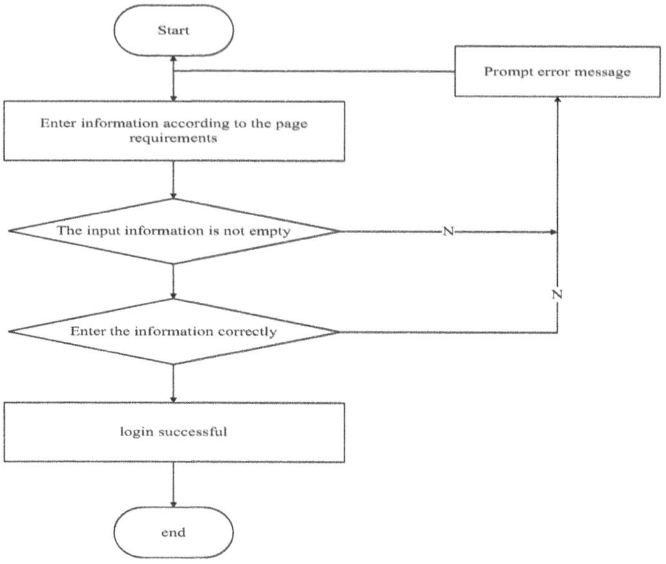

Fig. 1. User login flow chart

(2) User registration. This function is mainly aimed at new users of the system. After filling in, click the submit button, and the system will automatically check the validity of the input information. If the content is not empty and meets the quantity requirement, the registration is successful.

3.2 Data Information Collection Module

(1) Data collection management

The data collection management interface displays all collection records entered into the system, that is, a list of data collected and stored each time. Specific information includes data storage time, source type, total collection, etc. The module implements a complete data search function. Users can use data categories, names, keywords and other terms to filter the content they want to view.

(2) Data import management

The data entry management module can view and manage imported electronic data bills, such as user information, account information, personal call bills, etc. After entering the module, the list displays all the billing information entered by the system. Users can set data type, input volume, input time, keywords and other conditions to filter the content they want to see.

3.3 Data Management Module

The data management module stores the initial user information of the electronic bill data collection system, selects MySQL and REDIS databases as the storage technology of the bill data collection system, and stores the data electronically in the MySQL database. In order to demonstrate and adapt the electronic bill data collection method, it is convenient to call the results after the skill analysis, and save the user information, electronic bill information and other data into the database MySQL database. REDIS is a commonly used unstructured memory database. The data reading efficiency of REDIS is only related to memory. Saving frequently used data in the REDIS database can improve data storage efficiency.

3.4 System Test Method

(1) Integration test

Integration testing mainly uses manual testing methods to check whether the test results meet the expected results, and test the functions of the system according to the test cases; if it is satisfied, the test passes; otherwise, the system will be improved and the test will be executed again.

(2) Unit test

This unit uses manual testing to test the various functional modules of the test system, summarizes the test results statistics, points out the reasons for the irregularities, and uses this to further update the system.

(3) System test

The system uses manual testing to check the functionality and reliability of the system, and calculate the test results; if the test results are not satisfactory, please modify the system and retest.

4 Discussion

System performance testing is to test whether the system can have good performance during the use of the system. This text tests the safety, reliability, and response time of the system. Table 1 and Fig. 2 show the performance test results.

Table 1. System security test

Test content	Test case execution times	Passing rate
User authentication	30	100%
Role authentication	20	100%
Access control	20	100%
Data access control	20	100%

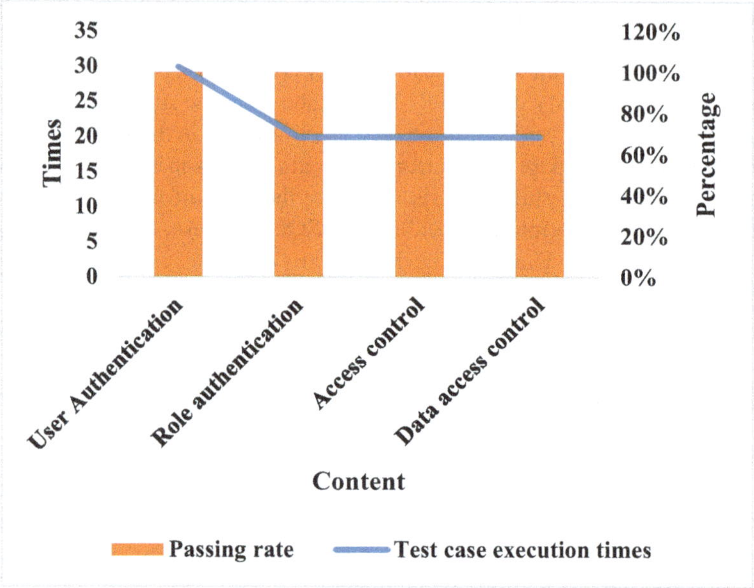

Fig. 2. System security test

As can be seen from Table 1 and Fig. 2, the user authentication test has been tested 30 times, and the test pass rate is 100%; the role authentication, authority control and data access control have been tested 20 times, and the test pass rate is also 100% .

Table 2. System response time test

	Estimated maximum time/s	Actual longest time/s	Estimated average time/s	Actual average time/s
Average visit time	3.4	2.8	2.3	1.7
User login access time	2.2	1.7	1.8	0.8
Data collection access time	3.2	2.9	1.8	1.5
Data import access time	3.2	2.8	2.7	1.5

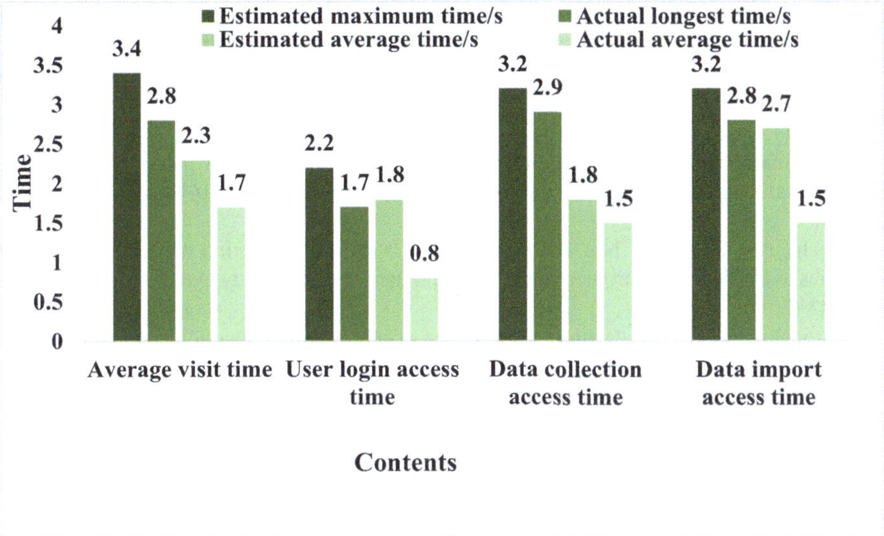

Fig. 3. System response time test

It can be seen from Table 2 and Fig. 3 that the expected longest access time for user login is 2.2 s, the actual longest time is 1.7 s, the expected average time is 1.8 s, and the actual average time is 0.8s; the expected longest access time for data collection is 3.2 s, the actual longest time is 2.9 s, the estimated average time is 1.8 s, and the actual average time is 1.5 s; the expected longest access time for user login is 3.2 s, the actual longest time is 2.8 s, and the estimated average time is 2.7 s. The actual average time is 1.5 s. By analyzing the performance test results of this system, we can see that the electronic bill data collection system designed in this paper has achieved good performance and meets the needs of this paper.

5 Conclusions

Data collection is the basic means of obtaining information. With the development of information technology, the Internet is constantly changing people's work and lifestyle, and people are paying more and more attention to the importance of protecting the environment. The elimination of paper bills and the promotion of electronic bills are essential products in this era, as well as business measures to reduce business costs and improve the efficiency of bill delivery.

References

1. Alali, H.: eFAWATEERcom as an electronic bill payment service: the case of Jordan. Int. J. Adv. Sci. Eng. Inf. Technol. **10**(4), 1430 (2020)
2. Choi, S.B.: Compliance of electronic bill of lading regulation in Korea with model law on electronic transferable records. J. Korea Trade **23**(3), 68–83 (2019)

3. Baloch, N., Jun, C.S., Kwon, B.I.: Design and performance evaluation of a modular linear induction machine for rotating electronic billboard. IEEE Access **PP**(99), 1 (2019)
4. Zhang, L., Kong, J., Cui, B., et al.: Safety effects of freeway roadside electronic billboards on visual properties of drivers: insights from field experiments. J. Transp. Eng. **146**(2), 04019071.1–04019071.11 (2020)
5. Lu, S., Sun, X.: Economic evaluation model of greenhouse soil information collection technology based on fuzzy comprehensive evaluation. J. Phys.: Conf. Ser. **1648**(4), 042108 (2020). (6pp)
6. Tenqchen, S., Su, Y.J., Chen, K.P.: Using the CVP traffic detection model at road-section applies to traffic information collection and monitor -the case study. Artif. Intell. Eng. **01**(2), 38–43 (2019)
7. Shestopalova, O.L.: Forecasting obsolescence of information collection and processing systems of space complexes according to the criterion of information support efficiency. Izvestiâ vysših učebnyh zavedenij Priborostroenie **64**(2), 81–89 (2021)
8. Phillips, J., Babcock, R.A., Orbinski, J.: The digital response to COVID-19: exploring the use of digital technology for information collection, dissemination and social control in a global pandemic. J. Bus. Contin. Emer. Plan. **14**(4), 333–353 (2021)
9. Ding, T., Li, J., Pan, J., et al.: Human remote mobile medical information collection method based on internet of things and intelligent algorithm. Rev. Brasileira Medicina Esporte **27**(spe), 28–30 (2021)
10. Wu, Y., Mu, X., Kahrizi, D.: Design of biological gene information collection system based on data mining technology. Cell. Mol. Biol. (Noisy-le-Grand, France) **66**(7), 93 (2020)
11. Flores-Luyo, L., Agra, A., et al.: Mixed integer formulations for a routing problem with information collection in wireless networks - ScienceDirect. Eur. J. Oper. Res. **280**(2), 621–638 (2020)
12. Lin, C., Xu, J., Ma, R., et al.: A mobile client oriented to subjective and objective information collection for township land resources surveying. Eco. Inform. **60**(840), 101173 (2020)

Marketing Strategy of Digital Transformation of Electricity Market in the Internet Era

Rui Xiao(✉) and Benquan Chen

Digital Grid Research Institute, China Southern Power Grid, Guangzhou 510700, Guangdong, China
xiaochongtougao2@163.com

Abstract. With the rapid development of market econoour and the continuous deepening of market-oriented reforms of the power system, power supply companies should actively integrate into the tide of "Internet +", build a marketing service system that meets the current market competition environment and the Internet background, and satisfy power customers demand, in order to achieve the goal of common development of power supply companies and power customers. At present, the marketing management of electric power enterprises seems to be an important work for enterprises, which is closely related to the development and growth of the company. Therefore, this article mainly studies the MS of the digital transformation of the electric power market in the Internet era. First of all, the electricity sales of company A in recent years and the average time limit for customers of various voltage levels comprehensively reflect the current status of EM (electricity marketing). The results show that the electricity sales from 2017 to 2020 are gradually increasing, and the electricity sales in 2020 will be 567.71 million kWh. 220 V customers are the main customers of company A, accounting for 72.85% of the total number of customers. Secondly, it briefly introduced the transformation effect of the digital MS of the electric power market in the Internet era. 41% of the people believed that the EM strategy after the digital transformation would help the electric power company provide more efficient power services. The rest of the people also pointed out that the power marketing system supported by the Internet can enable customers to control their electricity consumption well.

Keywords: Internet · Digital transformation · Marketing strategy · Electricity market

1 Introduction

Electric power companies take the initiative to build a marketing service system that is in line with the current market competition environment and the Internet background, to bring diversified, convenient, highly reliable and personalized power supply services to the majority of power customers, further improve customer satisfaction, and build a good power company Brand image. This article mainly studies the MS of the digital transformation of the EM in the Internet era. I hope to provide theoretical and practical

I. Ahmad et al. (Eds.): STSIoT 2021, LNDECT 122, pp. 271–279, 2023.
https://doi.org/10.1007/978-981-19-3632-6_34

reference and reference for the improvement of MS of our country's power companies under the background of "Internet +".

There are not a few researches on the MS of the electricity market in the Internet age. Acquila-Natale E believes that big data has brought tremendous changes to all walks of life, and people are entering the era of big data [1]. Grabchak E P pointed out that with the deepening of our country's power system reform, competition from other energy models is changing the power market pattern [2]. Based on the marketing status of our country's power companies, Nahar S used the 4C theory to analyze in detail the four parts of power company marketing in consumer demand and demand, cost, convenience and communication, and put forward product strategies, pricing strategies, and service power company marketing [3]. Although the research on power market marketing strategy is abundant, the digital transformation marketing strategy combined with the Internet background needs to be supplemented. Therefore, this article mainly studies the power market digital transformation marketing strategy in the Internet era.

This article mainly studies the MS of the digital transformation of the electricity market in the Internet era. The experiment is mainly divided into two parts. The first half comprehensively reflects the current situation of EM from the electricity sales of company A in recent years and the average time limit for customers at various voltage levels. The second half of the experiment analyzes the effect of digital marketing strategy transformation in the electricity market in the Internet era by designing a questionnaire experiment.

2 Theoretical Basis of the Digital Transformation Marketing Strategy of the Power Market in the Internet Era

2.1 Internet

The Internet + model means that under the new Internet model, various fields need to be deeply integrated with the Internet [4]. The Internet has many characteristics, one is cross-border integration. The so-called "+" means cross-industry, change, and openness [5]. Cross-industry talents make the foundation of innovation more solid and further create a more novel situation. The second is innovation-driven. The growth of traditional industries can no longer meet the requirements of market competition. The essence of competition has been transformed into a comparison between talents and technology. Technology represents the country's production and innovation capabilities, and talents represent the driving force for future development. As a new type of innovation platform, the Internet adopts an innovation-driven development path to gather innovation resources and exert innovative power [1]. The third is to connect everything. Connections are hierarchical, and the value of connections will be affected by the number of connected customers. Generally, the greater the number of connected customers, the greater the value generated, and the more interactive information, but connecting everything is the goal of "Internet +". The fourth is to reshape the structure. Internetization has broken the original social structure, economic structure, and cultural structure, causing unprecedented changes in the structure, politics, econoour, culture, and production of various countries. The "Internet +" model breaks through the traditional social governance model, changes the lives of ordinary people, and enables social governance to be reformed and improved. The fifth is to respect human nature. Maslow's theory of needs points out that people will become more and more subjective and pay attention to their own interests, needs, privacy, and rights. The Internet itself has strong tolerance and equality. Anyone can share their views on things on the Internet and enjoy the rich resources of the Internet on an equal footing. The Internet can respect human nature to the greatest extent and tolerate the free development of all people [6].

2.2 Electricity Marketing Strategy

There is no difference in attributes between the electricity market and the traditional market, and it is also a kind of general market. The difference is that the defining standard of the electricity market is electricity, a special commodity. The components of the electricity market consist of three parts: the commodity that can be exchanged, the seller and the buyer who provide the commodity [7]. The electricity market can be defined as the sum of various relationships in the process of electricity production, trading, and use. For the power supply company, the main business object of its electricity market is electric energy. The main exchange of the electricity market is electricity, and the object is the power supply company and the majority of electricity customers [8].

Power supply marketing refers to the power supply companies formulating MS to meet the needs of different consumers. The behavior of power supply companies with service as the core can promote the rapid growth of power supply companies' economic benefits. Economic benefits are always the ultimate goal of power supply companies, and only a certain market benefit can meet the company's development planning and competition goals [9]. There is no competition from the perspective of market laws among power companies. In order to maintain national security and social stability, the degree of administrative intervention in our country's electricity market is still relatively obvious. Compared with other industry markets, the degree of openness of the electricity market is relatively low. EM methods have particularities [10]. The marketing method of electric power is obviously different from ordinary products. The pricing of ordinary products can be priced according to the company's own situation, while the price of power products is often controlled by policies. Electric power companies cannot directly display power products when they carry out power marketing, and power users lack an intuitive experience of power products. As a circulation material directly intervened by the state, electric power commodities can play a very strong role in guiding the entire circulation market. Electric power companies pay more attention to terminal marketing. Electricity plays a supporting role in the development of the national econoour and the production and life of the people. Once electric energy fails, it will directly affect social development. The gray forecast model can predict the company's future electricity sales law. α is the development ash number, and β is the endogenous control ash number. The one-time cumulative prediction model is:

$$A_{j+1}^{1} = \left(A_{1}^{0} - \frac{\alpha}{\beta}\right) \times e^{\alpha * j} + \frac{\beta}{\alpha} \quad (j = 0, 1, 2 \cdots, n) \tag{1}$$

The cumulative prediction model of the original sequence is:

$$A_{j+1}^{0} = A_{j+1}^{1} - A_{j}^{1} = \left(1 - e^{\alpha}\right) \times \left(A_{1}^{0} - \frac{\alpha}{\beta}\right) \times e^{\alpha * j} \quad (j = 0, 1, 2 \cdots, n) \tag{2}$$

2.3 Electricity Marketing Theory

The marketing of power products also includes power prices, power products, sales channels, and power sales [11]. At present, the competition between electric energy and alternative energy is becoming more and more fierce, and the electric power market is in a state of oversupply of electric energy products. The power supply company should enhance the quality and market share of the supplied power, and reduce the cost of power supply [12]. Electricity price is a currency expression of the value of electricity products. It is also a unified retail price for users, but the pricing standards for electricity are not completely consistent. Electricity companies have formulated different charging standards for different market entities and different scales of electricity consumption. For example, the price of commercial electricity is different from that of residential electricity. Electricity in our country basically adopts a cost-oriented pricing method. It is necessary to improve the original price formation mechanism and transform to a demand-oriented pricing mechanism and a competition-oriented pricing mechanism. The sales of power products are mainly through the unified sales of power companies. When building a distribution channel, a power supply company needs to formulate its power distribution channel strategy based on the market that the power supply company expects to reach, and under the combined effect of users, products, competitors, policies, and market environment. Electricity promotion refers to the use of personnel or non-personnel means to customize sales strategies for different customers based on detailed market positioning, so as to promote customers' satisfaction with power companies and help increase electricity sales. The power promotion portfolio roughly includes advertising promotion, business promotion, and staff sales. Common methods in the promotion strategy include the promotion of the good image of electric power companies and the reliable characteristics of electric energy products through newspapers, radio and TV stations and other media [13].

3 Based on the Experiment of the Digital Transformation Marketing Strategy of the Electric Power Market in the Internet Era

3.1 Experimental Background

Electricity is a clean energy that the country vigorously promotes and develops, and it has a great impact on industrial production, people's lives and many other fields. Electricity sales are largely related to the marketing methods of power companies. Electricity sales are no longer restricted to government departments. Power companies can more accurately grasp market trends in terms of marketing. The Internet has become an extremely important marketing tool for power supply companies, and high-quality and efficient "Internet + power marketing" will benefit more and more power customers. With the current rapid development of the Internet, power customers demand higher and higher quality power supply services. Therefore, it becomes more meaningful to study the digital transformation marketing strategy of the electricity market in the Internet era.

3.2 Experimental Process Steps

After fully expounding the related concepts of Internet + and power marketing, this article designs experiments using field investigation method, literature survey method and questionnaire survey method. In the first half of the experiment, company A was taken as an example to analyze the electricity sales of company A in recent years and the average electricity service time of customers of various voltage levels, hoping to comprehensively reflect the current marketing situation of the electricity market. The second half of the experiment mainly used questionnaires to investigate the transformation effects of digital marketing strategies in the electricity market. The subjects of the survey are 100 front-line employees of power companies. The content of the questionnaire is the transformation effect of the digital MS of the power market.

4 Based on the Experimental Analysis of the Digital Transformation Marketing Strategy of the Power Market in the Internet Era

4.1 Status Quo of Electricity Marketing

Table 1. Development status of a power supply company

	2017	2018	2019	2020
Electricity sales (10,000 kWh)	41337	46583	47990	56771
The average unit price	0.60	0.63	0.71	0.75
Line loss rate	6.07%	5.00%	5.29%	5.1%
Voltage qualification rate	99.45%	99.74%	99.77%	99.91%

It can be seen from Table 1 and Fig. 1 that the development of Power Supply Company A is in good condition. The electricity sales from 2017 to 2020 are gradually increasing. In 2017, the electricity sales of A company were 413.37 million kWh, and the electricity sales in 2020 were 56,771 10,000 kWh, the electricity sales of Company A increased by about 150 million kWh in four years. From 2017 to 2020, the electricity price has increased slightly. The average unit price in 2017 is 0.6 yuan per kilowatt hour, and the average unit price in 2020 is 0.75 yuan per kilowatt hour. The line loss rate in the four years was below 6.1%, and the line loss rate in 2017 was 6.07%. Due to the influence of technology and other factors, the line loss rate in 2020 will be reduced to 5.1%. The line loss rate in the four years is at a low and reasonable level, and the line loss rate is declining year by year. The voltage qualification rate from 2017 to 2020 is higher than 99%, and the voltage qualification rate in 2020 reaches 99.91%.

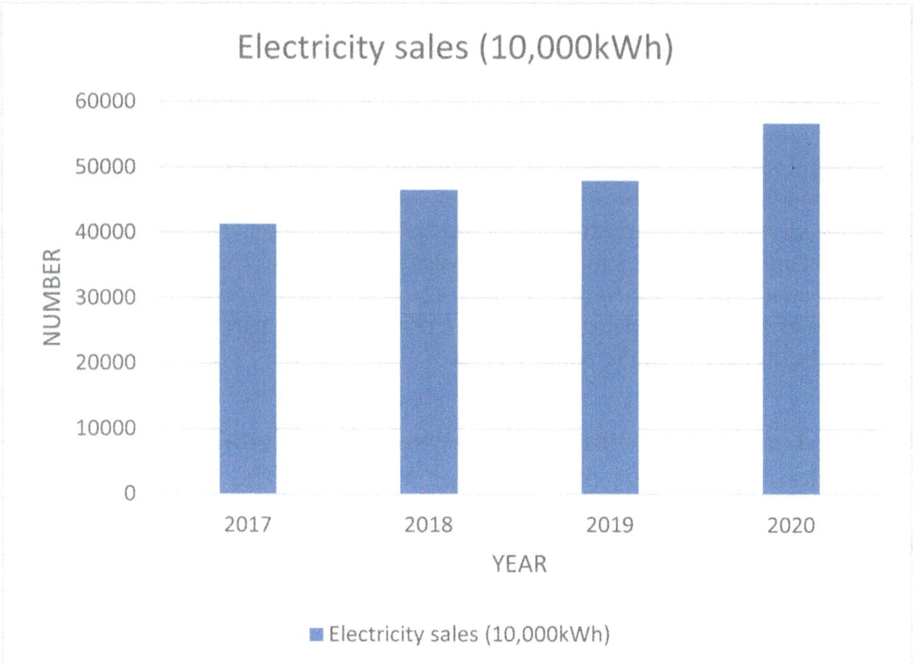

Fig. 1. Development status of a power supply company

Table 2. Average time limit for customers of various voltage levels

Customer voltage level	Number of customers doing electricity	Average time for customers
220 V	2831	3.82
380 V	977	7.97
10000 V	74	117.6
35000 V	4	134.77

It can be seen from Table 2 that in one working day, Company A served a total of 3886 customers who handle electricity. High-voltage customers above 10 kV have a longer time for powering up. The average time for 10 kV customers is 117.6 min, and the average time for 35 kV customers is as high as 134.77 min. At the same time, the number of high-voltage customers is much smaller than that of customers with voltage levels below 380 V. There are 74 households with 10 kV customers and 4 households with 35 kV customers. There are 2831 households with 220 V customers, accounting for 72.85% of the total number of customers. The average time for 220-V customers is 3.82 min, and the average for 380-V customers is 7.97 min.

4.2 Analysis of the Effect of Digital Marketing Strategy Transformation in the Electricity Market

Table 3. Transformation effect analysis

	Number	Proportion
Realize energy saving and consumption reduction	29	29%
Improve the convenience of the marketing process	21	21%
Provide efficient power services	41	41%
Other	9	9%

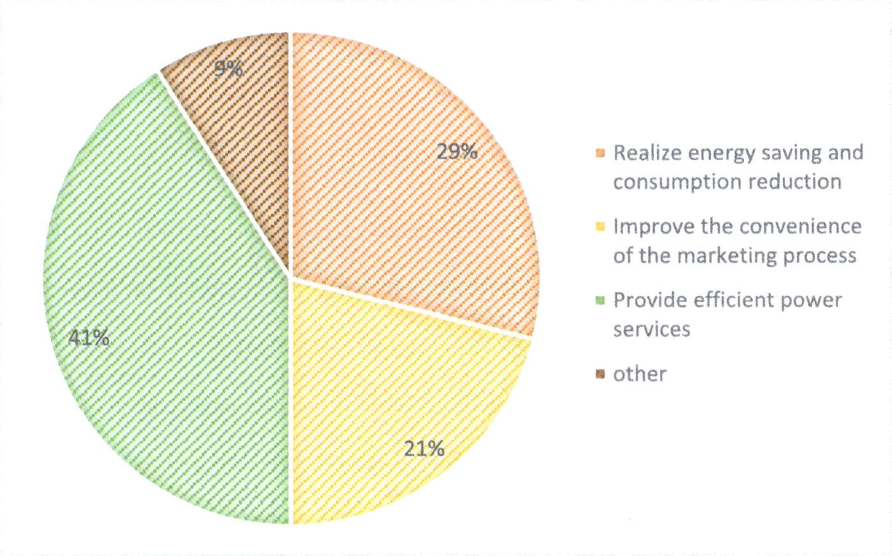

Fig. 2. Transformation effect analysis

It can be seen from Table 3 and Fig. 2 that 41% of people believe that the power marketing strategy after digital transformation will help power companies provide more efficient power services. In a fiercely competitive environment, power companies can only occupy a more advantageous position if they continuously improve their service level and quality. 29% of people think that the power marketing system supported by the Internet can enable customers to control their electricity consumption well. Electric power companies gradually use smart meters to eliminate traditional meters. On the one hand, it reduces the error in calculating and charging electricity charges, on the other hand, it reduces the time for consumers to handle business and improves customer satisfaction. 21% also pointed out that the Internet can improve the convenience of the marketing process. The Internet platform can effectively improve the fluency of information channels and improve communication efficiency. The remaining 9% believe that in addition to the above advantages, the power MS after digital transformation can also help power companies accurately predict market demand and promote corporate development.

5 Conclusion

With the country's new round of power system reform, our country's power distribution system has been completely broken. For power companies, opportunities and challenges coexist. At the same time, the demand for electric energy as a clean energy is increasing. The implementation of more effective MS is the prerequisite for the survival and development of power companies in the future. The Internet has become a necessary marketing channel for power supply companies, which will enable more and more power customers to experience the convenience and speed of power supply. Therefore, the research on the digital MS transformation of the power market in the Internet age has certain guiding significance, and also provides reference and reference for the marketing work of other power supply companies.

References

1. Chaparro-Peláez, J., Acquila-Natale, E., Hernández-García, N., et al.: The digital transformation of the retail electricity market in Spain. Energies **13**(8), 2085 (2020)
2. Grabchak, E.P., Loginov, E.L., Mischeryakov, S.V., et al.: Approaches to the integration of information about resource and financial flows in the fuel and power complex under the digital transformation of control systems. Upravlenie **8**(2), 13–19 (2020)
3. Nahar, S., Meero, A., Rahman, A.A.A., et al.: Analysis on the marketing strategy and competitive advantage of banking industry in Bangladesh: an entrepreneurial case study of HSBC bank. Acad. Entrepr. J. **27**(4), 1–7 (2021)
4. Hutomo, H.I., Wijaya, C.: Analysis of marketing strategy in increasing room occupation at GH universal hotel bandung during the COVID-19 pandemic. Int. J. Res. - GRANTHAALAYAH **9**(7), 61–74 (2021)
5. Haron, A.J.: Standardized versus localized strategy: the role of cultural patterns in society on consumption and market research. J. Account. Mark. **05**(01), 2–13 (2016)
6. Yuelin, X., Danyang, X., Li, H.: Research on E-commerce customer evaluation system in the context of big data: taking amazon as an example. J. Phys.: Conf. Ser. **1955**(1), 012024 (2021). (6pp)
7. Lageson, S.E., Maruna, S.: Digital degradation: stigma management in the internet age. Punishment Soc. **20**(1), 113–133 (2018)
8. Tang, W., Liu, Y.: University mobile employment network information system in the internet age. J. Phys.: Conf. Ser. **1881**(2), 022095 (2021). (6pp)
9. Zhang, L., Hooimeijer, P., Lin, Y., et al.: Strategies of the built-heritage stewardship movement in urban redevelopment in the internet age: the case of bell-drum towers controversy in Beijing, China. Geoforum **106**, 97–104 (2019)
10. Li, G.: The application and innovation of crowdsourcing in the internet age. Open J. Soc. Sci. **04**(3), 199–204 (2016)
11. Lopes, H., Watté, B.H., Gosling, M.: Business model change of printed newspapers in the internet age: the New York times case. Rev. Econ. Gestão **20**(55), 102–119 (2020)

12. Jiang, X., Luo, L.: Globalization of services in the internet age: a new engine, acceleration and major power competitiveness. Soc. Sci. China **41**(4), 5–23 (2020)
13. Jain, P.: Consumer management in internet age: a study of Indian car buyers digital communication adoption by applying competing models. FIIB Bus. Rev. **8**(1), 231971451988310 (2019)

Construction of Electricity Charge Information Management System Based on Network Microservice Technology

Haiyan Duan(✉) and Weifeng Dong

Digital Grid Research Institute, China Southern Power Grid, Guangzhou 510700, Guangdong, China
xiaochongtougao1@163.com

Abstract. With the development of society and the progress of science and technology, people's demand for electric energy is becoming stronger and stronger. In order to adapt to the increasing demand for electric energy, smart grids are gradually popularized. It not only changed the development mode of our country's power grid, but also brought new opportunities and challenges for our country's power companies. Among them, the electricity bill information management system based on network micro-service technology ensures the accuracy and stability of users' electricity consumption data. The purpose of this paper is to study the construction of electricity bill information management system based on network microservice technology. This article takes the electricity bill information management system as the research object, combined with the network micro-service technology, analyzes the functional characteristics of the system, and details the functional modules of the system's user management, meter reading management, electricity bill collection management, and electricity monitoring management. This article briefly introduces the system test environment, and conducts main functional module tests and stress tests on the system. The test results show that the shortest response time of each function of the system is 2.016 s, and the longest is 2.341 s. And the success rate of system function operation is above 98%. It can be seen that this system has good performance and meets the performance requirements of the system.

Keywords: Electricity bill management · Electricity consumption information collection · Management analysis · Network microservices

1 Introduction

With the advancement of social science and technology, all classes of society are increasingly dependent on the power industry, and our country's demand for power is also increasing year by year [1, 2]. With the continuous increase in electricity demand, it has become more and more difficult to manage the electricity bill information of power users across the country. The use of traditional manual methods to record and manage electricity bill data can no longer meet the actual work and business needs [3, 4]. With

I. Ahmad et al. (Eds.): STSIoT 2021, LNDECT 122, pp. 280–287, 2023.
https://doi.org/10.1007/978-981-19-3632-6_35

the development and application of network microservice technology, the electricity bill information management system has the advantages of automation, intelligence, informationization and sharing. It can realize the information and automation of electricity billing and related settlement, which is useful for improving the management of electricity costs for power supply companies. It has an important role in promoting [5, 6].

Regarding the research on electricity tariff management, many scholars at home and abroad have conducted in-depth and multi-faceted discussions on it. For example, Liu H proposed a simplified and highly cost-optimized method for renewable power grids [7]. Bushnell J analyzed Mexico the advantages of the liberalization of the electricity market have studied the relationship between transmission cost allocation and network efficiency [8]. Abushnaf J introduced the electricity tariff calculation method of time-of-use pricing, applied it to the user's home energy management system, and explored the impact of the electricity tariff calculation method on the user's electricity consumption [9]. It can be seen that the informatization of electricity bill management has long been an important task in the development of domestic and foreign power industries. The construction and research of electricity bill information management systems based on network microservices is of great significance.

This paper focuses on the research of electricity bill information management system based on network microservice technology. This article first analyzes the functions and characteristics of the system such as data information collection function, data information sorting function, data information control and so on. Then, the functional modules of the system, such as user management, meter reading management, electricity bill collection management, and electricity monitoring management, are described in detail. Finally, the system test environment is introduced, and the main function module test and stress test of the system are carried out. Through the analysis of the test results, it has been verified that the system has good performance.

2 Construction of Electricity Bill Information Management System Based on Network Microservice Technology

2.1 Functions and Characteristics of Electricity Bill Information Management System Based on Network Microservice Technology

(1) Power consumption data information collection function

The electricity bill information management system can collect and sort the user's electricity consumption, switches and the operating status of various components and other information, with good accuracy, reliability and completeness. At the same time, it can collect power grid analog current and voltage data information, prepaid electricity bill data information, electricity overrun data information, and terminal meter status information.

(2) Power consumption data information sorting function

The main function of this function is to ensure that the collected data is complete and accurate. Therefore, the system strictly checks the collected information, finds abnormal data in time through sorting, storing and querying the data, and automatically repairs missing data [10, 11]. In addition, it can also estimate the load situation, power quality, and line loss of the entire grid, thereby improving the ability to regulate and control the grid.

(3) Comprehensive application finishing function

The system also has functions such as automatic meter reading, operating status supervision, orderly power usage, abnormal power usage, accident handling, quality inspection, line loss management, and value-added operations.

(4) Electricity fee control function

The electricity charge information management system can perform charge management according to the electricity consumption status of users in a specific time period, and realize the regulation and management of electricity charges for users of the entire grid. The system can monitor the electricity consumption of power users in real time, and feedback the electricity consumption to the users. In addition, users can also inquire about their own electricity consumption through the mobile network. When the system finds that the user's balance is insufficient, the smart master station will issue a trip command to notify the user to pay the electricity bill in time.

Neural networks need to be used to calculate electricity consumption for the regulation and management of electricity charges for grid users. The functions used in this calculation are Sigmoid (excitation equation) equations and general linear equations. The calculation method of the output layer is shown in formula (1):

$$f(x) = \frac{1}{\sqrt{2\pi}\,\partial} exp(\frac{(x-a)^2}{2\partial^2}) \tag{1}$$

In the formula, ∂ represents the input amount.

The calculation method of Sigmoid function is shown in formula (2):

$$f(x) = \frac{1}{1 + exp(-x)} \tag{2}$$

The total calculation error is shown in formula (3):

$$e = \frac{1}{2 * \sum_{i=1}^{n} (y_i - \hat{y}_i)^2} \tag{3}$$

In the formula, i represents the sample number, n represents the number of samples; y_i represents the output of the i-th sample; $\hat{y}_i$ represents the actual output of the sample, so that the calculated target value e is the smallest [12].

2.2 Design of Electricity Bill Information Management System Based on Network Microservice Technology

(1) User management

The electricity bill information management system based on network microservice technology is divided into different types of users, such as ordinary electricity users, meter readers, toll collectors, business managers, and so on. The system administrator can manage the users of the system, including adding, deleting, modifying, viewing, and querying user information.

(2) Meter reading management

Meter reading management is mainly responsible for planning and executing the company's meter reading business. It is an important link and prerequisite for electricity bill management and an important part of the electricity bill management system. The main functions of meter reading management include: meter reading management, electricity and electricity bill calculation, electricity bill verification, charge management, parameter maintenance, summary statistics, and comprehensive query.

(3) Electricity fee verification and collection management

Electricity fee verification and collection mainly include four sub-functions: electricity fee calculation, electricity fee collection, special fee processing, and payment invoice printing functions.

When the user needs to pay the electricity bill, the toll collector gets the authority and enters the electricity bill collection management interface. First, check the power consumption and power consumption category of the user during the power consumption cycle. After confirming that they are correct, the system automatically calculates the amount to be paid. The toll collector updates the user's electricity bill status after charging the relevant fees in full. At the same time, toll collectors can refund overcharges and make up payments to users who owe fees. After all the charges are completed, the invoice for the user payment must be printed for the user.

(4) Power consumption supervision and management

The system can perform statistical analysis on the power consumption and power consumption category of the entire power grid users. Through the analysis of historical data, the user's electricity consumption trends in different years and months are obtained, which is convenient for the power supply company to plan scientifically in advance, cope with the peak electricity consumption in a specific period, and ensure the smooth and smooth electricity consumption of factories and citizens.

The system can analyze the user's power level, power consumption category, and power consumption time according to time, and can also compare power consumption in different regions and present it in the form of a chart. System administrators can detect the user's electricity consumption in real time based on this information, and predict future electricity consumption trends.

3 Research Design

3.1 System Test Environment

Hardware environment: Intel(R) Corei5, 2.4 GHz processor; 4 GB of system memory; hard disk space of 360 GB; network card is 10/100M adaptive network card;

Software environment: eclipse and Oracle database, the browser supports IE8.0 and above.

There are two servers in this experiment, one is a Web server, used to run Web programs; the other is a database server, used to run a database management system to achieve data persistence.

3.2 System Test Case

(1) User basic information query test

Enter the user profile management interface, enter the user basic information management interface through the menu; enter the query condition field; click the "query" button, and return to the list of records that meet the conditions; click details to open the detailed information of the selected record.

(2) User power consumption inquiry test

Enter the electricity bill management interface; enter the user power consumption interface; input conditions: available work code, user number, according to the category and month, click the "query" button; the system displays the query list.

(3) System stress test

By continuously increasing the number of users accessed by the system, test the maximum affordability of the system. This system stress test uses Load Runner test software to simulate the use environment of 50, 100, and 150 concurrent users in the test client host by means of test scripts, and read the relevant performance indicators from the Load Runner tool. The specific content of the test includes the number of concurrent users and response time of the system.

4 Data Analysis

4.1 Analysis of the Test Results of the Main Functional Modules of the System

The test results of the main modules of the system are shown in Table 1. The number of concurrent users is 50, and the average response time of user basic information query is 2.129 s; the average response time of user power consumption information query is 2.134 s; the average response of user electricity bill calculation is 2.016 s; the average response time for printing user bills is 2.341 s.

Table 1. Test data of main functional modules of the system

Business name	Serial number	Concurrency	Average response time (seconds)	Success rate (%)
User basic information query	1	50	2.129	98.27%
User power consumption information query	2	50	2.134	98.04%
User electricity bill calculation	3	50	2.016	98.43%
Print user bill	4	50	2.341	98.12%

According to Fig. 1, in the performance test process of this system, the shortest response time of each function of the system is 2.016 s, and the longest is 2.341 s. In

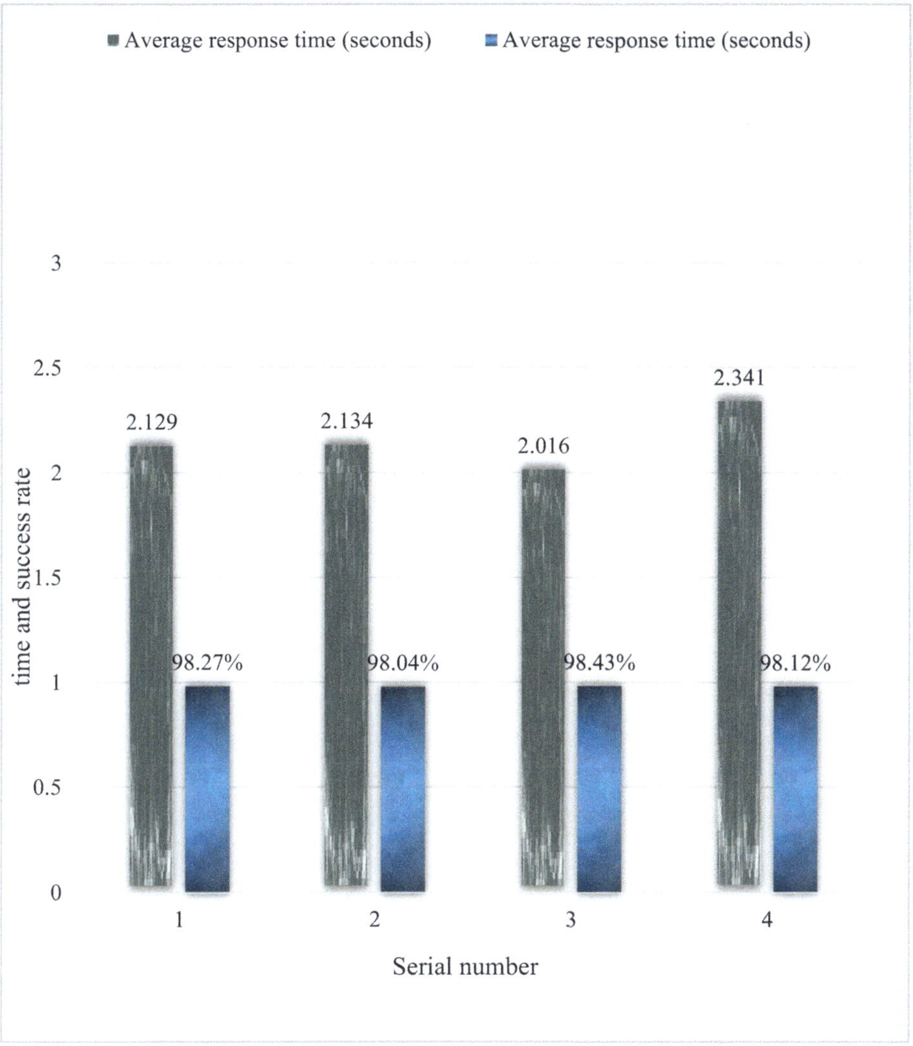

Fig. 1. Test data of main functional modules of the system

addition, the success rate of system function operation is above 98%, which shows that the performance of the system is good and meets the performance requirements of the system.

4.2 Analysis of System Stress Test Results

Load Runner software was used to carry out a stress test. After the concurrent operations of 50, 100, and 150 users were simulated, the performance and efficiency of the system were tested. The system stress test results are shown in Table 2: when the number of concurrent users is 150, the response time of user basic information query is 2.578 s;

the response time of user power consumption information query is 2.603 s; the response time of user electricity bill calculation is 2.341 s; the response time for printing user bills is 2.647 s.

Table 2. System stress test data

Business name	Serial number	50	100	150
User basic information query	1	2.129	2.940	2.578
User power consumption information query	2	2.130	3.120	2.603
User electricity bill calculation	3	2.017	3.106	2.341
Print user bill	4	2.334	3.224	2.647

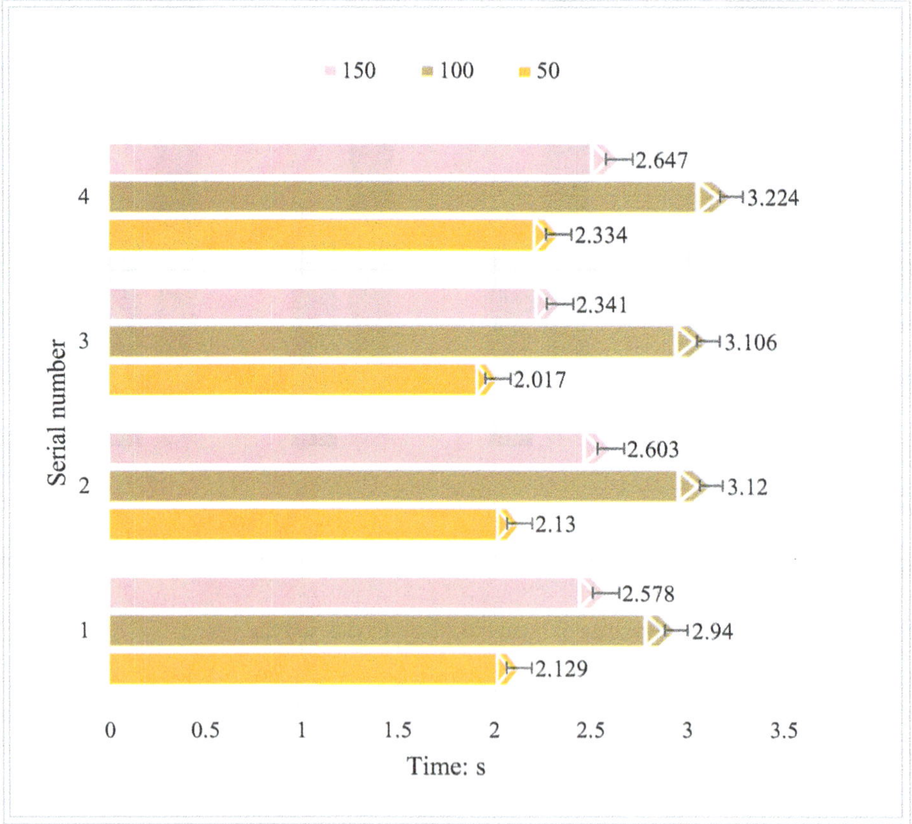

Fig. 2. System stress test data

As shown in Fig. 2, it shows the response time test results of the four business functions of the system. When 50, 100, and 150 users log in, the average response time for these services is less than 4 s, which meets the performance requirements.

5 Conclusion

The management of electricity charges is one of the most important businesses of a power company. It directly determines the quality of service to electricity customers and affects the economic benefits of the power company's marketing. The construction and application of electricity bill information management system can not only save a lot of economic costs for power companies, but also improve work efficiency and social benefits. This paper studies the construction of electricity fee information management system based on network micro-service technology. The functional modules of the system, such as user management, meter reading management, electricity bill checking management and electricity consumption monitoring management, are explained in detail. Through the main functional module test and pressure test of the system, it verifies that the system has good performance characteristics.

References

1. Tushar, M., Assi, C.: Optimal energy management and marginal-cost electricity pricing in microgrid network. IEEE Trans. Ind. Inform. **PP**(6), 1 (2017)
2. Zhao, X., Gao, W., Qian, F., et al.: Electricity cost comparison of dynamic pricing model based on load forecasting in home energy management system. Energy **229**(1), 120538 (2021)
3. Barja-Martinez, S., Rucker, F., Aragues-Penalba, M., et al.: A novel hybrid home energy management system considering electricity cost and greenhouse gas emissions minimization. IEEE Trans. Ind. Appl. **PP**(99), 1 (2021)
4. Bukhsh, R., Javed, M.U., Fatima, A., et al.: Cost efficient real time electricity management services for green community using fog. Energies **13**(12), 3164 (2020)
5. Mittapalli, R.K., et al.: FACTS devices cost recovery during congestion management in deregulated electricity markets. J. Inst. Eng. (India) Ser. B Electr. Eng. Electron. Telecommun. Eng. Comput. Eng. **97**(3), 339–354 (2016)
6. Abdalla, M., Wang, M., Abbaker, O.: Two-stage energy management strategy of EV and PV integrated smart home to minimize electricity cost and flatten power load profile. Energies **13**(23), 6387 (2020)
7. Liu, H., Andresen, G.B., Greiner, M.: Cost-optimal design of a simplified highly renewable Chinese electricity network. Energy **147**(MAR.15), 534–546 (2018)
8. Bushnell, J., Ibarra-Yunez, A., Pappas, N.: Electricity transmission cost allocation and network efficiency: implications for Mexico's liberalized power market. Utilit. Policy **59**(AUG.), 100932.1–100932.9 (2019)
9. Abushnaf, J., Rassau, A., Górnisiewicz, W.: Impact on electricity use of introducing time-of-use pricing to a multi-user home energy management system. Int.Trans. Electr. Energy Syst. **26**(5), 993–1005 (2016)
10. Iwafune, Y., Yagita, Y.: High-resolution determinant analysis of Japanese residential electricity consumption using home energy management system data. Energy Build. **116**(mar.), 274–284 (2016)
11. Baptiste Feron, I.D., Monti, A.: A market-based optimization approach for domestic thermal and electricity energy management system: formulation and assessment. Information **9**(5), 120 (2018)
12. Syamsiana, I.N., Wibowo, S.S., Hakim, M.F., et al.: Energy database management system (EDBMS)-based data acquisition audit for electricity savings analysis. In: IOP Conference Series: Materials Science and Engineering, vol. 1073, no. 1, p. 012036 (2021). (5pp)

Deep Dense Autoencoder Using Modulation Spectrogram for Machine Unsupervised Anomaly Detection

Shuai Zhao, Shuang Li, Ziqiang Bao, Guisong Jiang, Lifen Jiang, and Long Zhang(✉)

College of Computer and Information Engineering, Tianjin Normal University, Tianjin, China
zlwalkman@sohu.com

Abstract. The purpose of this paper is to design an abnormal sound detection system to detect abnormal sound during mechanical operation. The system uses the modulation spectrogram as a feature of the sound signal to train a dense autoencoder. Using the development set provided by DCASE2021 to compare 7 machine types with its baseline system, the result is better than the baseline system. Among the 7 machine types, the effects of Fan, Gearbox, Pump, Slider and Valve are significantly better than the baseline system provided by DCASE2021. By comparison, we believe that the method of feature extraction has an impact on the training of the neural network. In addition, the number of layers of the neural network should not be too large.

Keywords: Modulation spectrogram · Autoencoder · Unsupervised · Anomaly detection

1 Introduction

During the operation of the machine, abnormal sounds can be used to detect whether the machine is operating normally. In reality, abnormalities may be caused by various circumstances and cannot be predicted. Deliberately damaging valuable machines will bring unnecessary losses. Because abnormal sounds are difficult to collect, and the inappropriate ratio of normal sounds to abnormal sounds makes it difficult to develop a supervised learning dataset, so unsupervised anomaly detection is necessary.

In our proposed method, we use the modulation spectrogram to represent the features of the sound instead of using the Mel spectrogram, and send the features to the autoencoder that adapts the dimensions of the sound feature for training, the encoder and decoder to minimize the mean square error (MSE) between input and reconstruction. The baseline system [1] given by DCASE2021 uses Mel spectrogram to represent features, the encoder and decoder use 4 dense layers. Our method can exceed the performance of the baseline system.

I. Ahmad et al. (Eds.): STSIoT 2021, LNDECT 122, pp. 288–295, 2023.
https://doi.org/10.1007/978-981-19-3632-6_36

2 Method

2.1 Modulation Spectrogram

Our method uses the modulation spectrogram to represent the features of the sound instead of the commonly used Mel spectrogram. The method of extracting modulation spectrogram is proposed in [2].

Figure 4 shows the process of extracting the modulation spectrogram. First, we simulate the cochlea [3] and filter the voice signal through a filter bank composed of 60 gammatone filters. Figure 1 shows the gammatone filterbank. The center of the first filter is 125 Hz, and the center of the last filter is lower than half of the sampling rate. The filter bandwidth calculation follows the equivalent rectangular bandwidth (ERB). ERB is an approximate value similar to the bandwidth of the human auditory filter. The calculation method is formula 1.

$$ERB_j = \frac{f_i}{Q_{ear}} + B_{min} \tag{1}$$

where f_j represents the center frequency of the j-th filter, Q_{ear} and B_{min} are set to constants 9.265 and 24.7 respectively.

The original voice signal $x(n)$ is output as $\hat{x}_j(n)$ after filter by the j-th gammatone filter, where n represents the time variable, j = 1, ..., 60. The time envelope $e_j(n)$ is calculated by $\hat{x}_j(n)$ and Hibbert Transform:

$$e_j(n) = \sqrt{\hat{x}_j(n)^2 + H\{\hat{x}_j(n)\}^2} \tag{2}$$

Which $H\{.\}$ represents the Hilbert Transform. Use 256 ms Hamming window and 40 ms shifts for windowing. After that, the discrete Fourier transform (DFT) is performed on the windowed time envelope.

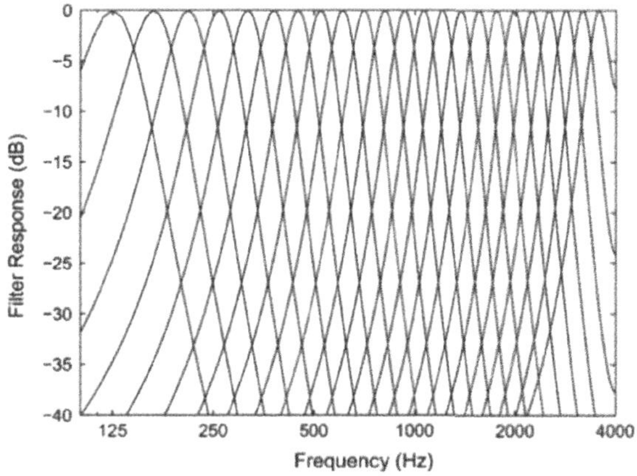

Fig. 1. Gammatone filterbank

$$E_j(m; f) = |F(e_j(m))| \tag{3}$$

where $e_j(m)$ represents the m-th frame after windowing, $F\{.\}$ represents the discrete Fourier transform, and f represents the modulation frequency. In [3], a modulated filter bank that simulates the human auditory system is proposed, and $E(j, k)(m)$ is obtained through an 8-channel modulator bank, j = 1, ..., 60, k = 1, ...8, where j represents a gammatone filter, and k represents a modulation filter. The final result is a 60 × 8-dimensional modulation spectrogram. Figure 2 shows the modulator filterbank.

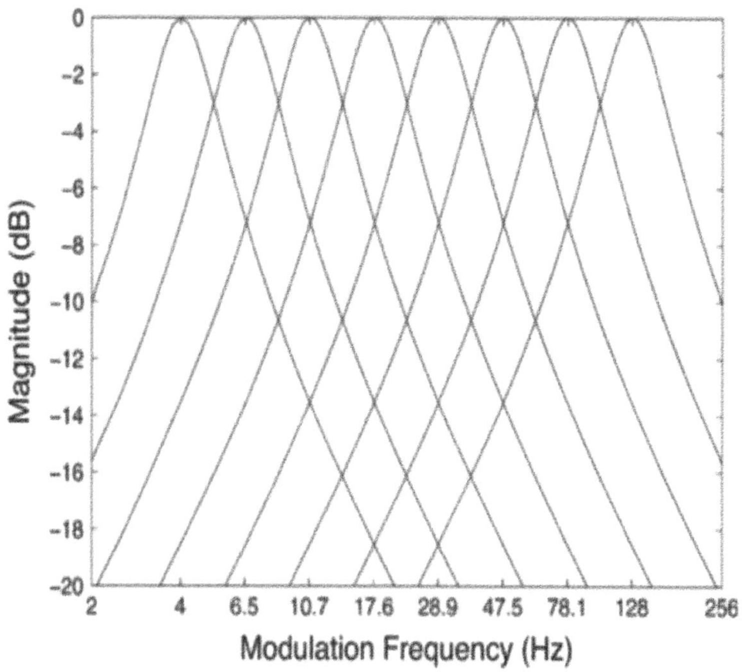

Fig. 2. Modulator filterbank

After getting the modulation spectrogram, we expanded it and finally got the 480-dimensional feature vector.

2.2 Dense Autoencoder

Many scholars have obtained good results in abnormal sound detection [4–7] by using autoencoders. The autoencoder is a neural network structure composed of an encoder, a decoder and a latent vector space. After the input through the encoder, a feature vector is mapped in the latent space, and then these feature vectors are used to decode and reconstruct a similar reconstructed output of the input. The difference between input and output called reconstruction error [8].

Since the training only uses normal sound data, the goal is to train a set of encoder and decoder parameters to minimize the reconstruction error, that is, to make the reconstruction error between the output and input small, and the reconstruction error between the abnormal sound and the input is large, so as to distinguish between normal and abnormal sound. Using a complex neural network structure is likely to train a model that cannot distinguish abnormal sounds and consumes a lot of time. Therefore, a simple and dense autoencoder is trained to detect abnormal sounds.

Experiment with the number of layers, activation functions, and hidden units of the model, use the modulation spectrogram to compare the autoencoders of 6 fully connected layers and the autoencoders of 8 fully connected layers, it is found that too many layers are not good for anomaly detection. Therefore, the encoder and decoder are composed of three fully connected layers, 128 hidden units, Batch Normalization and ReLu are used as activation functions, and the dimension of the latent space is 8 dimensions. Figure 3 shows the structure of the dense autoencoder.

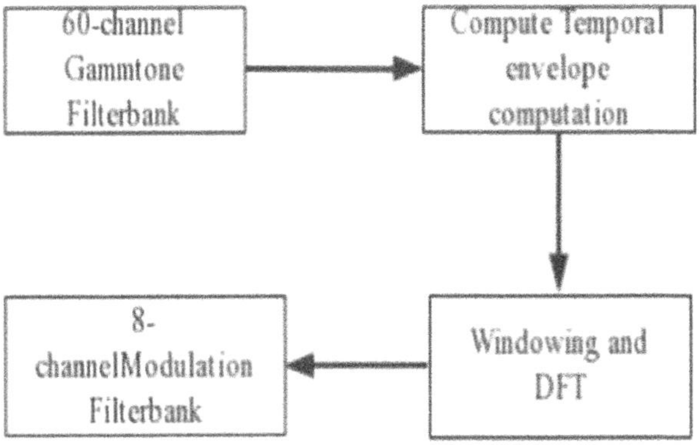

Fig. 3. Step for extract the modulatio spectrogram

3 Experiment and Result

3.1 Dataset

The dataset [9, 10] used for development includes normal and abnormal sounds. The training set only includes normal sounds, and the test set includes normal sounds and abnormal sounds.

All sounds are single channel and the duration is 10 s. The dataset includes different machine types such as Fan, Gearbox, Pump, Slider, ToyCar, ToyTrain, Valve. Each type of machine in the training set includes 3 different devices, providing 1003 normal operating sounds for each device. 3 out of 1003 sounds are under different environmental noise, load or running speed.

In addition, each machine includes normal sounds in three situations, and is distinguished by different id. For example, the sound of the fan type with id of 0 indicates that the wind intensity has changed, and the sound of id of 1 indicates that the size of the fan has changed.

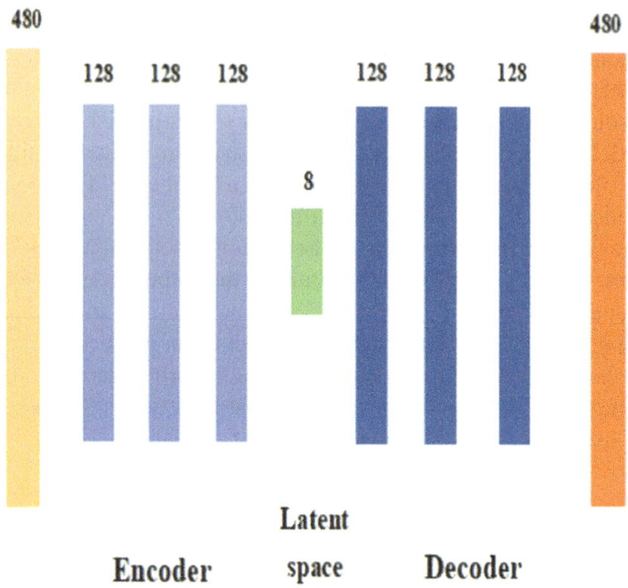

Fig. 4. Dense autoencoder network architectures

3.2 Training Settings and Result

Due to the small number of normal sounds provided by different noises, loads or operating speeds, simple mix up data augmentation is carried out.

$$x = \alpha x_s + (1 - \alpha)x_t \tag{4}$$

where x_s is the source domain, that is, the sound of the machine that works normally under normal conditions, and x_t is the target domain, that is, the sound of the load, the operating speed or the ambient noise is different. After many tests, the value of α is 0.8.

Before extracting the modulation spectrogram, we use the noise-reduce in Python reduce the noise of the signal. We averaged 200 normal training sounds, treated the averaged segments as noise segments and performed fast Fourier transform calculations. After listing the statistical data of each frequency band we calculate the threshold value of each frequency band based on these statistical data, and then perform fast Fourier transform on the signal.

After reduce the noise, we used the expanded data and the provided data to extract the modulation spectrogram, the method is mentioned in Sect. 2.1

We set the learning rate 0.001, the optimizer uses Adam, the training is repeated 500 epochs, the batch size is set to 128, and the training loss function is to minimize the mean square error (MSE) between the input and the reconstruction. The calculation method is formula 5.

$$Loss = \frac{1}{n}\sum_{i=1}^{n}(x^{(i)} - x_r^{(i)})^2 \tag{5}$$

Each machine trains an autoencoder, in order to train a general model for each machine, the id of the machine is not distinguished during training. The experiment uses AUC and pAUC indicators for evaluation, and the definition is given in the task description [1].

In order to compare the impact of different feature extraction methods on accuracy. We designed an 8-layer dense autoencoder using modulated spectrogram to compare with the 8-layer baseline system. In order to compare the impact of the number of network layers on accuracy, we designed a 6-layer dense autoencoder that uses a modulation spectrogram to compare with an 8-layer dense autoencoder that uses a modulation spectrogram.

Table 1 compares the results of the provided baseline dense encoder, 6-layer fully connected autoencoder, and 8-layer autoencoder in the development dataset. The best result is marked in bold. Except for ToyTrain, the results of dense encoders that use modulated spectrograms are better than baseline systems in other machine types. Among them, the effect of Fan, Pump, Valve on the 6-layer dense autoencoder using modulation spectrogram has been improved by 20%. Gearbox, Slider have been improved by 10%. Although the effect of ToyCar has also been improved, it is not as good as the 8-layer dense autoencoder. It can be compared that too many neural network layers will not have a good effect on abnormal sound detection. At the same time, modulating spectrogram to represent sound characteristics can improve the performance of anomaly detection.

Table 1. Performance comparison between the baseline system and the proposed method

Type	Baseline		8-layer dense AE		6-layer dense AE	
	AUC	pAUC	AUC	pAUC	AUC	pAUC
Fan	63.24%	53.38%	74.8%	64.8%	**85.50%**	**80.28%**
Gearbox	65.97%	52.76%	62.8%	55.6%	**76.69%**	**69.23%**
Pump	61.92%	54.41%	65.6%	60.3%	**82.50%**	**55.83%**
Slider	66.74%	55.94%	63.4%	61.1%	**69.14%**	**55.50%**
ToyCar	62.49%	52.36%	**79.5%**	**68.6%**	74.83%	62.25%
ToyTrain	**61.71%**	**53.81%**	50.7%	50.5%	56.33%	51.23%
Valve	53.41%	50.54%	66.7%	57.1&	**78.01%**	**77.61%**

4 Conclusions

In this paper, we propose a dense autoencoder model that uses modulation spectrograms as sound features. The 7 machine types in the data set provided by DCASE2021 are trained and tested and compared with the baseline system provided. On Fan, Gearbox, Pump, Slider, ToyCar, Valve, dense autoencoder using modulated spectrogram results better than baseline system.

We have used CNN with many parameters during training. It takes a long time to train, but we have obtained poor results. The characteristics of the reference data set are that the difference between abnormal and normal sounds is not very obvious. Therefore, inference that for the detection of small anomalies, complex neural networks is not necessarily the optimal solution. The comparison between the 8-layer dense autoencoder using the modulation spectrogram and the baseline system using the Mel spectrogram also shows that more reasonable feature extraction is beneficial to the detection of abnormal sounds.

This work has trained an autoencoder for each type of machine, but it can be seen from the results that each type of machine does not perform best in the same network. The experimental comparison may be because the neural network structure is too complex to detect subtle abnormal sounds. It may be that the method of feature extraction needs to be improved, or of course, it may also be the method of calculating error, and the calculation of loss needs to be optimized. Therefore, future work can consider how to optimize the network structure and calculation methods to train a general autoencoder to analyze all the machines through the same characteristics of different sounds.

Acknowledgements. This work was supported by the National Natural Science Foundation of China (Grants No. 61771173), the Tianjin General Plan of Art and Science Project (Grants No. E20016) and the Tianjin Natural Science Foundation of China (Grants No. 20JCZDJC00400).

References

1. Kawaguchi, Y., et al.: Description and discussion on DCASE 2021 challenge task 2: unsupervised anomalous sound detection for machine condition monitoring under domain shifted conditions. arXiv e-prints arXiv:2106.04492, pp. 1–5 (2021)
2. Falk, T., Chan, W.-Y.: Modulation spectral features for robust far-field speaker identification. IEEE Trans. Audio Speech Lang. Process. **18**, 90–100 (2009). https://doi.org/10.1109/TASL.2009.2023679
3. Slaney, M.: An efficient implementation of the Patterson-Holdsworth auditory filter bank (2000)
4. Koizumi, Y., Saito, S., Uematsu, H., Kawachi, Y., Harada, N.: Unsupervised detection of anomalous sound based on deep learning and the Neyman–Pearson lemma. IEEE/ACM Trans. Audio Speech Lang. Process. **27**, 212–224 (2019)
5. Suefusa, K., Nishida, T., Purohit, H., Tanabe, R., Endo, T., Kawaguchi, Y.: Anomalous sound detection based on interpolation deep neural network. In: ICASSP 2020 - 2020 IEEE International Conference on Acoustics, Speech and Signal Processing (ICASSP), pp. 271–275 (2020)
6. Perez-Castanos, S., Naranjo-Alcazar, J., Zuccarello, P., Cobos, M.: Anomalous sound detection using unsupervised and semi-supervised autoencoders and gammatone audio representation. ArXiv abs/2006.15321 (2020)
7. Koizumi Y., Kawaguchi Y., Imoto K., et al.: Description and discussion on DCASE2020 challenge task2: unsupervised anomalous sound detection for machine condition monitoring (2020)
8. An, J., Cho, S.: Variational autoencoder based anomaly detection using reconstruction probability (2015)

9. Tanabe, R., et al.: MIMII DUE: sound dataset for malfunctioning industrial machine investigation and inspection with domain shifts due to changes in operational and environmental conditions. arXiv e-prints arXiv:2006.05822, pp. 1–4 (2021)
10. Harada, N., Niizumi, D., Takeuchi, D., Ohishi, Y., Yasuda, M., Saito, S.: ToyADMOS2: another dataset of miniature-machine operating sounds for anomal

Multi-dimensional Convolutional Neural Network for Speech Emotion Recognition

Ziqiang Bao, Shuai Zhao, Shuang Li, Guisong Jiang, Huazhi Sun, and Long Zhang(✉)

College of Computer and Information Engineering, Tianjin Normal University, Tianjin, China
zlwalkman@sohu.com

Abstract. Speech Emotion Recognition (SER) is a difficulty of deep learning algorithms. The difficulty is that people's own understanding of emotions is not absolute. Different people may also have different judgments on the same speech. And speech emotion recognition plays a huge role in many real-time applications. With the continuous development of deep learning in recent years, many people use convolutional neural networks (CNN) to extract high-dimensional features in speech from speech spectrograms, thereby improving the accuracy of speech emotion recognition. In contrast, we propose a new model of speech emotion recognition. The model uses the eGeMAPS feature set extracted through the openSMILE toolkit to input into our model. The model learns the correlation and timing between features. In addition, we perform intra-class normalization on the input features to ensure more accurate recognition and faster data fitting. In our model, the key speech segments can be selected through the characteristics of convolutional neural network (CNN), so that the recognition accuracy of the model can achieve a better effect. Our model was evaluated experimentally in the IEMOCAP dataset. Experimental results show that our unweighted accuracy (UA) and weighted accuracy (WA) on the test set reached 60.9% and 63.0%.

Keywords: Speech emotion recognition · Convolutional neural networks · Attention mechanism

1 Introduction

With the continuous development of artificial intelligence, issues related to human-computer interaction are now getting more and more attention. Emotions play a very important role in people's communication. It enables the machine to successfully interpret the emotional state in the process of human communication. It enables the machine to successfully interpret the emotional state in the process of human communication. It can greatly improve the efficiency of human-computer interaction. It can also solve many problems that people encounter in the process of daily life. Some examples are mobile phone voice assistants, driverless cars, and companion robots. The emotional state that the machine extracts from people's voice can make the machine better provide better services to humans.

I. Ahmad et al. (Eds.): STSIoT 2021, LNDECT 122, pp. 296–303, 2023.
https://doi.org/10.1007/978-981-19-3632-6_37

The continuous improvement of deep learning technology has also prompted the speech emotion recognition (SER) model to be inspired by the framework of different fields, and it has been developed rapidly in the past few years. In the early stage, people did research on speech emotion recognition. First, it extracts some hand-designed low level descriptors (LLDs) from the original audio. Then do some statistical calculations on the basis of these low-level features to obtain high level statistics features (HSFs). Finally, the features are sent to machine learning algorithms (for example: Hidden Markov Model (HMM) [1], Gaussian Mixture Model (GMM) [2] and Support Vector Machine (SVM) [3]) for classification. In recent years, deep learning methods have been introduced into this field. Zhao J et al. [4] used a combination of deep neural network (DNN), long short-term memory network (LSTM) and convolutional neural network (CNN) to classify the original audio and the log-mel spectrograms. The experimental research results show that the original sound signal is The emotion classification ability is not as good as the log-mel spectrogram after the feature is extracted, indicating that we need to extract a good feature to have a good effect on the prediction of speech emotion. Subsequently, Zengwei Yao et al. [5] separately designed a special network to integrate the three characteristics of log-mel spectrograms, LLDs and HSFs to identify emotions, and achieved a good result.

In this paper, we propose a new feature extraction structure Multi-dimensional Convolutional Neural Network (MD-CNN). Compared with traditional CNN networks, this paper not only considers the relationship between adjacent features. It also considers the relationship between each voice feature and different time periods. Finally, Bidirectional long short-term memory network (BiLSTM) is used to extract the time information of the voice in different dimensions. Then use the attention mechanism to merge features. In order to obtain the generalized features extracted in different dimensions. We conducted experiments using the IEMOCAP dataset. Experiments show that extracting features in different dimensions is beneficial to the final emotion recognition. The main contributions of this research are as follows: (1) We propose a model that extracts features from different dimensions based on 1D CNN. (2) We fused the features of different dimensions based on the attention mechanism. So that the two models can be effectively merged.

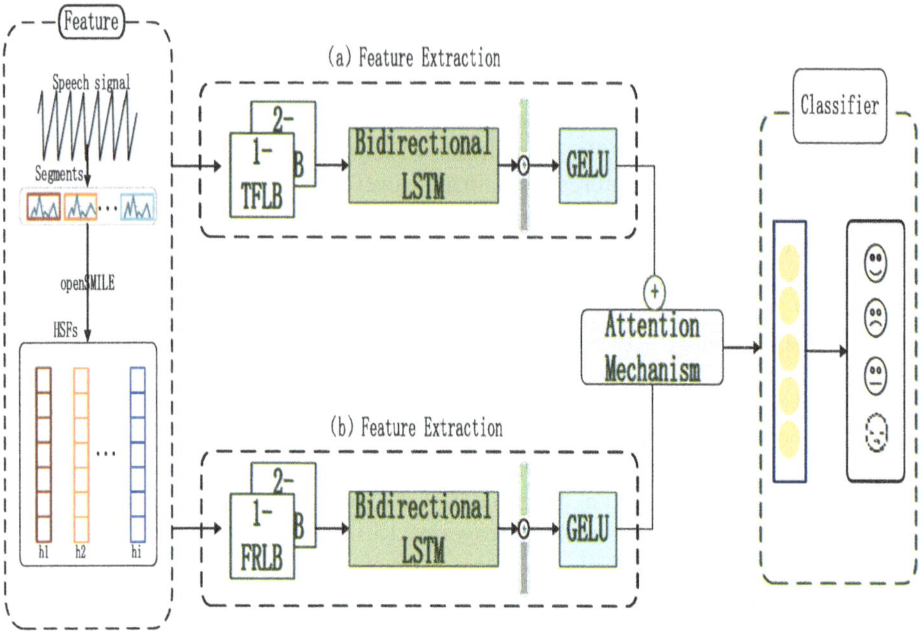

Fig. 1. The model structure of this paper. The input is 88-dimensional HSFs features. (a) The feature extraction block extracts the correlation between features (b) The feature extraction block extracts the temporal relationship between each segment. BiLSTM extracts the spliced features of forward propagation and reverse propagation of features.

2 Materials and Methods

2.1 Dataset

IEMOCAP is an emotional binary interactive English database. This database is collected by the Speech Analysis and Interpretation Laboratory (SAIL) of the University of Southern California (USC). The corpus contains approximately 12 h of data. This database was recorded by 10 actors. Contains 5 dialogues, each of which consists of a man and a woman. They not only perform pre-written emotional scripts, but also improvise imaginary scenes designed to trigger specific types of emotions (happiness, anger, sadness, depression and neutral states). In this paper, we have selected four main types of emotions: neutral, happy, sad, and anger. Among them, because the emotions of happiness and excitement are close, we classify them into one category. These four emotions have a total of 5531 utterances [6] (Table 1).

2.2 Methods

As shown in Fig. 1, we use two different types of feature selection blocks to extract the input features. Finally, the attention mechanism is used for feature fusion at the decision-making level to get the final prediction. Specifically, we separately send the extracted HSFs features into two temporal feature learning blocks (TFLB) and feature

Table 1. The detailed description of emotions distribution in different classes and each class data participation in percentage of the IEMOCAP dataset.

Class	Neutral	Happy	Sad	Anger
Total utterances	1708	1636	1084	1103
Participation in (%)	30.8%	29.6%	19.7%	19.9%

relationship learning blocks (FRLB), which together form a multi-dimensional learning framework. The following sections will introduce the function of each module in detail.

2.2.1 Input Features

For speech emotion recognition, the choice of the length of each speech in speech has always been an uncertain thing. Here we show according to (Mustaqeem et al., 2020) [7] research that fragments longer than 250 ms contain more emotional information. Therefore, after comprehensive consideration, we choose to divide each voice into several 500 ms voices. And each voice share the same label. The speech features are 88-dimensional HSFs extracted using the openSMILE toolkit [8] and eGeMAPS feature set. These features are calculated by some statistical functions based on the features of LLDs. For example, there are: mean, maximum, and standard deviation. HSFs is to make statistics on multi-frame speech on utterance, so we use it to represent a feature of utterance.

2.2.2 Temporal Feature Learning Blocks

For the corpus of IEMOCAP, the length of different utterances is not the same. Long speech fragments last as long as 30 s, while short speeches are even less than 1s. It has caused great difficulties for people studying emotion recognition. Dealing with the problem of different lengths of utterances will greatly improve the accuracy of emotion prediction. We use the deep learning method to intelligently select the speech segment of each time step by combining features through CNN.

2.2.3 Feature Relationship Learning Blocks

Convolutional neural networks can extract the correlation between adjacent features through local receptive fields and parameter sharing. In Fig. 1, the FRLB can extract the relationship between features in the sequence. Each of our features is a combination of a segment of utterance cut into multiple 500 ms HSFs. Therefore, FRLB is needed in this group of features to extract the relationship between different features.

2.2.4 Bidirectional Long Short-Term Memory Network

For continuous speech segments, we use BiLSTM to enable the network to fully learn context-related features. In BiLSTM, the output hidden layer of each BiLSTM unit is a splicing of the hidden layers of forward propagation and backward propagation. In order to select the comprehensive features that can represent forward propagation and backward propagation. We select the output of the last unit of forward propagation and the output of the last unit of backward propagation to be spliced together.

2.2.5 Attention Mechanism

The attention mechanism can make the neural network have the ability to focus on a certain feature. It can make certain important features get more weight. It can make certain important features get more weight.

$$e_{i,j} = \tanh(Linear(s_{i,j}^{S})) \tag{1}$$

$$\alpha_t = \frac{\exp(e_{i,j})}{\sum_{k=1}^{U}(\exp(e_{i,k}))} \tag{2}$$

$$H = s_t^T + \alpha_t s_t^S \tag{3}$$

where s_t represents the feature of the tth segment of utterance after the feature extraction block. Its superscripts T and S represent the features extracted through the feature extraction of blocks (a) and (b). $s_{i,j}$ represents the jth feature of the ith segment of utterance. U represents the number of features contained in a utterance. H represents the final embedding of utterance based on attention mechanism.

3 Experimental Setup and Results

The details of our model are as follows: the two BiLSTMs in the model are set as one layer, and each layer has 32 hidden dimensions. The dropout after BiLSTM is set to 0.25. The hidden dimension in the attention mechanism is set to 64. The learning rate of the entire model is 1e−3, and the batch size of the model is set to 256. Use Adam optimizer, cross entropy loss function. And use L2 regularization to optimize our model training.

In the next section, we will compare with the baseline model we designed and some excellent models in recent years. We will compare with some advanced models in recent years.

3.1 Result and Discussion

In this part, we further research and evaluate the performance of the proposed model. In order to better express the effectiveness and robustness of our model. The evaluation methods we use are the confusion matrix, weighted average accuracy, and unweighted average accuracy commonly used in the literature to evaluate the performance of the model. In addition, in order to show the effectiveness of extracting key speech segments using TFLB. We propose a baseline model. The configuration of the baseline model is as follows:

- Baseline model: The parameters of the input features and classification modules in our model remain unchanged. In the feature extraction module, we only use one layer of FRLB convolution block to extract features, and then pass through a layer of BiLSTM. Finally, enter the classification module to get the final prediction result.

Table 2. The table shows the unweighted accuracy (UA) and weighted accuracy (WA) of the baseline model and other mature models on the IEMOCAP database

Model	Accuary	
	WA	UA
Baseline	57.7	56.1
(Zengwei Yao) 2020 [5]	57.1	58.3
(Guo et al., 2019) [9]	–	57.1
(Latif S et al., 2019) [10]	–	60.2
Our method	**63.0**	**60.9**

Table 2 shows the comparison between our model and the model proposed in recent years on the IEMOCAP dataset. We can see that the effect of our proposed model is better than that of the baseline system. This proves that our model is effective. From the results in Fig. 3, our model achieves the best effect on the test set at the 37th epoch. Although the accuracy of the training set is still rising after the 37th epoch. But it is obvious that the training of the model has been over-fitting. Figure 2 shows the prediction distribution on the test set when the model is optimal.

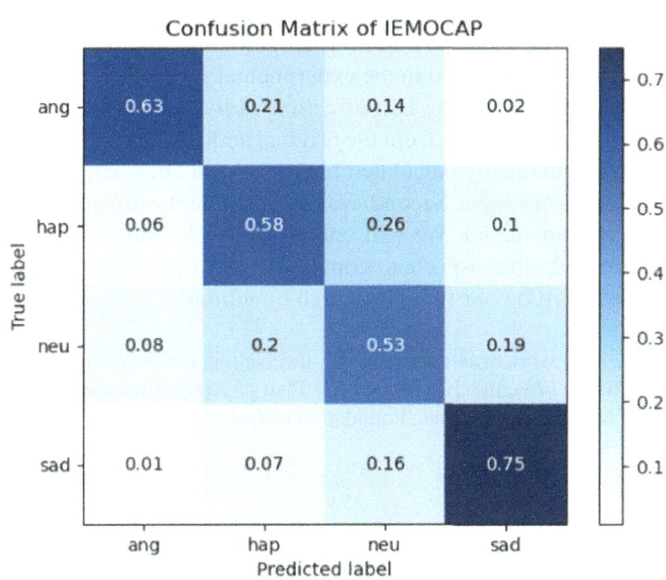

Fig. 2. Confusion matrix of the proposed model. It shows the prediction effect of the model on IEMOCAP, with an unweighted recognition effect of 60.9%.

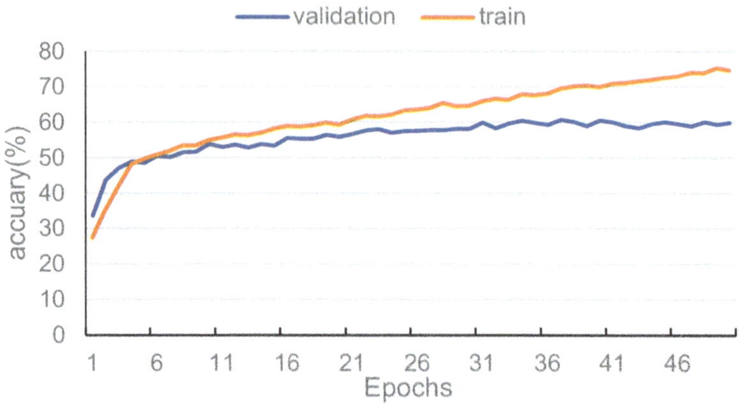

Fig. 3. Trend diagram of the accuracy of the proposed model.

4 Conclusions

This paper mainly introduces the network framework of 1D CNN and BiLSTM for speech emotion recognition. It studied how to use SFLB to extract the local correlation of features from the eGeMAPS feature set and how to use FRLB to extract important speech segments. In a word, we propose a framework that can effectively learn key information between features and features at different time steps.

With the gradual development of emotion recognition technology, combined with the experience gained and accumulated in the experimental process, our follow-up research will be carried out in the following two directions. First, we will consider using deep learning method to extract features from the original audio. The combination of machine extracted features and manually calculated high level statistics features should achieve better results in later experiments. Second, we will consider the use of multi-modal fusion methods to optimize our model. We will comprehensively analyze the information of voice, text, pictures and other aspects to complete the judgment of voice emotion from many aspects. These will be our future research directions.

Acknowledgment. This work was supported by the National Natural Science Foundation of China (Grants No. 61771173), the Tianjin General Plan of Art and Science Project (Grants No. E20016) and the Tianjin Natural Science Foundation of China (Grants No. 20JCZDJC00400).

References

1. Nwe, T.L., Foo, S.W., De Silva, L.C.: Speech emotion recognition using hidden Markov models. Speech Commun. **41**(4), 603–623 (2003)
2. Ayadi, M., Kamel, M.S., Karray, F.: Speech emotion recognition using Gaussian mixture vector autoregressive models. In: IEEE International Conference on Acoustics. IEEE (2007)
3. Samantaray, A.K., Mahapatra, K., Kabi, B., et al. A novel approach of speech emotion recognition with prosody, quality and derived features using SVM classifier for a class of North-Eastern Languages. In: IEEE International Conference on Recent Trends in Information Systems. IEEE (2015)

4. Zhao, J., Mao, X., Chen, L.: Speech emotion recognition using deep 1D & 2D CNN LSTM networks. Biomed. Signal Process. Control **47**(JAN.), 312–323 (2019)
5. Yao, Z., Wang, Z., Liu, W., et al.: Speech emotion recognition using fusion of three multi-task learning-based classifiers: HSF-DNN MS-CNN and LLD-RNN. Speech Commun. **120**, 11–19 (2020)
6. Busso, C., Bulut, M., Lee, C.C., et al.: IEMOCAP: interactive emotional dyadic motion capture database. Lang. Resour. Eval. **42**(4), 335–359 (2008)
7. Mustaqeem, Sajjad, M., Kwon, S.: Clustering-based speech emotion recognition by incorporating learned features and deep BiLSTM. IEEE Access **8**, 79861–79875 (2020)
8. Eyben, F., Wllmer, M., Schuller, B.: OpenSMILE: the Munich versatile and fast open-source audio feature extractor. In: ACM International Conference on Multimedia. ACM (2010)
9. Guo, L., Wang, L., Dang, J., et al.: Exploration of complementary features for speech emotion recognition based on kernel extreme learning machine. IEEE Access **7**, 75798–75809 (2019)
10. Latif, S., Rana, R., Khalifa, S., Jurdak, R., Epps, J.: Direct modelling of speech emotion from raw speech. In: Proceedings of Interspeech 2019, pp. 3920–3924 (2019). https://doi.org/10.21437/Interspeech.2019-3252

Discussion on the Development and Application of BIM Technology in Information Environment

Shujie Zhang(✉)

Shandong Huayu University of Technology, Dezhou, Shandong, China
wang201806324@126.com

Abstract. With the rapid development of information technology, computers have been applied to all walks of life. Computer technology has been widely used in the construction industry. If CAD is to free people from the heavy labor of hand-painting engineering drawings, then BIM technology is another major change in the construction industry, which is called a "revolution" technology. Now BIM technology has been applied in various fields of the construction industry. For example, it plays an irreplaceable role in heating engineering management, structural construction drawings and green building design. The application in the industry is introduced in detail.

Keywords: BIM technology · Computer · Information technology

1 BIM Technology Concept

Computer BIM technology refers to Building Information Modeling, which is translated as "Building Information Modeling" in China. It is a concept proposed by Dr. Chuck Eastman of Georgia Institute of Technology in the United States. The theoretical basis of BIM comes from the CIMS concept of a computer integrated manufacturing system that integrates CAD and CAM in the manufacturing industry, and the product information model based on product data management PDM and STEP standards. The international definition of BIM is: "under open industrial standards, the physical and functional characteristics of the facility and its related project life cycle information can be expressed in a computable form, so as to provide support for decision-making to better realize the value of the project" [1].

Different from two-dimensional models such as traditional CAD, BIM is a multi-dimensional design, which can be designed in three-dimensional, four-dimensional and multi-dimensional models, which can meet the different needs of the construction industry. BIM is of great importance to the construction industry. Compared with traditional drawing methods, BIM has many advantages:

(1) Project construction has become more convenient: The traditional construction work mode is generally done by the architect in charge of coordinating the construction of the entire unit, and various construction problems are often caused by poor communication, or the communication speed is slow and the uploading cycle

I. Ahmad et al. (Eds.): STSIoT 2021, LNDECT 122, pp. 304–313, 2023.
https://doi.org/10.1007/978-981-19-3632-6_38

is long and the delay is long. Construction Schedule. The BIM technology changes the work mode to a network relationship, which enables all parties involved in the project to work together, share information, and accelerate the construction progress [2].

(2) Project construction becomes visualized: In addition to designing construction models, BIM technology can also be used for management. The designer uses computers to complete all aspects from the design stage, construction stage to sales and investment, operation management, etc., covering the entire construction process. All parties involved in the project can share information, submit and review documents and drawings, and negotiate through the Internet to ensure orderly and safe construction [3].

And we also did a survey to find out which parties would choose to use BIM technology when a project is in progress. See the following table for details (Fig. 1):

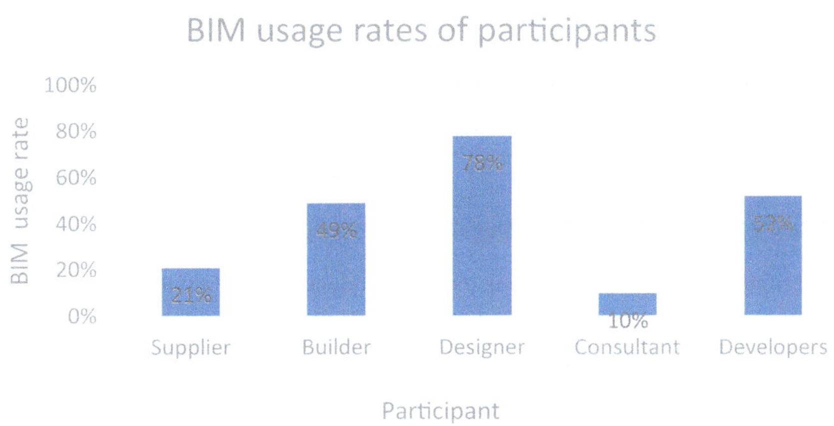

Fig. 1. The usage rate of BIM among the participating parties

It can be seen from the above table that all participants can use computer BIM technology, and the designer is the main one. This also shows that BIM can run through the entire project phase, avoid the attenuation of information transmission in the project construction process, and realize visual and intelligent management [4].

Realize the virtual construction of the project: Various problems often occur during the construction of the project, leading to delays in the construction period. BIM technology can allow the project to carry out construction simulation exercises in the computer design stage. The virtual model is basically the same as the actual model. Some problems in the construction process can be found through virtual construction in advance, and optimization and transformation can be carried out; risk assessment can also be carried out to eliminate the construction process some of the unsafe factors in. This advantage of BIM technology can greatly reduce the probability of problems during the construction process, and reduce costs and energy consumption [5] (Fig. 2).

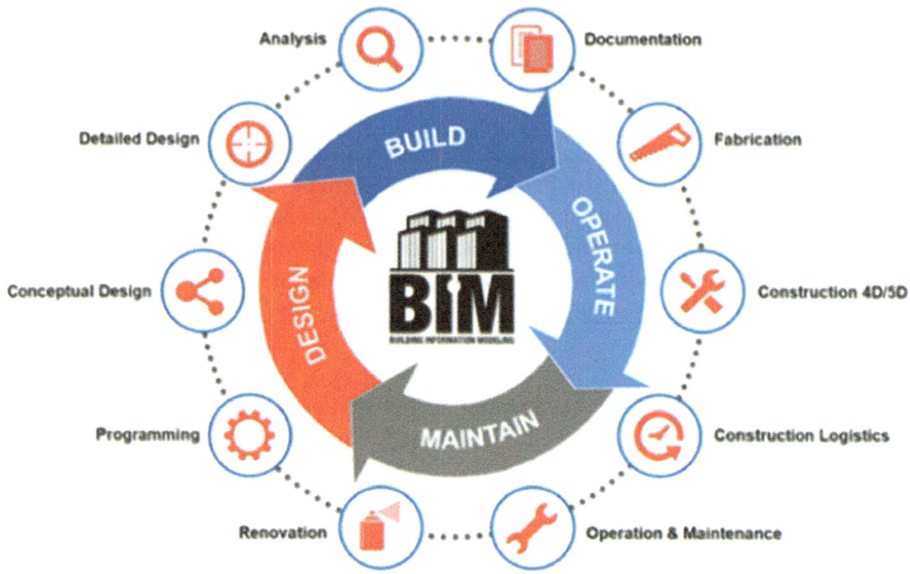

Fig. 2. The process of using BIM technology for project construction

2 Development Status of BIM Technology at Home and Abroad

2.1 The Current Status of BIM Development Abroad

At the beginning of the article, it has also been introduced that computer BIM technology originated in the United States, so the United States is the first country that has matured BIM technology. The BIM standard of the United States is a relatively advanced national standard recognized worldwide [6]. This standard also provides convenience for users. Everyone adopts a unified standard for project formulation and maximizes the economic benefits of all participants.

Driven by the United States, computer BIM technology is rapidly popularized in many European countries. The United Kingdom formulated and revised its own standards based on the standards of the United States, and the utilization rate of BIM technology has rapidly increased in China after its implementation. Moreover, the research and development of computer technology in European countries has always been relatively advanced. After more than ten years of exploration and research, they have formed a set of relatively mature information management tools from project design, transportation, and production, and they have also implemented BIM technical data. The system used in conjunction with other equipment maximizes the advantages of BIM technology.

Japan began to apply computer BIM technology in 2009, and there is a saying in Japan that "2009 is the first year of BIM". The Ministry of Transport of Japan selected a government construction project as a BIM technology pilot in 2010 [7]. Japan's computer software industry is relatively developed. Therefore, after BIM technology has become popular, many software companies have developed BIM technology software and used it in conjunction with other software to establish BIM data processing, design processes, engineering budgets, and building simulations. Wait for a series of completed application systems to meet the needs of the construction industry.

South Korea issued the "Guidelines for the Application of BIM in the Construction Field" in 2010. This guide provides precautions and guidelines when using BIM technology for all cities and many public institutions in South Korea, and establishes its own BIM standards [8]. The measures have prompted many leading Korean companies to adopt BIM technology, such as Samsung Construction, Daewoo Construction, and Hyundai Construction.

2.2 Domestic Development Status of BIM

(1) Status quo of BIM development in mainland china

Computer BIM technology was introduced to mainland China in 2003. Once it was introduced, it attracted the attention of people in the industry, and the country began to learn BIM technology. BIM technology has also received great attention from the country and the government. In the Twelfth Five-Year Development Program, BIM technology is included in the development focus of construction industry information technology, and the development of BIM information technology is emphasized [9].

The first BIM local standard was born in March 2013. The standard "Beijing Civil Building Information Model Design Standard (DB11/1063–2014)" also indicates that BIM technology will gradually get on track in China in the future [10, 11]. Design units, scientific research institutions, and construction companies have all begun to set up BIM technology departments, investing a lot of time and energy on BIM technology research. The research report shows that in 2011, the national awareness of BIM technology has increased to 87%, which shows that China the emphasis on BIM technology is extraordinary.

(2) Status quo of BIM technology development in Hong Kong

The Hong Kong Housing Department first used BIM technology in 2006 and applied BIM technology to the design of complex-shaped buildings. In 2009, BIM standards were issued, and the BIM Society was established. The Director of Hong Kong Housing Department Feng Yixuan pointed out that BIM technology will cover all projects of the Hong Kong Housing Department from 2014 to 2015. In recent years, construction companies have required the use of BIM technology for the design and development of construction projects, realizing the rapid development of the construction industry [12].

(3) Status quo of BIM development in Taiwan

In 2009, the National Taiwan University established the BIM Research Center to "engineering information simulation and management research" throughout the process. It is committed to developing a service platform for experience exchange, talent training and achievement sharing of BIM-related computer technology. The Taiwan government attaches great importance to BIM. It has held several BIM seminars from 2010 to 2012. It hopes to promote the promotion of BIM among enterprises. Completed with BIM technology.

However, due to the high cost of introducing BIM technology and software, many small companies are hesitant to invest a lot of money in the development of computer software and hardware, and the government has not given incentives for the introduction of BIM and hopes that companies can complete the introduction on their own, leading to the development of BIM technology in Taiwan Subject to certain restrictions [13].

3 Application of BIM Technology

3.1 Application of BIM in Structural Construction Drawing Design

As we mentioned earlier, BIM is a multi-dimensional model design through a computer. Compared with the traditional two-dimensional model, it is more three-dimensional and more intuitive. We use the Revit model to reduce it. The Revit model is one of the most commonly used computer software in BIM design. The Revit model can convert all drawings, views, and schedules into information representations in the same database, and can automatically collect different information representations and coordinate the information, that is to say, it can realize the conversion of two-dimensional views and three-dimensional views at any time (Fig. 3).

Fig. 3. Application of BIM technology in heating engineering

When designing in multiple majors at the same time, Revit can provide a platform to use the collaborative mechanism of the software to work together, and merge the design of each major into the same model, which is convenient for each major to view and use. This avoids the complicated linking of diagrams in the past, shortens the design time, realizes real-time synchronization of various professional operations on the computer, and reduces the occurrence of errors. Real-time synchronization is also a major advantage of this model [14].

For complex structures, the flat graphics are not intuitive, making it difficult to think of a three-dimensional space model. The model designed by Revit model can intuitively grasp the spatial relationship between the structures, and the model can also be automatically converted into a calculation and analysis model to reduce the workload in the design process is improved, and the accuracy rate is improved.

The Revit model can realize the expression of the flat method. The flat method of the traditional construction drawing design refers to the direct expression of the size and reinforcement of the structural components on the structural layout plan, and then integrate it with the detailed structural drawing to form a new and complete flat design As shown in the figure, the core of the flat method is to accurately express the steel bar information, and the expression of the flat method in the Revit model is through the flat method symbol family. The flat method symbol family can be associated with the steel bar information to complete the model Conversion of two-dimensional views.

However, Revit also has certain shortcomings. When the structure of the building model that needs to be established is complex, the performance requirements for the computer are relatively high, which requires a higher configuration to achieve; secondly, Revit is automatically converted into a computational analysis modeler, sometimes Errors occur, stability and accuracy need to be improved.

Taking assembly building engineering as an example, BIM technology is used to design according to the following parameter model, and reasonable data can be obtained according to the specified parameters:

$$r = \sqrt{\frac{\mathrm{q}\mathrm{t}_p}{2\pi b}} \tag{1}$$

Q represents the prefabricated building quality parameter, T_P represents the parameter change rate, and B represents the rationality parameter. According to the formula, the processing process can be expressed as:

$$\mu(r, t_p) = \mu\left(\sqrt{\mathrm{q}\mathrm{t}_p/2\pi b}\right) \tag{2}$$

μ indicates the dimension parameter, and the meanings of other parameters remain unchanged. According to the corresponding dimensions of different nodes, a prefabricated building with a general structure is simulated in BIM software, parameters of prefabricated building components are set, and the space-time distribution equation of an prefabricated building is constructed:

$$\frac{2\pi br^2}{q} t_p = r_{max} \tag{3}$$

where, Rmax represents the maximum rationality value of prefabricated building, and the meanings of other parameters remain unchanged.

3.2 Application of BIM Technology in Heating Engineering Management

The application of BIM technology in the design, implementation and use phases of heating projects has achieved good results. The design phase is an important beginning of a project. In the design phase, a project proposal and a project feasibility study report need to be formulated. Only in this way can it be determined whether a project is feasible. Computer BIM technology can just accomplish this task. For example, when the heating pipe network is in the engineering design, different areas will be completed by different design institutes. If the communication is not timely, there will be heating pipe collision problems, and BIM technology By displaying all the design drawings on the computer and performing collision detection, problems can be found and resolved in time, avoiding the problem of being discovered during construction, and the need to dismantle and rebuild the construction waste cost, which reflects the advantages of BIM technology visualization (Fig. 4).

Fig. 4. BIM technology construction effect drawing

In the later stage of project construction, the real-time update and transmission of construction information is particularly important. If the information is not transmitted in time, the construction rate will be reduced. The computer BIM technology can be used in conjunction with other software to update the construction progress and construction problems in time. And BIM technology can also be combined with mobile phone software. Managers can use the mobile phone software to know the progress of the construction anytime and anywhere, and they can also know the status of the construction at any time without a computer.

3.3 Application of BIM in Green Buildings

Green building refers to the construction of a type of building that saves resources, protects the environment, and reduces pollution in combination with the climate, environment, economy, and culture of the location. It is also the development trend of future architecture. According to the concept of green buildings, we can see that in order to achieve the purpose of saving resources and reducing pollution, the structure of green buildings will inevitably be more complex, which means that traditional two-dimensional models or paper graphics are difficult to structure well. Express it. This needs to be applied to computer BIM technology, using BIM technology to build a three-dimensional model on a computer with a configuration that meets the requirements, which can be rotated 360° to intuitively understand the structure of the building and the location of each building structure [15] (Fig. 5).

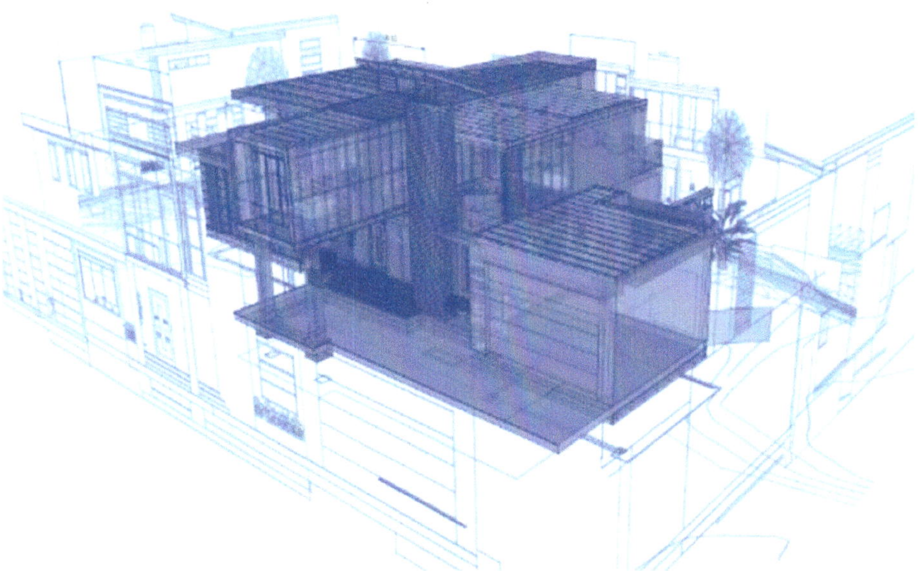

Fig. 5. BIM technology for 3d visualization design schematic diagram

In the process of constructing green buildings, it is necessary to realize the rational use of water and light resources. In the early stage, by collecting local water resources and climate conditions, and uploading the use data to the computer BIM design system, a low-carbon and environmentally friendly design system is simulated. Building model. For example, for water resources, BIM technology provides a three-dimensional pipeline layout through analysis and design of pipeline layout, pipe size and material, and then simulates and uses it to determine the feasibility. The daily maximum water consumption and maximum drainage of the building can be calculated, and the water usage data can be collected and analyzed to determine the peak water usage period. Based on this result, the designer can propose a water usage plan to avoid water cuts and water cuts., Small water pressure and other issues. In addition, water resources can be recycled through computer

BIM technology simulation. For example, solar energy is used to heat water to supply domestic water, and domestic water is processed for toilet flushing, road flushing, etc.; and for the use of natural light, it can be used Measures such as light wells to improve natural daylighting. In view of the problem of insufficient natural daylighting, solar photovoltaic power generation systems can be built to combine with them to reduce the use of coal-powered power generation and realize energy saving and emission reduction.

Computer BIM technology can also realize the saving and utilization of materials. Green buildings have extremely high requirements for the use of materials, and environmentally friendly materials must be used. Using BIM technology to simulate the construction model on the computer, through the calculation of the internal structure of the building, repeated simulation exercises to determine the minimum amount of materials used. And through data calculation, the construction materials can be recycled, and the materials are recycled.

The combination of green building and computer BIM technology has become an inevitable trend. In 2000, the Green Building Council of the United States formulated a standard system called "LEED 1.0 version" in order to realize the combination with BIM technology. This standard stipulates the combination of the two should be reflected in several aspects, namely: building location, water resource utilization, energy utilization, environmental utilization, material utilization, resource utilization, air quality, and comprehensive design, indicating that the United States attaches great importance to BIM technology. The combination of these will inevitably bring about tremendous changes in human life.

4 Disadvantages of Computer BIM Technology

(1) High requirements for computer performance: We all know that BIM technology is not only for the design stage, it includes the production and management of engineering data information sharing in the entire life cycle of an engineering project from planning, design, construction, operation management to demolition The platform, which means that its amount of information is very large, easily reaching tens or hundreds of GB. In addition, BIM technology also needs to work with other software and even realize remote operation. Therefore, it must be matched with high-performance computers or mobile workstations, and it is difficult to activate multiple BIM files at the same time, which limits the promotion of BIM technology.

(2) Information security cannot be guaranteed: Computer BIM technology requires the participation of multiple parties to achieve resource sharing, which means that the security of information will be reduced, and any party can perform license authorization, which is likely to cause technical leakage.

In summary, computer BIM technology has indeed subverted traditional modeling methods and has gradually been recognized by the industry. I believe that in the future, the award will establish a unified standard for BIM technology and actively encourage construction companies to use BIM technology. With continuous research and updates on BIM technology, the security of BIM technology will be better in the future. Moreover, computers are developing rapidly nowadays. It is believed that in the near future, ordinary computers can be used for BIM technology, and BIM technology will also bring greater convenience and value to the construction industry and gradually extend to other industries.

Acknowledgements. This work was supported by R&D Center of BIM Technology and Application in Engineering (2019-02) of Shandong Huayu University of Technology.

References

1. Malagnino, A., Montanaro, T., Lazoi, M., Sergi, I., Corallo, A., Patrono, L.: Building information modeling and internet of things integration for smart and sustainable environments: a review. J. Clean. Prod. **312**, 127716 (2021). [5]
2. Khodabakhshian, A., Toosi, H.: Residential real estate valuation framework based on life cycle cost by building information modeling. J. Archit. Eng. **27**(3), 04021020 (2021)
3. Dinis, F.M., Martins, J.P., Guimarães, A.S., Rangel, B.: BIM and semantic enrichment methods and applications: a review of recent developments. Arch. Comput. Methods Eng. (2021). [4] (prepublish)
4. Li, S., Zhang, D., Tian, C.: Research on the Application of BIM Technology Based on Information Technology in Construction Engineering Safety Management. J. Phys.: Conf. Ser. **1915**(2) (2021)
5. Liu, Z.: Application of BIM technology in a prefabricated complex project. In: IOP Conference Series: Earth and Environmental Science, vol. 783, no. 1 (2021)
6. Fu, F.: Evaluation of risk management in engineering construction based on BIM technology. Int. J. Comput. Eng. **6**(1) (2021)
7. Pan, Y., Zhang, L.: Roles of artificial intelligence in construction engineering and management: A critical review and future trends. Autom. Constr. **122**, 103517 (2021). [5]
8. Yang, S.: The practice of project teaching method in engineering cost teaching–the application of Bim technology. Front. Educ. Res. **3**(15) (2020)
9. Bassier, M., Vergauwen, M.: Unsupervised reconstruction of building information modeling wall objects from point cloud data. Autom. Constr. **120**, 103338 (2020). [5]
10. Gohil, S., Verma, S.: Evolution of quantitative effects of construction changes on labor productivity time and cost control using building information modeling. J. Trend Sci. Res. Dev. **5**(1) (2020)
11. Zhao, P.: Innovation and reform of talent training mode of higher vocational architectural engineering technology specialty based on BIM technology. Int. J. Educ. Teach. Res. **1**(3) (2020)
12. Gong, J.: Research on the refined management of engineering cost in the field of bim technology. Mod. Manage. Forum **4**(3) (2020)
13. Sun, F.: BIM technology application in civil engineering. Jo. Archit. Res. Dev. **4**(5) (2020)
14. MaskilLeitan, R., Gurevich, U., Reychav, I.: BIM management measure for an effective green building project. Buildings **10**(9), 147 (2020)
15. Sporr, A., Zucker, G., Hofmann, R.: Automatically creating HVAC control strategies based on building information modeling (BIM): heat provisioning and distribution. Energies **13**(17), 4403 (2020)

Design of Automatic Verification System for Evaporation Sensor

Jianyu Li[1](✉), Chen Chen[2], and Mursaha Abstueri[3]

[1] The Meteorological Equipment Center of Hunan, Changsha, Hunan, China
zbzxljy@163.com
[2] Key Laboratory of Hunan Meteorological Disaster Prevention and Reduction, Changsha, Hunan, China
[3] University of Delhi, Delhi, India

Abstract. In order to solve the problem that the evaporation sensor can only be verified manually at present, an automatic evaporation sensor verification equipment is developed. By simulating different liquid level heights and equipped with a high-precision acquisition module to collect the signal of the detected sensor, the liquid level height and the sensor output are compared to determine whether the accuracy of the sensor meets the requirements, so as to realize the automatic verification of the evaporation sensor and improve the verification efficiency.

Keywords: Evaporation · Sensor · Automatic verification · System design

1 Introduction

Evaporation is one of the main elements of meteorological observation and is also very important in agricultural production and hydrological work. As one of the necessary meteorological observation key elements, evaporation taken as a foreseen sign to forecast extreme weather situation, for instance rainstorm, hailstone, snowfall. To guarantee diversity of sample sizes, every meteorological observation stations owns a standard evaporation equipment to collect and measure evaporation data, then compare data in central server to ensure consistency and accuracy. During comparing program individual extreme value will be filter and reject, then all of data will be sent to national database to complete forecast of subsequent weather. At the same time, in order to ensure the accuracy of the observation data of evaporation sensors, verification is needed every two years by national meteorological administration. At present, the verification of evaporation sensors mainly relies on standard modules of different heights. In the verification process, it is necessary to manually replace the standard modules of different heights, and read the evaporation data on the automatic weather station and compare it with the nominal value of the standard module, which is very inconvenient to use. Moreover, the evaporation sensor is not allowed to operate with electricity, and improper operation is very easy to damage the evaporation sensor. Traditional evaporation observation requires a 12 square meters large area to set observation machine, considering manually operation able to reflect evaporation data more accurately and it is hard to repeat observations

I. Ahmad et al. (Eds.): STSIoT 2021, LNDECT 122, pp. 314–323, 2023.
https://doi.org/10.1007/978-981-19-3632-6_39

in a short period of time, automatic observation is necessary to take part in evaporation observation system satisfy needs of safety and verification.

In order to simplify the verification process of evaporation sensor, we develop an automatic verification system of evaporation sensor. By simulating different liquid level heights and equipping a high-precision acquisition module to collect the signal of the detected sensor, the simulated liquid level height and the sensor output are compared to determine whether the accuracy of the sensor meets the requirements, so as to realize the automatic verification of the evaporation sensor and improve the verification efficiency.

2 Related Work

The automatic verification system of evaporation sensor mainly uses STM32F single chip microcomputer, stepping motor and grating ruler. Hung Nguyen Tan Using STM32 single chip microcomputer to realize the control of multi-level inverter [1], argues single chip able to satisfy functions in complicated jobs, especially in workplaces required small mechanical equipment and highlight STM32F language to assemble complex functions. Hung Nguyen Tan designed a low-cost security control system [2] using single chip microcomputer, compares to other security control system developed by single chip microcomputer, it confirms security performances [3] can be greatly protected by specific logic programs and statements, and argues a complicated language command line can be merged in single chip system. Farzin Asadi and others have carried out product design and research by using STM32 single chip microcomputer [4], which demonstrates a logic chain and function list may meet needs of different scenarios [5], they highlight digital circuits and systems training project, demonstrates how single chip microcomputer used in various workplaces. Yan-Lin Huang designed a stepping motor torque sensor [6], that demonstrates potential of large mechanical equipment application, they finish a experiment of low cost compliant motion generation and prove it can be run by embedded system. Jafar Tavoosi achieved the control of stepping motor through analog neural network [7], and argues that single chip owns a huge potential to self-improve and complete complicated tasks by design a application of stepping motor [8], Yosra Miladi and others have carried out application design and Research on stepping motor [9], which aim to research a newly mechanical control method and develops a application satisfy needs of standardized control, they develops a chaotic switched system to simulation different workplace and find out whether a system able find the best way to solution problems. Seok-Kyoon Kim and others [10] adopts a similar approach to setting obstacles. Hisao Fukumoto and others [11] try to combine newly developed network technology and cloud technology to complete remote control function, which proved as a efficient way to improve study. John and others [12] designs a algorithm and embedded by program to reduce position error, which shows stepper motor drive waveform and actual performances can be further optimization by take advantage of various functions. Simin Li and others designed a real-time direction judgment system for sub nano grating ruler, and analyzes what a real-time direction judgement system program needs and how is the application of single-chip microcomputer system works [13].

At present, there is no automatic calibration equipment for evaporation sensor due to cost and many other reasons, simple logic of micro-controller is characteristic of its low cost and clear logical chain, which can be developed as a operation mode of automatic evaporation equipment. Besides, single chip microcomputer able to merge many integrated circuit such as clock circuit, timer, and et cl. Which may greatly reduce time and cost required to develop new system.

3 System Principle

3.1 Introduction to the Evaporation Sensor

The picture of an evaporation sensor is shown in Fig. 1, including evaporation barrel, shutter box, hydrosphere, communication tube, thermometer, ultrasonic sensor, stainless-steel measuring tube and other equipment [14, 15]. The evaporation barrel adopts the original large evaporator of the station, with a diameter of 61.8cm, exposed to the air and 30cm from the ground, and the change of its water level is often called evaporation. The water in the evaporation barrel enters through the connecting pipes in the stainless-steel measurement barrel placed in the thermometer shelters. The ultrasonic sensor is fixed on the measurement barrel to measure the change of the water level regularly.

Fig. 1. The evaporation sensor

3.2 Measurement Principle of the Evaporation Sensor

The main measuring components of the evaporation sensor are ultrasonic sensor and stainless-steel measuring tube [16], as shown in Fig. 2. They are placed in shelters to prevent the influence of meteorological factors such as wind and radiation. The stainless-steel measuring cylinder is connected with the evaporation barrel, and the change of the water level is consistent with the change of the water level of the evaporation barrel. The ultrasonic sensor is placed about 180cm above the water surface, and sends ultrasonic wave to the water surface regularly, and accurately measures the height of the water level according to the ultrasonic signal reflected from the water surface. The water surface height is measured by $H = C_W \cdot t/2$, C_W is the sound velocity in water, and t is the time that ultrasonic pulse travels back and forth to the water surface height H.The evaporation of the water surface can be obtained by calculating the difference between the two measurements. The output signal of ultrasonic evaporation sensor is a 4 ~ 20 mA current, and the corresponding water level height is 100 mm ~ 0 mm.

Fig. 2. The measurement part of the evaporation sensor

3.3 Verification of the Evaporation Sensor

The verification of the evaporation sensor is based on "JJG (Meteorological) 006–2011-Verification regulation of Evaporation Sensor of Automatic weather Station". According to the requirements, the maximum relative error of 6 evaporation water level points (0 mm, 20 mm, 40 mm, 60 mm, 80 mm, 90 mm) should be verified to determine whether the sensor is qualified. First, the 72mm module is put into the cylinder of the evaporation sensor, and the evaporation zero value is recorded from the collector. Then, the evaporation module group is used to form the standard height values of 10 mm, 30 mm, 50 mm, 80 m and 100 mm, which are put into the evaporation cylinder. The indication values of the evaporation sensor are read respectively, and the relative error values of each inspection point are calculated as shown in formula (1).

$$\Delta h = \frac{(h - h_0) - h_s}{h_s} \times 100\% \tag{1}$$

Where: h is the height value of each module, h_0 is the zero value, and h_s is the standard height value. When the relative error of each test point is less than ± 1.5%FS, it is qualified [17].

3.4 Design Principle of Automatic Verification Instrument for the Evaporation Sensor

The schematic diagram of automatic verification instrument for the evaporation sensor is shown in Fig. 3.

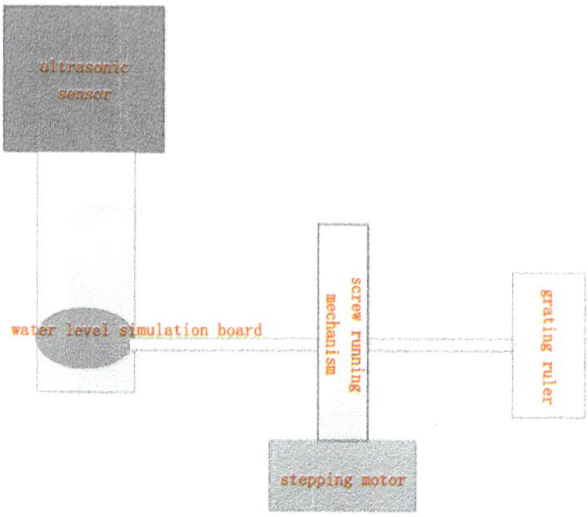

Fig. 3. Schematic diagram of automatic verification instrument for the evaporation sensor

The automatic verification instrument of the evaporation sensor uses stepping motor and water level simulation board to realize the simulation of standard evaporation. A water level simulation board is installed at one end of the stepper motor, and the laser emitted by the evaporation sensor shines on the water level simulation board and reflects back. By adjusting the height of the water level simulation board, the simulation of evaporation water level at different heights can be realized. According to the relevant technical indexes and verification regulations of evaporation sensors, the simulated water level height is 0–100 mm.

High precision grating ruler displacement sensor is used for automatic reading of evaporation. The grating ruler displacement sensor is a measuring feedback device based on the optical principle of grating. The grating ruler displacement sensor and the water level simulation board operate synchronously, and the position change of the water level simulation board can be feedback in real time.

During verification, controlling the stepper motor to rotate and raising or lowering the position of water level simulation board can simulate different water levels. The grating ruler displacement sensor connected with the water level simulation board outputs position information in real time, the control module reads the output data information of the grating ruler displacement sensor and the current signal of the evaporation sensor at the same time, and compares them to realize the verification of the sensor.

4 Hardware Design

The main components of evaporation sensor verification instrument include support, sensor support, stepping motor, screw running mechanism, water level simulation board installed on the screw sliding block, high-precision grating ruler and control module. The control module mainly includes stepper motor control unit, sensor signal acquisition unit, grating ruler signal acquisition unit, etc.

4.1 Design of the Main Controller

STM32F103CVT6 single-chip microcomputer is used in the main control system of the evaporation sensor verification instrument to realize the control of stepper motor, the acquisition of grating ruler data, screen display and data communication. STM32 microcontroller is a 32-bit microprocessor based on embedded ARM Cortex-M3 core, which works at the frequency of 72 MHz. It contains a built-in high-speed memory, rich enhanced I/O ports and peripherals connecting two APB buses, three 12-bit ADCs, four general 16-bit timers and two PWM timers, standard and advanced USART, USB, CAN and other communication interfaces.

4.2 Design of Stepper Motor Drive Circuit

The driving circuit of the stepper motor uses A4988 chip, as shown in Fig. 4. The chip is a DMOS micro stepper motor driver with converter and overcurrent protection, which can operate the bipolar stepper motor in full, half, 1/4, 1/8 and 1/16 stepping modes. During operation, through the control of STEP and DIR two ports, inputting a pulse in the STEP port can drive the motor to produce micro step, without phase sequence table or complex interface programming.

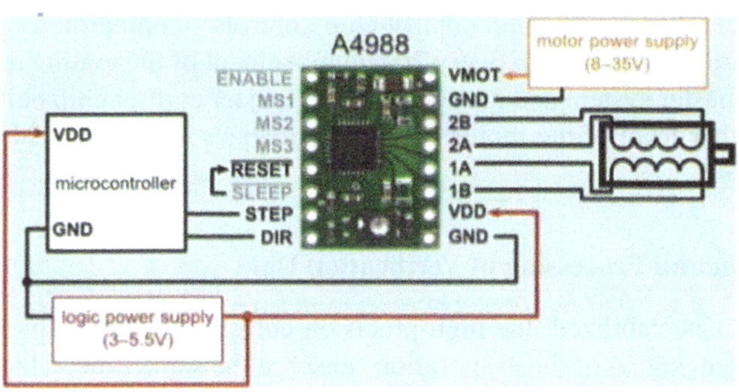

Fig. 4. Schematic diagram of stepper motor drive circuit

4.3 Data Acquisition Circuit of Water Level Simulation Board

The data acquisition of water level simulation board is realized by high-precision grating ruler. The high-precision grating ruler connected to the water level simulation board outputs TTL waveform, corresponding to water levels of different heights, and the grating ruler pulse is read by the master SCM. The simulated water level height is generally 10 mm, 20 mm, 30 mm, 40 mm, 50 mm, 60 mm, 70 mm, 80 mm, 90 mm, etc., and the error of each simulation point is not more than $\pm$ 0.15%.

5 Software Design

In order to realize the automatic control of evaporation sensor verification instrument by computer, the verification software corresponding to the product is designed. The main functions include communication parameters, instrument information input, system reset, manual single point verification, automatic whole verification, verification information chart display, verification report output, etc.

5.1 Software System Process

See Fig. 5 for the flowchart of the software system. After entering the system, the sensor information is input first, the communication port is opened, and the system resets the sensor. Then the automatic verification mode or manual single point verification mode can be selected. After the verification is completed, the verification results are checked out and the verification report is generated.

5.2 Control of Collection Calibration Instrument

The software architecture is controlled by double closed loop. The system starts into the initial state. After receiving the instruction of a given water level values, system comes into a testing condition, and main control chip controls stepping motor speed to drive linear grating ruler sliding head. Every 20 μm movement of the grating ruler feeds back a pulse to the master system and master chip, and master control chip built-in algorithm adjusts the higher level of the motor and finally realizes the grating ruler to achieve a given level value.

5.3 Collection and Processing of Verification Data

After the motor is stabilized, the high-precision current acquisition module measures the output current signal of the evaporation sensor at the same time, which is converted into the measured value of the water level height, and then the system error is finally obtained by making a difference with the standard value of the water level height output by the grating ruler. After verification, Word format report can be generated for storage and printing.

5.4 Measurement Mode Selection

The measurement mode is divided into automatic mode and manual mode two operation modes: manual mode can directly test the error rate of a water level; automatic mode can directly complete the requirements of all the inspection point measurement. The program automatically controls the stepper motor simulation of different water level values, reads the standard water level through the grating ruler and the sensor current calculation sensor water level at the same time, then calculates the error, and the results are transmitted to the outer screen display.

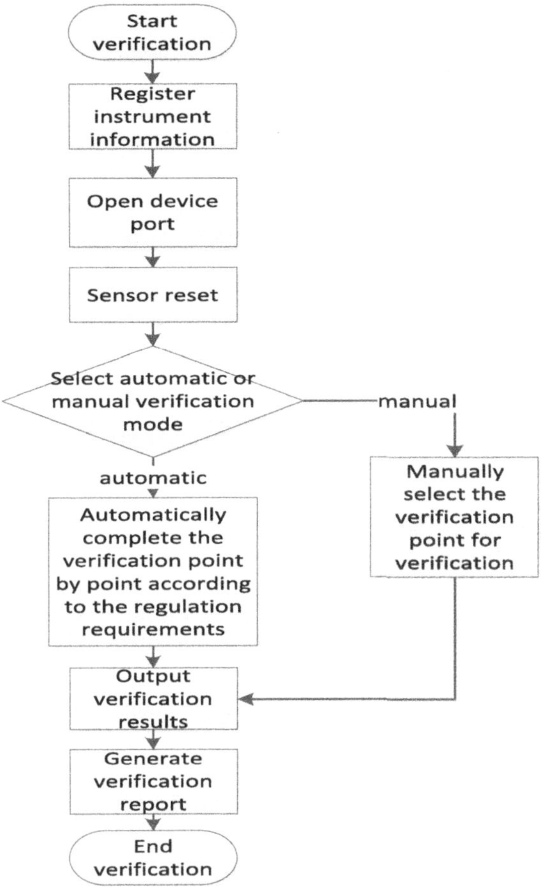

Fig. 5. Flowchart of main program

6 System Function Test

Wush-tv2 evaporation sensor from Wuxi was used for 5 tests, and 3 inspection points of 20mm, 40mm and 60mm were selected. The automatic and manual test results are shown in Table 1 and Table 2. By comparison, the average error of automatic and manual verification is not much different, and the variance is larger than that of manual verification, but it can also meet the requirements of verification business.

Table 1. Automatic verification result

Test point (mm)	1	2	3	4	5	Average	Variance
20	0.41	0.5	0.45	0.4	0.41	0.43	0.001384
40	0.63	0.8	0.7	0.53	0.63	0.66	0.007976
60	0.32	0.43	0.34	0.21	0.32	0.32	0.004904

Table 2. Manual verification result

Test point (mm)	1	2	3	4	5	Average	Variance
20	0.41	0.36	0.34	0.38	0.39	0.38	0.000584
40	0.55	0.48	0.48	0.52	0.55	0.52	0.000984
60	0.5	0.4	0.42	0.49	0.5	0.46	0.001856

7 Summary

The evaporation sensor automatic verification system solves the problem that the evaporation sensor must be verified manually in the past. The verification time is shortened by more than 60%, and the work efficiency of the evaporation sensor is greatly improved. Through the grating ruler and the signal acquisition module, the system can realize the magnitude traceability of the etalon, and the measurement accuracy can reach ± 0.02 mm, which can fully meet the requirements of the verification regulation.

This year, the system has been used to verify 22 sets of evaporation sensors in our province, and 20 sets are qualified, with a qualified rate of more than 90%. It can better test whether the performance of evaporation sensors meets the requirements.

References

1. Tan, H.N., Van, H.P., Duy, T.N., Duc, T.D.: STM32F407 Implementation of Unipolar SPWM for Three-phase 3 Level Inverter. In: ICSSE 2021, pp. 27–31 (2021)
2. Marie, T.F.B., Han, D., An, B.: Microcontroller design for security system: implementation of a microcontroller based on STM32F103 microchip. Int. J. Embed. Syst. **11**(5), 541–550 (2019)
3. Asadi, F., Pongswatd, S.: Programming the ARM® Cortex®-M4-based STM32F4 microcontrollers with simulink®. Synthesis Lectures on Digital Circuits and Systems, Morgan & Claypool Publishers, pp. 1–183 (2021)
4. Wang, Z.: Design of ship attitude angle detection system based on STM32F103ZET6. In: ICIT, pp. 262–266 (2020)
5. Zhou, Y., Shi, F., Chen, J.: Design and application of pocket experiment system based on STM32F4. In: ICIT 2020, pp. 40–45 (2020)
6. Huang, Y.-L., Liang, C.-H., Chen, B.-H., Lan, C.-C.: Torque-sensorless control of stepper motors for low-cost compliant motion generation. IEEE Access **9**, 94495–94504 (2021)

7. Tavoosi, J.: A novel recurrent type-2 fuzzy neural network for stepper motor control. Mechatron. Syst. Control. **49**(1) (2021)
8. Miladi, Y., Derbel, N., Feki, M.: Optimal control based on multiple models approach of chaotic switched systems, application to a stepper motor. Int. J. Autom. Control. **15**(2), 240–258 (2021)
9. Groenhuis, V., Rolff, G., Bosman, K., Abelmann, L., Stramigioli, S.: Multi-Axis electric stepper motor. IEEE Robotics Autom. Lett. **6**(4), 7201–7208 (2021)
10. Kim, S.-K., Ahn, C.K.: Variable-performance positioning law for hybrid-type stepper motors via active damping injection and disturbance observer. IEEE Trans. Circuit. Syst. II Express Briefs **68**(4), 1308–1312 (2021)
11. Fukumoto, H., Yamaguchi, T., Ishibashi, M., Furukawa, T.: Developing a remote laboratory system of stepper motor for learning support. IEEE Trans. Educ. **64**(3), 292–298 (2021)
12. Pillans, J.: Reducing position errors by vibration optimization of stepper motor drive waveforms. IEEE Trans. Ind. Electron. **68**(6), 5176–5183 (2021)
13. Li, S., Wang, J., Zhang, W., Hao, D., Xiong, X.: Real-time direction judgment system of sub-nanometer scale grating ruler. IEEE Access **9**, 74939–74948 (2021)
14. Alonso, S., Álvarez, A.M., Pérez, D., Prada, M.A., Fuertes, J.J., Domínguez, M.: Virtual sensor for probabilistic estimation of the evaporation in cooling towers. Integr. Comput. Aided Eng. **28**(4), 369–381 (2021)
15. Comprehensive observation Department of China Meteorological Administration, code for ground meteorological observation (2019)
16. Walczak, A., Lipinski, M., Janik, G.: Application of the TDR sensor and the parameters of injection irrigation for the estimation of soil evaporation intensity. Sensors **21**(7), 2309 (2021)
17. Pei, B., et al.: Oven-controlled MEMS oscillator with integrated micro-evaporation trimming. Sensors **20**(8), 2373 (2020)

Application of Intelligent Operation and Maintenance Platform for Rail Transit Power Supply System

Yangning Zheng[1,2](✉) and Xiaoyu Zhang[1]

[1] School of Electronics and Information, Nanchang Institute of Technology, Nanchang, Jiangxi, China
zyn@nut.edu.cn
[2] Bansomdejchaopraya Rajabhat University, Bangkok, Thailand

Abstract. The construction of urban rail transit is one of the important ways to solve traffic congestion, drive employment as well as drive economic development. By the end of 2020, China's rail transit has opened 1083 km of operation, while there are about 45 approved cities. The development of intelligent system and the construction of smart urban rail is the necessary way to realize the development of urban rail transportation from high speed to high quality, and it is also the premise and guarantee to promote the construction of a strong transportation country. In this paper, firstly, the system architecture and functional modules are designed according to the actual requirements. Secondly, the system database is designed based on the system development environment. The system provides intelligent inspection function and leaky cable monitoring function, realizing good equipment visualization effect, complete asset management data and high vehicle electrical system overhaul efficiency, etc. Finally, through the test of the platform, the results show that the accuracy of fault diagnosis module for turnout fault diagnosis reaches more than 95%, especially for the troubleshooting of uncommon faults, the effect of on-site guidance is better, which improves the efficiency of on-site troubleshooting and fault location.

Keywords: Rail transit · Intelligent operation and maintenance · Power supply system · Fault detection

1 Introduction

With the increasing scale of rail transit network construction, it has become a trend to establish a full-dimensional intelligent monitoring and maintenance platform centered on equipment control based on the application of equipment online monitoring technology and artificial intelligence technology in order to improve the production operation efficiency and equipment reliability [1, 2]. In terms of the development status of the rail transit system, the intelligent technology sensing technology and information technology carried by the trains have provided a benchmark technical support system for the

I. Ahmad et al. (Eds.): STSIoT 2021, LNDECT 122, pp. 324–332, 2023.
https://doi.org/10.1007/978-981-19-3632-6_40

construction of the whole intelligent system to ensure that the rail transit trains can operate intelligently in the system based on the synchronized transmission of information during operation [3, 4].

Regarding the study of intelligent operation and maintenance platform for RTPSS, many scholars at home and abroad have conducted research on it. Foreign research on O&M technology is more advanced, as early as 1951, the U.S. Westinghouse Company has begun to monitor research on generator failure, and then widely used in the field of military equipment and electrical industry [5, 6]. Domestic O&M technology research began in the 1980s, and after continuous development, the detection technology can be more maturely applied to power transformers and high-voltage circuit breakers, etc., however, the research development of condition maintenance is still relatively lagging behind [7]. In the construction of intelligent operation and maintenance system, Diao P H et al. introduced an expert system of big data in report statistics, information push and fault query and achieved good results [8]. It A, Yh A et al. described and analyzed the intelligent operation and maintenance of Beijing metro power supply equipment from two aspects, online monitoring system and substation intelligent inspection robot, respectively, which provided a reference for intelligent operation and maintenance scheme [9]. Yu J et al. described and analyzed the intelligent operation and maintenance of the subway signal system and gave a comprehensive plan design [10]. Although there are many studies on intelligent O&M platforms for rail transit power supply system (RTPSS)s, no solutions have been given so far that can solve some other problems; therefore, it is necessary to strengthen the research on them.

In this paper, through the analysis of the current situation of rail transit operation and maintenance, an intelligent operation and maintenance platform for RTPSS is proposed, which has two major functions, intelligent inspection function and leaky cable monitoring function; then the technology used in this platform is analyzed, deep learning and decision tree; and the implementation of this platform is introduced, and finally the platform is tested.

2 Research and Application of Intelligent Operation and Maintenance Platform for RTPSS

2.1 Current Situation of Rail Transit Operation and Maintenance

The equipment maintenance of RTPSS is complex, specialized and widely distributed, and the frequency, cost and requirements of maintenance are increasing. The traditional regular maintenance and after-action maintenance mode can hardly meet the demand of sustainable development, and the following problems exist.

1) There are manpower bottlenecks in operation and maintenance work.
2) The systems are not interoperable, unable to generate new quality capabilities and lacking a panoramic monitoring platform.
3) The overall scattering of operation and maintenance data and the lack of data analysis and application capabilities.
4) Equipment operation control fails to close the loop management.

At present, in practice, a large number of detection and monitoring devices are installed on various devices to obtain specific collection information, but there is a problem of scattered monitoring data and no comprehensive fault diagnosis function. From "planned fix" to "condition fix" or even "predictive fix" is not feasible. Planned maintenance requires more human resources to locate, troubleshoot and complete the repair, of which the human factor accounts for a larger proportion. In general, the real-time monitoring capability of traditional O&M is not strong and comprehensive enough, and the coverage is not enough, which consumes a lot of time and human resources.

2.2 Intelligent Operation and Maintenance Platform for RTPSS

In view of the current situation of rail transit cloud cataract maintenance mentioned above, this paper proposes a new system which takes production management as the main line, personnel management, equipment management, fault management, material management, work apparatus management and document management modules as auxiliary support, and constantly improves the card control and reminder of the main line module of production management. The system realizes the connection between the data of business modules, and then achieves the closed-loop management of business modules, and finally realizes the goal of whole life cycle management of facilities and equipment. Specifically, it is as follows: to solve the characteristics of the existing information systems which are many in number and scattered, independent of each other, lacking openness and having many data silos, etc., to establish an integrated platform for safe operation and maintenance and repair of equipment and facilities by means of multiple data access; to optimize data entry by establishing metadata standards, designing business associations. It realizes the optimization, monitoring and management of the whole process of the existing power supply maintenance work, so that the power supply facilities and equipment and related resources can play the best performance to achieve cost reduction, efficiency improvement and management improvement [11, 12].

On the basis of summarizing the industry experience, advanced technology concepts such as cloud computing, Internet of Things, big data, and artificial intelligence are used to realize the optimization, monitoring, and management of the whole process of existing power supply overhaul work by establishing metadata standards, designing business associations, and optimizing data entry to establish an intelligent operation and maintenance management platform for the power supply system, through which the equipment of the power supply system is managed in an all-round and whole-process manner. The data collected from the stereo sensing network and the analysis data of the business system are uploaded to the cloud platform in real time. Based on data processing technologies such as HDFS, Spark, HIVE and Kafka, the data uploaded to the cloud platform in real time are analyzed by using big data analysis and machine learning, and the real-time status data of the same type of equipment are compared horizontally and the historical data are compared vertically to precisely locate the equipment status under different environments and working conditions. The intelligent operation and maintenance platform of RTPSS mainly includes intelligent inspection and cable leakage monitoring.

(1) Intelligent inspection

The intelligent inspection system of equipment room can significantly reduce the manpower cost of daily inspection, and can realize remote real-time monitoring and alarm of dynamic environment and equipment status, and also provide real-time images of the scene for fault disposal at the first time.

The orbital inspection robot is set up to inspect the equipment room and cabinets according to the pre-set walking trajectory and points, and the inspection results are transmitted to the local server through data. If the inspection results are inconsistent with the normal state of the equipment, the location and status information of the faulty equipment is accurately provided. Under special circumstances, the robot can be manually controlled remotely to perform inspection tasks.

In the operation and maintenance site, contact network, track, bridge and tunnel, housing construction and other related professions usually have deployed their own professional monitoring, testing instruments and equipment, and have a large amount of data, such as contact network pull-out value, guide height, wear, combustion arc, track gauge, elevation, track direction, positive vector, triangle pit, wear, bridge and tunnel and housing construction settlement, horizontal shift, cracks, water leakage, etc.

(2) Leaky cable monitoring

Through the collection of leaky cable data and the establishment of data models, preventive monitoring of leaky cable status can be carried out, and the detection steps are as follows.

1) Real-time monitoring includes: bending and breaking trends of leaking cables, patch cords, loosening trends of joints and other hidden problems and other abnormalities.
2) The use of waveforms and other means to leak between the key connection points, joints and nodes such as combiners at the abnormal trend as real-time monitoring content.
3) Preventive analysis of the above-mentioned contents, setting early warning values and providing effective disposal basis for on-site fault treatment in the form of alarm information displayed at the terminal.

2.3 Research on Key Technologies for Intelligent Operation and Maintenance of RTPSS

(1) Prediction of remaining life and health based on deep learning

The system mainly adopts the Recurrent Neural Network (RNN) model based on the attention mechanism to model the remaining life of turnouts.

The original recurrent neural network structure, which is commonly characterized by gradient disappearance and gradient explosion, has been proposed as a Gated Recurrent Network (GRU) to better capture the long sequence dependence in time series. Concepts to solve the mentioned long time series dependence problem.

For the GRU network, the inputs to both the Reset Gate and Update Gate in the gating unit are the current state $X_t \in {}^{n\times d}$ (h is the number of hidden layer units of the neural network, n the number of samples of the current input) and the hidden state $H_{t-1} \in h^{n\times h}$ of the previous time step. The relationship between the reset gate $R_t \in h^{n\times h}$ and the update gate $Z_t \in h^{n \times h}$ is shown in the following equation.

$$R_t = \sigma(X_t W_{xr} + H_{t-1} W_{hr} + b_r) \tag{1}$$

$$Z_t = \sigma(X_t W_{xz} + H_{t-1} W_{hz} + b_z) \tag{2}$$

where $W_{xr} \in h^{d \times h}$, $W_{xz} \in h^{d \times h}$, $W_{hz} \in h^{d \times h}$ and $W_{hr} \in h^{d \times h}$ are the weight matrices, while b_r, $b_z \in h^{l \times h}$ is the bias matrix and $\sigma()$ is the sigmoid function that embeds the nonlinearity in the network.

The prediction of deep learning is mainly divided into the following steps.

1) Data cleaning: cleaning for the historically collected data to obtain higher quality current data.
2) Feature extraction and feature analysis: analyze the relevant factors affecting the final life for the turnout life prediction problem, and analyze the possible mechanisms.
3) Deep learning algorithm modeling: using deep neural network to construct life prediction model.
4) Residual life prediction: Deploy the relevant model on the actual line, iteratively optimize it and compare it with the life threshold to obtain the evaluation of healthiness.

(2) Fault diagnosis based on decision tree

In this paper, by studying the methods of fault diagnosis, it is found that the decision tree method in artificial intelligence technology has a better learning and reasoning ability, so this paper designs an integrated method for fault diagnosis, which consists of two major modules.

1) Pattern mining method based on integrated clustering: Through unsupervised learning algorithms, frequent patterns in the historical data collected by the equipment are mined, and normal patterns and abnormal patterns among them are distinguished. In the process, the proposed method is based on integrated clustering to obtain the frequent patterns in the data.
2) Building fault diagnosis model based on pattern library and historical data: After obtaining the fault patterns and normal patterns in the historical data, the fault detection algorithm based on abnormality detection and the fault diagnosis algorithm based on deep learning and template matching are used to synthesize the training model and obtain the trained fault diagnosis model.

3 Implementation of Intelligent Operation and Maintenance Platform for RTPSS

3.1 Database Implementation

The database is the main factor to decide the scalability and timeliness of the platform data layer. The database design of the main modules of the intelligent operation and maintenance platform is rooted, and the specific implementation of it is carried out. In the database layer, the table names are standardized in order to distinguish different data in the platform database.

3.2 Web-Side Management System Implementation

The intelligent operation and maintenance platform of railway power supply system is developed by ASP.NET MVC framework, and the framework of the platform is divided into: data storage layer, data access layer, Repository layer, Service layer, Controller layer, View layer and infrastructure layer on the basis of its ASP.NET MVC framework combined with CQRS software design pattern.

4 Application and Testing of Intelligent Operation and Maintenance Platform for RTPSS

4.1 Equipment Health Assessment

The key equipment health management provides the current equipment status of the whole network, and the health assessment and life prediction of various types of equipment are mainly done through the assessment and prediction algorithm, equipment status data and assessment and prediction process.

Take turnout system as an example, firstly, the health status of turnout is divided into three categories: healthy, sub-healthy and faulty. And the overall situation of all turnouts at line or line network level is displayed on the interface, including the number and percentage of normal turnouts, faulty turnouts and sub-healthy turnouts. The details are shown in Table 1.

Table 1. Health record of point machines

Site	Turnout	State	Health value
Site 1	Turnout 1	Healthy	100
Site 2	Turnout 2	Sub-healthy	75
Site 3	Turnout 5	Fault	0
Site 4	Turnout 6	Sub-healthy	75
Site 1	Turnout 1	Healthy	100
Site 2	Turnout 2	Sub-healthy	75
Site 3	Turnout 1	Healthy	100
Site 4	Turnout 5	Fault	0

According to the evaluation of health degree, the faulty turnouts and sub-healthy turnouts are listed according to the severity, so that the users can carry out the hidden troubleshooting of the rutters according to the listed sub-healthy and faulty conditions, as shown in Fig. 1. From Fig. 1, we can clearly see the type and number of faults in the turnouts, and the most number of faults is fault 2, which is 60 times, accounting for 21.1%. The deep learning algorithm can analyze and predict the trend of the data in the figure, and the corresponding maintenance strategy can be proposed based on these data.

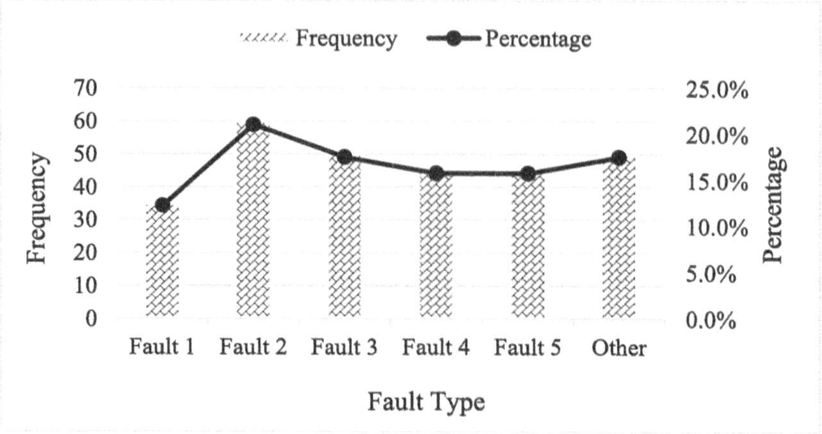

Fig. 1. Common fault distribution of turnouts

4.2 Equipment Fault Diagnosis Function

Intelligent fault diagnosis is realized based on real-time monitoring data and artificial intelligence technology, which no longer relies on human thinking to locate the cause of faults. It solves the contradiction between the rapid growth in lines and the lack of high-grade maintenance personnel.

Fault phenomenon: a metro vehicle returned to the depot at 10:30, the DDU did not show the fault, and the pantograph did not land.

By clicking on the intelligent fault diagnosis you can view the number and proportion of component failures that cause the DDU to be abnormal. Through the intelligent analysis module to calculate the a priori and a posteriori probability of the root node, to get a table of values, and then view the posteriori probability of failure of parts leading to DDU abnormal bar analysis chart, and can quickly determine the order of overhaul, as shown in Fig. 2.

According to the system display results, we can determine the inspection sequence as follows: LPTDRY relay, FBSK relay, lowering bow indicator, and position sensor, and the system prompts the post-test failure probability as [0.0854,0.1547], [0.0321,0.0411], [0.0125,0.0178], [0.0154,0.0151] respectively. Upon inspection, the FBSK relay is faulty.

The functional module of fault diagnosis aggregates 15 types of turnout fault problems, containing common problems such as relay faults and uncommon ones and other fault diagnosis models. After the online test of the fully automatic system and CBTC system, the accuracy of the fault diagnosis module for turnout fault diagnosis reaches more than 95%, especially for the troubleshooting of uncommon faults, the field guidance is better, and the efficiency of field troubleshooting and fault location is improved.

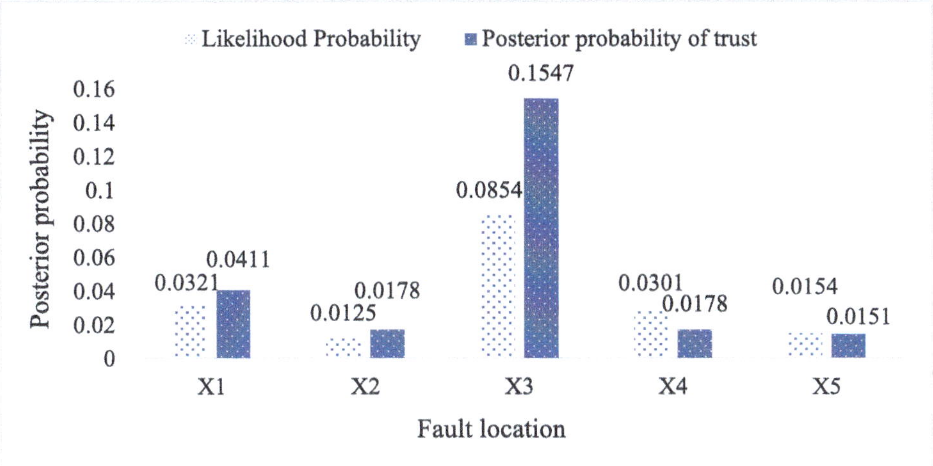

Fig. 2. DDU display anomaly analysis

5 Conclusions

The intelligent operation and maintenance platform can provide decision support for the maintenance of equipment of urban RTPSS and effectively improve the intelligent management of urban rail transit power supply equipment, making the operation and maintenance of on-site equipment more intelligent, real-time and accurate, and making the asset management of many equipments more economical and efficient. In view of the existing problems in the intelligent operation and maintenance mode of RTPSS, it is necessary to speed up the research on the application of intelligent operation and maintenance system and realize the management of the whole life cycle of equipment by promoting the construction of intelligent operation and maintenance system, improve the overall operation and maintenance level of equipment, prevent and reduce the occurrence of equipment failure, reduce the maintenance cost, and provide guarantee for the safety of metro operation, transportation efficiency and service quality. At the same time, it is also necessary to increase the breadth and depth of the application of intelligent operation and maintenance platform for RTPSS and gradually improve the level of intelligent operation and maintenance platform.

Acknowledgements. China's National University Student Innovation and Entrepreneurship Training Program-upgrading and Promotion of Intelligent Operation Inspection Platform for Urban Rail Transit Power Supply System (202112795003S).

References

1. Li, R., Shi, P.B., Li, M.K., et al.: Research on intelligent operation and maintenance platform based on automatic calibration system of electric energy meter. E3S Web Conf. **194**(2), 03001 (2020)

2. Xiao, B., Wang, W.: Intelligent network operation and maintenance system based on big data. J. Phys. Conf. Ser. **1744**(3), 032033 (2021)
3. Yu, S.D.: Research on cloud computing in the key technologies of railway intelligent operation and maintenance sharing platform. J. Phys. Conf. Ser. **1800**(1), 012010 (2021)
4. Hao, X., Li, B.: Research on intelligent system architecture of urban domestic sewage treatment. IOP Conf. Ser. Earth Environ. Sci. **450**(1), 012021 (2020)
5. Yu, F., Sun, M., Zhuang, W., et al.: Research on integrated operation and maintenance acquisition and monitoring technology for new generation dispatching control system. IOP Conf. Ser. Earth Environ. Sci. **467**(1), 012036 (2020)
6. Chen, X., Eder, M.A., Shihavuddin, A., et al.: A human-cyber-physical system toward intelligent wind turbine operation and maintenance. Sustainability **13**(2), 561 (2021)
7. Zhang, K., Tan, L., Chen, S., et al.: Research on intelligent operation and maintenance technology of primary equipment in substation. IOP Conf. Ser. Earth Environ. Sci. **769**(4), 042045 (2021)
8. Diao, P.H., Shih, N.J.: BIM-based AR maintenance system (BARMS) as an intelligent instruction platform for complex plumbing facilities. Appl. Sci. **9**(8), 1592 (2019)
9. Lt, A., Yh, A., Shuang, L.B., et al.: Application of photovoltaic power generation in RTPSS under the background of energy low carbon transformation. Alex. Eng. J. **60**(6), 5167–5174 (2021)
10. Yu, J., Wang, J., Tong, F.: Research and analysis of power supply load forecasting and self-healing control in urban rail transit system. IOP Conf. Ser. Earth Environ. Sci. **769**(4), 042093 (2021)
11. Wang, M., Yang, X., Zheng, T.Q., et al.: DC auto-transformer based traction power supply for urban transit rail potential and stray current mitigation. IEEE Trans. Transp. Electrification **6**(2), 762–773 (2020)
12. Li, E., Lu, Y., Mo, X., et al.: Research on non-catenary power supply technology. J. Phys. Conf. Ser. **1910**(1), 012037 (2021)

Innovation of Smart City Management System Based on Computer Application Technology

Xingfeng Liu[1(✉)], Guocheng Liu[1], and Hung-da Wan[2]

[1] International School of Education, Guangzhou Railway Polytechnic, Guangzhou, Guangdong, China
liuxingfeng@gtxy.edu.cn
[2] Trinity College Dublin, Dublin, Ireland

Abstract. In recent years, with the rapid development of information technology, the combination of mobile communication and Internet technology appears in all walks of life. The concept of smart city has gradually entered the daily life of community residents. Smart city is a new concept proposed by combining Internet of Things, cloud computing and mobile Internet technologies to achieve a safe, comfortable, convenient and efficient living experience. By investigating the current actual situation of smart city, this paper analyzes some specific requirements and overall development direction of smart city, and makes a detailed analysis of the functional requirements of the system. A smart city management system based on Java programming language and B/S architecture is designed. The design of the back-end management terminal of the system is based on B/S architecture and adopts the MVC framework. The whole system is divided into three basic levels: interface display layer, logical control layer, data access and data model layer. Through the lightweight Web container management work provided by the framework to achieve the management of smart city, greatly improve the management efficiency of managers. This paper conducts a pressure test on the smart city management system, and the test results show that both the system concurrency and the system resource occupation can meet the daily use requirements.

Keywords: Computer application · Smart city · Management system · MVC pattern

1 Introduction

With the construction of China's political system, community participation in urban management once became an important way to improve the level and efficiency of urban management [1]. With the advent of information and digital age, information technology has gradually become an important means of modern city management and governance. As the main means of urban management, the digital urban management mode provided a new solution to the dilemma faced by domestic urban management at that time, and gradually changed the inherent urban management mode and traditional ideas, which played a significant role in improving the efficiency of urban management and improving the level of public service [2]. At present, the community participation

I. Ahmad et al. (Eds.): STSIoT 2021, LNDECT 122, pp. 333–341, 2023.
https://doi.org/10.1007/978-981-19-3632-6_41

urban management mode and digital urban management mode are the most common in domestic cities. In order to solve the problem of urban development and realize the sustainable development of the city, it is bound to explore a new urban management model in accordance with the status quo of domestic cities. In recent years, the Internet, Internet of Things, cloud computing and other new-generation information technologies have emerged and developed rapidly. The "smart city" proposed by IBM has ushered in the construction of smart city [3]. With the upgrading of supporting facilities for smart city management, the inherent urban management mode is unable to meet the needs of the government and the public, and the transformation and upgrading of urban management mode under the background of smart city construction is poised for progress.

Foreign scholars started early in the study of city-related theories, with mature theoretical research on urban management, urban governance and urban management mode. After IBM put forward the concept of smart city, foreign scholars took the lead in the theoretical research on smart city construction and management [4]. Foreign scholars have studied and analyzed the urban management mode of France. The urban management of France focuses on the management of the urban environment, which is regulated by strict laws and standards and strong law enforcement means. Through the study of German urban drainage system, the importance of forward-looking scientific macro planning for urban construction and management is demonstrated [5]. Iranian urban management starts from strengthening infrastructure construction and improving public service level to improve the production and living environment of urban residents [6]. However, some foreign research contents are not applicable to China's national conditions. This paper will propose a smart city management system based on Java development according to China's actual situation.

This paper constructively integrates smart city construction and urban management, puts forward the view that technological innovation is the breakthrough point to promote the reform of urban management mode, and constructs a smart city management system based on Java technology, which provides a new perspective for urban management research in the new era.

2 Smart City Management System Based on Java Development

2.1 Smart City Management System Related Technology

(1) Java EE development technology

Java has the main advantages of object orientation, stability, robustness, cross-platform, etc. Compared with C++ language, Java has better memory processing mechanism and development convenience. For large enterprises or government departments, Web system development using Java language has become a mainstream trend [7].

SSH framework is the integration framework of Struts framework, Spring framework and Hibernate, and is widely used in the development of medium and large Web systems. Based on MVC hierarchical design pattern, Struts is responsible for the hierarchical division of the whole software system. Struts framework and Spring framework control business jump, and Hibernate framework is used for unified data management [8, 9]. The workflow of the SSH framework is as follows:

The presentation layer mainly completes the page display and user interaction. Through the configuration of Struts framework, the mapping management of front-end Request and control layer Action is realized by using Action Servlet.

The business logic is processed by using the Spring IoC container to interact with the Action class and provide data model components and data processing services.

Data persistence layer, through the use of Hibernate framework database entity object-oriented management, through the JDBC database driver encapsulation, complete the operation and reading of the data table, greatly simplifies the data management process.

(2) MVC design pattern

The hierarchical architecture and development mode of the system mostly adopt MVC mode, which is a mature development mode. MVC is mainly composed of three parts, M is the Model layer, which is the core content of the system and responsible for providing logical support for the system. V is the View layer, which is responsible for providing users with the interface display of the system and facilitating users to interact with the system; C is the control controller layer, which is mainly responsible for analyzing and calculating the logic of the system and controlling interface components and data forwarding [10, 11].

In essence, the MVC design pattern to maximize play the role of the controller, using the controller monitors the flow of business logic, and scheduling system resources, can be instantiated to specific processing business, so every business process essence is an implementation model aiming at specific problems, and according to the model logic to solve problems. At the same time, data interaction is also needed to solve problems. This structure has good functional scalability in various business processes, making business processing more flexible [12].

When using the MVC design pattern, display and the background model is independent of the interface, which is very suitable for the actual development mode, developers often focus on their work, workers of different content of developing in the whole development process are less dependent on the appearance of the work content, to ensure the whole system development orderly, each group have a clear division of responsibilities. It significantly reduces the development time of large-scale systems, improves the efficiency of system development, and ensures the quality of products. Figure 1 shows how the MVC design pattern works.

2.2 Smart City Management System Architecture Design

(1) System architecture

In the era of information sharing through the network, safe, flexible and elastic network data processing technology is gradually becoming the core of information. Application systems need to be constantly upgraded to increase functions and upgrade software and hardware to keep up with the needs of technological development. However, the traditional two-layer data processing method has many problems such as high maintenance cost, limited system expansion scope and weak data security, which can not meet the current needs. Therefore, the distributed Internet architecture must be adopted for system construction.

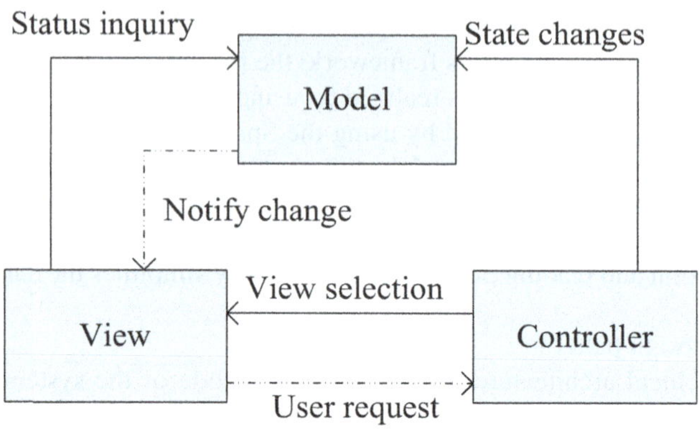

Fig. 1. MVC design pattern works

(2) Technical architecture

The system is mainly composed of basic platform and application system. The basic platform consists of urban management portal and many application systems, including wireless data acquisition subsystem, urban component online update subsystem, geocoding subsystem, video monitoring subsystem, vehicle positioning subsystem and so on. From the level of smart city management system, it is composed of data layer, platform layer and application layer.

The data layer means that the whole smart city management system is composed of countless data. The data forms the data layer through a certain sorting law. The data layer provides real data support for the system by providing data resources and managing data resources for the system. There are many kinds of data, usually including component data, inter data, various business data, basic geographic map data and other spatial and non spatial information. Its management function is mainly used to manage and maintain the above information, realize data stratification and data classification after expansion, and realize safe sharing on the system.

The platform layer refers to the basic information platform for digital urban management, which is composed of core module and expansion module. The main function is to support the operation of the system business application system as a platform. With the progress of science and technology and the upgrading of software and hardware, the platform layer also realizes the increase of system functions and the expansion of data.

The application layer belongs to an application system, which is mainly used to realize the expansion purpose of urban comprehensive management and comprehensive law enforcement management by relying on the platform layer, and to meet the needs of expanding urban management applications by relying on the development platform.

3 Smart City Management System Performance Test Experiment

3.1 Test Overview

The method used in this test is black box test. The test process is based on the user's use process, and the specific functions of different modules of the system are tested in detail for many times. In the specific test process, the background logic code is regarded as a black box, ignoring the internal implementation logic. Without considering the internal logic structure of the program, the tester tests according to the pre-designed test cases to check whether the function of the program is like the requirements analysis stage, and whether the program can give the correct results after obtaining the correct parameters. Then, according to the results, compare the records in the database and draw the test conclusion.

3.2 Performance Test

The purpose of this stress test is to test the performance of online teaching system so as to verify the server load. The project team simulates the number of concurrent users in the production environment as far as possible, conducts performance tests on the core business of the system, collects test results, and finally serves as the basis for the stable operation of the system and provides guidance for system tuning.

The relevant test formula is as follows:

Formula (1) is used to calculate the average number of concurrent users

$$C = nL/T \tag{1}$$

Formula (2) is used to calculate the peak value of the number of concurrent users

$$C' = C + 3\sqrt{C} \tag{2}$$

$$F = \frac{N_{PU} \times R}{T} \tag{3}$$

In Formula (1), C is the average number of concurrent users; N is the number of Login sessions; L is the average length of the Login Session; T refers to the length of time period investigated.

Formula (2) provides the calculation formula of the peak value of the number of concurrent users, where C refers to the peak value of the number of concurrent users, and C is the average number of concurrent users obtained in Formula (1). The formula is estimated by assuming that the user login session generation conforms to Poisson distribution.

4 Smart City Management System Performance Test Results

4.1 System Response Time Under Multiple Concurrent Users

Concurrent performance test simulates a situation in which a large number of users log in and submit a large amount of data at the same time, so as to verify the running status of the system under heavy load or even overload. By analyzing these running status, the bottleneck of the system can be found.

The concurrent tests were performed using Apache JMeter, a Java-based automated stress performance testing tool developed by the Apache organization. As shown in Table

Table 1. Multi-user concurrent test cases

	100	200	400	600	800
Login response time	0.21	0.28	0.42	0.75	0.83
Functional response time	0.27	0.33	0.46	0.79	0.95

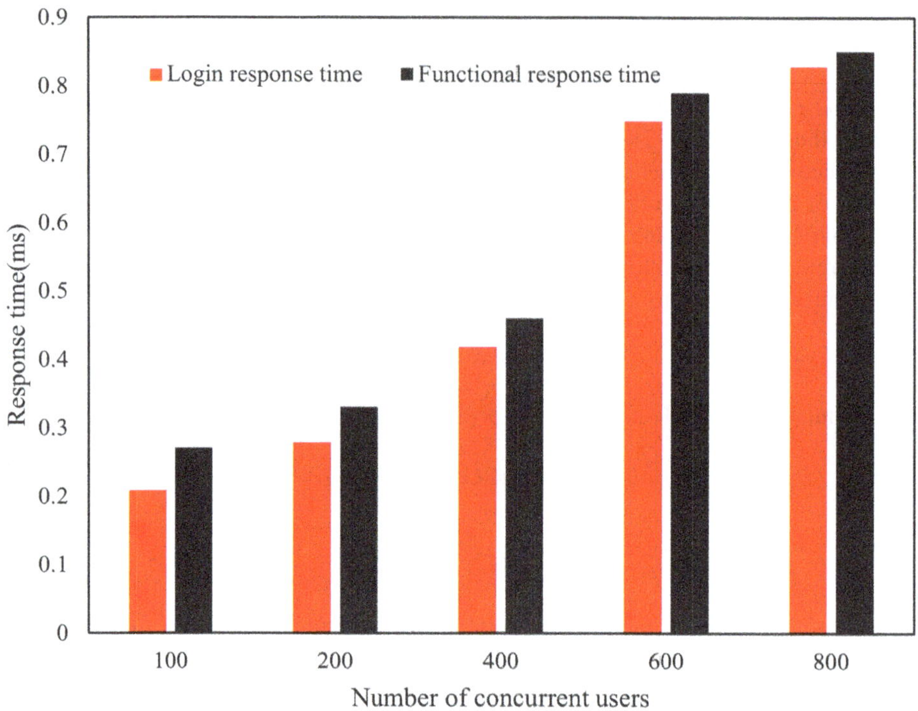

Fig. 2. Multi-user concurrent test cases

1 and Fig. 2, when the number of concurrent users is 100, the system login response time is 0.21 ms, and the functional response test time is 0.27 ms. When the number of concurrent users is 200, the system login response time is 0.28 ms, and the function response test time is 0.33 ms. When the number of concurrent users is 800, the system login response time is 0.83 ms, and the function response test time is 0.95 ms. The maximum response time of the system is more than 2 ms, which meets the requirements of daily use.

4.2 Multiple Concurrent Users Occupy System Resources

As shown in Fig. 3, when the number of concurrent users is 100, the CPU usage is only 7% and the memory usage is 11%. When the number of concurrent users is 200, the CPU usage is 16%, and the memory usage is 23%. When the number of concurrent users is 800, the CPU usage is 47% and the memory usage is 56%. In the case of multiple concurrent users, the CPU and memory usage of the system does not exceed 60%, indicating that the system designed in this document occupies less resources and conforms to the actual application situation (Table 2).

Table 2. System resource usage test

	100	200	400	600	800
CPU utilization	7%	16%	26%	35%	47%
Memory usage	11%	23%	29%	41%	56%

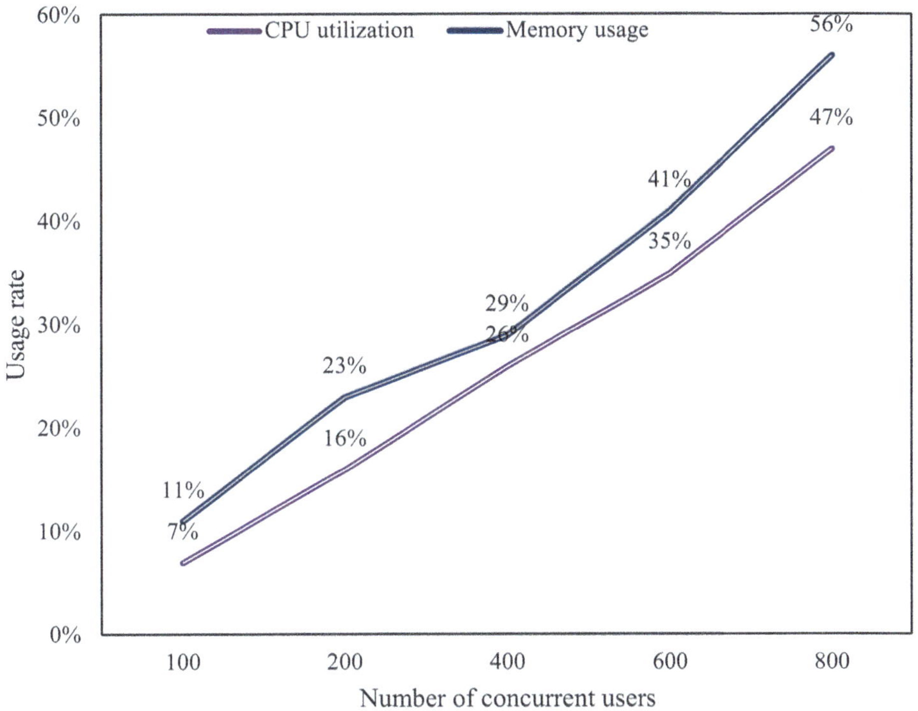

Fig. 3. System resource usage test

5 Conclusions

With the rapid development of information technology, the mature application of big data and cloud computing technology has promoted the process of smart city construction, and the construction of urban management innovation mode matching smart city construction has provided a good entry point for solving the current urban management problems. Based on the research and analysis of advanced urban management modes at home and abroad, this paper draws lessons from their successes and analyzes the shortcomings of traditional urban management modes at home and abroad, demonstrates each other, and proposes a smart city management system that can be practically applied to Chinese cities. The design of the back-end management terminal of the system is based on B/S architecture and adopts the MVC framework. The whole system is divided into three basic levels: interface display layer, logical control layer, data access and data model layer. Through the performance test of the system after construction, it can be seen that the system performance meets the requirements of daily use.

Funding. The authors have received financial support from the Programmes of Research and Practice Project of Innovation and Entrepreneurship Education for Computer Application Technology Major under Smart City Vision (2019KC213); Research on the Transformation Mode and Collaborative Innovation Mechanism of Scientific and Technological Achievements of Higher Vocational Colleges in Rail Transit under the Construction of "Guangdong-Hongkong-Macao Greater Bay Area" (2020GZQN62); Research on Key Technologies of Abnormal Behavior Recognition of Metro Crowd Based on Video Image Analysis (2017GKTSCX053).

References

1. Daely, P.T., Reda, H.T., Satrya, G.B., et al.: Design of smart LED streetlight system for smart city with web-based management system. IEEE Sens. J. **17**(18), 6100–6110 (2017)
2. Kovalev, D., Kulik, E.: The solution of environmental monitoring based on the integration of GIS technology in management system smart city. Interexpo GEO-Siberia **6**(1), 231–236 (2019)
3. Koukopoulos, Z., Koukopoulos, D., Jung, J.J.: Real-time crowd management for cultural heritage events: a case study on carnival parades. J. Ambient Intell. Smart Environ. **10**(3), 275–287 (2018)
4. Chen, L., Han, P.: The construction of a smart city energy efficiency management system oriented to the mobile data aggregation of the internet of things. Complexity **2021**(2), 1–13 (2021)
5. Yang, Z., Li, D.: WasNet: a neural network-based garbage collection management system. IEEE Access **8**, 103984–103993 (2020)
6. Ban, Y.: Energy decision making of steel company based on energy management system. IFAC-Pap. **53**(5), 608–613 (2020)
7. Luo, Q., Zuo, Z., Wei, H.: Application of a technology based on JavaEE in national rural environmental health monitoring information management system. Wei sheng yan jiu = J. Hyg. Res. **49**(6), 1014–1017 (2020)
8. Kronis, K., Uhanova, M.: Performance comparison of Java EE and ASP.NET core technologies for web API development. Appl. Comput. Syst. **23**(1), 37–44 (2018)
9. Sun, Z., Wang, Y., Zhang, L., et al.: Design and realization of intelligent service system for monitoring and warning of meteorological disasters in facility agriculture in North China. Nongye Gongcheng Xuebao/Trans. Chin. Soc. Agric. Eng. **34**(23), 149–156 (2018)
10. Sathyarajasekaran, K., Ganesan, R.: Effect of code generation in change impact analysis using MVC design pattern. J. Adv. Res. Dyn. Control Syst. **10**(10), 2417–2429 (2018)

11. Tremelling, A.M., Marley, R.A., Marley, M.B., et al.: Similarities between large animal-related and motor vehicle crash-related injuries. Wilderness Environ. Med. **28**(3), 213–218 (2017)
12. Juyuspan, F.A., Oktivasari, P.: Pengembangan website dinamis menggunakan ASP.NET MVC dan SQL server dengan metode RAD (studi kasus: PT X). InfoTekJar (Jurnal Nasional Informatika dan Teknologi Jaringan) **2**(1), 16–21 (2017)

Real-Time Vehicle Detection Based on YOLOv4 Neutral Network

Liping Lai[1(✉)], Han Wang[2], and Dashi Lin[2]

[1] School of Applied Technology, Hezhou University, Hezhou, Guangxi, China
laiIiping66@163.com

[2] School of Artificial Intelligence, Hezhou University, Hezhou, Guangxi, China

Abstract. With the rapid growth of people's living standards and national economic levels, the increase of per capita vehicles leads to the exponential boost in urban traffic congestion. To solve the existing problems of traffic congestion, a deep learning architecture based on yolov4 was proposed to realize monitoring of vehicles, which is used for the real-time detection and statistics of traffic stream information. The result shows that the mean average precision (mAP) of vehicle detection can reach 85% under different occasions of light, traffic flow and vehicle speed. The method has strong environmental adaptability and broad applicability.

Keywords: Deep learning · YOLOv4 · Vehicle detection

1 Introduction

At present, there are many problems in urban traffic management need to be solved urgently. For example, Traffic jams happened frequently in some sections with traffic accidents or improper management during the peak commuting period. To solve above problems, we can deal with them from two aspects: one is to speed up the construction of urban roads and improve the transportation efficiency. The other is to improve public awareness of traffic regulation. However, we may be limited by space and government financial allocation. Meanwhile, we cannot greatly improve public awareness of traffic regulation in a short time. Domestic and foreign researchers in artificial intelligence field have put forward a series of methods such as R-CNN and fast R-CNN series to detect vehicles in images. In recent years, YOLO series image vehicle detection methods have been put forward one after another. But different model architectures result in different accuracy and speed of detection. In this paper, we proposed the detection architecture based on YOLOv4 neutral network to detect vehicles in real time. The performance of detective speed and accuracy has been further verified.

2 Description of the Problem

2.1 Traditional Machine Learning

The steps of machine learning mainly include raw data acquisition, pre-processing, feature extraction, feature selection, recognition reasoning and prediction. The accuracy of results depends on pre-processing, feature extraction and feature selection.

I. Ahmad et al. (Eds.): STSIoT 2021, LNDECT 122, pp. 342–347, 2023.
https://doi.org/10.1007/978-981-19-3632-6_42

2.2 Overview of Deep Learning

With the further research of target detection, the current algorithms are mainly divided into two categories. One is the R-CNN series of Two-stage algorithms, including R-CNN [1], Fast R-CNN [2], Faster R-CNN [3], etc., the other is the one-stage algorithms, including SSD, YOLO series algorithms, etc.

2.3 R-CNN Target Detection Algorithm

R-CNN (Region-Convolutional Neural Networks) is an object detection algorithm based on sliding window. This algorithm transforms the problem of object detection scene into the problem of image classification. Through building windows with different sizes and proportions (the ratio of length, width, and height) to slide on the image from left to right, top to bottom in turn. According to the set step size, then using convolutional neural network directly to classify the image corresponding to each sliding window.

2.4 YOLO Target Detection Algorithm

The innovation point of YOLO (You Only Look Once) is that training and detection are both carried out in a single network. The specific horizontal, vertical coordinates and confidence rates can be inferred when the input image passes through one inference link. YOLO detection network is composed of 24 convolution layers for extracting features, two full connection layers for predicting image position and fully connected layer of attribute probability. YOLO network learns from GooGleNet classification network structure adopts 1×1 convolution layer, which has the advantages of edge lightening, cross-channel information blending, reducing parameter quantity and increasing model depth to improve nonlinear representation ability.

In train process, YOLOv1 [4] pre-trains the classification network on ImageNet. The pre-training classification network consists of the first 20 convolution layers, a pool layer and a full connection layer. Then, four convolution layers, two fully connected layers and random initial weights are added. The resolution of the input image is increased from 224*224 to 448*448 to meet the requirement of more fine-grained visual information during detection. YOLOv1's loss function is to optimize the output terminal with square sum loss.

Compared with YOLOv1, YOLOv2 [5] can detect more vehicles in different scenarios and expand the types of vehicles in urban roads by using ImageNet classification data set, which can detect up to 9000 types of targets; At the same time, YOLOv2's gradient increases and converges faster. Also, it can improve the generalization ability of the model effectively by using high-resolution training and learning from RPN's anchor boxes and prior boxes. The passthrough layer proposed by YOLOv2 samples and segments Feature Map and re-splices it, which is beneficial to the detection of small targets.

YOLOv3 [6] is a technology with constant speed and accuracy before YOLOv4, which has excellent detection performance. YOLOv3 is improved based on YOLOv2, in which sigmoid activation function is used to realize multi-classification detection, and feature maps of different convolution layers are fused for multi-level prediction.

The feature map sampled on the current layer and the feature map of the first layer are combined to get the combined feature map again, and then combined with the features extracted from other convolution layers, and so on, which can predict the target more accurately.

3 Vehicle Detection Based on YOLOv4

3.1 YOLOv4 Architecture

At present, YOLOv4 [7] target detection network model generally consists of four parts: Head, Neck, Backbone, and Input. Input refers to the input of the algorithm, including the whole image. Backbone is used to extract the partial structure of image features, which is mainly used to extract the shallow features of urban road vehicle images. The function of Neck module is to strengthen and train the whole features by processing the shallow features, so that the features learned by the model are the features I need. Head is the detection head, which is used to output the desired result. YOLOv4 model optimizes Backbone network to CSP-Darknet53 based on YOLOv3 and uses SPP as the feature fusion module of Neck, PANet as additional module of Head and swish activation function as the excitation layer.

3.2 Influence of Special Package on Training and Reasoning of Detection Framework

Bag of Freebies can be divided into three main types, Data augmentation, balance problems in object detection and GIoU loss. Firstly, the function of Data augmentation is to increase the diversity of sample training. The main ways of data enhancement are geometric enhancement and color enhancement. Geometric enhancement mainly includes rotation, stretching, random turning and random cutting. Color enhancement includes brightness enhancement, contrast enhancement and HSV spatial enhancement. Secondly, Imbalance problems in object detection is mainly used to solve the problem of data imbalance. There are two kinds of data imbalance, one is the imbalance between the identified object and the training background, the other is the imbalance between categories. GIoU loss is an improvement based on loss function. BBox of Glou loss is the overlapping part between BBox of predicted coverage and BBox of real coverage. This improved BBox replaces the denominator originally used in IOU loss. Glou loss has better accuracy (mAP) and convergence speed (FPS) in BBox regression. In addition, GIoU loss also involves the direction and shape of objects.

4 Experiment

The running experiment of the research was implemented under Windows 10 system which installed with OpenCV, Python, VS2019 and PyCharm. The system equipment was Nvidia Graphics card using CUDA, CUDNN.

The precision (P) is one of the performance evaluation systems, and its calculation method is shown in formula (1). In which TP (true positives) represents the number of instances (number of samples) that are positive examples and are classified as positive examples by the classifier; FP (False positives) is the number of samples that have not been successfully and accurately detected.

$$\mathbf{P} = \frac{TP}{TP + FP} \tag{1}$$

The mean average precision (mAP) refers to the average accuracy and is an index to measure the detection accuracy. MAP is equal to the sum of the average precision of all categories divided by all categories.

The Max P refers to the maximum accuracy and the Min P refers to the minimum accuracy.

Figure 1 shows the vehicle detection performance with good light, dim light, heavy traffic and fast vehicle speed. There are three categories in the inspection chart: car, bus and traffic light.

(a). Good light.

(b). Dim light.

(c). Heavy traffic.

(d). Fast vehicle speed.

Fig. 1. Detection of vehicles under four different conditions

Figure 1(a) shows that most vehicles were accurately detected under the good light condition, the accuracy rate was between 57% and 98%, with an average accuracy rate of 85%. Figure 1(b) shows that large amounts of vehicles were detected under the dim light condition, and the maximum accuracy was 99%, so the dim light has little influence on the vehicle detection method based on YOLOv4. Figure 1(c) shows that the traffic is heavy, and large numbers of vehicle bodies were blocked out, which led to the failure detection of vehicles with blocked bodies. Figure 1(d) shows that the speed of vehicles was relatively fast, which led to the complete deformation of some vehicles in the captured image. YOLOv4 model could accurately identify deformed vehicles with an accuracy rate of 90%. Which shows that the speed of vehicle has little influence on vehicle detection based on YOLOv4.

According to the results of vehicle detection under the above four different conditions, the experimental results showed that the greater the proportion of vehicles relative to captured images, the higher the accuracy. The unfavorable conditions such as dim light and too fast speed have little influence on the vehicle detection of YOLOv4. The accuracy of vehicle detection in four groups of captured images is shown in Table 1.

Table 1. Performance comparison between four conditions

Conditions	Serial number	Max P	Min P	mAP	Detection success rate
Good light	Fig. 1(a)	98%	55%	89%	75%
Dim light	Fig. 1(b)	99%	51%	88%	50%
Heavy traffic	Fig. 1(c)	95%	51%	60%	25%
Fast vehicle speed	Fig. 1(d)	94%	54%	62%	95%

5 Conclusions

In this paper, we used Real-time target detection based on YOLOv4 neutral network to detect vehicle. To verify YOLOv4's performance of detective speed and accuracy, we selected four groups of urban road vehicle images under different conditions for testing, such as different scenes, different angles, different road sections and different vehicle densities. The experimental results showed that the mAP of vehicle detection was 85%, which verified YOLOv4 has good applicability for vehicle detection in urban roads. Although this experiment can achieve better real-time detection of vehicle target, the performance of the algorithm is still affected by many factors, for example: the proportion of captured vehicle images is too small, and the part of vehicle body is blocked too much. In addition, how to use the backward adjustment of weights and gradients to reduce the loss function and increase the number of trained neural networks to achieve accurate detection of small targets are still something to be explored in future research.

Acknowledgements. This work was financially supported by the Hezhou University of scientific research project in 2019 (No. 2019ZZZK01), Undergraduate training programs for innovation and entrepreneurship (No. 201911838038 and 201911838122).

References

1. Girshick, R., Donahue, J., Darrell, T., et al.: Rich feature hierarchies for accurate object detection and semantic segmentation. In: IEEE Computer Society, pp. 580–587 (2014)
2. Girshick, R.: Fast R-CNN. arXiv e-prints (2015)
3. Ren, S., et al.: Faster R-CNN: towards real-time object detection with region proposal networks. IEEE Trans. Pattern Anal. Mach. Intell. **39**(6), 1137–1149 (2017)
4. He, K., Zhang, X., Ren, S., et al.: Deep residual learning for image recognition. In: IEEE Conference on Computer Vision and Pattern Recognition. IEEE Computer Society (2016)
5. Redmon, J., Farhadi, A.: YOLO9000: better, faster, stronger. In: IEEE Conference on Computer Vision and Pattern Recognition, pp. 6517–6525 (2017)
6. Redmon, J., Farhadi, A.: YOLOv3: an incremental improvement. Computer Vision and Pattern Recognition (2018)
7. Bochkovskiy, A., Wang, C.Y., Liao, H.: YOLOv4: optimal speed and accuracy of object detection. Computer Vision and Pattern Recognition (2020)

Analysis on the System of Single-Chip PWM Technology Controlling the Switching of Automobile Lighting Lamps

Long Yang(✉)

School of Electrical and Information Engineering, Anhui University of Technology, Maanshan 243032, Anhui, China
ynn0376@163.com

Abstract. In recent years, light pollution has become a very serious problem, and many nighttime traffic accidents are caused by rapid changes in the intensity of car headlights. As the rapid change of external light intensity can reduce people's ability to observe objects, cause human discomfort, and even damage the naked eye. Therefore, trying to achieve the brightness switching of car headlights by means of breathing lights is an important way to reduce traffic accidents. In this paper, we use microcontroller as the main control device and PWM technology to design a set of headlight state switching by breathing. Finally, the safety and security of night travelers are improved and the traffic accident rate is reduced.

Keywords: Microcontroller · PWM · Breathing light · Humanized design · Automotive lighting

1 Introduction

Automobile headlights are mainly divided into high beams and near lights. When the automobile is driving at night, the high beams of the headlights can illuminate the objects above 100 m in front of the automobile. The main role of high beam is to ensure that the driver can find the car in front of the distant pedestrians, vehicles and some special road conditions, to reserve enough reaction time for the driver, can let the driver take measures such as braking in time, to ensure driving safety. When the lighting condition is not good or when we pass the arch bridge or intersection, we can quickly switch the far and near light to remind the cars on the opposite side or at the intersection to ensure the safety of driving. Car headlamps dipped headlights lighting distance is not like high beam irradiation distance, low beam lights to ensure the car 30 to 40 m ahead in the distance you can see the road ahead and obstacles, and cannot make the lane to the drivers and pedestrians produces dazzle, ensure that the vehicle to change lanes, brakes and other safety operation. Traffic safety is a very important factor in people's daily travel, and in China, for example, more than 240,000 traffic accidents occurred in the year 2019, the vast majority of which occurred at night. At night, there are many factors that lead to traffic accidents. It has to be said that in the process of switching between the high beam

I. Ahmad et al. (Eds.): STSIoT 2021, LNDECT 122, pp. 348–357, 2023.
https://doi.org/10.1007/978-981-19-3632-6_43

and low beam of a vehicle, the rapid change in light intensity causes extreme discomfort to the naked eye of the person on the opposite side, or even brief blindness. Therefore, making the change of light intensity brought by the state switching of car lighting tends to be humanized and improve its comfort level, which becomes an important measure to improve traffic safety at night [1]. For most people, it is more willing to accept the gradual change of light intensity rather than rapid change. Therefore, the research in this paper is about the gradual change of light intensity achieved by the automotive lighting during state switching. In this regard, this paper investigates the possibility of using microcontroller to realize the lighting control system, and proposes a method to control the state switching of automobile lighting based on PWM technology of microcontroller. Finally, the Proteus simulation platform is built.

2 Microcontroller and PWM Dimming Principle Related

2.1 Introduction to Microcontroller

Microcontroller, also called monolithic microcontroller, integrates a computer system onto a chip and develops powerfully, mainly consisting of three parts, including memory, controller and operator, which can be regarded as a microcontroller. In actual development, programs are written in C, C++ and other program development languages to achieve different control functions according to the demand of control functions. All the written programs are burned and downloaded to the microcontroller, and finally stored in the memory, according to the command requirements to give control commands.

2.2 PWM Dimming Principle

PWM is Pulse Width Modulation, which uses a high-resolution timer/counter to encode the level of a specific analog signal by modulating the duty cycle of the square wave, which can make the light breathe effect. Different duty cycle produces different brightness, such as positive dimming: 10% very dark, 50% medium brightness, 90% very bright. Therefore, changing the duty cycle of PWM pulses can change the luminous brightness of the light [2]. And according to the human visual transient effect, the pulse frequency of PWM is set to 100 Hz, and the flicker is not perceptible to the vision, thus realizing the effect of breathing light.

3 Related Work

Bastien first proposed a vehicle-to-vehicle communication and ranging system using both headlights and taillights, the Visible light communication rangefinder. By exchanging the clock signals contained in the Manchester coded signal, the following vehicle and the car in front can share information and estimate the distance between them through phase shift measurements [3]. Bastien et al. first proposed a vehicle-to-vehicle communication and ranging system using both headlights and taillights – visible light communication rangefinder. By exchanging the clock signals contained in the Manchester coded signal, the following vehicle and the car in front can share information and estimate the distance between them through phase shift measurements [3]. Pavel et al. proposed a new average current control dc–dc LED driver suitable for automotive pixel (matrix) lighting. Current is sensed only on an integrated high side switch and peak current is controlled to achieve required average output current independent of tolerances of components and parameters other than the current sensing itself [4]. Jose-Luis studied motor drives. When decelerating or braking, swing the arm from the sides of the car through the switch on the brake pedal. Left or right turns are controlled individually by a switch on the clutch pedal [5]. Lee C S obtains a more reasonable deflection start time of the intelligent headlight system by analyzing the equal-illuminance light intensity envelope curve of the headlight, and determines the deflection start conditions of the system, so that the original light of the traditional headlight state can be more fully utilized, Avoid the problem of advance or lag in the deflection start time of the automobile AFS system [6]. In terms of control algorithm, Sevilgen G combines genetic algorithm to obtain a new improved particle swarm optimization algorithm. Through this algorithm, the parameters of PID control algorithm are determined, which effectively controls the overshoot and makes the system performance more stable. Finally, through the MATLAB/Simulink platform, the model of the automotive intelligent headlight system is established, and the simulation analysis and optimization are carried out [7]. Oh SK uses a combination of theoretical analysis, computer simulation and experimental testing to propose a feasible control scheme, which improves the performance of the intelligent headlamp system and is of great significance to improving nighttime driving safety. At the same time, it also provides a new idea for the control of the intelligent headlamp system, which promotes the technological progress of headlamps to a certain extent [8].

Although there are relatively abundant research results on the design of adaptive headlight systems in academia, there are relatively few studies that use the principle of breathing light to achieve gradual changes in intensity when the lighting status is switched and can be stabilized when normal brightness is reached.

4 System Functional Requirements and Chip Selection

4.1 System Function and Composition

In this paper, we use PWM technology to design a system based on a microcontroller to switch the headlights of a car in a breathing mode with the following functional design requirements.

Firstly, the external toggle switch generates a digital signal, which is used as the main basis for the working state of the car lighting; the digital signal is collected by the signal acquisition terminal and input to the designated microcontroller pin; the microcontroller analyzes the change of the signal to determine the state switching of the car lighting will be required; finally, the corresponding control is realized by the headlight control system, which requires that when the signal becomes high level, the low beam gradually goes off and the high beam gradually becomes brighter to achieve the effect of breathing light, and the final light tends to be stable; conversely, the low beam gradually lights up and the high beam gradually goes off, and the final light tends to be stable. In addition, a light intensity detection device needs to be added to the circuit to detect whether the lights are malfunctioning, so that the warning can be issued through the alarm device.

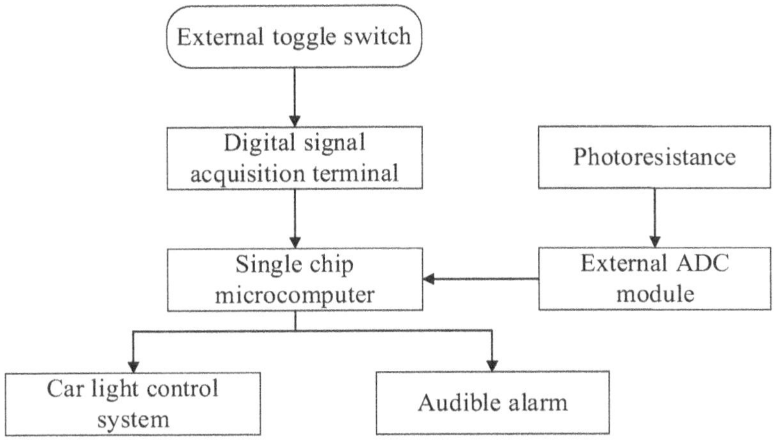

Fig. 1. Block diagram of the switching system scheme for PWM technology control of automotive lighting

According to the above requirements, the system scheme structure diagram is designed as shown in Fig. 1. In Fig. 1, the photoresistance and ADC module are responsible for detecting whether the light intensity is normal, and if it is, the microcontroller normally receives the digital signal brought by the externally operated toggle switch, and the headlight control system performs the corresponding function according to the instruction issued by the microcontroller to realize the breathing mode to switch the car lighting status.

4.2 Microcontroller Selection and Function Introduction

The STC89C52 single-chip microcomputer is a general-purpose single-chip microcomputer that not only has a relatively low development cost, but also is not susceptible to external interference. It consists of a serial interrupt, two external interrupts, three timing interrupts and four 8-bit parallel input ports. There is a clock circuit design inside the single chip microcomputer, which is composed of a quartz crystal and an external capacitor. The STC89C52 microcontroller mainly has the following features, as shown in Table 1.

In addition to the features in Table 1, there are some other features, such as low-power idle and power-down modes, the system can be woken up from idle mode during interrupts, fast programming features and power-down flags, and the most important thing is that it has a flexible in-system programming function, which is very useful for the design and development of the breathing lighting system. And this article chooses STC89C52 single-chip microcomputer as the control single-chip of the breathing switch car front lighting system, the following is the pin function.

Table 1. Main features of STC89C52 microcontroller

MCS-51 MCU instructions	Fully compatible
Erase cycle	1000 times
Voltage working range	4.0–5.5 V
Fully static working mode	0–33 MHz
Program memory lock	Level 3
Internal RAM	256 × 8 bits
Programmable I/O port	32
Timer/counter	Three 16 bits
Interrupt source	6

The pins of the single-chip microcomputer can be divided according to the different functions of the pins, which can be divided into 4 types in total, which are power, clock, controller, and I/O pins. The function of each pin of STC89C52 one-chip computer is shown in Table 2.

Table 2. Pin function table

Pin name	Function
VCC	Voltage
GND	Grounded
P0	Address/data bus multiplex port
P1	8-bit bidirectional I/O port with internal pull-up resistor
P2	8-bit bidirectional I/O port with internal pull-up resistor
P3	Input/output ports or other functions
RST	Reset function
ALE/PROG	Output pulse, external output clock or for timing purposes
PSEN	Read strobe signal of external program memory
EA/VPP	Control MCU to access internal and external program memory
XTAL1	Input terminal of oscillator inverting amplifier and internal clock generator
XTAL2	Output terminal of oscillator inverting amplifier

4.2.1 Power Supply

STC89C52 microcontroller has two power supplies, VCC and VSS.VCC refers to the chip power supply of the microprocessor, the operating voltage range is 4.0–5.5 V, and VSS is the grounding power supply.

4.2.2 Clock

The clock of the microcontroller is divided into two parts, including XTAL1 and XTAL2. XTAL1 refers to the input terminal of the oscillator inverting amplifier and the internal clock generator, and XTAL2 is the output terminal of the oscillator inverting amplifier.

4.2.3 Controller

STC89C52 microcontroller has 4 control lines, ALE/PROG, PSEN, RST/VPD, EA/VPP. ALE/PROG is the output pulse, external output clock or used for timing purposes; PSEN is the read strobe signal of the external program memory; RST/VPD is the reset function, in which RST is the input terminal for signal reset, and VPD is used to connect to standby power supply; EA/VPP is to control the microcontroller to access internal and external program memory.

4.2.4 I/O Pin

The microprocessor contains 32 I/O ports, divided into four types: P0, P1, P2, and P3. PO port is an address/data bus multiplexing port; P1 port is an 8-bit bidirectional I/O port with internal pull-up resistor; P2 port is an 8-bit bidirectional I/O port with internal pull-up resistor; P3 port is an input/output port or other functions.

5 System Design Specific Process

5.1 System Hardware Design

The 89C52 microcontroller is used to collect the signal brought by operating the toggle switch, and process the signal, and import the processed signal into the actuator. For the design of the chip hardware circuit connection, this paper is to use P2.2 as the input port of the control signal and P2.1 as the output of the control signal. The system circuit diagram is designed as shown in Fig. 2.

Based on the system circuit diagram in Fig. 2, the AT89C52 is used instead of the STC89C52 in the simulation diagram, considering the incomprehensive component library of the simulation software Proteus and the compatibility between each hardware. In addition to the reset and oscillation circuits in the microcontroller's minimal system, the circuit is divided into four main modules: photoresistor detection module, A/D conversion module, car light control module and alarm module.

5.1.1 Photoresistor Control Module

After connecting the two poles of the photoresistor to the voltage poles to establish a pathway, current flows in the circuit, and the output current changes when the light intensity is changed, and the output current is linearly proportional to the light intensity. Then the output current is converted to output voltage through a series resistor, and the output voltage can be used later as a check signal for the proper operation of the automotive lighting after A/D conversion.

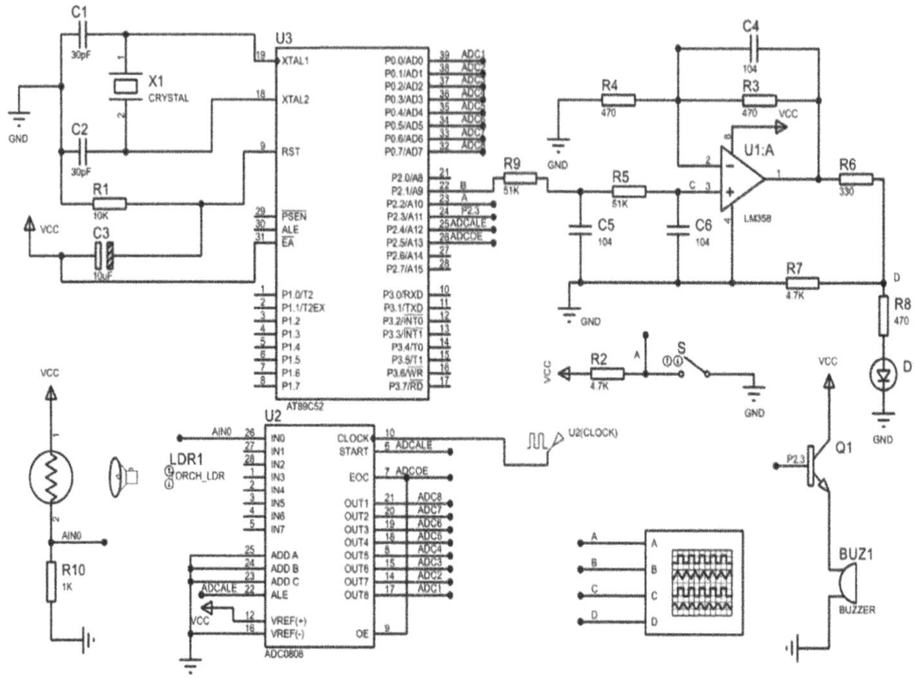

Fig. 2. System circuit design diagram

5.1.2 A/D Conversion Module

In this paper, the ADC0808 chip is selected as the A/D converter core, while the ADC0809 is commonly used in practical applications. As shown in Fig. 2, the main role of the chip is to convert the analog voltage value output from the photoresistor module into a digital quantity, which is received by the P0 ports of the microcontroller. The microcontroller analyzes whether the received digital quantity is within the normal range and thus determines whether the light intensity is normal.

5.1.3 Headlight Control Module

The microcontroller outputs PWM through port P2.1, which is filtered and amplified to eliminate external signal interference and obtain a stable PWM pulse waveform. This waveform drives the light-emitting diode through the D segment to simulate the car headlight [9]. Of course, usually, when switching the car headlights high beam and low beam light intensity changes in the opposite, which only need to add the opposite PWM pulse waveform can be. So the system circuit diagram designed in this paper may as well use a light-emitting diode to simulate the changing situation of the headlights, and finally realize the light intensity changes gradually and tends to be stable [10].

5.1.4 Alarm Signaling Module

When the microcontroller detects that the headlight intensity is not within the specified range, it indicates that the headlight control module is not working properly [11]. Using the P2.3 pin of the microcontroller to control the buzzer for alarm indication, a triode can be used to amplify the current and thus drive the buzzer. As shown in Fig. 2 shows the use of NPN-type triode connected to the buzzer, when the P2.3 port output high level, the triode conducts and the buzzer sounds, and vice versa the buzzer does not sound [12].

5.2 System Software Development

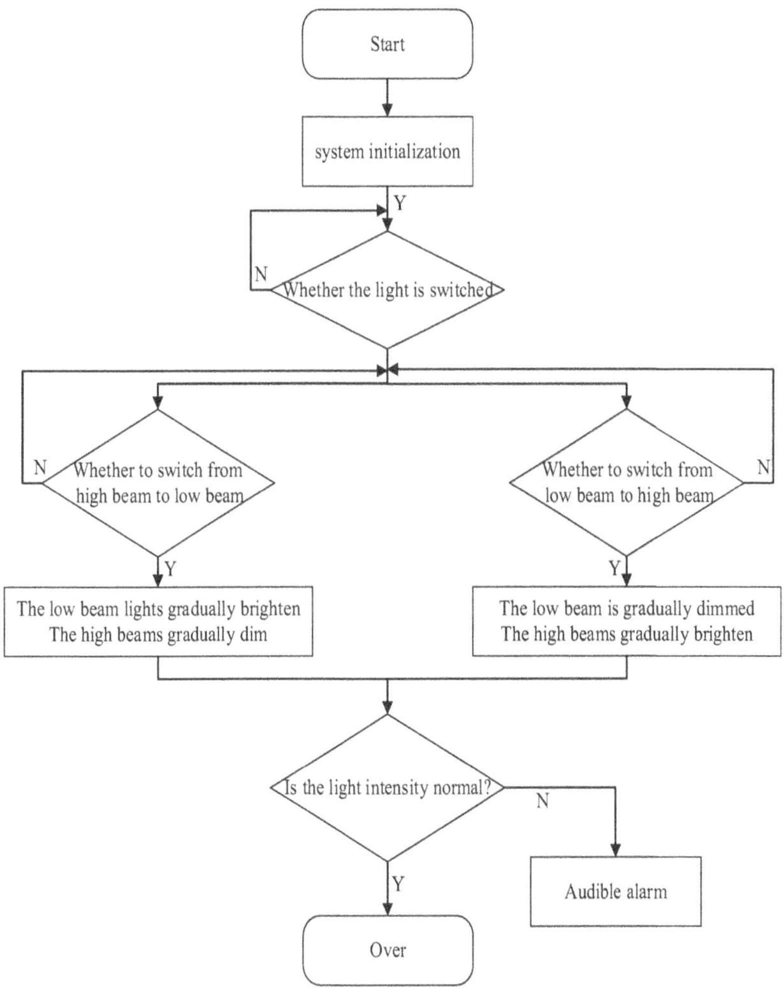

Fig. 3. System software flow chart

In this paper, the system software is developed using the upper PC. The microcontroller determines whether the signal generated by the external toggle switch turns on the high beam or the low beam, gives the headlight control command, and uses PWM technology so as to realize the state switching of the car lighting in the way of breathing light. The system software flow chart is shown in Fig. 3.

Step 1: System initialization settings.

Step 2: To determine whether the lighting occurs to switch, if there is a switch, the implementation of the third step; vice versa, continue to judge.

Step 3: Determine what state switching occurs.

Step 4: Wait for the headlight control system to complete the corresponding control.

Step 5: Detect whether the light intensity is normal, if normal, then end the current cycle and start to prepare for the next cycle; vice versa, the system issues a warning indication.

5.3 Related Algorithms

The calculation method of PWM frequency is shown in formula (1):

$$PWM = \frac{2T}{(psc + 1)(arr + 1)} \quad (1)$$

Among them, T is the clock period, PSC is the clock prescaler coefficient, and arr is the automatic reload value.

The formula of the light change form is as follows:

$$result = CCRX\, sin(param3.14/180) \quad (2)$$

Among them, result is the duty cycle value, and param is the radian value.

If there is a difference between the illuminance value and the expected illuminance value, it is expressed as e, and the rate of change between them is as follows:

$$\mathrm{ec} = de/dt \quad (3)$$

6 Discussion

This article will have a question here, since the goal of stabilizing the light intensity after gradually changing during the switching process between lighting states is achieved, how to solve the problem in some special situations that require the use of flashing lights? As we all know, flashing light is the driver manually switch back and forth between high beam and low beam at a certain frequency to achieve the effect of flashing light. A variable is set in the software development program, and the value of this variable is determined by the toggle switch switching frequency, and at the same time this value affects the PWM pulse working frequency of the microcontroller output. This operating frequency can determine how fast or slow the whole process of light intensity changes. When the time is too fast, there is no doubt that the naked eye is the effect of the flash. So when the owner uses the flash, the microcontroller will cooperate with the output of the PWM pulse needed to solve the above-mentioned problem.

7 Conclusion

In this paper, we have mainly studied the method of using microcontroller to output PWM pulse wave to achieve the state switching of car lighting in the way of breathing, so as to improve the comfort of human eyes when receiving strong light. In addition, this paper can be extended to automatically switch the high beam to low beam in the way of breathing light when the car turns, turns around and pedestrians appear in front. It is hoped that through the research of this paper, the state switching of car lighting will be more humanized; people travelling at night will have a sense of security; and traffic accidents will be reduced.

References

1. Gong, J., Liu, L., Zhou, Z., et al.: Design of infrared imaging system in front of vehicle. J. Phys. Conf. Ser. **1624**(5), 052006 (2020)
2. Dimming with PWM spotlights. Electronic Newspaper 25 July 2021
3. Béchadergue, B., Chassagne, L., Guan, H.: Simultaneous visible light communication and distance measurement based on the automotive lighting. IEEE Trans. Intell. Veh. **4**(4), 532–547 (2019)
4. Horsky, P., Plojhar, J., Daniel, J.: Adaptive peak average current control LED driver for automotive lighting. IEEE Solid-State Circuits Lett. **2**(9), 199–202 (2019)
5. Poza-Lujan, J.L., Sáenz-Peñafiel, J.J., Posadas-Yagüe, J.L., Conejero, J.A., Cano, J.C.: Use of receiver operating characteristic curve to evaluate a street lighting control system. IEEE Access **9**, 144660–144675 (2021)
6. Lee, C.S., Lee, J.H., Pak, H., et al.: Phantom array and stroboscopic effects of a time-modulated moving light source during saccadic eye movement. Light. Res. Technol. **50**(5), 772–786 (2018)
7. Sevilgen, G., Kiliç, M., Aktaş, M.: Dual-separated cooling channel performance evaluation for high-power LED PCB in automotive headlight. Case Stud. Therm. Eng. **25**(2021), 100985 (2021)
8. Oh, S.K., Lundh, J.S., Shervin, S., et al.: Thermal management and characterization of high-power wide-bandgap semiconductor electronic and photonic devices in automotive applications. J. Electron. Packag. **141**(2), 020801.1–020801.17 (2019)
9. Ozluk, B., Muslu, A.M., Arik, M.: A comparative study for the junction temperature of green light-emitting diodes. IEEE Trans. Compon. Packag. Manuf. Technol. **9**(10), 2024–2035 (2019)
10. Wu, S.-Y., Zeng, H.-J., Yang, L.-L., et al.: Research and design of automatic dimming system based on auto-dimming LED lamp. Big Technol. **000**(019), 241 (2019)
11. Yang, J., Wang, J., Wei, Y.H.: Design of vehicle light assist system based on Arduino microcontroller. South. Agric. Mach. **050**(022), 24 (2019)
12. Xiang, Y.: A new type of intelligent headlight design for automobiles based on microcontroller. Sci. Technol. Innov. Appl. **309**(17), 105–106 (2020)

Influence of Mobile Internet Based on Big Data Analysis on Integrated Marketing Communication Mode

Xia Hua[1](✉), Yan Bao[2], and Eleni Theodoraki[3]

[1] HaoJing College of Shaanxi University of Science and Technology, Shaanxi, China
huaxia822@126.com
[2] Xi'an Jiaotong University, Shaanxi, China
[3] Edinburgh Napier University, Edinburgh, UK

Abstract. The development of the mobile Internet is an important trend today, and the development of all areas of life is based on the mobile Internet. Integrated marketing is the integration of certain different resources into a powerful, competitive product package, which can produce synergy and aim to maximize the value of trade. Integrated marketing is a company that integrates independent marketing into a whole according to its own development needs, business goals, and its ability to achieve synergy and maximize the company's profits. The company's integrated marketing communication means that the company integrates all communication activities related to marketing. This article takes mobile internet as the theoretical basis of the research, and analyzes and studies the influence of its important content on the integrated marketing model. This article takes the classic mobile Internet as the research object and separately optimizes and improves the integrated marketing model. The technology in the mobile Internet can be used to construct a variety of integrated MLM models. The experimental results show that this research is very useful for the use of mobile Internet to influence the integrated marketing communication model, and it has better results in the study of the influence of integrated marketing communication methods observed from the mobile Internet.

Keywords: Mobile internet · Integrated marketing · Communication model · Model construction

1 Introduction

The rapid development of the mobile Internet has provided new opportunities and challenges for companies' integrated marketing communications. Products can become mainstream brands within months or even days. On the other hand, the rapid development of mobile e-commerce quickly conquered the traditional trading market. The impact on traditional trade is like broken bamboo [1]. A company or even an industry can be liquidated from the market in a short time. The development of the mobile Internet means that the era of mass communication in marketing is about to end, and modern marketing communication is entering the era of ultra-fast micro-communication [2].

I. Ahmad et al. (Eds.): STSIoT 2021, LNDECT 122, pp. 358–365, 2023.
https://doi.org/10.1007/978-981-19-3632-6_44

First, the mobile Internet implements a customer database based on the understanding of basic customer information and related mobile media usage patterns. In a sense, personal identification and blocking is accomplished through personal marketing [3]. Secondly, technological progress has made it possible for mobile terminal systems to integrate multiple communication methods. After all, mobile marketing can implement location-based push services at any time. In order to achieve the integration of mobile Internet marketing models, operators need to understand the characteristics of online store activities and create interactive product and service models [4]. To create a direct marketing model for the database, operators can use temporary advertisements to improve the accuracy of product promotion services. The development of the Internet shows that the traditional marketing communication model should be replaced by an integrated marketing communication model [5]. Therefore, the traditional marketing model must continue to innovate and develop new models. The integrated marketing model must protect itself from business risks. In the mobile network, the integrated marketing model is both an opportunity and a risk. Integrated marketing is a company that integrates independent marketing into a single program package according to its own development needs, business goals, and its ability to achieve synergy and maximize the company's profits [6]. The company's integrated marketing communication means that the company integrates all communication activities related to marketing. In the era of the development of mobile Internet, the market environment and consumer behavior are changing, and integrated marketing communication is given new content.

Under the development of mobile Internet, the mode of integrated marketing communication has changed a lot. This article mainly briefly describes the current situation of integrated marketing and the construction of integrated marketing mode under the development trend of mobile Internet [7]. It analyzes the communication methods and technologies in the mobile Internet platform, and summarizes its current social marketing methods to guide, support and prevent the company's integrated marketing communication. In the current Internet era, the target consumers of online marketing are changing dynamically [8]. You can actively select the information you need, view it in different locations at any time, and switch between different media or customers at different times. The same is true for synchronized media usage [9]. Their perceptions are selectively filtered, especially when switching from multi-screen to multi-screen, their needs and preferences will dynamically change. In short, the audience of online marketing ranges from passive to active, from one-way to multi-directional, from vague to clear. It is difficult for marketers to adapt to the current marketing communication environment with only one marketing method and traditional marketing methods. As consumers become an important part of marketing activities, integrated marketing communications have become mainstream. This article introduces the influence of the integrated marketing communication model in the mobile Internet era [10].

2 Method

2.1 Particle Swarm Algorithm

Use particle swarm algorithm to estimate the influence of mobile internet on integrated marketing communication mode, and define the objective function formula:

$$M(t) = m_{max}^{-\frac{t^2}{k}} \tag{1}$$

2.2 Select Samples for Estimation

The basic algorithm is a new product distribution model that considers both external and internal influences. It divides the users of the product into two groups: innovators and followers. They are only affected by the media and oral communication. The formula is as follows:

$$N(t) = M\left[\frac{1 - e^{-(p+q)}t}{1 + \frac{q}{p}}\right] \tag{2}$$

Among them, N(t) represents the total number of users from the issuance to time t, P is the degree of innovation, q is the speed of imitation, and M is the maximum market potential. In this article, he introduced the number of mobile Internet users when they are saturated. In the research, the sample t is randomly selected under infinite conditions, where i represents the internal attributes of the sample, and f represents the external activities (the number of iterations) of the sample, expressed by Formula 1:

$$v_t = \lambda\omega_t - I[y_{i_t}\{\omega_{t,}x_{i_i}\} < 1]y_{i_t}x_{i_i} \tag{3}$$

$$\omega_{t+1} \leq \omega_t - \beta_t v_t \tag{4}$$

Formula (3) is brought into formula (5):

$$\omega_{t+1} \leq \omega_t - \beta_\lambda\omega_t - I[y_{i_t}\{\omega_t, x_{i_i}\} < 1]y_{i_t}x_{i_i} \tag{5}$$

Simplified derivation:

$$\omega_{t+1} \leq (1 - \frac{1}{t})\omega_t + \beta_t I[y_{i_t}\{\omega_t, x_{i_i}\} < 1]y_{i_t}x_{i_i} \tag{6}$$

3 Methods and Experimental Research Design

3.1 Consumer Platform Marketing

The company needs to find the essence of the industry and the core consumption theme of users, and use the content of the communication as the basis of communication: in the era of the rise of the Internet, everyone is a medium of communication, resulting in a greater increase in the effect of communication, and the quality of content is important Has an obvious manifestation. Good-quality content and events are the main links to promote the brand. Marketing communication under the mobile Internet has many new development directions: First, the combination of products, mobile networks and economy. Use the mobile network to publish the fund-raising, let the designated masses participate in the product and project activities, and become a consortium of benefits, to maximize the positive interaction between the product and the target customer. Second, use the mobile Internet marketing platform. Marketing through WeChat, Weibo, APP-Client and other platforms is still a hidden marketing goal. The company disseminates its own information and product information to target audiences by updating the content of its platform. Finally, through various types of marketing communication platforms. The use of mobile device communication modes by enterprises should be based on business departments and user needs, and the content of messages should be used as the basis of communication. In the information age, everyone can be seen as a means of transportation, and the efficiency of information dissemination is rapidly increasing. The content of the message is very important and must be of high quality. There are many new models and technologies available for all mobile network marketing.

3.2 The Era of Fragmentation

In the era of mobile Internet, the all-round development of smart cities and smart communities allows consumers to choose multiple, repetitive and flexible ways. People's needs and social development have made the business district lose its main significance. Large-scale business districts have become small business districts, and they are about to evolve into the scale of units. The form of consumption is constantly changing. In the past, people's consumption may have traditional consumer business circles in fixed locations and regions. Nowadays, you can shop online, you can go to a store, you can go to a shopping place you own or have just appeared, or you can go to a small shop in the community that you aspire to. The same consumer can buy things through online stores, community stores, and large shopping malls. Business districts have become fragmented. Consumers' decision-making approach has undergone a fundamental change, not in unity, but in a mode of multiple choices. Marketing communication must be based on the level of consumer decisions, assuming, limiting, and choosing how to influence consumers, get close to shoppers, and touch new methods, new places, and new carriers to control the core of their consumption. Today's integrated marketing communication focuses on three types of people and does three types of things, that is, how to let the consumer group know, make shoppers approach the product, and let customers like the product. In the marketing situation of all channels, it must be on the consumer's decision channel. Through the effect of communication and interaction, everyone is not only a sales worker in the past, but also a carrier of brand communication today.

3.3 Pay Attention to Customer Discovery and Increase Loyalty

Internet finance providers must protect the rights of consumers. If there are any problems with product quality or customer service, they should actively correct them and accept comments and criticisms. Only by ensuring that most users have the right to know the truth can we gain the trust of more users and conduct fair online transactions. For silent customers who have used it before but gave up for a certain period of time or within a short period of time, the operator must conduct illegal customer acquisition marketing to retain customers and use online access to access these customers online. Be on time and greatly reduce the number of customer inquiries. You must keep abreast of new product releases, and you must carefully sell the company's flagship products to appease customers so that high-quality products can encourage consumers to buy. Active customers who need to use the store regularly and stable customers with high traffic and high business loyalty should repeat these users regularly. These customers are relatively active and loyal, and it is relatively easy to maintain the relationship between them. By encouraging these customers to independently share and test online store products with third parties, companies can develop in unexpected ways.

3.4 Experimental Investigation Objects

First of all, the investigation and research method is used to conduct detailed and in-depth investigation and research on the choice of each communication marketing method, research data, research rules, and refine and summarize the first-hand information. This paper selects a variety of Internet marketing communication platforms, and conducts a practical investigation and research on whether to choose to use mobile Internet technology from factors such as communication methods and marketing models.

Table 1. Questionnaire survey report

Questionnaire issuance and recovery	Marketing communication domestic platform	Marketing communication foreign platform	Total
Issue	88	66	154
Recycle	82	63	145
Effective	80	60	140
Efficient	90.1%	95.2%	90.1%

As shown in Table 1, in the investigation of the research report of the experiment, a total of 154 marketing communication platform experimental research questionnaires were reasonably selected and given, and a questionnaire survey was conducted on the 154 marketing communication platforms at home and abroad to evaluate each type of marketing. The dissemination of the platform is a factor of whether mobile internet technology is used. For this, we first need to study the various modes of integrating MLM in detail and thoroughly, and at the same time study the direct assimilation of data information in order to apply research and research methods, as well as to concretize and summarize specific cases. Secondly, it analyzes the use of case analysis method, which requires the communication mode of many marketing platforms in order to analyze the current situation of mobile Internet communication marketing mode through case studies.

4 Evaluation Results and Research

4.1 The Influence of Communication Mode on Online Marketing

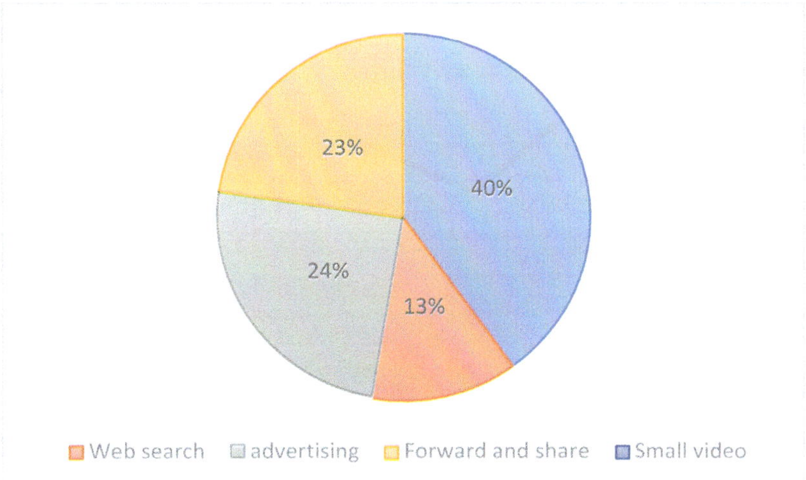

Fig. 1. Surveying the impact of communication models on network marketing

It can be seen from Fig. 1 that the mobile short video has a wide range of urban image dissemination, the arrival rate of information dissemination is more optimistic, and the user's contact rate is also good. Because this survey mobile phone end questionnaire is the majority, therefore has 40% proportion is the small video platform. Because this item is multi-topic, the viewing rate of its communication mode is relatively high. users like the elements with personalization and entertainment in mobile short video, and the urban image content is a pluralistic tendency about the content analysis of mobile short video users watching. this survey examines the recommendations of the most concerned people in mobile short video APP and the favorite content in mobile short video. The change of user's attitude and behavior is the last stage of communication and the most difficult and important stage. Whether the city image achieves good results in the process of mobile short video transmission.

Figure 2 shows that the proportion of Internet consumers has increased year by year, while the previous year has increased, but the growth rate in 2018 has been underestimated. As a result of the epidemic, the share of consumption rose sharply in 2020 and the level of consumption rose sharply, much higher than in 2019. On the other hand, our economy has reached a new level. The macroeconomic level and situation are consistent with the overall trend of the development of modern and innovative industries. Whether it is brand communication for customers, the marketing needs of different industries are appropriate and advertising accordingly. The diversification and commercialization of scenarios provide new opportunities for companies. The Internet has entered the second half of its development. As the demographic dividend falls, a further recovery in the stock

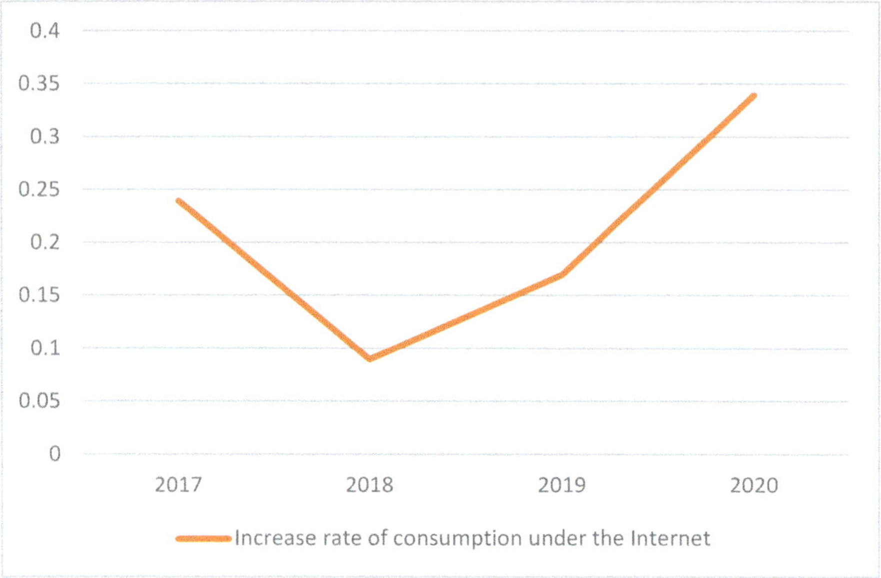

Fig. 2. Survey of annual consumption growth rates in mobile Internet traffic patterns

market is crucial. Internship is another breakthrough, internship marketing provides new development opportunities. At the same time, according to the technical conditions, the use of sensors, positioning systems are constantly accurate, wearable devices are popularized and offline infrastructure is constantly improved, which provides a new scene and data for scene marketing. In policy, the gradual improvement of laws and policies has created a benign development environment for Internet advertising.

5 Conclusion

Combined with the above, mobile media and application communication channels create a broad new communication experience by making communication and interaction more direct and decentralized. In the mobile Internet environment, traditional enterprise marketing has not been in line with the development of the times, integrated marketing communication has brought new models, new concepts, innovative communication mode, pay attention to the quality of communication content, pay attention to risk prevention, so that enterprises will improve market competitiveness and achieve long-term development. G mobile network and its technology, text, image, audio, video, website, email and audio-visual media features have been integrated into mobile media. By means of mesh communication mode and atomic separation, the product is combined with 3D mobile multimedia products, which can provide the public with improved and efficient information and services, and realize the comprehensive utilization of piecemeal viewing time. At the same time, it reconstructs the traditional understanding of social time, affects the structure and organization of people in their daily life, and stimulates new ways of communication and interaction. This proves that communication between target groups is reasonable and simple.

Acknowledgments. School-level Teaching Reform Project: Research on the Optimization of Practical Teaching System and Teaching Content Reform of Marketing Major Independent Colleges from the Perspective of Innovation and Entrepreneurship, Project No.: 19JGY02.

References

1. Saranya, N., Geetha, K., Rajan, C.: Data replication in mobile edge computing systems to reduce latency in internet of things. Wirel. Pers. Commun. **112**(4), 2643–2662 (2020)
2. Wu, H., Tian, H., Fan, S., et al.: Data age aware scheduling for wireless powered mobile-edge computing in industrial internet of things. IEEE Trans. Industr. Inform. **17**(1), 398–408 (2020)
3. Qian, L.P., Shi, B., Wu, Y., et al.: NOMA-enabled mobile edge computing for internet of things via joint communication and computation resource allocations. IEEE Internet Things J. **7**(1), 718–733 (2020)
4. Barolli, L., Xhafa, F., Javaid, N., Enokido, T. (eds.): [Advances in Intelligent Systems and Computing] Innovative Mobile and Internet Services in Ubiquitous Computing Volume 773 || Cognitive Informatics Approaches for Data Sharing and Management in Cloud Computing. IMIS 2018. AISC, vol. 773. Springer, Cham (2019). https://doi.org/10.1007/978-3-319-93554-6
5. Mogaji, E., Yoon, H.: Thematic analysis of marketing messages in UK universities' prospectuses. Int. J. Educ. Manag. **33**(7), 1561–1581 (2019)
6. Charoensukmongkol, P., Tarsakoo, P.: Dimensions of social media marketing capabilities and their contribution to business performance of firms in Thailand. J. Asia Bus. Stud. **14**(4), 441–461 (2019)
7. Tawafak, R.M., Romli, A.B.T., Arshah, R.B.A., Malik, S.I.: Framework design of university communication model (UCOM) to enhance continuous intentions in teaching and e-learning process. Educ. Inf. Technol. **25**(2), 817–843 (2019). https://doi.org/10.1007/s10639-019-09984-2
8. Sandiumenge, A., et al.: Donor referral from outside the intensive care unit: a multidisciplinary cooperation model using communication apps and redefining refereal criteria. Med. Intensiva (Engl. Ed.) **44**(3), 142–149 (2020)
9. Yang, Q., Hao, W., Ge, L., et al.: FARIMA model-based communication traffic anomaly detection in intelligent electric power substations. IET Cyber-Phys. Syst.: Theory Appl. **4**(1), 22–29 (2019)
10. Gu, X., Leng, B., Zhang, L., et al.: A stochastic geometry approach to model and analyze future vehicular communication networks. IEEE Access **8**, 14500–14512 (2020)

Design and Implementation of Intelligent Stadium System Based on RFID Technology

Zhong Wu[1] and Chuan Zhou[2(✉)]

[1] Physical Education School, Wuhan Business University, Wuhan, China
[2] Institute of Machanical Engineering, Wuhan Institute of Shipbuilding Technology, Wuhan, China
20150459@wbu.edu.cn

Abstract. With the emergence of intelligent buildings, as stadiums, the design of stadiums is also developing rapidly towards the direction of intelligence. Computer technology can not only help Venue Managers to complete their work, but also realize the information exchange between systems. Control technology can realize the automatic operation of various equipment. This paper discusses the design points of intelligent weak current system for stadiums. The results show that there are at most seven badminton fields and at least two volleyball fields in a city's functional core area.

Keywords: RFID technology · Intelligence · Stadiums · Sports

1 Introduction

With the development of information technology, intelligent system organically combines the structure, system, service and management of stadiums and gymnasiums to make their functions more reasonable. Modern intelligent stadium is more and more safe, efficient, comfortable and convenient, which reflects the characteristics of information society. The intelligent construction of gymnasium can not only meet the needs of physical education and sports activities, but also meet the needs of sports meeting, assembly and other large-scale activities.

With the continuous development of science and technology, many experts have studied the stadium. For example, some domestic teams have studied the stadium operation and maintenance management system, introduced the characteristics of the stadium, and designed the composition of the intelligent system according to the characteristics of the stadium. Combined with the distributed GIS special database, the stadium intelligent service system based on WebGIS is studied. Through the in-depth study of various traditional optimal path algorithms, an improved algorithm is proposed. Then, on the basis of in-depth analysis of users' travel mode, this paper designs a sports venue service system based on WebGIS to realize the optimal path of barrier free. A new crowd detection method based on gray correlation matrix (GLDM) is proposed. GLDM is a technique for measuring image texture. This paper proposes a new feature to describe GLDM, and applies AdaBoost and Bayesian classifier to the new feature, and tests it

I. Ahmad et al. (Eds.): STSIoT 2021, LNDECT 122, pp. 366–374, 2023.
https://doi.org/10.1007/978-981-19-3632-6_45

in the actual stadium scene [1]. Some experts have studied the intelligent management system of comprehensive stadiums and gymnasiums, analyzed the characteristics and advantages of intelligent lighting system of stadiums and Gymnasiums in detail, and focused on the glare problem that must be solved in the design of stadiums and gymnasiums. The influence mechanism of physical factors on human visual response was discussed. In the aspect of glare suppression, a series of countermeasures are put forward from the aspects of lighting design, lamp selection and installation. Based on OpenGL's 3D graphics construction ability and machine learning's powerful intelligent processing ability, this paper analyzes the stadium data, carries out 3D modeling and reconstruction, and constructs a 3D sports game automatic interpretation system. In the process of 3D reconstruction of stadium, candidate region extraction and pattern recognition are used to recognize the athletes and targets, and then the athletes are tracked by Calman filter. Finally, the information is used to reconstruct the stadium in OpenGL, and the corresponding interpretation system is added. The system composition, schematic diagram, hardware and software design of the measurement network are introduced. An intelligent wireless wind speed measurement network based on ZigBee technology is designed, which can measure the wind speed at any place in the stadium according to the predetermined measurement scheme [2]. Some experts have studied the development of stadiums under the trend of big data, and introduced the characteristics of stadium safety design. Combined with the requirements of current stadium operation management, this paper discusses the operation mode of stadium safety management. Through the comprehensive security design of the stadium, the unified management of video monitoring system, intruder alarm system, parking system and access control system is realized on the central platform of the security system. Through the combination of Internet, the operation and promotion of intelligent stadium can be realized, and the traffic and utilization rate of stadium can be improved. By mapping the input data to the appropriate bit pattern and creating a structure through which the data flows according to the results of various and/or masking operations, the data itself has a content function. According to the actual needs of stadium electronic accounting system, C language is used as programming language to realize the automation and intelligence of stadium electronic accounting system. A city stadium service system based on web is designed. Firstly, this paper introduces the key terms of system design, and uses MVC design pattern and four tier B/S structure to complete the system architecture design. According to the needs of users and the application of the system, this paper expounds the design of functional modules and subsystems, and realizes the intelligent development of urban sports facilities. The composition and construction steps of the National Indoor intelligent system are planned and analyzed. This paper introduces the design overview, design points and design functions of each system, which provides a model for the construction of intelligent system of sports facilities [3]. Although the research on stadiums and gymnasiums is fruitful, there are still some deficiencies in the design of intelligent stadium system based on RFID technology.

In order to study the design of intelligent stadium system based on RFID technology, this paper studies the intelligent stadium system based on RFID technology, and finds out the public stadium service quality evaluation index system. The results show that RFID technology is conducive to the design of intelligent stadium system.

2 Method

2.1 Intelligent Management System

(1) Intelligent motion management system
"Intelligent sports management system" is an application system which combines human intelligence with computer technology and applies it to sports management. Through computer simulation of human intelligence, it constantly realizes sports management activities on all kinds of equipment except human, so as to reduce human labor intensity and gradually realize the replacement of artificial computer [4]. The essence of intelligent management system is to process and exchange all kinds of information in daily work, and transfer it through known text, language, image and other carriers [5].

(2) System management
With the development of applications and the increasing amount of data information, how to ensure the security of data through the system maintenance function is a key problem [6]. When the computer software and hardware failure, through the backup way to restore the database, you can restore the damaged database files or control files or other files. Intelligent system management is to connect goods with the Internet through information sensing equipment according to the agreed protocol to realize information exchange and communication, so as to complete intelligent identification, positioning, tracking, monitoring and management of information [7, 8]. Intelligent system management is an extension based on Internet. Intelligent system management will realize the interconnection of people and things, things and things, information resources and intelligent services, and further realize the integration of the world and the virtual world [9]. The connection between people does not rely on the network to realize the communication between people; the connection between people and things is that people use a series of equipment and instruments to realize the connection between people and things [10]. Intelligent system management mainly realizes the connection between people, but does not realize the connection between objects. Therefore, the connection between objects is called intelligence.

(3) Research on the service quality evaluation index system of public sports venues
In order to simplify the analysis, the specific evaluation criteria should be energetic, the indicators should be concise, the respondents should make clear and concise judgments, some professional terms should be reasonably explained, and the quality of the questionnaire should not be affected by subjective judgments. In the evaluation of the service quality of public sports venues, due to many evaluation indexes, the weight of each index is different. Therefore, in order to reflect the service level of public stadiums more truly and accurately, each evaluation index should have a hierarchical structure.

2.2 RFID Technology

(1) RFID technology concept
RFID technology in English is called radio frequency identification, namely radio frequency identification, also known as electronic tags. RFID technology is a kind of non-contact automatic identification technology using radio wave to transmit data. It has the characteristics of fast recognition, non-contact, high degree of automation, multi-target recognition and so on. Electronic tags are mainly used to store the data of object marking information. The reader writer is used to write specific information to the label memory and read information from the label. The antenna is used to transmit and receive radio frequency signals to complete the data exchange between reader and tag.

(2) The role of RFID Technology
Using Internet of things technology to realize the perception and positioning of display objects is the current popular development direction. RFID technology is the key technology to achieve target positioning management. In the real world, we can track and manage the specific position changes of objects through 3D positioning in the real world, so as to realize the tracking and query of object state. At present, there are three kinds of positioning technology. One is GPS positioning, one is mobile base station positioning, the other is infrared positioning. The development of indoor positioning technology has always been an important part of positioning technology research. RFID technology is one of the hotspots of current research. Its basic principle is to use the uniqueness of RFID tag to realize the perception and positioning of objects in a specific location, including RF collector, tag and identification technology. The complexity of RFID related systems and the cost of data reading and writing, electronic tags and other components mainly depend on the working frequency of RFID related systems. This attribute can also determine the final RFID distance of the RFID system. In the whole RFID system, application software system is an indispensable part, which controls all behaviors of reader and tag. In a specific application, the application software system will actively send commands to the reader, and the reader will respond to the commands, including setting its own parameters or returning the corresponding data to the application software system. These data will be further processed by the application software system.

2.3 Evaluation Index System of Service Quality of Public Sports Venues

Firstly, the average value of each row of the standardized judgment matrix is calculated to obtain the eigenvector of the matrix, that is, the weight of each index. Then the maximum eigenvalue λ Max is calculated (1):

$$\lambda_{\max} = \frac{1}{n}\sum_{i=1}^{n} \lambda_1 \tag{1}$$

Due to the complexity of qualitative problems, it is difficult to guarantee the consistency of conclusions to a great extent. Therefore, the consistency judgment index (CR) is introduced into AHP to detect the consistency of judgment (2):

$$C_1 = \frac{\lambda_{\max} - 1}{n - 1} \tag{2}$$

Iaqi index of PM 2.5 is directly used as the evaluation index of air quality index. The calculation formula of iaqi is shown in (3):

$$S = \frac{\sum_{i=1}^{N} D}{N} \tag{3}$$

Carry out troubleshooting according to the abnormal condition of indoor air environment. Taking temperature as an example, the calculation formula of average temperature is shown in Eq. (4):

$$LAQI_p = \frac{LAQI_{Hi} - LAQI_{Lo}}{BP_{Hi} - BP_{Lo}} \tag{4}$$

3 Experience

3.1 Extraction of Experimental Objects

The system uses wireless module communication, if the communication module sends or receives data at the same time, there will be serious data conflict. Therefore, the system uses polling communication, so that at the same time, only one wireless module is working. The data base station sets the corresponding number of polling signals according to the number of data card swiping terminals adopted by the system. The data base station sends a polling signal (adjustable) to a terminal every 3S. After receiving and confirming the polling signal, the data acquisition terminal sends the stored student number information to the database, and then the database sends the student number information to the server for processing. The database layer depends on the server layer. The server sends the previous data request to the database layer (after the transformation of a database interface). The database carries out a series of read and write operations according to the data information, and feeds back the results to the service layer, which communicates with the browser layer.

3.2 Experimental Analysis

In the development of the system, we first write the code of the overall architecture of the system, and then create the functional modules of the system, including the front-end interface layer, business object processing layer and data layer. After the overall structure of the system is constructed, the specific functional modules of the system are coded. Using RFID technology, manufacturers, wholesalers and retailers only use the unique electronic tag card number corresponding to the product number to identify the identity of products on the basis of establishing strict product sales channels, and are compatible with the shared database system. The anti-counterfeiting system based on RFID technology will change the existing data management system to ensure the legitimacy of products.

4 Discussion

4.1 A Sports Events of Public Stadiums in the Functional Core Area of the City

According to the statistics of the Sports Bureau of a certain city, there are 31 Public Stadiums in the functional core area of a city. In the investigation stage, Yuetan gymnasium in a city is not open to the public because of its comprehensive transformation. Through the summary of the sports items of the public stadiums in the functional core area, the sports items provided by the public stadiums in the urban functional core area are divided into seven categories: badminton, basketball, swimming, tennis, table tennis, volleyball and billiards. As shown in Table 1.

Table 1. A summary of sports events in public stadiums and Gymnasiums in functional core area of a city

Sport event	Number of venues
Badminton	7
Basketball	6
Swimming	3
Table tennis	4
Billiards	4
Volleyball	2
Tennis	5

It can be seen from the above that there are 7 public sports venues providing badminton, 6 basketball, 3 swimming, 4 table tennis, 4 billiards, 2 volleyball and 5 tennis in the functional core area of city A. The results are shown in Fig. 1.

It can be seen from the above that the public sports venues in the functional core area of city a provide badminton sports at most 7 seats and volleyball sports at least 2 seats.

4.2 Statistics of the Number of People who Often Participate in Sports

With the rapid development of economy and the continuous improvement of people's living standards, people pay more and more attention to sports and health. In addition, the State Council and the State General Administration of sports have issued many documents to advocate and strongly encourage national sports fitness, which shows that the sports industry has been booming. Public information shows that from 2012 to 2018, the number of people who often participate in sports in China has been increasing, as shown in Table 2.

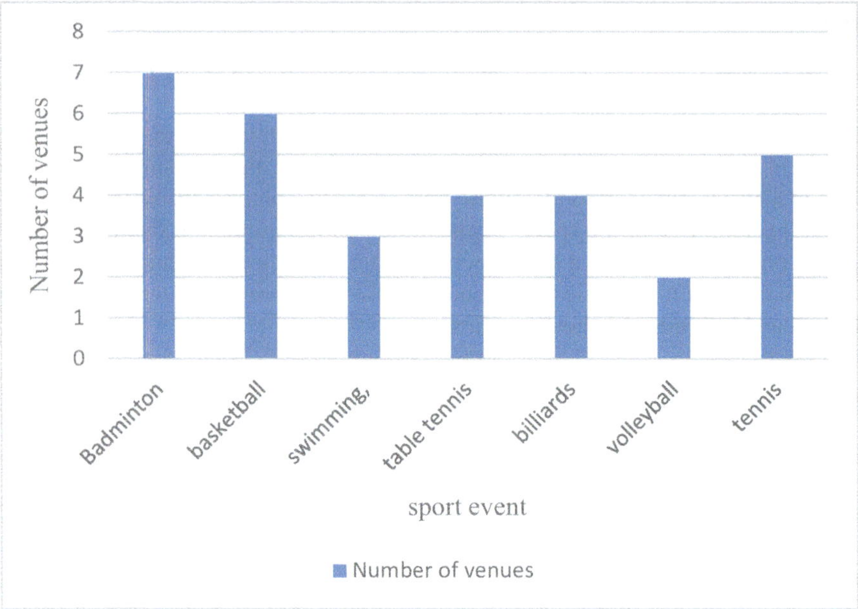

Fig. 1. A summary of sports events in public stadiums and Gymnasiums in functional core area of a city

Table 2. The number of people who often participate in sports in China from 2017 to 2020 (10000 people)

Particular year	2017	2018	2019	2020
Number of participants	3237	4356	5347	5579

It can be seen from the above that the number of people who often participate in sports in 2017 is 32.37 million, the number of people who often participate in sports in 2018 is 43.56 million, the number of people who often participate in sports in 2019 is 53.47 million, and the number of people who often participate in sports in 2020 is 55.79 million. The results are shown in Fig. 2.

It can be seen from the above that the development trend of the number of people who often participate in sports increases with the increase of years.

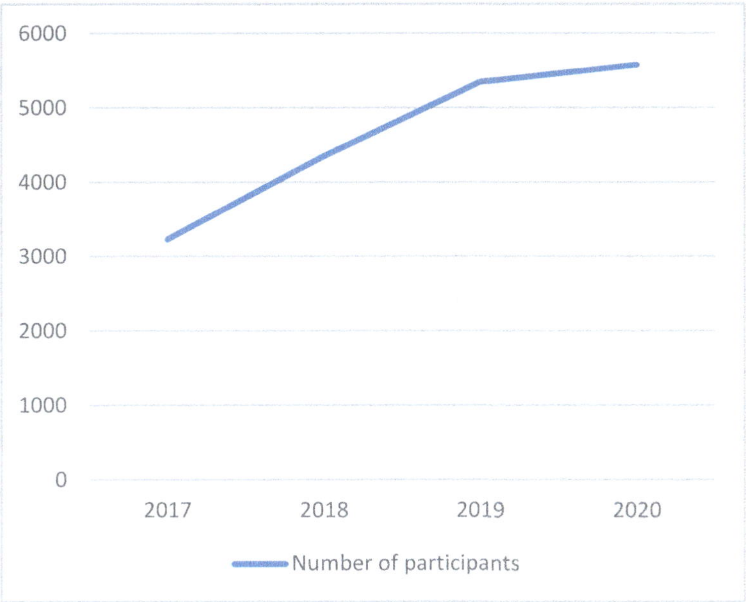

Fig. 2. The number of people who often participate in sports in China from 2017 to 2020 (10000 people)

5 Conclusion

With the development of China's economy, intelligent system is becoming the main body of more and more construction projects. The key technologies of intelligent monitoring system for stadium engineering are introduced. This paper expounds the quality control scheme of the intelligent system of stadium construction engineering in detail, which ensures that the system can meet the design requirements and industry standards in the aspects of construction, system debugging and testing. In this paper, it has become a main trend to optimize the design and development of stadium intelligent system. This paper analyzes the design requirements of stadium intelligent system based on Internet of things technology, and expounds the design and development of stadium intelligent system from the aspects of system overall structure, system function and Internet of things architecture. The results confirm the feasibility of developing and designing intelligent stadium system based on Internet of things. This will help to realize the intelligent management of stadiums and improve the management efficiency of intelligent stadiums.

Acknowledgements. This work was supported by Wuhan Application Foundation Frontier Project (2020010601012294) and the School-level Academic Team Project of Wuhan Business University (2018TD011).

References

1. Zhou, F., Cao, L., Ma, Y.: Design of intelligent stadium management system based on ASP.NET. Bol. Tec. Tech. Bull. **55**(4), 340–346 (2017)
2. Dinham, P.: ZTE, China mobile claim first 5g sporting event broadcast. Exchange, pp. 8–8, 13 August 2019
3. Schmidt, M., Schülke, A., Venturi, A., et al.: Cyber-physical system for energy-efficient stadium operation: methodology and experimental validation. ACM Trans. Cyber-Phys. Syst. **2**(4), 1–26 (2018)
4. Zhang, Q., Yang, Z., Xiao, Z.: The 3D reconstruction of the football stadium and automatic game commentary system based on opengl. Bol. Tec. Tech. Bull. **55**(6), 722–729 (2017)
5. Guo, Q., Li, B.: Role of AI physical education based on application of functional sports training. J. Intell. Fuzzy Syst. **2**, 1–9 (2020)
6. Consoli, F.A., et al.: Smart event traffic management: impact on the central Florida regional transportation network and lessons learned. Transp. Res. Rec. **2396**(1), 107–116 (2018)
7. Tallman, T.N., Hassan, H.: A computational exploration of the effect of alignment and aspect ratio on alternating current conductivity in carbon nanofiber-modified epoxy. J. Intell. Mater. Syst. Struct. **31**(5), 756–770 (2020)
8. Guo, C.: Research on pre-competition emotion recognition of student athletes based on improved machine learning. J. Intell. Fuzzy Syst. **39**(4), 5687–5698 (2020)
9. Mischa, S., Anett, S., Alberto, V., et al.: Cyber-physical system for energy-efficient stadium operation. Acm Trans. Cyber Phys. Syst. **2**(4), 1–26 (2018)
10. Beato, M., Jamil, M., Devereux, G.: The reliability of technical and tactical tagging analysis conducted by a semi-automatic VTS in Soccer. J. Hum. Kinet. **62**(1), 103–110 (2018)

Application of Computer Augmented Reality (AR) Technology in Landscape Architecture Design

Jing Zhao[1(✉)] and Charith Perera[2]

[1] Xi'an Jiaotong University City College, Xi'an 710018, Shaanxi, China
c18227598420@163.com
[2] University of Bologna, Bologna, Italy

Abstract. With the fast advancement of science and technology, internet and information technology are increasingly being integrated into people's professional and personal lives, bringing ease. The environment, or landscape, is a vital link in the urbanization process; it not only creates a pleasant living environment for people, but it also contributes considerably to the restoration of ecological balance, which is necessary for future urban expansion. This article examines the features and growth trend of computer augmented reality (AR) technology, as well as its potential use in landscape architectural design, with the goal of providing other perspectives for China's urban development.

Keywords: Computer Augmented Reality (AR) technology · Landscape architecture · Design practice

1 Introduction

With the popularization of the Internet and the advent of electronic information age, virtual technology has been applied to all walks of life, but with the continuous development of virtual technology, its disadvantages are gradually found by people. People just for the purpose of application of the technology to meet their own needs, want to have a real environment is more detailed and comprehensive observation, avoid because of some objective factors lead to observation of the environment have gaps, and is not want to take advantage of the virtual environment as a substitute for the real environment, but due to some technical problems, such as equipment running speed and the quality of the modeling, Not only can not restore the real scene, but also will make the image distortion, resulting in information loss, leading to the user's experience is not ideal, limit the user's observation and cognition of the environment. Therefore, in order to improve this problem, people invented a new technology–augmented reality technology. Augmented reality (AR), as an extension of VR technology, is to superposition virtual information formed by computers to real scenes, so that people can enhance their experience of the real world in visual, auditory, tactile and other aspects. It has the characteristics of virtual and real combination, interaction and infiltration.Computer augmented reality

I. Ahmad et al. (Eds.): STSIoT 2021, LNDECT 122, pp. 375–381, 2023.
https://doi.org/10.1007/978-981-19-3632-6_46

(AR) has been studied and debated extensively by scholars for a long period of time, but the technological circumstances at the time were insufficiently advanced, and hence the discussion is purely technical. he continuous advancement of software and hardware technology in recent years has created an excellent opportunity for AR technology. Because AR technology simultaneously enhances the real - world environment and virtual elements, Augmented reality has proven to have great potential in learning due to its high level of visibility and interaction. Therefore, it has gained widespread interest in a variety of fields. The application of computer augmented reality technology in garden design builds a bridge between reality and virtual environment. It's not only the plant characteristics for optimal collocation that make a landscape design unique; there are more to it. as long as aesthetics are considered. However, since landscape engineering consumes resources and cash, it must prioritize early design and scheme investigation, which is why AR technology expands the possibilities for landscape design by enabling the development of multi-angle environmental space.

2 Related Work

Numerous local and international specialists have successfully completed research on augmented reality technology in architecture and landscape design.

Juan Manuel Davila Delgado examined the potentials for AR and VR applications in architecture, engineering, and construction by combining qualitative and quantitative data collection and analysis. The findings indicate that, while AR and VR technologies have not been fully adopted by the construction industry, there is great interest within and between construction companies in investing in these technologies [1]. Francis Baek analyzed the user's interior location and orientation to create an AR system for facility management by comparing the user's perspective to deep learning-based building information modeling [2]. Philipp R.W. Urech created the landscape with a laser scanning point cloud model, opening up new avenues for scenario generation and testing using place-based metrics [3]. Nathalie Wergles conducted a preliminary research to determine the use and market for computer visualization for architecture and landscape design, which enables construction workers to have a better understanding of the spatial structure of the given buildings [4].

In his design research, Baranidharan Raman made extensive use of machine learning methods. The algorithm's application in pedestrian landscape design was proved by demonstrating its usage in the construction of pedestrian landscapes and testing it in a traffic simulator [5]. Adam R used augmented reality to construct a map-based flood visualization tool. Flood risk assessment and early action were made possible via the use of 3D visualization tools, minimizing the possibility of flood-related catastrophes [6]. Kelly Hamilton investigated and described the use of three-dimensional visualization design for contemporary landscape design via the use of mapping linkages specified by the principle of projection [7].

Despite substantial research into augmented reality technology in architecture and landscape design, there would be a need to apply science and technology to complex interior design difficulties in architectural settings.

3 Computer Augmented Reality: Characteristics and Development

3.1 The Technical Characteristics of Computer Augmented Reality

As the name implies, AR technology is used to supplement the physical world by rebuilding it. A virtual architecture may be built in response to user requirements throughout the expansion process. Simultaneously, it should collaborate with the relating photography and 3D modeling technologies in order to accomplish the goal of integrating the virtual and physical worlds, which is more visible at the visual level. AR technology is a subset of VR technology; the primary difference is that AR enables interaction between the physical and virtual worlds, simultaneously displays two different types of information from the two worlds, and can improve and tweak based on the information, allowing for more "creation" and "design" on the basis of the real world.

3.2 The Development of Computer Augmented Reality (Ar) Technology

Landscape design employing 3D image processing techniques may include preprocessing 3D landscape pictures to reduce noise and unnecessary information and then enhancing the clarity of landscape design images using 3D landscape image feature improvement approaches [8]. Furthermore, 3D-GIS allows the study of past human landscapes and the representation of extinct human landscapes through virtual models [9].

American scientists developed a robot to imitate street view in the late 1950s. However, it falls within the category of virtual reality technology, which is primarily concerned with simulating the actual world and is unable to "make extra" in accordance with the real world. However, it created quite a stir at the time. Although technology and associated goods are under continual iteration as part of the process of computer updating and iteration, they have not yet crossed the "threshold" of virtual reality. The term "augmented reality" first appeared in the mid-1990s, although it was still in the process of aggregating the region of the fundamental layer. It was not until around 2018 that augmented reality technology started to gain traction, owing to significant improvements in the performance of computer software and hardware. Simultaneously, the widespread usage of high-frequency wireless networks (5g) enabled the development of augmented reality technology, which became extensively employed in corporate and civilian applications. While the use of computer augmented reality technology in landscape design is yet experimental and requires more in-depth study and debate, there is little question that it has a bright future.

3.3 The Advantages of Computer Augmented Reality (AR) Technology

A digital landscape refers to a high-fidelity simulation model of a system that is coupled with PC hardware and software systems; it may be used to control rainfall, store water (recycling), and design CAD green space. New artificial wetland systems, Innovative space design, and water-saving garden rainwater (inlet) storage system design are all technical measures for plazas that can contribute to the long-term advancement of urban landscape environment and new plaza rainwater regulation (inlet) storage system [10].

Fig. 1. Landscape produced by AR technology

At the moment, many shopping malls include virtual experience halls where customers may play games in the virtual world using visual gadgets. Computer AR technology is the "merger" of the virtual and real worlds; this "merger" may provide a greater feeling of space creation in the experience. The benefits are mostly represented in the following areas. To begin, we may merge the actual and virtual worlds and apply 3D registration tracking technology to make the two seamless, allowing for information interchange and interconnection. Second, AR has moved away from the traditional "keyboard + mouse" setup in favor of a 3D system that emphasizes a more natural connection between machine and human. Finally, computer augmented reality technology provides the experimenter with a better sensation of realism and the benefit of mental immersion, which is also the most important element of AV technology in achieving the aim of "confusing the real with the false." AR technology, as seen in Fig. 1, may be utilized to create a more realistic simulation of landscape design.

4 Application of Computer Augmented Reality Technology in Landscape Design

4.1 Traditional VR Technology's Drawbacks

Traditional VR Technology's Drawbacks

In other words, in the actual world, virtual reality technology is unable to simulate 3D space. Instead, information must be gathered and organized in the real world before being modeled in the virtual one. The use of 3D computer and graphics technology, as well as 3D reality technology, is required for this strategy. In other words, it is essential to put at a lot of effort and time in the beginning phases, but modeling in the later stages

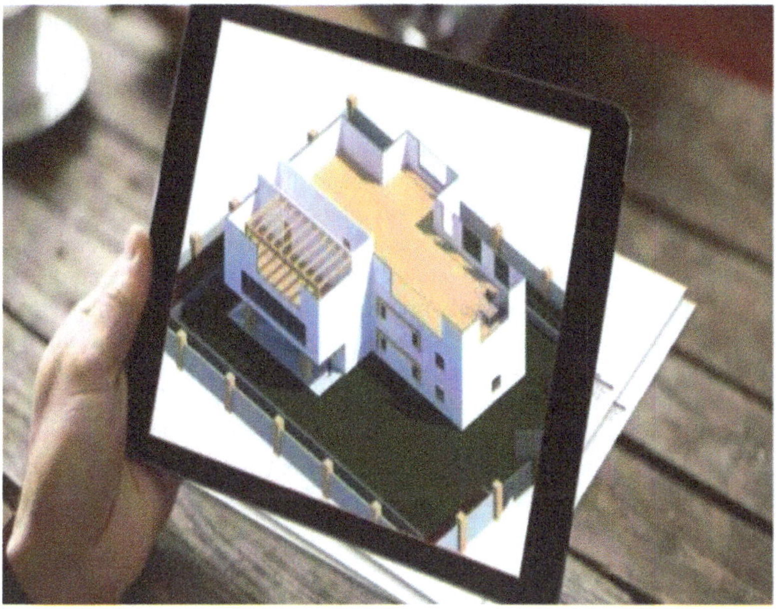

Fig. 2. The field landscape produced by AR technology

is also crucial. Because of the early modeling, it's important to note that the picture's refresh rate is vital. A complicated virtual environment with high software and hardware requirements needs constant refinement and modification in order to achieve adequate image quality. Figure 2 shows how landscape architecture design may be simulated using AR technology.

4.2 AR Technology Application Characteristics in Augmented Reality

In comparison to virtual reality, augmented reality technology does not need prior modeling and offers a better feeling of realism. It can mirror all things in the actual world and is unaffected by their quantity or size. Additionally, landscape design often incorporates sophisticated elements, such as not just architectural designs, but also a variety of flora. If unified digital modeling is performed, the task is enormous, and the time required is likewise lengthy. With the blessing of augmented reality technology, all types of plants and structures can be genuinely presented, and the integration of actual and virtual data can be accomplished efficiently, enhancing the design of landscape architecture while decreasing burden and the cost of time.

4.3 The Trend of AR Technology in Landscape Architecture Design

Landscape architecture is inextricably related to the growth and planning of cities. It is not only the city's finest "card," but also a critical livelihood initiative aimed at improving the people' surroundings. China is through a critical era of societal change. In the midst of fast urbanization, we should also consider the pace with which landscape architecture is constructed. However, since the project includes a large number of people, resolving the issue is challenging. This requires continuation of the preliminary plan discussion, and computer augmented reality (AR) that utilizes human-machine interaction to fuse the virtual and real worlds, thus bringing a large number of people together in an environment of virtual reality with creation, and discusses detailed issues. This has a profound effect on the industry's growth. Simultaneously, as compared to virtual reality technology, it not only saves modeling time and severe hardware requirements, but also successfully prevents any details deviations during conventional scheme discussion. It can dynamically collect and effectively preserve visitor behavior fragments through the spatial interface, enabling multi-party conversation schemes to be shown simultaneously in the creative space, so achieving effect synchronization and landscape reconstruction. Furthermore, with the advancement of current technology and science, AR for computer augmented reality is reliant on the incorporation of additional technologies, such as data gloves, panoramic scanning, and mobile scene. This can enhance the experience and will eventually bring in more people to participate.

5 Conclusion

To summarize, computer technology is a significant social development trend, while landscape architecture is a building project that aims to enhance the living environment. Relevant staff should pay close attention to the benefits of AR technology, investigate, invent, and use AR technology to landscape design in order to generate chances for the growth and development of our nation, and continually seek methods to penetrate the Internet industrial avenue.

References

1. Delgado, J.M.D.: et al. A research agenda for augmented and virtual reality in architecture, engineering and construction. Adv. Eng. Inf., **45** (2020). https://doi.org/10.1016/j.aei.2020.101122
2. Baek, F., Ha, I., Kim, H.: Augmented reality system for facility management using image-based indoor localization. Autom. Constr. **99**, 18–26 (2019). https://doi.org/10.1016/j.autcon.2018.11.034
3. Urech, P.R.W., et al.: Point cloud modeling as a bridge between landscape design and planning. Landscape Urban Plann., **203** (2020). https://doi.org/10.1016/j.landurbplan.2020.103903
4. Wergles, N., Muhar, A.: The role of computer visualization in the communication of urban design—a comparison of viewer responses to visualizations versus on-site visits. Landscape Urban Plan. **91**(4), 171–182 (2008). https://doi.org/10.1016/j.landurbplan.2008.12.010
5. Raman, B., Naderi, J R.: Computer based pedestrian landscape design using decision tree templates, Adv. Eng. Inform., **20**(1), 23–30, ISSN 1474–0346 (2006). https://doi.org/10.1016/j.aei.2005.08.002
6. Tomkins, A., Lange, E.: Interactive landscape design and flood visualisation in augmented reality. Multimodal Technol. Interact. **3**(2), 43 (2019). https://doi.org/10.3390/mti3020043
7. Hamilton, K.: Mapping spaces: three-dimensional visualisation and design in landscape architecture. https://doi.org/10.14236/ewic/EVA2017.56

8. Shan, P., Sun, W.: Research on landscape design system based on 3D virtual reality and image processing technology, Ecol. Inform., **63**, 101287, ISSN 1574–9541 (2021). https://doi.org/10.1016/j.ecoinf.2021.101287
9. Stewart, E.E.R.: 'The Royaltie of Sight': A 3D-GIS recreation of 'prospects' and 'perspectives' within an English designed landscape, c.1550–1660. DH (2017)
10. Yang, L.: Computer simulation of urban garden landscape design based on FPGA and neural network, Microprocess. Microsyst., **83**, 103988, ISSN 0141-9331 (2021). https://doi.org/10.1016/j.micpro.2021.103988

Protocol Adaptive Conversion Method of Power Transmission Internet of Things Terminal Based on Protocol Matching

Cheng Chen[1], Kang Jiao[1], Letao Ling[1], Zhenhua Wang[1], Yuan Liu[1], and Jie Zheng[2](✉)

[1] Shenzhen Comtop Information Technology Co., Ltd., Shenzhen 518000, Guangdong, China
[2] China Southern Power Grid Shenzhen Digital Grid Research Institute Co., Ltd., Guangdong 518000 Shenzhen, China
284421644@qq.com

Abstract. With the development of power transmission Internet of Things communication technology, higher requirements are put forward for data transmission speed, and various terminal protocols are proposed according to different needs. There is an increasing demand for communication between devices using different protocols. Therefore, it is necessary to seek an efficient and universal protocol adaptive conversion method to achieve this goal. The purpose of this article is to study the protocol adaptive conversion method of power transmission IoT terminal based on protocol matching. This article starts from the protocol grammar matching, gives a multi-pattern matching algorithm, and gives a detailed description of the overall design scheme and interface design of the adaptive conversion system. Finally, this paper tests the methods proposed in the research, introduces the implementation of simulation software NS-2, NS-2 and its simulation process, and uses adaptive mechanism, PCF mechanism and DCF mechanism to perform delay and packet loss rate performance for different services. The simulation results show that the delay time of the conversion service under the PCF mechanism is within 50 ms; before the simulation time is 55 s, the delay time of the conversion service under the DCF mechanism is within 50 ms and within 60 s; the conversion service delay time of the adaptive mechanism is all within 20 ms. It can be seen that the adaptive conversion method proposed in this paper shows better QoS performance than the standard PCF and DCF mechanisms.

Keywords: Power transmission and transformation equipment · Internet of Things · Communication protocol · Adaptive conversion

1 Introduction

With the rapid development of computer network technology, both industry and life scenes are no longer just connecting things with things, nor just connecting people with things [1, 2]. The Internet of Everything is already a world development trend, and the way of connection is also complicated depending on the form, that is, the terminal

I. Ahmad et al. (Eds.): STSIoT 2021, LNDECT 122, pp. 382–390, 2023.
https://doi.org/10.1007/978-981-19-3632-6_47

protocol of the Internet of Things [3, 4]. The combination of power transmission and transformation equipment and the Internet of Things technology can realize the needs of comprehensive, real-time and accurate information acquisition of power transmission and transformation equipment, and provide technical support for equipment intelligence and power grid equipment modernization [5, 6]. To achieve the interconnection of all things in the power transmission Internet of Things, it is bound to be the integration of many protocols, that is, protocol conversion, and the protocols used in different scenarios are naturally different. Research on an adaptive conversion based on protocol matching and flexibly configured to adapt to any protocol the method becomes the task that must be completed [7, 8].

In the research on the Internet of Things and protocol conversion, many scholars have discussed it. For example, Lee SW believes that the correct use of relevant Internet of Things technology can effectively improve economic benefits and further reduce operating costs [9]; Greene E pointed out that individuals can connect with users and communicate with society and the entire environment in the entire smart space through the Internet of Things [10]; Xu Z pointed out that because the Internet of Things contains thousands of objects, it not only involves sensors, radio frequency identification and software development, but also involves many communication networks, such as radio and television, telecommunication networks, etc. In actual use, further analysis of related applications can be more effective in adapting to the entire society and life [11].

The purpose of this article is to study the protocol adaptive conversion method of power transmission IoT terminal based on protocol matching. This article starts from the protocol grammar matching, gives a multi-pattern matching algorithm, and gives a detailed description of the overall design scheme and interface design of the adaptive conversion system. Finally, this paper tests the methods proposed in the research, introduces the implementation of simulation software NS-2, NS-2 and its simulation process, and uses adaptive mechanism, PCF mechanism and DCF mechanism to perform delay and packet loss rate performance for different services.

2 Protocol Adaptive Conversion Method of Power Transmission Internet of Things Terminal Based on Protocol Matching

2.1 Architecture of the Protocol Adaptive Conversion of the Power Transmission Iot Terminal Based on Protocol Matching

(1) Overall system architecture

The specific protocol conversion task is performed in the data conversion module in the power transmission Internet of Things terminal, and the conversion module can be configured by the microprocessor to make it suitable for the corresponding environment. The content of the configuration includes the type of conversion protocol and the priority corresponding to each protocol.

The architecture is roughly composed of 4 parts, configuration system, protocol priority FIFO, data conversion, load balancing data output. Because the logic of the system is solidified into hardware, the conversion system can be configured flexibly through a PC or embedded micro-processing system before the system runs. Tell the protocol priority FIFOG which protocols in the scene need to be opened, and which port is the corresponding one. According to the characteristics of each protocol data, the priority is set according to the use requirements. The data conversion module then obtains the final protocol type that needs to be output, and processes the data of each port in an orderly manner under the setting of priority polling.

(2) Data input buffer

The main function of this module is to prioritize the signal data frames of different protocols. The first is the ordering between different protocols, and then the ordering between the same protocols. The ordering of the same protocol will be set to 4 levels. Priority FIFO then parses and sorts each data frame according to the configured information [12]. Finally, output according to priority.

(3) Protocol conversion based on five-level pipeline hard core

The main function of the data coding and decoding module is to extract the valid bits of several protocol data and to frame the data to be sent. The main function of the rate matching processing unit is to match communication protocols of different rates. The data conversion module is mainly to convert specific data formats. When the data formats of the two different protocols are converted between each other, the communication rate must be consistent.

The conversion architecture is based on the configuration of the system to complete the protocol conversion between zigbee protocol, Bluetooth protocol, WIFI protocol, infrared protocol, RS 485 protocol, NB-IOT protocol, and Ethernet protocol. The rate of zigbee protocol is 2 Mbit/s; the rate of NB-IOT protocol and WIFI protocol can be expressed as N*64 kbits/s (N is a positive integer); the rate of infrared is 128 kbit/s; the rate of Bluetooth protocol is 560 kbit/s; the rate of RS 485 agreement is 1200 baud, 9600 baud, 38400 baud. The rate matching processing unit performs rate processing on the configured protocol to be transferred, so that the rates can be kept the same between the protocols.

2.2 Implementation Method of Protocol Conversion Rate Matching

(1) Protocol conversion system

Both ends are data devices, and the middle is a multi-protocol conversion system. If there are two devices with inconsistent transmission protocols, and device 1 is connected to a high-rate interface, then device 2 is connected to the low-rate interface. To achieve a high-speed interface, invalid data bits must be added through a vulgar protocol to ensure rate matching. Protocol conversion and discarding invalid data are ports with a low de-match rate, so the low rate of data sent by device 1 cannot exceed the frame threshold of protocol conversion. In device 2, when sending data, when sending data, the invalid data bits are matched with the high-rate interface, and the invalid data is discarded in the data terminal 1, thereby reducing valid data to achieve the purpose of recovering valid data.

(2) Protocol priority cache

The main function of the protocol priority buffer module is to prioritize and output the data streams of each protocol by setting different priorities and configure the priority of data frames in the same priority.

1) Data input integrated module

   It contains dozens of small modules. Priorx_data configures the module for each protocol frame format. The packetx_priior module is the priority definition module, and its receiving register accepts configuration information to mark the priority of each protocol. packetx_speed is the rate matching module for each protocol. Priorx_data_valid is the priority definition of the same protocol. Priorx_leng is the output strategy module for each priority. Each module is connected with data link through wire type and reg type. Each module is executed in parallel. Prior_switchUprior_switch is a data input comprehensive module. This module comprehensively analyzes the information of each module and finally completes the priority ordering of each input protocol frame data. After compiling and synthesizing the above modules, the RTL level circuit diagram will be obtained.

2) Read the forwarding control module

   If it detects that there is data in the frame information FIFO, it means that at least one packet of Ethernet frame data has been stored in the dual-port data RAM. The judgment is based on the IP packet type: if the type is 0800, the received data is read from the dual-port RAM And forward it to the received data FIFO, and finally store the frame length information in the frame length FIFO; if the type is 0806, the corresponding processing is performed, that is, the exchange destination and source MAC generate a response packet, and forward the response packet to in the sending FIFO, it is finally multiplexed with the downstream data of the protocol conversion system and sent to the MAC for the response of the ARP packet; if it is of other types, just do not read the data in the dual-port RAM, that is discard the packet data.

(3) Data conversion

The function of the data conversion module is to receive the data frame from the output strategy module and convert the data frame according to the pre-configured finite state machine. After the protocol signal is pre-processed in the early stage, the data frame enters the data conversion module with an 8-bit wide bus. And according to the configured finite state machine for automatic conversion.

1) Finite state machine module

   The finite state machine is composed of a large number of registers, data bit comparators, multiplex selectors, etc. The registers complete the transfer flag bit and state storage, the data bit comparator completes the status flag bit output according to the protocol frame, and the multiplex selector completes the state Transfer.

2.3 Pattern Specification Matching Algorithm

Pattern matching algorithm is an important research field in computer science. It has been widely used in the field of telecommunications network. Pattern matching algorithm plays a vital role in the security of telecommunications network data transmission.

1) Bad character rules

Suppose the pattern string is P and the bad character is x. According to whether x is included in P, the calculation method of the skip value skip(x) can be divided into two cases, and the corresponding calculation formula is formula (1).

$$skip(x) = \begin{cases} m, x \neq P[j](1 \leq j \leq m) \\ m - \max(x), \{k|P[k] = x, 1 \leq k \leq m\} \end{cases} \quad (1)$$

2) Good suffix rules

Suppose the pattern string is P and the good suffix is suffix. According to whether suffi (x) appears again in P, the jump value is calculated in two cases. Let shift (j) be the right shift distance calculated according to the good suffix rule, j is the position of the good suffix, and s is the distance between t' and t or the distance between y P''.

$$\text{shift}(j) = \min\{s|P[j + 1...m]= P[j - s + 1...m - s])\&\&(P[j] \neq P[j - s])(j \succ s) \quad (2)$$

$$P[s + 1... + m]= P[1...m](j \leq s) \quad (3)$$

3 Simulation of Protocol Adaptive Conversion Method Based on Protocol Matching for Power Transmission Iot Terminal

3.1 Software Simulation Experiment Tool

(1) NS-2

The simulation realization of NS-2 adopts C++ and OTcl language for programming. C++ is a very fast programming language, easy to accurately implement complex algorithms, and has strong scalability. Therefore, in NS-2 simulation, C++ is mainly used to implement the compilation layer of NS. Compared with C++, although the running speed of OTcl is much slower, the OTcl language can be easily implemented and modified, which can effectively solve the second task that needs to be completed in NS-2 simulation. Therefore, in NS-2 simulation, OTcl is mainly used to realize the interpretation layer of NS-2. That is, when the simulation environment needs to be established, configured, and modified, OTcl is used to implement operations on C++ objects.

(2) Data processing tool TRPR

The TRPR tool is used to process the data packets received by MGEN, and can count a variety of communication performance indicators in real time.

(3) Integrated flow generator MGEN

MGEN (Multi-Generator) is an open source comprehensive traffic generator, which can generate TCP or UDP data streams, supports generating data streams at a constant rate, and Poisson data streams and other traffic modes. You need to specify the data stream when using it. Destination address and port number, sending rate and other information. This experiment uses MGEN to generate a constant rate data stream, and then tests the business performance under different conversion mechanisms.

3.2 Parameter Setting

There are 4 wireless nodes, 2 for real-time services (CBR) and 2 for non-real-time services (FTP).

There are 20 real-time services (CBR) and non-real-time services (FTP) in the service stream.

Each data stream is of wireless type, and the transmission rate of the wireless channel is 2 Mbps. The data packet sizes of the first type of real-time service CBR application and the second type of application are 16 Bytes and 100 Bytes respectively, and the third type of application is the non-real-time service FTP data packet. The size is 500Bytes, and the super frame length is 16 ms.

3.3 Establish a Simulation Environment for Ns-2 Network Simulation

First, write the required components according to the specific application needs, that is, add the specific C++ and OTcl classes required, such as business flow, link, and node models. Then start writing OTcl scripts. Script writing is mainly divided into the following steps:

(1) Configure the network topology and set the basic characteristics of the link and other parameters;
(2) Establish protocol agents, configure applications and business flows;
(3) Set the trace object, that is, record the simulation time;
(4) Set the start and end time of the simulation;
(5) Start to execute the simulation script;
(6) After the simulation process is over, use Nam and Xgraph to view the running process of the network simulation, analyze the generated trace file, and extract useful data information from it.

4 Simulation Analysis of Protocol Adaptive Conversion of Power Transmission Iot Terminal Based on Protocol Matching

4.1 Comparison of Conversion Service Delays of Different Mechanisms

The simulation results of the conversion service delay of different mechanisms are shown in Table 1: Before the simulation time is 80 s, the delay time of the conversion service under the PCF mechanism is within 50 ms; before the simulation time is 55 s, the delay time of the conversion service under the DCF mechanism is 50 ms within 60 s; the conversion service delay time of the adaptive mechanism is all within 20 ms.

It can be seen from Fig. 1 that despite the increase in network load, the conversion service can control the delay time within 20 ms under the proposed adaptive mechanism. Before the simulation time is 80 s, although the delay of the control service under the PCF mechanism can be controlled at about 35 ms, as the number of network services increases, that is, when the number of nodes in the polling list continues to increase, the delay of the control service under the PCF mechanism also increases. Therefore, it cannot meet the real-time requirements of control services. However, the delay performance of the control

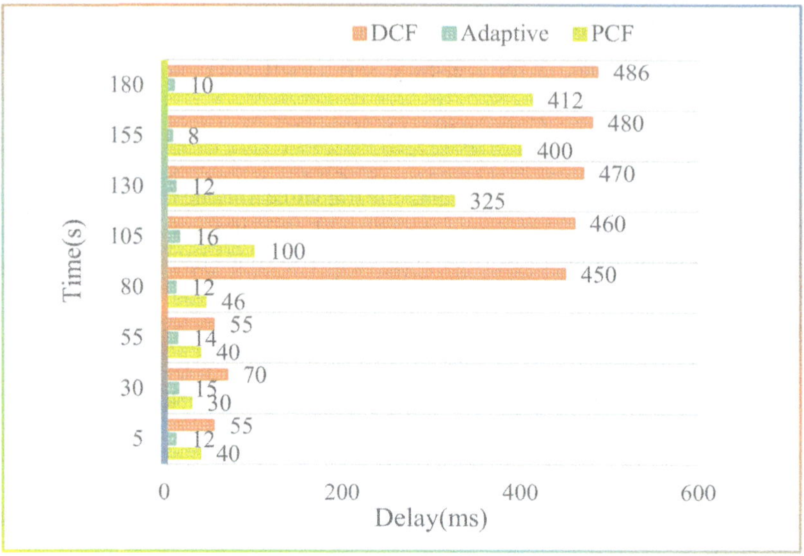

Fig. 1. Delays of control services under different mechanisms

service under the DCF mechanism is very poor, and the network performance of the control service cannot be guaranteed. This shows that the adaptive mechanism conversion method proposed in this paper shows superior performance than the traditional DCF mechanism and PCF mechanism.

Table 1. Delays of control services under different mechanisms

Time(s)	PCF	Adaptive	DCF
5	40	12	55
30	30	15	70
55	40	14	55
80	46	12	450
105	100	16	460
130	325	12	470
155	400	8	480
180	412	10	486

4.2 Packet Loss Rate Under Different Mechanisms

Calculating the packet loss rate of the converted data under different mechanisms, the results are shown in Table 2: The packet loss rate of the control service under the adaptive mechanism is less than 2.5%, and the packet loss rate under the PCF mechanism can be controlled at 3% in the first 80 s However, the packet loss rate gradually increases after 100 s. The packet loss rate under the DCF mechanism increases sharply, and the packet loss rate of the protocol conversion under the adaptive mechanism is the most stable and the smallest.

Table 2. Packet loss rate of control services under different mechanisms

Unit: %	PCF	Adaptive	DCF
5	1.8	0.5	3.6
30	1.7	0.6	3
55	1.8	0.4	12
80	2	0.6	20
105	4	2	35
130	5.8	2.2	36
155	12	2.4	42
180	16	2.3	49
205	20	2.5	51

It can be concluded from Fig. 2 that the packet loss rate of the protocol conversion under the adaptive mechanism proposed in this article is more stable than the traditional

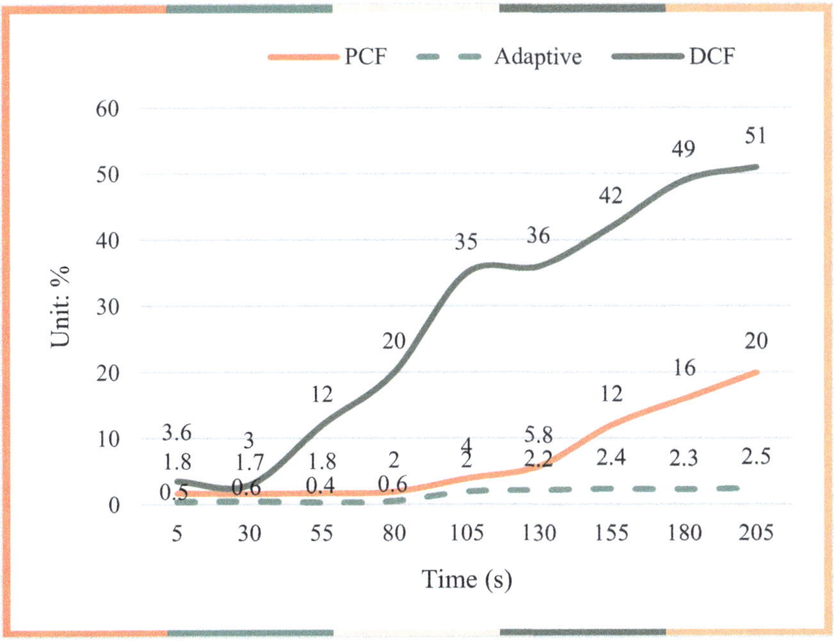

Fig. 2. Packet loss rate of control services under different mechanisms

DCF mechanism and PCF mechanism, and can be controlled within the acceptable range of the business, thus verifying the superiority of the adaptive conversion method.

5 Conclusion

This article is based on the research of protocol adaptive conversion method of power transmission IoT terminal based on protocol matching. According to the design goal, this article introduces the architecture of the protocol adaptive conversion system, and creatively designs the methods to realize the functions of the three modules: data receiving, data conversion, and data output. Through simulation, it is concluded that the protocol adaptive conversion system architecture can be more flexibly configured and applicable to the protocol space of various complex power transmission IoT terminals, so that the protocols can be converted to each other efficiently, and the data signal processing is complete during the conversion process. Relying on hardware completion, it has the advantages of small delay and high stability.

References

1. Cui, J., Zhao, X.: Study of IoT terminal interface platform based on embedded technology and zigbee protocol. Int. J. Future Gener. Commun.Netw. **9**(6), 55–64 (2016)
2. Sun, H., Chen, L., Hao, X., et al.: An energy-efficient and fast scheme for hybrid storage class memory in an AIoT terminal system. Electronics **9**(6), 1013 (2020)
3. Song, K., Wang, Q., Peng, L., et al.: Secrecy energy efficiency optimization for DF relaying IoT systems with passive eavesdropping terminal. Phys. Commun. **44**(5), 101254 (2020)
4. Sarwesh, P., Shet, N., Chandrasekaran, K.: ETRT–cross layer model for optimizing transmission range of nodes in low power wireless networks–an internet of things perspective. Phys. Commun. **29**, 307–318 (2018)
5. Zhai, X., Cheng, X.: Environmental protection requirements of power transmission and transformation project construction based on internet of things technology and its change trend analysis. IOP Conf. Ser.: Mater. Sci. Eng. **740**(1), 012144 (2020)
6. Li, D., Li, B., Qin, N., et al.: The research of NOMA-MEC network based on untrusted relay-assisted transmission in power internet of things. IOP Conf. Ser.: Earth Environ. Sci. **634**(1), 012052 (2021)
7. Dong, C.: Internet of things application in satellite communication in power transmission, transform and distribution. Int. J. Adv. Netw. Monit. Controls **4**(4), 38–46 (2019)
8. Gawanavar, M., Dinakaran, M.: Wireless power transmission (Wipow) for internet of things (IOT). Int. J. Pharm. Technol. **8**(4), 25584–25595 (2016)
9. Lee, S.W., Lim, S.H.: Analytic model for the power-optimal data transmission interval of wireless sensors in internet of things. J. Digit. Contents Soc. **19**(7), 1373–1379 (2018)
10. Greene, E., Hautus, M.J.: Demonstrating invariant encoding of shapes using a matching judgment protocol. AIMS Neurosci. **4**(3), 120–146 (2017)
11. Xu, Z., Zhou, F., Li, Y., et al.: Privacy-preserving subgraph matching protocol for two parties. Int. J. Found. Comput. Sci. **30**(4), 571–588 (2019)
12. Zeng, P., Tan, Q., Meng, X., et al.: A multi-pattern hash-binary hybrid algorithm for URL matching in the HTTP protocol. PLoS ONE **12**(4), e0175500 (2017)

Simulation Model of UHV AC Transmission System Based on PSCAD

Mingjiu Pan[1], Zhou Lan[1], Kai Yang[1], Zhifang Yu[1], Yuqian She[2], Di Zheng[2](✉), and Balestrieri Absier[3]

[1] Economic and Technical Research Institute of Zhejiang Electric Power Corporation, Hangzhou, China

[2] China Ji Liang University, Hangzhou, China
dizheng0715@163.com

[3] Georgetown University, Washington, DC, USA

Abstract. In this paper, the model of two 1000 kV UHV AC transmission lines of B - J is built on PSCAD, and set the type of fault on the model. By observing the voltage and current at the bus and the voltage change at the fault point, the fault point can be located quickly, and the reliability of the model can be verified by simulation. At the same time, this paper introduces in detail the calculation method of basic electrical parameters of power system transmission line, the function of PSCAD and the modeling method. In addition, the influence of the installation location of shunt reactance and the distribution parameters of transmission line is considered in the model, which reduces the influence of shunt reactance and line parameters on the model accuracy.

Keywords: AC Transmission System · Ultra high voltage · PSCAD · Fault location

1 Introduction

Generally, the voltage of the high-voltage AC transmission can reach 1000 kV, which can achieve a remote transmission of more than 500 km or more, and the transmission line is low, and the transmission capacity is large. The electric transmission process is in the environment of a small field, inevitably encountering some problems, such as bad weather, birds and nearby trees, etc., can affect the normal operation of the line, so that part of the system is working properly. The state was destroyed. Both faults and abnormal operating conditions have an impact on power systems, leading to decline in electric energy, and even cause economic losses, personal injury and equipment damage. Therefore, when the line is fails, it is necessary to remove the fault or perform manual fix quickly and selectively. In fact, such a transmission system is very strict for failover requirements, and it must ensure its safety reliability. In addition, because the distance of the whole line is too long and some transmission lines are in remote places with few people, it is difficult for people to find the fault point quickly and accurately for manual repair in such a long transmission line once the non-self-recovery fault occurs in reality. Therefore, it is necessary to carry out fault simulation of transmission system to verify the safety and reliability of the system and to locate the fault quickly [1–3].

I. Ahmad et al. (Eds.): STSIoT 2021, LNDECT 122, pp. 391–399, 2023.
https://doi.org/10.1007/978-981-19-3632-6_48

2 Model Parameter Calculation of UHV AC Transmission System

2.1 Equivalent Parameters of Transmission Lines

The electrical parameters of a transmission line include resistance, conductance, inductance and capacitance. Among them, inductance is calculated in the form of reactance, capacitance is calculated in the form of susceptance, these four basic parameters are evenly distributed along the length of the road. Under normal weather conditions, the corona power loss and insulator leakage loss of transmission lines are very small, almost negligible, and the electrical conductance can be assumed $g = 0$ [4–7].

The equivalent resistance of split wire per unit length is calculated by:

$$R = \frac{\rho}{NS} \tag{1}$$

The calculation formula of equivalent inductance of split conductor per unit length is:

$$l_1 = 2 \times 10^{-7} \ln\left[\frac{Deq}{\left(ND_sA^{N-1}\right)^{\frac{1}{N}}}\right] \tag{2}$$

$$Deq = \sqrt[3]{D_{ab}D_{bc}D_{ca}} \tag{3}$$

$$Ds = re^{-\mu_r/4} \tag{4}$$

According to wire parameters, the basic parameters of overhead line can be calculated, as shown in Table 1.

Table 1. Basic parameters of UHV transmission lines

R1/(Ω/km)	X1/(Ω/km)	B1/(Ω/km)	R0/(Ω/km)	X0/(Ω/km)	B0/(Ω/km)
6.109×10^{-3}	213.1×10^{-3}	4.66×10^{-6}	$0.134.5 \times 10^{-3}$	732.3×10^{-3}	2.59×10^{-6}

2.2 Equivalent Parameters of Transmission Lines

In practice, it is difficult to obtain detailed load parameters of a large transmission system due to the wide variety of loads and complex transient changes, and it is difficult to establish a model. In order to simplify the modeling process and data analysis, constant power equivalence method of load equivalence is adopted to calculate equivalent point loads according to the load rate of 50% and power factor of 0.8. Equivalence load parameters of lines are shown in Table 2.

Table 2. Load parameters of B - J

Station	Transformer/(MVA)	Load/(MVA)	P/(MW)	Q/(Mvar)
B	2 × 3000	3000	2400	1800
J	2 × 3000	3000	2400	1800

3 PSCAD Model of UHV AC Transmission System

3.1 The Power Supply Model

Set the power supply voltage as 1050 kV, the power supply capacity as 3000 MVA, the frequency as 50 Hz and the initial phase Angle as 0°.

3.2 Overhead Line Model

In the UHV system, the influence of distribution parameters needs to be considered, and the Bergeron model of overhead lines can be selected to reduce the influence. Bergeron model is a kind of π model, but the "S" model adopts centralized parameters, while the "Bergeron" model is a model with distributed parameters (L, C distributed parameters, R centralized parameters), which can be regarded as the sum of infinite π models.

3.3 The Fault Model

By using the external control function of the fault element, the different fault types can be switched. The fault occurrence time is set at 0.5 s and the duration is set at 0.5 s. The transition resistance is 0 Ω or 10 Ω.

3.4 The System Model

Establish B - J respectively using PSCAD segment transmission engineering simulation model, considering the transmission lines can produce capacitive reactive power, and the failure happened during online power frequency overvoltage occur on the road, in order to reduce this part of the factors, in the simulation model is set up, Shunt reactors need to be added to the system to suppress the influence of power frequency overvoltage.

In order to analyze the influence of the location of the shunt reactance on the model accuracy, the shunt reactance is installed on the bus and transmission lines. Figure 1 shows shunt reactance mounted on the bus and Fig. 2 shows shunt reactance mounted on the transmission line [8–10].

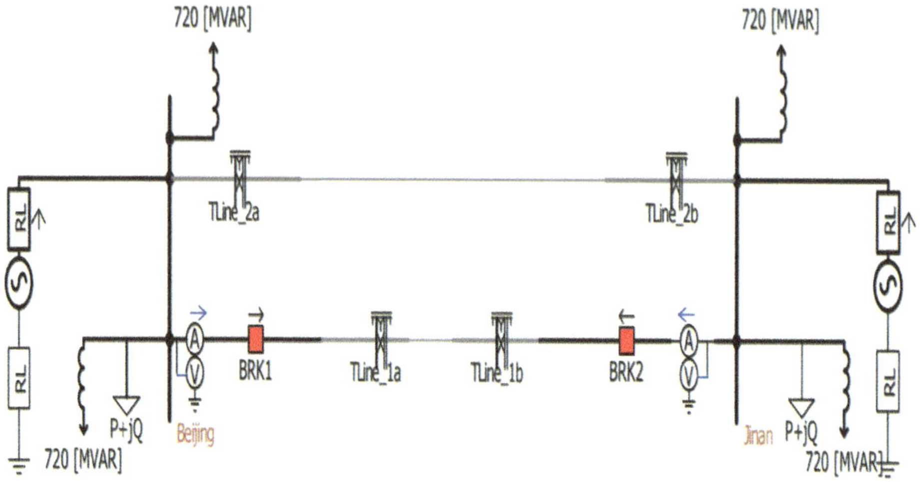

Fig. 1. Simulation model of B - J Sect. 1

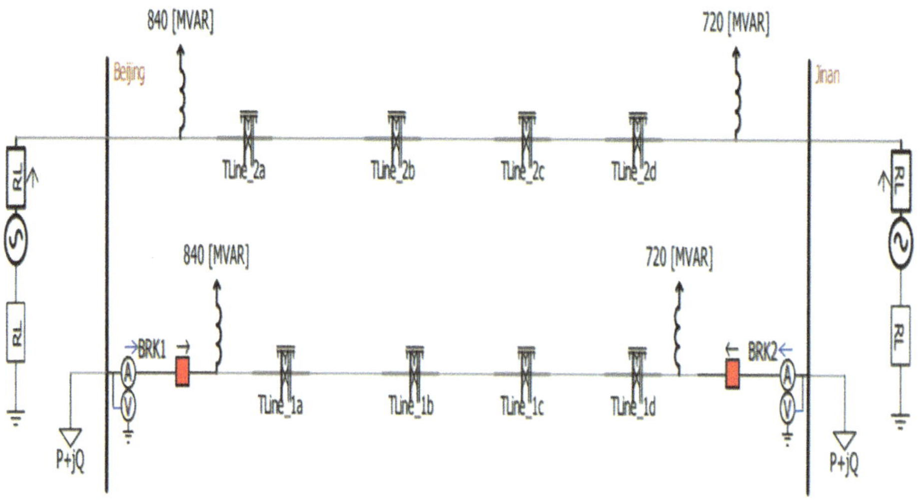

Fig. 2. Simulation model of B - J n Sect. 2

4 Case Analysis

In the simulation model of B - J transmission line, four fault types are set at 150 km from the east of B, and the fault transition resistance value is changed for simulation.

4.1 The Shunt Reactance is Installed Outside the Measuring Point

The waveform shown in Fig. 3 can be obtained by connecting A ground fault and changing the fault transition resistance value. shows shunt reactance mounted on the transmission line.

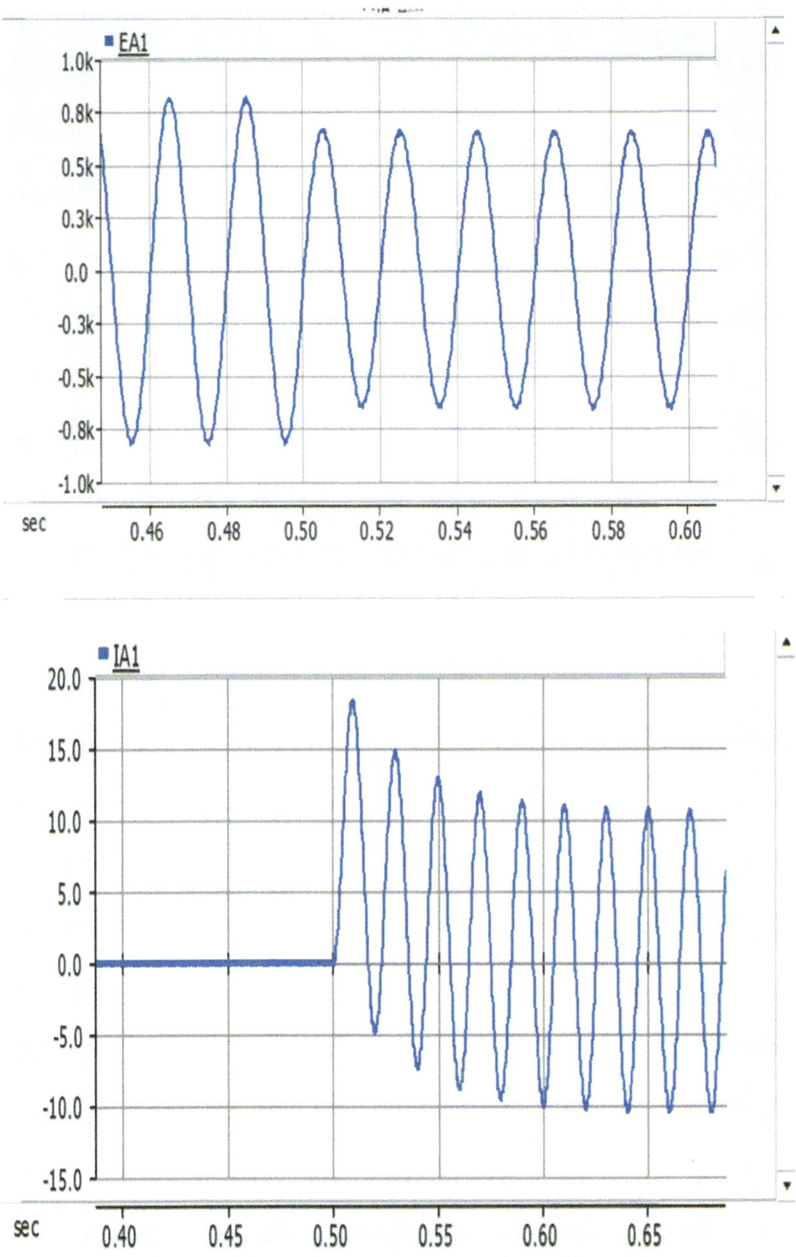

(a)R=0Ω

Fig. 3. Current and voltage waveforms of Phase A

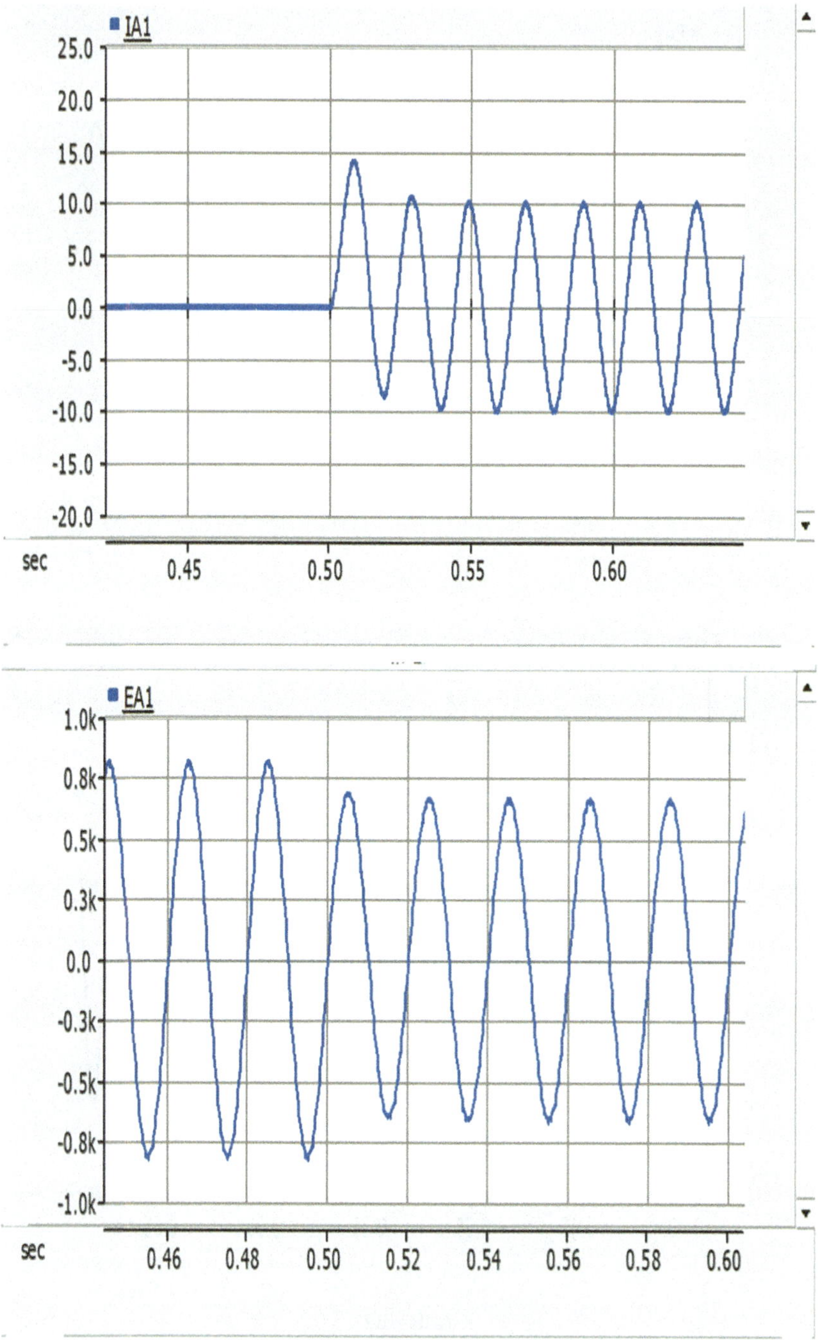

(b)R=10Ω

Fig. 3. continued

4.2 The Shunt Reactance is Installed Inside the Measuring Point

According to the simulation results, when the transition resistance is 0 Ω, the effective value of the fifth cycle is Um = 652.42 kV, Im = 17.62 kA, and the effective value of the A-phase voltage at the fault point is 0 kV after the A-phase voltage and current failure at the eastern end of B. The fault distance L = 173.83 km.

When the transition resistance is 10 Ω, the effective value of the fifth cycle is Um = 661.41 kV, Im = 173.40 kA, and the effective value of the A-phase voltage at the fault point is 168.7 kV after the A-phase voltage and current failure at the eastern end of B. The fault distance L = 172.62 km.

4.3 Analysis of Simulation Results

Table 3 is the simulation results of the simulation model of parallel reactance installed outside the measuring point for the transmission line from B-J. Table 4 is the simulation results of the simulation model of parallel reactance installed in the measuring point of the transmission line from B to J.

Table 3. Simulation results of PSCAD model 1 of B-J

R/Ω	Fault type	Fault location/km	Computing location/km	Relative error
0	AG	150	173.83	0.1589
	ABG	150	153.67	0.0245
	AB	150	147.62	0.0159
	ABC	150	146.03	0.0265
10	AG	150	172.62	0.1508
	ABG	150	151.48	0.0099
	AB	150	147.14	0.0191
	ABC	150	150.77	0.0051

The simulation results show that the accuracy of the two models is in line with the requirements, and the error is basically within 0.03. The installation position of shunt reactor has a certain influence on the accuracy of the simulation model. The accuracy of shunt reactor installed in the measuring point is lower than that of shunt reactor installed outside the measuring point. The accuracy of a transition resistor of 10 Ω is lower than that of a transition resistor of 0 Ω.

Table 4. Simulation results of PSCAD model 2 of B-J

R/Ω	Fault type	Fault location/km	Computing location/km	Relative error
0	AG	150	169.11	0.1274
	ABG	150	152.97	0.0198
	AB	150	150.96	0.0064
	ABC	150	147.90	0.0140
10	AG	150	167.28	0.1152
	ABG	150	153.45	0.0230
	AB	150	153.81	0.0254
	ABC	150	152.41	0.0161

5 Conclusions

(1) Through the simulation results, it can be seen that the voltage and current of the established model change normally when different types of faults occur. For example, in the case of single-phase grounding fault, the voltage of the fault phase at the short-circuit point decreases and the current of the non-fault phase is equal to zero. When a two-phase short circuit fault occurs, the fault phase voltage at the short circuit point is equal in magnitude and phase, and the current is equal in magnitude and opposite in direction.

(2) The installation of shunt reactor is mainly to eliminate the error caused by overvoltage of power frequency. Due to the existence of transition resistance, the measured voltage and current will change, resulting in changes in the measured impedance. The simulation results show that when the shunt reactor is installed outside the measuring point, that is, on the bus, the accuracy is higher than that of the shunt reactor installed inside the measuring point. As the transition resistance is pure resistance, it will affect the value of the measured voltage and current, thus affecting the calculation of the fault distance. It can be seen from the simulation results that the simulation accuracy is higher when the transition resistance is 0Ω.

Acknowledgements. This work was supported by SGZJJY00JJJS2100058.

References

1. Louie, K.W., Wang, A., Wilson, P., Buchanan, P., et al.: A newly built power flow program in PSCAD/EMTDC for electric power system studies. In: 2004 International Conference on Power System Technology, vol. 1, pp. 1502–1507 (2004)
2. Ishak, N.H., Isa, I.S., Abdullah, S., et al.: Performance comparison of electric power flow solutions using PSCAD. IEEE (2010)

3. Gopakumar, P., Mallikajuna, B., Reddy, M.J.B., Mohanta, D.K.: Remote monitoring system for real time detection and classification of transmission line faults in a power grid using PMU measurements. Prot. Control Mod. Power Syst. **3**(2), 159–168 (2018)
4. Song, G., Hou, J., Guo, B., Chen, Z.: Pilot protection of hybrid MMC DC grid based on active detection. Prot. Control Mod. Power Syst. **5**(1), 1–15 (2020). https://doi.org/10.1186/s41601-020-0152-2
5. Mishra, S.K., Tripathy, L.N.: A critical fault detection analysis & fault time in a UPFC transmission line. Prot. Control Mod. Power Syst. **4**(1), 24–33 (2019). https://doi.org/10.1186/s41601-019-0117-5
6. Huang, S., Cao, K., Luo, L.: A method of impedance measurement to eliminate influence of transition resistance. Autom. Electr. Power Syst. **37**(23), 108–113 (2013)
7. Rajaraman, P., Sundaravaradan, N.A., Mallikarjuna, B., Reddy, M.J.B., Mohanta, D.K.: Robust fault analysis in transmission lines using synchrophasor measurements. Prot. Control Mod. Power Syst. **3**(1), 108–110 (2018)
8. Das, S., Navalpakkam Ananthan, S., Santoso, S.: Relay performance verification using fault event records. Prot. Control Mod. Power Syst. **3**(1), 1–10 (2018). https://doi.org/10.1186/s41601-018-0094-0
9. Musa, M., He, Z., Fu, L., Deng, Y.: A cumulative standard deviation sum based method for high resistance fault identification and classification in power transmission lines. Prot. Control Mod. Power Syst. **3**(1), 1–12 (2018). https://doi.org/10.1186/s41601-018-0102-4
10. Castillo, R., Ramon Ramirez, J., Alonso, G., et al.: Prony's method application for BWR instabilities characterization. Nucl. Eng. Des. **284**(4), 67–73 (2015)

Intelligent Distribution Automation Fault Location System Based on Cloud Computing

Ke Ren(✉)

Qingdao University of Science and Technology (QUST), Qingdao 266000, China
rk13562832493@163.com

Abstract. As an important part of the smart grid, the smart distribution network is an important link between power companies and power users. Cloud computing technology has the advantages of distributed computing, strong fault tolerance and easy expansion, which can solve the problem of processing and storing massive data in smart distribution networks. The purpose of this paper is to study the intelligent distribution automation fault location system based on cloud computing. This article mainly uses cloud computing technology to build an intelligent power distribution network based on cloud computing, designs a distribution automation fault location system, and finally tests the system. The system test results show that the average value of the avalanche test results is 0.52, and the ideal value is 0.5. No matter how many digits of the plaintext are changed, even if one bit changes the smallest change, nearly half of the ciphertext digits change. It is very close to the ideal avalanche effect. That is, the AES encryption algorithm designed in this paper has good avalanche characteristics and can effectively resist linear analysis attacks launched by criminals.

Keywords: Cloud computing · Smart power distribution · Fault location · Power distribution automation

1 Introduction

The construction of smart grid is an inevitable trend of the development of the times, and it will bring great significance to the society [1, 2]. First, smart grids can guarantee energy security to a higher degree; second, smart grids consume much less energy than traditional grids, and are more adaptable to uncertain weather conditions [3, 4]. Finally, the smart grid is also an important measure to implement the energy sustainable development strategy [5, 6].

Regarding the systematic research on fault location of intelligent distribution automation, many scholars have conducted research on it [7]. By the 1990s, with the improvement of computer software and hardware levels and the successful development of primary power distribution equipment, the development of power distribution automation has achieved unprecedented achievements, especially in Japan and the United States, where power distribution automation has developed to a higher stage, and research technology It is being gradually promoted, and the reliability index of power supply in some

I. Ahmad et al. (Eds.): STSIoT 2021, LNDECT 122, pp. 400–409, 2023.
https://doi.org/10.1007/978-981-19-3632-6_49

areas has reached more than 99.99% [8]. Although the academic community has relatively rich research results on intelligent distribution automation fault location systems, there are relatively few intelligent distribution automation fault location systems based on cloud computing. Therefore, the research on the intelligent distribution automation fault location system based on cloud computing is of great significance.

This article first introduces CC technology and data transmission security, and then designs the CC-based smart power distribution network hardware architecture and power distribution automation fault location system. The smart power distribution network hardware architecture mainly includes three parts, the overall system framework the design and implementation of the main hardware equipment of the smart power distribution network and the smart power distribution network based on CC, and finally the system is tested.

2 Intelligent Distribution Automation Fault Location System Based on CC

2.1 CC Technology and Data Transmission Security

(1) CC technology.

1) HDFS distributed file system.
   HDFS has the function of distributing and storing data in block sequence.
2) Map Reduce distributed computing framework.
   The essence of Map Reduce is a programming model that can run large-scale data sets in parallel. The key to its framework is the map function and reduce function. Multiple map functions and reduce functions can be run simultaneously on a Map Reduce [9, 10].

(2) Data transmission security technology in CC environment.
   In the process of data transmission of smart distribution network to CC server, due to the existence of various security risks, its confidentiality may be destroyed. In this paper, the AES algorithm is selected to ensure the security and confidentiality of data transmission of smart distribution network in the CC environment.

The encryption round function of AES consists of 4 different modules, namely byte substitution transformation S-Byte(), row shift transformation S-Row(), column hybrid transformation M-Column() and round key plus transformation ARoundKey() [11, 12].

Byte substitution S-Byte() is a non-linear transformation of bytes. Each byte in the state is transformed into another byte by looking up a known table using the S-box. The S-box is based on a matrix in a finite field. Operation, there are 256 elements in total, each element is one byte in size, and all elements are different.

S-Row() cyclically shifts each row of each state to the left by a different amount of displacement. The first line remains unchanged, the second line is circularly shifted to the left by 1 byte, the third line is circularly shifted to the left by 2 bytes, and the fourth line is circularly shifted to the left by 3 bytes.

Let $s_j(x)$ and $s_j^{'}(x)$ as shown in Eqs. (1) and (2):

$$s_j(x) = s_{3j}x^3 + s_{2j}x^2 + s_{1j}x + s_{0j}, 0 \le j \le 3 \tag{1}$$

$$s_j^{'}(x) = s_{3j}^{'}x^3 + s_{2j}^{'}x^2 + s_{1j}^{'}x + s_{0j}^{'}, 0 \le j \le 3 \tag{2}$$

Then there is the following formula (3):

$$s_j^{'}(x) = a(x) \otimes s_j(x) \tag{3}$$

Among them, $a(x) = \{03\}x^3 + \{01\}x^2 + \{01\}x + \{02\}$, $\otimes$ means $x^4 + 1$ multiplication.

2.2 Hardware Architecture of Smart Power Distribution Network Based on CC

(1) Overall system framework.

The system adopts a hierarchical overall structure, which is divided into three layers, namely the execution layer, the link layer and the platform layer. The hierarchical overall architecture has the characteristics of high cohesion and low coupling.

The execution layer is mainly controlled by the area, and the intelligent control element is the most basic execution layer of the system.

The link layer provides connection functions for use by the management layer. The link layer is in a very important position. It is the main artery of the entire system and the transmission channel of information services. Through the services of the link layer, the system can realize various functions.

The platform layer is mainly composed of various middleware, service components and interfaces, and is the supporting platform for system business applications. The platform layer is the center of the realization of the entire system function. It is composed of a large amount of basic data. The platform layer can ensure the seamless combination of the hardware system and the software and management system.

The intelligent distribution network has different requirements for the real-time data transmission of the above-mentioned different services. Table 1 shows the technical indicators of different transmission data responses in the intelligent distribution network.

(2) Main hardware equipment of intelligent power distribution network

1) Sensor unit

The data of the sensor unit will be a global resource, not the private of a single component. Distributed and collaborative sensors will have greater significance for intelligent networks. They can share data to improve the reliability of the overall data and speed up the speed of decision-making.

2) Communication Technology

The combination of advanced communication technology, information technology, sensor technology, automation technology and power grid technology will provide a solid platform foundation and flexible access for the

Table 1. Technical indicators of transmission information response

	Type of data	Data priority	Data length	Real-time requirements
State quantity	Switch position			
	Switch remote			Less than 5S
	Terminal equipment status			
	Communication network status	A	1–3 bytes	Less than 5S
	Fault and action signal			

application of smart technology in the power grid, and truly connect the various neurons of the smart grid to together, it is a must for smart power distribution networks.

3) Power generation end

With the development of smart power grids, power generation institutions supplemented by distributed power generation will be connected to the power distribution system through smart power distribution networks, thereby improving the economy, flexibility and environmental performance of the overall power grid.

(3) Design and implementation of smart power distribution network based on CC.

1) Architecture design of smart distribution network.

Information security protection is the prerequisite for the practical application of the smart grid cloud model, which mainly includes the security protection mechanism of the computing platform and application modules in the cloud environment. With the support of the intelligent infrastructure of the cloud system, the power grid has realized the distributed allocation of resources, and the realized uncertain resource path physically constitutes the first line of defense against intrusion.

The characteristics of the smart distribution network are: highly integrated resources, adapt to business changes as needed, seamlessly fit business and green, environmentally friendly, and efficient power grid system, safety and reasonable operation and maintenance costs.

2) The realization of the function of the intelligent distribution network.

The intelligent distribution network can provide basic resources, application platforms, and resources as services to the system for distribution. Resource integration meets the needs of intensive management of the power grid. Through the support of the new system, the purchase and installation of new equipment adopt a unified standard, which greatly simplifies the process and provides the basis and prerequisite for the unified operation and maintenance of the system, which hardly takes up system downtime. It can realize the seamless and non-destructive installation and commissioning of the new equipment.

The design of intelligent power distribution network needs to break away from the traditional model and establish a fast, safe and effective information system construction process. The function of the cloud system provides a simple, intuitive and effective platform for system development and testing by integrating basic resources, thereby providing a more direct way for the design and realization of functions and requirements.

The intelligent power distribution network structure after cloud theory virtualization is shown in Fig. 1:

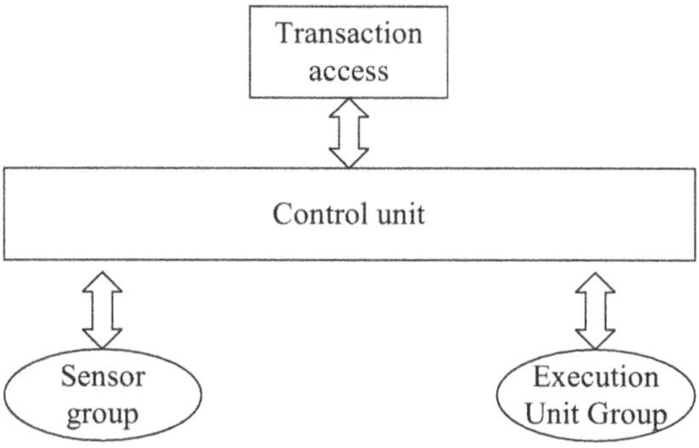

Fig. 1. Smart power distribution network structure after cloud theory virtualization

2.3 Distribution Automation Fault Location System

The distribution automation fault location system includes a fault indicator (FI) with communication function, a communication terminals (CT), a central station (CS), a monitoring master station and a communication system.

(1) FI

When the line fails, the indicator on the fault branch is triggered after the failure, and emits a red luminous indication. At the same time, the digital coded signal transmitting unit or optical fiber communication unit is sent out by radio wave or light signal.

(2) CT

CT is generally installed at the branch point of the online branch, which can accept the coding information of the FI installed on multiple branch lines. After the received action is processed, it is sent to the CS through address coding and timing control.

(3) CS

CS mainly accepts the information sent by CT through the communication yellow paper, decodes it, then processes the data, completes the protocol conversion, and sends it to the master station to complete the communication between the master station and FI.

(4) Monitoring master station

The main function of the monitoring master station is to collect the fault address information forwarded by the central station, and then after correcting it, find out the fault location and fault path through analysis and calculation.

The short-circuit fault indicator is an indicating device that can be directly installed on the power distribution line. It usually includes current and voltage, fault identification, fault indication drive, fault status indication, signal output and automatic delay reset control, etc., as shown in Fig. 2 shown.

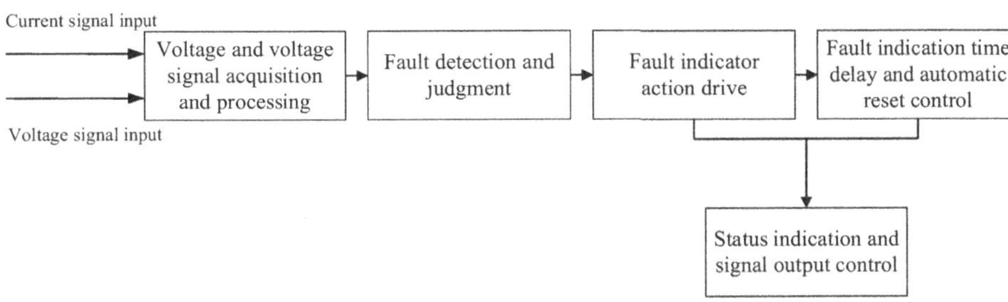

Fig. 2. Block diagram of the fault indicator

3 Realization of Intelligent Distribution Automation Fault Location System Based on CC

3.1 Hardware Environment

To build an intelligent power distribution automation fault location system based on CC, at least 3 computers are required to build a hardware environment, of which 2 computers are used as smart power distribution network terminal equipment. The main task is to process the data to be transmitted by the distribution network terminal. It interacts with the server to complete the secure data transmission; a computer is used as a physical machine built on the CC platform, and virtualization technology is used in the machine. One switch is mainly responsible for the information exchange between the intelligent power distribution terminal equipment and the CC platform.

3.2 System Environment Configuration

Since the Java language is more compatible and compatible with the operation of the Hadoop platform, the construction of the Hadoop CC platform needs to be developed in the Java language, so the JDK of the CC platform needs to be established and configured. Secondly, the process daemon of the slave node DataNode in the Hadoop cluster relies on the support of SSH. The orderly operation of the cluster also needs to control SSH. Therefore, SSH needs to be configured correctly to ensure the efficient operation of the Hadoop platform.

4 System Test Experiment

4.1 Linear Analysis Test

The principle of using linear cryptanalysis to decipher a cryptographic system is that an attacker launches an attack by analyzing the probability of plaintext, key and ciphertext, and finding a linear relationship between ciphertext and plaintext. The prerequisite for an attacker to launch a linear attack is that the algorithm has linear properties.

In order to detect whether the algorithm has an avalanche effect, this article tampered with 128 bits of plaintext data, and counted the number of bits changed in the corresponding ciphertext after tampering, and then detected the ability of the algorithm to resist linear attacks. Randomly generate a group of 128 bits plaintext and change it 128 times, and count the changes of each ciphertext. The obtained plaintext avalanche test results are shown in Fig. 1.

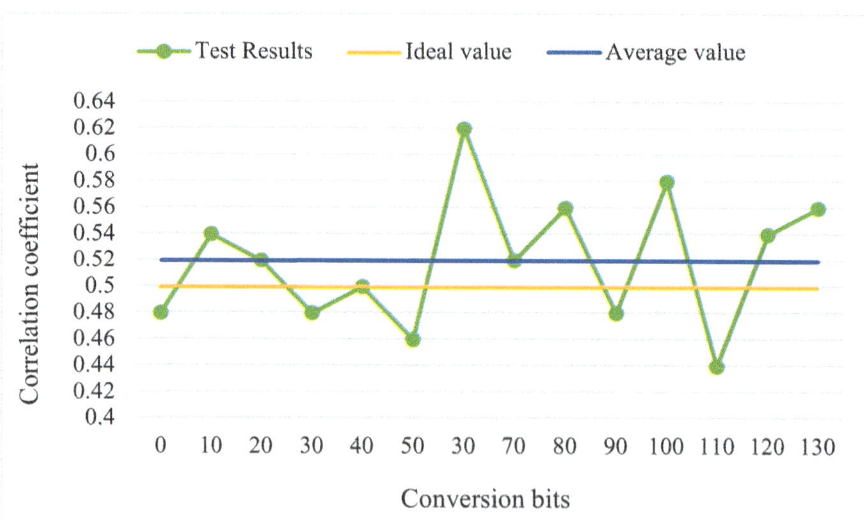

Fig. 3. Avalanche test result

It can be seen from Fig. 1 that the average value of the avalanche test results is 0.52, the ideal value is 0.5, and the actual value fluctuates around the ideal value. Through analysis, it can be known that no matter how many bits of the plaintext are changed, even if one bit changes the smallest change, nearly half of the ciphertext bits change. This situation is very similar to the ideal avalanche effect. It shows that the AES encryption algorithm designed in this paper has good avalanche characteristics and can effectively resist linear analysis attacks launched by criminals.

4.2 Correlation Analysis Test

Correlation analysis refers to the analysis of two or more related variable elements, and the correlation coefficients obtained from the analysis are used to determine the degree of correlation between the variable elements.

In this paper, 10 different sets of plaintext data are used to generate ciphertexts and message digests, respectively, to simulate an attacker launching a tampering attack on the ciphertext generated by each set of plaintext data, and to count the changes in the authentication code. The specific operation is: tampering with 10 groups of ciphertexts of 128 bits, 2 MB, 4 MB, 5 MB, 8 MB, 10 MB, 12 MB, 14 MB, 16 MB, 18 MB, each group of ciphertexts is tampered 20 times, the average coefficient is calculated, and the ciphertexts are counted. Close relationship between message digests. Use correlation analysis software to analyze the change digits of the message summary and the 128 bits ciphertext, and the results are shown in Table 1.

According to mathematical statistics, the value range of the correlation coefficient r is between −1 and 1. It is generally believed that the absolute value of r is less than or equal to 0.3 if it is correlated, and it is low correlation between 0.3 and 0.5, 0.5 and 0.8 There is a significant correlation, and between 0.8 and 1 is a high correlation. It can be seen from Table 1 that the correlation coefficient of 0.114 is less than 0.3, so it is a weak correlation. Therefore, it can be concluded that the relationship between the ciphertext generated by 128bits of plaintext data and the message digest is weakly correlated. The phase relationship analysis results of 10 groups of ciphertexts are shown in Table 2.

It can be seen from Fig. 4 that the correlation coefficients of the 10 groups of ciphertexts are all less than 0.3. Therefore, the correlation between message digests and ciphertext blocks of different sizes is poor. When criminals intercept both ciphertext and message digest at the same time, it can effectively prevent data from being attacked by correlation analysis.

Table 2. 128bits ciphertext correlation analysis results.

		Plaintext	Authentication code
Plaintext	Person correlation	0.9	−1.14
	Sig. (2-tailed)		.515
	N	25	25
Authentication code	Person correlation	−.114	0.9
	Sig. (2-tailed)	.515	
	N	25	25

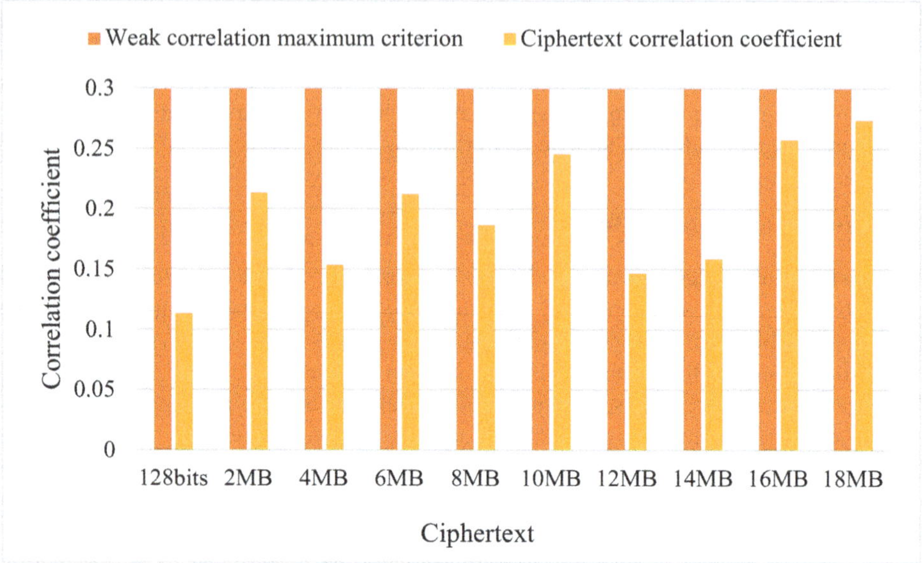

Fig. 4. 10 groups of ciphertext correlation analysis results.

5 Conclusions

CC technology can solve the problem of processing and storing massive amounts of data in smart distribution networks. The data transmission system of the intelligent distribution network bears the major responsibility of the data transmission of the intelligent distribution network, and is the basis for the normal operation of the intelligent distribution system. After a fault occurs in the distribution network, fault location, isolation, and recovery are required. The location accuracy, speed, and recovery effect directly affect the power supply safety and reliability of the system. Fault handling automatic control is one of the important components of the distribution system automation, and it is also the key to the reliable operation of the distribution system.

References

1. Li, Z., Dong, Y., Fu, L., et al.: Integrated research on power distribution intelligent switching equipment. Int. Core J. Eng. **6**(1), 48–54 (2020)
2. Hosseini, M.M., Umunnakwe, A., Parvania, M., et al.: Intelligent damage classification and estimation in power distribution poles using unmanned aerial vehicles and convolutional neural networks. IEEE Trans. Smart Grid **11**(4), 3325–3333 (2020)
3. Hou, Z., Sun, X., Ju, Z., et al.: Design of the monitoring system for power distribution area by clean energy intelligent integration terminals. In: IOP Conference Series: Earth and Environmental Science, vol. 804, no. (3), p. 032051 (2021). (9pp)
4. Gueddouche, R., Boudour, M.: Intelligent nonlinear dynamic equivalent approach applied to Algerian power system. IET Gener. Transm. Distrib. **13**(14), 2919–2929 (2019)
5. Nwazor, N.O., Orisadare, W.S., Phan, E.: Intelligent substation automatic load shedding system for power distribution. Int. J. Sci. Eng. Res. **10**(2), 1296–1301 (2019)

6. Samanlioglu, F., Aya, Z.: An intelligent approach for the evaluation of transformers in a power distribution project. J. Intell. Fuzzy Syst. **39**(3), 4133–4145 (2020)
7. Ikechukwu, U.K., Ude, K., Ngang, N.B., et al.: Improving power system stability in distribution network with intelligent distributed generation scheme. Am. J. Eng. Res. **10**(6), 64–76 (2021)
8. Wang, Z., Ni, Y., Zhang, Z., et al.: Intelligent distribution network information processing based on power data virtual plane. Appl. Sci. **10**(3), 736 (2020)
9. Yellagoud, S.K., Talluri, P.R.: A comparative evaluation of AI based fault location tools for electric distribution networks. Int. J. Power Energy Syst **39**(4), 177–183 (2019)
10. Ananthan, S.N., Santoso, S.: Universal model-based fault location for improved system integrity. Gener. Transm. Distrib. IET **13**(8), 1212–1219 (2019)
11. Sun, Y., Sun, J., Quan, X., et al.: A novel transmission line fault location method based on time frequency correlation. In: IOP Conference Series: Earth and Environmental Science, vol. 696, no. (1), p. 012047 (2021). (8pp)
12. Dashtdar, M., Esmaeilbeig, M., Najafi, M., Bushehri, M.E.N.: Fault location in the transmission network using artificial neural network. Autom. Control. Comput. Sci. **54**(1), 39–51 (2020). https://doi.org/10.3103/S0146411620010022

Technology of Radial Fluid Enhanced Diffusion Based on Machine Learning

Deyu Zhang(✉)

University of Manchester, Manchester City M139PL, UK
zhangdeyuessay82@163.com

Abstract. In the 1990s, it became easier for people to obtain digital information and to spread the information through the Internet. Against this background, machine learning has begun to develop vigorously, focusing more on solving practical problems. The phenomenon of radial fluid enhanced diffusion has always been the focus of attention in the field of fluid mechanics. However, due to the limitations of various actual physical conditions, convection dominates the radial fluid enhanced diffusion problem. The classic solution method will cause serious non-physical oscillations when solving the problem, and no more accurate numerical results can be obtained. Based on this, this paper proposes a radial fluid-enhanced diffusion technology based on machine learning. The logarithmic increment method is proposed to improve the non-parametric estimation of the drift coefficient. Experiments show that in the two typical models, the mean value of the logarithmic increment method fluctuates between 0.4–0.6, which is closer to the actual value of 0.495, indicating that the mean value of the logarithmic increment method is closer to the true value and the variance is smaller. The effect is better than that of the direct incremental method.

Keywords: Machine learning · Radial fluid · Diffusion enhancement · Drift-diffusion

1 Introduction

The convection diffusion model can be applied to many fields closely related to our actual life, such as river health protection, air quality monitoring and management, the transfer of water in the soil, the exploration and exploitation of oil and natural gas in the underground or ocean, wind energy, solar energy, hydropower, the rational development and utilization of biomass energy, geothermal energy, ocean energy, and nuclear energy, etc. [1, 2]. Therefore, the solution of the convection-diffusion equation (also known as advection-diffusion equation, drift-diffusion equation or scalar transport equation) has been widely concerned by many experts and scholars at home and abroad [3, 4]. However, due to the limitations of many practical factors, it is often extremely difficult to obtain the solution of the convection-diffusion equation. Therefore, it is necessary to find a precise, effective and relatively stable numerical calculation method for solving the convection-diffusion equation [5, 6].

I. Ahmad et al. (Eds.): STSIoT 2021, LNDECT 122, pp. 410–418, 2023.
https://doi.org/10.1007/978-981-19-3632-6_50

Many scholars at home and abroad have conducted research on the technology of radial fluid enhanced diffusion based on machine learning. For example, Li R, Deng Q, Tian D, etc. have proposed a pollution-free and high-order upwind method with high-order calculation accuracy to solve the singularly perturbed convection-diffusion problem [7]. Although there are many related researches on the technology of radial fluid enhanced diffusion, so far there is no solution that can solve some other problems. Therefore, it is necessary to strengthen the research on it.

This article first introduces several machine learning algorithms, t-SNE algorithm, FCN algorithm and CNN algorithm, then compares the two diffusions equation drift coefficient estimation methods, and finally, the technology of radial fluid enhanced diffusion based on machine learning is proposed, and the two parameter estimation methods are compared.

2 Technology of Radial Fluid Enhanced Diffusion Based on Machine Learning

2.1 Machine Learning

With the development of modern technology, data has become more and more complex, and the dimensions of the second data are also increasing, which makes the effective extraction and utilization of data more difficult. Computer technology can effectively help us process these data, and the effective features of the data can be extracted through these algorithms. Next, mainly introduce the FCN algorithm of machine learning [8, 9].

One way to understand a machine learning model is to start with a linear model, design the mechanism and parameters of the model, and try to overcome its limitations. For example, linear regression and logistic regression are widely used linear models. They can be efficiently and reliably applied to the solution of closed-form convex optimization problems. The feedforward neural network diagram used to simulate XNOR is shown in Fig. 1.

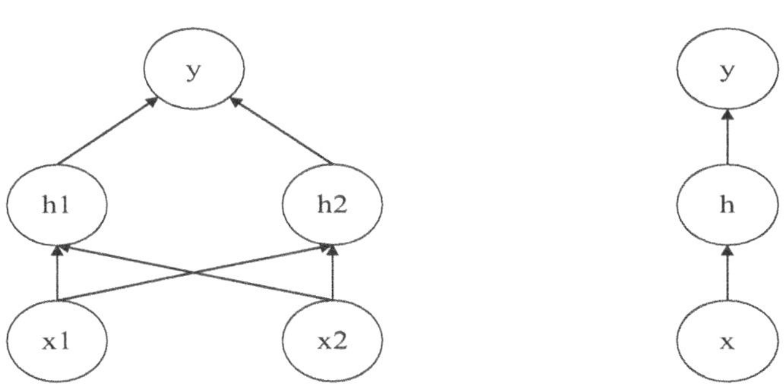

Fig. 1. Feedforward neural network diagram used to simulate XNOR

(1) FCN algorithm (full convolutional network algorithm)

The FCN algorithm mainly includes three layers: input layer, hidden layer and output layer. The neurons in each layer will be connected to the neurons in the previous or the next layer, and the algorithm can be defined by setting the weight coefficient and threshold. As shown in formula (1).

$$\sigma(z) = \frac{1}{1 + e^{-z}} \tag{1}$$

There is also a softmax function, as shown in formula (2), which combines the vector of z_1 components to obtain another vector combination.

$$a_j = \frac{e^{z_j}}{\sum k^{e^{z_k}}} \tag{2}$$

This function can be close to any given function, making machine learning more powerful.

(2) Input feature design

The idea of this article is to screen the appropriate input amount by predicting the output amount.

Therefore, in the process of designing candidate input quantities, the idea of controlling variables should be adopted so that the prediction difference between working conditions depends only on the relative change between working conditions—the difference in input quantities. Therefore, under all working conditions, the basic structure of the neural network remains unchanged, that is, the total number of layers of the network and the number of neurons in each layer are constant; the working ability of the neural network remains unchanged, that is, the loss function, activation function, and training algorithm are constant; the network expects that the target is unchanged, that is, the output is constant; the initial state of the neural network is unchanged, that is, the initial value of the parameter is constant in the process of training the neural network [10, 11].

2.2 Estimation Method of Drift Coefficient of Diffusion Equation

(1) The structure of the diffusion coefficient estimator

Consider the diffusion equation:

$$dX_t = \mu(X_t)dt + \sigma(X_t)dB_t \tag{3}$$

Among them, B_1 is the standard motion, and the bivariate functions $\mu(X_t)$ and $\sigma(X_t)$ are called process X_t drift coefficient and diffusion coefficient respectively [12].

(2) Logarithmic increment method and direct increment method to estimate drift coefficient

Suppose $X_{t_i}, i = 1, 2, \ldots, n$ the observed during [0, T]. There are two ways to construct samples of drift coefficients according to $X_{t_i}, i = 1, 2, \ldots, n$. One is the traditional method (direct increment method): $X_{t_i} - X_{t_i}, i = 1, 2, \ldots, n - 1$ is used as the sample of $\mu(X_t)$, based on the non-parametric estimation of $\mu(X_t)$, constructed by $\left(X_{t_i}, X_{t_i} - X_{t_i}, i = 1, 2, \ldots, n\right)$. This estimator is only compatible when $n \to \infty, T \to$

∞, and empirical analysis shows that it is only suitable for a limited sample. Using $\left(X_{t_i} - X_{t_i}\right)^2$, $i = 1, 2, \ldots, n-1$ as the sample, the non-parametric estimation of $\sigma^2(X_t)$ has better convergence, and Its non-parametric estimation is only compatible with the fixed T $n \to \infty$.

When X_t satisfies the condition of $P\{X_t \leq 0\} = 0$, another method of constructing drift coefficient samples (logarithmic increment method) is to rewrite formula (3) as:

$$dX_t = \mu_t X_t dt + \sigma_t X_t dB_t \tag{4}$$

Among them $\mu_t = \frac{\mu(X_t)}{X_t}$, $\sigma_t = \frac{\sigma(X_t)}{X_t}$, its discretization form is:

$$InX_{t_{i+1}} - InX_{t_i} = \int_{t_i}^{t_{i+1}} \tilde{\mu}_t dt + \int_{t_i}^{t_{i+1}} \sigma_t dB_t \tag{5}$$

Among them $E\left[\int_{t_i}^{t_{i+1}} \sigma_t dB_t \middle| X_{t_i}\right] = 0$, then, $InX_{t_{i+1}} - InX_{t_i}$, $i = 1, 2, \ldots, n$ can be used as the realization of $\int_{t_i}^{t_{i+1}} \tilde{\mu}_t dt$ to construct a non-parametric estimator of.

2.3 Radial Fluid-Enhanced Diffusion Technology Based on Machine Learning

This article is the first to deal with the enhanced dissipation problem of the entire R2 space. There are two main difficulties here: On the one hand, we do not have Poincare's inequality, so in general, the estimate of exponential decay is far from trivial. On the other hand, the possibility of increasing the uncertainty of the solution forces us to work in a weighted space (hence the definition of X-norm), adding some technical problems at the end of the estimation.

Combining the views of this article and some scholars, it is possible to deal with the more general case of radial flow, in which r^q is replaced by an arbitrary smooth function u(r), and in the case of a bounded domain (disk), an appropriate no flux is applied to f Boundary conditions. In this case, the weight of X-norm becomes redundant, but it is expected that the rate will not change substantially, except possibly in the case of q = 1.

2.4 Boundary Processing

Due to the limitations of computing resources and storage resources, when simulating fluids, we usually only focus on the area of interest. At this time, the simulated area and the surrounding area will form a boundary. When simulating the interaction between a solid and a fluid, the solid area can be used as a boundary so that the fluid cannot pass through the solid.

The simplest aspect of the boundary processing problem is the velocity. Because it is necessary to ensure that the fluid particles do not flow into or penetrate the solid, when the solid boundary is fixed, the normal component of the velocity should be 0; and when the boundary is also moving, generally, the normal velocity of fluid particles is required to be 0 relative to the normal velocity of the boundary. This is also called the non-viscous condition, because here only the normal component of the velocity is limited. For the pressure at the solid boundary, the terms in the Navier-Stokes equation can also be applied, because the normal velocity of the fluid particles at the boundary is to be controlled.

2.5 Density Correction of Fluid Particles at the Boundary

The fluid particles near the boundary also have the problem of insufficient neighboring particles. Therefore, the density approximated by smoothed-particle hydrodynamics (SPH) interpolation will be lower than the actual situation. This situation can be effectively improved by simple correction, but care must be taken to ensure that the density field gradient calculated after the correction is still continuous near the boundary. In addition, the particles near the boundary are not spherical, so the force exerted on these particles restricts their movement toward the boundary and causes adhesion. To avoid this problem, when calculating the density and force of fluid particles, the neighboring boundary particles should be considered.

Since the interaction between the non-deformable rigid body without melting effect and the fluid is considered in this work, it is only necessary to generate particles on the surface of the solid boundary. So only need to create a layer of particles on the solid surface. This saves space and storage space, and creates a very realistic effect. In this case, the particle representation of the rigid body can be calculated directly or from the network representation, so that the particles can be placed at any distance from the surface network and create a relatively uniform set of samples.

2.6 Radial Fluid Simulation

This article uses the V-Ray renderer, which can render special effects, such as focal dispersion, global illumination, and underground scattering. It can be used in many fields, such as architectural design, animation rendering and lighting design.

The following are the basic steps to render an image on a V-ray diagram.

(1) Create or open a scene.
(2) Specify the V-Ray renderer.
(3) Set up materials.
(4) Set the "Renderer" tab to "Fixed Mode".
(5) Adjust the lighting to suit the scene.

    1) If the ambient light is not ideal, you can adjust the intensity of the skylight or increase the shading multiple in the exposure method until it is suitable.
    2) Turn on reflection and refraction to match the material.

(6) Re-adjust the lights and materials in the scene to meet the actual situation.
(7) Render and save the photo file.

    1) Save the settings of the photon file.

2) Set the light imaging mode, minimum and maximum sampling to form a small image.

(8) Official rendering.

1) Increase the anti-aliasing value.
2) Set the size of the outgoing image.
3) Enter the photo file to render the big picture.

3 Examination of Typical Models

In order to study whether the logarithmic increment method can improve the limited sampling characteristics of the non-parametric estimator of the drift coefficient, this paper uses Matlab simulation to generate sample paths for the two typical models of geometric Brownian motion and OU model. The interval is logarithmic increment method or direct increment method, n = 10, 20, 30, 40, 50 five sample sizes, the kernel estimation of the drift coefficient is constructed and the sample limitation of the two estimation methods is compared.

The measurement used for the investigation is the mean and variance of the non-parametric estimates of the drift coefficient; the closer the mean is to the true value, the smaller the variance, the better the method, and vice versa.

4 Analysis of Typical Models

4.1 Geometric Brownian Motion

Samples of different volumes are first obtained by measuring the interval along the route, for example, a sample with a volume of 50 is obtained every 10 points. After obtaining the sample, use the direct increment method to estimate the parameters of the drift coefficient of the obtained sample and compare it with the actual value of 0.5X; also use the logarithmic increment method to estimate the sample and compare it with the actual value at this time 0.495 for comparison. It can be clearly seen from Fig. 1 that the logarithmic increment method chosen in this paper has a better effect.

Next, calculate the variance of the estimated parameters, and the results are shown in Table 1.

It can be seen from Table 1 that the variance of the drift coefficient estimated by the logarithmic increment method is smaller than that of the direct increment method, and it converges to zero faster.

Combining the data in Table 1 and Fig. 2, it can be seen that the average value of the logarithmic increment method fluctuates between 0.4–0.6, which is closer to the actual value of 0.495; while the average value of the direct increment method is higher than the actual value, at 0.6-Fluctuating between 1.2. From the point of view of the variance value, the variance of the logarithmic increment method is also smaller, indicating that the logarithmic increment method is more accurate and stable in estimating the drift coefficient than the direct increment method.

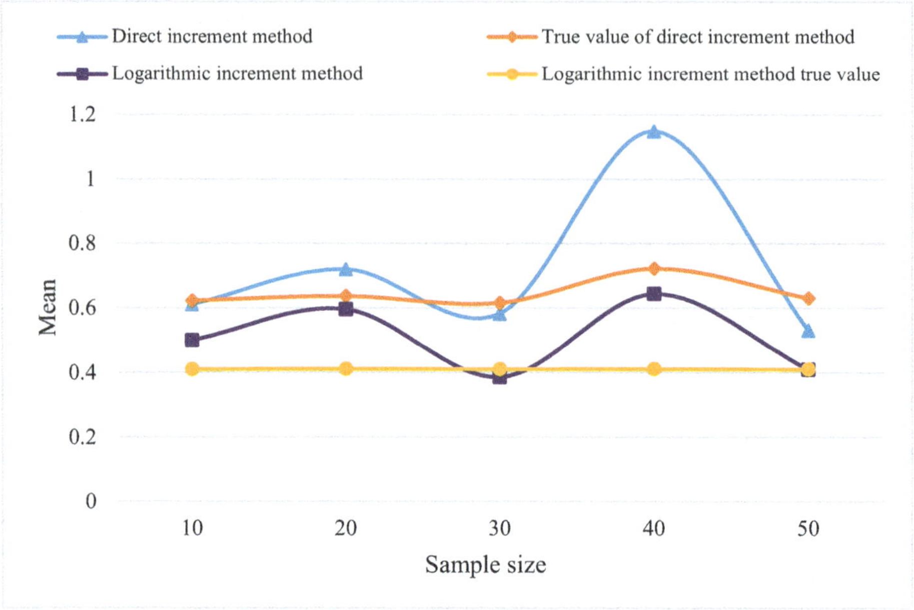

Fig. 2. Geometric brownian motion model drift coefficient nonparametric estimated mean value

Table 1. Comparison of drift coefficient variance of geometric brownian motion model

Sample size	10	20	30	40	50
Direct increment method	0.001	0.06	0.005	0.05	0.007
Logarithmic increment method	0.003	0.02	0.002	0.021	0.002

4.2 OU Stochastic Volatility Model

The method is the same as above, as shown in Fig. 3.

Then we did a statistical analysis as shown in Table 2.

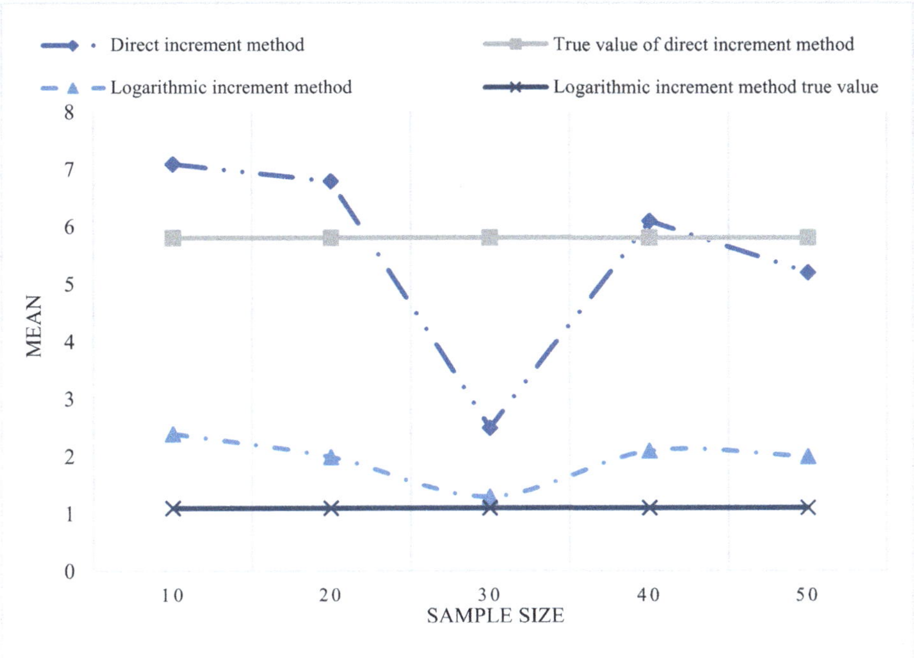

Fig. 3. OU random volatility model mean value

Table 2. Basic statistical indicators of exchange rate data

Average	1.97	Standard deviation	0.72
Skewness	0.24	Kurtosis	−0.53
Min	0.79	Max	3.93

It can be seen from Fig. 3 that the drift coefficient estimated by the logarithmic incremental method is closer, the variance of the drift coefficient estimated by the logarithmic incremental method fluctuates between [0, 20], and the variance estimated by the direct incremental method is in [40, 270], in contrast, the logarithmic increment method is more effective.

5 Conclusions

This article introduces machine learning algorithms, t-SNE, FCN and CNN algorithms, and then gives two diffusion equation drift coefficient estimation methods, namely logarithmic increment method and direct increment method, and proposes a path based on machine learning. The technology of enhancing diffusion into the fluid, and then using two typical models of geometric Brown motion and OU model to test which of the two coefficient estimation methods is more accurate and stable. Experimental results show that the logarithmic increment method is indeed better than the direct increment method.

References

1. Dang, S., Sondergeld, C., Rai, C.: Novel technique to measure mutual bulk fluid diffusion using NMR 1-D gradient. In: E3S Web of Conferences, vol. 146, no. (1–2), p. 03007 (2020)
2. Macdonald, T., Pr Ice, W.S., Astumian, R.D., et al.: Enhanced diffusion of molecular catalysts is due to convection. Angew. Chem. **131**(52), 19040–19043 (2019)
3. Kuo, D.P., Kuo, P.C., Chen, Y.C., et al.: Machine learning-based segmentation of ischemic penumbra by using diffusion tensor metrics in a rat model. J. Biomed. Sci. **27**(1), 80 (2020). https://doi.org/10.1186/s12929-020-00672-9
4. Dagdanpurev, S., Abe, S., Sun, G., et al.: A novel machine-learning-based infection screening system via 2013–2017 seasonal influenza patients' vital signs as training datasets. J. Infect. **78**(5), 409–421 (2019)
5. Boidi, G., Silva, M., Profito, F.J., et al.: Using machine learning radial basis function (RBF) method for predicting lubricated friction on textured and porous surfaces. Surf. Topogr. Metrol. Prop. **8**(4), 044002 (2020). (12p)
6. Yafei, L., Wanlu, J., Hongjie, N., et al.: Fault diagnosis of axial piston pump based on extreme-point symmetric mode decomposition and random forests. Shock. Vib. **2021**(4), 1–16 (2021)
7. Li, R., Deng, Q., Tian, D., et al.: Predicting perovskite performance with multiple machine-learning algorithms. Crystals **11**(7), 818 (2021)
8. Alhamadani, A.A., Al-Faiz, M.Z.: Inverse kinematic based brain computer interface to control humanoid robotic arm. Int. J. Mech. Mechatron. Eng. **20**(1), 15–24 (2020)
9. Dankwa, S., Zheng, W.: Special issue on using machine learning algorithms in the prediction of kyphosis disease: a comparative study. Appl. Sci. **9**(16), 3322 (2019)
10. Chen, Y., Huang, Q., Zhao, J., et al.: Unsupervised machine learning on domes in the lunar gardner region: implications for dome classification and local magmatic activities on the moon. Remote Sens. **13**(5), 845 (2021)
11. Koziarski, M., Krawczyk, B., Wozniak, M.: Radial-based undersampling for imbalanced data classification. Neurocomputing, **343**(MAY 28), 19–33 (2019)
12. Kronberg, E.A., Gastaldello, F., Haaland, S., et al.: Prediction and understanding of soft proton contamination in XMM-Newton: a machine learning approach. Astrophys. J. **903**(2), 89 (2020). (15p)

Risk Model and Decision Support System of State Grid Operation Management Based on Big Data

Siyu Zhang[1], Wei Ou[1], Guanghai Ren[2], Hongyan Wang[2], Pingfei Zhu[2], and Wei Zhang[2](✉)

[1] State Grid Customer Service Center, Tianjin, China
[2] Beijing China Power Information Technology Co. Ltd., Beijing, China
jingjing20200825@163.com

Abstract. With the rapid development of our country's national economic system, the further improvement of domestic power reforms, the improvement of corporate social responsibility, the increase in environmental pollution and natural disasters, power grid companies are facing huge impacts from economic, social, technological, natural and other fields and challenges. In order to reduce the impact of energy and environmental factors on the national economy, it is necessary to improve energy utilization efficiency, strengthen environmental protection, and create a model for scientific and technological development. State Grid operation risk management is critical to whether the State Grid can operate safely and stably. Therefore, this article aims to study the State Grid operation management risk model and decision support system based on big data. Based on the model construction and system design principles, the functional modules of the decision support system based on the State Grid operation management risk model were designed, including the system management module, project module, risk evaluation module and risk control module, and finally tested the performance of the system. The test results show that the maximum response time of the system is 140 ms, which meets the requirements of this article.

Keywords: Big data · State grid corporation · Operational risk · Decision support system

1 Introduction

Electric power companies occupy an important position in Chinese society. Electrical appliances manufacturing products are deeply rooted in the hearts of people in modern society. They are essential products in people's work and life in modern society. They are also an important mainstream industry for national development. It is also used to measure a country's and One of the weights of the overall economic growth rate of the region [1, 2]. With the acceleration and deepening of the current global energy market economic integration development trend, the development of the energy market is also

I. Ahmad et al. (Eds.): STSIoT 2021, LNDECT 122, pp. 419–427, 2023.
https://doi.org/10.1007/978-981-19-3632-6_51

facing potential contradictions, which are related to the healthy growth and sustainable survival of energy companies in the future [3, 4].

In recent years, many scholars have conducted research on the State Grid operation management risk model and decision support system, and have achieved good results. Some scholars have proposed grey relational analysis based on entropy weight and improvement. First, use entropy weight to determine the weight of the power grid enterprise's operating risk index system. Secondly, determine the gray-scale weighted correlation degree of the ideal objects of each power grid company, and then evaluate the business risk of each power grid company. Finally, an example is used to prove the effectiveness and practicability of the method [5]. As more and more variable resources and demand response resources are integrated into the grid, the uncertainty of grid operation becomes higher and higher, and the decision-making process in this environment becomes more challenging. A researcher investigated current risk management practices and proposed a risk management framework for future grid operations. Under this framework, three risk management procedures are proposed. The first process is a robust unit commitment to reliability, the second process is a forward-looking commitment of a scenario-based quick start unit, and the third process is economic dispatch with corrective measures [6]. An expert and scholar designed and compiled a set of CIM-based power grid planning information collection and management decision-making support system for the problem of low efficiency in data collection and management of multiple departments and multi-level power grid planning information. This time the system designed a large number of databases based on the CIM model as a bottom support of the system, and realized data sharing through the gis system, marketing management system and a large number of externally updated databases that were regularly updated [7]. In summary, through the research of domestic and foreign scholars, the risk management and decision support system has been developed very maturely and has been developed by leaps and bounds.

In this paper, combining the characteristics of State Grid's operation, constructs a mathematical model of State Grid's operation and management risk, and designs a decision support system based on the State Grid's operation and management risk model on the basis of following the system design principles, and finally performs a performance test on it.

2 Research on the Risk Model and Decision Support System of State Grid Operation Management Based on Big Data

2.1 Characteristics of State Grid Operations

(1) Difference

Different types of power users in this system have obvious differences in power consumption and usage. Their existence supports the fine division of relevant markets, and at the same time better meets the requirements, adaptation and changing trends of all aspects of the market, and on this basis, the individual needs of relevant users can be met [8, 9].

(2) Chance

The potential for operational risks of the State Grid means that it is not clear when and where the operational risks of the State Grid will occur. The danger of the power system during operation is certain, but due to certain risk factors such as manager negligence and natural disasters, the spread of dangerous accidents is also accidental and inevitable. Such accidents include the uncertainty when the danger occurs, the uncertainty of which unit operates the national grid, the uncertainty of the risk caused by man-made or physical factors, and the uncertainty of the loss of the power grid caused by the danger [10].

(3) Integrity

Both supply and demand parties need to use their own supply facilities during the electricity transaction process, and there is a close relationship between these activities. Although the energy market is connected through the grid as a carrier, although energy companies have the consumption and time of energy users, they provide comprehensive services for all users in a specific area [11, 12].

2.2 Construction of State Grid Operation Management Risk Model

Using genetic neural network to optimize the research on the hierarchical risk management model of smart grid investment projects, taking construction projects as the research object, obtain the fund distribution chain between the various projects of the State Grid investment project. Its mathematical model is shown in formulas (1) and (2):

$$min\,R = \mu_1 R_1(W, X) + \mu_2 R_2(W, X) + \ldots + \mu_n R_n(W, X) \tag{1}$$

$$s.t \begin{array}{c} R_i(W, X) = f\left(\sum WX_i - \theta\right) \\ \theta_i \in (0, 1) \end{array} \tag{2}$$

$$s.t \begin{array}{c} \mu_1 + \mu_2 + \ldots + \mu_n = 1 \\ 0 < \mu_1 < 1 \end{array} \tag{3}$$

Among them: R is the overall risk level of the State Grid investment project; W is the transfer rate of the impact of the risk element on the project; R_i is the risk level of the i-th sub-project; μ_i is the contribution rate of the i-th project to the entire project.

2.3 Design Principles of Decision Support System

(1) Safety and reliability

The system must encrypt the database, apply strict principles for management, and set different levels of user authority for each module and system data. In addition, you need to be able to create Word files, monitor and record detailed information about all user interactions on the system, and create a regular data backup mechanism. It must also have debugging, debugging functions, virus protection and data corruption recovery functions.

(2) Economy

When the corresponding requirements are met, the system needs to save costs as much as possible. In the process of selecting materials, it is necessary to consider its applicability, rather than blindly pursuing technological progress. The design process in this area should try to avoid unnecessary complication, ensure that each part is concise, easy to analyze and process, and the corresponding cost is low.

(3) Openness and scalability

This system is based on the J2EE standard and supports data exchange and sharing with other systems. At the same time, the system's software and hardware have strong scalability, which is conducive to later business customization and system expansion.

(4) Practicality and advancement

The State Grid decision support system needs to be highly practical, combined with cutting-edge computer technologies such as the Internet of Things and big data, to ensure the advanced nature of the system and meet the needs of energy optimization implementation.

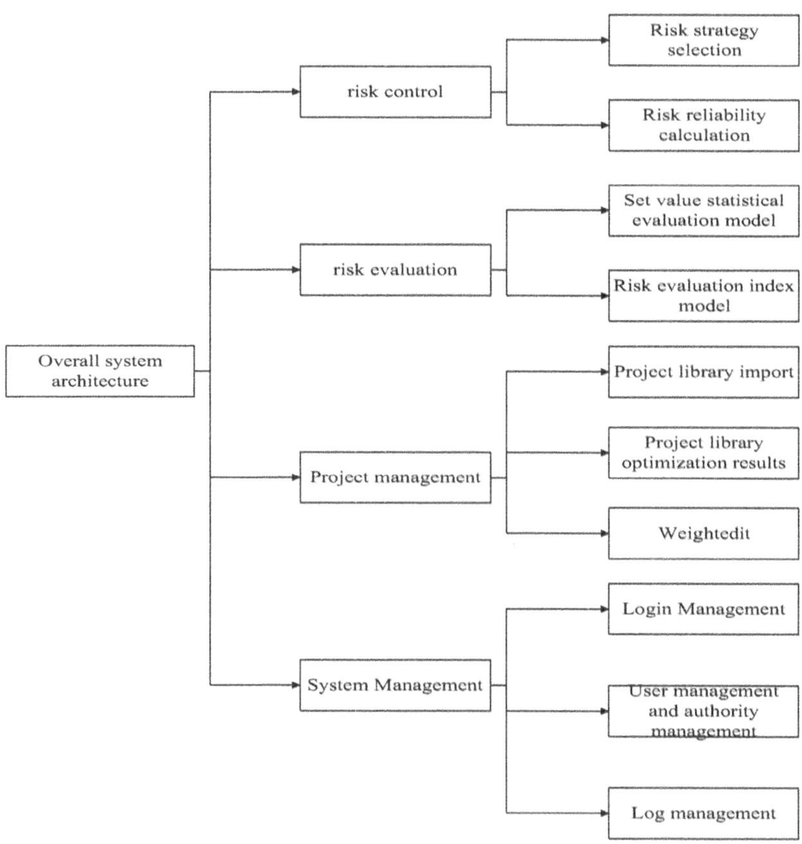

Fig. 1. System overall design

3 Experiment

3.1 System Overall Design

The functional modules of the decision support system based on the State Grid operation management risk model designed in this paper are shown in Fig. 1:

3.2 System Management

(1) Login management

When using this decision support system for the first time, you must register. You can connect after registration. Online users are divided into ordinary users and administrators. Ordinary users can perform query and data entry tasks. The administrator is responsible for managing user information and maintaining the database, and can perform functions

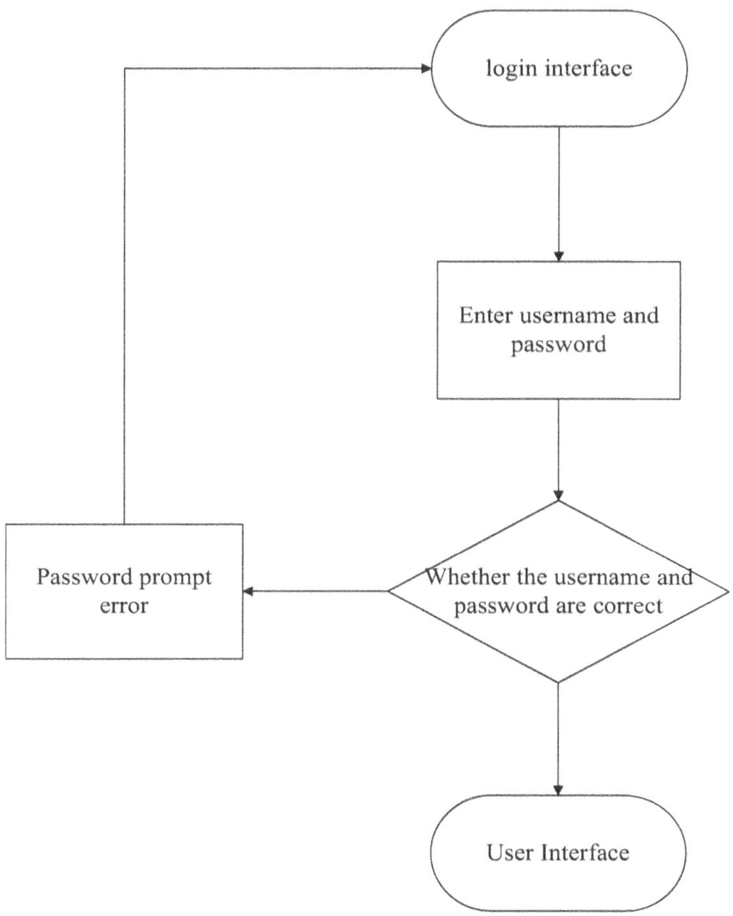

Fig. 2. System login process

such as backup and modification of related database data. The login process is shown in Fig. 2:

(2) User management and authority management

After the user is registered, the system administrator assigns permissions to the user, and users with different permissions have different permissions to the system. The administrator can access user information in the user management unit. This section contains the user name, password, level, registration time, and last login time of users of different levels. The user information manager can add, change or delete this information. When the system is designed, all accesses are subdivided and intercepted, and judged according to authority, to judge whether the user has authority for this function. If you can perform this function at your permission level, you will be redirected to the login page. If not, the system will prompt you "Unauthorized" and take you to the homepage.

(3) Log management

Log management is also an important part of the security system. It is used to record system operations and user usage, provide error messages, and help system administrators maintain the system. Only system administrators and maintenance personnel have the right to use it. The system provides a good basic log support tool. Application-level logs can usually be divided into user logs, system logs and user operating system logs, and a log entry needs to be added to display the basic functions of the user.

3.3 Project Management

The main function of the project library unit is to activate the measurement unit after the project is entered, select the project, send statistics and results into the system dynamics model model, and serve the calculation unit.

Project library import: Enter 5 years of the main distribution network project in the main distribution network project library to view all detailed information and information.

Project selection results: display the project selection results, and rank the projects according to the total score. The displayed content includes all the content in the project, the weight of each indicator, the total score, the number of problems solved, the statistical results of specific indicators, etc. The results can be exported to Excel.

Weight editing: View the weight indicators of the main distribution network and edit the weights to optimize auxiliary components.

3.4 Risk Assessment

Project risk assessment can be implemented using a variety of models, including risk measurement models and fixed-value statistical assessment models. Through the project risk assessment, the risk level of each stage of the project can be reached, providing a basis for risk management.

3.5 Risk Control

Risk management is an important step in the implementation of project risk management. According to the project risk assessment results and risk reliability calculations, reasonable risk strategies are selected to achieve the best project risk management.

4 Discussion

Performance control is mainly to control the performance of the system, such as speed, response time and system performance. Mainly to control system performance and ensure system quality. If the developed system only meets the operational requirements but not the performance requirements, the system will also fail. System test performance points are shown in Table 1.

Table 1. System test performance point table

Test point	Test content	Steps	Explanation of expected results	Actual test result
1	System runtime test	The system is always running	The system runs for a long time without any failure	Consistent with expected results
2	System response time test	Enter different pages, record page conversion time	The maximum response time of the page should be within 3 s	Consistent with expected results
3	System load test	Use LoadRunner to simulate a large number of users accessing the system at the same time	The system is operating normally	Consistent with expected results

The system performance test process uses LoadRunner as a load control tool. Its main function is to predict the behavior and performance of the system. It simulates the behavior of real users in the form of virtual users, simulating multiple users who can access the system, and makes the system respond to different numbers of user access times. The following chart shows the system concurrent response test results.

It can be seen from Table 2 and Fig. 3 that as the number of users accessing the system increases, the response time of the system also increases. When the number of users is 50, the expected response time of the system is 25 ms and the actual response time is 19 ms; when the number of users is 100, the expected response time of the system is 47 ms, and the actual response time is 41 ms; when the number of users is 150, the expected response time of the system is 62 ms and the actual response time is 58 ms; when the number of users is 200, the expected response time of the system is 71ms,

Table 2. The system concurrently responds to test results

	Expected response time	Actual response time
50	25	19
100	47	41
150	62	58
200	71	63
250	92	84
300	111	105
350	147	140

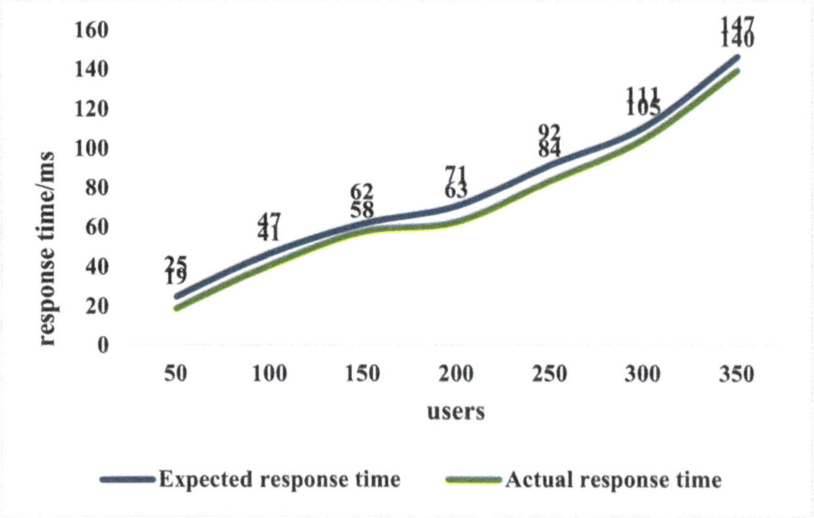

Fig. 3. System concurrent response time test results

and the actual response time is 63 ms; when the number of users is 250, the expected response time of the system is 92 ms, and the actual response time is 84 ms; when the number of users is 300, the expected response time of the system is 111 ms, and the actual response time is 105 ms; when the number of users is 350, the expected response time of the system is 147 ms, and the actual response time is 140 ms. It can be seen that the system meets the design requirements.

5 Conclusions

With the development of society and economy, the demand for electricity continues to increase, and the problem of low energy consumption is becoming more and more serious. It has promoted energy conservation, emission reduction and consumption, and improved efficiency in part of the energy demand. It has become the energy system and all energy-using enterprises. Predicting and responding to future energy usage, helping companies collect, process, and export energy data, examine their energy optimization capabilities, and improve the scientific and intelligent level of energy usage decision-making has far-reaching research value and practical significance.

References

1. Attia, S., Shafik, Z., Ibrahim, A.: New Cities and Community Extensions in Egypt and the Middle East Operational Risk Model Managing Urban Safety in Real State, pp. 199-216 (2019). https://doi.org/10.1007/978-3-319-77875-4 (Chapter 11)
2. Baldwin, S.: Business continuity management as an operational risk service provider: an approach to organisational resilience. J. Bus. Contin. Emer. Plan. **13**(2), 102–110 (2019)
3. Ssa, B., Nq, A., Msma, B., et al.: Integrating flare gas with cogeneration systems: operational risk assessment. J. Loss Prev. Process Ind. **72**(3), 104571 (2021)
4. Boyer, M.M., Cowins, E.P., Reddic, W.D.: Operational risk management and regulatory investment constraints on portfolio allocation: evidence from property and casualty insurers. J. Regul. Econ. **57**(1), 20–52 (2019). https://doi.org/10.1007/s11149-019-09396-7
5. Juri, T., et al.: Error grid analysis for risk management in the difference between invasive and noninvasive blood pressure measurements. J. Anesth. **35**(2), 189–196 (2021). https://doi.org/10.1007/s00540-020-02893-6
6. Veeramany, A., Coles, G.A., Unwin, S.D., et al.: Trial implementation of a multihazard risk assessment framework for high-impact low-frequency power grid events. IEEE Syst. J. **12**(4), 3807–3815 (2018)
7. Wawrzyniak, K., Padrón, E.U., Gomulski, K., et al.: Methodology of risk assessment and decomposition in power grid applications. IET Gener. Trans. Distrib. **12**(15), 3666–3672 (2018)
8. Magnus, D.S., Schindler, M.B., Marlow, R.D., et al.: A service evaluation of a hospital child death review process to elucidate understanding of contributory factors to child mortality and inform practice in the English national health service. BMJ Open **8**(3), e015802 (2018)
9. Ding, Z., Xie, L., Lu, Y., et al.: Emission-aware stochastic resource planning scheme for data center microgrid considering batch workload scheduling and risk management. IEEE Trans. Ind. Appl. **54**(6), 5599–5608 (2018)
10. Lundstrom, C., Lindblom, J.: Considering farmers' situated knowledge of using agricultural decision support systems (AgriDSS) to Foster farming practices: the case of CropSAT. Agric. Syst. **159**(1), 9–20 (2018)
11. Damoiseaux-Volman, B.A., Medlock, S., Eijk, M., et al.: Falls and delirium in older inpatients: work-as-imagined, work-as-done and preferences for clinical decision support systems. Saf. Sci. **142**(6), 105355 (2021)
12. Hosni, M., García-Mateos, G., Carrillo-de-Gea, J.M., et al.: A mapping study of ensemble classification methods in lung cancer decision support systems. Med. Biol. Eng. Comput. **58**(10), 1-17 (2020). https://doi.org/10.1007/s11517-020-02223-8

WSN Data Fusion Algorithm Based on Improved ARMA Prediction Model and Compressed Sensing

Mufang Hu(✉), Wenchao Yang, and Deming Jiang

School of Artificial Intelligence, Hezhou University, Hezhou, Guangxi, China
humufang2021@163.com

Abstract. In order to reduce the energy consumption of WSN data fusion and improve the reliability of data fusion, we propose a hybrid data fusion algorithm based on improved data ARMA prediction model and compressed sensing technology (HDFAC). By analyzing the characteristics of the temporal and spatial correlation of monitoring data, the excess valued elimination mechanism is used to remove redundant invalid data, and then a prediction model is established to estimate the monitoring value. The predicted value of high credibility is uploaded to the cluster head, the cluster head node compresses the data, and then the original data is reconstructed at the Sink node to reduce the overall energy consumption of the network. Experimental results show that the HDFAC algorithm can effectively balance the load between nodes, reduce the amount of nodes sent, and extend the network life.

Keywords: Wireless sensor networks · Data fusion · Improved ARMA model · Compressed sensing

1 Introduction

WSNs is a special wireless network composed of many sensor nodes for monitoring, collecting and processing environmental data. At present, it is mainly used in environmental monitoring, disaster prevention and target tracking [1–3]. However, most WSN nodes are limited by cost and volume, usually powered by batteries with limited energy. Therefore, when transmitting data, the influence of data transmission volume on energy consumption of nodes should be fully considered on the premise of ensuring data reliability. Among many technologies of WSNs, data fusion technology occupies a very important position [2, 4]. Data fusion technology eliminates redundant information and reduces the amount of data needed to be transmitted in the network by processing the signals collected by sensor nodes, thus reducing the energy consumption of the network [5].

Because the method of data fusion can effectively reduce energy consumption, many scholars have studied data fusion. Literature [6] proposed a data fusion algorithm for wireless sensor networks based on compressed sensing theory. The algorithm can reduce

I. Ahmad et al. (Eds.): STSIoT 2021, LNDECT 122, pp. 428–436, 2023.
https://doi.org/10.1007/978-981-19-3632-6_52

the data transmission of cluster head to a certain extent, but the algorithm has no abnormal data processing mechanism and does not consider the reliability of data. Reference [7] proposed an intra-cluster data fusion algorithm. Based on the statistical characteristics of node data distribution reflected by information entropy, the upper and lower thresholds of data fusion are determined by using the maximum optimization of one-dimensional to two-dimensional information entropy for local fusion and redundant data filtering.

This paper adopts different data processing methods for different levels of data. At the source node, an outlier removal mechanism is used to remove redundant invalid data, and an improved ARMA prediction model is used to estimate the monitoring value. The nodes in the cluster upload the data with high credibility to the cluster head, and the cluster head node data is compressed and uploaded to the Sink node using the compression sensing algorithm. After accepting the compressed data, the Sink node uses the reconstruction algorithm to restore the original data and upload it to the user, reducing the energy consumption of the network data and improving the reliability of the data.

2 Time Series Prediction Model and Compressive Sensing Theory

According to the time correlation and spatial correlation of monitoring data, an autoregressive moving average mixed model (ARMA model) is used in the data prediction stage [8–11]. In the ARMA prediction model, the time series of the monitoring data is set to {Xt} = {X1,X2,...,Xn}, and the time series has a certain degree of stationarity, normality, and zero mean value, and then the {Xt} It must be possible to fit the stochastic difference equation as shown in Eq. (1):

$$S_t = \lambda_1 S_{\mathrm{t-1}} + \lambda_2 S_{\mathrm{t-2}} + \ldots + \lambda_n S_{\mathrm{t-n}} - \gamma_1 \alpha_{\mathrm{t-1}} - \gamma_2 \alpha_{\mathrm{t-2}} - \ldots - \gamma_m \alpha_{\mathrm{t-m}} + \alpha_t \quad (1)$$

In the formula, St Is the data value of time series X at t; λi(i = 1,2,...,n) is the parameter of the autoregressive (AR) part; The sequence {αt} is the residual sequence between the real and predicted values;{αt} Satisfy αt ~ NID (0, σa2)) If formula (1) can better reflect the real situation of the system, then the sequence {αt} is white noise. The order of the autoregressive part and the moving average part of the model is n and m. Predictive value expressed as follows.

$$\hat{S}_t = \lambda_1 S_{\mathrm{t-1}} + \lambda_2 S_{\mathrm{t-2}} + \ldots + \lambda_n S_{\mathrm{t-n}} - \gamma_1 \alpha_{\mathrm{t-1}} - \gamma_2 \alpha_{\mathrm{t-2}} - \ldots - \gamma_m \alpha_{\mathrm{t-m}} \quad (2)$$

Confidence level α The range of true values at time t falls in the range: $M_t \in \left[\hat{S}_t - Z_{1-\mathrm{a}/2}\sigma_a, \hat{S}_t + Z_{1-a/2}\sigma_a\right]$; The standard deviation of the model residual.

$$\sigma_a^2 = \sqrt{\frac{1}{N}\sum\nolimits_{i=1}^{N} (a_i - \overline{a})^2} \quad (3)$$

where N is the length of the model data. $\overline{a}$ is the average of error sequences.

Considering the seasonality of monitoring data requirements. When using the ARMA model to predict data, seasonal influence factors can be added to the original model. We should know the monitoring data at the same time in the previous cycle.

The improved ARMA prediction model is:

$$Y = AX + BX' + CE \tag{4}$$

In the formula B is the seasonal factor coefficient matrix B = [b1,b2,…,bp];Where X is the preperiod measurement matrix X' = [x(k-24) x(k-25)….x(k-24-(p-1)]T;p is the seasonal factor order for this model; Recorded as ARMA (n, p, m) seasonal model.

After the source nodes establishe the ARMA (n,p,m) model, then use the model to predict the next monitoring data. Given confidence level α; Obtaining monitored data falls in the interval Mt in the normal state, if the node monitoring value is in the interval Mt, it means that the prediction model is normal and there is no need to update the model. Otherwise, judge whether the detection value is abnormal, if it is abnormal, discard the value, otherwise use the latest S data to re-model.

Compressed sensing theory compresses the monitored data according to the temporal and spatial correlation of the monitored data. [12, 13] To reduce the amount of data transmission and improve the utilization of the network. For a vector of length N, y = (y1,y2,…,yN)T, an invertible orthogonal transform matrix MN*N = [M1,…,MN], which can be a column vector in M the linear combination means that:

$$y = M\theta = \sum\nolimits_{n=1}^{N} \theta_n m_n = \sum_{l=1}^{j} \theta_{nl} m_{nl} \tag{5}$$

where y is a j sparse signal at base M, and MN*N is a sparse transform base. The signal y is j sparse signal at base M, then, the high-dimensional signal y may be projected onto the low-dimensional space XM through a measurement matrix ON*N that is irrelevant to the sparsely-transformed base MN*N, and high-probability reconstruction of the data is guaranteed. That is $x_M = O_{M*N} y_N$ combined with formula (5).

$$x_M = O_{M*N} y_{N*1} = O_{M*N} M_{N*N} \theta_{N*1} = A_{M*N} \theta_{N*1} \tag{6}$$

where OM*N is the number of measurements of the measurement matrix, and its value is related to the sparsity j, where M = cjlg(N/k), where c is a constant.

3 Data Fusion Algorithm

Based on the LEACH clustering routing mechanism, a data fusion algorithm HDFAC (Improved Forecast Method and Compressed Sensing in Data Fusion) is proposed. Increase the seasonal influence factor, and use the outlier elimination mechanism at the source node to remove redundant invalid data. The nodes in the cluster upload the data with high credibility to the cluster head, and then compress the data using compression-aware algorithm and upload it to the Sink node, and then use the reconstruction algorithm to obtain an algorithm for the original data number.

Assume that the sensor nodes are broadcast unevenly in the network monitoring area, and the cluster head node is elected through a routing algorithm, and there is only one Sink node in the monitoring area. Nodes in the network have the following properties:

①Normal nodes, aggregation nodes, and base stations are fixed once they are deployed, and all nodes have the same initial energy. Sink node energy is provided by the external power supply.
②All nodes are homogeneous and have one and only one ID.
③All nodes can adjust the transmit power according to the distance.
④Ordinary nodes can complete data prediction and outlier elimination.

The source node uses the above-mentioned abnormal data detection mechanism to filter out the invalid data, and then introduces the improved ARMA prediction model to predict the data, and uploads the highly reliable data to the cluster head node. The cluster head node uses the sparse matrix to compress the uploaded data group and upload it to the Sink node. The Sink node reconstructs the data using the newly proposed threshold segmentation algorithm. The algorithm has low computational complexity and is suitable for wireless sensor networks. Starting from X0 = 0, the reconstruction algorithm calculates the sparse solution of Eq. (7) according to the iterative equation of Eq. (8).

$$Z = \Phi X = \Phi\Psi S = AS, A = \Phi\Psi \tag{7}$$

In the formula, the observation matrix is $\Phi = [\Phi_1, \Phi_2, \ldots, \Phi_n]^T$, the observation value is $Z = [z_1, z_2, \ldots, z_n]^T$. After the observation matrix projection, the N-dimensional original signal X is compressed into the M-dimensional observation value Z.

$$x^{n+1} = H_k\left(x^n + \Phi^T\left(y - \Phi x^n\right)\right) \tag{8}$$

In the formula, it is a non-linear operator, except for the largest parameter, other values are set to 0. If $||\Phi_2|| < 1$, then the algorithm will definitely converge to the local minimum of the cost function $||Z - \Phi x||_2$.This iteration gives the most sparse approximation.

4 Simulation and Result Analysis

In MATLAB environment, the performance of the algorithm was simulated. The parameter settings are shown in Table 1.

The experimental data are from Intel Berkeley research lab dataset. The data have two different time periods (Time1, Time2) and predicted using the improved ARMA prediction model. The prediction results are shown in Fig. 1 below. The predicted value and the measured value are the average values of the same area within the same time point.

From the analysis of Fig. 1. There is a maximum relative error of 6.7% over the Time1 period, with an average relative error of 2.7%. The maximum relative error in the Time2 period was 6.4% with an average relative error of 2.1%. The average relative error of the two sets of data was 2.4%. The calculated relative error results are all less than 10%. It can be seen from Fig. 1 that the monitoring value is similar to the prediction. Within the

Table 1. Simulation parameter settings

Parameter	Values and units	Parameter	Values and units
Number of nodes	100	Transmit power	50 nJ/bit
Scene range	200 m × 200 m	Fusion unit data energy consumption	5 nJ/(bit.signal)
Initial energy	0.6 J	threshold ε	0.05
Communication radius	50 m		

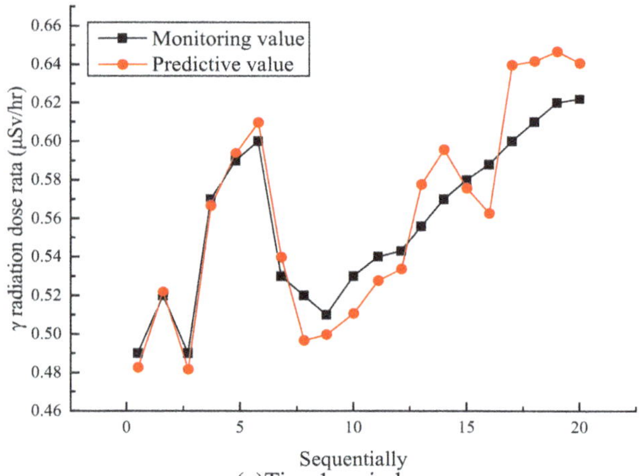

(a)Time1 period

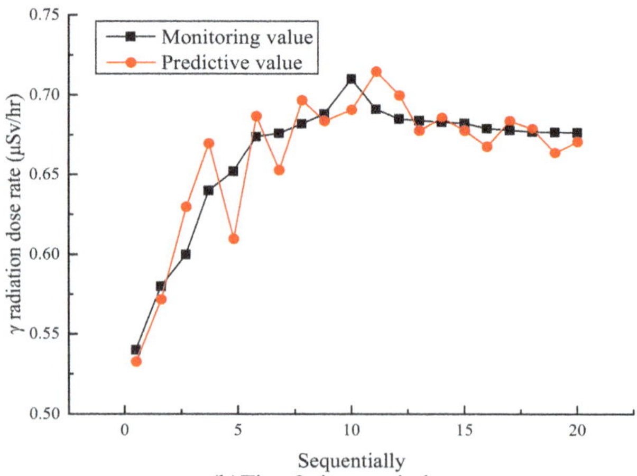

(b)Time2 time period

Fig. 1. Comparison of improved ARMA predictions

general requirements, the improved model can be used to predict the monitoring value well.

In the simulation experiment, the sparse base adopts DCT base and adopts the general compression rate signal as the test signal respectively. The conversion coefficient is shown in the following formula (9).

$$|\theta|_{(i)} = i^{-\frac{1}{p}} \tag{9}$$

The compressive sensing technology is used to observe, project and reconstruct the data collected by sensors to verify the application of compressive sensing technology in practical wireless sensor networks [14]. The quality of data reconstruction uses the reconstruction error as an evaluation criterion, as shown in the following formula (10).

$$\lambda = E\left(\frac{||\overline{x} - x||x_F}{||x||_F}\right) \tag{10}$$

where $\overline{x}$ for the reconstructed data, x for the original data, E express average value.

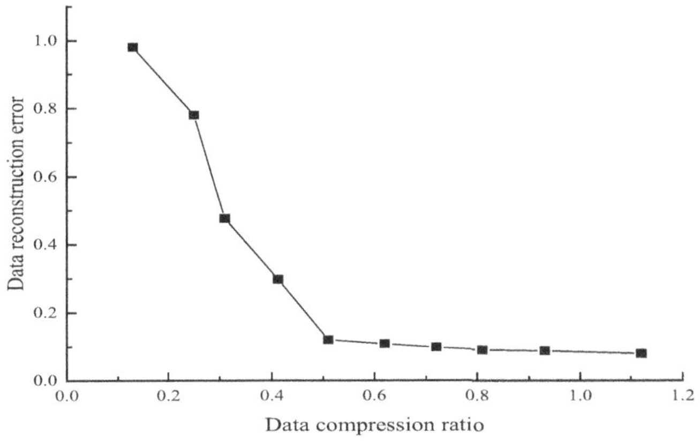

Fig. 2. Reconstruction error and compression ratio

Figure 2 showing the variation of the reconstruction error with the data compression ratio when there are 40 cluster nodes in the simulation area.With the increase of data compression ratio, the error of data reconstruction is correspondingly decreasing. When the data compression ratio is 0.1−0.5, the data reconstruction error decreases most rapidly. When the compression ratio exceeds 0.5, the data reconstruction error is Within a certain range, there is no major change. When the data compression ratio reaches 1, the reconstruction error of the data is close to 0. This shows that the HDFAC fusion algorithm can effectively reduce the cluster head data transmission, saving network energy consumption.

Figure 3 showing the compares the HDFAC algorithm with the LEACH algorithm and the literature [6]. From the perspective of the overall trend, the average remaining energy of the three algorithm nodes decreases with the increase in the number of network

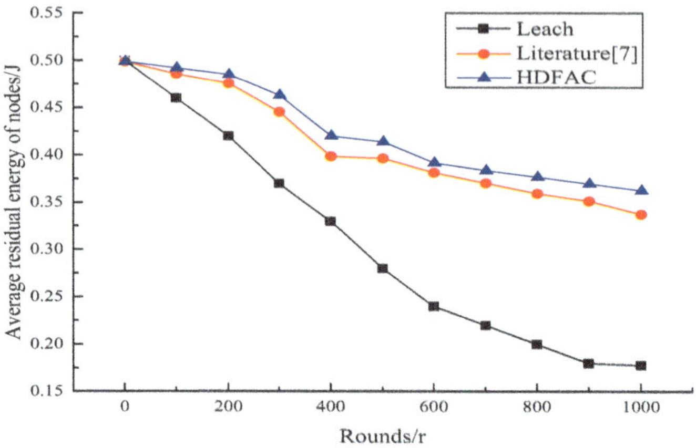

Fig. 3. Comparison of residual energy of three algorithm

operations.The energy consumption of the LEACH algorithm is the fastest. The node consumes the fastest energy in the 200 to 600 rounds of the network operation. However, in the literature [6] and the HDFAC algorithm, the energy consumption of the node is relatively stable. The latter two algorithms play a greater advantage in the later period. The HDFAC algorithm has significant advantages over the LEACH algorithm and literature [6] in the average residual energy of nodes, and the average energy consumption of the nodes in the network is the smallest. The average energy consumption of the nodes of the HDFAC algorithm is small, which can improve the utilization of the nodes in the entire network area and extend the life cycle of the nodes, thereby prolonging the life of the network.

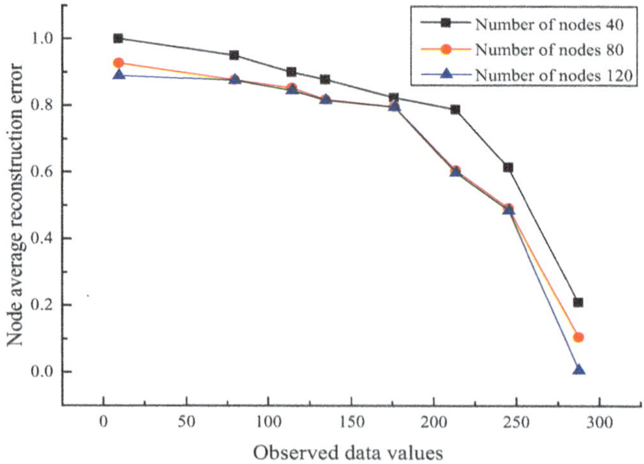

Fig. 4. Change of data with average reconstruction error at different node ratios

Figure 4 showing the average reconstruction error of nodes 40, 80, and 120 in the cluster as a function of the number of nodes observed data. The signal length N of the nodes in the network is set to 300. From the analysis in Fig. 4, we can see that when the network observation value is close to the node signal length, the average reconstruction error of the node is close to 0; When the number of observations is constant, when the intra-cluster node is 120, the reconstruction error value is the smallest, and the fewer nodes in the cluster, the larger the error. It shows that the HDFAC algorithm has higher precision in reconstructing data when the number of nodes in the cluster is large, which is suitable for large-scale networks.

5 Conclusion

Aiming at the problem of energy consumption and data loss in data fusion, a new data fusion algorithm based on improved prediction model compressive sensing (HDFAC) was proposed. First, use the improved ARMA prediction model at the source node to predict data and reject outliers. Upload high-reliability monitoring values to the cluster head. Then use the spatio-temporal correlation of the data collected between WSN nodes and the joint sparse model. Data is compressed and reconstructed. The simulation verifies and analyzes the relationship between the reconstruction error and the number of observed data and the energy efficiency of the algorithm. It shows that the HDFAC algorithm can eliminate data redundancy, reduce energy consumption, balance the network load between nodes, and can effectively improve the network lifetime.

Acknowledgements. This work was supported by Hezhou University Natural Science Foundation 2020ZZK01.

References

1. Chen, Z., Tian, L., Lin, C.: Trust model of wireless sensor networks and its application in data fusion. Sensors **17**(4), 703 (2017)
2. Chen, I.Z.: An algorithm of mobile sensors data fusion tracking for wireless sensor networks. Wirel. Pers. Commun. **58**(2), 197–214 (2011). https://doi.org/10.1007/s11277-009-9888-8
3. Khaleghi, B., Khamis, A., Karray, F.O., Razavi, S.N.: Multisensor data fusion: a review of the state-of-the-art. Inf. Fusion **14**, 28–44 (2013)
4. Roy, S., Conti, M., Setia, S., Jajodia, S.: Secure data aggregation in wireless sensor networks. IEEE Trans. Inf. Forensics Secur. **7**, 1040–1052 (2012)
5. He, H., Zhu, Z., Mäkinen, E.: Task-oriented distributed data fusion in autonomous wireless sensor networks. Soft. Comput. **19**(8), 2305–2319 (2014). https://doi.org/10.1007/s00500-014-1421-7
6. Xianju, F., Xiaofang, L.: Wireless sensor network data fusion algorithm based on compressed sensing theory. J. Jilin Univ. (Science Edition), **54**, pp. 575–579 (2016). (in Chinese)
7. Li, X.L., Kang, H., Cao, J.N.: Coordinated workload scheduling in hierarchical sensor networks for data fusion applications. J. Comput. Sci. Technol. **23**(3), 355–364 (2008). https://doi.org/10.1007/s11390-008-9138-7

8. Nezhad, S.M.T., Nazari, M., Gharavol, E.A.: A novel DoS and DDoS attacks detection algorithm using ARIMA time series model and chaotic system in computer networks. IEEE Commun. Lett. **20**(4), 700–703 (2016)
9. Zhang, X.G., Zhu, S.F., Zheng, Q.S., et al.: Lower order data fusion algorithm based on ARMA model for wireless sensor networks. J. Chin. Comput. Syst. **35**(8), 1771–1774 (2014)
10. Xiguang, Z., Sifeng, Z., Qiusheng, Z., et al.: Lower order data fusion algorithm based on ARMA model for wireless sensor networks. J. Chin. Comput. Syst. **35**(8), 1771–1774 (2014)
11. Xia, Y., Zhao, Z., Zhang, H.: Distributed anomaly event detection in wireless networks using compressed sensing. In: Proceedings of the 2011 11th International Symposium on Communications and Information Technologies (ISCIT), Hangzhou, China, pp.12–14 (2011)
12. Xu, X., Ansari, R., Khokhar, A., et al.: Hierarchical data aggregation using compressive sensing (HDACS) in WSNs. ACM Trans. Sens. Networks **11**(3), 1–25 (2015)
13. Singh, V.K., Singh, V.K., Kumar, M.: In-network data processing based on compressed sensing in WSN: a survey. Wirel. Pers. Commun. **95**(5), 1–38 (2017)
14. Caione, C., Brunelli, D., Benini, L.: Compressive sensing optimization for signal ensembles in WSNs. IEEE Trans. Ind. Inform. **10**(1), 382–392 (2013)

Navigation System in Space Environment Under Internet of Things Era

Ming Lv and Aimeng Wang(✉)

Shenyang Jianzhu University, Shenyang, Liaoning, China
1020617982@qq.com

Abstract. With the increasing development of life aesthetics, the forms of guide system design are becoming more and more diversified. It is no longer a single design and production project defined by materials, shapes, or processing types, but a "people-oriented" system design that integrates into People's Daily life to meet the aesthetic requirements and combines with the space environment. In modern city life, the information between people and the environment is more and more closely related to its requirements are also more and more high, small to an office building, large to a region or even a city are required to have a scientific and humane guide system design.

Keywords: Intelligent interconnection · Guide system · Environmental space · Design · People-oriented · Internt of Things

1 Introduction

The Internet of Things is the key to making any system smart [4]. The development of the society, technological progress, product updates, the accelerating rhythm of life, and so on a series of socia land physical factors, make people in the enjoyment of material life at the same time, pay more attention to the product in the "convenient", "comfort", "reliable", "value", "safe" and "efficiency" and so on, which is mentioned in the product design humanization design problems. In the rapidly developing city, the guide system can be seen everywhere, infiltrating into everyone around us, it is no longer a simple traffic guide and life guide, but gradually become a society where people's spiritual culture development, and reflects the harmonious development of social civilization. Is a part of the public environment, from the design concept to the actual implementation, must follow the "people-oriented" design purpose, combined with the space environment.

2 Visual Design is Applied to Environmental and Spatial Traceability

The design and application of visual elements in environmental space has a long history, which can be traced back to the cuneiform script used by people over 2000 BC. The use of cuneiform script marked the beginning of the use of abstract symbols to record

I. Ahmad et al. (Eds.): STSIoT 2021, LNDECT 122, pp. 437–442, 2023.
https://doi.org/10.1007/978-981-19-3632-6_53

events, and provided the basis for the development of abstract visual symbols in the future. The design of guiding system is not a sign or a road sign in the traditional sense. It needs to combine the content of symbols, colors, architecture, landscape, space and other professional fields to form a systematic design that integrates with the environment and space. It is an organic combination of artistic creation and rational conception.

The guide system is composed of the sign and the sign system. The sign is the medium of information The guide system is composed of the sign and the sign system. The sign is the medium of information transmission, which is conveyed in the form of symbols and conveyed by the carrier of materials. Such indicative symbols are widely used in public space, closely related to People's Daily activities, and are not limited to visual expression, but also include auditory, tactile and other sensory systems. Sign system design is the overall systematic design research of sign, its constituent elements include symbol, text, color and so on. The content information, the form of expression and the environment position are the important contents to be considered in the design of the navigation system.

The advertising system design cannot leave it in the space environment, visual guide system design includes indicator, signs, route guidance, etc., its color, pattern, visual should match the space environment. We often have this experience, their direction is difficult to find the destination in accordance with the instruction of the advertising, advertising font is not clear, it is difficult to identify, is beyond the scope of best visual sign position, is truly reflected in people's life of some of the advertising system design in space is not really from the perspective of "user" to design.

3 Understanding the Design of an Internet-of-Things

A good first step is to think abou what we mean when we talk about the loT [1]. Internet of Things is an important part of the new generation of information technology, is the extensionand expansion of the Internet based on the network, will be a variety of information sensor equipmentand network combined to form a huge network, to achieve any time anywhere, the interconnection ofpeople, machines and things. He Internet of Things platform can process massive data user resources, etc. [3].

4 The Role of Navigation System in Space Environment in the Era of Internet of Things

4.1 Environmental Factors

The vision of the Internet of Things and its impact on product design and manufacturing has been shaping future implementations for years [2].

Land is the basis of human activities, but the development and expansion of urban economy has brought a lot of ecological problems [5]. In contemporary art, visual art and environmental art have a particularly profound influence on the design form of visual guide system. The development of diversified forms of visual art has a direct impact on the visual perception of the design of the guiding system. The transformation from standardized and popular style to personalized and artistic style also elevates the two-dimensional and three-dimensional visual experience to a new height. The design of the guiding system, as the guiding road sign of coordinating the city, is born in and serves the environment. Environmental art has a deep influence on the development of the guiding system design for the relationship between people and the environment and environmental factors.

City guide system design needs to understand the culture and customs of each urban area, understand the cultural characteristics and differences of the nation, only to the city culture and ethnic customs in-depth understanding, in order to better for people's life and urban construction.. This kind of urban guide system with regional natural style is designed to become the unique image of this region or city. When designing a visual guide system in a city of different cultures, we should abstract out the elements such as lines, colors and culture according to the urban environment and architectural style, and make them into recognizable forms of beauty. Enrich the connotation of urban culture, for the city's beautiful environment to add highlights.

4.2 Human Factors

The Internet of Things offers companies the opportunity to design new technologies [6]. The design concept based on "people" aims to adjust the psychological state of the public through the visual design of the environment, and to play the bridging role between man and nature, man and the emotional environment of space through the external expression form of design. Humanized visual guide system design starts from the study of human needs. Because different groups of people have different behaviors and psychological conditions, different needs of various groups should be studied and visual images designed to meet their needs should be designed according to their activity characteristics. Aesthetic psychology refers to the psychological aesthetic feeling obtained in the experience and feeling, including pleasure and enjoyment. In life, people's psychological characteristics of "discovering beauty, creating beauty and needing beauty" are also constantly affecting the design field, making the design form of guiding system change. Any design is the result of the internal environment adapting to the external environment. Internal environment refers to people's own thinking ability to analyze and judge the environment formed by external natural conditions so as to find the best solution. Survival of the fittest, design is for people to survive better, in the city visual guidance system, such as road signs, signal lights, station signs, so that our life becomes more convenient, reflecting the humanistic care in the guide design.

5 The Role of Navigation System in Space Environment in the Era of Internet of Things

Now, the Internet of Things generally refers to a ubiquitous network of smart everyday objects [8].

5.1 The Fuctional

By melding the digital and physical worlds, the Internet of Things will generate vast amounts of real-world data, enabling new ways of creating value based on data [7]

Indicating function refers to the imperative visual communication effects of visual guide system, through studies of semiotics, extract the abstract symbols graphics to represent the role of information description and explanation, to be accurate to convey command information, in the selection of elements such as form, color symbolism contact region has a lot of space, To analyze the space environment to develop a unique guide system to maximize the indicative function. The Internet of Things is made up of billions of connected knowledge sharing and useful data devices, people and services [9] Directional function refers to the guiding function of the visual guidance system, which conveys the region and characteristics of the system to people through indicative symbols. With the acceleration of globalization, the competition among cities in different countries includes the competition in economy, quality of life and culture. The important criterion to measure the livability of a city is the characteristics and convenience of the city.

The ideographic function refers to its specific function, which shapes the strategic image and regional characteristics of a city or region through visual guidance. For example, in the process of driving on the highway, entering the entrance of a city, you will see some pointing signs, mainly to promote the city's main characteristic industries and tourism image. These greatly enhance the image of the city and highlight the regional cultural connotation of the specific space, which is the core of distinguishing the different urban styles. At the same time has the appearance, mainly refers to the visual guidance system has a specific indication, pointing role.

Media-centric applications and services in the Internet of Things can provide the interpretation of multimedia content delivered by media devices such as cameras and microphones [10].

In the design of visual guiding system, the form of guiding system design is judged according to the characteristics and personality of the city image. Human vision is very sensitive to images. The information combined with graphics and colors can be integrated into a specific space environment to play a good role in guiding signs. From the perspective of visual communication, the focus of the design of the guiding system is to use simple and eye-catching graphic symbols to express accurate meaning, that is, to construct visual language with symbol language, and to form a unique guiding logo with unique visual language.

5.2 The Artistry

In the design, we follow the principle of formal beauty, use the reasonable expression form of art, and give people visual impact and appeal through the expression form of art. Art design of each era is related to the material environment of each era, appropriate, green, circular, sustainable, is the direction of development we should focus on. The ability of observation determines the level and depth of a person's perception, which is closely related to the rationality and creativity of design.

6 Conclusion

Guide system every day can be easily found in our living environment, and even sometimes rely on public space guide system to continue our next step of work or life. It can be said that the public space guide system is closely related to our life. To build a convenient and unique city, we need to take the ergonomics theory as the basis and guide design as the carrier to deliver the cultural core of the city. The most important part of the public space guide system lies in the interactive information communication with people. It clearly conveys the accurate location information to people in the obvious visual area through color, design and material, so that people can understand the key information and reach the destination in a short time. In addition to the basic function of information transmission, the public information system also affects people's visual psychological feelings in the aspects of environmental atmosphere and design beauty. In terms of emphasizing the harmony of environment and aesthetic value of design, public information system also influences people's aesthetic feeling. The guiding system of public space not only influences the way of guiding people's direction and conveying the guiding information, but also has an important aesthetic influence on the level of environmental design and design.

As a part of the environmental space, the visual guide system has strong spatial expression power. In the design, we should make good use of the environmental conditions to increase the visual focus. The new artistic expression and aesthetic way are integrated into the design of the visual guiding device, so that the guiding system not only has the cognitive guiding function, but also can fully display the spatial experience, and to the maximum extent to meet people's psychological and spiritual needs. Different environmental space has different characteristics, in the design process of visual guide system should consider the connection between humanistic care and environmental space, on the basis of adapting to environmental conditions, people-oriented, fully consider the psychological characteristics of the public, into the humanistic care; On the basis of paying attention to the design of humanistic care, the use of modern scientific concepts and advanced technology, constantly carry out innovative design, green design, so that the guide system gradually standardized and rationalized. In the design of continuous innovation, the visual guide system in the environmental space is no longer limited to the original positioning, but is integrated with the environmental space and becomes a symbol of social civilization.

References

1. Schuck, T.M.: Cybernetics, complexity, and the challenges to the realization of the Internet-of-Things. Procedia Comput. Sci. **185**, 45–54 (2021)
2. Bradley, R., Jawahir, I.S., Murrell, N., Whitney, J.: Parallel design of a product and Internet of Things (IoT) architecture to minimize the cost of utilizing Big Data (BD) for sustainable value creation. Procedia CIRP **61**, 58–62 (2017)
3. Almagrabi, A.O.: A pervasive controlled access with privacy delegation design for smart internet of things applications. Measurement **172**, 108875 (2021)
4. Ahmed, M.S.: Designing of internet of things for real time system. Mater. today Proceedings (2021)
5. Cheng, F., Wang, Y.: Research and application of 3D visualization and Internet of Things technology in urban land use efficiency management. Displays **69**, 102050 (2021)
6. Hughes-Lartey, K., Li, M., Botchey, F.E., Qin, Z.: Human factor, a critical weak point in the information security of an organization's Internet of things. Heliyon **7**(3), e06522 (2021)

7. Nitschke, P., Williams, S.P.: Conceptualizing the Internet of Things data supply. Procedia Comput. Sci. **181**, 642–649 (2021)
8. Hu, W., Ma, X.: Research on design and development of home virtual system based on internet of things system. Microprocess. Microsyst. **82**, 103857 (2021)
9. Srinadh, V., Rao, M.S., Sahoo, M.R., Rameshchandra, K.: An analytical study on security and future research of Internet of Things. Mater. Today Proceedings (2021)
10. Allouche, M., Mitrea, M., Moreaux, A., Kim, S.-K.: Automatic smart contract generation for Internet of Media Things. ICT Express **7**, 274–277 (2021)

Intelligent Grid Operation and Maintenance Management and Command Platform Based on Computer Distributed Network

Pengyu Zhang[1](✉), Xiaochun Li[1], Ying Li[1], Banghua Jin[1], and Ryota Hinami[2]

[1] Guizhou Power Grid Co., Ltd. Power Transmission Yunxing Repair Branch, Guiyang 550002, Guizhou, China
zhangpy866@126.com
[2] University of Vienna, Vienna, Austria

Abstract. With the leap-forward development of power grid, the number of transmission line equipment has increased significantly. Traditional operation and maintenance methods can no longer meet the needs of comprehensively controlling equipment status and accurately assessing power grid risks. It is urgent to accelerate the gradual integration of modern information and communication technology and traditional inspection technology. This paper mainly studies the design of smart grid operation and maintenance management command platform based on computer distributed network. The docker-based distributed Web platform designed and implemented in this paper will explore continuous integration, performance monitoring, gray publishing and log retrieval. The platform developed in this paper realizes the functions of remote monitoring and analysis, fault prediction, intelligent personnel scheduling and so on. It can greatly shorten the emergency response time and solve the problems such as low efficiency of transmission line operation and maintenance management and lack of intelligent level.

Keywords: Distributed network · Smart grid · Power grid operation and maintenance · Operation and maintenance platform

1 Introduction

In the power system, transmission line is a very important infrastructure, which mainly undertakes the whole process of business data interaction function from power energy production to power consumption terminal of the power system, and is an important technical foundation support for power safety production management [1]. In recent years, with the continuous progress of the smart grid construction project of China Southern Power Grid Corporation, China's power system has gradually stepped into a new development period of "three sets and five" from the original "four modernizations", which requires local power companies to follow the overall development plan of the power industry proposed in the "fourteenth Five-year Plan". Optimize and adjust the organization and management mode of production data information and communication in the

I. Ahmad et al. (Eds.): STSIoT 2021, LNDECT 122, pp. 443–451, 2023.
https://doi.org/10.1007/978-981-19-3632-6_54

company, establish the integrated management and maintenance mode of production data communication in the power system, realize the intensification and integration of production data communication and management, and take efficient power communication network as the smooth realization of smart grid construction project. To achieve the automation, digitalization, informatization and interactivity of power grid management to provide necessary infrastructure support and technical support [2]. In recent years as power supply company around the southern power grid company for the improving of the inter-district grid service ability request, the electric power communication network management model and system there exist obvious layered management, information isolated island and adverse factors, limiting the interregional power grid communication and the construction of network resource intensive management system, It has become the main factor restricting the information communication network management ability of the current domestic electric power industry [3].

In electric power communication network management information software research, at present the world electric power department of electric power communication network management software is mainly divided into two types: the first type is based on the network protocol communication hardware and communication channel maintenance management software platform, the second type is based on the general network management information platform of [4]. At the moment, typical products, better known abroad electric power communication network operations management software platform mainly include general electric research and development of IIP communication network management tools, research and development of German electricity group GESP - X series network management tools, in the typical software products are usually includes the daily operational management of the electric power communication network tools, In addition, management and maintenance tool interfaces are provided for operation and maintenance managers in a customized way, and customized deployment can be made according to the management needs of power communication networks in different regions [5, 6].

The smart grid operation and maintenance management command platform developed in this paper breaks through the traditional power grid operation and management mode, improves the management level of power grid operation and maintenance based on the actual work status of power grid operation and maintenance, and makes the power grid operation and maintenance management more scientific, standardized and reasonable.

2 Construction of a Distributed Platform for Smart Grid Operation and Maintenance Management and Command

2.1 System Requirements and Distributed Technologies

(1) Demand analysis of distributed system

The core points of the construction of distributed Web platform are as follows: providing a reliable release scheduling platform that ensures the consistency of all environments, that is, continuous integration and stable release of application platform versions; Provide strategies to ensure balanced utilization of platform resources and comprehensive monitoring, that is, effective resource monitoring, effective and timely grasp of the global application running state; To ensure the efficient, flexible, stable, isolated and uninterrupted resource scheduling of each application platform, that is, stable gray scale release, to achieve business continuity and online elastic expansion mechanism; In addition, the log of each node is timely obtained, that is, the efficient log retrieval function, and stable audit and log services are provided [7, 8].

In order to meet the high performance requirements of test points, the platform resources are utilized to the maximum extent. Firstly, cloud computing virtualization technology is considered. However, the traditional IaaS layer virtualization still has the problems of complex implementation and slow scheduling. In contrast, Docker, as an emerging paravirtualization product, is relatively easy to enter and has a higher utilization of resources. Many domestic companies have started to develop Docker into products and customized products, and Docker technology research has gradually matured. Therefore, Docker virtualization is adopted as the main structure of this test [9].

(2) Docker engine

Docker project is a lightweight virtualization solution, which is further encapsulated on the basis of Linux kernel virtualization. Different from traditional virtualization at the hardware level, Docker mainly virtualizes at the operating system level, greatly reusing the resources of the host operating system [10].

Docke is a architecture pattern based on C/S. It sends a request to the Docker Daemon Daemon through the Client Docker Client or RESTAPI, and the Docker Daemon will return to the Client after processing. The main components of Docker are as follows.

Mirroring: Docker images are static templates that the container runs on. Each image starts with a base image, each containing a series of layers that Docker mirrors using UnionFS (federated file system).

Containers: The Docker container is created from the Docker image and contains the operating system, user files, and metadata. Container runtime isolation of namespace and storage relies on Namespaces and UnionFS, while resource isolation is achieved by means of Linux's underlying cgroups.

Repository: Docker repositories are used to store Docker images, which can be divided into public and private.

(3) Docker deployment application

The distribution version of Docker is divided into CE (Community Edition) Community Edition and EE(Enterprise Edition) Enterprise Edition. CE is maintained and supported by the community and is a free version. EE version is a paid version, supported by the after-sales team and the technical team. It supports not only the mainstream Linux operating system, but also Mac OS(Apple operating system) and Microsoft Windows 10 [11, 12]. The simple deployment and application are as follows.

Take the example of deploying the CE version of Docker under Centos7. Since the Docker source is already built in, enter the following command to install it directly (make sure the host is networked).

yum install -y docker.

After the installation, if Docker service is not started successfully, you need to start Docker service through systemctl.

systemctl start docker.

After Docker startup is complete, you can view the installed version of Docker through Docker version. Create a simple container to apply the hello-world command as follows.

docker run hello-world.

2.2 Smart Grid Operation and Maintenance Management and Command Platform Architecture

(1) Platform design principles
Considering the disadvantages brought by multiple business systems, the research and development of integrated operation and maintenance platform of power grid put forward four principles pertinently: "system simplification, information standardization, unified interface and business collaboration".

System simplification: the integrated operation and maintenance platform of the power grid should be simplified through computers, so as to reduce the number of equipment terminals that the power grid needs to pay attention to and simplify the configuration of the station control layer.

Information specification: through the unified coding and modeling of all kinds of information, the unified specification of information is realized, and the eight types of data are standardized to realize the information integration.

Unified interface: according to the technical principles of modularization of function, labelling of interface, simplicity of content and clarity of task, unified the three screens of power grid monitoring, equipment monitoring and operation management to achieve unified interface. The three screens correspond to monitoring host/operator station, operation and maintenance management host/operator station and video and environment host/operator station respectively.

Business collaboration: Through breaking down the business barriers of various professional systems in the power grid, cross-professional business collaboration can be realized, and trend analysis, intelligent judgment and other business collaboration applications can be carried out.

(2) Platform service architecture
Database: it is used to store various types of data information in documents, which can be structured data or unstructured data, and can meet the information storage requirements of both types of structures.

Common component layer: it provides functional support for the underlying system architecture in the form of components by using abstract methods. Its main work is to handle various types of requests submitted by the service layer, such as some timing tasks or data-related operations.

Service layer: It can generally be divided into two layers, the core service types at the bottom and the specialized service types at the top. The former mainly refers to general service types such as integration services, while the latter has more specific service types such as reading and evaluation.

Service bus: provides a channel for information transfer and business call execution at the lower level, which can not only realize method call within the system, but also support mutual call scenarios between systems.

Application layer: Implement various types of business functions in the form of modularization and partitioning of subsystems. Using the resources of service layer as the basis of function realization, it provides services for all kinds of business applications on the premise of keeping the attributes of service layer. In addition, it can not only be a good integration of existing applications management. It also has good extensibility, which provides the possibility for future business application update.The system framework components are listed as shown in Fig. 1

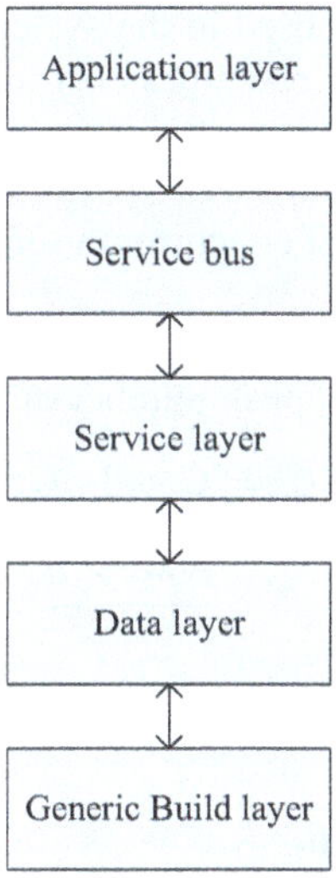

Fig. 1. System framework composition

3 System Test

3.1 System Function Test

The function of this system test using automated manner, all functions in LoadRunner test case execution is completed, recording and summarizing the function of the detected bugs, through the analysis of bugs, code positioning, check and modification, use cases for the function of the bugs iterative testing, until not BUG check out function.

In the actual test, the number of functional test iterations was 5 times. Functional bugs were classified, recorded and counted according to their types. In the system test, functional bugs were classified according to their severity:

Very serious BUG: the system cannot provide services normally, such as running crash, unresponsive page, GIS publishing failure, etc.

Serious bugs: logical anomalies or errors occur in system functional services, such as data errors, substandard GIS visualization styles, abnormal operation responses on GIS maps, etc.

General bugs: errors that do not affect the logical correctness of the system, such as Web page layout error, incomplete prompt information, incomplete display information, etc.

3.2 System Performance Test

For the system performance test, in the actual test, the LoadRunner tool is used to test the system performance in the way of automatic simulation. Divided into the system in the performance test with the target of concurrent ability and the key indicators such as response time, at the same time to ensure that the system's overall resources occupancy rate can achieve a good level, also used in the system server host I3 monitoring tools, the system for the server hardware resources cost in the process of running condition, quantitative monitoring.

The relevant test formula is as follows:

Calculate the average number of concurrent users using Eq. (1)

$$C = nL/T \tag{1}$$

Use formula (2) to calculate the peak number of concurrent users

$$C' = C + 3\sqrt{C} \tag{2}$$

$$F = \frac{N_{PU} \times R}{T} \tag{3}$$

4 System Test Results

4.1 System Function Test Results

Table 1. Functional test results

	1	2	3	4	5
General BUG	75	56	22	7	0
Serious BUG	12	10	5	3	0
Very serious BUG	4	2	1	0	0

As shown in Table 1 and Fig. 2, 6 very serious bugs were detected in the function test of the system, among which 2 BUG2 were repeated, mainly concentrated in the network communication function of the system and GIS service release, due to the wrong configuration of network communication, firewall and release parameters. Serious bugs are mainly concentrated in the image scheduling operation function of the system. General bugs mainly focus on the visual layout of Web pages and browser compatible display.

Through 5 rounds of iterative testing, code inspection and modification, all the above functional bugs were finally solved. In the last round of testing, all the bugs were not repeated, and no new functional bugs were detected, so the system functional test passed.

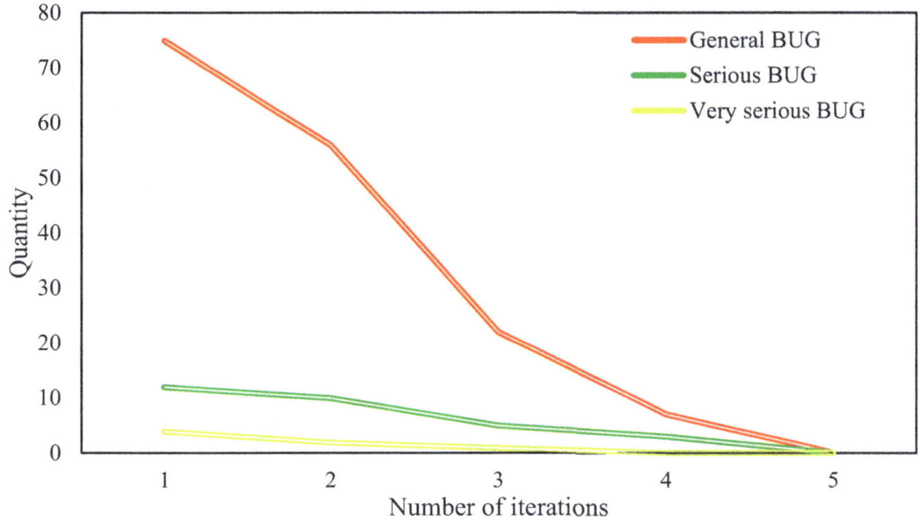

Fig. 2. Functional test results

4.2 System Performance Test Results

Table 2. System concurrency ability and response time test results

	50	100	150	200
Sercice layer	1.08	1.29	1.74	2.06
Sercice bus	1.17	1.32	1.95	2.17
Application layer	1.24	1.36	1.83	2.09

As shown in Fig. 3, the response time of the internal logical operations of the system did not exceed 3 s under the expected concurrency stress. At the same time, according to the background service log of cross-platform data interaction of the system, a total of 276 interactive operations were carried out in the whole performance test process of the system, and the maximum interaction time was 2.93 s, less than the expected standard of 4 s. So in terms of concurrency capability and response time, the system test passed (Table 2).

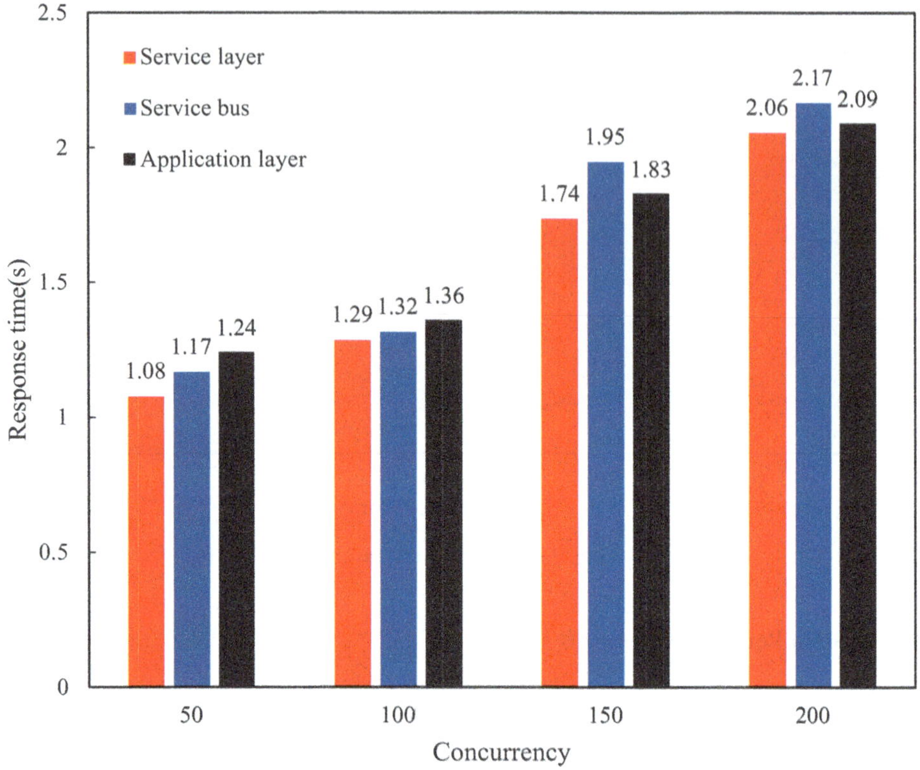

Fig. 3. System concurrency ability and response time test results

5 Conclusions

With the rapid growth of power grid equipment, the scale of transmission lines keeps expanding, and the equipment network becomes more and more complex, the traditional operation and maintenance mode can no longer meet the operation and maintenance requirements. In order to further improve the quality and efficiency of transmission line operation and maintenance, support transmission line operation and maintenance technology development to digital, lean, intelligent and automatic transformation. In this paper, Java Web development technology, Spring MVC framework technology and distributed technology are used to build an intelligent transmission line operation and maintenance management system. The system can scientifically manage the operation and maintenance tasks and operation and maintenance personnel, conduct real-time statistics and analysis of the data uploaded outdoors on the transmission line, and make fault prediction according to the analysis results. For operation and maintenance management personnel can be efficient scheduling; For operation and maintenance workers, the operation and maintenance effect is optimized; For residents, enterprises is to ensure the safety and stability of electricity and high quality.

References

1. Zhen, M., Muzaffar, H.K.T., et al.: Intelligent fusion algorithm for multi-sensor information in integrated power grid operation system. J. Intell. Fuzzy Syst. **35**(4), 1–11 (2018)
2. Xu, C., Wu, J., Feng, H., et al.: Operation of off-grid power supply system using IoT monitoring platform for oil and gas pipeline based on RESOC. CSEE J. Power Energy Syst. **6**(1), 12–21 (2020)
3. Mohammad, H.H., et al.: Optimum islanded microgrid reconfiguration based on maximization of system loadability and minimization of power losses. Int. J. Electr. Power Energy Syst. **78**(4), 343–355 (2016)
4. Tavakoli, M., Shokridehaki, F., Akorede, M.F., et al.: CVaR-based energy management scheme for optimal resilience and operational cost in commercial building microgrids. Int. J. Electr. Power Energy Syst. **100**(SEP.), 1–9 (2018)
5. Javanbakht, P., Mohagheghi, S.: A risk-aware generation dispatch including wind power for a power grid subjected to hurricanes. Int. Trans. Electr. Energy Syst. **25**(11), 2982–3003 (2016)
6. Ajami, W.A., Arief, A., Nappu, M.B.: Optimal power flow for power system interconnection considering wind power plants intermittency. Int. J. Smart Grid Clean Energy **8**(3), 372–376 (2019)
7. ChandyKM, L.: Distributed snapshots: determining global states of a distributed system. ACM Trans. Comput. Syst. **3**(1), 63–75 (2016)
8. Bal, H., Epema, D., De Laat, C., et al.: A medium-scale distributed system for computer science research: infrastructure for the long term. Computer **49**(5), 54–63 (2016)
9. Amudhavel, J., Kathavate, P., Reddy, L., et al.: Assessment on authentication mechanisms in distributed system: a case study. J. Adv. Res. Dyn. Control Syst. **9**(12), 1437–1448 (2017)
10. Siami, M., Skaf, J.: Structural analysis and optimal design of distributed system Throttlers. IEEE Trans. Autom. Control, (99), 1 (2017)
11. Ovseevich, A.I., et al.: Asymptotically optimal control for a simplest distributed system. Doklady Math. **95**(2), 194–197 (2017)
12. Nesterov, R.A., Lomazova, I.A.: Using interface patterns for compositional discovery of distributed system models. Proceedings Inst. Syst. Program. RAS **29**(4), 21–38 (2017)

Electric Automation Control System Based on Improved KMP Algorithm

Youliang Yuan(✉) and Yu Qiu

Tiandi (Changzhou) Automation Co., Ltd., Changzhou 213015, Jiangsu, China
35596737@qq.com

Abstract. Electrical automation technology is a comprehensive technology that optimizes the production process through computers, information technology and control theory, and realizes the improvement of production efficiency. Hardware automation system and software system are the prerequisites for realizing electrical automation technology. In addition to the hardware part of this platform, electrical automation systems are also used in it. Man-machine interface refers to industrial control computers and operating screens. Touch screens are widely used devices in recent years. They have the characteristics of instant information interaction, easy operation, and high reliability, and are widely used in the field of industrial control. The advantage of this automatic control function is that it is separated from the traditional manual control pendulum, which greatly reduces the cost of supervision and improves the quality of supervision at the same time. From this perspective, the study of electrical automation control systems based on computer algorithms has certain practical significance. The purpose of this paper is to study the electrical automation control system based on the improved KMP algorithm. For electrical automation control systems, the kmp mode distribution algorithm is used on the basis of the bf mode distribution algorithm, and the kmp algorithm is further improved on the basis of the kmp mode distribution algorithm. In practical applications, the feasibility of the improved kmp algorithm is studied. The PID control model algorithm and the improved kmp algorithm control experiment verify the efficiency and practicability of the improved kmp calculation and transmission, and the experimental results are in line with expectations. The experimental results show that the improved algorithm proposed in this paper is based on the kmp algorithm, and the error does not exceed 0.1 in actual experiments compared with the experimental results of the standard control group, which proves that the algorithm is feasible in practical applications.

Keywords: Kmp algorithm · Automation control · Electrical system · Computer control system

1 Introduction

In recent years, computer control systems have participated in many fields. Its essence is a logical way of thinking. In terms of mechanical control, commands can be issued by simulating the human brain, which significantly improves the efficiency and functions of

I. Ahmad et al. (Eds.): STSIoT 2021, LNDECT 122, pp. 452–460, 2023.
https://doi.org/10.1007/978-981-19-3632-6_55

the machine [1]. Nowadays, the pace of industrialization in countries all over the world is accelerating, and intelligence is becoming more and more important in the research of industrial projects. In order to further improve the efficiency and mechanical reliability of the manufacturing industry, many researchers have tried to apply the concept of automated mechanical control to mechanical equipment. In order to keep up with the pace of the times and meet the needs of the current society, a technological innovation in traditional industries is imminent [2]. In the process of studying the kmp algorithm, many researchers have conducted research on it and have achieved many research results. Pandey G first proposed to use Brute-Force algorithm to control electrical automation [3]. On the basis of the first one, Bargshady N proposed pattern matching using an improved kmp algorithm. In the Brute-Force algorithm, there are many characters in the pattern string, and several consecutive characters in the main string are equal, but the last character is the right, and the comparison position of the main string must be folded [4]. The kmp algorithm does not need to give way to the position of the main chord in the above state, which can significantly improve performance. Okokpujie K O is further enhanced by using an improved kmp algorithm and applied to electrical automation control, which significantly improves efficiency and reduces potential safety hazards [5]. These researchers have done a lot of research on the electrical automation control system based on the improved KMP algorithm, which provides a good theoretical and research basis for this article.

This article first conducted experiments on the PID model algorithm and kmp algorithm, and then compared the experimental results, and found that they were not satisfactory. Then the kmp algorithm is improved to a certain extent. Comparison of the results after the experiment with the standard control group: the improved kmp algorithm has a smooth broken line trajectory and good stability, and the error with the control group is within the standard range. Confirm the reliability of the algorithm.

2 Method of Electrical Automation Control System Based on Improved Kmp Algorithm

2.1 BF Pattern Matching Algorithm

The idea of the BF pattern matching algorithm is to match the first character in the content string with the first character in the keyword string one by one. If they are equal, the second character in the two strings is compared, and then both strings are moved one bit backward. If the two points do not match during character comparison, the first character in the keyword string and the second character in the content string will be the comparison point, and then re-compare, and when they match, go backwards Move bit [6–8].

2.2 Kmp Pattern Matching Algorithm

The Kmp algorithm is an improved algorithm based on the bf algorithm. The reason why this algorithm is more efficient than the bf algorithm is that there is no need to re-allocate the target character sequence after the match fails, which reduces the number of matches and improves the efficiency. The core of the algorithm application is based on a navigation table, which stores the characteristic information of the keyword string.

2.3 Plc Hardware System Configuration

The bus is used for the communication between the PLC host and the inverter. Through the communication with the man-machine touch screen interface. Each slave uses a different address on the bus, and the PLC sends or receives data through scanning.

(1) Motor control hardware design

1) Inverter

Generally, the frequency converter is installed on the PLC main controller as auxiliary equipment, and the PLC output port is used for parameter input to the frequency converter. In order to increase the control speed, increase the communication distance, and enhance the interference, the communication between the inverter and the PLC is selected as serial communication based on RS485. In this paper, PLC is used as the host computer, and the horizontal seal motor inverter, the drive motor inverter and the vertical seal motor inverter are connected through a serial bus to control multiple inverter channels.

2) Photoelectric encoder

Use photoelectric encoder to detect the running speed of each motor. Photoelectric encoders are mainly divided into incremental and absolute rotary encoders. The functions of the two encoders are different. In this system, the additional type is selected as the initial measurement of engine speed.

3) Hardware design of heating component detection

PLC installation and wiring In order to make the entire control system as cool as possible, all components of the fully automatic must be installed vertically. The input/output rack is installed under or near the host. Because a powerful power supply generates high-frequency and high-voltage signals, although the optical isolation and other components inside the PLC filter these signals to prevent them, they will be installed as far away as possible from these interference sources.

3 Experiment of Electrical Automation Control System Based on Improved Kmp Algorithm

3.1 Pid Control Model Experiment

PID is the abbreviation of control strategy, which contains a combination of three functions: integrated link, differential link and analog link. The PID system is actually a negative feedback system, which balances the system through feedback signals [9, 10]. Digital PID uses a computer to perform PID control calculations. It is a unique PID. It belongs to a method in modern control theory. Whether it is mathematical modeling or actual circuit, PID customization can be seen everywhere [11].

3.2 Classic Pid Control and Adjustment

In the classic initial PID control chart, there is an initial input value. This input is combined through analog, integrated, and differential connections to control the controlled object. The current state of the test object is sent back as feedback, and then the expected value and feedback value are combined as a new system input, and the controller is continuously rotated until the system is stable, that is, the current state and the expected value of the controlled object. Objects tend to be equal. In this way, even if there is not much prior knowledge of the dynamic process, control can be achieved [12].

In the continuous control system, the input and output of the general PID control algorithm have the following relationship, as shown in formula (1):

$$u(t) = k_p\left[e(t) + \frac{1}{T_i}\int_0^t e(t)dt + T_d\frac{de(t)}{dt}\right] \tag{1}$$

In formula (1), K_p is the proportional coefficient, T_i is the integral coefficient, and T_d is the differential coefficient.

(1) Proportional part

When there is a difference between the initial value input by the system and the feedback value, the system will give feedback in time. However, the system sometimes does not respond sensitively to this difference, resulting in a long response time and cannot be used in actual projects. Therefore, an analog link is added to the system to enhance the system's sensitivity to deviations. The scale factor is actually a constant, and the specific value is determined by the system. The magnitude of this constant is proportional to its influence. The larger the value, the stronger the effect, the faster the transition process, and the smaller the corresponding static deviation in the control process. But when the value is too high, the system will oscillate, which is not conducive to control. Therefore, the scale factor must be selected reasonably.

(2) Points part

The function of the integral part is to eliminate the deviation. When the deviation is eliminated, the integral is stable, and the overall action is stable at this time. However, the negative impact of this component on the system is the delay in system response time. When the coefficient is high, the error elimination time is long, and there will be no oscillation during the transition period, otherwise there will be oscillation, but the error elimination process will speed up. Therefore, the size of the integer constant must also be determined according to the actual situation.

(3) Differential part

The system must also judge the development direction of the deviation and correct it in time. Performing this process can not only reduce overruns, but also speed up the monitoring system. The differential link achieves this goal. In the control system, if there is a deviation, it is necessary not only to respond immediately, but also to understand its dynamics, so that the deviation can be reduced faster and more accurately. However, when the input contains a lot of noise, the system cannot add the generated energy. The effect of this link is also determined by its coefficient, too large or too small a coefficient is not conducive to system control. In short, differentiation plays a great role in the stability of the system, and appropriate differential coefficients must be selected.

3.3 Digital Pid Control Experiment

Due to the development of computer technology, the current PID controller design mostly adopts the digital PID customization method. First, the signals must be distinguished, and then the computer is used to calculate the PID. The number type of PID is:

$$u(n) = K_p\left\{e(n) + \frac{T}{T_i}\sum\nolimits_{i=0}^{n} e(t) + \frac{T_d}{T}[e(n) - e(n-1)]\right\} \tag{2}$$

Among them, T is the sampling period, n is the sampling number, e(n) is the deviation of the sampling, and u(n) is the control quantity.

3.4 PID Regulator Control Model

In the design of the electrical automation control system in this article, the drive motor adopts the closed-loop control of the motor to realize the precision of the motor control. The closed loop system consists of PLC, frequency converter, AC motor and speed encoder. The PID regulator model formula for motor control is:

$$M(t) = K_p \times \int_0^t edt + K_1 + M + K_D \times \frac{de}{dt} \tag{3}$$

Among them, KP is the loop gain, e is the PID loop deviation, M is the initial value of the loop output, and M(t) is the PID loop output.

4 Experimental Analysis of Electrical Automation Control System Based on Improved Kmp Algorithm

4.1 Pid Algorithm Module Simulation

In the actual calculation, each variable may be changed, so before the calculation, each parameter in the project needs to be converted into a value according to the rules, and then sent to the controller for calculation. Similarly, the output data of the controller will be different from the actual situation and must be used after conversion. The experimental results are shown in Table 1:

The trajectory of each joint is shown in Fig. 1. It can be seen that the numerical change is not smooth enough to meet the conditions of some performance indicators, and the robustness is poor. On this basis, this article recommends that you use the improved kmp algorithm to experiment again.

Table 1. Comparison table of algorithm experiment results

Number of experiments	J1PID	J1KMP
1	1	1
2	1.4	1.3
3	1.6	1.5
4	1.8	1.8
5	1.9	2
6	1.8	1.9
7	1.9	2.1

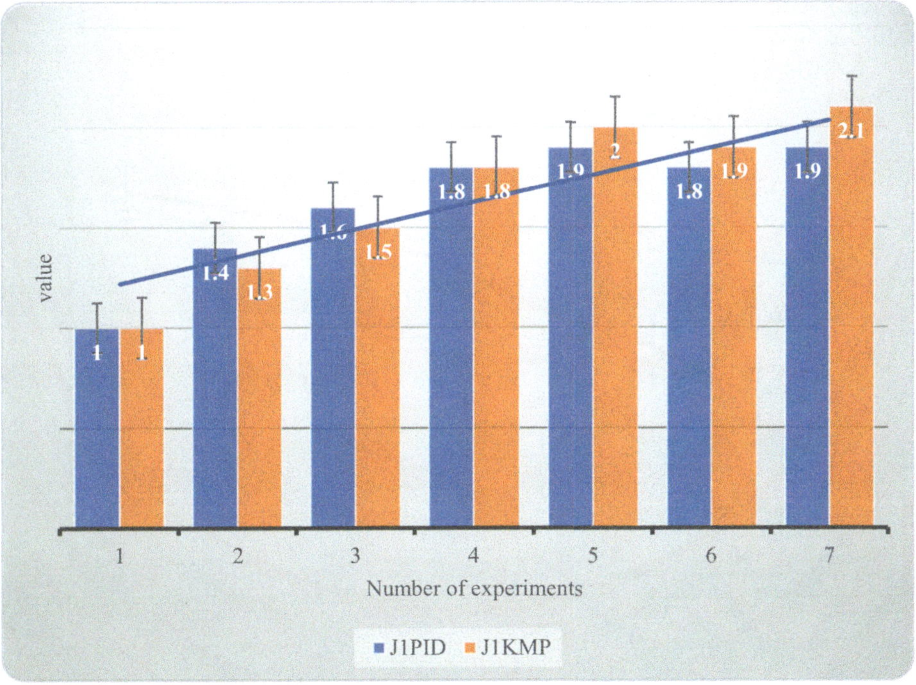

Fig. 1. Comparison chart of algorithm experiment results

4.2 Kmp Improved Algorithm Experiment

From the previous experiments, the smoothness of PID algorithm and kmp algorithm will become unstable as the number of experiments increases, which obviously does not meet the expectations of the experiment. Therefore, this article improves the kmp algorithm on the basis of its experiments, and uses the improved kmp algorithm to conduct experiments. The experimental results are shown in Table 2:

Table 2. Comparison table of experimental results

Number of experiments	Control group	Improve kmp
1	1	1
2	1.3	1.2
3	1.5	1.4
4	1.8	1.7
5	2.0	1.9
6	2.1	2.0
7	2.2	2.1

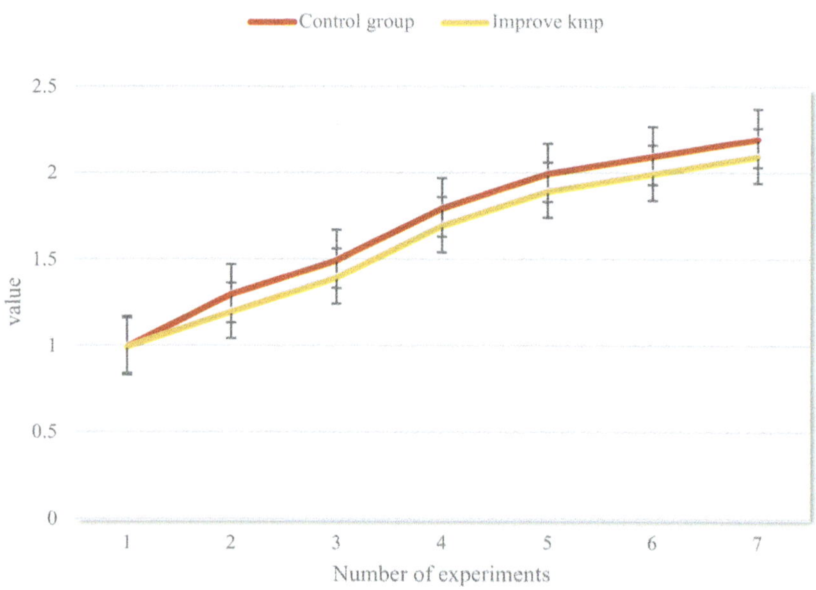

Fig. 2. Comparison of experimental results

It can be seen from Fig. 2 that when the kmp algorithm is improved and used in the experiment, the trajectory is relatively smooth without obvious fluctuations, similar to the dotted line of the control group, and has good stability. The error with the control group is only 0.1. The experimental results are in line with expectations, which proves that the improvement of kmp algorithm is effective and practical in practical applications.Based on the introduction of the improved kmp algorithm, this article also conducted experiments on the dynamic change of the error value. The starting point of each experiment was set to 0.85, and the trend of the error value was studied. The experimental results are shown in Table 3:

Table 3. Error value change table

Number of experiments	0.25 s	0.5 s	0.75 s	1 s	1.25 s	1.5 s
1	0.85	0.69	0.52	0.38	0.19	0.08
2	0.85	0.72	0.61	0.49	0.22	0.13
3	0.85	0.7	0.59	0.42	0.19	0.1
4	0.85	0.68	0.56	0.4	0.21	0.09

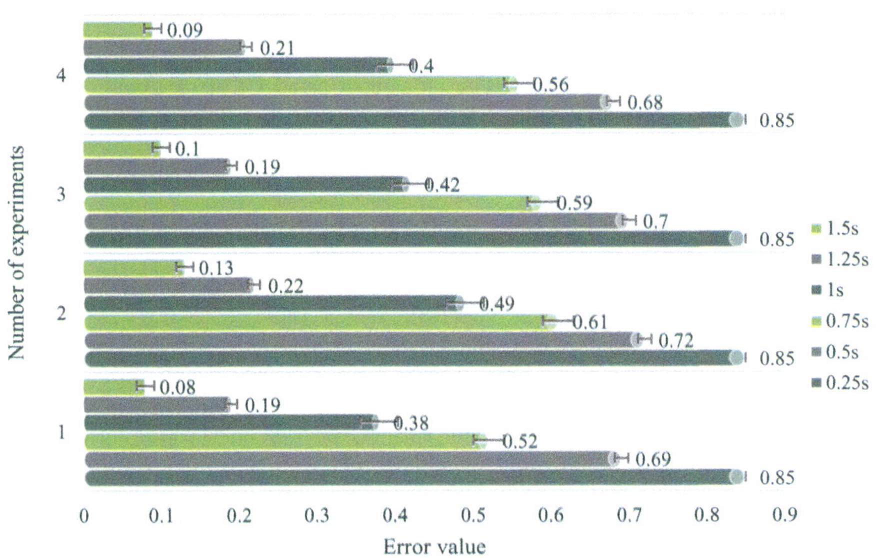

Fig. 3. Error trend analysis chart

It can be seen from Fig. 3 that in the error trend, with the change of time, the error value will gradually tend to 0.1, which proves that the stability and reliability of the system after the introduction of the improved KMP algorithm mentioned above have been improved to a certain extent. It has certain reference value for the design of electric automation control system in the future.

5 Conclusions

This article has made preliminary results in the research of electrical automation control system based on the improved KMP algorithm, but there are still some shortcomings, such as no further analysis of the control of programmable controllers, frequency converters and motion controllers. The experiment is actually carried out in an ideal environment. All conditions are good by default. However, the conditions in the actual engineering project cannot be guaranteed to be stable. There will be many unexpected factors that affect the variables of the system, so the research involved in this article There are still some shortcomings, and further improvements are still needed. It is necessary to conduct a more in-depth study on adjusting the PID proportional amount and entering high-speed counting. Although the proposed KMP improved algorithm can meet the requirements of the control system in theory, it still needs continuous improvement and optimization in practical applications.

References

1. Park, S., Kim, D., Lee, M., et al.: High throughput parallel KMP algorithm considering CPU-GPU memory hierarchy. Trans. Korean Inst. Electr. Eng. **67**(5), 656–662 (2018)
2. Bernet, L., Navas-Guzmán, F., et al.: The effect of cloud liquid water on tropospheric temperature retrievals from microwave measurements. Atmos. Measur. Tech. Discuss. **19**(11), 1–35 (2017)
3. Pandey, G., Martolia, M., Arora, N.: A novel string matching algorithm and comparison with KMP algorithm. Int. J. Comput. Appl. **179**(3), 6–8 (2017)
4. Bargshady, N., Garza, G., Pahlavan, K.: Precise tracking of things via hybrid 3-D fingerprint database and kernel method particle filter. IEEE Sens. J. **16**(24), 8963–8971 (2016)
5. Okokpujie, K.O., Modulo, O., Okokpujie, I.P., et al.: A model for automatic control of home appliances using DTMF technique. Res. J. Appl. Sci. **12**(2), 266–272 (2017)
6. Jia, C.B., Du, Y.P., Wu, C.S., et al.: Design of an automatic control system for magnetic controlled narrow-gap tig arc welding of thick plates. Huanan Ligong Daxue Xuebao/J. South China Univ. Technol. (Nat. Sci.) **45**(9), 40–46 (2017)
7. Guangqing, N.: Application of automatic control technology of drilling machine based on PLC. Acta Technica CSAV (Ceskoslovensk Akademie Ved) **62**(1), 531–540 (2017)
8. Hou, P., Hu, W., Chen, C., et al.: Overall optimization for offshore wind farm electrical system. Wind Energ. **20**(6), 1017–1032 (2017)
9. Adail, A.S., Hassan, H.M., Shaat, M.K.: Safety and reliability of the electrical system based on optimal FACTS allocation for a research reactor. Prog. Nucl. Energ. **104**, 143–149 (2018)
10. Muzzammel, R., Khail, I., Tariq, M.H., et al.: Design and power flow analysis of electrical system using electrical transient and program software. Energ. Power Eng. **11**(4), 186–199 (2019)
11. Ngan Kee, W.D., Khaw, K.S., Tam, Y.-H., Ng, F.F., Lee, S.W.: Performance of a closed-loop feedback computer-controlled infusion system for maintaining blood pressure during spinal anaesthesia for caesarean section: a randomized controlled comparison of norepinephrine versus phenylephrine. J. Clin. Monit. Comput. **31**(3), 617–623 (2016). https://doi.org/10.1007/s10877-016-9883-z
12. Imran, T., Hussain, M., Figueira, G.: Computer controlled multi-shot frequency-resolved optical gating diagnostic system for femtosecond optical pulse measurement. Microw. Opt. Technol. Lett. **59**(12), 3155–3160 (2017)

Construction Cost Prediction of Transmission Line Engineering Under the Background of Big Data

Junqiang Sha, Huiting Dong(✉), Hongping Xie, Bo Yang, Xiao Shang, and Yuchen Ling

State Grid Jiangsu Electric Power Engineering Consulting Co., LTD., Nanjing, Jiangsu, China
donghuiting1994@163.com

Abstract. Transmission line engineering have a wide range of regions and many uncertain factors, which have brought great difficulties to the design, construction and operation of transmission lines. Cost management of transmission line engineering is crucial. The cost prediction is a key link in the cost management. This paper takes the transmission line construction cost data as the research object, analyzes it by using the multiple regression analysis theory. By selecting the research index, a reasonable multiple linear regression equation was established and the statistical test was carried out. The results show that the equation is in line with the reality and can be used to predict the transmission line construction cost. It provides ideas and suggestions for cost prediction of power enterprises.

Keywords: Transmission line engineering · Cost prediction · Multiple regression analysis

1 Introduction

Toffler first mentioned the word "big data" in the third wave in 1980. In 2008, the journal Nature published a special issue on big data, which introduced the concept of "big data". Since then, "big data" has been rapidly applied to various industries. The power industry has a large amount of high-precision, diversified and heterogeneous structured and unstructured data. These data sets combined with public information will produce great value. In 2013, "The White Paper on China's Power Big Data Development" first proposed the definition of power big data. Since then, the academic circles have conducted relevant research from different angles such as generation, transmission, distribution and sales.

Reference [1] mentioned that big data technology could establish an accurate prediction model of renewable energy in the prediction of new energy power generation, improve energy efficiency and intelligence, and achieve the goal of clean energy. Reference [2] proposed using big data for weather prediction. cloud movement mode to analyze regional light intensity, and neural network algorithm to predict photovoltaic power generation related power data; Reference [3] combines k-clustering and neural

I. Ahmad et al. (Eds.): STSIoT 2021, LNDECT 122, pp. 461–468, 2023.
https://doi.org/10.1007/978-981-19-3632-6_56

network algorithm to predict wind power generation through the processing of atmospheric temperature, wind speed and other data through big data, so as to further help wind power consumption.

In terms of power transmission and transformation, with the help of large transmission data sets and SVM algorithm collected by wide area data measurement and collection system, references [4] and [5] combined with random matrix theory quantitatively analyzes the impact of different degrees of disturbance on real-time transient stability of power grid. Reference [6] further combined with big data mining to analyze the transient impact of transmission line trip and lightning trip on transmission line voltage and current within a certain range, and analyze the trip factors, which make the safety and robustness of transmission lines have been greatly improved.

In distribution network, references [7] and [8] use big data technology combined with classifier model and artificial neural network to predict load trend and locate fault. On this basis, reference [9] proposed the specific application of big data technology in active distribution network.

In terms of power sales, reference [10] constructs a circular neural network with the help of big data technology to predict the price of the power market, so that all participants in the dynamic power market can maximize the economic and environmental value objectives while meeting the power supply and demand. Reference [11] constructs a power marketing system based on collaborative filtering for customer power consumption data and historical transaction big data, carries out precision marketing, reduces users' energy expenditure and improves customers' power consumption experience. References [12] and [13] proposed to use optimal system clustering, fuzzy C clustering, BIRCH algorithm and CLIOUB algorithm to analyze users' electricity consumption behavior patterns in combination with big data technology.

2 Multiple Regression Analysis

Multiple linear regression is to study whether there is an interdependent relationship (linear relationship) between two or more independent variables and a dependent variable. This relationship can usually be expressed by multiple regression equation, which depicts the relationship between a dependent variable and multiple independent variables. The linear regression model with two or more independent variables in the equation is called multiple linear regression model. In this model, the dependent variable Y is a linear function of multiple independent variables $X_1, X_2, \cdots, X_k$ and error terms, and the expression is as follows:

$$Y = \beta_0 + \beta_1 X_1 + \beta_2 X_2 + \cdots + \beta_k X_k$$

For random error term ε, $E(\varepsilon) = 0, Var(\varepsilon) = \sigma^2$ is often assumed, and: $E(\varepsilon) = 0, Var(\varepsilon) = \sigma^2$, and the theoretical regression equation is:

$$E(Y) = \beta_0 + \beta_1 X_1 + \beta_2 X_2 + \cdots + \beta_k X_k$$

In practical application, if n groups of observation data $(X_{i1}, X_{i2}, \cdots, X_{ik}; Y_i), i = 1, 2, \cdots, n$ are obtained, then:

$$Y_i = \beta_0 + \beta_1 X_{i1} + \beta_2 X_{i2} + \cdots + \beta_k X_{ik} + \varepsilon_i$$

The corresponding matrix expression is: $Y = X\beta + \varepsilon$.

In order to facilitate the parameter estimation of the model, the following basic assumptions are made for the above equation:

(1) The explanatory variable $X_1, X_2, \cdots, X_k$ is a deterministic variable, and X is a full rank matrix;
(2) The random error term has zero mean and equal variance, that is, it satisfies the Gauss-Markov condition:

$$\begin{cases} E(\varepsilon_i) = 0, i = 1, 2, \cdots, n \\ Cov(\varepsilon_i, \varepsilon_j) = \begin{cases} \sigma^2, i = j(i, j = 1, 2, & \cdots, n) \\ 0, & i \neq j \end{cases} \end{cases}$$

(3) Random error obeys normal distribution.

$$\varepsilon_i \sim N(0, \sigma^2), i = 1, 2, \cdots, n$$

For the matrix form of multiple linear regression, this condition is:

$$\varepsilon \sim N(0, \sigma^2 I_n)$$

According to the above assumptions and the properties of multivariate normal distribution, Y obeys n-dimensional normal distribution:

$$Y \sim N(X\beta, \sigma^2 I_n)$$

3 Case Analysis

3.1 Variable Selection

Transmission line engineering includes foundation construction, tower construction, stringing construction, etc. In this paper, we choose the tower construction cost as the research object (Fig. 1).

During the construction of the tower, a large number of cost basic data will be generated. The change of cost is affected by many factors. Integrating the existing research results and merging some influencing factors with relevant information, this paper selects three influencing factors: tower type, tower height and tower weight for regression analysis. The above three influencing factors are recorded as X1, X2 and X3 respectively as explanatory variables of multiple regression model, and the construction cost is recorded as Y. Take the actual cost data of a power grid company in 2020. These data have removed all kinds of irrelevant information that is easy to affect the prediction results, and study the above three influencing factors of tower construction cost. The actual data are shown in the table below (Table 1):

Fig. 1. Tower construction drawing

Table 1. Sample actual data sheet

Tower type (X1)	Tower height (X2)	Tower weight (X3)	Construction cost (Y)
Angle steel tower	110	88	153109
Angle steel tower	110	132	208212
Angle steel tower	110	176	250359
Angle steel tower	110	264	364738
Angle steel tower	110	396	534301
Angle steel tower	110	462	615494
Angle steel tower	120	192	300172
Angle steel tower	120	288	435677
Angle steel tower	120	432	642081
Angle steel tower	130	208	385418
Angle steel tower	130	312	568569
Angle steel tower	130	468	824209
Angle steel tower	140	224	496564
Angle steel tower	140	336	724557
Angle steel tower	140	504	1073015
Angle steel tower	150	360	838886
Angle steel tower	170	408	1047657
Angle steel tower	190	456	1277302
Angle steel tower	220	528	1579092
Steel tube tower	110	88	168741
Steel tube tower	110	132	230008
Steel tube tower	110	176	275365

(continued)

Table 1. (*continued*)

Tower type (X1)	Tower height (X2)	Tower weight (X3)	Construction cost (Y)
Steel tube tower	110	264	401003
Steel tube tower	110	396	587886
Steel tube tower	110	462	677313
Steel tube tower	120	192	330111
Steel tube tower	120	288	478970
Steel tube tower	120	432	705929
Steel tube tower	130	208	424019
Steel tube tower	130	312	625570
Steel tube tower	130	468	906856
Steel tube tower	140	224	546338
Steel tube tower	140	336	797134
Steel tube tower	140	504	1180543
Steel tube tower	150	360	923592
Steel tube tower	170	408	1153463
Steel tube tower	190	456	1412565
Steel tube tower	220	528	1754657

3.2 Establish Mathematical Model to Solve

This paper will use the statistical analysis tool EViews to fit the regression relationship between construction cost and influence factors, and evaluate the fitting of the model and the reasonable reliability of the results according to the obtained results.

Firstly, the correlation between the obtained data is analyzed to ensure the rationality of the linear model, and the results are shown in the table below. Judging from the data of the analysis results, it can be seen that there is a strong correlation between tower type, tower height, tower weight and cost. Through the significance test of confidence level $\alpha = 0.01$, it shows that the eternal linear model is more appropriate to explain their relationship similarly (Table 2).

Table 2. Correlation analysis result table between various factors

		X1	X2	X3
Y	Pearson correlation	0.0763	0.0802	0.0778
	Covariance	0.084	0.089	0.084

Multicollinearity between independent variables would distort the model estimation. To avoid this situation, multicollinearity test is carried out on the data, and the results are shown in the table below. The variance ratio obtained in the table does not have a number close to 1, and there is no multicollinearity between the data (Table 3).

Table 3. Multiple collinearity test results table

	X1	X2	X3
X1	1	0	0
X2	0	1	0.558
X3	0	0.558	1

A linear relationship between independent and dependent variables is evident. The prediction model of transmission line construction cost obtained by running EViews10 software is as follows: Y = 66350.05X1 + 8217.803X2 + 1555.010X3 – 969057.7

3.3 Statistical Test

The goodness of fit and significance test of the regression equation is essential in the application of multiple linear regression models for predicting, which can judge whether the research of the model is valuable. The steps of statistical test of the model are as follows:

(1) Test of goodness of fit. We usually use the fitting degree to test the fitting of regression equation to the value of independent variable. Generally, the coefficient of determination R is in the range of 0.8–1. The closer R^2 is to 1, the higher fitting degree. It means that the independent variable has a strong correlation with the dependent variable. Through it, we can judge the correlation between the predicted data and the sample data. The model $R^2 = 0.970$, which is close to 1, proving that there is strong fit between the actual value and the predicted value of the construction cost of the transmission line.
(2) F-test. F-test is also called joint hypotheses test, used to test whether the coefficient of the global equation is significantly different from zero. Given the significance level of 0.01, the F-inspection Sig = 0.00, which is far less than 0.01, showing that the overall effect of the regression equation is significant, and the regression model has significant significance.
(3) DW-inspection. A method commonly used in statistical analysis to test the autocorrelation of sequences. Explanatory variables are not related to random items, that is, there is no heteroscedasticity. The DW value of output test heteroscedasticity is d = 0.97, which indicates that there is no heteroscedasticity problem in the model.
(4) The predicted results of the model are compared with the real values. The following table shows the relationship between the predicted value, the real value and the difference obtained by substituting 38 data samples. From the analysis of nearly 38 samples, the total cost of tower assembly shows an upward trend, and the relative error between the real value and the predicted value changes by about 0.082. Further analysis shows that this is consistent with the actual situation and the economic significance test. In conclusion,

the multiple linear regression models are effective in predicting the construction cost of transmission line (Table 4).

Table 4. Table of the difference between the predicted value and the true value.

Number	Construction cost	Predictive value	Difference	Number	Construction cost	Predictive value	Difference
1	153109	71742	81367	20	168741	138092	30649
2	208212	140162	68050	21	230008	206512	23496
3	250359	208582	41777	22	275365	274932	433
4	364738	345423	19315	23	401003	411773	10770
5	534301	550685	16384	24	587886	617035	29149
6	615494	653315	37822	25	677313	719665	42352
7	300172	315641	15469	26	330111	381991	51880
8	435677	464922	29244	27	478970	531272	52301
9	642081	688843	46762	28	705929	755193	49264
10	385418	422699	37281	29	424019	489049	65030
11	568569	584420	15851	30	625570	650770	25200
12	824209	827001	2792	31	906856	893351	13504
13	496564	529757	33193	32	546338	596107	49769
14	724557	703918	20639	33	797134	770268	26866
15	1073015	965160	107856	34	1180543	1031510	149033
16	838886	823416	15470	35	923592	889766	33826
17	1047657	1062413	14756	36	1153463	1128763	24700
18	1277302	1301409	24107	37	1412565	1367759	44806
19	1579092	1659904	80812	38	1754657	1726254	28403

4 Conclusion

By using multiple linear regression analysis theory and EViews statistical analysis tool, this paper analyzes the relevant indicators affecting the construction cost of transmission line, discards some theoretical parameters that have no practical contribution to the calculation accuracy of the model, and obtains a good prediction model. Among them, the tower height has a very significant correlation with the total cost, which is an important index to forecast the construction cost of transmission line to a great extent. As one of many methods to forecast the construction cost of transmission line, the cost prediction method based on multiple linear regression mathematical model has obvious characteristics. This method has clear theory, simple structure, simple calculation, strong practicability and good fitting. The prediction of construction cost in this paper can help power enterprises to control costs reasonably and improve the accuracy of management, which has a certain practical significance.

References

1. Priya, S.: Big data analysis is used in renewable energy power generation. Int. J. Comput. Appl. **174**(2), 37–39 (2017)
2. Zhou, H., Zhang, Y., Yang, L., et al.: Short term photovoltaic power forecasting based on long short term memory neural network and attention mechanism. IEEE Access **7**, 78063–78074 (2019)
3. Wu, W., Peng, M.: A data mining approach combining K-means clustering with bagging neural network for short-term wind power forecasting. IEEE Internet Things J. **4**(4), 23274662 (2017)
4. Bo, W., Jiali, W., Dachen, L., et al.: Power grid weak point evaluation method based on high-dimensional random matrix theory. Chin. J. Electr. Eng. **39**(6), 1682–1691 (2019)
5. HU, W., et al.: Real-time transient stability assessment in power system based on improved SVM. J. Mod. Power Syst. Clean Energ. **7**(1), 26–37 (2018). https://doi.org/10.1007/s40565-018-0453-x
6. Yuquan, L., Yuanjun, G., Guojun, L., et al.: Transmission line trip analysis and correlation mining of related factors in Guangzhou power grid based on big data. China South. Power Grid Technol. **11**(04), 38–44 (2017)
7. Sophi, S.G., Hazlie, M.H., Azil, I.: Fault location using mathematical analysis and database approach. Int. J. Comput. Math. Electr. Electron. Eng. **39**(1), 415–430 (2019)
8. Xin, M., Dongxia, Z., Dedong, S., et al.: Opportunities and challenges of applying big data in distribution network. Power Grid Technol. **39**(11), 3122–3127 (2015)
9. Lin, Z., Duan, D., Yang, Qi., et al.: Data-driven fault localization in distribution systems with distributed energy resources. In: 2019 IEEE Sustainable Power and Energy Conference, pp. 1021-1026, Beijing, China (2019)
10. Dehghan, B.A., Taufik, T., Feliachi, A.: Big data analytics in a day-head electricity price forecasting using tensor flow in restructured power systems. In: 2018 International Conference on Computational Science and Computational Intelligence, pp. 1065–1068, Las Vegas,USA (2018)
11. Zhang, Y., Kong, W., Zhao, Y., et al.: Big data driven electricity plan recommender system. In: 2018 IEFE Power & Energy Society General Meeting, pp. 1944-9933, Portland, USA (2018)
12. Singh, S., Yassine, A.: Mining energy consumption behavior patterns for households in smart grid. IEEE Trans. Emerg. Top. Comput. **7**(3), 2168–6750 (2019)
13. Pan, S., Qiao, J., Zhu, L.: Application of parallel clustering algorithm based on in power customer classification. In: IEEE 4th International Conference on Cloud Computing and Big Data Analysis, pp. 165–169, Chengdu, China (2019)

Data Security Detection and Location Technology Based on DLP Network

Wei Zhan[1], Mingyang Yu[1], Bo Jin[1], Feng Guo[1], Guoru Deng[1], Rongtao Liao[1], Jinhui Zhao[1], Geng Wu[1], Hanghan Liang[1], Ruixue Li[2], and Xin He[1](✉)

[1] State Grid Hubei Information and Telecommunication Company Limited, Wuhan, Hubei, China
2271639122@qq.com

[2] Electric Power Evaluation and Consulting Institue, China Electricity Council, Beijing 100055, China

Abstract. Faced with a complex network environment, network security issues are getting more and more serious. Cyber attacks will not only leak user privacy, but also cause huge economic losses. In the face of massive network data, decision trees have become an effective method for detecting abnormal network data. The decision tree method trains a model on a large amount of data, classifies normal data and abnormal data, and detects network attacks more efficiently and accurately. This article aims to study DLP network data security detection and positioning technology. Based on the analysis of DLP trends, the development direction of intrusion detection, abnormal data classification algorithms and positioning technology, the KDD CUP1999 data set is selected as the experimental data set. These three methods, namely, decision tree, support vector machine, are used to detect the data set. The detection results show that the data detection rate and false alarm rate of the decision tree algorithm perform better among the three algorithms, and are suitable for network data security detection.

Keywords: DLP · Data security detection · Security positioning · Decision tree

1 Introduction

With the rapid development of information technology, data has become an important resource in people's daily life and work [1, 2]. Data security issues related to data resources, especially the theft of corporate information resources, corporate confidentiality, privacy, and other illegal and criminal activities initiated by mobile devices and information networks are gradually attracting people's attention [3, 4].

In recent years, many scholars have conducted research on network data security detection and have achieved good results. Some scholars have proposed a method of real-time monitoring of files, by modifying the Linux kernel to monitor files, recording file access, modification, opening, closing, creating and deleting operations, analyzing and checking file security to check files, and proposing a real-time monitoring method [5]. Some scholars have proposed to implement a privacy leak detection system based

I. Ahmad et al. (Eds.): STSIoT 2021, LNDECT 122, pp. 469–477, 2023.
https://doi.org/10.1007/978-981-19-3632-6_57

on dynamic pollution tracking, marking the privacy data read by the application as pollution, tracking the spread of the data, and recording the behavior when the data transmission is detected, and analyzing the log information, to determine whether there is a hidden danger of privacy data leakage [6]. In addition, some researchers are inspired by computer system intrusion detection systems to design and implement an Android intrusion detection system to process data collected using anomaly detection algorithms [7]. Other researchers designed a security tracking system based on network monitoring. The mobile terminal can operate with ultra-low power consumption, can monitor the input/output of network data for a long time, display the flow of network data in real time, detect the data sent to the remote server via WLAN or 4G network, and set the server data and threshold parameters to pass text or electronic email messages notify users of network abnormalities [8]. The research results of predecessors provide a theoretical basis for the research of this article.

In this paper, by consulting related literature and in-depth analysis of DLP, intrusion detection technology, etc., the data set is trained and classified, and the performance of the three classification algorithms of naive Bayes, decision tree, and support vector machine in detecting abnormal data is compared.

2 Research on Data Security Detection and Location Technology Based on DLP Network

2.1 DLP Trends

(1) Cloud technology needs to expand the scope of leakage protection.

With the popularization of cloud operations such as cloud computing, cloud storage, cloud platforms, and cloud services, companies have started to compete on the cloud. Some business applications are also developed on the cloud platform. Competition in the cloud is becoming increasingly fierce. Data leakage prevention has also put forward new requirements, and the scope of data leakage prevention needs to be completely expanded [9, 10].

(2) Virtualization needs to change leakage prevention technology.

The implementation and development of virtualization technologies such as servers, storage, networks, desktops, and terminals have urged companies to use multiple products to connect virtualization and cloud services. Therefore, new data leakage prevention requirements that require data leakage prevention technology have emerged, and changes have been made to meet the application requirements of virtualization technology [11, 12].

(3) Mobile equipment needs to improve leakage protection technology.

With the application and popularization of mobile devices such as smartphones, tablets, smart watches, and smart bracelets, attacks on mobile devices are becoming more and more targeted. With the rapid development of mobile networks, data leakage incidents in mobile devices have increased exponentially, which requires a more comprehensive improvement in leakage prevention technology.

2.2 The Development Direction of Intrusion Detection

(1) Distributed intrusion detection system

Many network applications are currently distributed. For example, Hadoop technology in big data analysis and processing implements a distributed file system. For such distributed applications, traditional intrusion detection systems cannot adapt to this change, and intrusion detection systems need to be used for actual application functions. Distributed is mainly manifested in two aspects, one is intrusion detection and protection for distributed networks, and the other is to use distributed detection methods. In other words, the intrusion detection method itself is distributed. In a distributed network, it is necessary to coordinate the overall situation and complete the collective processing of information.

(2) Establish an intrusion detection and evaluation system

Intrusion Detection System (IDS) is an important technical tool to ensure network security. Experts and researchers at home and abroad have conducted detailed studies on this and proposed various types of IDS. Choosing the right type for a large number of IDS is very difficult and requires detailed and comprehensive system evaluation.

(3) Collaborative intrusion detection

Coordinated intrusion detection system is a basic part of distributed intrusion detection, and collaboration has different meanings in different scenarios. For example, the collaboration between heterogeneous platforms is to enable intrusion detection systems to work powerfully on distributed platforms, and the collaboration between different detection systems to ensure system stability.

2.3 Anomaly Data Classification Algorithm

(1) Support vector machine

Support vector machine is a very commonly used sorting algorithm. This classification algorithm is considered a very effective machine learning. It is based on traditional statistical theory and has been proven by many modern practical tools (such as face recognition, text classification, etc.). This is how the research results are obtained. Among them, the algorithm has achieved good results in dealing with disproportionate data.

Support vector machine is based on the learning accuracy limit of training samples to achieve the best classification results. Some data sets are difficult to distinguish in low-dimensional spaces. The support vector machine method solves this problem by using the kernel function method to map them to a large space. Therefore, for the difficult problems of nonlinear or large data samples, the support vector machine method can be well separated, but the calculation amount and time complexity are increased, and the kernel function can solve this problem.

(2) Decision tree

The decision tree is output in the form of a decision tree, and the decision tree method is widely used in analyzing and solving problem classification. In addition, it is also very responsive to noise and large amounts of data. The functional classification structure of the decision tree is very similar to the process diagram.

All nodes in the tree that are not leaves are considered as a functional test, and the test output is all branches of the decision tree. The distribution or type of classes are considered as leaf nodes. The smallest scale can have higher classification accuracy. Figure 1 is a schematic diagram of a simple decision tree.

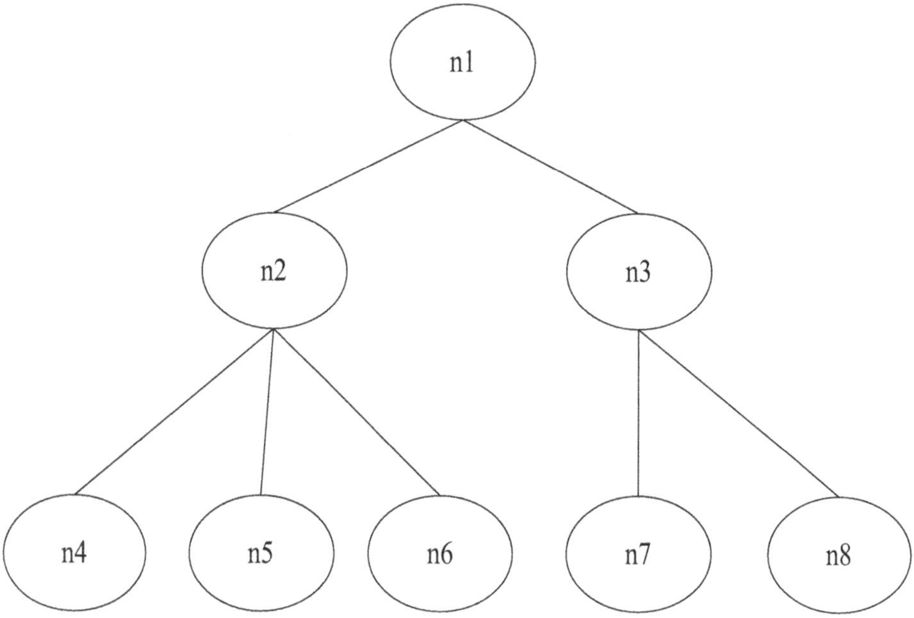

Fig. 1. Schematic diagram of decision tree model

ID3 algorithm is one of the most commonly used and important decision tree classification algorithms. The more information you get, the more functions you have. By selecting an attribute with the highest information gain as the tested sample, and dividing the sample, the goal of classifying a tested sample data set with the least information gain to obtain the highest accuracy is achieved.

Suppose S is a collection of s data samples. Assuming that there are m different class label attribute values, n different classes $C_i(i = 1, ..., n)$ are defined. Let s_i be the number of samples in class c_i. The following formula is the expected information of a specific sample:

$$I(s_1, ...s_m) = -\sum\nolimits_m p_i \log_2(p_i) \tag{1}$$

where p_i is the probability of any sample c_i.

$$E(A) = \sum\nolimits_{j=1}^{v} \frac{S_{1j} + ... + S_{mj}}{S} I\left(S_{1j}, ...S_{mj}\right) \tag{2}$$

Among them, item $\frac{S_{1j}+...+S_{mj}}{S}$ is the weight of the jth subset.

(3) Naive Bayes

To construct a classification algorithm, it is necessary to summarize the rules or functional relationships between class indicators and attributes according to a set of training data, and to classify the sorted samples according to the values of existing variables. The basis of the classification process is the first step in defining rules. Nowadays, there are many different models to classify. Each model has its own advantages, disadvantages and shortcomings, and is applied to different fields in order to solve practical problems. Random classification algorithm is an important type of various algorithms. If you want to use a random classifier, you must first calculate the common probability, and then classify and distinguish it based on the maximum probability value using the principle of minimum probability. Due to the different basic structures of various probability classification algorithms, and the different calculation methods of decomposition and probability, various probability classification algorithms have been formed.

Bayesian classification algorithm is considered as the basic classification algorithm of random classification algorithm. The superiority of this algorithm has been proved in statistical theory. Although this is generally considered to be the best classifier, it also has obvious shortcomings. Bayesian classification algorithms need to calculate probabilities together immediately, which reduces performance, has a more complex structure, and may also lead to changes in the data set.

2.4 Positioning Technology

(1) Distance-independent positioning algorithm

1) Centroid positioning algorithm

The basic idea of the centroid positioning algorithm is to use the node signal connected between two nodes to calculate the position and information of an unknown node. All these algorithms are layout algorithms that do not require amplitude. In these algorithms, the anchor node periodically sends its coordinate information to the neighboring unknown node, which contains the identifier and location information. If the location information received and sent from an unknown node in a certain period of time exceeds a certain limit, we can consider that an unknown node and an anchor node are interconnected.

2) Energy attenuation positioning algorithm

When the unknown node is covered by the transmission signal of the anchor node, the signal amplitude of the anchor node decreases as the transmission distance increases, and the signal energy received by the beacon node from the unknown node also decreases. The energy attenuation algorithm calculates the distance from the unknown node to the anchor node through the attenuation energy of the beacon signal.

(2) Ranging and positioning algorithm

The placement algorithm based on distance measurement has high placement accuracy. However, measuring the distance between the unknown node and the anchor node usually requires additional expensive hardware equipment.

Typical distance measurement techniques include RSSI, TOA, TDOA, AOA and other methods. The location algorithm based on distance measurement uses the above-mentioned various distance measurement techniques to locate unknown nodes.

3 Experiment

3.1 Experimental Data

This article uses the KDD CUP1999 data set in the field of network anomaly detection as the experimental data set. The data set contains a total of 21 attack types, 494,114 log entries, 97,365 regular log entries, each entry has 40 different configurations, of which 32 are numbers and 8 are character attributes; test data set contains 312041 log entries and 35 attack types, 15 of which are not shown in the training samples.

3.2 Evaluation Indicators

In this paper, the sample data is classified into two categories, and the data samples are divided into two categories: safety and abnormality. The detection rate and false alarm rate are used to evaluate the detection performance of the model.

The detection rate indicates the probability that the system correctly detects malicious data. The false alarm rate is the probability that normal data is detected as malicious data. The calculation formula for detection rate and false alarm rate is as follows:

$$TP_{rate} = TP/(TP + FP) \tag{3}$$

$$FP_{rate} = FP/(FP + TN) \tag{4}$$

Among them, TP and TN indicate that the data category is correctly detected, and FP indicates false alarms.

4 Discussion

4.1 The Impact of the Classifier on the Classification Results

In this paper, three classification algorithms: Naive Bayes, Decision Tree, and Support Vector Machine are used to train sample data and establish classification models. The evaluation results of each classification model are shown in Table 1:

According to Fig. 2, the classification performance of the naive Bayes classification model is very poor, the detection rate is 44.6%, and the false alarm rate is 11%; the classification performance of the decision tree and the support vector machine classification model is good, and the decision tree has a good classification performance. The detection rate is 92.5%, the false alarm rate is 4.6%, the support vector machine detection rate is 92.2%, and the false alarm rate is 8.2%. Among the three algorithms, decision trees are the most suitable for data classification.

Table 1. Model performance evaluation results

	Detection rate	False alarm rate
Naive Bayes	0.446	0.110
Decision tree	0.925	0.046
Support Vector Machines	0.922	0.082

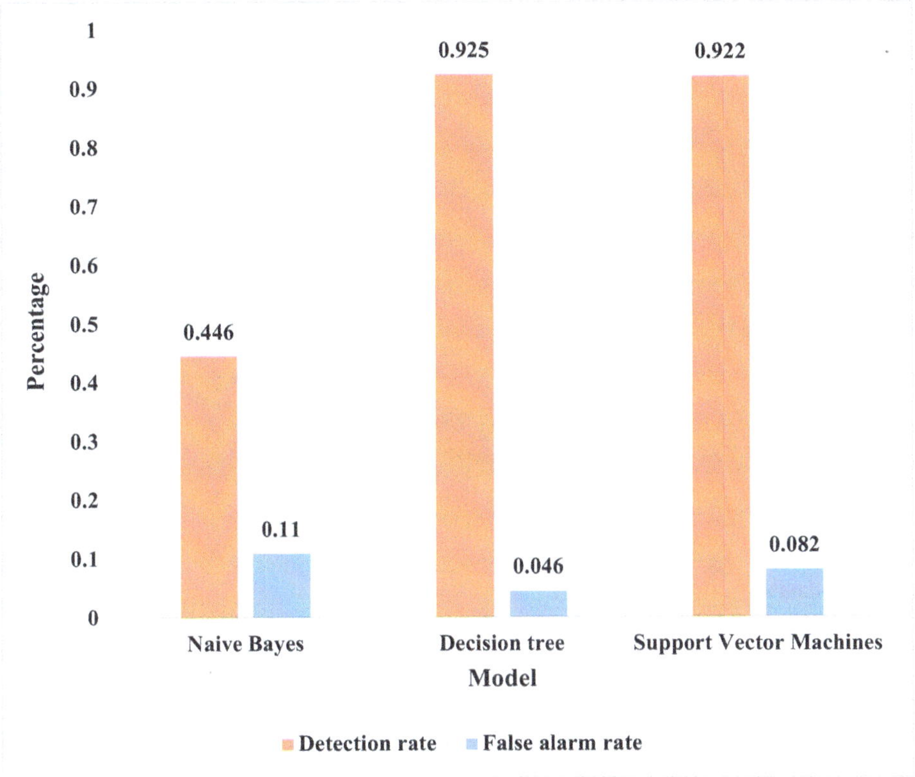

Fig. 2. Model performance evaluation results

4.2 The Impact of Data Size on Classification Results

In fact, the size of anomalous data samples is often relatively small. In order to simulate the performance of the classifier when the abnormal sample data is small, this paper first disrupts the abnormal sample data set, and then selects a specific percentage of abnormal samples to train the classification model based on the original abnormal data samples. Due to the better classification performance of the decision tree classification model, this paper chooses a decision tree classification model to further test the impact of the abnormal sample data set size on the classification performance of the classifier.

The following figure shows the performance test results of the model under different abnormal sample data set sizes as shown in Table 2 and Fig. 3:

Table 2. The impact of data size on classification results

	Detection rate	False alarm rate
1%	38.70%	0.00%
3%	62.00%	0.00%
5%	72.40%	0.00%
10%	75.00%	0.10%
20%	78.20%	0.20%
50%	84.30%	2.10%
80%	90.10%	4.50%
100%	91.60%	4.90%

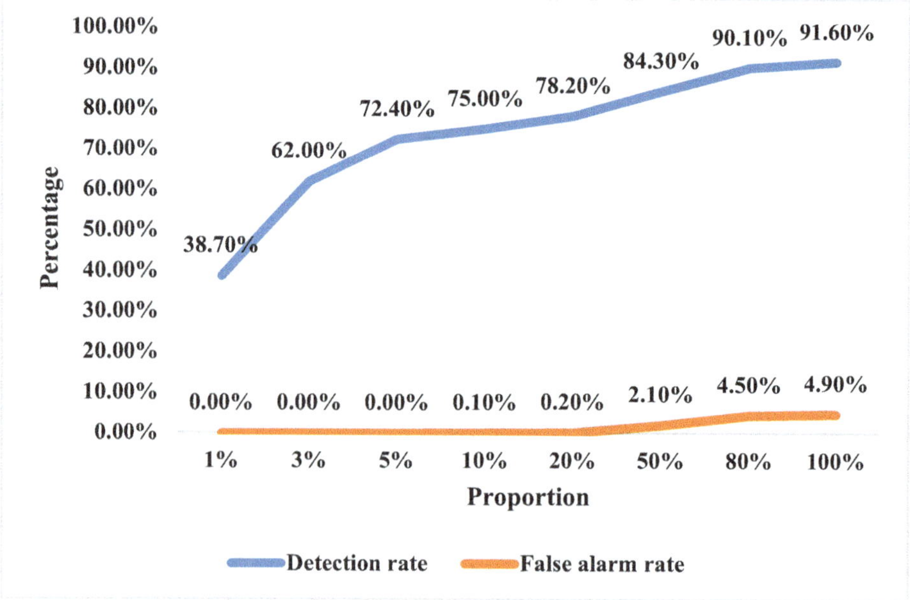

Fig. 3. The impact of data size on classification results

It can be seen from Table 2 and Fig. 3 that when the abnormal samples in the training set account for 5% or less of the original abnormal samples, the detection rate of the model is 72.4%, the false alarm rate is 0%, and the performance is very impressive. With the increase of, the detection rate and false alarm rate of the model are slowly increasing, but the false alarm rate is also below 5%.

5 Conclusions

As the core technology of intrusion detection technology, network abnormal data detection technology has attracted much attention. However, with the popularity of the network and the continuous expansion of network scale, network security problems still exist. Although the traditional network anomaly detection technology can ensure network security to a certain extent, it still has the problem of false positives and high false positives. The combination of classification algorithm and abnormal network data detection technology provides a shortcut to solve these problems.

References

1. Sujendran, R., Arunachalam, M.: Design and development of suginer filter for intrusion detection using real time network data. Int. Arab J. Inf. Technol. **15**(4), 633–638 (2018)
2. Alagrash, Y., Drebee, A., Zirjawi, N.: Comparing the area of data mining algorithms in network intrusion detection. J. Inf. Secur. **11**(1), 1–18 (2020)
3. Yuan, X., Wang, S., Peng, C., et al.: Efficient near-duplicate document detection using consistent weighted sampling filter. Int. J. Netw. Secur. **21**(6), 947–956 (2019)
4. Alkasassbeh, M.: A novel hybrid method for network anomaly detection based on traffic prediction and change point detection. J. Comput. Sci. **14**(2), 153–162 (2018)
5. Zhao, H., Liu, H., Wenjing, H., Yan, X.: Anomaly detection and fault analysis of wind turbine components based on deep learning network. Renew. Energ. **127**, 825–834 (2018)
6. Carvalho, L.F., Abrao, T., Mendes, L., et al.: An ecosystem for anomaly detection and mitigation in software-defined networking. Expert Syst. Appl. **104**, 121–133 (2018)
7. Burgueo, J., De-La-Bandera, I., Mendoza, J., et al.: Online anomaly detection system for mobile networks. Sensors **20**(24), 7232 (2020)
8. Callegari, C., Donatini, L., Giordano, S., et al.: Improving stability of PCA-based network anomaly detection by means of kernel-PCA. Int. J. Comput. Sci. Eng. **16**(1), 9 (2018)
9. Nevat, I., Divakaran, D.M., Nagarajan, S.G., et al.: Anomaly detection and attribution in networks with temporally correlated traffic. IEEE/ACM Trans. Networking **26**(1), 131–144 (2018)
10. Kaur, G., Jain, M.: A comparison of two blending-based ensemble techniques for network anomaly detection in Spark distributed environment. Int. J. Ad Hoc Ubiquitous Comput. **35**(2), 71 (2020)
11. Jain, M., Kaur, G.: A study of feature reduction techniques and classification for network anomaly detection. J. Comput. Inf. Technol. **27**(4), 1–16 (2020)
12. Nkenyereye, L., Tama, B.A., Lim, S.: A stacking-based deep neural network approach for effective network anomaly detection. Comput. Mater. Continua **66**(2), 2217–2227 (2021)

Construction of Network Data Security Detection System Based on Data Mining Algorithm

Wei Zhan, Zhiyong Zha, Bo Jin, Rongtao Liao, Feng Guo, Guoru Deng, Zheng Yu, Liang Dong, Jinhui Zhao, Chenxi Dong, and Xin He(✉)

State Grid Hubei Information and Telecommunication Company Limited, Wuhan, Hubei, China
2271639122@qq.com

Abstract. With the rapid popularization and development of the mobile Internet, the issue of network data security has received more and more attention. In order to effectively guarantee Internet data security, an advanced, efficient and reliable intrusion monitoring and detection system is bound to be indispensable. This article mainly focuses on the research of the Internet data security detection management system based on data mining algorithms. Based on the collection of relevant literature materials, it summarizes the actual needs of the Internet data security detection management system, and then analyzes the Internet data mining in the Internet data security detection application research in the management system. Based on these technologies, the Internet data security detection management system of the Internet data mining algorithm is designed, and the designed system is tested. The detection results are obtained. The system after the algorithm is improved. It is lower than the traditional system, and as the amount of data increases, the system saves more time. The detection rate of the improved system is 4.44% higher than the traditional detection rate.

Keywords: Data mining · Network data · Security detection · System design

1 Introductions

While the rapid development of Internet technology provides convenience for people, it also poses many threats to network security [1, 2]. In the field of network security, there is an urgent need for accurate and effective detection of abnormal network behaviors [3, 4]. Data mining is a data processing method that can dig out a lot of useful information based on facts and rules [5, 6]. Data mining algorithms are usually combined with anomaly detection technology, because they can generate supervised detection models in real time and detect abnormal data, and algorithms can eliminate the complex redundant interference features of abnormal data through feature selection, so as to avoid dimensional damage and improve anomaly detection speed, reduce the cost of inspection operation. This makes data mining algorithms widely used in the field of anomaly detection [7, 8].

I. Ahmad et al. (Eds.): STSIoT 2021, LNDECT 122, pp. 478–486, 2023.
https://doi.org/10.1007/978-981-19-3632-6_58

Regarding the research on network data security detection, some researchers believe that static network intrusion protection based on firewalls alone is no longer sufficient to deal with various attack methods. As an active monitoring tool, network intrusion detection technology can effectively fill the gaps in firewalls. Monitoring objects can be divided into hosts and network intrusion detection systems. The host-based anomaly detection system detects abnormal network activities by monitoring and analyzing server system logs, and there is a problem with interactivity. Network intrusion detection system is currently a research hotspot that uses data mining technology to quickly detect intrusive network behaviors. The system uses a grouping algorithm to divide clusters into normal behaviors and extreme behaviors into abnormal behaviors. The detection performance is significantly better than server-based detection systems. However, with the rapid development of the network, many new large-scale attack methods continue to emerge, which have certain problems in flexibility and adaptability, and are not sufficient to deal with multiple attacks [9]. Some researchers believe that when using data mining techniques to process real-world data sets, cluster analysis and detection techniques are complementary. When performing data processing, it is important to consider how to analyze and deal with the redundancy in the data set. One idea is to cluster the data set before analyzing extreme values. This can also significantly reduce the size of the data set and reduce the complexity of the algorithm [10]. In summary, there are still many research results on network data security detection, but there are few researches on the design of security detection systems and the improvement of data mining algorithms.

This paper studies the network data security detection system based on data mining algorithms. After a general understanding of related theories, the network data security detection system based on data mining algorithms is designed, and the designed system is tested, and finally passed. The test results draw relevant conclusions.

2 Research on Network Data Security Detection System

2.1 Demand Analysis of Network Data Security Detection System

(1) The system intrusion detection function can monitor the behavior of the user database, that is, detect the intrusion behavior in the database. Collect database control logs, conduct behavior analysis, detect possible abnormal behavior, and use it as evidence of intrusion. The intrusion detection function needs to analyze and monitor the files in the running database in order to detect intrusions in the database in time, and through training data, identify the characteristics of abnormal user behavior patterns, or analyze the intrusion behavior to determine the root cause, and avoid using it to attack the database again, to ensure the safe operation of the database system.

(2) The functional requirements of the database intrusion detection system of the response module are mainly to perform the function of immediately notifying the database administrator of the detection of an intrusion and passing it to the database administrator to process unidentified data. The alarm content includes the user name of the intruder, the intrusion time, the intrusion server, the intrusion operation, etc. For unidentified data, the administrator must determine whether this is a new intrusion.

(3) The basic functional requirements of the database intrusion detection system are mainly to realize user management and secure data management of the database system. Users include administrators and general users. Different users have different permissions to manage their own permissions, preventing unauthorized users from breaking into the system and performing malicious functions. System administrators need to manage security logs, use training history data to create rules, delete control logs that consume memory for a long time, and free up server disk space to speed up the server.

2.2 Application of Data Mining in Network Data Security Detection System

(1) With the passage of technology and time, the data changes, and the information and basic knowledge included in the data will also change (especially, for example, in the system for intrusion and detection of the network, new intrusion methods and behaviors continue to emerge and change [11]. The model needs to be updated, and the update process requires the use of a new data mining system to recreate a new model with new data, and then apply it to the system.
(2) Use data development and mining technology to solve network performance to evaluate from the perspective of data. Undefined useful information is derived from databases including data. For the data set of the monitored event, the technology of classification and comprehensive analysis of the results is used to classify the event, analyze the relationship between the event and the data, and evaluate the data set and results controlled by the event, and finally achieve the goal of intrusion detection.
(3) Because the data attributes of different information sources are different, different data mining algorithms need to be used to detect their hidden laws, and different data mining algorithms will also set corresponding data attributes according to their own different needs [12]. Therefore, different types of intrusion retrieval data should be compared with different types of data mining algorithms to study the corresponding data and discover the rules.

2.3 Data Mining Algorithm

This paper designs a hybrid feature selection method based on adaptive genetic algorithm, which fully combines some advantages of filtering and encapsulated feature selection methods based on adaptive genetic algorithm. First of all, the data set that has been detected by intrusion in the input network is cleared and transformed into a format suitable for processing data mining algorithms. Second, use cs-reg to filter irrelevant data in the data set. Finally, it is mainly to filter the characteristics. The lightgbm classifier and adaptive genetics are closely combined with each other. The ranking algorithm is a method for comprehensive selection of intrusion detection features, finding a subset of features with good ranking results, improving the results of intrusion detection, and reducing ranking the sorting time of the algorithm in the network.

After preprocessing the data, use cs to filter the attributed data. In view of the existing problems of each intrusion detection, the independence between each attribute and category in χ^2 is checked, and the difference of each intrusion detection attribute is descended and sorted according to the statistics of χ^2. Here, the χ^2 statistic received by each attribute is higher. This attribute can ensure that our classification effect on intrusion detection results and the correlation with other types are higher. Therefore, the χ^2 expression between each attribute and category c is expressed as:

$$\chi^2(f, c) = \sum\nolimits_{f \notin 0,1} \sum\nolimits_{c \in 0,1} \frac{(N_{fc} - E_{fc})}{E_{fc}} \tag{1}$$

$$E_{fc} = R_f \times \frac{I_c}{N} \tag{2}$$

In the formula, C-a category, normal or abnormal; f–a feature in intrusion detection data; N-observed value; E-expected value.

Given that the observation set X = {x}, i = 1, 2 …n, each observation value is a d-dimensional real vector, and k-means clustering is to divide the observation set X of the observation set X into the k set C = {c}, k = 1, 2, ..K, and kSn, so that the sum of squares in the group is the smallest. Let μh be the mean value of c, then the variance of each sample data and the mean value in each category is defined as:

$$J(c_k) = \sum\nolimits_{x_i \in c_k} \|x_i - \mu_k\|^2 \tag{3}$$

3 Design of Network Data Security Detection System Based on Data Mining Algorithm

3.1 The Overall Architecture of the Network Data Security Detection System

Based on the above analysis of the requirements of the network data security detection system, the overall architecture of the network data security detection system based on the data mining algorithm designed in this paper includes user management, data acquisition, data processing, data mining, and visualization of anomaly detection.

3.2 User Management

User id mainly completes adding and deleting all users of the entire system. This system adopts a role-based access control mode, which manages the user's identity information. Its main functions are two points: create a new character and delete a character. After selecting a role, you can directly change the role (change the name and character description of a role at the same time), according to the settings of the control rules. A control rule is determined by selecting a strategy, a new strategy corresponding to the role is edited and written into a database, and executed by the security monitor agent.

3.3 Data Collection

The Wincap data package is a free public network access architecture on the Windows platform. It has a set of standard data packet download interface, compatible with Libcap. Wincap is composed of three main parts: acket Filter Driver: Packet.dll, a basic dynamic link library, the underlying API contained in it can be used to directly access driver functions. Wpcap.dll has a powerful set of advanced packet logging libraries. The library allows data packets to be independent of the operating system and network hardware. Wincap can record data packets on the network and filter the data packets according to the configured rules.

3.4 Data Processing

The received original network data packet cannot be used directly, and can be used after processing. This is because the network data packet is split into data with unique numbers before being sent. However, because the Internet is based on packet switching, there is no relationship between the order in which data arrives at the receiver and the serial number. Therefore, when the data piece arrives at the sink machine, it appears that the data flow is interrupted. To process the data, you need to reintegrate the data. The process of each protocol is different. UDP and ICMP protocols are offline, and their data packets can be regarded as independent connections. For TCP protocol messages, information such as flag bits will reset the message data belonging to the same TCP connection based on the source address, destination address, sequence number.

3.5 Data Mining

The purpose of correlation analysis is to discover the characteristics of unknown or hidden intrusion behavior, and the extracted data source is the packet information stored in the system log. The intrusion detection features are written in a format that meets the requirements and added to the Snort rule library, so that the detection system has the ability to self-learn, and new intrusions can also be detected. On the one hand, this module allows administrators to greatly reduce their workload and eliminate a large number of manual registration of new intrusion detection rules. On the other hand, management strategies that can be adjusted for different situations.

3.6 Visualization of Anomaly Detection

The server usually uses node as the development language. At the same time, node is a time-based asynchronous programming model. This computing model allows a node-based system to quickly push a large number of data sets to its associated data sets, so that the server-side efficiency is higher and the latency is lower. When multiple clients connect to node and service, they can execute the corresponding service efficiently and communicate data. In addition, the library that can guarantee a continuous connection between the client and the server is called Socket.io. This is a js library that provides a persistent link to the server and supports multiple browsers. Socket.io and Node.js are used in combination, so if the server updates some service data, these data will be active on the client side, instead of manually refreshing the page on the client side, it will be forwarded to it. In the above architecture diagram, data information is stored in the Kafka distributed message queue. When other modules are applied and data is received, the data is put into Kafka. Node.js can track registration issues and set events of interest. Node.js forwards events of interest to clients through the persistent connection to Socket.io mentioned above. Therefore, this article chooses a big data real-time visualization architecture based on Kafka, Node.js and Socket.js.

4 System Detection

4.1 Testing Time Test

The detection time is mainly performed using a set of known data. If you want to protect the same amount of crawled data, first do not pay attention to the accuracy of the crawling results, but mainly record the time spent crawling before and after the improvement. In the experimental simulation of this article, the test algorithm detection time uses kddcup.newtestdata_10_percent__unlabeled.gz. The test subset data of the sample comes from 10% of the experimental KDD Cup99 data set. The experimental results are shown in Table 1:

Table 1. Detection time test result

	The algorithm of this article	Traditional algorithm	Time saving rate (%)
1500	17	15	0.10
2500	29	25	0.13
5500	97	91	1.09
15000	358	344	2.23
55000	869	839	3.12
150000	2267	2175	3.34
250000	5087	4878	3.56
550000	13999	13467	3.78

It can be seen from the above Fig. 1 and Fig. 2 that the improved algorithm system is first compared with the conventional database, eliminating most of the normally undetected data, and also adding related analysis units. Therefore, the more data you are processing, the more time you will save. This improved method will improve the system detection time to a certain extent.

4.2 Comparison Test of Detection Accuracy

The accuracy of algorithm test detection before and after the improvement should be from 10% of data sampling to experiment, mainly using data training subset (kddcup.data_10_cent.gz) and test subset data (kddcup.newtestdata_10_cent_unlabeled.gz).

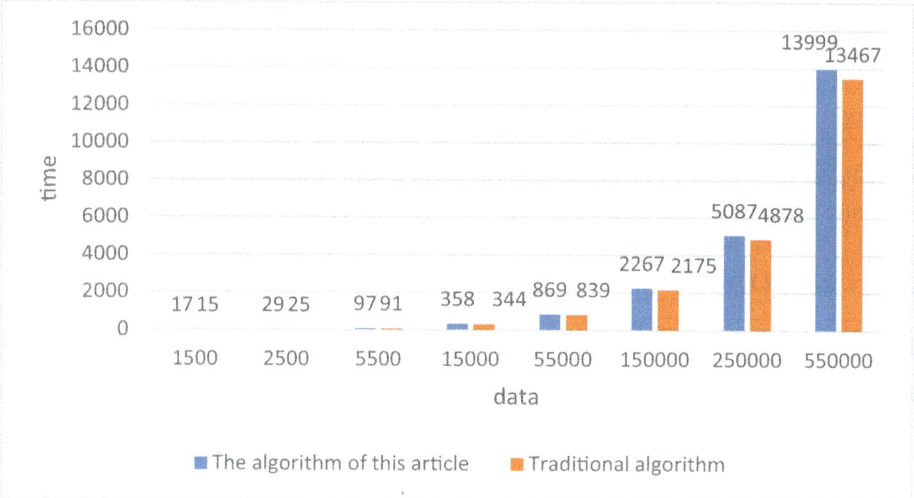

Fig. 1. Detection time test result

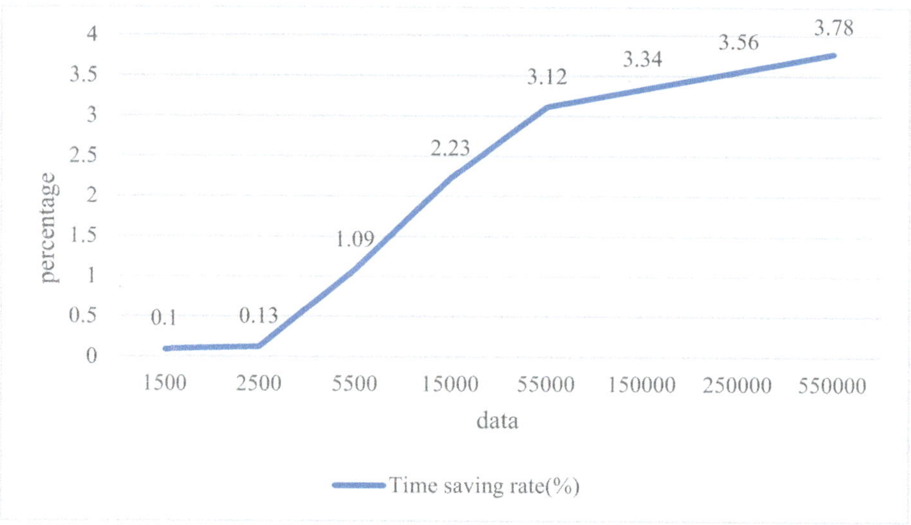

Fig. 2. Time saving rate

Without paying attention to the crawling time, first use the training subset data to train the required data in the regular and rules database, and then use the test subset data to test the results. The experimental data results are shown in Table 2:

As can be seen from Fig. 3, the system after the improved algorithm has increased the detection rate by 4.44% compared to the traditional system due to the decrease in packet loss rate and other reasons. The system is reduced by 2.01% and 3.09% respectively.

Table 2. Comparison test of detection accuracy

	The algorithm of this article	Traditional algorithm
Detection rate	89.12%	93.56%
False alarm rate	3.02%	1.01%
False negative rate	7.43%	4.34%

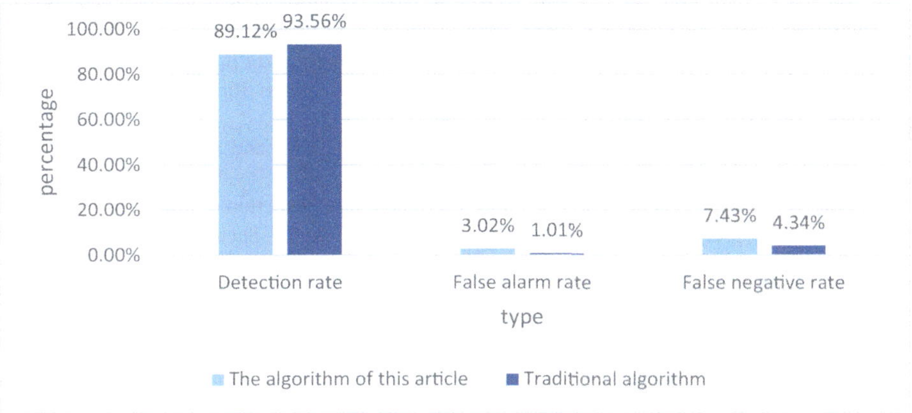

Fig. 3. Comparison test of detection accuracy

5 Conclusions

This paper studies the network data security detection system based on data mining algorithms. After analyzing the system requirements, the system is designed, and the designed system is tested, and the solution is obtained through the test. After the algorithm is improved, the system is tested. The time has been reduced and the detection rate has increased by 4.44%.

References

1. Mao, W., Cai, Z., Yuan, Y., et al.: From big data to knowledge: a spatio-temporal approach to malware detection. Comput. Secur. **74**(MAY), 167–183 (2017)
2. Alagrash, Y., Drebee, A., Zirjawi, N.: Comparing the area of data mining algorithms in network intrusion detection. J. Inf. Secur. **11**(1), 1–18 (2020)
3. Camacho, J., García-Giménez, J.M., Fuentes-García, N.M., et al.: Multivariate big data analysis for intrusion detection: 5 steps from the haystack to the needle. Comput. Secur. **87**(Nov.), 101603.1–101603.11 (2019)
4. Nogueira, R.F., et al.: Fingerprint liveness detection using convolutional neural networks. IEEE Trans. Inf. Forensics Secur. **11**(6), 1206–1213 (2016)
5. Pierazzi, F., Casolari, S., Colajanni, M., et al.: Exploratory security analytics for anomaly detection. Comput. Secur. **56**(FEB.), 28–49 (2016)

6. Shitharth, S., Prince, W.D.: An enhanced optimization based algorithm for intrusion detection in SCADA network. Comput. Secur. **70**(sep.), 16–26 (2017)
7. Zhao, X., Lin, S., Chen, X., Ou, C., Liao, C.: Application of face image detection based on deep learning in privacy security of intelligent cloud platform. Multimed. Tools Appl. **79**(23–24), 16707–16718 (2019). https://doi.org/10.1007/s11042-019-08014-0
8. Jafarian, T., Masdari, M., Ghaffari, A., Majidzadeh, K.: A survey and classification of the security anomaly detection mechanisms in software defined networks. Clust. Comput. **24**(2), 1235–1253 (2020). https://doi.org/10.1007/s10586-020-03184-1
9. Larriva-Novo, X., Vega-Barbas, M., Villagra, V., et al.: Evaluation of cybersecurity data set characteristics for their applicability to neural networks algorithms detecting cybersecurity anomalies. IEEE Access **8**(99), 9000–9014 (2020)
10. George, A., Mostaani, Z., Geissenbuhler, D., et al.: Biometric face presentation attack detection with multi-channel convolutional neural network. IEEE Trans. Inf. Forensics Secur. **15**(1), 42–55 (2020)
11. Campfield, M.: The problem with (most) network detection and response. Netw. Secur. **2020**(9), 6–9 (2020)
12. Tundis, A., Bck, L., Stanilescu, V., et al.: Experiencing the detection of radicalized criminals on Facebook social network and data-related issues. J. Cyber Secur. Mob. **9**(2), 203–236 (2020)

Security Protection Technology Based on Intelligent Semantic Analysis

Ning Xu, Liang Dong, Cheng Zhang, Yue Guo, Guoru Deng, Yan Zhuang, Liang He, Jie Wang, Zheng Yu, Yuan Liang, Hao Xu, and Ping Zhang(✉)

State Grid Hubei Information and Telecommunication Company Limited, Wuhan, Hubei, China
3025531580@qq.com

Abstract. Intelligent semantic analysis is based on the user's security protection needs, according to the current system application system problems, using artificial intelligence technology and information theory and other related knowledge to realize data mining, screening and sorting. Intelligent semantic retrieval technology is used to realize the identification and judgment of the dangerous factors in the user's environment and the information that may cause accident consequences. This is a new safety protection technology. The significance of studying this technology in this paper is to optimize the safety system through the understanding of the technology. This article mainly uses experimental method, data method and data analysis method to gain an in-depth understanding of intelligent semantic analysis technology and security protection technology, and conduct experiments. The experimental results show that the accuracy rate of the safety protection system designed in this paper can reach 99%, and the real-time performance is relatively strong.

Keywords: Intelligent semantics · Semantic analysis · Security protection · Technical research

1 Introduction

With the development of information construction, there is an urgent need to build appropriate information security infrastructure at home and abroad to ensure the normal operation of network and information system. In the process of information construction, the complexity of the network system and the dynamics of the application environment cause a sharp increase in security risks, but the attack technology is improving and becoming more and more popular. Various attacks are becoming more and more frequent, and the consequences of security problems are becoming more and more serious. Therefore, users' demand for network security products is becoming more and more urgent, and the technical requirements for security products are becoming higher and higher. The development of an intelligent and integrated network security protection system is of great strategic significance for adapting to and meeting the requirements of national information security construction and building China's information security infrastructure with independent intellectual property rights. However, in the past

I. Ahmad et al. (Eds.): STSIoT 2021, LNDECT 122, pp. 487–496, 2023.
https://doi.org/10.1007/978-981-19-3632-6_59

IDS system, technicians paid more attention to how to design an effective system and its detection performance. The research on intrusion response, effective processing of alarm events, auxiliary decision-making process and tracking of intrusion behavior is not very sufficient and lack of corresponding means. On the other hand, a prominent problem of the current IDS system is the high rate of false positives, which can not accurately find all intrusion and attack behaviors. From a global perspective, in the information age, information system has become the base of national security, no matter what.

The national infrastructure systems related to the national economy and people's life, such as finance, electricity and telecommunications, are also various other industrial networks, and their operation depends on the information network system. Once these systems have problems, they will lead to economic losses and inconvenience to social life, or paralysis of the whole country's politics, economy or military, and out of control of social order.

Intelligent security protection system is a new type of protection object based on computer network technology and centered on information processing. It can be realized through functions such as intrusion detection, monitoring and alarm, and auxiliary control. During the construction process, the system will automatically filter, store and process information according to functional requirements. The association between this part of the data feature and the environment is realized through intelligent semantic retrieval technology.

2 Related Work

There are many researches on security protection technology based on intelligent semantic analysis.

For example, Marios Kountouris and Nikolaos Pappas said, wireless connections have traditionally been viewed as opaque data pipes. However, in emerging cyber-physical and autonomous network systems using semantic-enhanced sampling and communication strategies, the number of non-informative samples generated is significantly reduced [1]. Gerard Deepak et al. proposed that extracting relevant information from networks is a major challenge due to the diversity of network information and its dynamically changing content. As the World Wide Web transforms into a more organized Semantic Web, it is necessary to incorporate semantic techniques to retrieve relevant information [2]. According to Y. S. Mehanna and M. B. Mahmuddin, emotions can be implicit or explicit in the text. Therefore, identifying hidden emotions is a major challenge for current sentiment analysis (SA) methods. Other common challenges are misclassification of opinion words, ignorance of contextual information and misinterpretation of text leading to mishandling of short texts, and lack of sufficient data for analysis tasks [3]. According to M. Hina et al., artificial intelligence (AI) combined with the Internet of Things (IoT), known as (AIoT), is an emerging trend in industrial applications that enables intelligent decision-making through autonomous analysis. The widespread use of IoT devices in various scenarios generates massive amounts of data designed by attackers to disrupt normal operations and services. Therefore, active data analysis is necessary to prevent cyber-attacks and crimes [4]. According to Chiara Renso et al., location-based applications and devices generate large amounts of spatiotemporal data and other useful information. This data can form a trajectory with a series of time-stamped locations. Analyze the lifecycle around the complete semantic trajectory [5]. D. Amitrano et al. propose a new method for unsupervised detection of land cover changes. It overcomes these limiting layers by leveraging multi-temporal, geographic, and object-based image analysis to reveal the intrinsic semantics of the data, and directly identifying the precise categories of variation through dictionary-based pre-classification and fuzzy combinations of category-specific data [6].

The application of intelligent semantic analysis technology in the field of security protection can effectively solve the problems existing in traditional security protection. Therefore, starting from intelligent semantic analysis, this paper studies intelligent security protection technology and its system.

3 Security Protection Technology Based on Intelligent Semantic Analysis

3.1 Smart Search

The semantic-based intelligent retrieval system is the need of human-computer natural communication, and the need of users to use network information more conveniently and effectively. The intelligent retrieval system can provide a more humanized user interface, which has strong practical significance for improving retrieval accuracy and enabling users to share network resources more efficiently and conveniently [7].

The key technology to realize intelligent information retrieval is to understand the natural language of the machine. The goal of natural language understanding research has not yet reached a very high level due to its difficulty. Therefore, further research and investigation are needed [8].

(1) Main methods of intelligent retrieval

1) Vector space model. By calculating the similarity between the recovery vector and each document vector, any document that is more similar to the recovery vector will be marked to the user. 2) Probabilistic retrieval strategy. The probability model is based on the principles of Bayesian probability theory and is different from the vector space model. The disadvantage of the probabilistic model is that the probability of calculating the correlation is not clearly defined. 3) Reasoning network. The method is to create a network diagram between all documents and the retrieved information through the relationship between user retrieval information and document information. 4) Neural network. A neural network consists of nodes and connections between nodes [9].

3.2 Semantic Analysis Technology

In the development of artificial intelligence, the security protection system is combined with intelligent semantic analysis technology to achieve effective protection of user information and improve the security and reliability [10].

There are some problems with the safety protection system. (1) Misreporting and underreporting issues. The data processing and identification process in the intelligent security protection system is very complicated. In actual application, misreading errors and missing information are often caused due to defects in the operating environment or the system itself. (2) The network topology is unstable and the intrusion detection is difficult. For different users, the corresponding functions are also different [11].

Semantic analysis is a computational theory and a method for acquiring and displaying knowledge. It uses statistical calculation methods to analyze a large number of text sets, extracts the potential semantic structure between words, and uses this potential semantic structure to remove the correlation between words, simplify the text vector, and achieve the purpose of dimensionality reduction. Semantic analysis can analyze the overall structure of the safety protection system well, and can clearly detect the technology and faults in the safety protection system, and can better extract the faults of the system. For semantic analysis, the speed of sentence segmentation has an impact on the speed of the final result.

If intelligent semantic analysis only decomposes matrix A, it is not enough to completely display the relationship between words and documents. It is necessary to define a weight function W (a, b), perform weighted transformation on it, and obtain a weighted matrix A*, and then perform operations on it.

(1) Boolean weights are the simplest and most typical weight assignment method in the field of information retrieval.

$$W_{ab} = \begin{cases} 1 \; if \; g_{ab} \succ 0 \\ 0 \; if \; g_{ab} = 0 \end{cases} \tag{1}$$

If a word appears in the document, the weight of the word is recorded as 1, and if it does not appear, the weight is 0.

(2) Word frequency weight method

Unlike Boolean weights, term frequency weights use the number of times the word appears in the document as the weight.

$$W_{ab} = sg_{ab} \tag{2}$$

The word frequency weight takes into account the influence that different words should be given different weights.

(3) Probability weight method

We define the local weight of a word by its importance in a particular document. This method of assigning weights has certain defects. The logarithmic function can slow down the excessive growth of the function value caused by the growth of the independent variable without changing the monotonicity of the original function, so the logarithmic function of the frequency function can effectively solve this problem. Specifically as formula (3):

$$WQ(a, b) = log_2(sg_{ab} + 1) \tag{3}$$

Among them, sg_{ab} is the frequency of the vocabulary. The purpose of adding 1 here is to make the weight not appear negative.

3.3 Design of Security Protection System for Big Data Platform Based on Intelligent Semantics

(1) Safety protection technology

1) The firewall is the first barrier to protect the local network. Through the establishment of security policies, illegal users are prevented from accessing the local network and the loss of local network information. At the same time, the firewall has a powerful audit function, as long as the access is recorded by the firewall. A firewall is a widely used security mechanism today, and large local area networks are usually equipped with a firewall. 2) IDS is the second layer of defense behind the firewall and a powerful supplement to security protection. IDS is a defense-oriented monitoring firewall, and access through the firewall is regarded as normal access. As a proactive defense system, IDS collects and analyzes IT information and alerts network administrators when abnormalities are detected. 3) WAF recognizes user access requests and provides good protection for web applications. Currently, there are thousands of sites on the network and many users are faced, so it is very necessary to install WAF. 4) Digital certificates are used to ensure the reliability of the identities of the communicating parties, and to encrypt and decrypt the transmitted information. Even if the information is intercepted by hackers, it can be guaranteed that the information is not leaked, and the digital certificate can also prevent malicious data transmission. Its specific composition and process are shown in Fig. 1:

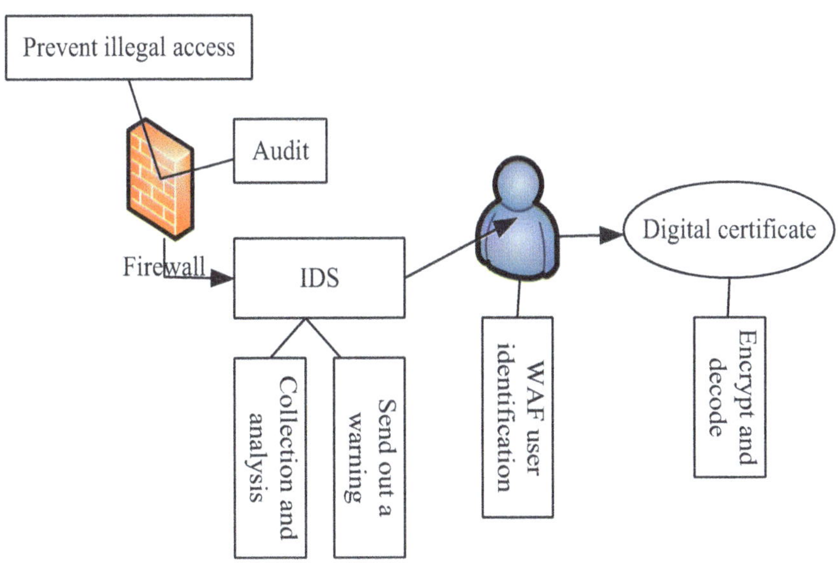

Fig. 1. The specific composition and process of safety protection technology

(2) The big data platform security protection model mainly realizes the following functions:

1) Provide a unified authentication mechanism for each component in the big data platform, and provide a strong authentication strategy. 2) Provide user single sign-on function and provide centralized authorization mechanism. 3) Provide data protection functions on the platform. 4) Provide a fine-grained authorization mechanism, which can provide different access control methods for different big data platforms. 5) It can shield the vulnerabilities of the big data platform itself and provide a unified interface to access different components in the big data platform. 6) Control the data returned to users, and be able to filter sensitive data.

(3) Key points of the security protection model

The security protection model of the big data platform should include six functions, namely authentication, authorization, security audit, account management, service agent and data protection.

1) Certification model design. The authentication module should be composed of two sub-modules, an external authentication module and an internal authentication module. The external authentication module is responsible for the authentication between the external application and the security protection system, and the internal authentication module is mainly responsible for the authentication between the security protection system and the big data platform. The external authentication module performs authentication based on the Apache Shiro authentication framework, ensuring that users only need to authenticate once for the big data authentication authorization control system. 2) Authorization model design. The authorization module is mainly responsible for performing corresponding authorization operations on the legal identities in the system. The authorization module mainly includes two sub-modules: service layer authorization and fine-grained authorization. The service layer authorization is mainly based on the four-tuple to determine and formulate the corresponding authorization strategy. The security protection model of the big data platform uses a gateway to proxy user access requests. Therefore, each user's identity in the big data platform is the same, and it is the identity of the proxy gateway. However, this makes it impossible to distinguish between authority management and control. The permissions of different users. 3) The goal of the security audit module is to detect abnormal behaviors in the system in time or when a security incident occurs, to trace the root cause of the problem through the audit log. 4) The account management module is mainly to manage the user's account. 5) Service agent algorithm design. The service proxy module is mainly responsible for proxying requests, sending access requests to the internal protected big data platform, and returning the execution results to the user. 6) Data protection algorithm design. There are two main aspects of data protection in a big data platform. One is encryption to protect data, and the other is sensitive data protection.

(4) The service agent module processing request process is:

The client sends an HTTP request to the gateway through the REST API, and the gateway provides services through Jetty.

The Jetty service will query the corresponding filter chain according to the requested URL to determine the responsibility chain for processing the request.

The filter chain is called, and the request will be passed in the responsibility chain.

The request is called by the objects in the responsibility chain in turn, until the end of the responsibility chain.

After the request is processed by the object in the responsibility chain, the service proxy module will map the request to an internal URL and send the request to the internal big data platform.

The internal big data platform returns the result of the request through JSON.

The returned results are processed through the chain of responsibility in turn.

When the objects in the responsibility chain have processed the returned result data, they will send the data to the Jetty service.

The Jetty service returns the requested result to the customer.

(5) System architecture design

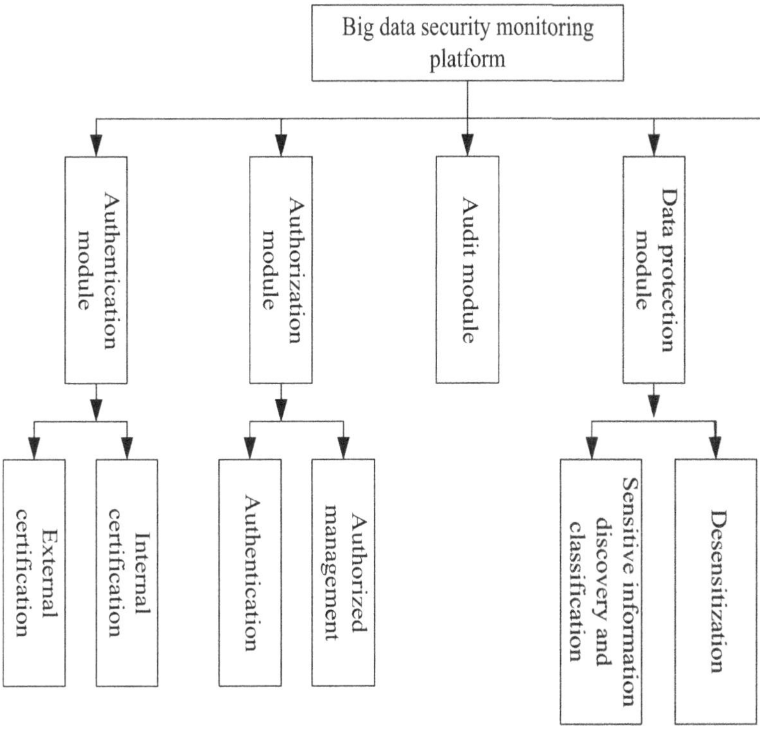

Fig. 2. Big data security protection system composition

As shown in Fig. 2, the big data security monitoring platform is mainly composed of five major modules: authentication module, authorization module, audit module, data protection module and service agent module.

4 Experiment

4.1 System Experimental Environment Deployment

The big data security monitoring platform is mainly used to protect the big data platform, and realize the big data platform's authentication, centralized authorization, access audit and data protection functions. The experimental environment contains a total of 3 servers. The server where the big data security monitoring platform is located usually includes two IP addresses. The user accesses the big data security monitoring platform through this IP. The intranet IP is one of the big data security monitoring platform and the big data platform. The specific experimental configuration is shown in Table 1:

Table 1. Test environment specific configuration

Hardware environment		Software environment	
Project	Configuration	Project	Configuration
Server	PowerEdge R730XD	Operating system	CentOS 9 server
processor	E5-2630 v3	CDH software package	CDH6.9.0
RAM	64 GB RDIMM	Cloudera Manager version	6.9
Network daughter card	I360 QP 2 GB	Knox	0.9.0

Table 2. Security protection domain name framework test results

	Number of domain names (basic analysis)	Number of domain names (page detection)	Accuracy
Take over	956	32	99.6
Transmission	979	16	99.5
Encryption	964	20	99.7
Enter	997	8	98.6
Warning	986	12	99.8

4.2 Safety Protection System Verification

First, develop a program that sends Modbus communication messages to simulate the Modbus master station, and realize the function of sending Modbus/TCP communication messages through this program. Then, use ModbusSlave to simulate the Modbus slave station to receive the communication messages sent by the Modbus master station. Test the security protection performance and recognition effect of the big data platform. The DNS servers of receiving, transmitting and encrypting, input and three parts are selected respectively, and 1000 domain names are randomly selected from each DNS server, and the domain names are used for warning detection.

5 Result Analysis

5.1 Analysis of Safety Protection Test Results

The experiment's testing of the security protection domain name detection framework includes two aspects. On the one hand, it tests the accuracy of the domain name detection of the entire framework. On the other hand, the test can directly determine whether the domain name is the number of domain names in the basic analysis stage (Table 2).

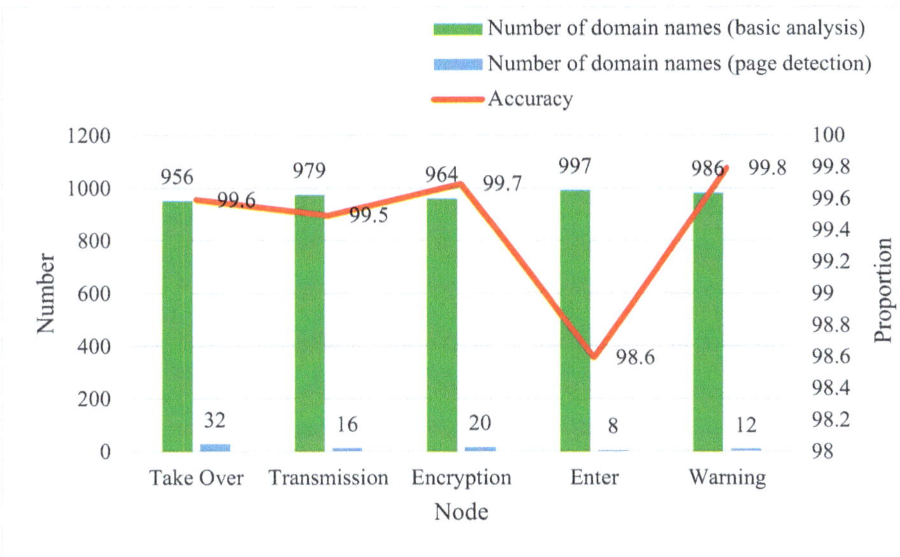

Fig. 3. Security protection domain name framework test results

As shown in Fig. 3, the entire detection framework has a high accuracy of detecting domain names under the DNS server, basically above 99%, and most domain names can be directly detected through the basic analysis stage.

6 Conclusion

Intelligent security protection technology is a new type of system. It plays an important role in protecting personal privacy and preventing information leakage. With the development of network communication, users have doubts about the security and reliability of network applications. The application of intelligent semantic analysis technology in the field of security protection can effectively solve the problems existing in traditional security protection. Through the integration of system data information, interactive access control between users and devices is realized. This paper studies the application of the security protection system to intelligent semantic analysis, and conducts experiments on the security protection system of the big data platform. Experiments show that the system designed in this paper has high accuracy, can perform real-time detection and improve safety.

References

1. Kountouris, M., Pappas, N.: Semantics-empowered communication for networked intelligent systems. IEEE Commun. Mag. **59**(6), 96–102 (2021)
2. Deepak, G., Ahmed, A., Skanda, B.: An intelligent inventive system for personalised webpage recommendation based on ontology semantics. Int. J. Intell. Syst. Technol. Appl. **18**(1/2), 115–132 (2019)
3. Mehanna, Y.S., Mahmuddin, M.B.: A semantic conceptualization using tagged bag-of-concepts for sentiment analysis. IEEE Access **9**, 118736–118756 (2021). https://doi.org/10.1109/ACCESS.2021.3107237
4. Hina, M., Ali, M., Javed, A.R., Ghabban, F., Khan, L.A., Jalil, Z.: SeFACED: semantic-based forensic analysis and classification of e-mail data using deep learning. IEEE Access **9**, 98398–98411 (2021). https://doi.org/10.1109/ACCESS.2021.3095730
5. Renso, C., Bogorny, V., Tserpes, K., Matwin, S., de Macêdo, J.A.F.: Multiple-aspect analysis of semantic trajectories (MASTER). Int. J. Geogr. Inf. Sci. **35**(4), 763–766 (2021)
6. Amitrano, D., Guida, R., Iervolino, P.: Semantic unsupervised change detection of natural land cover with multitemporal object-based analysis on SAR images. IEEE Trans. Geosci. Remote Sens. **59**(7), 5494–5514 (2021). https://doi.org/10.1109/TGRS.2020.3029841
7. Mahmoud, A., Zrigui, M.: Semantic similarity analysis for corpus development and paraphrase detection in Arabic. Int. Arab J. Inf. Technol. **18**(1), 1–7 (2021)
8. Chan, L., Hosseini, M.S., Plataniotis, K.N.: A comprehensive analysis of weakly-supervised semantic segmentation in different image domains. Int. J. Comput. Vis. **129**(2), 361–384 (2021)
9. Essel, D.D., Benuwa, B.-B., Ghansah, B.: Video semantic analysis: the sparsity based locality-sensitive discriminative dictionary learning factor. Int. J. Comput. Vis. Image Process. **11**(2), 1–21 (2021)
10. Ghansah, B., Benuwa, B.-B., Monney, A.: A discriminative locality-sensitive dictionary learning with kernel weighted KNN classification for video semantic concepts analysis. Int. J. Intell. Inf. Technol. **17**(1), 68–91 (2021)
11. Jain, S., Seeja, K.R., Jindal, R.: Computing semantic relatedness using latent semantic analysis and fuzzy formal concept analysis. Int. J. Reason. based Intell. Syst. **13**(2), 92–100 (2021)

Intelligent Algorithm of Semantic Analysis Based on BP Neural Network

Yan Zhuang, Cheng Zhang, Jie Xu, Liang Dong, Ning Xu, Liang He, Jie Wang, Feng Guo, Bo Jin, Zheng Yu, Wangsong Ke, Yaodong Hu, and Ping Zhang(✉)

State Grid Hubei Information and Telecommunication Company Limited, Wuhan, Hubei, China
3025531580@qq.com

Abstract. With the growth of scientific and technological information technology and the rapid popularization of the Internet, network big data and information technology are also growing rapidly. Information technology provides people with more information, and it also significantly increases the operating and management costs of my country's Internet companies. In order to solve this problem thoroughly, people propose a new type of it development, research and business model, namely BP neural network technology. At present, BP neural network technology has been widely used in various application fields such as network storage, search engines, distributed computers, e-commerce, social networks, and has achieved rapid growth. This article mainly adopts the method of organically combining theoretical exploration and empirical research, and systematically analyzes the data collected through research based on the views and research contents of some scholars in recent years. Combining with the analysis of the data of intelligent semantic analysis algorithm, some relevant characteristics of BP neural network are summarized. This article mainly focuses on the research of an intelligent algorithm for image semantic analysis for image processing. The semantic analysis intelligent algorithm can well change the situation of target detection difficulties. This article uses an intelligent algorithm based on BP neural network to automatically analyze and distinguish differences. The final results of the research show that this paper uses the attention model and proposes a semantic analysis algorithm combined with graphic target detection through a multi-scale segmentation network. The experiment shows that the three performances of attention are 71.6, 56.5 and 49.3, which can be learned this algorithm is better than the same comparison algorithm in terms of three performance evaluation indexes.

Keywords: Neural network · Data analysis · Semantic analysis · Target detection

1 Introduction

With the development of Internet technology step by step to the present, computer technology has gradually matured. Using computer technology can solve some tedious and complicated problems. Computer technology is a field that can cover many fields, such as using computer technology to process digital images and videos and gain a higher level of understanding. In terms of mechanical manufacturing and application, it can automate

I. Ahmad et al. (Eds.): STSIoT 2021, LNDECT 122, pp. 497–504, 2023.
https://doi.org/10.1007/978-981-19-3632-6_60

humans to complete tasks that can be accomplished by visual systems. Computer vision includes the following methods: capturing, editing, identifying, and processing digital images. In addition, if digital information and symbolic expressions are to be generated, large-scale data must be drawn from real-world networks [1]. Semantic image analysis is mainly a basic analysis method based on computer vision processing technology. It involves semantically divided regions belonging to different objects. Semantic analysis and object detection technology play a vital role in the understanding of images, and they also have an indispensable position in our real life.

In recent years, many researchers have conducted in-depth research on the semantic analysis intelligent algorithm of BP neural network and achieved good practical application results. For example, ConjetiS believes that the BP neural network system is a parallel and distributed processing structure, composed of the interconnection between two processing units and a combination of directions called the connection signal path. These processing units have local memory and perform local operations. Each processing unit contains a single input and link. This input link can rely on one or more different inputs and branch into multiple parallel links according to its needs [2]. CarboJ believes that the BP neural intelligent network algorithm is very helpful for semantic analysis. The algorithm has strong adaptability and adaptability. It has brought very fast and effective advanced algorithms for the development of human science. It should be popularized in semantic analysis. In application [3]. At present, there are many researches on BP neural network semantic analysis algorithm. These previous theories and experimental results provide a theoretical basis for the research of this article.

This paper analyzes image semantic analysis combined with BP neural network. BP neural network not only improves the performance of semantic image analysis and classification, but also makes good progress in part of the work of structured production. This progress includes determining demarcation targets and forecasting key points. The next step from rough to subtle inference is of course to predict each pixel. Therefore, the BP network once again promotes the development of semantic image analysis, especially in more complex scenes and large-scale data sets, the performance of the analysis far exceeds the traditional methods [4]. Although the BP neural network has the ability to automatically process deep features, due to its numerous features, the existing semantic analysis methods based on the BP neural network still have some challenges.

2 Related Characteristics of BP Neural Network and Network Model Analysis

2.1 Basic Characteristics of BP Neural Network

(1) Non-linear mapping capability

BP neural network technology is essentially a mapping function that performs input and output at the same time. Relevant theories have confirmed that BP neural network technology can approach any nonlinear continuous function under different accuracy [5]. In terms of calculation and mathematics, the traditional BP neural network is actually a method of local search and optimization. It needs to deal with and study how to solve complicated nonlinear related problems. As the weight of the Internet gradually adjusted to local changes. When he falls to the end of the local minimum, the weight will quickly converge to the local minimum, causing pure training to fail. In addition, the neural network BP is also very sensitive to the initial weight of the network. Networks with different weights tend to converge at different low positions.

(2) The convergence speed of the BP neural network algorithm is slow

The original algorithm of the intelligent algorithm in the BP neural network is the italicity algorithm. Therefore, "sawtooth effect" will inevitably occur, resulting in lower efficiency of the BP algorithm. And because the optimized objective function is so complicated [6, 7]. In the BP neural network model, the weight and error change very little, which leads to the BP neural network model in the training. In order to enable the entire network to run the BP algorithm, the traditional one-dimensional search method cannot use the step size of each iteration, but has given the network step size update rules. This method will also lead to inefficient algorithm, all of which lead to BP The neural network algorithm converges slowly [8, 9].

2.2 Fusion of BP Neural Network Model

The merging of BP neural network models is to merge the results of multiple models using a specific algorithm, so that the merged result has the advantages of multiple models, and at the same time eliminates the shortcomings of each model. Using a more appropriate model fusion algorithm, the final performance will be better than that of any sub-model [10, 11]. In this paper, three BP neural network model algorithms are applied, and the division performance of the three algorithms is compared through experiments. BP neural network model fusion methods can be simply divided into learning methods and non-learning methods. The learning method is to linearly weight the feature score-cards of all sub-models through the learned weights. The difference from the above linear combination method is that this method is not needed, just in the test phase, after normalizing the feature score map of all models, each element takes the maximum value or the average value [12].

3 Research on Experimental Preparation of Intelligent Algorithms for Image Semantic Analysis

3.1 Experimental Method

The main method used in this article is to study the BP neural network model. In actual economic activities, the activation of BP neural network parameters plays a decisive role in the training speed of the model. If the initial parameters are close to the target parameters, the training model is easier to converge. The parameters are very different from the target parameters, so training the model will be more difficult. Take the two-channel BP neural network with attention network as the main structure proposed in this paper as an example. This article uses a two-stage training method, that is, the training task is divided into two stages. The results obtained from the first stage training serve the second stage. Experiments verify that this method can effectively improve the training accuracy of the model and reduce the training of the model.

(1) Differential network semantic segmentation algorithm

The perception of BP neural network refers to the size of the area corresponding to the original image of the element in the output feature map of each layer. The size of the receptive field is determined by the size of the nucleus and the step length at the same time, so the size of the receptive field of the first layer of pixels is R^l:

$$R^l = R^{L-1} + \left(K^l - 1\right)S^l \tag{1}$$

(2) Where S^l and S^l are the core size and step length of the first layer, respectively. According to the formula from back to front, the receptive field of any neuron in the final classification layer relative to the input image is called the receptive field of the entire FCN network. If the network has L layers, then

$$R^{FCN} = R^L \tag{2}$$

Assuming that the size of the input image of the FCN network is h_0 (that is, the width or height of the image), then the size of the feature map output by the first layer (the width or height of the feature map)

$$h_l = \left[\left(h_0 - R^l\right)/S^l + 1\right] \tag{3}$$

3.2 Experimental Data Collection

This paper establishes a semantic model under BP neural network to achieve balanced results. First, the relevant elements of BP neural network and the important links that must be considered in the research process of semantic analysis model configuration are explained. Several commonly used objective functions are analyzed to facilitate the selection of the best operation objective. In order to facilitate the establishment of the model and the research of the problem, combined with the analysis of the objective function and the theoretical summary of the BP neural network management, the model assumptions are made. Finally determine the objective function of this article, the test object is the performance comparison of different methods in the new test set.

4 Experimental Study on Image Semantic Analysis of BP Neural Network

4.1 Performance Comparative Analysis of Different Methods Used in the Test Set

Table 1 shows the three evaluation indicators of average pixel accuracy (pixel accuracy), class average pixel accuracy (mean accuracy) and average iou (region intersection over union) to show the different accuracy splits in the newly released test set of pascalvoc 2012 Algorithm to judge.

According to the data shown in Fig. 1, the evaluation results of different segmentation algorithms under the MSCOCO2014 test set are shown. The segmentation algorithms include: BP neural network (FCN), cavity pyramid module (ASPP) neural network,

Table 1. Performance comparison analysis of different methods used in the test set (%)

Algorithm	Pixel accuracy	Mean accuracy	Region intersection over union
FCN	68.5	52.7	42.1
ASPP	68.2	51.6	52.5
Attention	71.6	56.5	49.3
Ours	73.5	60.5	51.1

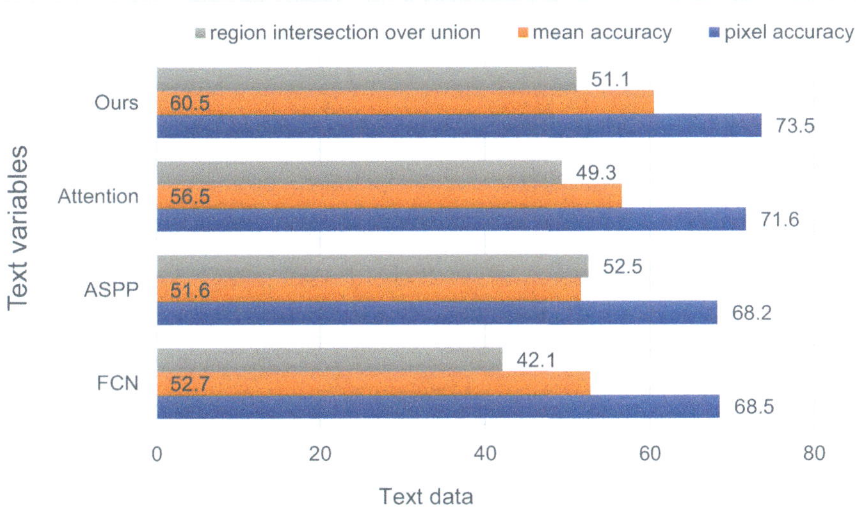

Fig. 1. Performance comparison analysis of different methods used in the test set (%)

using Attention model combined with multi-scale The segmentation network and the semantic analysis algorithm combined with graphic object detection proposed in this article, the three performances of FCN are 68.5, 52.7 and 42.1, the three performances of ASPP are 68.2, 51.6 and 52.5, and the three performances of Attention are 71.6, 56.5 And 49.3 It can be seen from the table that the algorithm in this chapter is better than the comparison algorithm in the three evaluation indicators.

4.2 Attention Performance Comparison Analysis

By comparing the performance advantages of Attention, it is compared with traditional algorithms for performance testing and analysis, and testing from the three aspects of propagation speed, memory usage and model size (Table 2).

According to the data in Fig. 2, it can be concluded that Attention has achieved a good balance between accuracy and speed. Attention's operating speed is obviously beyond the classic methods such as FCN, ASPP, Deeplab, etc. The forward and backward propagation speeds are 485ms and 612ms respectively, and the memory footprint and model size are relatively small, only 1732M and 82M. Attention The feature re-use is completed by feature addition, which is similar to the FCN approach. In addition, this method is also conducive to the propagation of the loss value in the network. The calculated loss value can be directly transferred from the back layer to the front layer, thereby avoiding the problem of gradient disappearance, which allows the network to train more BP layer.

Table 2. Attention performance comparison analysis table

Algorithm	Forward propagation (ms)	Backpropagation (ms)	Memory footprint (M)	Model size (M)
FCN	412.5	498.71	1123	123
ASPP	115.06	172.73	1985	165
Attention	485	612	1732	82
Ours	325	454	1885	886
DeepLab	47.1	71.91	1853	87

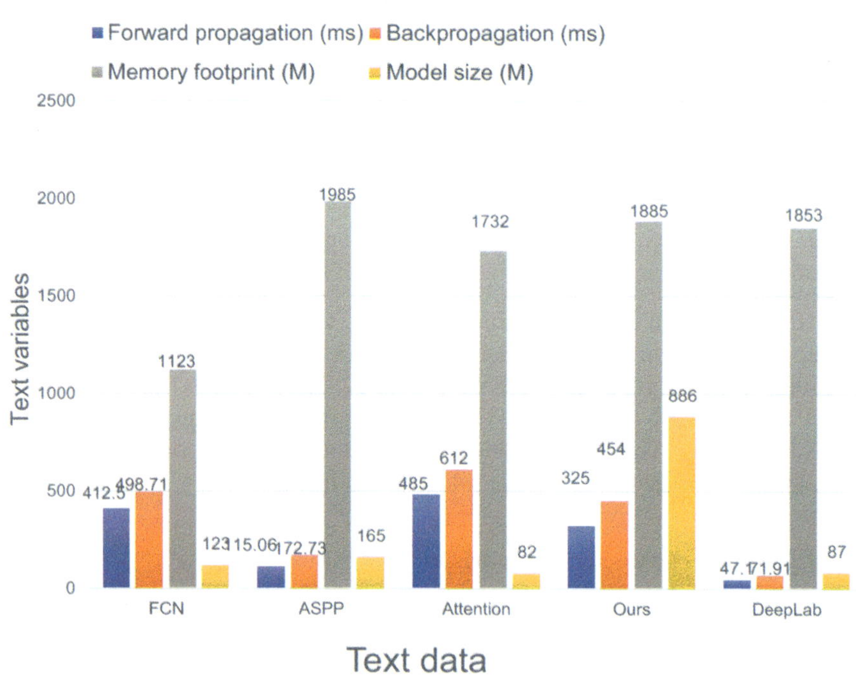

Fig. 2. Attention performance comparison analysis

4.3 Comparison and Analysis of the Accuracy of Different Methods Used in the Test Set

This article collects data on the results of 400 questions answered, and is screened according to the answers, and there are 388 valid sentences remaining. The effective sentences are analyzed by FCN, ASPP, Attention, Ours and DeepLab algorithms respectively. The analysis results are shown in Table 3.

Table 3. Accuracy comparison analysis table

Algorithm	testing amount	Correct quantity	Correct rate (%)
FCN	388	302	77.84
ASPP	388	265	68.3
Attention	388	200	51.55
Ours	388	261	67.27
DeepLab	388	190	49

As shown in Table 3, Through experimental comparison, it is found that the accuracy of the BP network algorithm (FCN) is the highest, and the accuracy of 388 sentence semantic analysis is 77.84%, which is 68.30%, 51.55%, and 67.27% compared to ASPP, Attention, Ours, and Deeplab respectively. Compared with 49%, the BP network algorithm (FCN) has certain advantages over other algorithms in the processing of semantic analysis through comparative analysis of the accuracy rate.

5 Conclusions

In this paper, the semantic image analysis and the analysis of the BP neural network model are combined with each other. The experimental verification shows that the algorithm has a better positioning effect in locating microscopic targets in objects in complex scenes. As a high-tech technology, computer vision has been developed for decades, and its low price, cost, intelligence and other advantages make it irreplaceable and important in certain areas of the commercial market. In the next few years, with the relatively rapid development of big data and the Internet market, this provides a good opportunity for computer vision technology. The BP neural intelligence algorithm and its improved model have solved the adaptation, other problems are difficult to optimize, and are far superior to other traditional resource extraction algorithms. Due to these external and inherent environmental influences, computer vision technology has achieved obvious and rapid development. This research work uses BP neural network as the main basic framework to solve the problem of semantic analysis of targets in complex scenarios. However, deep learning technology includes not only excellent models such as BP neural network, but also other networks suitable for semantic segmentation such as GAN. Therefore, I hope that in future work, other deep learning techniques can be explored to meet the existing challenges of semantic image analysis.

References

1. Wu, Y.: Product form evolutionary design system construction based on neural network model and multi-objective optimization. J. Intell. Fuzzy Syst. **39**(5), 1–15 (2020)
2. Singh, R.R., Conjeti, S., Banerjee, R.: A comparative evaluation of neural network classifiers for stress level analysis of automotive drivers using physiological signals. Biomed. Sig. Process. Control **8**(6), 740–754 (2013)
3. Conjeti, S., Carbo, J., Molina, J.M.: A statistical simulation technique to develop and evaluate conversational agents. AI Commun. **26**(4), 355–371 (2013)
4. Naruniec, J.: A survey on facial features detection. Int. J. Electron. Telecommun. **56**(3), 267–272 (2010)
5. Gobet, F.: Chunk hierarchies and retrieval structures: comments on Saariluoma and Laine. Scand. J. Psychol. **42**(2), 149–155 (2010)
6. Dospinescu, A.S.: Local environment analysis and rules inferring procedure in an agent-based model – applications in economics. J. Econ. Forecast. **15**(1), 128–143 (2012)
7. Gu, D.X., Liang, C.Y., Li, X.G., et al.: Intelligent technique for knowledge reuse of dental medical records based on case-based reasoning. J. Med. Syst. **34**(2), 213–222 (2010)
8. Zhang, C., Liu, X.: Feature extraction of ancient Chinese characters based on deep convolution neural network and big data analysis. Comput. Intell. Neurosci. **2021**(3), 1–10 (2021)
9. Bernardini, F.C., Garcia, A.C.B., Ferraz, I.N.: Artificial intelligence based methods to support motor pump multi-failure diagnostic. Int. J. Eng. Intell. Syst. Electr. Eng. Commun. **17**(2), 71–84 (2009)
10. Zhu, L.: Computer vision-driven evaluation system for assisted decision-making in sports training. Wirel. Commun. Mob. Comput. **2021**(7), 1–7 (2021)
11. Zhang, H., Ji, P., Wang, J.-Q., Chen, X.-H.: A neutrosophic normal cloud and its application in decision-making. Cogn. Comput. **8**(4), 649–669 (2016). https://doi.org/10.1007/s12559-016-9394-8
12. Pellejero, N.F., Grinblat, G., Uzal, L.: Semantic analysis on faces using deep neural networks. Intell. Artif. **21**(61), 14–29 (2018)

The Creation and Dissemination of Popular Science Animation Based on Computer Technology

Xiaoyu Liu[1(✉)] and Constable Edwin[2]

[1] Department of Art, Tianjin Ren'ai College, Tianjin 301636, China
liuxiaoyu62007@163.com
[2] University of London, London, UK

Abstract. Science popularization is the popularization of science and technology. In the post-epidemic era, in response to the popular science issues exposed during the epidemic, the creation and dissemination of popular science animations urgently require in-depth reflection and research. The research object of this topic is the creation and dissemination of popular science animation in the post-epidemic era, and the purpose is to use popular science animation to conduct scientific and rational publicity and guidance to the public. The main methods used in this article are questionnaire survey and interview method. The survey results show that about 45% of people think that dynamic design is very important, and 30% think that animation duration is also the key. Therefore, the innovation of popular science animation should focus on these two aspects.

Keywords: Popular science animation · MG animation · Post-epidemic era · Computer technology

1 Introduction

Facing the urgent public demand for popular science knowledge, driven by the Internet, a large number of popular science works appeared in a short period of time in the early stage of the epidemic. Popular science animation is an animation work that disseminates scientific knowledge. Under the background of "big animation" education, animation majors in colleges and universities serve as animation producers and training places for animation talents. The interdisciplinary characteristics of the talent training model should adapt to the development of the times. Therefore, this article studies the methods and teaching directions of popular science animation creation and dissemination under computer technology.

There are many researches on the creation and dissemination of popular science animation under computer technology. For example, Su Chang pointed out that how to deliver scientific and technological achievements to the public vividly, intuitively and quickly requires a communication method that constantly adapts to the development of the times. With its unique advantages, animation has become one of the important

I. Ahmad et al. (Eds.): STSIoT 2021, LNDECT 122, pp. 505–513, 2023.
https://doi.org/10.1007/978-981-19-3632-6_61

forms of popularization of science. Zhu Kaiwen said that with the rapid development of three-dimensional animation 3D, its technology has been paid more and more attention. Not only have relevant courses opened in colleges and universities, but also occupy a certain position in the interest science education of young people. Wang Rui said that popular science animation plays an extremely important role in disseminating scientific knowledge. In the new media environment, popular science animation reinterprets the creative form, transmission path and penetration platform field of popular science animation through the characteristics of new media. This article also attaches importance to the role of popular science animation in the post-epidemic era, so computer technology is used to study the creation and spread of popular science animation.

The innovation of this article is to use the Internet as a platform to promote the interaction between science and the public, and to provide new ways to improve the scientific quality of the public. The research on this subject reflects the creative integration of science and humanity, and establishes a public-centered awareness of popular science services. Finally, this article drives the curriculum reform according to the needs of the society, and provides an effective way of thinking for the interdisciplinary and interprofessional development of the profession.

2 Related Work

There are many researches on the creation and dissemination of popular science animation under computer technology [1].

Wolfgang introduced a hybrid animation method, which combines the example based animation method and neural animation method to create a simple but powerful animation system for the face [2]. Remi et al. present a history and recent summary of algorithmic film director research by characterizations of applications that require film directors, and identify promising paths and hot topics for future research [3]. Prashant et al. propose that modern hardware developments have also made possible the development of technologies that enable cloud display and animation at interactive frame rates, review outstanding work in this area and summarize the evolution of this research over time [4]. Artem et al. noted the importance of visual support for educational activities, provided the results of an analysis of computer animation production tools, and described the authors' findings [5].

Under the premise of ensuring scientific and professional, reasonable scientific animation programs should be developed for different groups and regions to meet the personalized needs of part of the public [6].

The popularization and application of computer technology in the field of animation in developed countries has pushed the animation industry to a new stage. In particular, the DIGITAL technology in the United States has reached a mature stage. A variety of international popular production of two-dimensional, three-dimensional and synthetic software is also available, complete functions, and excellent compatibility, suitable for running in a variety of platforms. Chinese computer animation technology closely follows the development of foreign countries and has no lack of independent innovation. However, due to the lack of good script and preliminary design, it can only survive in the middle and late animation processing [7].

The impact of foreign market and the limitation of its own conditions make Chinese animation fall into an awkward situation. Chinese animation has always been in a weak position in the world market, but its development potential should not be underestimated. Only by thoroughly studying the computer animation technology and applying the new thinking and technology in practice can the national animation of our country be revitalized.

3 Creation and Dissemination of Popular Science Animation Under Computer Technology

3.1 Computer Technology

The development of computer technology has brought mankind into a whole new era. As a high-tech industry, popular science animation not only transforms some simple and abstract concepts into intuitive and vivid concrete visual images. In the design of popular science animation, computers are often used to achieve certain effects. The recognition of computer vision technology in images is a basic ability. Human real vision is a process of multi-resolution analysis from coarse to fine. The basic function relationship of wavelet transform can be expressed as:

$$RS_a(x, l) = \frac{1}{\sqrt{x}} \int_{-\infty}^{+\infty} a(s)\beta\left(\frac{s-l}{x}\right) vs = \langle a(s),\ \beta_{xl}(s)\rangle \tag{1}$$

Among them, a(s) is the square integrable function, $RS_a(x, l)$ is the wavelet transform of a(s) and:

$$\beta_{\text{xl}}(s) = \frac{1}{\sqrt{x}}\beta\left(\frac{s-l}{x}\right) \tag{2}$$

Formula (2) is the displacement and scale expansion of the basic wavelet. In addition, in the formula, x is the scale factor, and l reflects the displacement. The equivalent frequency domain representation is:

$$RS_a(x, l) = \frac{\sqrt{x}}{2\partial} \int_{-\infty}^{+\infty} A(\theta)\beta(x\theta)f^{k\theta\partial} c\theta \tag{3}$$

In the formula, $A(\theta)$ and $C(\theta)$ are the fourier transform of a(s) and c(s) respectively.

3.2 Popular Science Animation

Popular science animation is a way to popularize science through animation. It integrates a variety of art forms, and then uses digital technology to produce videos that can be used for scientific communication [8, 9]. Popular science animation is adapted from popular science books, bringing a brand new way of popularizing science for children [10, 11].

(1) Forms of expression

The expression form of popular science animation is logical. In the creative process, the content is presented to the audience in a variety of artistic languages through a reasonable structure. Combine fun, entertainment and life [12].

Its implementation methods mainly include: 1) Video playback. Use words, sounds or images as carriers to express opinions and thoughts and emotions. 2) The audio commentary system plays the content of the program and completes the publicity task. 3) The interactive communication platform publishes news messages to the audience and other related auxiliary function modules to achieve the effect of popular science animation.

(2) Principle

The creation and dissemination of popular science animation is a gradual process. In the epidemic stage, it is necessary to take into account that people in different countries, regions, and different ethnic cultural backgrounds have great differences in scientific awareness and aesthetic concepts. Therefore, when designing popular science cartoons, we must first understand what content can attract the public's attention. The second is to choose appropriate themes to express the theme according to the background of the epidemic. The final step is to apply it to the animation.

(3) Advantages

Popular science dissemination in the new media environment is derived from the combination of traditional media and new technologies. Its greatest advantage is the ability to deliver information to a broad audience. 1) Strong interaction. 2) Aesthetic advantage. 3) Production advantage. The development of new media technology provides convenience for people to obtain news. In this case, traditional media can better deliver news to the audience.

3.3 MG Animation in the New Media Environment

(1) MG animation in the era of new media

MG animation is a language that combines film and graphic design, which is mainly used in the fields of film, TV program design and advertising. In the era of new media, with the development of new media, the efficient features and demands of new media have created more space for dynamic graphics. Animation has also made new developments in the wave of the new media era. MG animation has rich information carrying capacity and diversified visual expressive power.

(2) Features of MG animation

Simplified functions. In the creative process, the creator exaggerated and simplified the performance object, and created the performance object based on points, lines and fundamentals. 2) Plane features. Flat design also represents the mainstream of design in the context of the new media era. In the graphic design of MG Animation, the emphasis is on simple shapes, pure colors and flat outlines.

(3) Technical advantages

MG animation has the characteristics of comprehensive nature, creating MG animation, diversified software cooperation is an indispensable path. Due to the rapid release of digital media technology, the variety of software, and the speed of update are greatly improved, the software and technical environment involved in the production of MG animation are relatively rich.

4 Questionnaire Survey on the Creation and Dissemination of Popular Science Animation

4.1 Investigation Background

Since the outbreak of the new crown epidemic in 2019, people have gone from panic and turmoil to calmness, and it has taken a long time to achieve this state. In the face of the new crown epidemic and the face of abnormal toxins, our country has taken a calm attitude and proactively dealt with it. The severity of the epidemic at that time and the knowledge of how to prevent it were urgently needed to be understood and followed. Therefore, the popularization of knowledge about the epidemic was necessary at that time and bear the brunt. For science popularization, you need to choose an appropriate method. Therefore, the choice of popular science animation is to attract the public's attention to epidemic prevention with vivid and interesting images.

4.2 Questionnaire Design

The specific survey subjects include students and teachers majoring in art, design, journalism and animation. The content of the survey centered on the related issues of popular science animation and the training methods of animation majors in colleges and universities.

The questionnaire in this article mainly consists of the following questions:

1) The current audience and dissemination of popular science animation. 2) The science and story of popular science animation. 3) How to create popular science animations with high public acceptance in the media environment. 4) The direction in which the teaching design of animation can be improved.

4.3 Questionnaire Process

This survey selected 100 students majoring in fine arts, design and animation to fill in the offline questionnaire. A total of 100 questionnaires were distributed. While the questionnaire was distributed, interviews were conducted with college students. Ask students to fill in while asking, and effectively collect 100 questionnaires. The questionnaire process lasted 2 weeks and was divided into three times. Finally, organize the data.

5 Questionnaire Analysis

5.1 Survey on the Status Quo of Popular Science Animation

According to the survey, there are some problems in popular science animation nowadays. For example, due to the update of the way, the audience has become singular and more suitable for young people. Moreover, the scope of popular science animation is generally small. The specific situation is shown in Table 1:

Table 1. Survey on the status quo of popular science animation

	Audience is monotonous	Small range	Publicity is not in place
Fine arts	10	6	6
Design	12	7	5
Animation	12	10	8
Journalism	13	9	10

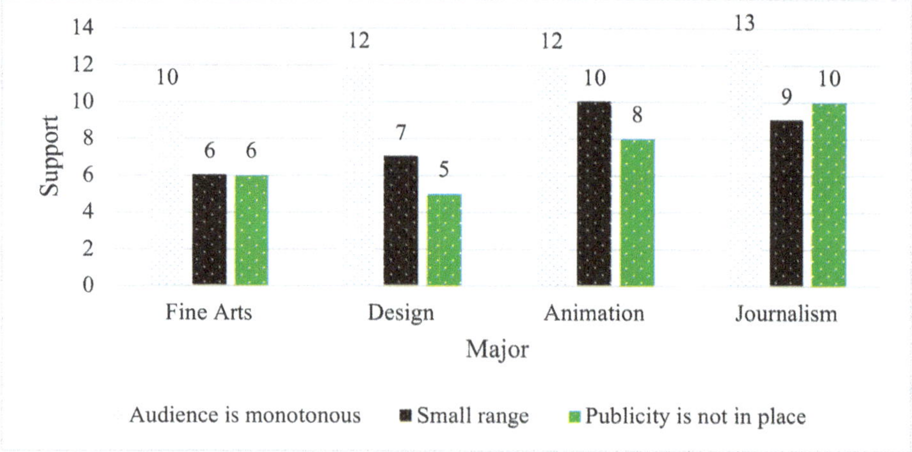

Fig. 1. Survey on the status quo of popular science animation

As shown in Fig. 1, there are 13 people majoring in journalism, and 12 people each in animation and design think that popular science animation audiences are relatively monotonous.

5.2 The Creation and Dissemination Method of Popular Science Animation

According to the survey results, it is learned that the creation of popular science animation can start from the aspects of video, audio, interactive platform and graphic design. Specific opinions are shown in Table 2:

Table 2. The creation and dissemination method of popular science animation

	Smart phone	Computer	Television
Video	10	8	13
Audio	7	5	5
Interactive platform	9	7	9
Graphic design	13	8	6

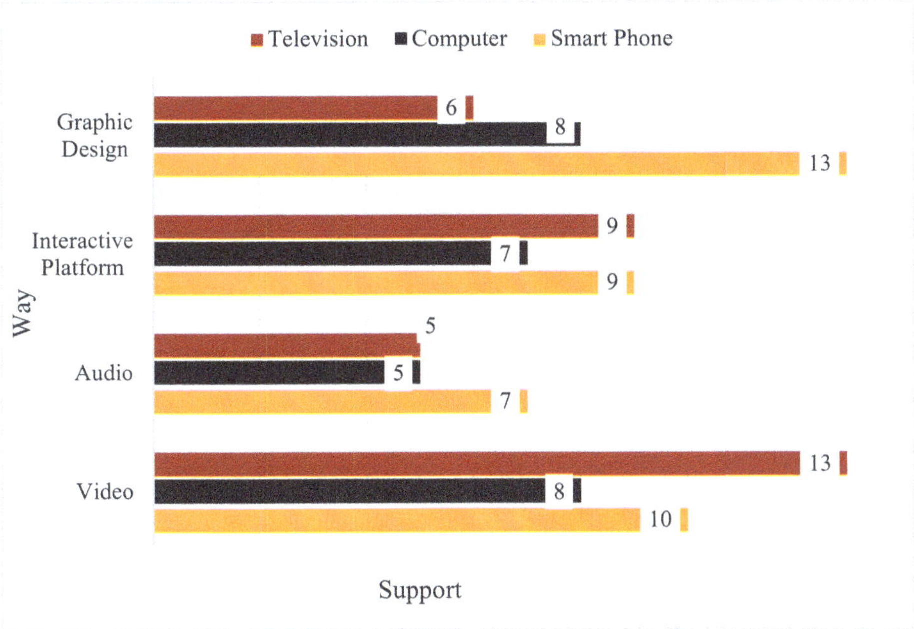

Fig. 2. The creation and dissemination method of popular science animation

As shown in Fig. 2, we can see that most people now learn the knowledge of popular science animation through interior design and TV. As for video and graphic design, they are the two most popular aspects of animation creation.

5.3 Training Direction of Animation Major in Colleges and Universities

The four majors have different views on the cultivation of animation majors in popular science animation in colleges and universities. They discussed in terms of vision, copywriting, and animation duration, as shown in Table 3:

Table 3. Training direction of animation major in colleges and universities

	Visually	Copywriting	Animation duration
Fine arts	43%	24%	33%
Design	40%	31%	29%
Animation	50%	27%	23%
Journalism	34%	36%	30%

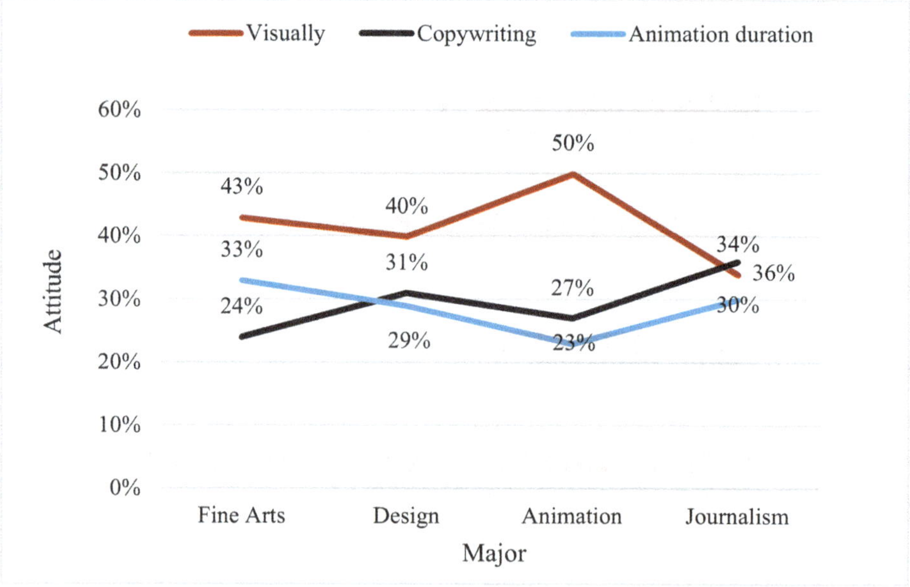

Fig. 3. Training direction of animation major in colleges and universities

As shown in Fig. 3, we can see that different majors agree that the teaching design of popular science animation should pay attention to dynamics. This tells teachers and students from the side that the importance of dynamics in animation design. As for the length of copywriting and animation, the art majors that value time most, and journalism the most valued copywriting.

6 Conclusion

Through a series of research and design of this article, this article provides development ideas for the creation of popular science animation. Affected by the new crown epidemic, network technology, and media changes, social media and self-media facilitate the rapid spread of popular science knowledge. On the premise of ensuring scientific and professionalism, in the face of different groups and regions, formulate reasonable science animation programs to meet the individual needs of some of the public. This article also provides a reference for the training of talents in animation majors in colleges and universities. Aiming at the comprehensive characteristics of popular science animation, it can provide a certain reference value for the training mode of animation professionals.

Acknowledgements. This work was supported by XX20008.

References

1. Carlson, W.E., Hackathorn, R., Parent, R.E.: Computer graphics and animation at the Ohio state university. IEEE Comput. Graph. Appl. **41**(3), 8–17 (2021)
2. Paier, W., Hilsmann, A., Eisert, P.: Example-based facial animation of virtual reality avatars using auto-regressive neural networks. IEEE Comput. Graph. Appl. **41**(4), 52–63 (2021)
3. Ronfard, R.: Film directing for computer games and animation. Comput. Graph. Forum **40**(2), 713–730 (2021)
4. Goswami, P.: A survey of modeling, rendering and animation of clouds in computer graphics. Vis. Comput. **37**(7), 1931–1948 (2020). https://doi.org/10.1007/s00371-020-01953-y
5. Yurchenko, A., Shamonia, V., Udovychenko, O., Momot, R., Semenikhina, O.: Improvement of teacher qualification in the field of computer animation: training or master class? In: MIPRO 2021, pp. 631–635 (2021)
6. Albinski, C., Salwin, M.: @Julian Tuwim: to everyman. SIGGRAPH Comput. Anim. Festiv. **11**, 1 (2021)
7. Ascher, U.M., Larionov, E., Sheen, S.H., Pai, D.K.: Simulating deformable objects for computer animation: a numerical perspective. CoRR abs/2103.01891 (2021)
8. Izdebski, Ł, Kopiecki, R., Sawicki, D.: Bézier curve as a generalization of the easing function in computer animation. In: Magnenat-Thalmann, N., Stephanidis, C., Wu, E., Thalmann, D., Sheng, B., Kim, J., Papagiannakis, G., Gavrilova, M. (eds.) CGI 2020. LNCS, vol. 12221, pp. 382–393. Springer, Cham (2020). https://doi.org/10.1007/978-3-030-61864-3_32
9. Kolling, C., Araujo, V., Barros, R.C., Musse, S.R.: How does computer animation affect our perception of emotions in video summarization? In: Bebis, G., et al. (eds.) ISVC 2020. LNCS, vol. 12510, pp. 374–385. Springer, Cham (2020). https://doi.org/10.1007/978-3-030-64559-5_29
10. Michels, D.L.: 19th ACM SIGGRAPH/Eurographics Symposium on Computer Animation 2020, SCA 2020 - Posters, Online, 6–9 October 2020. Eurographics Association (2020). ISBN 978-3-03868-119-9 [contents]
11. Eckert, M.L., Um, K., Thuerey, N.: ScalarFlow: a large-scale volumetric data set of real-world scalar transport flows for computer animation and machine learning. ACM Trans. Graph. **38**(6), 239:1–239:16 (2019)
12. Anjum, B.: A conversation with Santiago Montesdeoca: how current 3-D computer animations are constraining individual creative expression. Ubiquity **2019**(July), 1–6 (2019)

Fault Recovery and Reconfiguration of Distribution Network Based on Artificial Intelligence Algorithm

Ruibang Gong[1], Yuchen Wang[1], Junwen Cheng[1](✉), Xiao Sun[1], and Kazuyoshi Yoshii[2]

[1] State Grid Xinjiang Electric Power Co. Ltd. Electric Power Research Institute, Urumqi 830001, China
17690807467@163.com
[2] University of Sydney, Sydney, Australia

Abstract. Distribution network is an important part of power system and the last link of power system facing customers. Distribution lines mostly cover urban and rural areas with complex operation environment, which are vulnerable to different types of faults such as bad weather, vegetation growth and equipment failure. With the rapid development of distribution automation technology, people have higher and higher requirements for the reliability of distribution network. The purpose of this paper is to study an efficient distribution network fault recovery and reconstruction platform to solve the problems of time-consuming, labor-consuming, inconvenient management and low degree of automation in the current distribution network fault recovery and reconstruction. In this paper, the distribution network automation is not high in the area as the research object, when the distribution network fault occurs, using the available measurement information of the distribution network, using the method of state estimation to solve the problem of distribution network fault location, put forward the distribution network fault recovery model based on differential evolution algorithm to speed up the location search ability. In this paper, a mathematical model is proposed to solve the problem of fault recovery and reconfiguration. By studying the collaborative work mechanism and process optimization of distribution network fault repair, the demand analysis and design of functional modules of distribution network fault platform are carried out. After the application of the platform, the fault processing efficiency of the distribution network is effectively improved, the fault location time is shortened by 19 min, and the repair time is shortened by 15 min; simplify the personnel allocation of emergency repair and reduce the staffing by 31%; the operation reliability of distribution network is improved by 1.02%.

Keywords: Distribution network · Distribution automation · State estimation · Fault recovery

1 Introduction

Distribution network is directly connected with users, with wide distribution range, numerous branches and complex network structure [1, 2]. Because the distribution line

I. Ahmad et al. (Eds.): STSIoT 2021, LNDECT 122, pp. 514–524, 2023.
https://doi.org/10.1007/978-981-19-3632-6_62

is usually located in the area with dense buildings, the fault frequency is often high in the distribution network, so the location is very difficult [3, 4]. The safety, stability and normal operation management of power grid is an important driving force of our party and national economic and social development. The safety, stability and normal operation management of the power grid is an important driving force for the development of our party and the national economy and society. When our power grid equipment fails, it will cause great damage to our enterprises and users. Research on efficient and practical fault diagnosis and recovery solutions for distribution network equipment plays an important role in improving and enhancing the reliability of our grid power supply system [5, 6].

There are many research results on the problem of fault recovery and reconfiguration of distribution network. For example, some scholars proposed a WSN based cable grounding fault location scheme, designed the structure of the WSN based cable grounding fault location system, used the sensor nodes to collect the zero sequence current in the cable, and transmitted the real-time zero sequence current information of each node to the fault information processing center through the communication between the sink node and the gateway [7]. Some scholars also study the design and implementation of distribution network fault location system based on WSN, and propose an improved differential evolution algorithm for distribution network fault location [8].

The main work of this paper is as follows: firstly, the mathematical model of distribution network fault recovery and reconfiguration is constructed, including objective function and constraints. Secondly, taking the areas with low degree of automation of distribution network as the research object, this paper uses the available measurement information of distribution network to solve the problem of fault location of distribution network by using the method of state estimation when the distribution network fails. On the basis of this, this paper proposes the differential evolution algorithm to speed up the fault location. Thirdly, the requirement of the fault recovery platform is analyzed and designed by modules. Finally, an application example is given to verify whether the fault handling efficiency has been improved after using the distribution network fault recovery and reconstruction platform.

2 Fault Recovery and Reconfiguration of Distribution Network

2.1 Mathematical Model of Distribution Network Fault Recovery and Reconfiguration

Fault recovery and reconfiguration is to solve a kind of NP hard problem. Its mathematical model includes objective function and related constraints.

2.1.1 Objective Function

1) Reconfiguration model with the goal of maximizing power loss load recovery

The objective function is as follows

$$\max f_1(x) = \sum\nolimits_{i=1}^{n} P_i s_i \tag{1}$$

$$\max f_2(x) = \sum\nolimits_{i=1}^{n} Q_i s_i \tag{2}$$

where I is the fault node and N is the total number of fault nodes.

2) Reconfiguration model for minimizing the number of switches

The objective function is as follows:

$$\min f_3(x) = \sum\nolimits_{i=1}^{n} k_i - z_i \tag{3}$$

where k_i is the opening and closing state of the restoration, "1" is the closed state, "0" is the open state.

3) Reconfiguration Model Aiming at Reducing Network Loss

The network loss of distribution network should be reduced as much as possible after reconstruction. The objective function is as follows:

$$\min \sum\nolimits_{j=1}^{n_1} I_j^2 R_j \Rightarrow \max f_4(x) = \frac{1}{\sum_{j=1}^{n_1} I_j^2 R_j} \tag{4}$$

4) Reconfiguration Model for Improving Power Quality

The influence of voltage quality on equipment operation should be considered in fault recovery and reconstruction. The objective function is as follows:

$$\delta U = \frac{U_{re} - U_N}{U_N} \times 100\% \Rightarrow \max f_5(X) = \max(U_{rel} - U_N, \frac{U_N}{U_{rel} - U_N}, \ldots, U_{ren_2} - U_N) \tag{5}$$

5) Reconfiguration Model Aiming at Load Balancing

The objective function of evenly distributing the power loss load to each line during fault recovery and reconstruction is as follows:

$$\min \sum\nolimits_{i=1}^{n_i} \frac{S_i^2}{S_i^{max2}} \Rightarrow \max f_6(x) = \frac{1}{\sum_{i=1}^{n_i} \frac{S_i^2}{S_i^{max}}} \tag{6}$$

S_i^{max} 为 is the maximum allowable transfer power of line I; S_i is the amplitude of the complex power of the power loss load at the I sending end of the line.

2.1.2 Constraints

1) Radial structure

$$g_k \in G_k \tag{7}$$

2) Capacity constraints of distribution network lines

$$I_l \leq I_{lmax} \tag{8}$$

3) Node voltage constraint

$$U_{imin} \leq U_i \leq U_{imax} \tag{9}$$

2.2 Distribution Network Fault Recovery Based on Differential Evolution Algorithm

2.2.1 Basic Idea of Differential Evolution Algorithm

The essence of differential evolution algorithm is similar to other evolutionary algorithms. The core idea of intelligent optimization algorithm based on population evolutionary search is originated from Darwin's idea of natural selection, mainly including excessive reproduction, survival competition, heredity and mutation, and survival of the fittest [9, 10].

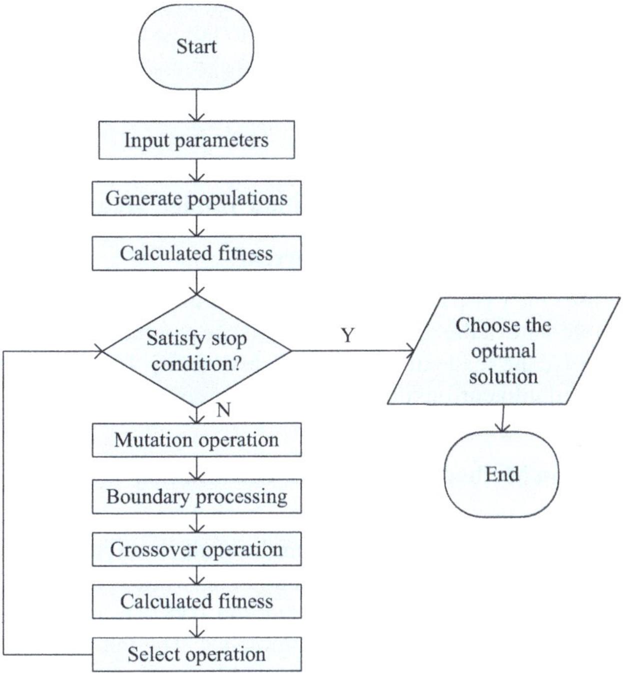

Fig. 1. Flow chart of differential evolution algorithm

2.2.2 Distribution Network Fault Recovery Based on Differential Evolution Algorithm

Differential evolution algorithm has the characteristics of simple principle and less coding difficulty. Relevant scholars have applied it to solve power system related problems, and proved that DE algorithm has good convergence and optimization ability [11, 12].

1) Binary and coding rules of DE algorithm

In order to improve the efficiency of the algorithm, reduce the infeasible solutions and shorten the coding dimension, the following two principles are usually adopted:

A. It is assumed that all the switches of the system are closed to form a ring network structure. The operation of connecting other switches not related to the circuit to the power grid must be turned off, so it is not necessary to consider such switches in restoration and reconstruction. B. In order to achieve better optimization effect and speed up the search speed, the switch connected to the power side should also be closed, so this kind of switch cannot be considered.

2) Parameter setting in binary de

The population size NP is usually 2 to 5 times of the individual vector dimension. The scaling factor f ranges from 0 to 2. The size of F can play the role of local search. If f is large, the accuracy of the algorithm will be reduced; if f is small, it will make it difficult for the algorithm to jump out of the local extremum and find the global optimum. Therefore, formula (7) can be used to adjust the size of zoom F.

$$\mathrm{F} = F_d - F_x * \frac{T_{max} - t}{T_{max}} + F_x \tag{10}$$

2.3 Fault Location Based on State Estimation Voltage Matching

2.3.1 State Estimation Based on Load Current

State estimation uses the existing power grid measurement and mathematical model to estimate the possible state of the power grid. In the state estimation based on load current, it is necessary to solve the load current as the state quantity. Firstly, other measurement needs to be converted into current quantity by measurement conversion.

2.3.2 Fault Location Method Based on State Estimation Voltage Matching

Through the analysis of the method of distribution network state estimation, it can be found that the state estimation can be used in the fault location method in this paper, when the measurement information of distribution network is not complete. First, the input data is detected to identify and eliminate the bad data in measurement. Then, the fault area is determined as a set of adjacent nodes. The fault model is transformed into a special load temporarily connected to the suspicious node. Then, the fault node is determined by using the state estimation algorithm based on load current. The calculated voltage of the state estimation is matched with the voltage at the known measuring point. Finally, the location of the fault is determined by the fault node and the data calculated in the state estimation.

2.4 Overall Framework Design of the Platform

2.4.1 Integrated Framework Design

Based on SOA structure, the bus provides the corresponding services, and the model and interaction are processed in strict accordance with IEC61970/61968-cim and sg-cim specifications.

2.4.2 Functional Framework Design

The operation and fault recovery platform of distribution network carries out information interaction through interactive bus and marketing system, pms1.0, gis1.5, 95598, distribution automation master station system and other related systems, which determines the production and emergency repair control as the basis, so that all kinds of information can be processed uniformly and give full play to the application value.

3 Design and Implementation of the Operation Fault Recovery Platform of Distribution Network

3.1 Visualization Module

The visualization module is the geographic map module, which combines gis1.5 geographic map with production repair to realize the visualization of production information.

The interface of the main interface of geographical map is measured as equipment navigation and layer management on the left. The equipment navigation deployment can query specific feeders, as well as the distribution transformer, switchgear and feeder line equipment under the feeder. The toolbar is below the map. The basic function of geographic map can be realized by the buttons in the toolbar.

3.2 Planned Power Failure Management

The planned outage management module realizes the record management of power outage plan, mainly maintains the pre-arranged outage information proposed according to the operation requirements of users and distribution network, and preliminarily generates monthly blackout plan by referring to monthly reliability index, historical power outage times of users and power conservation plan information; during the implementation, risk management shall be completed according to monthly blackout plan, alarm prompt work shall be done and power failure message shall be issued in time; we should fully supervise the blackout plan today, and at the same time, we need to let the coverage area and users know the power outage situation accurately.

3.3 Edit Plan Power Failure

In the edit planned outage interface, further supplement and modification of planned blackout information can be realized, and the actual operation of planned outage can be maintained. When planned power failure starts, fill in actual time of blackout and change the status to being implemented; When the actual power failure is finished, maintain the actual recovery time and change the status of planned power failure to completed, then the planned power outage will be completed.

3.4 Power Failure of Release Plan

Select an unpublished planned power failure in the main maintenance interface, and use the publish planned power off button to release the planned outage successfully. After the planned power outage is released, the system sends the planned outage information to the marketing management system, which is convenient for 95598 to inform the user of the power failure information.

3.5 Check the Power Failure Scope

In the main interface of planned blackout, select a planned outage event, and use the view blackout scope button to open the influence scope window of planned blackout.

3.6 Failure and Power Failure Management

The main control interface can perform basic maintenance for the failure and power failure, list all the failure and power failure information, or display the number of users affected by the current failure and the details of the power failure equipment.

After the switch is changed in the distribution automation system, the power loss equipment is sent to the emergency repair command platform, and the fault power failure event is automatically pushed out in the emergency repair platform. Meanwhile, the power failure scope affected by the medium voltage fault is automatically pushed out in the geographical map.

3.7 Manual Maintenance of Failure and Power Failure

If the distribution automation information is not pushed to the fault recovery platform, it can be manually maintained in the main control interface and the new fault is cut off. The new fault power failure interface will pop up, which is the same as the planned power failure. In addition, it can delete the fault power failure, view the power failure range and query the failure power failure.

4 Application and Analysis of Fault Recovery Platform for Distribution Network Operation

4.1 Application Examples

4.1.1 Examples Before Platform Application

Before the application of the fault recovery platform for distribution network operation, the control personnel and operation inspection personnel of the electric power department all handled the work orders submitted by users in the 95598 system, the responsibility interface was not clear, the fault diagnosis could not be carried out, and the efficiency of receiving orders and orders was low. For example, at 19:23 on December 30, 2020, Mr. Li of the city called for repair through the 95598 telephone repair platform, the customer service center enters the fault information and assigns the work order to the company according to the user's area at 19:24. The company sent the specific fault information to the fault repair team by fax, and team a of the distribution fault repair team rushed to the scene after knowing it. The fault was not found until 20:25, and it was fixed at 20:41.

4.1.2 Examples After Platform Application

Through the use of the distribution network operation fault recovery platform for five months, combined with the distribution operation, fault repair and other management mechanisms, the fault repair, command, decision-making and other work has been effectively managed. Based on the distribution network production command visualization technology, the fault repair and inspection work has more scientific auxiliary technology. At 9:24 on March 26, 2020, Mr. Yang reported for repair by telephone. The customer service sent the repair order to the distribution network operation fault recovery platform through the marketing system. At 9:44, the emergency repair team arrived at the accident point and determined that one distribution load switch in No. 3 FCF box tripped. Through emergency repair, the power supply was normal at 10:11, and the emergency repair time was within 30 min. Compared with the past, the efficiency was significantly faster.

4.2 Application Effect Analysis

After the distribution network operation fault recovery platform is online, human resources are saved, and the total time of fault repair and the time to reach the site are greatly improved. The specific improvement is shown in the Table 1:

Table 1. Comparison of manpower and time effect after platform application

Index name		Before November 2020	November 2020 to present	Rate of change
Number of planning staff		11	9	18% reduction
Planned outage time per household		1.28	1.14	11% reduction
Failure repair time (minutes)	Total time	123	95	23% reduction
	Arrival time	30	26	13% reduction
	Fault location time	41	22	41% reduction
	Troubleshooting time	50	35	24% reduction
Number of maintenance personnel (person)		592	421	31% reduction
Reliability index of distribution network (%)		99.02	99.98	Increased by 1.02%

The time comparison chart according to the above table is shown in Fig. 1.

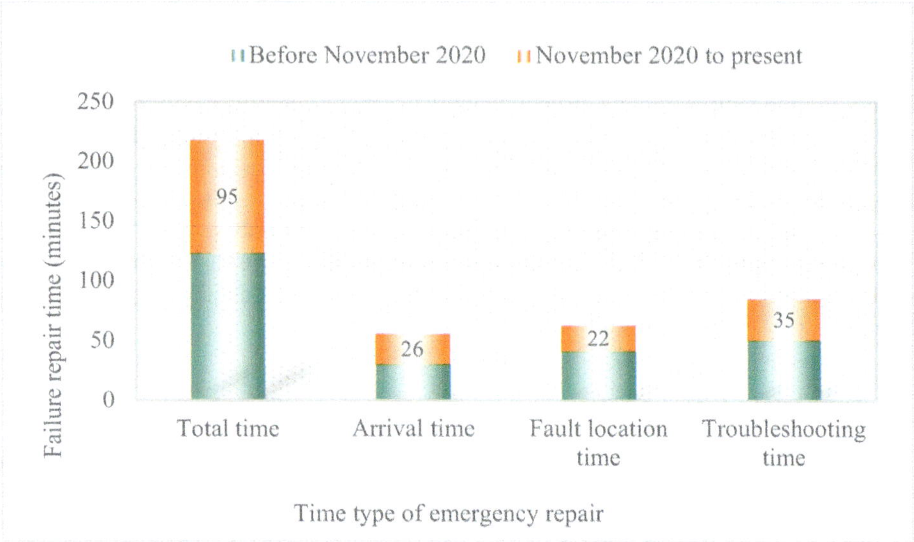

Fig. 2. Time dimension effect improvement chart

As can be seen from Fig. 2, the fault repair time has been significantly reduced after the application of the platform, and the fault location time has been reduced by 41%, which shows that the application effect of the platform is better.

According to Table 1, the comparison chart of human resources is shown in Fig. 3.

Table 2. Effect improvement chart of human dimension

	Number of planning staff	Number of maintenance personnel
Before November 2020	11	592
November 2020 to present	9	421

It can be seen from Table 2 and Fig. 3 that the number of maintenance personnel and the number of planning personnel are reduced after using the distribution network fault recovery and reconstruction platform, which shows that the use of the platform increases the efficiency of the staff.

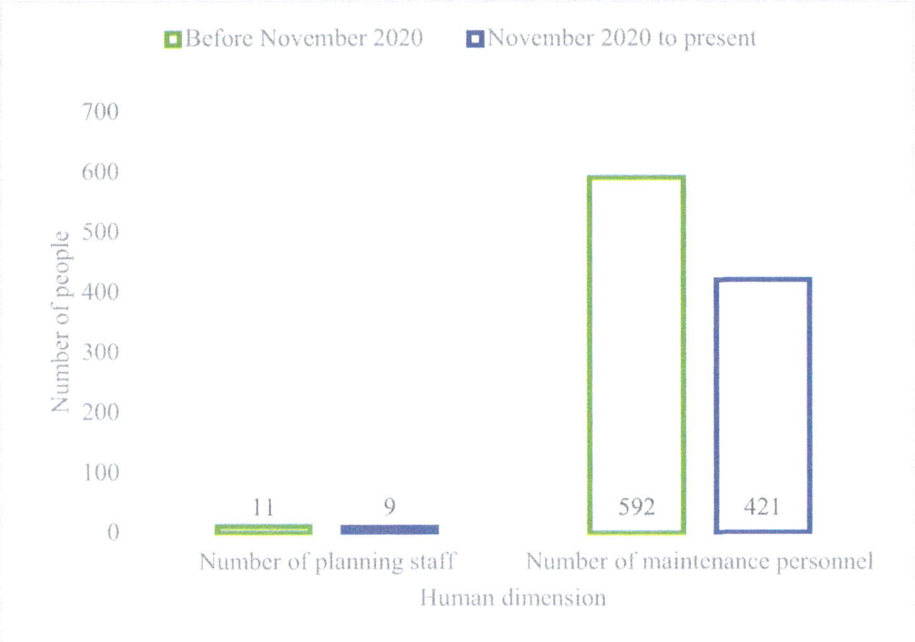

Fig. 3. Effect improvement chart of human dimension

5 Conclusion

The development and application of fault recovery platform for distribution network operation reflect the great advantages of distribution network automation technology, and it is also an important achievement of the integration of informatization and industrialization. The platform fully reflects the role of the command organization. Under the guidance of the command department, it completes the work of information collection, coordination and deployment, and ensures the timely and effective exchange of various information. It not only enhances the management ability of distribution network emergency repair, ensures the reliability of power supply, but also plays an important role in improving the service quality.

References

1. Wen, J., Tan, Y., Jiang, L., et al.: Dynamic reconfiguration of distribution networks considering the real-time topology variation. IET Gener. Transm. Distrib. **12**(7), 1509–1517 (2017)
2. Garau, M., Ghiani, E., Celli, G., et al.: Co-simulation of smart distribution network fault management and reconfiguration with LTE communication. Energies **11**(6), 1332 (2018)
3. Jain, T., Ghosh, D., Mohanta, D.K.: Augmentation of situational awareness by fault passage indicators in distribution network incorporating network reconfiguration. Protect. Control Mod. Power Syst. **4**(1), 1–14 (2019). https://doi.org/10.1186/s41601-019-0140-6

4. Ke, S., Lin, T., Chen, R., et al.: A novel self-healing strategy for distribution network with distributed generators considering uncertain power-quality constraints. Appl. Sci. **10**(4), 1469 (2020)
5. Golub, I., Voitov, O., Boloev, E., et al.: Reconfiguration of primary distribution network with several independent power sources. IFAC-PapersOnLine **52**(4), 437–442 (2019)
6. Ravikumar, T., Rao, G.K.: Reconfiguration of power distribution network and allocation of DG's for loss reduction and enhanced voltage profile. J. Adv. Res. Dyn. Control Syst. **9**(15), 815–822 (2017)
7. Saleh, O.A., Elshahed, M., Elsayed, M.: Enhancement of radial distribution network with distributed generation and system reconfiguration. J. Electr. Syst. **14**(3), 36–50 (2018)
8. Rajesh, R., Elango, K., Rajanbabu, S.: Optimum location and sizing of DG with network reconfiguration in distribution system. Int. J. Innov. Technol. Explor. Eng. **9**(6), 2278–3075 (2020)
9. Mosbah, M., Zine, R., Arif, S., et al.: Optimum distribution network reconfiguration in presence DG unit using BBO algorithm. J. Electr. Syst. **14**(4), 180–189 (2018)
10. Huang, S., Venkata, D.: Fast distribution network reconfiguration with graph theory. IET Gener. Transm. Distrib. **12**(13), 3286–3295 (2018)
11. Liu, J., Srikantha, P.: Decentralized topology reconfiguration in multiphase distribution networks. IEEE Trans. Sig. Inf. Process. Netw. **5**(3), 598–610 (2019)
12. Syahputra, R.: ANFIS Approach for distribution network reconfiguration. Int. J. Appl. Eng. Res. **12**(18), 7775–7782 (2017)

Large-Capacity Data Processing of Main Distribution Network Based on Information Processing Cluster Framework

Hongbo Wei[1(✉)], Guinan Ye[1], Jiancheng Wei[1], and Hu Xie[2]

[1] Power Dispatch and Control Center of Guangxi Power Grid Co., Ltd., Nanning, Guangxi, China
xiaochongtougao5@163.com

[2] Digital Grid Research Institute, China Southern Power Grid, Guangzhou, Guangdong, China

Abstract. Due to the continuous development and in-depth promotion of smart grid construction in China, the amount of information accumulated has increased exponentially. The technical method of extracting "treasure" from these important historical materials has gradually become an urgent need for building a powerful intelligent power grid, and the rise of big data storage and processing technology has also opened up a new road for data mining. This paper studies the high-capacity data processing of main distribution network based on information processing cluster framework. After understanding the relevant theories, a high-capacity data processing system of main distribution network based on information processing cluster framework is designed and tested. According to the accuracy test results of the system, the accuracy of the system is about 70%, which basically meets the needs of the system, and then the system management efficiency is good. The effectiveness test results show that the parallel test time of the system is greatly reduced compared with the serial time, so the system has good parallel processing efficiency.

Keywords: Information processing · Large-capacity data · Data processing · Smart grid

1 Introductions

At present, the data of power companies is structured and semi-structured, and the growth rate changes from TB level to PB level. It has already bid farewell to the era of relatively uniform data types and relatively slow growth [1, 2]. Existing platform functions: data storage capabilities, data display capabilities, and data processing capabilities may not meet the data analysis needs [3, 4]. In addition, most data processing platforms now also use expensive mainframe computers, which have low scalability and high cost [5, 6], and disk arrays are mainly used to store a large amount of data, and one or several servers need to be upgraded. To reduce the storage time, these are not well enough for the processing and analysis of big data. Therefore, the data analysis platform must have good scalability and high fault tolerance. In order to provide better value-added services

I. Ahmad et al. (Eds.): STSIoT 2021, LNDECT 122, pp. 525–534, 2023.
https://doi.org/10.1007/978-981-19-3632-6_63

to our country's power companies, we can make full use of emerging technologies such as big data processing technology and massive data mining, comprehensively analyze most of the data resources, and find valuable information among them [7, 8].

Regarding the research of large-capacity data processing, some researchers pointed out that in recent years, data storage control systems have become more and more important, and storage testing technology is also constantly developing. High requirements are put forward on storage test technology and require test equipment reliability, high anti-interference, high data measurement accuracy, etc. [9]. Some researchers also pointed out that processing massive amounts of data is a major challenge facing all industries in this era, not only in the field of defense, but also in various fields of massive data. Data analysis can help companies make wise decisions based on this information, and decisions play a leading role in the rise and fall of enterprises. Therefore, research on massive data can bring great value to enterprises, which also promotes the continuous development and innovation of massive data processing technology, and drives the arrival of the era of big data [10]. Regarding the research on the information processing cluster framework, some scholars pointed out that there are still many problems in the information processing cluster framework. For example, JobTracker will increase the overall memory resources [11]. Some designers have used threads as a task management framework to solve the above problems, and redesigned the Map-Reduce framework with a new version of the information processing complex architecture, decomposing the old JobTracker into two independent components. One is dedicated to resource scheduling, and the other is dedicated to task control. In this way, tasks can be assigned to various resource filling states to reduce resource consumption on a single host and further optimize the resources on each node. This makes Hadoop a very large, powerful and reliable data processing framework [12]. In summary, there are still many research results on big data processing technology, but there are relatively few researches on active network data processing.

This paper studies the large-capacity data processing of the main distribution network based on the information processing cluster framework, summarizes the characteristics of the active network data on the basis of relevant documents, and then analyzes the application of the information processing cluster framework in data processing. Based on these, design a large-capacity data processing system for the main distribution network based on the information processing cluster framework, test the designed system, and draw relevant conclusions through the test results.

2 Research on Large-Capacity Data of Main Distribution Network

2.1 Main Distribution Network Data Characteristics

With the digital transformation of power generation intelligence, power transmission monitoring networks, smart substations, smart meters, etc., the scale and types of power data have increased dramatically, and huge power grids have generated a lot of big data. From power generation to electricity consumption, many data collection sources pay great attention to collecting specific data information, especially image formats. The final smart grid data is huge and diverse, the value density is very low, and the speed is very fast, so the characteristics of the main distribution network data are summarized as follows:

(1) Large amounts of data. With the construction of smart grids, grid data has increased from GB and TB to PB, EB, and ZB.
(2) There are many types of data. There are many types of smart grid big data, including structured, unstructured and semi-structured. With the growth of video applications, the proportion of unstructured multimedia data in the power network is gradually increasing. In addition, data applications require correlation analysis of various data types such as meteorological data and non-industrial energy data, adding data types and complicating data processing.
(3) High speed. The high-power data processing speed needs to be very high, which can reach the processing speed of microseconds, and can quickly analyze the data in a short time and support the database, so as to make accurate decisions on the reliable operation of the power grid. Compared with offline data, the requirements for online data processing are higher, and the analysis and extraction of Web data streams are very different from traditional data mining techniques.
(4) The value density is low. Take video viewing as an example, useful data is only 1–2 s. In addition, most of the information collected by the equipment is normal, with very few abnormalities. This happens to be an important basis for the operation and maintenance of the power grid.

2.2 Application of Information Processing Cluster Framework in Large-Capacity Data Processing of Main Distribution Network

(1) The distributed file system (HDFS) in the information processing cluster framework is designed to run on general-purpose hardware. Based on the cloud platform of the information processing cluster framework, the HDFS distributed file index is created, distributed massive data processing, and real-time search is provided. HDFS ensures that very large files can be stored on the machine, and each file can be stored as a series of data blocks. The files are written at the same time, and there can only be one writer at a time. This ensures reliable data storage. When the client receives the file, it checks whether it is suitable. Based on HDFS, large data blocks are decomposed into a large number of small data blocks with complete fault tolerance.
(2) The information processing cluster framework first sorts the distribution network data, then divides the map job according to the information processing module, processes multiple calculation tasks in parallel, and finally reduces the calculation results of each map. Effectively solve the problem of excessive data calculation delay caused by the excessive amount of existing data in the SQL database.

2.3 Data Processing Algorithm

(1) The MapReduce data processing model

The MapReduce data processing model is currently the most common data processing model in big data processing. Therefore, many big data processing systems based on the MapReduce data processing model have been implemented, such as Hadoop, Pig, and AsterData. In order to calculate and process large and small graph data, this section mainly introduces the Network TopValue PageRank algorithm.

The user randomly selects a specific network node (web page) as the default node. After the user initiates an access operation, the user randomly selects a directed edge to enter the next network node with a specific probability d according to the existing network graph G architecture (the probability d ranges from 0.1 to 0.2, usually 0.15). According to the above rules, in the directed graph G(V, E), the PageRank value of node u is obtained by formula (1). Where V represents the network node, and E represents the network connection (directed edge).

$$R(u) = (1 - d) + d \times \sum\nolimits_{v \in B} \frac{R(v)}{N} \tag{1}$$

where B is the set of all neighbors that have internal relations with node u, and N is the externality of different nodes v.

The PageRank value of each network node in Figure G is determined by the score, except for all nodes pointing to it and the corresponding PageRank value. If you want the final starting value of different nodes in the network, you need to repeat Eq. (1) until the starting value of each node does not change. The iterative expression is shown in (2).

$$R_i(u) = (1 - d) + d \times \sum\nolimits_{v \in B} \frac{R_{i-1}(v)}{N} \tag{2}$$

(2) Hadoop computing cost

The biggest advantage of the Hadoop big data processing system lies in the map mapping function of MapReduce, that is, the basic data processing model, that is, it can map input data to the local computer as much as possible. However, due to the limited hardware resources of each computing node, the local calculation of input data during the map processing stage also depends on the distribution of the data. If the number of data fragments on a particular node exceeds the maximum number of data segments that a node can handle, this means that some data nodes in the node must be used by other remote nodes, but cannot perform local map functions. Therefore, the number of map functions started by the data node in the processing phase of the map mapping function depends on the amount of data that the data node needs to edit. The cost of reading the Map function is shown in formula (3).

$$\text{cost}M_{in} = \lambda * \overset{s}{D_r} + (1 - \lambda) * \underset{m}{\overset{s}{N_s}} + Cost_{map\text{-}calls}(num) \tag{3}$$

3 The Large-Capacity Data Processing System of the Main Distribution Network Based on the Information Processing Cluster Framework

3.1 Overall Framework of Large-Capacity Data Processing System

This paper proposes a large-capacity data processing system for the main distribution network based on the data characteristics of the main distribution network. The functional framework of the system is shown in Fig. 1, including data collection, data storage, data processing, and data management.

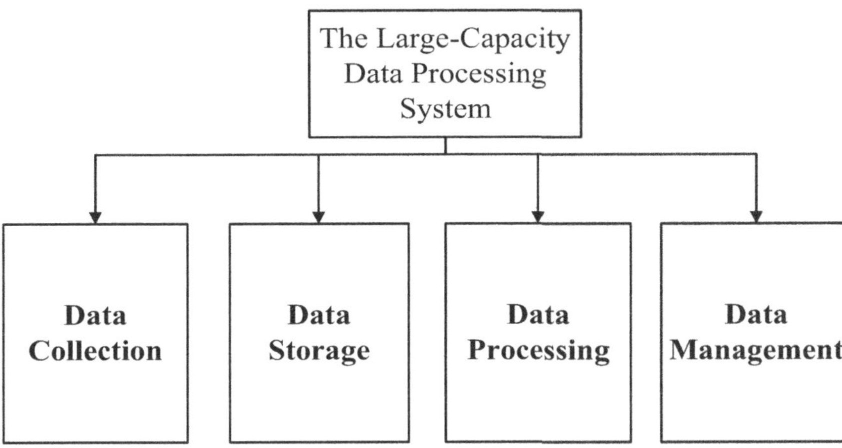

Fig. 1. The functional framework of the system

3.2 Data Collection

A large number of distribution network data are obtained from various monitoring terminals. According to the grid index system, it is divided into safety, reliability and power quality. The specific parameters mainly include power supply reliability, average number of power outages, distribution transformer operating life, switch operating life, average trunk length, etc. The monitoring terminal can select the appropriate collection frequency according to the characteristics of the index, generally once every 3 to 5 min. At this time, the monitoring terminal automatically collects data and chooses to send the data to the top layer for processing by the terminal. With the development of the power system and the increase in monitoring frequency requirements, the collection frequency will inevitably increase. Assuming that the acquisition frequency is set to the acquisition time of 2 min, each channel acquires 3 values, the amount of data collected per hour is 90, and the amount of data collected per day is 2160. Because each data has its own characteristics and different acquisition characteristics, some for data, such as harmonics, a sequence of at least 40 waves is collected.

3.3 Data Storage

At present, the power data warehouse can only meet the static statistical requirements, and the function is T+1. This means that big data in the power grid is statically stored, and its statistical data can only measure data in previous time units. The relationship of heterogeneous big data is complex, and information operators cannot build efficient big data warehouses above the XB level, which brings major problems to the data processing of information operators. Currently, it is mainly supported by various distributed technologies. According to the location of the data, the data will be distributed and stored in the local database. This mashup architecture is usually a parallel new-generation MPP+Hadoop database and some computing technologies. At present, this model can barely support, as the amount of data grows, this technology will soon become difficult to use. So this article applies new database technology.

The basic technology of the new database is very different from the traditional database. Efficiently and technically process PB data is to solve data storage problems for industry users. The new database will gradually integrate with Hadoop to provide rich SQL support for semi-structured and unstructured data processing. In this way, the processing needs of complex data are also met.

3.4 Data Processing

In order to standardize the complete production data chain of the system, this paper chooses to simulate an innovative model, which is a full-process data processing technology system based on data set delivery of labeled data. Focus on enhanced data calculation methods and production data business capabilities. In order to improve the efficiency of query and data calculation, and then improve and innovate applications, lay the foundation for subsequent big data processing and applications.

As the user's business needs change, the system will automatically recommend a suitable data label for the user, and automatically identify the logic between the physical structure in the data table and the historical data. Utilizing the powerful scalability, fast search efficiency and the powerful functions of highly accurate mark-based analysis technology, the search speed of massive telemetry data is significantly enhanced. Based on the tag center, an integrated logic model is created on the big data resources. The "tag" model log view can provide a variety of data service modules for users in a variety of different business scenarios, including drawing image analysis, rule prompts, text mining, personalized recommendations, relational networks, etc. The combination of interfaces can realize rapid analysis and application program construction. Interactive data metadata is created through a set of tags, and the data structure is centrally managed in the system. The system automatically creates a data source list while performing interactive data source security management and control.

3.5 Data Management

Data management includes a data management mechanism, which consists of three parts: technology, tools, and systems. It usually supports data governance to ensure efficient and effective work related to data governance.

3.6 Construction of Information Processing Cluster Framework

Hadoop is based on the Java language, so it has strong cross-platform capabilities. It can be well adapted to the system environment of Windows, Linux and Mac systems.

(1) Hardware description
    The Hadoop platform uses four computers, one of which is the master node and the other three are child nodes. The main computer parameters are: CPU: dual-core processor, 2.40 GHz, memory: 8.0 GB, system type: 64-bit operating system, hard disk: 2 T.
(2) Software description

    1) Linux system selection
        Linux systems can use virtual machines or dual-system installations together with Windows, so they will not affect other computer application requirements during the testing phase. Therefore, this experiment uses the Linux Ubuntu operating system for corresponding experimental operations. For this experiment, choose Ubuntu LTS 15.04.
    2) Hadoop version selection
        This article uses Hadoop 3.6.0 (stable) version.

4 Test of the Large-Capacity Data Processing System of the Main Distribution Network Based on the Information Processing Cluster Framework

4.1 Algorithm and Accuracy of the System

In this experiment, in order to test the accuracy of the algorithm and the system, the test process is as follows: (1) the collected data is divided into five data subsets, which is convenient for comparison and verification of the experiment. (2) A total of 3 experiments were carried out. In each experiment, three sets of data subsets are selected as the training set, and one set is used as the validation set. The experimental results are shown in Table 1.

Table 1. Algorithm and accuracy of the system

	1	2	3
Subset 1	67%	68%	71%
Subset 2	71%	70%	72%
Subset 3	70%	71%	69%
Subset 4	73%	72%	70%
Subset 5	72%	71%	72%

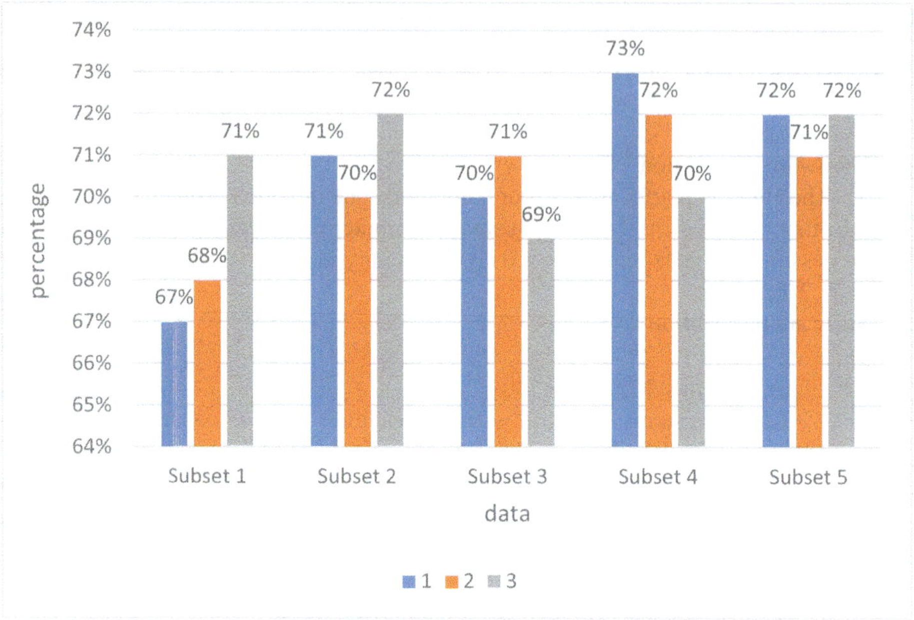

Fig. 2. Algorithm and accuracy of the system

It can be seen from Fig. 2 that the data processing accuracy rate of the system is about 70%. It can be seen that the accuracy of the parallel algorithm implemented by the system is within an acceptable range.

4.2 Validity Test

In this experiment, in order to test the efficiency of the system and algorithm, a total of three data sets are created, and the serial algorithm (the node is set to 1, which is equivalent to a serial function) is compared with the parallel algorithm. The experiment is based on the use of Data for predicting the wrong category. The five data sets are (10 million, 20 million, 30 million, 40 million, and 50 million data volumes). The experimental results are shown in Table 2.

Table 2. System effectiveness test

	Serial time (ms)	Parallel time (ms)
1	150006	224571
2	191220	306576
3	233453	343255
4	312452	421341
5	432145	456325

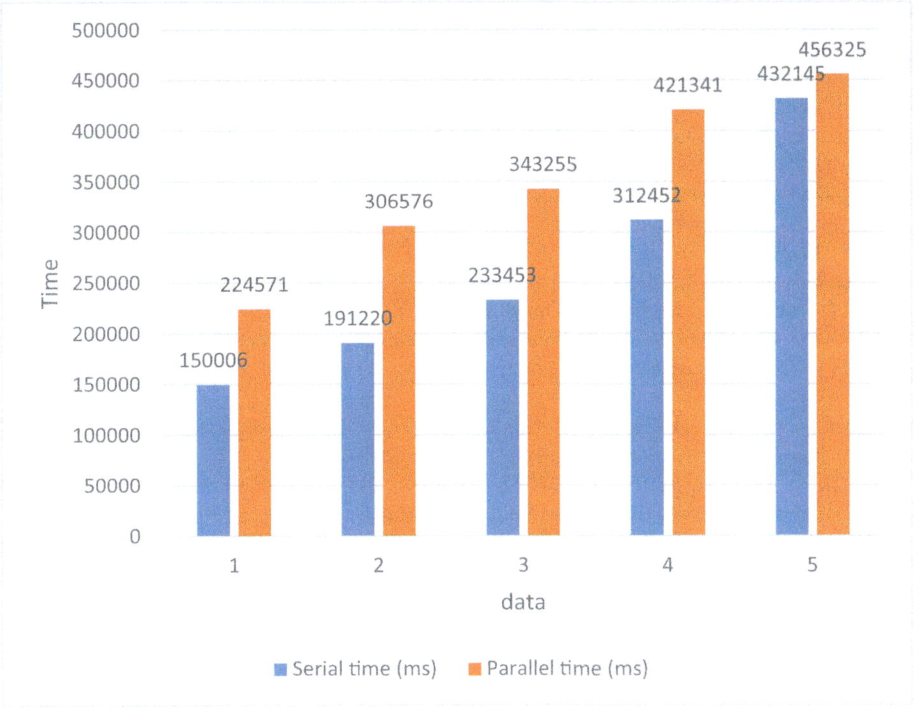

Fig. 3. System effectiveness test

It can be seen from Fig. 3 that compared with serial time, parallel test time is significantly reduced. This experiment shows that the parallel algorithm based on Hadoop has been improved more successfully, and the parallel processing performance of the platform is better.

5 Conclusions

This paper studies the large-capacity data processing of the main distribution network based on the information processing cluster framework. After analyzing the characteristics of the main power grid data, according to the situation, a large-capacity data processing system for the main distribution network based on the information processing cluster framework is designed. Then the designed system is tested. According to the accuracy test of the system, the test result shows that the data processing accuracy of the system is about 70%, which basically meets the system requirements.

Acknowledgements. "(Fund Project: China Southern Power Grid Corporation Key Technology Project Funding (R&D and Engineering Demonstration of Edge Computing Cluster Technology for Provincial Dispatching Main Station of Smart Grid, GXKJXM20190619)".

References

1. Wang, Z., Ng, D., Wong, V., et al.: Robust beamforming design in C-RAN with sigmoidal utility and capacity-limited Backhaul. IEEE Trans. Wirel. Commun. **16**(9), 5583–5598 (2017)
2. Silva, D.A.N.S., Souza, L.C., Motta, G.H.M.B.: An instance selection method for large datasets based on Markov Geometric Diffusion. Data Knowl. Eng. **101**(Jan.), 24–41 (2016)
3. Ferreira, R.S., Bentes, C., Costa, G., et al.: A set of methods to support object-based distributed analysis of large volumes of earth observation data. IEEE J. Sel. Top. Appl. Earth Obs. Remote Sens. **10**(2), 681–690 (2017)
4. Saleem, A., Khan, A., Malik, S., et al.: FESDA: fog-enabled secure data aggregation in smart grid IoT network. IEEE Internet Things J. **7**(7), 6132–6142 (2020)
5. Lowik, S., Kraaijenbrink, J., Groen, A.: The team absorptive capacity triad: a configurational study of individual, enabling, and motivating factors. J. Knowl. Manag. **20**(5), 1083–1103 (2016)
6. Yan, C.G., Wang, X.D., Zuo, X.N., et al.: DPABI: data processing & analysis for (resting-state) brain imaging. Neuroinformatics **14**(3), 339–351 (2016)
7. Jindal, A., Dua, A., Kaur, K., et al.: Decision tree and SVM-based data analytics for theft detection in smart grid. IEEE Trans. Industr. Inf. **12**(3), 1005–1016 (2016)
8. Hsieh, K., Ebrahimi, E., Kim, G., et al.: Transparent offloading and mapping (TOM): enabling programmer-transparent near-data processing in GPU systems. Comput. Archit. News **44**(3), 204–216 (2016)
9. Mehenni, A., Alimazighi, Z., Bouktir, T., Ahmed-Nacer, M.: An optimal big data processing for smart grid based on hybrid MDM/R architecture to strengthening RE integration and EE in datacenter. J. Ambient Intell. Humaniz. Comput. **10**(9), 3709–3722 (2018). https://doi.org/10.1007/s12652-018-1097-4
10. Vanfretti, L., Olsen, S.H., Arava, V., et al.: An open data repository and a data processing software toolset of an equivalent Nordic grid model matched to historical electricity market data. Data Brief **11**(C), 349–357 (2017)
11. Al-Rubaye, S., Kadhum, E., Ni, Q., et al.: Industrial internet of things driven by SDN platform for smart grid resiliency. IEEE Internet Things J. **6**(1), 267–277 (2019)
12. Hu, J., Yang, K., et al.: Guest editorial: smart grid inspired data sensing, processing and networking technologies. Mob. Netw. Appl. **24**(5), 1699–1700 (2019)

Modeling Method of Power Grid CIM Model Based on Graph Data Model

Yini He[1(✉)], Wei Cao[1], Changfu Wei[1], and Hu Xie[2]

[1] Power Dispatch and Control Center of Guangxi Power Grid Co., Ltd., Nanning, Guangxi, China
xiaochongtougao4@163.com
[2] Digital Grid Research Institute, China Southern Power Grid, Guangzhou, Guangdong, China

Abstract. With the expansion of power system scale and more frequent operation adjustments, people are increasing the need for real-time analysis and calculations. Graph data model is a new type of data model derived from parallel and analytical processing of a large number of data information on the mobile Internet that has appeared in recent years. Its data model provides users with an intuitive expression of grid topology, and can easily realize the parallel query of the data. This paper aims to study the power grid CIM model modeling method based on the graph data model. On the basis of analyzing the classes and relationships in the graph data model, CIM model and CIM model, a power grid CIM model based on the graph data model is constructed, which mainly includes the power grid equipment asset model and the power grid topology model are two parts, and then the parallel network topology analysis algorithm is proposed and implemented. Finally, the test verification shows that the model and algorithm in this paper can speed up the network topology analysis speed and improve the power grid calculation efficiency.

Keywords: Graph data model · CIM model · Network topology · Power grid · Information modeling

1 Introduction

With the continuous advancement of my country's smart grid, the requirements for information and communication between different systems are getting higher and higher, and the information and communication interaction between different systems is particularly obvious. Nowadays, different systems in the power grid must follow different standards, and they cannot exchange information. The problem of "information islands" has become increasingly prominent" [1, 2]. One of the main reasons for this problem is that manufacturers develop their own. There is no international standard for products. The general trend of self-developed products is to follow a single international standard because of the poor data exchange and systemicity between different systems [3, 4].

My country has also made rapid progress in the field of CIM model research, mainly to expand the original standards and carry out related experiments and test applications.

I. Ahmad et al. (Eds.): STSIoT 2021, LNDECT 122, pp. 535–543, 2023.
https://doi.org/10.1007/978-981-19-3632-6_64

In the field of CIM grid modeling, many researchers in our country have conducted research on this. For example, some researchers have studied the application of CIM-based distributed network models in specific modeling tools, explained the correlation between object-oriented models and relational database design, realized universal data input and output, and designed a effective resource descriptor frame analyzer [5]; some researchers discussed the extension of the distribution line model based on IEC 61970 CIM and the extended IEC 61968 model, analyzed the creation process of the distribution network topology model, and matched the mapping problem of power grid model to relational database [6]. In addition, some researchers have proposed a CIM-based distribution network model, which uses the power flow calculation of the three-phase distribution network for application [7]. Similarly, from the perspective of power asset information modeling, due to the continuous emergence of power equipment for new systems and asset life cycle management, many documents have been investigated and expanded. For example, some researchers have extended the CIM model to two levels of distribution network equipment containers and physical equipment, and based on the actual distribution network structure, established CIM models of distribution boards, network ring cabinets and reclosers [8]. In general, the current research on CIM models at home and abroad still needs to be expanded and improved.

This paper deeply studies the concepts of graph data model and CIM model, fully understands the definition of classes and the relationship between classes, and constructs the power grid CIM model based on graph data model. This model has the characteristics of scalability, follows the logical and strict expansion rules, and improves the power grid information model by adding new content according to the actual needs.

2 Research on Modeling Method of Power Grid CIM Model Based on Graph Data Model

2.1 Graph Data Model

(1) Data model

Different from other traditional databases, graph databases are based on graph theory and use nodes and sides to represent a data model. The attributes of data are stored on a node and edge, and the relationship between data is described by nodes and edges. Moreover, the graph database and the traditional database have some similarities in the way of physical description. For example, a relational database, hierarchical database, or a branch compensator description in an object-oriented database can directly use feature sets such as rated voltage, rated power capacitance, and reactive power measurement. The difference lies in the difference between these entities Relationship description and its treatment [9, 10].

(2) Data query access

Query includes a set of program interfaces for accessing large amounts of data and related support services, which are all database management mechanisms in the system. In the database model in the graph, each vertex or edge has a computing power that can be included as both a data storage unit and a dynamic data calculation unit. The data set for querying the parallel graph structure is also referred to as parallel graph channel or partial calculation. According to each vertex of the parallel graph, a local calculation is performed independently, and information is exchanged through both sides, so that the parallel graph calculation is performed locally it's easier [11, 12].

2.2 CIM Model

(1) IEC 61970

IEC 61970 is a set of standards developed by IEC TC57 WG13 to define the EMS application interface (API), also known as the EMSAPI standard. It mainly defines PSR (Power System Resource) as a base class. This and the main power resource category inherited from PSR are the main nodes of all power systems, and there are complex interactions between the nodes that make up the grid structure. In general, IEC 61790 focuses on the electrical operation information of the power supply. The IEC 61970-301 sub-package diagram includes core, domain, measurement, circuit, topology, generation, protection), power outage package, load model (LoadModel), SCADA package.

(2) IEC 61968

The IEC 61968 sub-package diagram includes asset package, core extension package (Core2), user package (consumer), work package (work) and document package (documents). In addition to IEC 61970, IEC 61968 mainly extends to various aspects, such as asset, business and user models, as well as monitoring and recording of user models. Similar to the PSR defined in IEC 61970, the asset class can be regarded as the center of IEC 61968. It focuses on simulating real-world power installations and is the basis for energy companies to manage asset life cycle projects.

The dependencies among the packages are shown in Fig. 1. The dashed line indicates the dependency relationship, and the arrow points from the dependent package to the package it depends on.

2.3 Classes and Relationships in the CIM Model

(1) Classes in the CIM model

Each CIM package contains many classes. The uml class diagram clearly shows all the classes in the package and the relationships between them. If these classes are closely related to all classes in other packages, the classes in these packages will automatically display comments indicating which package they belong to. Classes and objects mainly describe all ems applications that need to express power system information in a common form. These categories are the descriptions of the objects of various substances or events in the real world, such as transformers, generators, loads, etc. They should be represented in the ems system as an integral part of the entire power system model. The ems system needs to process, analyze and save other types of objects, such as design and measurement. These objects must also be designed according to various general expressions required to use ems-api. They must have plug-in and interoperability. Those special objects with unique identifiers in the power system can directly pass through their classes.

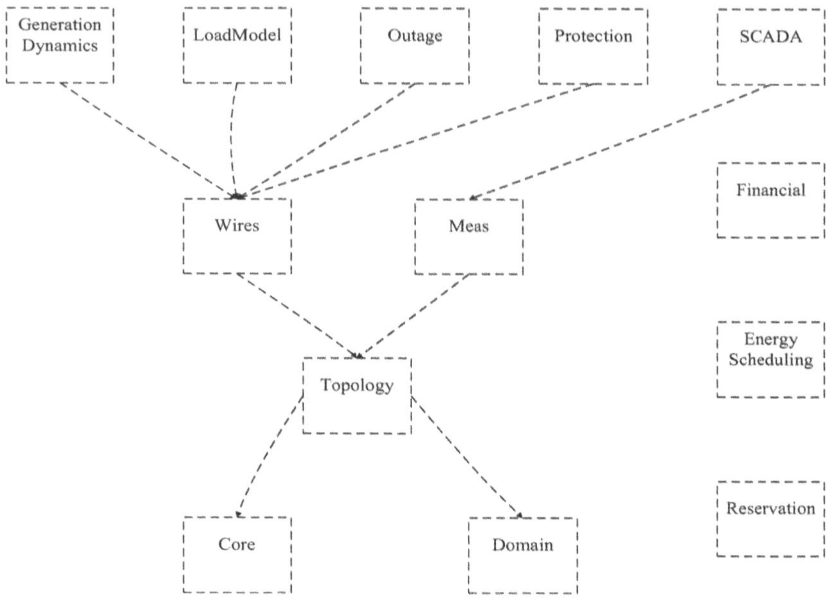

Fig. 1. CIM system main block diagram

(2) The relationship between the classes in the CIM model

1) Simple association
   Correlation is the conceptual association that exists between various classes. Each association should have two goals and roles. Each goal and role represents an interrelated goal and direction, and describes the status and role of the two goals and source categories in the inter-related process. The characters are named according to the target category, with or without verb phrases, and each character also has diversity and collection characteristics to determine the relationship that can participate in the set project. In cim, the number of associations does not need to be named.
2) Aggregation
   Aggregation is a special case formed by the basis of the binding relationship. Through the cumulative relationship, the relationship between the entire category and part of the grid system can be clarified. The whole-class construction of the power grid monitoring system in each region is composed of local classes, and the optimization of tracking technology allows the identification and retrieval of the tracking information content required by the local class. The local category of each monitored area network is part of the overall monitoring of the power network. These local categories are not left over from the grid system, but various monitoring information generated through intelligent technology, such as images, files, videos, and so on. With the help of the CIM model, local and overall collections are collected to realize intelligent monitoring information collection and collection, and information exchange and collection at a

platform point. Finally, the composite aggregation is concentrated in the overall relationship of the grid system.

3) Generalization

The generalization relationship is the relationship established by general data information and specific data information. Special categories include various information data. For example, grid system resource information identifies information data through specific categories that may have a generalized relationship. Figure 2 is an example of generalization. Generalization is a simplified graph technology relationship created on the basis of genetic grid monitoring technology.

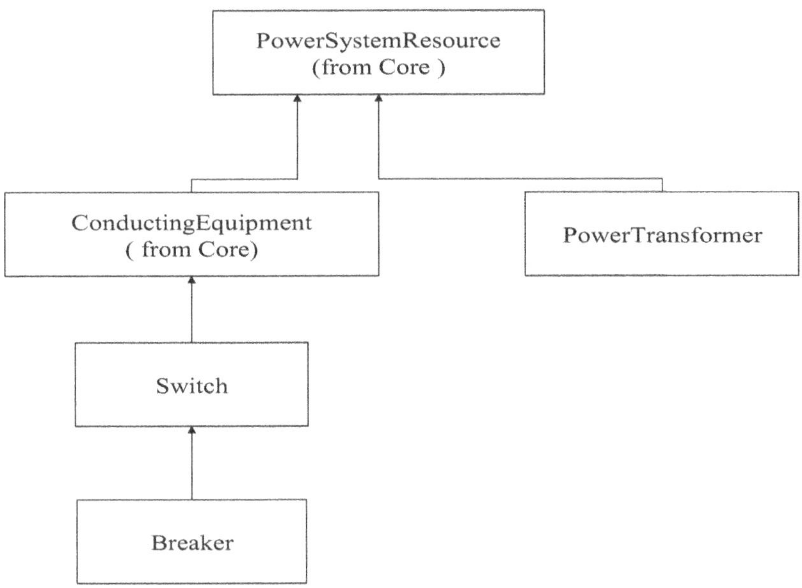

Fig. 2. A generalized example

3 Experiment

3.1 Power Grid CIM Modeling Based on Graph Data Model

With the continuous development of modern power business information analysis and calculation in China, an efficient and integrated data analysis and calculation model of the Internet information system can significantly improve and enhance the quality of information between power business systems and promote business data exchange. In the process of modeling it, the depth and level of the two models need to be considered in turn: the asset model of the power grid and the topology model of the power grid.

(1) Grid asset model

The asset model is the most basic part of the complete radio network information model, similar to the data dictionary of all devices. Grid asset requirements describe the entire power equipment and life cycle, including the life process of energy equipment, from bidding, production, commissioning, maintenance, exit to real-time query, activity log, status traceability, and mobile Internet asset management. The power grid can contain many production control devices. There are as many as 384 kinds of production control devices exported in the production control system. They need to contain various data and assets of the entire enterprise life cycle. There are hundreds of very systematic characteristic fields. You need to create a very abstract data and asset model that can cover basic information.

(2) Grid topology model

1) Topological model storage structure

Fast DB storage topology. When using the topology for analysis, it can be found that it is necessary to avoid the negative impact on Fast DB. The topology analysis mainly used in this article is applied through an object-oriented method, and then Fast DB is upgraded. The node-branch connection needs to be confirmed, so it needs to be stored in a specific way. Therefore, both are stored in a two-dimensional array. At present, the relationship between the two must be described: choosing the neighborhood table can solve this requirement.

2) Topology model interface design

As far as the system is concerned, the control mechanism, database interface, computing elements and Fast DB database work together to configure topology calculations. Four can perform topology calculations called subsystems. The first is the control element, which is mainly responsible for the management process. Now all topology calculations need to be started. The other is the computing unit, which is mainly responsible for static and dynamic topology analysis. With the help of calculations, the corresponding analysis covers the whole, not just the parts. Its basic structure is topology calculation and its interface and power grid topology. All operations of the Fast DB database can be completed through the damage interface, and the static topology information can be obtained instantly, and the analysis results can also be obtained. At the same time, device and topology information can be saved. The basic architecture can be understood with reference to Fig. 3 of the subsystem topology architecture.

3) Parallel local topology analysis

In the actual operation stage, the switching state of the power grid changes frequently, but only partially affects it. Transitions during handover are not common. Therefore, there is no need to re-create the topology of all networks, just re-create the factory to be adjusted. Therefore, once the global topology is established, the pressure and workload of local topology analysis are reduced, and efficiency is further improved.

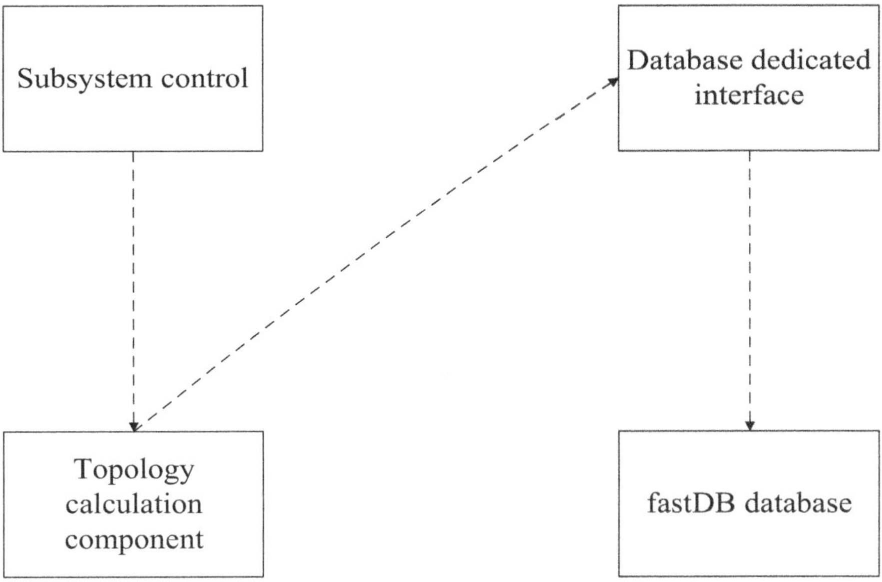

Fig. 3. The structure of the topology

3.2 Analysis of Network Topology Example

Analyzing the power network is actually to find out the connection relationship between any two nodes in the network. This is the connectivity analysis of the graph. The specific meaning is to traverse all the nodes to determine whether it is a connected graph, if not, find out several different connected domains formed by all the nodes.

(1) Node-node connection matrix

$$C = C_{ij} = \begin{cases} 1, \text{ i and j connected} \\ 0, \ others \end{cases} \tag{1}$$

(2) Matrix multiplication operation

For a network with m nodes and n branches,

$$C = A \cdot B \tag{2}$$

$$C_{ij} = \cup_{k=1}^{n} \left(a_{ik} \cap b_{kj} \right) \tag{3}$$

$\cap$ and $\cup$ respectively represent "and" and "or" operations. Obviously C is a symmetric matrix.

4 Discussion

Run a real-time network topology analysis software based on graph data model. The computer network topology analysis software mainly refers to a computer system running on a single network node of Linux multi-core multi-processor. The overall structure of software and hardware is as follows Table 1 shows.

Table 1. Software and hardware operating environment

Software and hardware	
CPU	Intel Xeon E5-26202
RAM	64 GB
operating system	CentOS 6.8 (linux)
Graph computing platform	TigerGraph

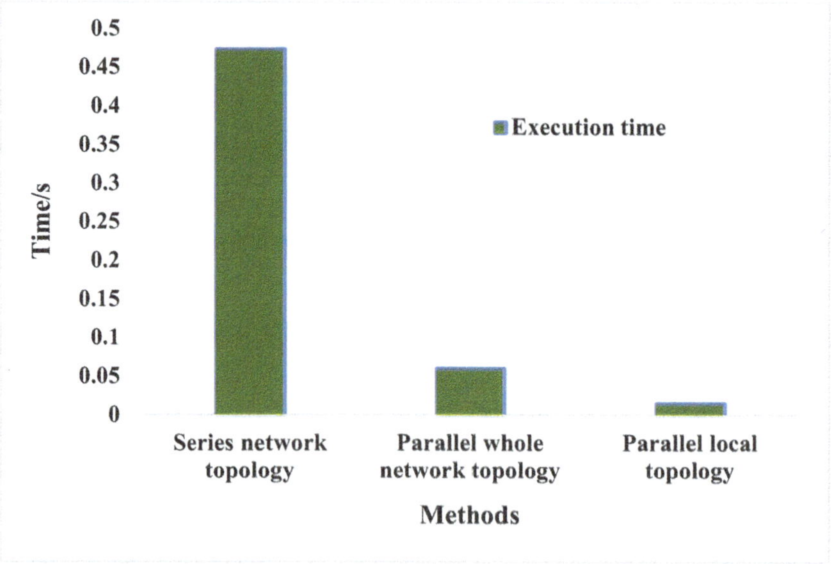

Fig. 4. Execution time of network topology analysis

Among them, the serial whole network topology technology is used as a topology analysis technology for the existing scheduling and control system. Figure 4 shows a parallel network topology based on online data model, which can significantly speed up the online analysis and calculation of network topology, and provides a second level basic data calculation model for online data analysis and calculation of various network topology models such as state evaluation and dispatcher.

5 Conclusions

With the rapid development of our country's modern market economy, energy demand is also increasing, and the scale of power grids is also increasing. The stability and safety of the power grid and the operation of universities are inseparable from the data support of the model and its operating data of the power grid. The information interaction between different application systems and the integration between applications have become problems to be solved urgently. Therefore, the analysis of the power grid model has important research significance for the exchange of model data in the power system and the improvement of system functions.

Acknowledgements. "(Fund Project: China Southern Power Grid Corporation Key Technology Project Funding (R&D and Engineering Demonstration of Edge Computing Cluster Technology for Provincial Dispatching Main Station of Smart Grid, GXKJXM20190619)".

References

1. Zhong, Z., Duan, Y., et al.: Empirical study on the evaluation model of public satisfaction with typhoon disaster information disclosure a case from China. Kybernetes: Int. J. Syst. Cybern. **47**(9), 1704–1720 (2018)
2. Wang, J., Wang, X., Fu, L.: Evolutionary game model of public opinion information propagation in online social networks. IEEE Access **8**, 127732–127747 (2020)
3. Nazir, M., Hussain, I., Tian, J., et al.: A multidimensional model of public health approaches against COVID-19. Int. J. Environ. Res. Public Health **17**(11), 3780 (2020)
4. Cavraro, G., Arghandeh, R.: Power distribution network topology detection with time-series signature verification method. IEEE Trans. Power Syst. **33**(4), 3500–3509 (2018)
5. Shine, J.M., Brink, R., Hernaus, D., et al.: Catecholaminergic manipulation alters dynamic network topology across behavioral states. Netw. Neurosci. **2**(10), 1–41 (2018)
6. Ledwoch, A., Yasarcan, H., Brintrup, A.: The moderating impact of supply network topology on the effectiveness of risk management. Int. J. Prod. Econ. **197**(MAR.), 13–26 (2018)
7. Baliki, M.N., Babbitt, E.M., Cherney, L.R., et al.: Brain network topology influences response to intensive comprehensive aphasia treatment. NeuroRehabilitation **43**(1), 63–76 (2018)
8. Wang, C., Huang, N., Bai, Y., et al.: A method of network topology optimization design considering application process characteristic. Mod. Phys. Lett. B **32**(07), 1850091 (2018)
9. Jiménez-Grande, D., Atashzar, S.F., Martinez-Valdes, E., et al.: Muscle network topology analysis for the classification of chronic neck pain based on EMG biomarkers extracted during walking. PLoS ONE **16**(6), e0252657 (2021)
10. Liu, Z., Wang, L.: Leveraging network topology optimization to strengthen power grid resilience against cyber-physical attacks. IEEE Trans. Smart Grid **12**, 1552–1564 (2020)
11. Yamamoto, K., Yoshida, J., Miyagi, S., et al.: Mobilities in network topology and simulation reproducibility of sightseeing vehicle detected by low-power wide-area positioning system. Electronics **9**(1), 116 (2020)
12. Zhang, M., Wang, L., Cui, Z., et al.: Fast nondominated sorting genetic algorithm II with Lévy distribution for network topology optimization. Math. Probl. Eng. **2020**(2), 1–12 (2020)

Design of DTU for Adaptive Information Collection in Internet of Things

Peixue Liu[1(✉)], Juan Song[1], Mingcheng Sun[1], and Trumone Sims[2]

[1] Qingdao Huanghai University, Qingdao 266427, Shandong, China
hhxylpx@163.com
[2] Ecole Normale Superieure de Lyon, Lyon, France

Abstract. In recent years, with the vigorous development of wireless communication technology, devices such as information acquisition DTU are becoming more and more common in our daily life. DTU is widely favored by the public because of its small size, low cost, high transmission accuracy and convenient networking. Aiming at the problems of single transmission mode, multi scene and poor applicability of traditional DTU, a multi-protocol transmission information acquisition DTU is designed, which takes STM32 single chip microcomputer as the core and integrates WiFi, Lora, 433 M and other wireless transmission modules as well as the corresponding receiver and LCD module. The DTU can complete the functions of information acquisition, wireless transmission mode switching, data receiving and display, etc.; The software and hardware design of information acquisition adaptive multi-protocol transmission DTU is introduced in detail, and the system test is carried out. The test results show that the DTU can collect information in real time and switch a variety of wireless transmission modes, can meet the needs of a variety of scenarios, and has great market application value.

Keywords: STM32F103 · Wireless transmission · Serial communication · Self-adaption

1 Introduction

Relying on the innovation and progress of mobile communication technology, wireless data transmission technology began to gradually replace the traditional wired transmission in the fields of remote information acquisition and system control [1]. Relatively speaking, wireless transmission does not have various restrictions to be considered in the wiring of traditional wired transmission. It only needs to match the transmitting end with the receiving end, which makes the installation more convenient, more flexible and easy to adjust. Therefore, it is more and more applied to the fields of data monitoring, system remote control and so on, and it is also the top priority of military communication technology. As a typical application of wireless transmission technology, DTU itself supports a variety of transmission protocols, which can not only transmit data transparently, but also meet the transmission of user-defined protocols. It supports various communication modes, including communication between independent stations, communication

I. Ahmad et al. (Eds.): STSIoT 2021, LNDECT 122, pp. 544–554, 2023.
https://doi.org/10.1007/978-981-19-3632-6_65

between multiple stations corresponding to a single station at the same time, etc. The technology of modern DTU has been relatively mature and various configurations are complete. Therefore, in practical use, users can send and receive data remotely without specifically understanding various complex communication protocols [2]. To sum up, with the rapid development of wireless data transmission technology, its application scenarios in all walks of life are becoming more and more diverse and complex [3]. Therefore, it is urgent to explore a way of adaptive transmission of information using a variety of transmission protocols to expand the application scenario of wireless data acquisition and transmission module.

With the demand of domestic modernization and economic development, more and more industries realize the importance of wireless transmission technology. The market scale of DTU in China is expanding rapidly and gradually applied to all walks of life. Then, with the development of wireless network technology, DTU began to transmit wirelessly through GPRS. At the same time, the vigorous development of short-range communication technology with low cost and convenient networking also led to the emergence of DTU using WiFi, Lora, Bluetooth and other transmission methods. Most common DTUs adopt a single transmission mode, which cannot continue to work when the signal is disturbed or the transmission module fails. However, this design adopts a variety of wireless transmission modes, which can switch adaptively according to the signal strength to solve this kind of problem.

2 Related Work

Many scholars have studied information acquisition DTU and information acquisition. Dinesh r proposed the museum collection environmental monitoring system, applied the Internet of things information acquisition to museum monitoring, and introduced the corresponding software and hardware design [4]. Sarah Tonello proposed an improved portable information collection device, which is characterized by small volume and easy to use [5]. Relevant experts also proposed multi-point information acquisition device and energy-saving information acquisition device, which organized multi-point information acquisition into a network [6, 7]. Prontheip pipitsunthonsan designed a stomatal transpiration sensor, which can collect temperature and humidity information. It is mainly improved from the sensor [8]. Imanol Picallo applied wireless sensor networks to the collection of athletes' physical state signals and organized the Internet of things network [9]. Washington Velasquez designed a real-time sensor network simulator to simulate the sensor network, which can easily simulate the sensor network [10]. Work A. Gomez et al. Designed a plug and play sensor based on the Internet of things, which can be used for environmental monitoring [11]. Andreas tsatsaris and anagha Rajput respectively applied the information collection of the Internet of things to geographical environment and underwater monitoring, and designed appropriate sensors [12, 13]. Andrea abrardo analyzed the application of data fusion in Internet of things information collection, which improved the accuracy of information collection [14]. Tanvir Alam shifat introduced deep learning into Internet of things information collection, further improving the intelligence of information collection [15].

To sum up, although many scholars have optimized the sensor design and Internet of things network respectively, there is a lack of a DTU capable of multi-protocol transmission in general, and information cannot be transmitted freely between different networks. Therefore, it is of great significance to design a DTU with multi-protocol transmission.

3 Overall System Design

The overall design of the system is shown in Fig. 1. According to the functional requirements of DTU for multi environment and long-term stable transmission of collected data, the following circuit control part is designed; The system hardware is divided into two parts: signal transmitter and receiver. The minimum system of the main control chip is the core of the circuit, which is responsible for processing the collected data and detecting the signal strength of the wireless module; The required data is detected by the temperature and humidity detection module at the transmitting end, processed by the main control chip and transmitted to the receiving end by the wireless module; The receiving end has a wireless receiving module corresponding to the transmitting end for receiving data. The received data is processed by the main control chip of the receiving end and displayed by the LCD display screen of the receiving end.

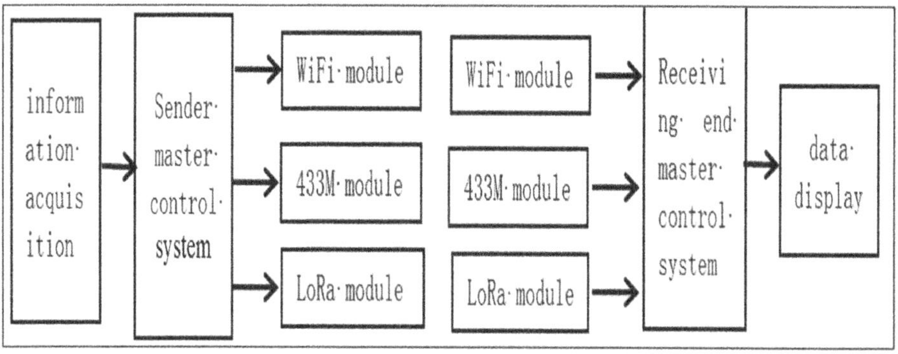

Fig. 1. Overall system design

Taking Stm32f103c8t6 as the main control chip of the transmitter, three different wireless transmission modes are designed, namely WiFi, 433M and Lora; The information transmission range, signal strength and power consumption of signals between corresponding wireless modules are different in different environments. When the design starts to run, the signal strength of each wireless module in the current environment is compared through the main control chip to enable it to independently select the transmission mode with higher signal strength for connection. If the signal strength of the transmission module changes due to changes in the environment, such as distance or obstacles, the main control chip will change the transmission mode again according to the signal strength, So as to realize adaptive wireless transmission. The main control chip used at the receiving end is stm32f103zet6. Compared with c8t6, zet6 has larger ram and flash storage capacity and more interfaces, which can access multiple wireless modules and display modules at the same time. The main control chip at the receiving end can display the received temperature and humidity information, the current signal transmission mode and the comparison value of signal strength of each wireless module on the screen.

4 Hardware Design

4.1 Control Core Circuit Design

In this design, the single chip microcomputer used at the information sending end is c8t6 of STM32 series. The Cortex-M3 of the single chip microcomputer has the advantages of high performance, low power consumption and low cost. The speed of serial peripheral interface reaches 18 MB/s, SRAM is 20 K and flash memory is 64 K. It has two debugging modes: SWD interface and JTAG interface. The whole circuit consists of the core chip of single chip microcomputer and peripheral circuit. The basic peripheral circuit is divided into three parts, which are power supply circuit, crystal oscillator circuit and reset circuit. Power supply circuit: when the STM32 single chip microcomputer used in the design works, it needs to input a voltage of more than 2 V and less than 3.6 V. The input voltage can provide power to the single chip microcomputer only after it is internally adjusted. After the system is powered down, the Vbat pin can obtain power from VDD or backup battery to supply power to RTC and backup register; Crystal oscillator circuit: the crystal oscillator circuit provides the clock frequency for the system, and the clock circuit is the key to the normal operation of the system. The function of the clock source is to drive the system clock. The single chip microcomputer used in this design has three built-in clock sources: high-speed internal oscillator HSI, low-speed internal oscillator LSI and PLL; Reset circuit: the circuit includes a resistor, a capacitor and a switch, which can control the reset of the single chip microcomputer through the key switch. The circuit of single chip microcomputer is shown in Fig. 2.

4.2 WiFi Module Design

Esp8266 module supports LVTTL serial port, and its working voltage is compatible with the voltage of single chip microcomputer, which is easy to connect. The module can be used not only as the central node of a network, but also as a station accessing the network. Therefore, it is more convenient to establish a serial port communication scheme.

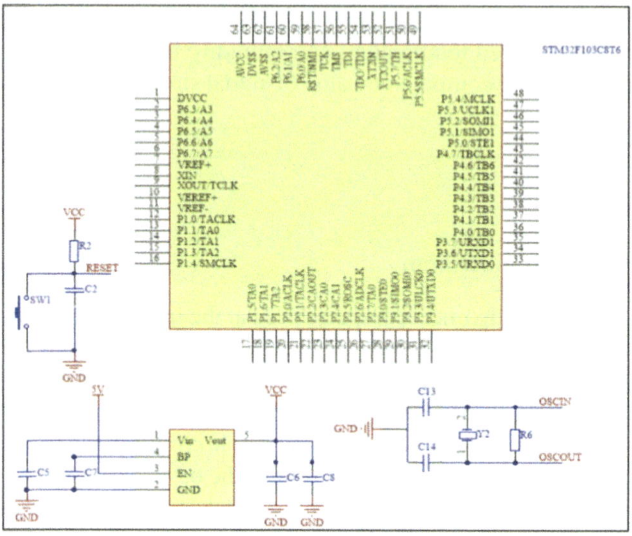

Fig. 2. Structure diagram of single chip microcomputer

The working frequency of the module is 2.4 GHz, the voltage of 3.3 v–5 v can be connected, the transmission power is 11–18 dbm, and the power consumption is 70 mW. This design uses the serial port 3 of single chip microcomputer to communicate with esp8266 module. The pin allocation is usart3_ Rx is connected to the sending pin of the module, and its corresponding MCU pins are pb11 and usart3_ TX is connected with the receiving pin of the module, and the corresponding MCU pin is pb10 (in Fig. 3).

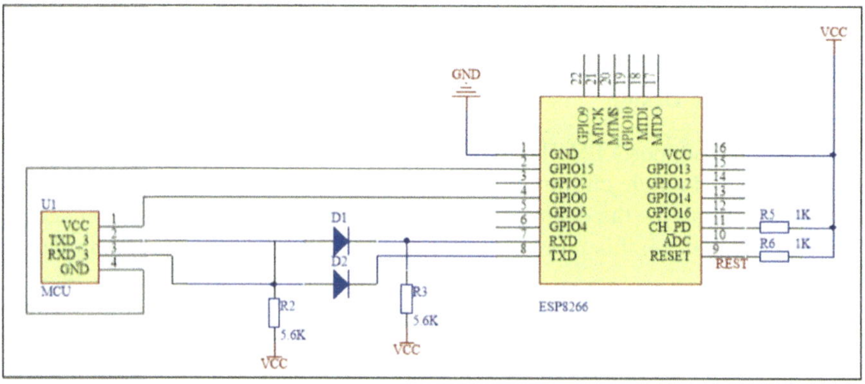

Fig. 3. Esp8266 circuit diagram

4.3 433M Module Design

The 433M wireless module used in this design is hc-12 wireless serial communication module, and the communication distance of this module can reach one kilometer. The module communicates with the MCU through the serial port. Various parameters and working modes of the module can be configured by using at instructions. The pins of hc-12 are: VCC pin is connected to the power supply, GND pin is grounded, RXD pin is data receiving, TXD pin is data transmitting, set parameter setting control pin, and ant is the antenna pin of 433 M (in Fig. 4).

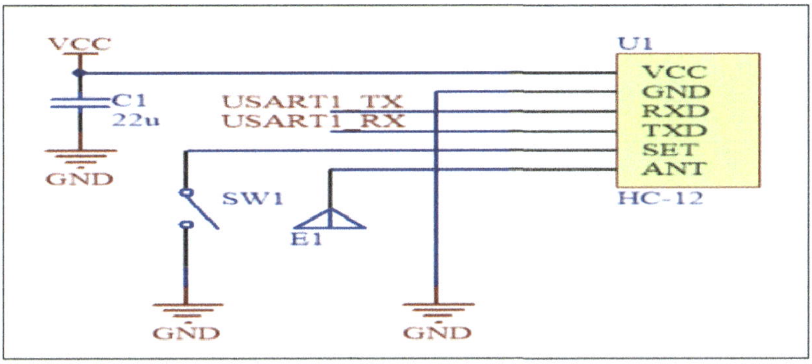

Fig. 4. 433M circuit connection diagram

4.4 Lora Module Design

The working voltage of the module is 3.3 v–5 v. Pay attention to cross access when connecting with the serial port of the single chip microcomputer. In this design, the module communicates with the serial port 1 of the single chip microcomputer. The signal input of the module corresponds to the signal output (PA9) of the single chip microcomputer, and the signal output corresponds to the signal input (PA10) of the single chip microcomputer. Generally, the module does not need series protection resistance.

Atk-lora-01 is connected to the outside through 1 * 6 row pins. The function introduction and detailed connection of each pin are as follows: md0 pin is connected to PC5 pin of MCU as configuration parameters; Aux pin is used to indicate the status and firmware upgrade during module debugging. It cooperates with md0 pin and will not be connected after debugging; In the design, the serial port 2 of single chip microcomputer is used to communicate with this module; The data receiving pin of the module is connected with the transmitting pin of MCU serial port 2, and the corresponding MCU pin is pa2,; The data sending pin of the module is connected with the receiving pin of the MCU serial port 2, and the corresponding MCU pin is PA3; And ensure the correct connection of the power supply pin. The wireless module is at TTL level. Pay attention when connecting the MCU, and md0 and aux pins are at low level in the suspended state. The module circuit and hardware connection are shown in Fig. 5.

5 Software Design

Starting from the overall operation of the system, initialize first. After the configuration of each module is completed, the system starts to read the data collected by the sensor and send it to the receiving end through the wireless transmission module after processing. After the wireless module is configured, it will select the transmission mode with high signal strength for connection. When the receiving end receives low signal strength or signal interruption, it means that the distance between the transmitting end and the receiving end is far. In order to ensure the reliability of information transmission, the system will switch to a wireless module more suitable for long-distance transmission. The temperature and humidity information collected by the transmitting end and the currently applied wireless transmission form are displayed on the LCD screen of the receiving end.

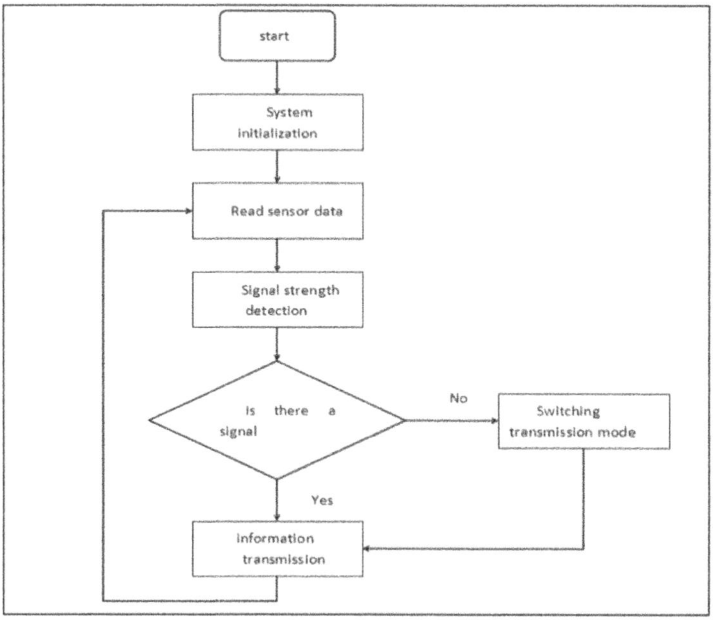

Fig. 5. Overall system flow chart

This design uses three different wireless transmission modules, namely WiFi module, Lora module and 433 M module. The working mode of the three modules is realized through serial communication. Before transmitting the collected information, configure the baud rate and working mode of the wireless module to ensure normal communication between modules. The sensor transmits the detected data to the single-chip microcomputer, and then the single-chip microcomputer initializes the serial port. The wireless module should send the data transmitted by the single-chip microcomputer. At this time, the relevant subroutines executed by the system are used for data sending and receiving.

Using the while cycle, read and send the data every time delay, so as to achieve real-time data transmission (Fig. 6).

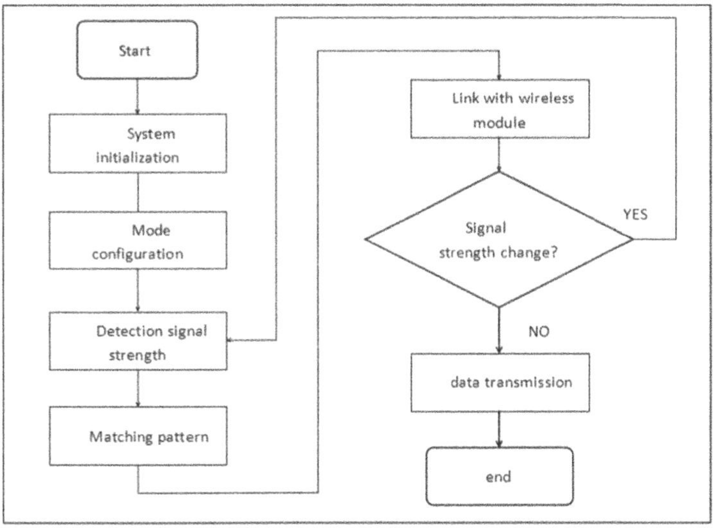

Fig. 6. Flow chart of mode switching

The transmission mode switching in this design is realized by judging the current RSSI value of the wireless module. The RSSI value of the module simply represents the signal strength in the development, which is usually negative. The greater the value, the stronger the signal strength. When each module obtains the current signal strength value, each interval can be defined as different signal strength levels by dividing the interval of signal strength value, and corresponding to the transmission module that should be used in this case, so as to realize the switching of wireless transmission. The wireless transmission module passes through at + cwjap in sta mode\The corresponding signal strength level can be obtained by the R\n command.

In this design, the signal strength is divided into four levels, level 1: $-100 < \text{RSSI} < -80$; Grade 2: $-80 < \text{RSSI} < -70$; Grade 3: $-70 < \text{RSSI} < -50$; Grade 4: $-50 < \text{RSSI} < -5$; The higher the level, the higher the priority. In the process of signal transmission, the current signal strength level of the module should be based first. At the same time, the transmission mode should be further selected according to the actual transmission efficiency, anti-interference and power consumption of each module. For example, the WiFi module has short transmission distance but high transmission efficiency. When the distance is close and the signal strength of each module is similar, the WiFi module is preferred for transmission.

In this design, the collected information can be processed by Kalman filter to make the information more accurate. The formula of Kalman filter is as follows:

$$X(k+1) = FX(k) + w(k) \tag{1}$$

$$Z(k+1) = HX(k+1) + v(k+1) \tag{2}$$

$$\widehat{X}(k+1|k) = F\widehat{X}(k|k) \tag{3}$$

$$P(k+1|k) = FP(k|k)F^{T} + \boldsymbol{Q}(k) \tag{4}$$

$X(\mathrm{k}+1)$ Is the observed value at K + 1 time, F Represents the state transition matrix, $w(k)$, $\mathrm{v}(k+1)$ are process noise and measurement noise respectively.

Status update:

$$\begin{aligned} k(k+1) &= P(k+1|k) \bullet \\ & H^{T}\left[HP(k+1|k)H^{T} + \boldsymbol{R}\right]^{-1} \end{aligned} \tag{5}$$

The estimated value is:

$$\begin{aligned} &\widehat{X}(k+1|k+1) = \\ &\widehat{X}(k+1|k) + k(k+1)\left[Z(k+1) - H\widehat{X}(k+1|k)\right] \end{aligned} \tag{6}$$

6 System Test

After the power on and lighting is successful, the transmission switching test of the physical object is started. Firstly, when the WiFi mode with short transmission distance is used by default, the distance between the transmitting end and the receiving end is extended to 15 m, and there is a wall barrier in the middle. At this time, due to the limitation of transmission distance, the WiFi signal is lost, and the display screen is expressed in the form of "XXX". In this case, the system can re judge the signal strength level of each module, It can be seen from the display screen that the signal strength level of 433 M module is 4 and that of Lora module is 3. Therefore, the system selects 433 M module with higher signal strength level to continue transmission. At this time, the interface is shown in Fig. 7.

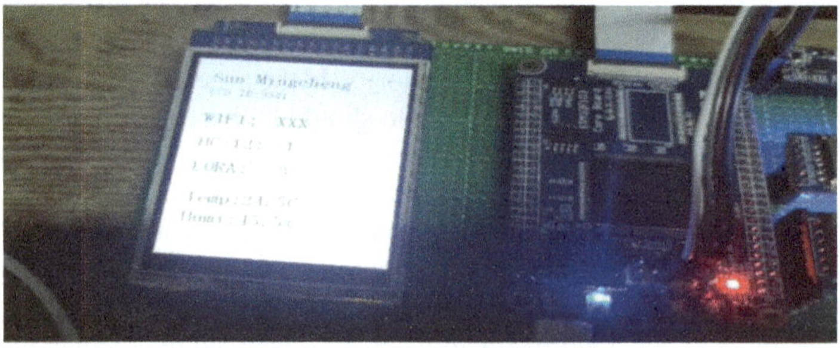

Fig. 7. 433M module switching test

However, when the distance is more than 500 m, due to the limitation of transmission distance, the signals of WiFi module and 433 M module are lost, and they are expressed in the form of "XXX" on the display screen. Only Lora module has signals, and the signal strength level is 3. In this case, the system successfully switches the transmission mode to Lora with the farthest transmission distance to continue the real-time transmission of information, The interface is shown in Fig. 8.

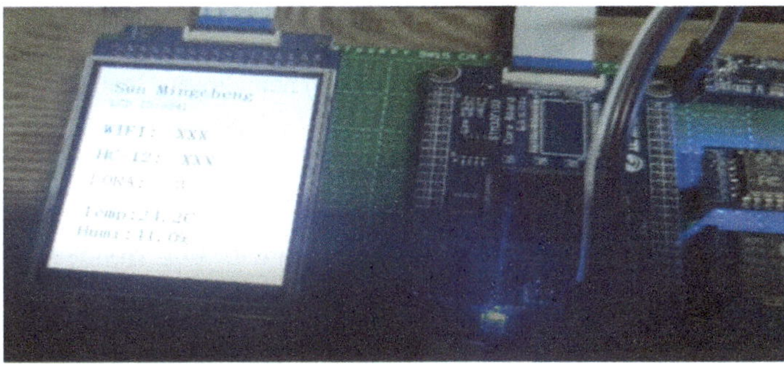

Fig. 8. Lora module switching test

7 Conclusion

Information acquisition adaptive multi-protocol transmission DTU is a device with large data acquisition range, high precision and wide applicability. Various transmission modules adopted by the equipment match the transmission mode by comparing the signal strength, so as to achieve the stable transmission of information under various distances and scenarios, and ensure the continuous transmission of data. This paper expounds the current situation and future progress trend of information acquisition adaptive multi-protocol transmission DTU at home and abroad, introduces the DTU from the aspects of the overall scheme of the system, the modules used in the design and the software content, and completes the operation and commissioning of the physical object.

Acknowledgements. This research was financially supported by Shandong Province key research and development programs of China (2019GGX105001) and the College Science and technology project of Shandong Province of China (J15LN59, J16LN75, and J16LN78).

References

1. Gaber, M.I., Khalaf, A., Mahmoud, I., et al.: Development of information collection scheme in internet of things environment for intelligent radiation monitoring systems. Telecommun. Syst. (12), 132–141 (2020).https://doi.org/10.1007/s11235-020-00697-3
2. Sagar, G.R., Jayapandian, N.: Internet of Things: service-oriented architecture opportunities and challenges. In: Ranganathan, G., Chen, J., Rocha, Á. (eds.) Inventive Communication and Computational Technologies. LNNS, vol. 89, pp. 71–78. Springer, Singapore (2020). https://doi.org/10.1007/978-981-15-0146-3_8

3. Sun, T.: Mining and utilization of special information for archives management based on 5G network and Internet of Things. Microprocess. Microsyst. **1**, 103410–103416 (2020)
4. Dinesh, R., et al.: A smart archive box for museum artifact monitoring using battery-less temperature and humidity sensing. Sensors **21**(14), 4903 (2021)
5. Tonello, S., et al.: Improved portable measuring device for real-time humidity and temperature monitoring in intensive care unit. IEEE Instrum. Meas. Mag. **23**(4), 79–86 (2020)
6. Zbiec, M., et al.: Multipoint wireless humidity and temperature monitoring network for HVAC systems validation. In: MIXDES, pp. 260–265 (2020)
7. Giacomo Fornari etal. Applying frugal innovation to humidity and temperature monitoring. FRUGALTHINGS@MOBICOM 2020, pp. 12–17 (2020)
8. Pipitsunthonsan, P., et al.: Leaf sensor for stomata transpiration monitoring using temperature and humidity. In: WPMC 2018, pp. 252–255 (2018)
9. Guembe, I.P., et al.: Basketball player on-body biophysical and environmental parameter monitoring based on wireless sensor network integration. IEEE Access **9**, 27051–27066 (2021)
10. Velasquez, W.: Sensor network simulator prototype with real-time environmental data monitoring to build smart application. IEEE Access **9**, 144530–144539 (2021)
11. Gomes, J.B.A., et al.: A novel Internet of things-based plug-and-play multigas sensor for environmental monitoring. Trans. Emerg. Telecommun. Technol. **32**(6), e3967 (2021)
12. Tsatsaris, A., et al.: Geoinformation technologies in support of environmental hazards monitoring under climate change: an extensive review. ISPRS Int. J. Geo Inf. **10**(2), 94 (2021)
13. Rajput, A., et al.: An IoT general-purpose sensor board for enabling remote aquatic environmental monitoring. Internet Things **16**, 100429 (2021)
14. Abrardo, A., et al.: Information Fusion in Distributed Sensor Networks with Byzantines, pp. 1–104. Springer, Singapore (2021). https://doi.org/10.1007/978-981-32-9001-3, ISBN 978–981–32–9000–6
15. Shifat, T.A., et al.: ANN assisted multi sensor information fusion for BLDC motor fault diagnosis. IEEE Access **9**, 9429–9441 (2021)

Application of Computer Trajectory Planning Algorithm in UAVs Power Line Patrolling System

Mingxi Jiang[1,2](✉), Jieyin Nan[1,2], Wentao Zhou[1,2], Zhenhui Chen[1,2], An Chang[1,2], and Amar Jain[1,2]

[1] State Grid Tianjin Maintenance Company, Tianjin 300000, China
jmx1986@163.com
[2] Department of Civil Engineering, Faculty of Engineering and Technology, Madhyanchal Professional University, Bhopal, India

Abstract. In order to complete the transmission circuit inorganic and automatic search, shrink deviation hand control, this paper is devoted to the complete type of uav autonomous cruise system based on laser point precise positioning, through the high precision 3 d laser spot time data to complete the course of independent planning, automatically generated, and then complete the inorganic and the whole flow of automatic cruise work. Experiment results shows that, given the precise positioning of the laser point cloud drones in autonomous cruise phase, with space collision testing and automatic blocking function, high efficiency to ensure the safety of the unmanned aerial vehicle (uav) navigation, reduce the latent threat to power grid, improve power transmission cable inspection results and the safety of the operation, provide strategies for the development of power transmission cable inspection to explore in the late.

Keywords: Trajectory movement · Unmanned aerial vehicle (uav) · Cruise detection · Power line inspection · Examples demonstrate

1 Introduction

In recent years, the continuous growth of power demand for the development of all walks of life has accelerated the intelligent transformation and upgrading of the power industry. Among them, the performance in inspection is the application of intelligent devices such as UAV. In the future, with the accelerated implementation of new infrastructure and the further popularization of 5g network and artificial intelligence, the industry is very optimistic about the future of electric patrol UAV. In this context, relying on technical means to promote industrial change has become the key for the industry to seek a breakthrough. In addition, UAV inspection also helps to collect and analyze data management and decision-making, so as to promote the development of precision intelligence in the industry. Therefore, the power patrol of UAV usually plays an important role in promoting the transformation of intelligent manufacturing industry. The further improvement

I. Ahmad et al. (Eds.): STSIoT 2021, LNDECT 122, pp. 555–563, 2023.
https://doi.org/10.1007/978-981-19-3632-6_66

and development of power grid will undoubtedly further improve the inspection demand and difficulty, and will also increase the dependence on UAVs [1].

With the steady development of the two technologies, the support of electric patrol UAV is also increasing. Among them, artificial intelligence further enhances the intelligent interaction, information processing, analysis and decision-making ability of UAV. 5g improves the ability of UAV Communication and data transmission, and jointly promotes the continuous release of UAV potential [2].

Based on this, Chinese companies need to continuously accelerate the research and development of UAV core technology and strive to solve the problems of product battery life, stability and function. At the same time, the industry also needs to establish and improve product standards, improve the threshold of the industry and improve the quality of products. At the same time, behind the UAV RD, production, repair and maintenance, a large number of talents are needed as the foundation. The implementation of Power Patrol can not only effectively reduce the dependence of traditional inspection on manpower, but also ensure high efficiency, high quality and safety. In the future, the potential of electric patrol UAV and blue ocean will be further released and expanded. At the same time, the Chinese government has also issued many preferential policies, which has laid a good foundation for the development trend and future potential release of China's power inspection UAVs [3]. Today, China has initially formed a complete industrial chain of electric patrol UAV, and the market development has gradually entered a good state. The construction of UHV and smart grid continues to advance, and the development of electric patrol UAV has ushered in a broader space. After all, at this stage, as an emerging subdivided industry, its application in the field of power inspection is still naive and faces more challenges and obstacles, such as technology, talents and supporting facilities. The follow-up work still needs to be further improved and effectively solved. In recent years, with the continuous maturity of technology and the development of important manufacturing countries and UAV demand market in the world, the development of UAV in China has achieved remarkable results [4].

After entering voltai in mid July, the peak of power consumption in summer arrived as scheduled. The burden of power grid system continues to rise, and the workload of substation inspection, control and detection also increases. Measure the infrared temperature of the whole station equipment and equipment joints, find hidden dangers in advance and have the function of high-altitude detection. This inspection method has high labor intensity and difficult working conditions, so it is unable to feed back the operation of transmission line in time. It has the advantages of fast, high work efficiency, no regional influence, high quality and high security.

The paper included five parts. The first part is introduction. The second part is related work. The third is data analysis. The example analysis is the fifth part. The last part is the conclusion.

2 Related Work

This section reviews the most relevant works related to the application of UAV technology. The reference [5] used the trajectory planning algorithm for the UAV, and the author proposed the algorithm by network controlled, the result is better and for the most UAV, the algorithm is applicable. Navigation of autonomous light vehicles using an optimal trajectory planning algorithm for UAV, according the experiment, the author get the algorithm is available and do some comparing the experiment for the algorithm and verify the algorithm was better than the other algorithm [6]. In addition, a noval algorithm was proposed by the Asim M, and applying the algorithm to the navigation of AUV, the algorithm extended the scope of the application [7].

Different algorithms have different architectures. So the reference [8] refered to the different architectures, the main purpose is to improve the accuracy of navigation for different controlled objectives [8], such as the references [9, 10]. Due to the real time problem in the UAV, the reference [11] is mainly concerned with real-time problems for the UAV, thus we can reduce the delayed, according to the experiment, this delay problem has been well solved.

Difficulties and challenges of target tracking algorithm research: actual complex application environment, background similar interference, changes in lighting conditions, occlusion and other external factors, as well as changes in target attitude, appearance deformation, scale change, out of plane rotation, in-plane rotation, out of field of view, fast motion and motion blur. Moreover, when the target tracking algorithm is put into practical application, an inevitable problem - real-time problem is also very important. It is these problems that make the algorithm research full of difficulties and challenges.

The advantage of infrared thermal imaging technology in the application of power detection technology is that infrared thermal imaging technology has good advantages. Infrared thermal imaging technology can directly display the temperature field of the object surface. The infrared thermal imager not affected by strong light can also measure the temperature level of each point on the object surface at the same time.

2.1 Isolation and Acquisition of Power Lines, Poles and Towers in Laser-Point Clouds

(1) Separation of laser point cloud features

For the separation of point cloud features, filtering technology is the core point of the overall process. First of all, road points and non-road points should be isolated. The author refers to the automatic filtering mode of airborne laser point cloud in the case of terrain fluctuation to isolate features. The mode is based on a linear predictor filtering algorithm to implement automatic filtering data set according to the landscape, and this data is the specification of the grid in the filtering, which in turn is cemented to separate the laser points (see Fig. 1).

(2) acquisition of power transmission lines and electric poles and towers

The shadow cast by the transmission line points at the horizontal position is smaller than that of the poles and towers, and the focus of the horizontal projection of the poles is much greater than that of the transmission line. Therefore, the position of the tower can be obtained easily according to the distribution characteristics of the laser points. The power transmission line appears sag between two poles and towers. The author firstly screens the positions of poles and towers, and then carries out curve simulation on the complete power transmission line points (See Fig. 2).

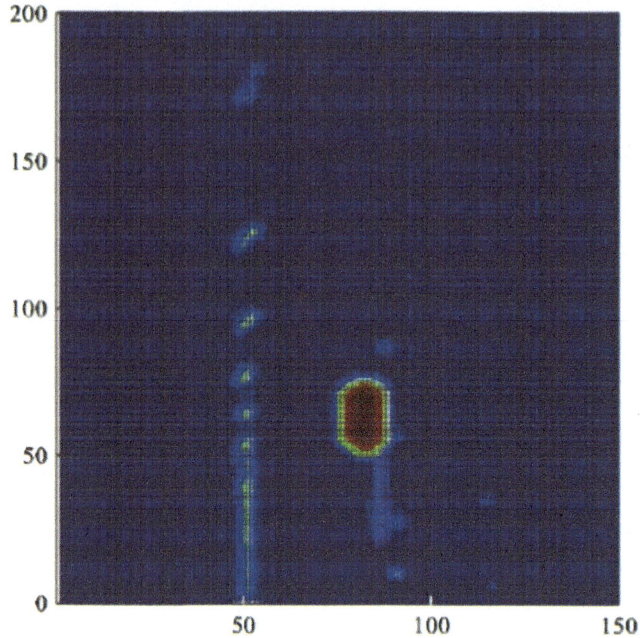

Fig. 1. Diagram of laser point cloud

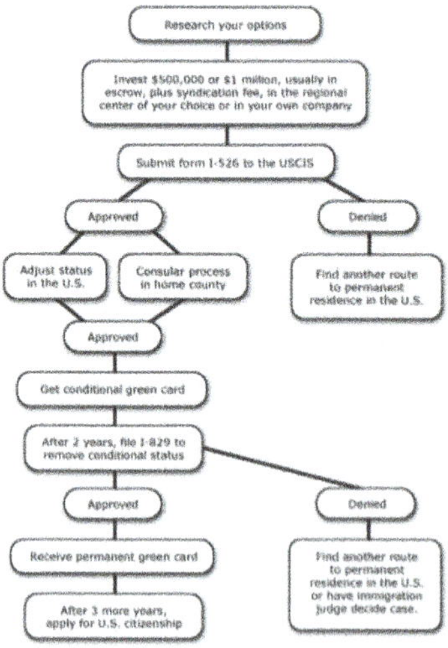

Fig. 2. Three-point perspective diagram of the tower

2.2 Lidar Patrol Inspection Based on Laser Point Cloud

Light Detection And Ranging (Lidar) works by using laser Ranging theory to get the rangy space between the laser radiator And the road point, And by visualizing the GPS And Inertial Measurement Unit, which are built synchronously. (IMU) parameters to complete the positioning and orientation of the laser point cloud, restore the 3D spatial coordinates, and then make the system more active, non-contact, penetration and high precision characteristics (See Fig. 3).

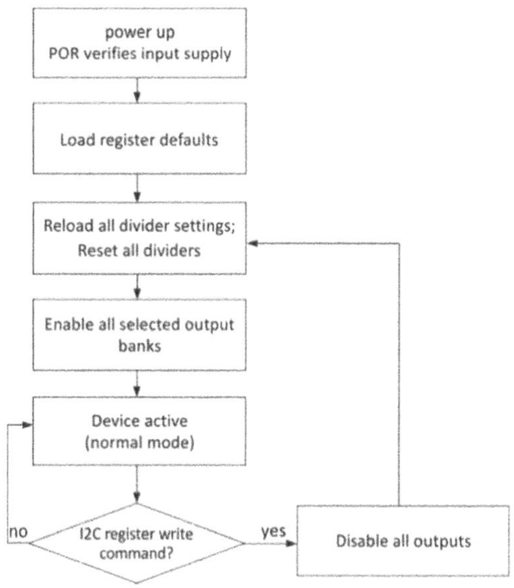

Fig. 3. Schematic diagram of UAV operation

(1) Automatic planning of cruise routes

Figure 4 is the high precision point cloud parameter route design and effect diagram. Tower model parameters according to the type of high precision complex track planning implementation, with the study algorithm to assist tower body of fine sia video automation precision correction, form continuous video point trajectory, which embodies the navigation ideas of what you see is what you have both continuous cable tower the sailing route of the default, more power for unmanned aerial vehicle (uav) navigation provide safe bearing locking; Both space collision test and automatic evasive function, efficient to ensure the reliability of navigation.

In addition, route data is set according to different tower types and terrain. The initial data preset for route planning of UAV Jingwei M210 RTK series with Z30 lens as an example.

(2) The unmanned aerial vehicle is fully self-operated

The UAV flight management system receives the patrol task information within the point cloud planning system. The system completes flight safety monitoring

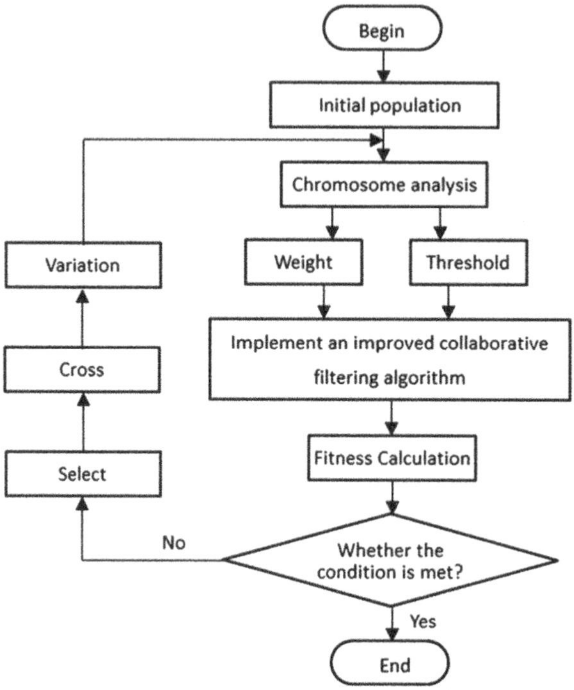

Fig. 4. Route planning and effect diagram with high accuracy point cloud parameters

before takeoff in accordance with the task characteristics of flight path planning, and automatically records the flight trajectory in the stage of fully autonomous flight. In case of emergency, one-key return is required to ensure unmanned flight in the whole stage.

3 Data Analysis

TOn March 3, in order to comprehensively improve the accuracy of transmission line outage maintenance, State Grid Taiwan power supply company conducted a detailed investigation on the body defects and channel hidden dangers before transmission line outage in combination with UAV inspection mode. The auxiliary facilities of transmission lines and towers are exposed outdoors for a long time, and are vulnerable to various weather, mechanical and man-made effects, resulting in collapse, strand breakage, wear and other damage, which poses a threat to the normal operation of transmission lines. In order to timely understand the current situation of power supply of the power grid, master the line load operation, and make preparations for autumn inspection and autumn investigation in advance, Pu Yang company used "UAV" to patrol 10 UHV lines of 500 kV and above in the Yang City. Measure the infrared temperature of the whole station equipment and equipment joints, find hidden dangers in advance and have the function of high-altitude detection. It has the advantages of fast, high work efficiency, no regional influence, high quality and high security. The efficiency and quality of equipment body inspection have been significantly improved, which greatly reduces the labor intensity, improves the inspection efficiency, and ensures the state operation and maintenance ability of power equipment.

Infrared thermal imaging technology can directly display the temperature field of the object surface. The infrared thermal imager not affected by strong light can also measure the temperature level of each point on the object surface at the same time. UAV patrol improves the speed and efficiency of power maintenance and overhaul, so that many tasks can be completed quickly when fully charged, which is more effective than manual patrol. UAV can test the temperature of high voltage line in close range. Therefore, the transmission and transportation inspection center proposes to use the UHV patrol mode of "UAV + manual" cooperative patrol, and use the manual UAV navigation system to reduce the cost of line patrol and improve the precision and accuracy of patrol. Open the intelligent patrol mode by autonomous flight of fixed wing UAV, patrol and take photos of special high-voltage lines one by one, find defects and hidden dangers as soon as possible, and ensure the operation safety of the line.

At the same time, it also has the largest power grid data collection library in China, accurate tower spatial data, image data, installation data, etc., which has the advantages of first mover and data. For the latter, the team uses the deep learning technology of artificial intelligence to build the algorithm model and train their own unique massive patrol photos to realize the automatic analysis of the photos taken by the computer. He is good at the R D and integration of technologies such as UAV, remote graphics transmission, online monitoring, intelligent interconnection and precise positioning. The technology can customize and quickly build advanced and practical application systems to help customers quickly improve their ability of intelligence, automation and comprehensive management.

4 Example Analysis

On the line, a UAV completed the patrol of high-voltage lines such as "station to station" automatic inspection and data image return according to the established route. By the end of last year, the transmission and transportation inspection center of Long South power supply company had carried out work in the fields of line acceptance, fault patrol, on-site survey and auxiliary point operation. The work is more standardized and safe, which greatly improves the inspection efficiency.

The official launch of the electric patrol UAV driving training base established by the State Grid Qinghai electric power company will effectively make up for the shortage of using UAVs to carry out power grid inspection and comprehensively create an intelligent inspection system in Qinghai. In order to solve the increasingly prominent problems such as low efficiency of manual inspection and serious shortage of personnel, since 2013, as a pilot unit of "helicopter + UAV + manual" joint patrol, Qinghai power of State Grid has continuously explored and vigorously promoted the patrol application of UAV, and gradually established an intelligent transmission line patrol operation system, The quality and efficiency of transportation maintenance are greatly improved.

It is reported that with the completion and opening of the first training base in Qinghai, the State Grid Qinghai electric power will rely on the base to regularly carry out large-scale personnel skill training to accelerate the transformation of traditional manual patrol methods. Meng Wei, special technical director of transmission and transportation inspection of Golmud operation and maintenance division of State Grid Qinghai maintenance company, introduced: "In the past, I was a hiker, and the only two sets of equipment were binoculars and cameras. Now our yunvita is relatively high, especially the Chaida line, which is basically between 30 and 40 m. The height can reach more than 50 m. There is jitter during use. It is difficult to find the defects in some key parts of the connection between binoculars and cameras." In recent years, with the progress of science and technology and the continuous innovation of inspection methods, the transmission professional UAV patrol of Qinghai maintenance company of State Grid has been comprehensively promoted. Through the close-up shooting and inspection from a special angle at high altitude, the parts above the tower and the overall construction condition of the line have been "scanned" in an all-round way Diagnosis and circuit inspection gradually stand out from manual inspection. In the past, we needed to climb the mountain. Now you can complete the inspection at the foot of the mountain. Regularly patrol the ground with personnel. The inspection quality and defect detection rate will be greatly improved, "Meng Wei said "Today, the efficiency of UAV Patrol has been greatly improved. We can operate UAVs in two groups at the same time. We can carry out two directions at the same time. In this case, when we check the first level tower, we can complete the shooting process of the whole tower in four to five minutes." as the main energy artery connecting the two provinces of Qinghai Tibet, this "power road" It has been in operation for more than nine years.

$$
\begin{aligned}
&E(t)\dot{x}_{d+1}(t) - E(t)\dot{x}_{k+1}(t) = E(t)\Delta\dot{x}_{k+1}(t) = \\
&f(t, x_d(t)) + B(t)u_d(t) - f(t, x_k(t)) - B(t)u_k(t) - d_k(t) = \\
&f(t, x_d(t)) - f(t, x_{k+1}(t)) + B(t)\Delta u_{k+1}(t) - d_{k+1}(t)
\end{aligned} \tag{1}
$$

$$
\begin{aligned}
&\Delta x_{k+1}(t) = \textstyle\int_0^t P^{-1}(\tau)(f(t, x_d(\tau)) - f(t, x_{k+1}(\tau)))d\tau + \\
&\textstyle\int_0^t P^{-1}(\tau)B(\tau)\Delta u_k(\tau)d\tau - \int_0^t P^{-1}(\tau)\big(B(\tau)L(\tau)\dot{C}(\tau) + \\
&B(\tau)L(\tau)C(\tau)\big)\Delta x_{k+1}(\tau)d\tau - \textstyle\int_0^t P^{-1}(\tau)d_{k+1}(\tau)
\end{aligned} \tag{2}
$$

5 Conclusions

Track planning algorithm has always been a very important and hot research direction in the field of Power Patrol UAV. In order to design a flight control system for electric patrol six rotor UAV and test the performance of the flight control system, an exploration scheme of UAV fully autonomous detection system for laser point cloud accurate positioning is proposed in this paper. Aiming at the high-precision three-dimensional light point cloud parameters, this study completes the autonomous planning and formation of flight route, and then completes the automatic flight navigation. UAV autonomous Patrol has the functions of space collision test and automatic shielding during navigation, which provides safe azimuth guidance for UAV unpowered navigation. Considering the quality and quantity of images taken by the UAV and the electromagnetic field on the high-voltage cable, the safe distance between the UAV and the tower must be controlled in the planning and wiring stage.

Acknowledgements. State Grid Tianjin electric power company science and technology project KJ20-1-01: Research and application of all weather autonomous inspection technology of UAV for transmission line.

References

1. Gao, X., He, G., et al.: Wavelet theory and its application in traffic incident detection. Highw. Traffic Branch **116**(13), 63–65 (2021)
2. Jiang, G., Wen, H., Yang, Z.: Design of automatic detection system and algorithm for expressway traffic incident. J. Transp. Eng. **1**(1), 77–81 (2020)
3. Zhou, W., Luo, S.: Event detection algorithm based on fuzzy comprehensive recognition. J. Xi'an Highw. Jiaotong Univ. **21**(2), 70–73 (2020)
4. Zhou, X.: automatic traffic incident detection based on wavelet analysis and neural network. Highw. Automob. Transp. **91**, 17–19 (2020)
5. Scordamaglia, V., Nardi, V.A.: A set-based trajectory planning algorithm for a network controlled skid-steered tracked mobile robot subject to skid and slip phenomena. J. Intell. Rob. Syst. **101**(1), 124–134 (2021)
6. Valera, N., Valero, F., Vallés, M., et al.: Navigation of autonomous light vehicles using an optimal trajectory planning algorithm. Sustainability **13**, 223–245 (2021)
7. Asim, M., Khan, W., Belhaouari, S.B.: An evolutionary trajectory planning algorithm for multi-UAV-assisted MEC system. Int. J. Highseed 225–246 (2021)
8. Li, Y., Wang, L., Wu, Y., et al.: A dynamic lane-change trajectory planning algorithm based on minimum safe spacing. Autom. Tech. **44**(07), 1108–1112 (2021)
9. Li, X., Zhang, J., Han, J.: Trajectory planning of load transportation with multi-quadrotors based on reinforcement learning algorithm. Aerosp. Sci. Technol. **10**(11), 124–134 (2021)
10. Kose, O.: Real-world application of various trajectory planning algorithms on MIT RACECAR. J. Control Theory **32**(01), 221–232 (2021)
11. Xu, Z., Deng, D., Dong, Y., et al.: DPMPC-planner: a real-time UAV trajectory planning framework for complex static environments with dynamic obstacles. J. Inf. Syst. **23**(03), 334–356 (2021)

Application of Blockchain Technology in the Construction of MOOC Digital Communication Platform

Jing Zhang[1,2](✉) and Sulaima Haleem[1,2]

[1] School of Art and Design, Modern College of Northwest University, Xi'an 710130, Shaanxi, China
zhangjingbsjs@163.com
[2] South Eastern University of Sri Lanka (SEUSL), Oluvil, Sri Lanka

Abstract. The rapid development of blockchain technology and the construction of MOOC communication platform have a positive role in promoting higher education, the problems we have to think about and solve up to now are how to effectively build a MOOC communication platform in higher vocational education and how to use blockchain technology to build a MOOC curriculum system, and how to deal with the development of various MOOC platform contradictions, the preliminary proposed corresponding solutions. In accordance with the trend of education reform, this paper focuses on the sustainable development of MOOC education production, teaching, learning, research and use integrated block chain ecosystem, and the development and implementation of MOOC communication platform, so as to evaluate its development prospects.

Keywords: Blockchain · MOOC · Communication platform in the picture classification

1 Introduction

Due to the lack of verification of the practicality and scientific degree of MOOC, there is no MOOC platform developed directly from blockchain technology at home and abroad, but from the MOOC construction how to promote the Chinese education, the block chain Electronic Archives Trust Guarantee, the block chain security detection model and so on has had the certain research. puts forward the advantages and application ways of blockchain in the platform of E-commerce, which can be used for reference in the application of MOOC communication platform. there are corresponding suggestions for the application mode of blockchain technology in education but there is no effective link with MOOC. Reference based on dual factor module and blockchain technology, some research results have been made on the identity authentication of MOOC learners, but there is still a lack of analysis, research and practical test from the full-scale operation and diversified management of MOOC. The construction and use analysis of different dimensions of MOOC platform and the influencing factors of blockchain technology are the key points that universities should pay attention to in the future.

I. Ahmad et al. (Eds.): STSIoT 2021, LNDECT 122, pp. 564–573, 2023.
https://doi.org/10.1007/978-981-19-3632-6_67

In response to the national policy and strategy, the construction of MOOC communication platform in the context of blockchain needs to be combined with the sharing of big data, authentication and security system, optimization of energy saving, mutual communication and integration, and play an important role. This paper consists of the following parts. The first part introduces the relevant background and significance of this paper, the second part is the related work of this paper, and the third part is data analysis. The fourth part is example analysis. The fifth part is conclusion.

2 Related Work

To solve these problems Zhang et. al provide five summarized ways [1]. Le et. al provide evaluation of these courses, comparing a manual feature engineering approach to an automatic feature learning approach using neural networks [2]. Based on this, Shao discussed the influence of MOOC teaching on the reform of ideological and political teaching in colleges and universities [3]. The MOOC model tries hard to be "student-centered", mainly in the following two aspects [4]. Gamage et. al find overall collaboration opportunities in platforms are significantly less than the opportunities to interact [5]. Ali et. al reveal that highest number of courses available through SWAYAM platform is in the field of Engineering with 654 courses, followed by Arts and Recreation with 127 courses and Science with 93 courses [6]. The research reported here was conducted to further understand learners social engagement on a MOOC platform and the impact of engagement on course completion [7]. The purpose of Sharov et. al is to analyze the possibilities provided by the Ukrainian online platform of the Open University of Maidan (OUM), and to compare it with other Ukrainian platforms (Prometheus, EdEra) according to certain criteria [8]. Febrianti et. al examine to acknowledge the learning model development of MOOC-based BIPA for Business Communication [9]. In order to improve the effective management of MOOC platform resources based on traditional methods, Tian et. al. implement a deep neural network algorithm, and recommends related services [10]. MOOC communication platform is an important operational medium connecting academic institutions, institutions, experts and scholars, teaching teams, enterprises, students and so on, creative thinking mode of digital art is more needed. Blockchain technology can help colleges and universities to build interoperable, safe and stable platform blockchain, and reduce the workload of MOOC development, to provide innovative business and facilitation in massive open online course's vision, deployment, development, management, operation and application of technology.

2.1 Introduction to Blockchain Technology

It is a data storage computing system, which can calculate on it. At the same time, the distributed network based on Internet is composed of many nodes. In other words, blockchain is a data computing and storage system under Internet distribution. It can be understood as a public database.

(1) Principle of blockchain technology. In the blockchain, each block in the blockchain contains data, the hash value of the current block and the hash value of the previous

block. Different types of data are stored according to different blockchain types. Each block contains the hash value of the current block and the hash value of the previous block, which makes there a chain structure between blocks. In this chain structure, the latter block points to the previous block, while the first block does not point to it. We call it the creation block. In this level-by-level directional structure, if a block in the middle is changed, the blocks of all subsequent nodes will be changed. Therefore, when someone wants to add a new node or tamper with the data, the hash values of all blocks must be recalculated, which is also called work proof.

(2) Decentralization. The blockchain uses P2P peer-to-peer network, and its point-to-point transmission technology is the basis for the blockchain to break away from the control of centralized server and form a public ledger. There is no central server in this network. When someone joins the network, he will get a complete copy of the blockchain, which can be used to verify whether the blockchain is normal and orderly.

2.2 Characteristics of Blockchain

(1) Consensus mechanism with high cheating cost. Blockchain is a distributed ledger without a central structure, so a consensus mechanism is needed to replace the central structure to assume the role of the decision maker, that is, consensus algorithm. Common consensus algorithms include pow, POS, dpos, pbft, etc. different blockchains are based on different consensus mechanisms. For example: in a POW based blockchain, when a new block is added, the new block will be sent to all nodes in the network. Each node user can use the consensus algorithm to verify whether the current block is normal. If it is a normal block, the user node will add the block to its own blockchain copy. When more than 50% of the nodes in the network join the block, This block can exist effectively in the whole network. If you want to add an abnormal block, you must control more than 50% of the user nodes in the blockchain and obtain sufficient workload proof. This is almost impossible.

(2) It exists on the Internet and is open to all users. The blockchain is decentralized. Each node in the chain has complete data of the whole chain, and all information is fully disclosed to users. The general architecture of the blockchain is shown in Fig. 1.

(3) Very secure encryption technology. The blockchain uses the hash value as the password for block data and authentication. The hash value is a string of non repetitive and irregular strings generated by a specific calculation formula according to the block data. Once the data changes, the hash value changes, and the hash value cannot deduce the data. The blockchain generates a public key and a private key according to the hash algorithm for authentication. The public key is used as the user address, and the private key is privately held by the user. Authentication requires both public and private keys. However, the overall design of the client also has some general concepts, as shown in Fig. 2.

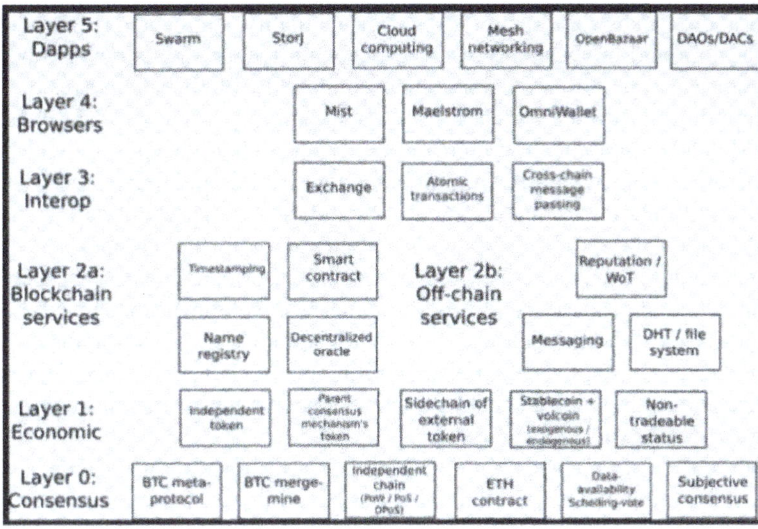

Fig. 1. Blockchain architecture

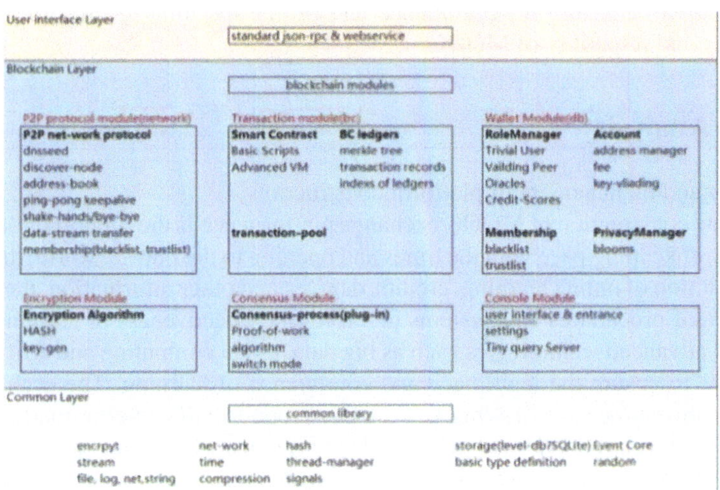

Fig. 2. Overall client architecture

2.3 Current Situation Analysis of MOOC Communication Platform

MOOC communication platform has the characteristics of big data, rich courses, sharing, large-scale, open, personalized and advanced technology. According to its teaching methods, it can break through the advantages of time and space constraints, the current popular non-contact network education. According to the current economic and technological development level, technologies such as big data, 5g, cloud computing, artificial intelligence and interactive experience are becoming more and more mature, and technologies such as Internet of things, virtual reality, augmented reality and hybrid reality are more and more widely used. Blockchain technology is developing rapidly with the improvement of these technologies, combined with the advantages of big data, cloud computing and artificial intelligence, A new pattern has emerged. The construction space of MOOC communication platform under blockchain technology will be larger and larger, which will subvert the tradition and make it fair and transparent.

According to the environment, the cross regional, Cross School and cross professional application mode of MOOC course needs to be further studied and explored, including the management and operation of the market-oriented platform, especially the issue of intellectual property rights and sustainable development. Due to the different backgrounds of universities, cultures, disciplines and majors, the communication channels of MOOC are limited. There are some problems, such as unbalanced development of MOOC content, single form, chaotic content, low completion rate of MOOC courses, weak learner management and so on.

The bottleneck of MOOC communication platform construction also lies in the lack of time accumulation and trust. All universities have relevant management systems and pay much attention to the introduction of MOOC. MOOC advocates a free learning model across time and space. Compared with face-to-face teaching, MOOC teaching in Colleges and universities lacks time accumulation and trust. Strengthen the synchronous communication between online guidance and offline teaching to improve the safety, authenticity and reliability of MOOC.

3 Data Analysis

(1) Policy and management of platform construction

The construction of MOOC exchange platform needs the support of the national policy of keeping pace with the times and opening to the outside world, the national recognition of online learning, credits, degrees and other information, the standardization of procedures and systems to ensure the effectiveness of learning, and the use of advanced technologies such as big data, cloud computing and artificial intelligence to ensure the legitimacy and compliance of learning, The evaluation and verification process will bring new business opportunities to the market after the opening of relevant national policies. Therefore, it will promote the rapid development of MOOC communication platform technology and inter enterprise art. The learning law is as follows:

$$u_{k+1}(t) = u_k(t) + L(t)(\dot{e}_{k+1}(t) + e_{k+1}(t)) \tag{1}$$

For example, in the current struggle against COVID-19 epidemic, the State Education Department is particularly necessary to introduce some quality online courses on relevant platforms according to its own learning characteristics for unified management of universities and credit certification. Teachers from offline professional schools should help themselves with the individual differences between schools and students by means of self media and communication platform. At present, some colleges and universities try to avoid blind catalog micro courses and do not directly organize, manage and choose online teaching. Due to immature technology, limited platform operation and unequal teaching level of teachers, it will cause negative effects such as poor teaching quality, chaotic management and unclear responsibility.

(2) Platform mode innovation

The ideal mode of MOOC communication platform construction needs to pay attention to different disciplines of different universities in the blockchain environment, pay attention to advanced technical functions, and create a MOOC communication platform with distinctive characteristics and innovative artistic forms.

1) MOOC communication platform technology innovation

The construction of MOOC communication platform should keep pace with the times and introduce more advanced technical functions, such as face recognition, virtual interaction, live communication, background big data synchronous analysis, etc. even student achievement and credit recognition should be linked to the blockchain. Timely and gradually improve the service consciousness of platform teaching. After MOOC enters the study, it should establish links such as teacher-student evaluation, teacher-student interaction and joint learning, so as to adapt to the new needs and differences of teaching in Colleges and universities, give full play to the advantages of first-class professional teaching resources, and then develop and improve the resources and management of MOOC platform, so that each individual and link can form a benign operation mechanism, It is necessary to treat the normal face-to-face teaching management of schools differently and resolve the operation conflict. For the platform system, its learning law is as follows:

$$u_{k+1}(t) = u_k(t) + \Gamma_{l1}\dot{e}_k(t) + \Gamma_{l2}\dot{e}_{k+1}(t) + \Gamma_{p1}\Delta\dot{e}_k(t) + \Gamma_{p2}\Delta\dot{e}_{k+1}(t) \quad (2)$$

2) The construction of MOOC communication platform should be carried out at different levels, paying attention to the differences of different disciplines in Colleges and universities

The education administrative department shall support, build, guide, operate and manage the MOOC communication platform at different levels, and formulate different standards and competition levels according to the school's hardware facilities, teachers, student level, management ability, systems and policies. If the unified standard can not play an important role in different institutions in different regions, or even form a teaching burden, the final MOOC communication is likely to become a formalism or image project, thus losing the significance of education and teaching.

MOOC communication platforms need not be too many, as shown in Fig. 3. The execution platform provides operation rules, resource sharing and service guarantee. The specific initiative and teaching work will be left to the institutions, managers or teachers operating on the platform. The school will operate specific MOOC courses according to the direction of each discipline, the advantages of professional resources and the characteristics of talent training. The fees charged also need to be combined with the University's tuition management policy. The education administrative department can choose incentive policies and certification standards to better play the role of the curriculum.

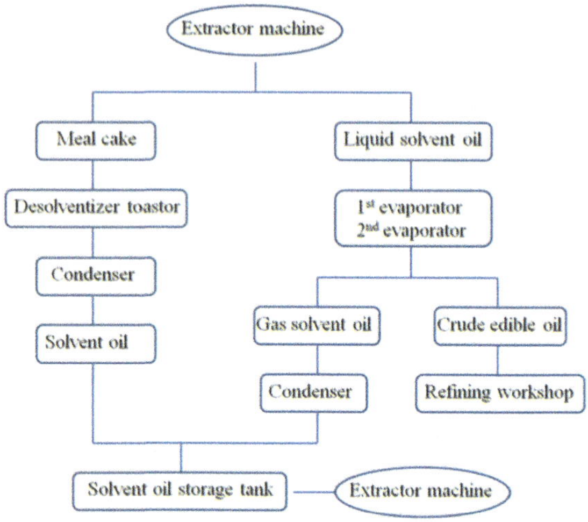

Fig. 3. MOOC digital communication platform process

3) Promote the integration of blockchain teaching and MOOC

Introducing blockchain technology into MOOC communication platform will bring high data and information security. College users will have strong collective trust in MOOC teaching resources, student information, teaching results and interactive information data, embed blockchain information content into various professional fields for classified teaching, and accelerate new technologies and education such as virtual experience, artificial intelligence and synchronous analysis of big data The in-depth integration of information and art will promote artistic innovation and information sharing and integration of industry and education. The key to promoting the integration of blockchain technology teaching and MOOC platform construction is to cultivate the innovative thinking of teachers and students in blockchain teaching, which is a comprehensive innovative thinking mode of technology, mathematics, industry and artistic logic. The innovative achievements should be tested in practice.

4 Example Analysis

(1) Strengthening the construction and management team of moocs

MOOC construction is an extremely complex systems engineering of computing, design, development, construction, technology, statistics, management, application, evaluation, feedback, and so on. In view of the current form of MOOC presentation, topics that need to be technically optimized for presentation, tend to block chain technology expansion material analysis, dynamic demonstration, case analysis, exercise evaluation, graphic animation, virtual experience, forum exchange, note sharing, data evaluation, results analysis, group interaction and other forms. Based on the background of the Internet of things, the introduction of advanced technology, the characteristics of multimedia, multi-dimensional perspective, the characteristics of subject direction, the changes of course contents, the diversified arts, and the personality of students, teachers decide the form of knowledge construction.

MOOC construction needs a strong, responsible, technical hard, strong sense of responsibility, dare to be the first, constantly enterprising, brave to try the production and management team. In order to ensure the smooth progress of each link, it is necessary to have appropriate and professional work management personnel when the MOOC course is completed. The Management Team should adhere to the principle of integration of MOOC construction and application, insist on the combination of its own construction platform and foreign introduction to expand the advantages of MOOC, and base on the ability and management level of teachers and students of the school, give full play to the universities' advantages of multi-media, multi-disciplinary, multi-disciplinary, multi-modal, multi-lingual and multi-cultural resources, and encourage them to independently develop MOOC courses, to promote MOOC construction, operation, management, evaluation and other processes more humane, systematic, safe, digital, standardized.

(2) Setting up the MOOC mixed model and implementing the resource sharing

According to the students' strong curiosity, they are very adaptable when they accept new things, new models, new media, new environment and new structure. Using the advantages of online MOOC resources to achieve resource sharing, through the block chain technology can better serve offline teaching, not only safe and effective, but also to improve students' autonomous initiative.The introduction of rollover classroom teaching features under blockchain technology, the autonomy of teachers, according to the characteristics of colleges and students, the selection and self-made learning resources, and then link to the resources of MOOC platform, in this way, it is convenient for online and offline moocs to integrate various teaching methods. Based on the establishment of this mixed model, the teaching resources are more diversified, rich and interesting. Teachers and students can work together to develop their strengths and avoid their weaknesses, and give full play to their respective advantages, in order to improve the teaching and learning efficiency of college teachers and students, and further improve the level and quality of education and teaching.

(3) The establishment of production, teaching, learning, research, and use of integrated regional blockchain ecosystem

MOOC construction under blockchain technology can enable teachers and students to absorb the latest, most professional, the most cutting-edge knowledge at home and abroad, so that students can improve their professional, technical, entrepreneurial capabilities, and thus enhance their competitiveness in the workplace. Combining with their own professional background and disciplinary advantages, colleges and universities have set up their own distinctive mooc communication platform to create an integrated block chain ecosystem of production, teaching, learning, research and use, it mainly includes building an innovative MOOC team platform, upgrading the content and management model of MOOC, implementing an innovative mooc drive system, gathering talents from various fields in the professional field, and cultivating research and development teams with strong technology, to guide the strength of all sectors to support moocs certification credits, stimulate the enthusiasm of teachers and students to create and so on, as shown in Table 1.

Table 1. MOOC team builds framework

Production, teaching, learning, research, and use of integrated block chain ecosystem			
Building a MOOC team			
Management Team	R & D team	Operation team	Teaching team
Improve the management system	Technology, art	Operation and maintenance	Senior experts, professionals, art optimization

(4) Supporting moocs and strengthening the policy of credit conversion
The need for national education authorities to encourage the promotion of online moocs and to establish real-time standards for teaching quality and credit recognition, taking into account the individual talent development goals and needs of each school, as well as different teachers, technical support, subject characteristics and management system, in order to ensure the quality of professional teaching, according to the blockchain technology, to develop a safe, reliable, rigorous, decentralized and transparent teaching method that integrates mooc online learning with offline classroom teaching, multiple methods are used to carry out credit verification, achievement assessment, credit conversion, question-answering test and learning process verification. Optimize the content and form of MOOC, turn to the specialized, systematic course, let students develop good habits, a sense of responsibility, put more time and energy into learning.

5 Conclusion

In the educational exploration of MOOC exchange platform construction, blockchain innovation technology is not suitable to tamper with in massive open online course management, learning record, course content, knowledge base, exchange record, academic information, authentication information, etc., the Factor of safety is high and has a wide range of applications. Blockchain makes it possible for universities all over the country to start from a starting line, which can be in advance of the layout of colleges and universities, do a good job in the construction of MOOC, will be able to preempt.It is hoped that the current research results of this topic can provide some research ideas and practical value for the application of blockchain technology in MOOC. Although the idea of building a MOOC communication platform based on blockchain technology has been put forward, as well as the related course production, due to the constraints of time, technology, production environment, cognitive level, and the strength of teachers, there are some deficiencies in curriculum design, resource sharing management, credit recognition and so on. In the future work will focus on technology, knowledge, platform construction and other aspects.

References

1. Zhang, J., Lu, L.: Practical dilemma and elaboration of MOOC out of the existence of internet. J. Inf. Teach. **218**(7), 3539–3552 (2018)
2. Le, C.V., Pardos, Z.A., Meyer, S.D., Thorp, R.: Communication at scale in a MOOC using predictive engagement analytics. J. Teach. Reform. **64**, 78–88 (2018)
3. Shao, M.: Research on the influence of MOOC teaching on the reform of ideological and political education in colleges and universities. Teach. Forum **37**(7), 4867–4880 (2018)
4. Zheng, J.: Discussion on the reform of mixed teaching mode of ideological and political courses in colleges and universities from the perspective of MOOC. J. Educ. Teach. **31**(5), 460–470 (2020)
5. Gamage, D., Perera, I., Fernando, S.: MOOCs lack interactivity and collaborativeness: evaluating MOOC platforms. Int. J. Eng. Pedagog. **63**, 226–235 (2020)
6. Ali, A., Bhat, M.A., Ganaie, S.A.: India on the cyber learning platform: an insight about SWAYAM MOOC platform. INT. J. Inf. Platf. **31**(5), 460–470 (2020)
7. Sunar, A.S., Abbasi, R.A., Davis, H.C., White, S., Aljohani, N.R.: Modelling MOOC learners' social behaviours. Comput. Hum. Behav. **96**(8), 925–932 (2020)
8. Sharov, S., Kolmakova, V., Sharova, T., Kamyshova, T.: Possibilities of the Ukrainian online platform OUM. Int. J. Inf. Educ. Technol. **28**(1), 131–138 (2021)
9. Febrianti, L.Y., Devina, Ningsih, R.Y.: BIPA for business communication learning model development through MOOC. J. Earth Environ. Sci. **35**(4), 309–316 (2021)
10. Tian, Y., Sun, Y., Zhang, L., Qi, W.: Research on MOOC teaching mode in higher education based on deep learning. Comput. Intell. Neurosci. **36**(2), 165–171 (2022)

Analysis and Comparison of Automatic Image Focusing Algorithms in Digital Image Processing

Hong Xiao[1,2(✉)] and Eric Rosales[1,2]

[1] Department of Information Technology, Guangzhou Sontan Polytechnic College, Guangzhou 511300, China
xhongong@126.com

[2] Universidad Autonoma de Zacatacas, Centro Historico, Jardin Juarez 147, 98000 Zacatecas, ZA, Mexico

Abstract. The selection of image focus discrimination function is the basis for obtaining high-quality images in automatic image scene measurement. The performance of several digital image processing algorithms for automatic image focus discrimination is compared comprehensively, and the calculation speed, uniqueness, accuracy and sensitivity of different algorithms are analyzed quantitatively. Firstly, this paper briefly summarizes the imaging principle and focusing principle of digital image processing automatic image focusing, and then from the image information entropy function, gray gradient function, frequency domain evaluation function, and other evaluation functions. Finally, the area selection and focus search algorithm of digital image processing window are described from the aspects of depth of field and focal depth, algorithm selection and algorithm improvement direction. The above analysis results of the characteristics of image focusing discriminant function have guiding significance for the automatic focusing control required by image automatic measurement.

Keywords: Digital image processing · Auto focus · Automatic image measurement · Evaluation function · Focus function

1 Introduction

Automatic image measurement technology can replace the human eye to realize the measurement of target parameters with image as the carrier, and automatic focusing technology is the guarantee to automatically complete the image measurement and make the measurement results accurate and reliable. As the key content of visual instrument research, it is valued by researchers in many fields at home and abroad. Image auto focusing technology is an important technology in digital image processing. If the level of image auto focusing technology is low, the function of digital image processing system will be greatly reduced. Only by improving the sensitivity of image auto focusing technology and reducing its complexity and dispersion, can we really play the role of digital image processing system.

The focus evaluation operation using the digital image processing method is based on the clarity of the object in the captured image. An ideal focus evaluation function should

I. Ahmad et al. (Eds.): STSIoT 2021, LNDECT 122, pp. 574–583, 2023.
https://doi.org/10.1007/978-981-19-3632-6_68

have the characteristics of high sensitivity, single value, no deviation, small amount of calculation and high signal-to-noise ratio. The so-called high sensitivity means that the data to be judged should have obvious numerical changes near the focus; Single value means that the extreme point to be judged should be single; No deviation means that the calculated focus point position is consistent with the actual measured position; Small amount of calculation means simple operation or short operation time; High SNR means that the result of function operation is less affected by irrelevant content. The characteristics of focus evaluation function are related to the selected function. In this paper, several common image focus evaluation algorithms using image processing algorithms are analyzed and compared, and their corresponding discrimination time, sensitivity and single value results are given. This paper consists of the following parts. The first part introduces the relevant background and significance of this paper, the second part is the related work of this paper, and the third part is data analysis. The fourth part is example analysis. The fifth part is conclusion.

2 Related Work

The soot particles produced from diffusion flames burning biodiesel fuel were thermophoretically sampled and the carbon nanostructure of soot particles were imaged using a high resolution transmission electron microscopy (HRTEM) [1]. Lei et. al report a new quantitative measurement method of polarization direction based on the polarization axis finder (PAF) and digital image processing [2]. Choodowicz et. al present an application of a hybrid algorithm for detection and recognition of railway signaling [3]. Various applications of the firefly algorithm in image analysis are also discussed [4]. A digital image processing algorithm based on sampling aerosol inhomogeneities was developed in the applied problem of laser remote sensing for measuring the velocity of wind [5]. Charu et. al focus on the outstanding features of FPGA technology [6]. This algorithm serves to identify the line/edge of the image object to highlight the boundary lines of the image information [7]. Tan et.al [8] shared the calculation method for image processing with new algorithm, according to the algorithm, the image can be calculated by the math algorithm, so as to we can only use the algorithm to handle many images. However for some image analysis, if we use the different algorithm to handle it. As in some images, there are many noises in the image, so the functional analysis is necessary [9]. Because this design focuses on using the opening and closing operation to denoise the binary image, and the edge extraction function to obtain some image edge feature data, it focuses on the theoretical knowledge involved in the opening and closing operation and edge extraction. Automatic image focusing can be realized by ranging method, that is, measuring the distance between the measured object and the imaging surface, or by image gray contrast analysis method. The former is called active mode and the latter is called passive mode. The judgment of image gray contrast can be realized by optical method or by focusing evaluation function method of digital image processing. The automatic focusing operation using ranging method and optical contrast judgment method has been quite mature, but the camera structure using these two methods is complex, and can not be used in some special automatic image measurement occasions, such as short-range image measurement beyond the resolution of ranging method and image analysis of small objects.

2.1 Related Theories and Tools of Image Processing

Firstly, the structural elements are introduced. The structural elements can be regarded as a small graph much smaller than an image to be processed. They are usually widely used in morphological operations, mainly including expansion, corrosion and opening and closing operations. The expression of inflation is defined as follows:

$$D = X \oplus B = \{(x, y)|B_{xy} \cap X \neq \phi\} \tag{1}$$

It should be explained that B is the structural element for expansion operation, X is the image to be expanded, D is the binary image obtained by image x under the expansion of structural element B, and all points in d meet one requirement: for structural element B, when the coordinate point (x, y) coincides with its origin, The intersection of the set of points covered by the structural element B and the binary image x is not empty.

In more popular terms, expansion is to replace the pixel value of the origin of the structural element with the value of the pixel with the largest value in the set of defined pixels. Since the image designed for expansion operation is a binary image, that is, it only includes black-and-white images, that is, the pixel value is only 0 and 255. Therefore, if a white pixel appears in an area covered by the binary image, the value of the pixel at the origin will be replaced by 255, that is, the color will be changed to white. Then, since the edge position of the object is usually a black 0 value, corrosion is equivalent to a contraction of the edge position.

So the role of expansion is obvious. The processing effect of expansion on binary images is that a small part of the black areas in contact with large black areas can be merged into large black areas, which can make the boundary expand outward, and fill the holes in large black areas.

The so-called corrosion, in terms of a relatively easy to understand description, is to replace the pixel value of the origin of the structural element with the value of the pixel with the smallest value in the set of defined pixels. Since the image designed for corrosion operation is a binary image, that is, it only includes black-and-white images, that is, the pixel value is only 0 and 255. Therefore, if a black pixel appears in an area covered by the binary image, the value of the pixel at the origin will be replaced by 0, that is, the color will be changed to black, Otherwise, the value will be replaced by 255, that is, the color will be changed to white. Then, since the edge position of the object is usually a black 0 value, corrosion is equivalent to a contraction of the edge position. Then the effect of corrosion is obvious, that is, to eliminate some scattered points on the edge, so that the edge position shrinks to the interior of the object, so as to eliminate some small and meaningless scattered small objects.

To sum up, expansion and corrosion are mutually inverse processes. The general algorithm flow chart is shown in Fig. 1 below.

2.2 Digital Image Focusing Principle

According to Fourier optics theory, the degree of image clarity or focus is mainly determined by the number of high-frequency components in the light intensity distribution.

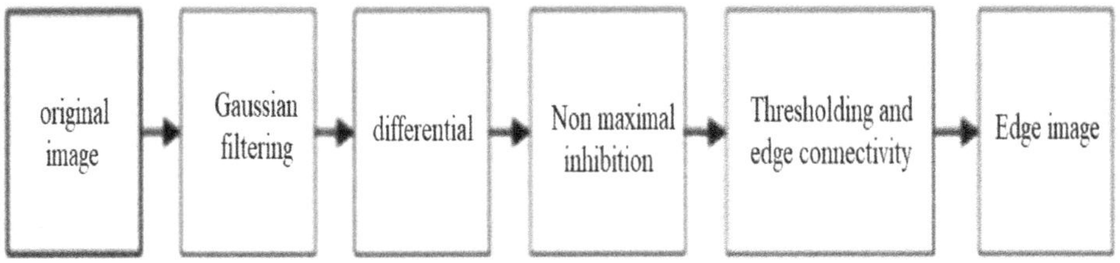

Fig. 1. Candy edge detection algorithm flow chart

If the high-frequency components are small, the image will be blurred, and if the high-frequency components are rich, the image will be clear. Therefore, the content of high-frequency components in the image light intensity distribution can be used as the main basis for the image definition evaluation function. Because the image has edge parts, when the image is fully focused, the image is clear, and the high-frequency components containing edge information are the most: when out of focus, the image is blurred, and the high-frequency components are less. Therefore, whether the image is focused can be determined by the number of high-frequency components of image edge information.

Any optical imaging system can be equivalent to an ideal Gaussian imaging system. According to Newton's imaging formula, the optical system can realize the conjugate between the object plane and the image plane by adjusting any one or more parameters in the object distance, image distance or focal length, that is, imaging. The better the conjugate relation is satisfied, the clearer the image will be, otherwise it is the opposite. Only in the case of correct focusing, the gray contrast everywhere in the image is the strongest, which is the theoretical basis for focusing judgment. In automatic image measurement, the object plane and imaging plane are generally fixed, so conjugate imaging can be realized by adjusting the position of imaging lens. The optical imaging optical path structure is shown in Fig. 2.

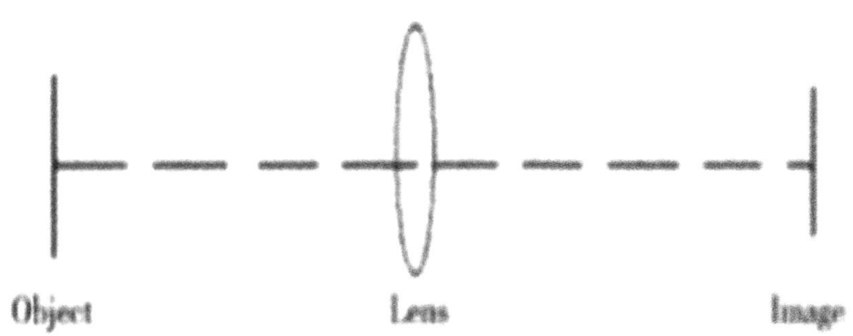

Fig. 2. Sketch of optical imaging

The gray variance of the image represents the degree of dispersion of the gray distribution of image pixels. When the gray value of image pixels in the calculation area changes greatly, the gray variance is also large, and when the gray values of all image pixels in the calculation area are equal, the gray variance is the smallest. When the image is completely blurred, the gray value distribution dispersion of pixels is small and the gray variance is small; When the image is sharp, the dispersion of pixel gray value distribution is large, so the gray variance is large. The gray variance function takes the gray average value of all pixels in the image window as the reference, calculates the difference of the gray value of each pixel, takes the sum of squares, and then standardizes the number of pixels. It represents the average degree of image gray change in the calculation area. Therefore, the gray variance function can characterize the sharpness of the image to a certain extent.

3 Data Analysis

3.1 Principle of Automatic Image Focusing for Digital Image Processing

(1) Principle of imaging
Although automatic image focusing technology is advanced, its imaging principle is basically the same as that of convex lens imaging. The formula of convex lens imaging principle is as follows:

$$\frac{1}{u} + \frac{1}{v} = \frac{1}{f} \tag{2}$$

The meaning of each index in formula (1) is as follows: the meaning of u is the distance between convex lens and object; the meaning of v is the distance between convex lens and imaging plane; the meaning of f is the focal length of convex lens. According to the principle of convex lens imaging, the convex lens imaging model can be obtained as shown in Fig. 3:

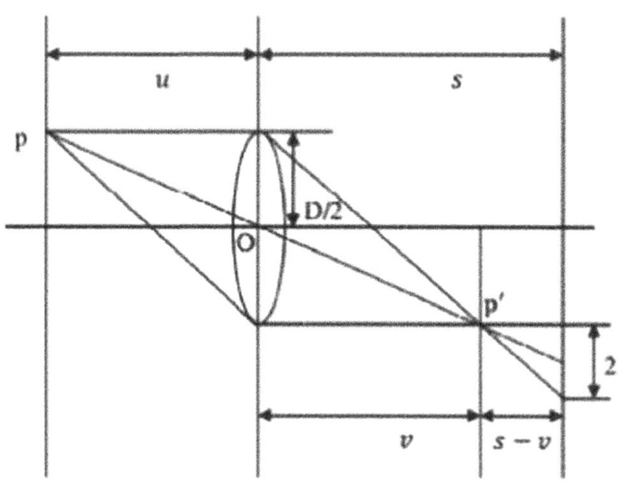

Fig. 3. Image model of convex lens

Figure 3 u, v, f, D, p, R refer to object distance, image distance, focus, convex lens diameter, object position, imaging radius, respectively. When the digital image processing system defocuss, the distance between the imaging and the convex lens will gradually decrease from s to v, object imaging will leave a fuzzy image on the image detector. The distance between the focus plane and the convex lens is s-v.. The distance between the focus plane and the convex lens is If the value of the s-v continues to increase, the image on the image detector will be more blurred. According to the similar triangles in Fig. 1, the imaging scaling factor formula can be obtained as follows:

$$q = \frac{2R}{D} = \frac{s-v}{v} = \left(\frac{1}{v} - \frac{1}{s}\right) \tag{3}$$

Formula (3) q refers to the imaging scaling factor. The following formula can be obtained from the convex lens imaging formula and the imaging scaling factor formula:

$$R = q\frac{D}{2} = S\frac{D}{2}\left(\frac{1}{f} - \frac{1}{u} - \frac{1}{s}\right) \tag{4}$$

Rs > v 0, and when the Rs > vq > is 0 and Rs > vq > q > v, the imaging surface is 0, The formula (4) shows that when the q > is 0 and s > v, the imaging surface is and the imaging surface is in front of the positive focus position. Therefore, digital image processing can realize auto-focusing according to formula (4) principle.

(2) Principle of focusing
The development process of digital image processing is divided into two stages. The first stage mainly adopts the traditional image automatic focusing system, and the second stage mainly adopts the automatic image focusing system. The traditional image automatic focusing principle first adjusts the lens to include the target and then enters the PC machine or embedded system by the CCD/CMOS camera. The embedded system determines whether the lens is readjusted through the motor control module according to the image definition. The auto-image focusing system is divided into two situations: focusing depth and defocusing depth. In focusing depth, the search algorithm is used to focus, then the image processing module is used to determine whether the image is clear or not. Finally, the defocusing depth is calculated by collecting defocusing image parameter information or defocusing image degradation model and fuzzy graphics. Finally, the image definition can be adjusted to the best。

3.2 Evaluation function of automatic image focusing for digital image processing

(1) Information entropy function for images
The formula of image information entropy function is as follows:

$$F = -\sum p_i log_b(pi) \tag{5}$$

The meaning of each index in formula (5) is as follows: the meaning of pi is the probability of characterizing information; the value of b is 2. In digital image processing, the gray level of auto-focusing image is independent, so the probability of representation information of each gray value is different. Based on this, the probability of gray value in gray histogram can be calculated.

(2) Grayscale gradient function
The change of gray scale fluctuation and absolute change of gray scale have a certain function relation with the gray value of a certain point in the image and the pixel of image scale, while the gradient vector mode square function is also related to the change of gray scale fluctuation and absolute change of gray scale. Therefore, the gray gradient vector mode function can be obtained according to gray fluctuation and absolute change:

$$F = \sum_{x}^{M} \sum_{y}^{N} \left\{ [g(x+1, y) - g(x, y)]^2 + [g(x, y+1) - g(x, y)]^2 \right\}^{1/2} \tag{6}$$

The meaning of each index in formula (6) is as follows: the meaning of M*N is image scale pixel;(x,y) is a point in image; the meaning of g (x,y) is the gray value of a point in image.

(3) Frequency domain evaluation function
Based on Fourier transform, the frequency domain evaluation function can be obtained as follows:

$$F = \sum_{X}^{M} \sum_{Y}^{N} \left(\sum_{X}^{M} \sum_{Y}^{N} g(x, y) W_{MN}^{xyXY} \right) - \varphi \tag{7}$$

The meanings of each index in formula (7) are as follows:(x,y) means the spatial coordinate vector of the image;(X,Y) means the coordinate vector of the image in the corresponding spatial frequency domain; and (x,y) means the two-dimensional Fourier transform matrix of the g.W_{MN}^{xyXY}

(4) Function analysis and comparison
After comparing the sensitivity, precision, deviation, complexity, signal-to-noise ratio, time and other parameters of each function, the following conclusions can be obtained: the image information entropy function has long focus time, poor focus position, short focus time, high focus dispersion, and strong focus sensitivity of frequency domain evaluation function. Therefore, the most suitable function for automatic focusing of digital image processing is the frequency domain evaluation function, but the function obtained from Fourier transform is not good in terms of the complexity of the function. Frequency domain evaluation functions need to be further optimized or explored for other functions, such as wavelet analysis, which are also obtained by Fourier transform.

4 Example analysis

4.1 Area Selection and Focus Search Algorithm for Digital Image Processing Window

(1) Effects of depth of field and focal depth
The deeper the depth of field and focal depth of the digital image processing window, the more blurred the image is, the larger the depth of field of the camera window is, the smaller the aperture is, the distance, focusing parameters and imaging clarity will be affected.

(2) Algorithm selection

a. Blind Mountain Climbing Algorithm
The principle of blind mountain climbing algorithm is to judge the position of mountain peak during mountain climbing, which can determine the best focus position of image definition. The algorithm can optimize the automatic image focusing evaluation function of digital image processing, improve the image focusing speed and reduce the deviation of focusing imaging.
b. curve fitting algorithm
The principle of curve fitting algorithm is to synthesize the original complex curve function into the simplest clustering evaluation function by simple function, and then the extreme point of the original curve function can be obtained by the extreme point of the near fitting function. The algorithm can improve the accuracy of the image, but it has certain requirements for the maximum value of the image data.
c. Fibonacci Search Algorithm
Fibonacci search algorithm is a search algorithm, which can use the hypothesis principle to analyze the most suitable points in the process of auto-focusing, and then determine the best auto-focusing interval by theoretical calculation. Although the algorithm can improve the focusing speed, it is easy to appear larger focusing deviation in the process of moving direction change.

(3) Direction of algorithm improvement

a. improve accuracy
Accuracy is one of the criteria for evaluating whether the automatic image focusing evaluation function conforms to the digital image processing. If the imaging image is fuzzy, the minimum gradient value can be adjusted according to the numerical change of gradient value. This can reduce the impact of minimum gradient image definition and improve the accuracy of evaluation function.
b. Improved SNR
Signal-to-noise ratio (SNR) represents the anti-noise interference ability of digital image processing auto-focusing algorithm, and the increase of SNR can reduce the probability of auto-focusing algorithm. Therefore, the image processing can directly take out most of the gradient values, and then use a simple algorithm to bring them into the gradient matrix to calculate, so that the SNR can be improved.

4.2 Analysis of Computational Complexity of Focus Function

The time required for various kinds of processing and operation is different on different levels of computers, but on the same computer, the calculation time required for relatively complex operation must be longer than that for simple operation. In theory, in order to compare the complexity of various operations, we use the addition times corresponding to the corresponding operation to express the complexity of the operation. The operation time involved in the formula is in ascending order: addition/subtraction, multiplication/division, square operation, square operation, logarithm operation and image gray value sorting and comparison operation. In this way, the operation complexity of the above focus evaluation function expressions is arranged in ascending order: F1, F2, F3, F4, F5, F6, F7 and F8, which is also confirmed by the operation experiments of specific images. Because the computer has random external and internal interrupt requests, even the calculation of the same content takes different time each time. In this paper, the average operation time of five operations is taken as the evaluation parameter of algorithm speed. Figure 4 is a corresponding graphical representation.

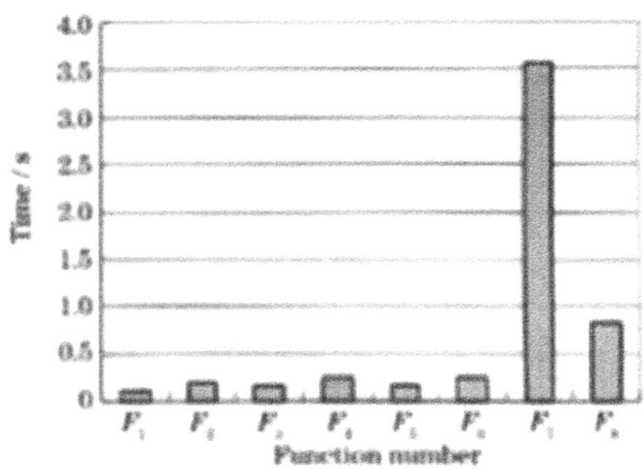

Fig. 4. Image model of convex lens

5 Conclusion

To sum up, this paper mainly analyzes and compares the image information entropy function, gray gradient function, frequency domain evaluation function, other evaluation functions and other digital image processing automatic image focusing algorithms. The image information entropy function has the disadvantages of inaccurate focus position and long focus time. Therefore, the image information entropy function and gray gradient function are not suitable for the automatic image focusing algorithm of digital image processing. Although the frequency domain evaluation function has some advantages, there are still some shortcomings in the automatic focusing time. These digital image processing automatic image focusing evaluation functions can be further improved.

References

1. Choi, S., Park, S.: Quantitative image analysis of carbon nanostructure of particles produced from combustion process. J. Nanosci. Nanotechnol. **26**(6), 1269–1278 (2018)
2. Lei, B., Liu, S.: Efficient polarization direction measurement by utilizing the polarization axis finder and digital image processing. Opt. Lett. **40**(4), 1468–1479 (2018)

3. Choodowicz, E., Lisiecki, P., Lech, P.: Hybrid algorithm for the detection and recognition of railway signs. In: Burduk, R., Kurzynski, M., Wozniak, M. (eds.) Progress in Computer Recognition Systems, pp. 337–347. Springer International Publishing, Cham (2020). https://doi.org/10.1007/978-3-030-19738-4_34
4. Dey, N., Chaki, J., Moraru, L., Fong, S., Yang, XS.: Firefly algorithm and its variants in digital image processing: a comprehensive review. In: Dey, N., (eds.) Applications of Firefly Algorithm and its Variants. Springer Tracts in Nature-Inspired Computing. Springer, Singapore, vol. 17 no(2), pp. 346–355 (2019) https://doi.org/10.1007/978-981-15-0306-1_1
5. Filimonov, P.A., Belov, M.L., Ivanov, S.E., Gorodnichev, V.A., Fedotov, Y.: An algorithm for measuring wind speed based on sampling aerosol inhomogeneities. Comput. Opt. **125**, 116–128 (2020)
6. Kumar, P., Singh, K.: Hardware model for efficient edge detection in images. In: 2020 IEEE International Conference On Computing, Power and Communication Technologies (GUCON), IEEE, pp.1-6 (2020)
7. Putra, A., Sihombing, V., Munandar, M.H.: Rancang bangun aplikasi deteksi tepi citra digital menggunakan algoritma prewitt. J. Digit. Inf. **113**, 189–199 (2021)
8. Tan, Y.L., Wang, H.L., Wang, Y.R.: Calculation of effective mode field area of photonic crystal fiber with digital image processing algorithm. Comput. Opt. **112**, 163–176 (2018)
9. Ibrahim, A., Mohamed, M., Ghoneim, M., Shawkly, M.: Digital image processing algorithm for functional analysis of renal perfusion. Int. Digit. Process. **38**, 15–26 (2021)

Algorithm Design Based on Intelligent Transportation Nonlinear Dynamic Control and Automatic Accident Detection Algorithm

Xudong Yang(✉)

City College, Southwest University of Science and Technology, Mianyang 621000, China
Donyt1689@163.com

Abstract. With the rapid development of social economy and the acceleration of urbanization in China, the scale of urban road network is expanding day by day, and the total mileage of all levels of roads is growing rapidly. As the road with the highest level of technology and service in the urban road network, urban expressway plays the role of the skeleton road network in the urban road system, bearing the traffic demand of high flow and high speed. Once a traffic incident occurs, it is easy to produce traffic congestion. If it is not handled in time, it will also cause secondary accidents, which will have a serious impact on the operation of the whole urban expressway network. In this paper, an algorithm design based on Intelligent Transportation nonlinear dynamic control and automatic accident detection algorithm is proposed to improve the coverage and uncertainty of expressway incident detection.

Keywords: Intelligent transportation · Nonlinear dynamic control · Automatic accident detection

1 Introduction

With the rapid development of China's economy and the continuous acceleration of urbanization, urban expressway has developed rapidly in various cities in China. The construction of urban expressway plays an increasingly important role in alleviating urban traffic congestion, reducing environmental pollution and promoting economic development.

Automatic traffic event detection (AID) is one of the important functions of traffic management and control system. In the past 40 years, many experts and scholars both at home and abroad have been devoted to the research of various traffic event detection methods. The traditional automatic detection algorithm of traffic events is mostly based on fixed detector data (such as the data of loop detector), and the development is relatively perfect. But the fixed detector can only collect the traffic data (such as occupancy, speed and flow) of the point, and it is difficult to ensure whether the data based on the point represents the real traffic condition; On the other hand, the layout distance, location and communication of fixed detector are also the key factors affecting the performance

I. Ahmad et al. (Eds.): STSIoT 2021, LNDECT 122, pp. 584–593, 2023.
https://doi.org/10.1007/978-981-19-3632-6_69

of the algorithm. In practical application, the fixed detector data often lacks data for a long time due to communication failure and equipment failure, which can not guarantee the smooth progress of event detection. In addition, the installation and maintenance of detectors will hinder the normal operation of road traffic, sometimes it is necessary to close the road, which will cause inconvenience to the giver. How to detect and confirm the time, place and nature of the event accurately and timely is the key technology for the successful operation of traffic management and control system. The performance of the automatic detection algorithm for traffic events is the core of traffic event management system, and also one of the important evaluation indexes for the successful operation of intelligent transportation system.

2 Related Work

Guerrero-Ibáñez et al. discuss how sensor technology can be integrated with the transportation infrastructure to achieve a sustainable Intelligent Transportation System (ITS) and how safety, traffic control and infotainment applications can benefit from multiple sensors deployed in different elements of an ITS [1]. Several case studies of big data analytics applications in intelligent transportation systems, including road traffic accidents analysis, road traffic flow prediction, public transportation service plan, personal travel route plan, rail transportation management and control, and assets maintenance are introduced [2]. Intelligent transportation is an emerging technology that integrates advanced sensors, network communication, data processing, and automatic control technologies to provide great convenience for the daily lives [3]. Patel et al. survey a set of solutions available in the literature to design of an ITS system using IoT along with challenges and future scope for the improvement of the existing solutions [4]. Veres et al. present a survey that highlights the role modeling techniques within the realm of deep learning have played within ITS [5]. In order to meet the demands of the intelligent transportation big data processing, this paper puts forward a high performance computing architecture of large-scale transportation video data management based on cloud computing, designs a parallel computing model containing the distributed file system and distributed computing system to solve the problems such as flexible server increase or decrease, load balancing and flexible dynamic storage increase or decrease, computing power and great improvement of storage efficiency [7]. Other influential work includes Refs [7–10].

This paper consists of the following parts. The first part introduces the related background and significance of this paper, the second part is the related work of this paper, and the third part is data analysis. The fourth part is example analysis. The fifth part is conclusion.

Nonlinear dynamic control is a method to study the feedback linearization design of general nonlinear control systems through the concept of "inverse" of dynamic systems. The application research shows that dynamic inversion is a more effective method in nonlinear control, and has good tracking performance for nonlinear rigid spacecraft system. However, the dynamic inverse method is sensitive to the modeling error, and how to improve the robustness of the controller has always been a difficult problem to solve.

2.1 Traditional Traffic Incident Detection Technology

At present, China is actively carrying out its research, hoping to realize the systematization and informatization of expressway management. According to the classification provided in the professional magazine path published by the IT center of the University of California, Berkeley, the road traffic flow parameters can be divided into: vehicle counting, vehicle type identification and accident detection. Automatic incident detection (AID) is one of the main research directions of intelligent transportation. It mainly focuses on the macro road traffic flow information, such as road flow, vehicle occupancy, vehicle flow density, etc. the micro phenomenon of traffic accidents can be detected indirectly.

According to the different sources of traffic flow information, the methods widely studied at present mainly include the following: aid based on ground induction coil; Aid based on global positioning system (GPS) signal; Aid based on video signal. The ground induction coil detects the traffic flow information according to the principle of the change of magnetic field intensity. When the vehicle passes through, the magnetic field intensity of the coil changes, and the parameters such as vehicle speed and flow can be obtained. These parameters can be used to identify traffic events such as vehicle suspension and traffic flow reduction, so as to indirectly detect traffic accidents. During the installation and maintenance of ground induction coil, the road must be excavated to block the traffic, which is very inconvenient and the detection parameters are limited.

GPS uses the triangulation principle of multiple satellites to work, and can accurately locate the geographical coordinates of GPS receiving terminal. With the progress of GPS technology, it has been more and more applied to the field of transportation. By installing GPS receiving terminal on the vehicle, the monitoring center can obtain the vehicle position information in real time, and calculate the vehicle motion information according to the position information at different times, so as to realize the detection of traffic events. Video detection is more and more widely used in traffic monitoring. Through the background modeling, moving target extraction and classification, moving target tracking and so on, the vehicle information, including vehicle shape, size, driving condition and traffic flow, can be detected. Finally, the vehicle behavior is analyzed by the methods of pattern matching and state estimation. However, because video detection is easily affected by difficulties such as weather conditions, object occlusion and color similarity between the target and the environment, the detection rate is not high. The analysis of vehicle behavior can only be aimed at simple situations and the accuracy is not high. Video based traffic flow detection technology is relatively mature, so the current video based aid technology still uses the macro information of traffic flow for indirect detection.

2.2 Overview of Linear Dynamic Control

In the local sense, a. Isidori and others consider the nonlinear regulation problem of nonlinear system under the assumption that the external system is Poisson stable, and the internal system dynamic index can stabilize the assumption of nonlinear regulation. The necessary conditions for the general local tracking of unrestricted external signals are obtained by J w. grizle, It is worth noting that local nonlinear regulation cannot track the unbounded external signals, which is essentially different from linear systems. In order to realize the tracking of unstable (unbounded) external signals, the adjustment problem in the global sense must be considered. In the global sense, the models considered by M. D. dayawansa and a. r.tee are as follows:

$$\begin{cases} x = \varphi(x, y) \\ y_1 = y_2 \\ \dots \\ y_m = \alpha(x, y) + \beta(x, y)u \end{cases} \tag{1}$$

In recent years, topological Photonics and non Hermite optics have become the two most active emerging research fields of photonics. The concept of topology originally came from mathematics and was used to study the properties of geometric shapes that remain unchanged under continuous deformation. For example, if a doughnut is not torn, no matter how it expands, rubs or contracts, it cannot be equivalent to a solid ball. The most famous topological invariant in topology is the "Chen number" named after Mr. Chen Shengshi of Nankai University. The development of topological photonics originated from the study of topological states in condensed matter physics. At first, the concept of topology was introduced into physical science to explain the famous quantum Hall effect. Therefore, the 2016 Nobel Prize in physics was also awarded to pioneer scientists in topological materials research.

Subsequently, the concept of topology was extended to the fields of optics, acoustics, metamaterials and cold atomic systems, which greatly promoted the development of topological physics. Especially in the field of optics. Topological photonics has gradually become an important frontier and cross field in optics and related scientific fields from the initial unidirectional transmission electromagnetic wave topological state experiment to the recent topological laser. On the other hand, the concept of non Hermite comes from quantum mechanics. It is generally believed that non Hermite systems have no physical meaning, and the introduction of parity time Pt symmetry has changed people's traditional understanding of non Ö Mi open systems. When the concept of Pt symmetry in non Hermite quantum mechanics is introduced into the optical field, the carefully designed IPS symmetry with reciprocal loss and easy to control system continues to bring new discoveries. The development of non Hermite optics also brings new prospects for a series of application technologies, such as sensing and detection, wireless transmission energy and single-mode laser.

Due to the difficulties in experiment and theory, most studies of topological Photonics and non Hermite optics in the past were carried out by an Shengye, almost focusing on the on-line effect. However, nonlinear effects can be found everywhere in both the classical world and the quantum world. The diversity of the natural world also promotes the development of Applied Science. For example, nonlinear response is the key to the powerful function of digital electronic technology. It is the fundamental reason why artificial neural network can perform complex operations and the basis for the development of many new photonics technologies. Until recently, it has been found that there are many interesting phenomena when considering nonlinearity in optical topological systems, such as topological optical solitons, topological lasers and nonlinear topological insulators. However, the "marriage" between topology and non Hermite has just begun. For complex systems with topological and non Hermitian characteristics, the research on nonlinear effects is almost blank. Even in the field of optics, it has not found or built an adjustable nonlinear non Hermite topology photonics experimental platform.

2.3 Global Adjustment of Nonlinear Dynamic Control System

Considering that the external dynamic control system $w = r(w)$ is free and uncontrolled, if the solution flow of the system is $\varphi_t^r(w)$, $\Omega = \{w|\varphi_t^r(w_0)$ is a limit point and $w_0 \in R^r\}$, then q is the invariant manifold of the external dynamic control system, and any bounded solution w(T) of the system has $w(t) \to \Omega(t \to \infty)$, so the following proposition can be obtained.

The necessary conditions for global adjustment problem to be solvable by state feedback are: existence of maps $c(w) \in L$ and $S(w) \in C'$, which makes

$$\begin{cases} \frac{\partial S(w)}{\partial w} r(w) = f(S(w)), w, c(w)) \\ h(S(w), w) = 0 \end{cases} \tag{2}$$

Due to the difficulties in experiment and theory, most studies of topological Photonics and non Hermite optics in the past were carried out by an Shengye, almost focusing on the on-line effect. However, nonlinear effects can be found everywhere in both the classical world and the quantum world. The diversity of the natural world also promotes the development of Applied Science. For example, nonlinear response is the key to the powerful function of digital electronic technology. It is the fundamental reason why artificial neural network can perform complex operations and the basis for the development of many new photonics technologies. Until recently, it has been found that there are many interesting phenomena when considering nonlinearity in optical topological systems, such as topological optical solitons, topological lasers and nonlinear topological insulators. However, the "marriage" between topology and non Hermite has just begun. For complex systems with topological and non Hermitian characteristics, the research on nonlinear effects is almost blank. Even in the field of optics, it has not found or built an adjustable nonlinear non Hermite topology photonics experimental platform.

To address this gap, researchers at Nankai University used the self-developed continuous laser direct writing technology to prepare non Hermite topological photonic lattices in nonlinear crystals for the first time, realizing the regulation of parity time and non Hermite topology. The influence of nonlinear effect and the resistance of sensitivity and robustness between outliers and singular points further reveal the nonlinear effect. The results show that the local nonlinear effect can affect and change the overall Pt symmetry of the system, resulting in the emergence and disappearance of topology and abnormal non Hermitian singularity dynamic control. This result changes people's understanding of the interaction of multiple characteristics in nonlinear complex systems, and provides a new research direction for topological Photonics and non Hermite optics.

3 Data Analysis

3.1 Data Analysis and Selection of Input Parameters

According to the data base of Sect. 3, the microwave detector data on Beijing Expressway can collect the traffic, speed, occupancy and traffic volume of a certain point on the road.

Among them, the flow, speed and occupancy can be used to describe the characteristics of traffic flow. The changing law of these parameters can reflect the operation state of traffic flow. When the traffic flow is in the normal and stable state, the change of traffic parameters is relatively stable or not obvious; When traffic events occur and affect the upstream detector, the traffic parameters detected by the upstream detector change obviously, and the traffic parameters detected by the downstream detector are not obvious. Therefore, the event detection can be carried out by considering the change rate of traffic parameters with time and the change rate of upstream and downstream traffic parameters.

3.2 Bayesian Algorithm

In the method of statistical analysis, the method of discriminant analysis is used to establish a better discriminant function according to a batch of samples with clear classification, so that the cases of misjudgment are the least. Then, for a given new sample, it can be judged which population it comes from. The main methods include Fisher, Bayesian, distance and so on. Among them, Bayesian discriminant thought is to calculate the posterior probability according to the prior probability and make statistical inference based on the distribution of posterior probability. The so-called prior probability is to describe the degree of people's understanding of the object studied in advance by probability; The so-called posterior probability is the probability calculated according to the specific data, prior probability and specific discrimination rules. It is the result of the correction of prior probability. Because Bayesian discriminant method considers the loss after misjudgment, it has certain superiority. Here, Bayesian discriminant analysis method is adopted.

$$
\begin{aligned}
OCCRDF &= \frac{OCC(i,t) - OCC(i+1,t)}{OCC(i,t)} \\
VOLRDF &= \frac{VOL(i+1,t) - VOL(i,t)}{VOL(i,t)}
\end{aligned} \tag{3}
$$

3.3 Improved Algorithm Based on Multi Parameters

California algorithm is the most classic and practical algorithm based on Discriminant recognition. It has been used as a comparison algorithm of other newly developed algorithms. The only disadvantage of this algorithm is that it only uses one traffic parameter of occupancy rate, and only one parameter is easy to cause high misjudgment rate, In this study, the occupancy rate, the change rate of vehicle speed with time and the change rate of upstream and downstream are used as the judgment conditions. This paper uses the improved California algorithm based on multi parameters to judge the relative difference of the upstream and downstream occupancy rate, the relative difference of the upstream and downstream speed, the change rate of the upstream occupancy rate with time, and whether the change rate of the upstream speed with time is greater than the specified threshold to give an event alarm. The flow chart of the algorithm is shown in Fig. 1

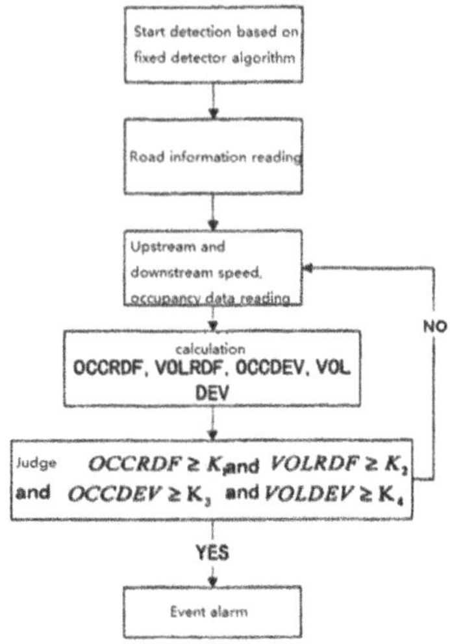

Fig. 1. Flow chart of multi parameter discrimination algorithm based on fixed detector

4 Example Analysis

4.1 Algorithm

This paper first introduces the relatively mature normal deviation method (SND). The normal deviation method uses the arithmetic mean of the traffic parameter values of the N sampling periods before the time t as the prediction value of the traffic parameter at the time t, and then uses the standard normal deviation to measure the change degree of the parameter in time. When it exceeds the corresponding threshold, the alarm will be triggered. Based on the analysis of the travel speed of expressways in the previous section, if only considering the change of traffic parameters in time dimension, it will cause false alarm in morning and evening peak, resulting in a high false alarm rate. Therefore, the space-time two-dimensional discrimination algorithm based on floating car proposed in this paper is as follows:

(1) From the time dimension, we first judge the speed value, then use the arithmetic mean of the driving speed of the N sampling periods before the time t to predict the traffic parameter value at the time t, and then use the standard normal deviation to measure the change degree of the driving speed relative to its previous average value. Algorithm is shown in Fig. 2.

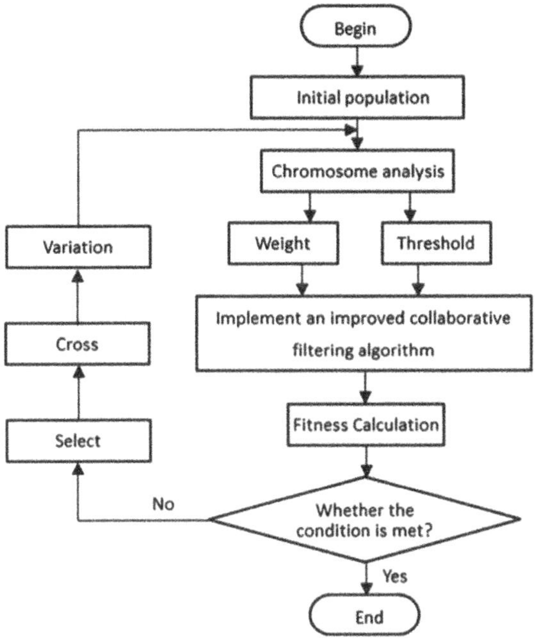

Fig. 2. Algorithm diagram

Let the actual value of the driving speed at time t be $v\,(t)$, and the actual values of the traffic parameters in the n sampling periods before time t are $v(t-n)$, $v(t-n+1)$,… $v\,(t-1)$.

$$\begin{aligned} & v(t) \le K_1 \\ & SND(t) = \frac{\bar{v}(t) - v(t)}{S} \ge K_2 \end{aligned} \tag{4}$$

(2) From the spatial dimension, based on the drastic change of the upstream and downstream driving speed when the event occurs, the following formula is used for discrimination;

$$VRDF(t)\frac{V(i+1,t) - V(i,t)}{V(i,t)} \ge K_3 \tag{5}$$

$$\begin{aligned} E(t)\dot{x}_{d+1}(t) - E(t)\dot{x}_{k+1}(t) &= E(t)\Delta\dot{x}_{k+1}(t) = f(t, x_d(t)) + B(t)u_d(t) - f(t, x_k(t)) \\ &- B(t)u_k(t) = f(t, x_d(t)) - f(t, x_{k+1}(t)) + B(t)\Delta u_{k+1}(t) \end{aligned} \tag{6}$$

$$\|\Delta x_{k+1}(t)\| \le \left(pk_f + m_2 + m_3\right)\int_0^t \Delta x_{k+1}(\tau)d\tau + \int_0^t (m_1\|\Delta u_k(\tau)\| + pd)d\tau \tag{7}$$

4.2 Algorithm Effectiveness Analysis

There are two methods to test and verify the event detection algorithm, one is based on simulation data, the other is based on actual data. However, under the simulation condition, the traffic condition is ideal, which is far from the real situation. Therefore, the algorithm verification based on the measured data is carried out, that is, the event detection algorithm is verified by collecting and processing the floating car detection data, fixed detector data and real event information in Beijing, and different threshold combinations are used to detect the algorithm, On the premise of ensuring a certain error rate, improve the detection rate, so as to determine the national value of the algorithm, and obtain the detection effect of the multi parameter discrimination algorithm based on fixed detector and the spatiotemporal two-dimensional discrimination algorithm based on floating car.

In 1996, through a survey of the traffic management center of the United States, abdulhai proposed that the acceptable average indicators of incident detection were Dr Z 88% and far s 1.8%. This index is also called TMC acceptable index. Here, this paper takes this as the index of the effectiveness analysis of the algorithm. If the algorithm meets this requirement, the algorithm can meet the needs of practical application (Fig. 3).

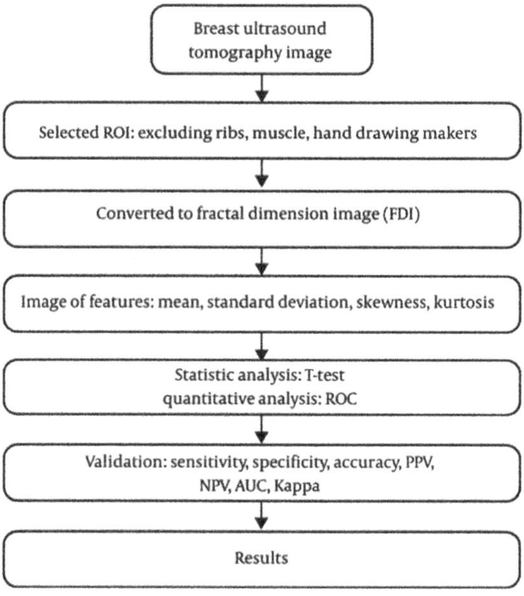

Fig. 3. Algorithm effectiveness analysis

5 Conclusions

Due to the fixed detector spacing, communication failure caused by the serious lack of data and the number of floating cars and other issues, if the simple use of a single data source for expressway incident detection, the effect is not ideal, often resulting in long-term interruption of detection, can not smoothly carry out detection and other issues. Therefore, in this paper, the fixed detector data and floating car data are effectively combined, and a good detection algorithm is proposed respectively. Finally, the D-S theory is applied to the fusion of algorithm results, which can effectively solve the problems of low coverage and reliability of single data source event detection algorithm. Finally, the effectiveness of the algorithm is analyzed with the data of Beijing Expressway. It is expected that the algorithm proposed in this study can provide some reference for automatic event detection of urban expressway.

References

1. Sumalee, A., Ho, H.W.: Smarter and more connected: future intelligent transportation system. IATSS Res. **42**(2), 67–71 (2018)
2. Guerrero-Ibáñez, J., Zeadally, S., Contreras-Castillo, J.: Sensor technologies for intelligent transportation systems. Sensors **18**(4), 1212 (2018)
3. Zhu, L., Yu, F.R., Wang, Y., Ning, B., Tang, T.: Big data analytics in intelligent transportation systems: a survey. IEEE Trans. Intell. Transp. Syst. **20**(1), 383–398 (2018)
4. Gao, H., Huang, W., Yang, X.: Applying probabilistic model checking to path planning in an intelligent transportation system using mobility trajectories and their statistical data. Intell. Automat. Soft Comput. **25**(3), 547–559 (2019)
5. Li, Q., Wang, F., Wang, J., Li, W.: LSTM-based SQL injection detection method for intelligent transportation system. IEEE Trans. Veh. Technol. **68**(5), 4182–4191 (2019)
6. Patel, P., Narmawala, Z., Thakkar, A.: A survey on intelligent transportation system using internet of things. In: Shetty, N.R., Patnaik, L.M., Nagaraj, H.C., Hamsavath, P.N., Nalini, N. (eds.) Emerging Research in Computing, Information, Communication and Applications. AISC, vol. 882, pp. 231–240. Springer, Singapore (2019). https://doi.org/10.1007/978-981-13-5953-8_20
7. Veres, M., Moussa, M.: Deep learning for intelligent transportation systems: a survey of emerging trends. IEEE Trans. Intell. Transp. Syst. **21**(8), 3152–3168 (2019)
8. Hao, Q., Qin, L.: The design of intelligent transportation video processing system in big data environment. IEEE Access **8**, 13769–13780 (2020)
9. Guevara, L., Auat Cheein, F.: The role of 5G technologies: challenges in smart cities and intelligent transportation systems. Sustainability **12**(16), 6469 (2020)
10. Zhang, D., Wang, J., Fan, H., Zhang, T., Gao, J., Yang, P.: New method of traffic flow forecasting based on quantum particle swarm optimization strategy for intelligent transportation system. Int. J. Commun. Syst. **34**(1), e4647 (2021)

Application of Big Data in Management Information System

Xudong Yang[1,2](✉) and Chadi Altrjman[1,2]

[1] City College, Southwest University of Science and Technology, Mianyang 621000, China
Donyt1689@163.com
[2] Near East University, North Nicosia, Turkey

Abstract. With the rapid development of social economy and the acceleration of urbanization in China, the scale of urban road network is expanding day by day, and the total mileage of all levels of roads is growing rapidly. As the road with the highest level of technology and service in the urban road network, urban expressway plays the role of the skeleton road network in the urban road system, bearing the traffic demand of high flow and high speed. Once a traffic incident occurs, it is easy to produce traffic congestion. If it is not handled in time, it will also cause secondary accidents, which will have a serious impact on the operation of the whole urban expressway network. In this paper, an algorithm design based on Intelligent Transportation nonlinear dynamic control and automatic accident detection algorithm is proposed to improve the coverage and uncertainty of expressway incident detection.

Keywords: Intelligent transportation · Nonlinear dynamic control · Automatic accident detection

1 Introduction

The anomaly detection method of ad system is to judge whether there are abnormal events by the algorithm according to the collected data information. The performance of automatic traffic incident detection algorithm is the core of traffic incident management system and one of the important evaluation indexes for the successful operation of intelligent transportation system. Common traffic j anomaly detection algorithms include state recognition algorithm, statistical prediction algorithm, catastrophe theory algorithm and high-level event detection technology, It depends on the preset threshold of relevant key parameters, and identifies the abnormal changes of traffic state through the abnormal mutation of key parameters beyond the threshold, so as to finally find the traffic accidents. For example, California algorithm gives an alarm when the three eigenvalues calculated by the upstream and downstream occupancy exceed the threshold at the same time. As a parameter statistical prediction algorithm for comparing and evaluating the new model, California model generally makes a short-term prediction of the future traffic conditions according to the historical data, and then compares the real-time collected data with the predicted value. If there are large changes between the two, it is considered that there

I. Ahmad et al. (Eds.): STSIoT 2021, LNDECT 122, pp. 594–603, 2023.
https://doi.org/10.1007/978-981-19-3632-6_70

is an accident. The characterization of traffic parameters to traffic state depends on the speed flow (or occupancy flow) relationship diagram. Such as McMaster algorithm. This paper consists of the following parts. The first part introduces the relevant background and significance of this paper, the second part is the related work of this paper, and the third part is data analysis. The fourth part is example analysis. The fifth part is conclusions.

2 Related Work

The article is devoted to the description of the developed software system of risk management [1]. The subject of Ref [2] was to determine how much the application of the industrial revolution 4.0 management information system influenced innovation. Tarigan et al. examine the impact of information system management implementation on the company performance with the mediating role of process innovation and process innovation [3]. Muslih et al. discuss the effectiveness of marriage services through Marriage Management Information System (SIMKAH) at Palu city religious court [4]. The process of supply chain management information system and the key technology of block chain are analysed, and the collaborative mechanism of supply chain management information system from the perspective of block chain is proposed, including the process and consensus collaborative management mechanism, which optimizes the transaction process management and block chain system consensus, accounting and so on [5]. Maqsudov et al. present the general results obtained during the automation of document management (workflow of documents) in university based on a management information system of Khujand Polytechnic institute of Tajik technical university (KPITTU) [6]. In order to solve the problems of low efficiency of management information systems and low utilization rate of the information resources, Hu et al. propose the designs of an exhibition management information system by using a B/S structure (Browser/Server mode) [7]. The personnel management information system designed and implemented integrates the functions of batch import of information, instant update of personnel information, position transfer, information inquiry and statistical analysis [8]. Other influential work includes [9, 10].

The algorithm mainly includes time series method, standard normal deviation method, double exponential smoothing method, filtering model method and Bayesian catastrophe theory algorithm. It is based on the sudden change of the traffic state represented by the traffic parameters when the traffic accident occurs. At the same time, the accident is finally confirmed by comparing the different traffic States of adjacent sections of the road at the same time.

2.1 Analysis of Big Data Characteristics in Enterprises

(1) Object of enterprise big data analysis

According to the analysis of business activities within the enterprise, the big data formats managed within the enterprise generally include structured data and unstructured data. Structured data generally refers to the data with unified format, fixed fields and can be defined based on relational database. These data have been well managed in their respective information systems. The other is unstructured data. This kind of data has no fixed format, or its format often changes. It can not be formatted with some fixed methods. These data are generally random and easy to be ignored by enterprises in general. Unstructured data generally has the characteristics of wide distribution, diversified formats and large amount of data, which poses a challenge to the collection, processing and storage of big data.

(2) Significance of unstructured data analysis

Although these unstructured data have no direct relationship with the daily operation and specific business of the enterprise, the analysis of these unstructured data may bring indirect value to the enterprise. For example, we can get some information that reflects the user status, operation and management status or employee behavior of the enterprise. Through this information, we can provide reference for the operation and management decision-making of the enterprise. The value of these data comes from the generation point of the data, which is the key point of the enterprise information system. For example, users access the enterprise's product information by accessing the enterprise portal website, and these access behaviors are recorded in the log file of the enterprise website. The enterprise obtains the information of the user's access behavior by analyzing the above files, so as to obtain the specific data of different product concerns. For example, if an enterprise finds that the attention level of a product suddenly decreases in a certain period, it can analyze and deal with it accordingly, such as reducing the production plan of the product, otherwise it can improve the production plan of the product. This shows that unstructured data has reference value for enterprise production and operation decision-making, but at present, most of the data are ignored or even discarded.

2.2 Overview of Linear Dynamic Control

In the local sense, a. Isidori and others consider the nonlinear regulation problem of nonlinear system under the assumption that the external system is Poisson stable, and the internal system dynamic index can stabilize the assumption of nonlinear regulation. The necessary conditions for the general local tracking of unrestricted external signals are obtained by J w. grizle, It is worth noting that local nonlinear regulation cannot track the unbounded external signals, which is essentially different from linear systems. In order to realize the tracking of unstable (unbounded) external signals, the adjustment problem in the global sense must be considered. In the global sense, the models considered by M. D. dayawansa and a. r.tee are as follows:

$$\begin{cases} x = \varphi(x, y) \\ y_1 = y_2 \\ \cdots \\ y_m = \alpha(x, y) + \beta(x, y)u \end{cases} \tag{1}$$

In recent years, topological Photonics and non Hermite optics have become the two most active emerging research fields of photonics. The concept of topology originally came from mathematics and was used to study the properties of geometric shapes that remain unchanged under continuous deformation. For example, if a doughnut is not torn, no matter how it expands, rubs or contracts, it cannot be equivalent to a solid ball. The most famous topological invariant in topology is the "Chen number" named after Mr. Chen Shengshi of Nankai University. The development of topological photonics originated from the study of topological states in condensed matter physics. At first, the concept of topology was introduced into physical science to explain the famous quantum Hall effect. Therefore, the 2016 Nobel Prize in physics was also awarded to pioneer scientists in topological materials research.

Subsequently, the concept of topology was extended to the fields of optics, acoustics, metamaterials and cold atomic systems, which greatly promoted the development of topological physics. Especially in the field of optics. Topological photonics has gradually become an important frontier and cross field in optics and related scientific fields from the initial unidirectional transmission electromagnetic wave topological state experiment to the recent topological laser.. On the other hand, the concept of non Hermite comes from quantum mechanics. It is generally believed that non Hermite systems have no physical meaning, and the introduction of parity time Pt symmetry has changed people's traditional understanding of non Ö Mi open systems. When the concept of Pt symmetry in non Hermite quantum mechanics is introduced into the optical field, the carefully designed IPS symmetry with reciprocal loss and easy to control system continues to bring new discoveries. The development of non Hermite optics also brings new prospects for a series of application technologies, such as sensing and detection, wireless transmission energy and single-mode laser.

Due to the difficulties in experiment and theory, most studies of topological Photonics and non Hermite optics in the past were carried out by an Shengye, almost focusing on the on-line effect. However, nonlinear effects can be found everywhere in both the classical world and the quantum world. The diversity of the natural world also promotes the development of Applied Science. For example, nonlinear response is the key to the powerful function of digital electronic technology. It is the fundamental reason why artificial neural network can perform complex operations and the basis for the development of many new photonics technologies. Until recently, it has been found that there are many interesting phenomena when considering nonlinearity in optical topological systems, such as topological optical solitons, topological lasers and nonlinear topological insulators. However, the "marriage" between topology and non Hermite has just begun. For complex systems with topological and non Hermitian characteristics, the research on nonlinear effects is almost blank. Even in the field of optics, it has not found or built an adjustable nonlinear non Hermite topology photonics experimental platform.

2.3 Global Adjustment of Nonlinear Dynamic Control System

Considering that the external dynamic control system $w = r\ (w)$ is free and uncontrolled, if the solution flow of the system is $\varphi_t^r(w)$, $\Omega = \left\{\mathrm{w}|\varphi_\mathrm{t}^\mathrm{r}(\mathrm{w}_0)\text{is a limit point and } \mathrm{w}_0 \in \mathrm{R}^\mathrm{r}\right\}$, then q is the invariant manifold of the external dynamic control system, and any bounded solution w (T) of the system has $w\ (t) \rightarrow \Omega\ (t \rightarrow \infty)$, so the following proposition can be obtained.

The necessary conditions for global adjustment problem to be solvable by state feedback are: existence of maps $c\ (w) \in L$ and $S\ (w) \in C'$, which makes

$$\begin{cases} \frac{\partial S(w)}{\partial w} r(w) = f\,(S(w)), w, c(w)) \\ h(S(w), w) = 0 \end{cases} \tag{2}$$

Due to the difficulties in experiment and theory, most studies of topological Photonics and non Hermite optics in the past were carried out by an Shengye, almost focusing on the on-line effect. However, nonlinear effects can be found everywhere in both the classical world and the quantum world. The diversity of the natural world also promotes the development of Applied Science. For example, nonlinear response is the key to the powerful function of digital electronic technology. It is the fundamental reason why artificial neural network can perform complex operations and the basis for the development of many new photonics technologies. Until recently, it has been found that there are many interesting phenomena when considering nonlinearity in optical topological systems, such as topological optical solitons, topological lasers and nonlinear topological insulators. However, the "marriage" between topology and non Hermite has just begun. For complex systems with topological and non Hermitian characteristics, the research on nonlinear effects is almost blank. Even in the field of optics, it has not found or built an adjustable nonlinear non Hermite topology photonics experimental platform.

To address this gap, researchers at Nankai University used the self-developed continuous laser direct writing technology to prepare non Hermite topological photonic lattices in nonlinear crystals for the first time, realizing the regulation of parity time and non Hermite topology. The influence of nonlinear effect and the resistance of sensitivity and robustness between outliers and singular points further reveal the nonlinear effect. The results show that the local nonlinear effect can affect and change the overall Pt symmetry of the system, resulting in the emergence and disappearance of topology and abnormal non Hermitian singularity dynamic control. This result changes people's understanding of the interaction of multiple characteristics in nonlinear complex systems, and provides a new research direction for topological Photonics and non Hermite optics.

3 Data Analysis

3.1 Data Analysis and Selection of Input Parameters

According to the data base of Chapter 3, the microwave detector data on Beijing Expressway can collect the traffic, speed, occupancy and traffic volume of a certain point on the road. Among them, the flow, speed and occupancy can be used to describe the characteristics of traffic flow. The changing law of these parameters can reflect the operation state of traffic flow. When the traffic flow is in the normal and stable state, the change of traffic parameters is relatively stable or not obvious; When traffic events occur and affect the upstream detector, the traffic parameters detected by the upstream detector change obviously, and the traffic parameters detected by the downstream detector are not obvious. Therefore, the event detection can be carried out by considering the change rate of traffic parameters with time and the change rate of upstream and downstream traffic parameters.

3.2 Bayesian Algorithm

In the method of statistical analysis, the method of discriminant analysis is used to establish a better discriminant function according to a batch of samples with clear classification, so that the cases of misjudgment are the least. Then, for a given new sample, it can be judged which population it comes from. The main methods include Fisher, Bayesian, distance and so on. Among them, Bayesian discriminant thought is to calculate the posterior probability according to the prior probability and make statistical inference based on the distribution of posterior probability. The so-called prior probability is to describe the degree of people's understanding of the object studied in advance by probability; The so-called posterior probability is the probability calculated according to the specific data, prior probability and specific discrimination rules. It is the result of the correction of prior probability. Because Bayesian discriminant method considers the loss after misjudgment, it has certain superiority. Here, Bayesian discriminant analysis method is adopted.

$$\begin{aligned} OCCRDF &= \frac{OCC(i,t) - OCC(i+1,t)}{OCC(i,t)} \\ VOLRDF &= \frac{VOL(i+1,t) - VOL(i,t)}{VOL(i,t)} \end{aligned} \tag{3}$$

3.3 Improved Algorithm Based on Multi Parameters

California algorithm is the most classic and practical algorithm based on Discriminant recognition. It has been used as a comparison algorithm of other newly developed algorithms. The only disadvantage of this algorithm is that it only uses one traffic parameter of occupancy rate, and only one parameter is easy to cause high misjudgment rate, In this study, the occupancy rate, the change rate of vehicle speed with time and the change rate of upstream and downstream are used as the judgment conditions. This paper uses the improved California algorithm based on multi parameters to judge the relative difference of the upstream and downstream occupancy rate, the relative difference of the upstream and downstream speed, the change rate of the upstream occupancy rate with time, and whether the change rate of the upstream speed with time is greater than the specified threshold to give an event alarm. The flow chart of the algorithm is shown in Fig. 1.

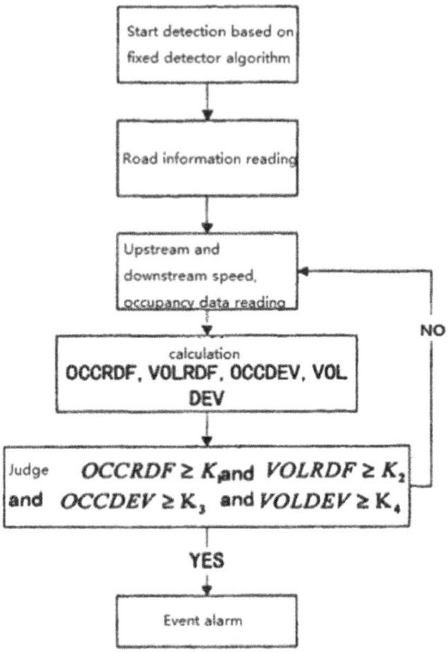

Fig. 1. Flow chart of multi parameter discrimination algorithm based on fixed detector

4 Example Analysis

4.1 Algorithm

This paper first introduces the relatively mature normal deviation method (SND). The normal deviation method uses the arithmetic mean of the traffic parameter values of the N sampling periods before the time t as the prediction value of the traffic parameter at the time t, and then uses the standard normal deviation to measure the change degree of the parameter in time. When it exceeds the corresponding threshold, the alarm will be triggered. Based on the analysis of the travel speed of expressways in the previous section, if only considering the change of traffic parameters in time dimension, it will cause false alarm in morning and evening peak, resulting in a high false alarm rate. Therefore, the space-time two-dimensional discrimination algorithm based on floating car proposed in this paper is as follows:

(1) From the time dimension, we first judge the speed value, then use the arithmetic mean of the driving speed of the N sampling periods before the time t to predict the traffic parameter value at the time t, and then use the standard normal deviation to measure the change degree of the driving speed relative to its previous average value. Algorithm is shown in Fig. 2.

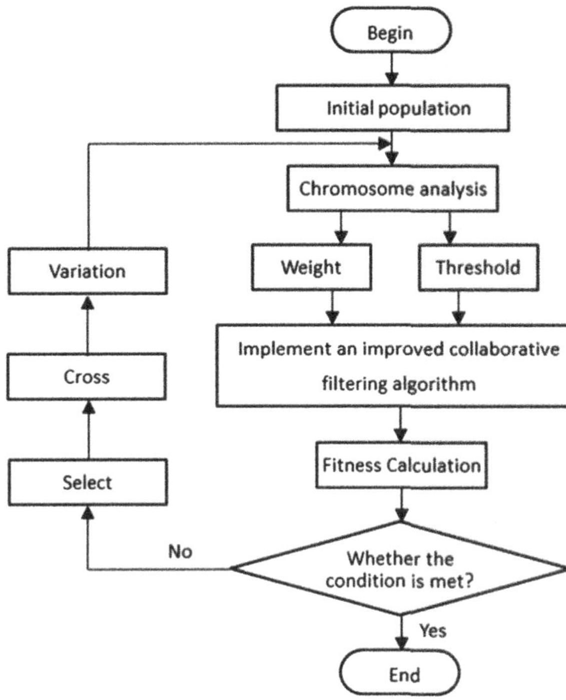

Fig. 2. Algorithm diagram

Let the actual value of the driving speed at time t be $v\ (t)$, and the actual values of the traffic parameters in the n sampling periods before time t are $v(t-n)$, $v(t-n+1)$,... $v\ (t-1)$.

$$\begin{array}{l} v(t) \leq K_1 \\ SND(t) = \frac{\bar{v}(t)-v(t)}{S} \geq K_2 \end{array} \tag{4}$$

(2) From the spatial dimension, based on the drastic change of the upstream and downstream driving speed when the event occurs, the following formula is used for discrimination;

$$VRDF(t)\frac{V(i+1,t)-V(i,t)}{V(i,t)} \geq K_3 \tag{5}$$

$$\begin{array}{l} E(t)\dot{x}_{d+1}(t) - E(t)\dot{x}_{k+1}(t) = E(t)\Delta\dot{x}_{k+1}(t) = f(t, x_d(t)) + B(t)u_d(t) - \\ f(t, x_k(t)) - B(t)u_k(t) = f(t, x_d(t)) - f(t, x_{k+1}(t)) + B(t)\Delta u_{k+1}(t) \end{array} \tag{6}$$

$$\|\Delta x_{k+1}(t)\| \leq \left(pk_f + m_2 + m_3\right)\int_0^t \Delta x_{k+1}(\tau)d\tau + \int_0^t (m_1\|\Delta u_k(\tau)\| + pd)d\tau \tag{7}$$

4.2 Algorithm Effectiveness Analysis

There are two methods to test and verify the event detection algorithm, one is based on simulation data, the other is based on actual data. However, under the simulation condition, the traffic condition is ideal, which is far from the real situation. Therefore, the algorithm verification based on the measured data is carried out, that is, the event detection algorithm is verified by collecting and processing the floating car detection data, fixed detector data and real event information in Beijing, and different threshold combinations are used to detect the algorithm, On the premise of ensuring a certain error rate, improve the detection rate, so as to determine the national value of the algorithm, and obtain the detection effect of the multi parameter discrimination algorithm based on fixed detector and the spatiotemporal two-dimensional discrimination algorithm based on floating car.

In 1996, through a survey of the traffic management center of the United States, abdulhai proposed that the acceptable average indicators of incident detection were Dr Z 88% and far s 1.8%. This index is also called TMC acceptable index. Here, this paper takes this as the index of the effectiveness analysis of the algorithm. If the algorithm meets this requirement, the algorithm can meet the needs of practical application (Fig. 3).

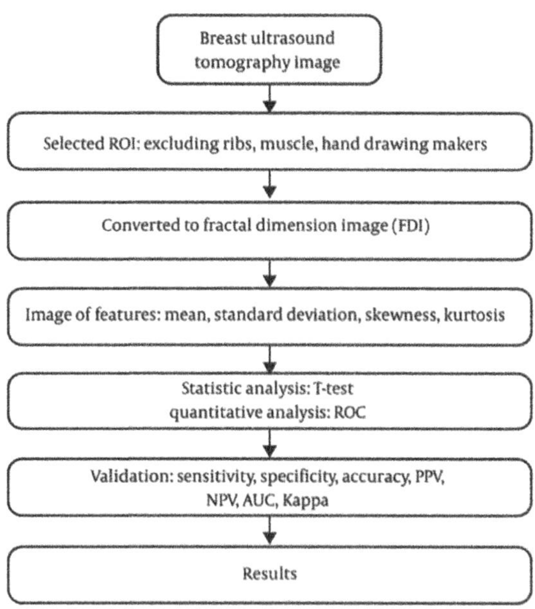

Fig. 3. Algorithm effectiveness analysis

5 Conclusions

Due to the fixed detector spacing, communication failure caused by the serious lack of data and the number of floating cars and other issues, if the simple use of a single data source for expressway incident detection, the effect is not ideal, often resulting in long-term interruption of detection, can not smoothly carry out detection and other issues. Therefore, in this paper, the fixed detector data and floating car data are effectively combined, and a good detection algorithm is proposed respectively. Finally, the D-S theory is applied to the fusion of algorithm results, which can effectively solve the problems of low coverage and reliability of single data source event detection algorithm. Finally, the effectiveness of the algorithm is analyzed with the data of Beijing Expressway. It is expected that the algorithm proposed in this study can provide some reference for automatic event detection of urban expressway.

References

1. Boranbayev, A., Boranbayev, S., Nurusheva, A., Yersakhanov, K., Seitkulov, Y.: A software system for risk management of information systems*, In: 2018 IEEE 12th International Conference On Application Of Information and Communication Technologies (AICT), pp. 211–220 (2018)
2. Hertati, L., Syafarudin, A.: How the implementation of the industrial revolution 4.0 management information system influenced innovation: the case of small and medium enterprises in Indonesia. J. Asian Bus. Strategy **125**, 106–116 (2018)
3. Tarigan, Z.J.H., Siagian, H., Bua, R.R.: The impact of information system implementation to the integrated system for increasing the supply chain performance of manufacturing companies. In: IOP Conference Series: Materials Science and Engineering, pp. 11–25 (2019)
4. Muslih, I., Nurdin, N., Marzuki, M.: Effectiveness of Marriage Services Through Information System Management (SIMKAH) at Palu city religious court **17**(2), 346–355 (2020)
5. Yuan, H., Qiu, H., Bi, Y., Chang, S.-H., Lam, A.: Analysis of coordination mechanism of supply chain management information system from the perspective of block chain. Inf. Syst. E-Bus. Manage. **44**(2), 215–224 (2020). https://doi.org/10.1007/s10257-018-0391-1
6. Maqsudov, K.T., Khudoyberdiev, K.A., Soliev, P.A.: Experience in development and implementation of an information management system in a technical university. ITM Web Conf. 40(4), 1468–1479 (2020)
7. Hu, B., Shi, W.: Design of a flexible exhibition management information system for exhibition services. Sci. Program. **26**(6), 1269–1278 (2021)
8. Sun, Q.-L.: Design and implementation of personnel management information system based on SSM. Destech Trans. Econ. Bus. Manage. **213**, 104–114 (2021)
9. Sui, Y., Ding, R., Wang, H.: An integrated management system for occupational health and safety and environment in an operating nuclear power plant in east china and its management information system. J. Cleaner Prod. **2672**(12), 137–147 (2018)
10. Kenneth, C.L., Jane, P.L.: Desing of the management information with the university teaching. Manag. Inf. Syst. **96**(7), 739–747 (2018)

Application of Blockchain Technology and Data Mining Technology in Public Utilities Management

Cirenlajie Ci(✉)

Fujian Agricultural and Forestry University, Fuzhou, Fujian, China
c13375000971@163.com

Abstract. As a government management evaluation system, it must be relatively independent and integrated. However, the existing research results lack the research on the technical platform of performance, which makes the design limitations and the repair of the index system, can not connect the index design with the performance evaluation, and can not make full use of the existing relevant data and information, resulting in the weak practicability of the designed index. This paper mainly introduces the econometric analysis method of extracting conceptual data from a large number of data by using the decision tree method of data mining technology in the process of public utility management performance evaluation, so as to improve the reliability and effectiveness of evaluation and save evaluation time.

Keywords: Decision tree · Data mining · Public utilities management · Performance evaluation

1 Introduction

Public utilities are social public affairs that take the common interests and living standards of all the public as the basic content, and include the activities and results of necessary economic affairs, mainly including education, science and technology, culture, health, sports, social security, environmental protection and so on. Public utilities management is a process of adjusting and controlling public affairs according to law under the guidance of the government, so as to promote the coordinated development of the overall interests of the society. The research on the performance of public utilities management began from the perspective of human resource management. Campbell, McCloy, Borman and other foreign scholars have made outstanding contributions to the research on performance.

With the development of information technology and the implementation of transaction automation and e-government, a large amount of data has been accumulated in the field of public utilities management. Traditional query and data processing technology can not solve the problem of real information explosion. The decision-making errors and efficiency caused by information redundancy perplex managers. This requires making

I. Ahmad et al. (Eds.): STSIoT 2021, LNDECT 122, pp. 604–613, 2023.
https://doi.org/10.1007/978-981-19-3632-6_71

full use of the information processing capacity of the computer in order to automatically and efficiently obtain useful knowledge and information from the vast amount of data. This is KDD (knowledge discovery in database) technology. KDD is an automatic process of discovering useful, new and available information from a data set. Generally speaking, it includes three stages: data preparation, data mining (DM) and knowledge expression, interpretation and verification. DM is the core process of KDD. It mainly includes classification, clustering, regression analysis, generalization, construction of dependent patterns, change and deviation analysis, pattern discovery and path discovery, using decision tree method, neural network method, statistical method and so on. Among them, decision tree method has the advantages of high speed, high precision and simple generation mode, which is widely popular in data mining. The evaluation of job performance is not a new concept. A considerable number of experts and scholars at home and abroad have studied and explored it. The research contents mainly include the following aspects: first, study the variables, indicators and definitions of performance; Second, research performance evaluation; The third is to study the structure of performance; The fourth is to study the input and output of performance. Different schools have different perspectives and different analysis tools, and the conclusions are often different or even contradictory. Therefore, it is very necessary to strengthen the quantitative refinement of the evaluation indicators and integrate accurate quantitative analysis methods in the analysis. It can be divided into the following steps: (1) query the data sets related to the task to generate a multidimensional data view; (2) Summarize and summarize the original data to a higher level of abstraction; (3) Further summarize and standardize; (4) Describe the knowledge found. This paper consists of the following parts. The first part introduces the relevant background and significance of this paper, the second part is the related work of this paper, and the third part is data analysis. The fourth part is example analysis. The fifth part is conclusion.

2 Related Work

The authors propose an algorithm of actions to follow when introducing occupational standards in a higher education institution; examine the problem of applying occupational standards in the context of training bachelors and masters in economics and housing and public utilities management; and show that the HPU industry suffers from a lack of methodical literature on using the requirements of the occupational standard in working training programmes [1]. Aim of Bresciani et al. investigate the governance of Italian public utilities whose top management is engaged in balancing the conflicting pressures of the business model and the social functions [2]. This research study analyses the project management of agricultural facilities that protect agricultural food and public health from associated toxic hazardous landfill emissions and risks [3]. Lukmanova et al. present the needs to improve the controlling systems in public utilities [4]. Golladay et al. consider PES to include a range of knowledge necessary to develop a resilient and sustainable water management strategy based on ecological, environmental, and engineering principles [5]. Liu used pattern discriminant analysis technology and BP neural network model to build a local public utility management performance prediction model and uses 11 regions in the east, middle, and west of the country as samples to predict the local public utility management performance [6]. Based on the above background, the purpose of Ref. [7] is the application of big data information system in the field of public utilities management. At present, there are many evaluation systems for public health, but their function is different, so this paper provide a new evaluation system for public health, its mainly function is emergence management [8]. However what is the fundamentals of public utilities management? The paper shared the answer for us [9]. Other influential work includes Refs. [10]. The core process of public health management is artificial intelligence, which is realized through machine learning, statistical calculation and logical derivation. However, public health management itself is not a simple combination of multiple technologies, but a complete whole. It needs to connect various technical means and information resources to complete a series of tasks such as data collection, preprocessing, information analysis and result description, and finally present the results.

2.1 Data Mining Concepts

Data mining is to find the "knowledge Nuggets" hidden in the "data mine", so as to help enterprises reduce unnecessary investment and improve capital return at the same time. Practice has proved that the potential return on investment brought by data mining to enterprises is almost endless. Innovative companies around the world have begun to use data mining technology to determine which customers are their most valuable customers, so as to re formulate their product promotion strategy (promote products to the people who need them most) and get the best sales with the least cost. For another example, telecom companies can use data mining technology to find customers with high probability of loss (no longer using the company's services), possible fraud (no payment after call), potential large customers and so on. In short, data mining is to extract knowledge or mine knowledge from a large amount of data. Data mining is a basic step in the process of database knowledge discovery. The knowledge discovery process includes the next step, such as Fig. 1.

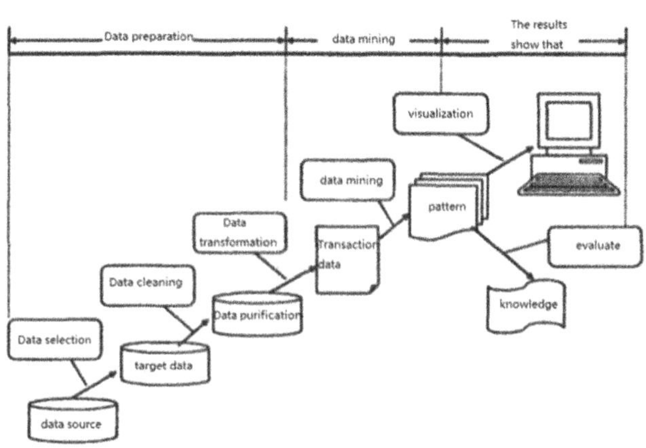

Fig. 1. Data mining and knowledge discovery

Data mining is related to the integration of multiple disciplines and technologies. This includes database technology, statistics, machine learning, high-performance computing, model recognition, neural network, data visualization, information extraction, image and signal processing, and spatial data analysis. Through data mining, interesting knowledge, rules or advanced information can be extracted from the database for observation or query from different angles. The discovered knowledge can be used in decision-making, process control, information management, query processing and so on. Data mining is one of the most important and promising subjects in the information industry.

The purpose factor of performance evaluation if we want to establish the performance evaluation index system of local government public utilities management, we must first determine its establishment basis, which is the basis of performance evaluation. The selection of performance evaluation indicators is determined by the purpose of performance evaluation. The fundamental indicators of performance evaluation are different in different regions. If the government wants to improve the performance of public utilities, the index system should pay more attention to the analysis of potential factors affecting the performance evaluation than the simple performance evaluation, and have an in-depth discussion on how to improve it. However, in real life, there are many kinds of performance evaluation indicators for local government public utilities management, and the number can not be ignored. In the evaluation, we can not be comprehensive and specific, so we can only evaluate the government management performance according to the evaluation purpose. Otherwise, not only the performance evaluation is meaningless, but also the operability is not strong, and the due value of evaluation is lost.

2.2 Data Mining System Structure

The essence of local government's function factor performance is the performance quality and degree of government's public utilities management function. Based on the government's management function, the local government's public utilities management performance evaluation index system is established φ The content of the indicators shall be carefully marked according to the management functions, so as to fully and accurately understand the main functions of local government management. From a macro perspective, the main functions of local governments include: education, scientific and technological development, cultural communication, health supervision, sports training, infrastructure construction, environmental protection, etc. θ The indicators to evaluate these functions need to establish corresponding performance evaluation systems, such as education performance evaluation system, science and technology development performance evaluation system, cultural communication performance evaluation system, health supervision performance evaluation system, sports training performance evaluation system, equipment construction performance evaluation system, social security performance evaluation system, environmental protection performance evaluation system, etc. Formulate corresponding performance evaluation indicators according to different functions. China is rich in land resources. Different regions have different historical factors and natural conditions. Therefore, when establishing indicators, we should fully consider that different functions affect the formulation of indicators. It is necessary to formulate not only a special index system, but also a general index system, which is analyzed and formulated by using quantity, so as to minimize the functional differences in different regions and the errors and negative effects caused by the differences. 3. The subjective and objective conditions of local governments will affect the performance evaluation system due to regional differences. This mainly includes three aspects: first, according to the requirements of the superior leaders and the general objectives put forward by the superior leaders, convey and complete the task allocation, refer to the relevant provisions, laws and regulations of the plan, etc.; Second, the quality of local personnel, economic and social development, the quality and degree of achievement of objectives, government management system, etc.; Third, the direct impact of the external environment, that is, the development quality and speed of relevant enterprises, regional characteristics, departments, etc. Take it as the basis of establishing the system, and establish the management performance evaluation index system of local government public utilities. Only by truly and comprehensively understanding, mastering, researching, analyzing and making use of these conditions can we reasonably formulate the management performance evaluation indicators of local government public utilities.

A typical data mining system has the following main components, and its system structure is shown in Fig. 2:

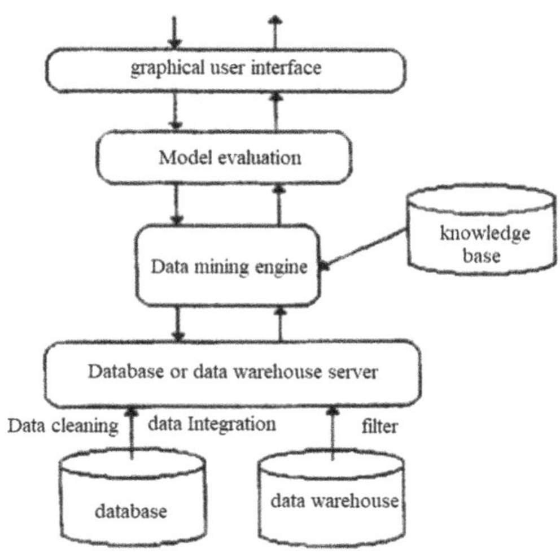

Fig. 2. Typical data mining system structure

2.3 Research on Decision Tree Algorithm

In many methods of data mining, classification algorithm is an important topic, which is the most widely used and studied by scholars. It can be used for prediction and decision-making. By learning and analyzing the experience data that have been classified in the past, the classification algorithm establishes the corresponding algorithm model to describe the differences of each category in the training data, and can classify the new data of unknown categories. By predicting the class labels of these new data, we can have a better understanding of the categories in the data and obtain the category knowledge of the new data, even though they are unknown.

Recursive top-down method is a common way in the learning process of decision tree. The general process is as follows: firstly, the attribute values of different variables to be compared are compared in the inner nodes of the tree, and then the branch is decided based on the different attribute values of different variables, According to this step, the classification of the variable attribute is obtained in the leaf node of the tree. Therefore, from the root node to the leaf node of the tree, each path in the middle is essentially a disjunctive rule. To sum up, a decision tree is essentially a set of disjunctive rules.

According to the learning process of decision tree, the algorithm of generating decision tree is divided into two steps: the first step is the generation of decision tree, the root node stores all the data to be classified during initialization, and then the data is segmented by recursion. The second step is to prune the decision tree. The essence of pruning is to eliminate noise or abnormal data similar to noise. When the data in an internal node belongs to a category or the attribute cannot be further divided, the data segmentation can be stopped.

3 Data Analysis

ID3 uses greedy method and recursively uses top-down divide and conquer method to generate decision tree. Whenever the branch node of the tree is selected, the information gain is measured, that is, the amount of information required to classify the current data is calculated before division. Then, after the segmentation according to the segmentation, the amount of information required to classify the current data is calculated again. The difference between the amount of information before and after splitting is the information gain of the splitting attribute. After calculating the information gain of all the current split attributes, the split attribute with the largest information gain is finally selected, which can make the amount of classification information of data decrease the fastest. Most of the decision tree induction algorithms follow this top-down classification method. The decision tree is generated from the training tuple set and their associated class labels. The decision tree consists of three types of nodes: root node, internal node (decision node) and leaf node. The root node and the inner node correspond to an attribute in the training tuple set, while the leaf node is the set of class label attributes in the classification. As the tree grows, the training set is recursively divided into smaller training subsets. The growth process stops until the leaf node, when the entropy is zero, and the instances in the corresponding instance set of each leaf node belong to the same class.

Assuming that the number of samples belonging to category s is, the amount of information needed to classify a given data object is Formula 1:

$$AOI(C_1, C_2, \cdots, C_m) = -\sum\nolimits_{i=1}^{m} p_i \log_2 p_i \tag{1}$$

The faster the entropy decreases, the better. In this way, we can get a decision tree with the smallest height. The smaller the value is, the purer the result is. For a given subset, the information quantity formula 2:

$$AOI\left(C_{1j}, C_{2j}, \cdots, C_{mj}\right) = -\sum\nolimits_{i=1}^{m} p_{ij} \log_2 p_{ij} \tag{2}$$

Split information degree spiltinfo is used to measure the breadth and evenness of an attribute's split data. The calculation formula is 3:

$$IGR(A) = \frac{IG(A)}{SpiltInfo(A)} \tag{3}$$

Information gain rate is a kind of compensation for the number of attribute values. The importance of attributes will decrease with the increase of split information degree. When the number of branches generated by data set splitting is small, the gain rate will increase accordingly, which effectively solves the problem of attribute multi value bias.

For numerical data, ID3 algorithm can't deal with continuous attributes, but can only deal with discrete attributes, so it discretizes the data set; for descriptive data, every node or concept of spanning tree is actually a range or interval, so we use grid method to quantify the statistical indicators.

4 Example Analysis

4.1 Generation Algorithm of Public Utility Management Performance Decision Tree

For numerical data, ID3 algorithm can not deal with continuous attributes, but only discrete attributes, so the data set is discretized; For descriptive data, each node or concept of the spanning tree is actually a range or interval. We use the grid method to quantify the statistical indicators. Here, taking the education management in public utilities as an example, we select three indicators for evaluation: the proportion of education in GDP (g), the number of full-time teachers per 100 students (s), and the proportion of college students in the population (P). The data are from cities in Hunan Province (Zhangjiajie, Changsha, Shaoyang, Zhuzhou, Xiangtan, Hengyang, Chenzhou, Xiangxi, Yiyang, Loudi and Yueyang), in order to avoid unnecessary impact caused by the evaluation, the specific unit name is hidden in the text, supplemented by a, B, C… Instead. Interested readers can refer to relevant literature or the local statistical yearbook of that year.

The virtual government education management performance indicators are constructed, and the statistical indicators of each region are weighted and averaged to form a virtual evaluation index. According to the given index weight, the proportion in the education management performance evaluation is 0.5391, 0.2971 and 0.1638 respectively through factor conversion; the classification comparison results are obtained by comparing the actual indicators with the virtual indicators.

The principle of data preprocessing and sorting is to compare the regional indicators with the virtual indicators and convert them into a unified format suitable for data mining. The corresponding value range of each type is that if 20% of the performance of the virtual government is excellent and 10% is good, generally less than - 10% is poor and - 20% is very poor (expressed in 1–5 respectively). It can be seen that "education accounts for the proportion of GDP" "The information value of this attribute for training set classification is the largest, so this index is selected as the basic attribute for division, and the decision tree is generated by analogy, as shown in Fig. 3.

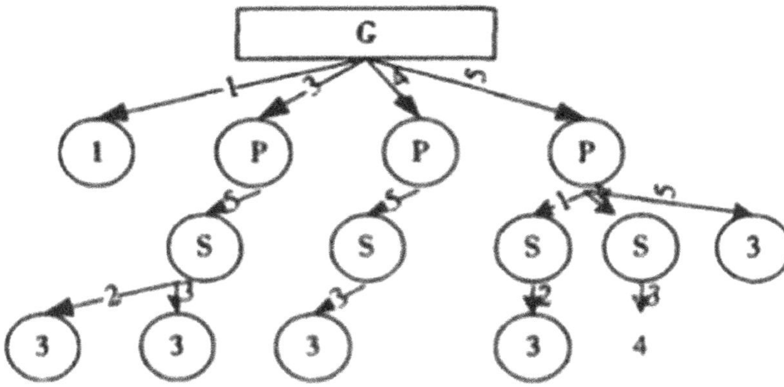

Fig. 3. Education management performance index evaluation decision tree

4.2 Optimization and Pruning of Decision Tree

Decision tree is one of the most important classification methods in data mining. The training data sets are combined repeatedly to construct the decision tree. If the data of the training data set can accurately reflect the character of the analysis object, the decision tree of the training data set can correctly classify the problem. However, there are many uncertain factors in practical problems. By using the decision tree structure algorithm to classify these data, the obtained decision tree becomes huge and complex, and the generated knowledge rule set becomes huge and complex. Therefore, it is necessary to prune the decision tree. The purpose of pruning is to reduce the fluctuation caused by the noise in the training set. Because many branches may reflect the noise or isolated points in the training data, pruning can improve the accuracy of classification of unknown data sets, avoid the systematic error caused by such points, and improve the mining accuracy. There are two kinds of pruning methods:

(1) Pruning afterwards. It allows the decision tree to get the most full growth, and then according to certain rules, it cuts out the leaf nodes or branches which are not generally representative in the decision tree. After pruning, the pruned branch node becomes a leaf node, and it is marked as the class with the largest number of categories in the sample it contains. This is a process of pruning while checking. Of course, this process may be at the expense of prediction accuracy. When the standard deviation of classification data increases rapidly, pruning should be abandoned, otherwise the classification results will be affected.
(2) This deliberative method determines whether the point generation process should be stopped as soon as possible and continue to provide the training sample set of the current node. If the blanking key stops, the current node becomes an inversion node. Leaf nodes can include different types of training samples. Hiccups are formed in front of the brand, so the general way to hiccup in advance is to set the maximum height (layer) of the crystal tree to limit the growth of the tree. Another method is to set the minimum number of records to include in each node. If the number of records of the node is less than this value, the segment stops. But it is often difficult

to determine such a reasonable threshold. If the threshold is too large, the decision tree will be too simple, and if the threshold is too small, the redundant branches will not be pruned. The decision tree generated in this paper adopts this kind of method. Of course, pre pruning and post pruning can be combined with each other. Using the advantages of the two methods, it can save time and improve the accuracy of the tree, thus forming a hybrid pruning method to obtain a more reliable decision tree.

5 Conclusion

Based on the three indicators of educational management performance, this paper puts forward a kind of data mining method for public management performance evaluation. Its practicability lies in that it can analyze the results of management performance by substituting the corresponding case data under its generated decision tree evaluation rules. At the same time, by calculating the information entropy of each performance index, find out the main factors that affect performance, so as to find measures to improve performance.

Of course, the performance of public utilities management is affected by various factors. The structure of evaluation index is complex, non market, non discrete and non deterministic, which makes the performance itself fuzzy. Although there is a great correlation between the satisfaction degree of service objects and the administrative efforts of government organs, in fact, it is not entirely determined by the administrative efforts of the government, but the result of the comprehensive action of many factors, among which the social environment factors have an important influence on the satisfaction degree. In addition, the relationship between government cost and management performance of public utilities is not completely linear. Therefore, as a basic data mining technology, decision tree is used to describe the performance evaluation of public utilities with a large number of data. It is a very useful fuzzy mathematical evaluation tool to provide a wide range of evaluation and reasonable quantitative space.

References

1. Astratova, G.V., Shklyaeva, N.A.: On application of occupational standards in training bachelors and masters in economics and housing and public utilities management. J. Manag. Syst. **158**, 107704 (2018)
2. Bresciani, S., Del Giudice, M., Papa, A.: Public control and strategic governance in state-owned public utilities: empirical evidence from Italian listed firms. Sinergie Ital. J. Manag. **74**(18), 3823–3831 (2018)
3. Koliopoulos, T., et al.: A project management utility for agricultural facilities and public health. J. Eng. Stud. Res. **155**, 23–36 (2018)
4. Lukmanova, O., Volkova, E., Zabolotnyi, A., Gorelik, A.: Blockchain technology for public utilities. In: 2019 IEEE Conference of Russian Young Researchers in Electrical and Electronic Engineering (EIConRus), pp. 9–16 (2019)
5. Golladay, S.W., Craig, L.S., DePalma-Dow, A.D., Emanuel, B.N., Rogers, S.G.: Building resilience into water management through public engagement. Freshw. Sci. **29**(2), 300–312 (2020)

6. Liu, X.: Forecasting utilities management performance based on discriminant technology and BP neural network. J. Intell. Fuzzy Syst. **20**(4), 609–616 (2021)
7. Yu, Q.: Application of big data information system in the field of public utilities management. J. Inf. 785–788 (2021)
8. Xie, X., Li, H.: Analysis on the evaluation system of emergency response ability to public health emergencies. J. Health Manag. **20**(4), 609–616 (2020)
9. Spellman, F.R.: Fundamentals of public utilities management. Int. Public Util. Manag. **220**, 106918 (2020)
10. Spellman, F.R.: What is public utility management? Int. Public Util. Manag. **220**, 106877 (2020)

Integrated as a Service in the Construction of Small and Micro Enterprise Financial Management Platform System

Hualing Huang[1,2(✉)], Noura Metawa[1,2], and G. Rajendran[1,2]

[1] Nanfang College of Sun Yat-Sen University, Guangzhou 510900, Guangdong, China
hhlx2019@163.com
[2] Karpagam College of Engineering, Othakalmandapam, Coimbatore, India

Abstract. The technology of the system is .Net. It has the advantages of low requirement for client and good expansibility. The whole system is flexible and efficient. The main work of this paper is the system analysis of the financial management system, and based on the system analysis, puts forward the overall business objectives of the financial management system; gives the overall design of the system, as well as the detailed design of the main functional modules. It includes voucher management, salary management, fixed assets management, account book management, statement management, period end management, business transaction management, cashier management and financial analysis. On the basis of detailed analysis of each functional module of the system, coding is carried out to achieve the expected function.

Keywords: Construction system · Multi-objective optimization · Ant colony algorithm

1 Introduction

At present, there are still many defects in the domestic financial management system, for example, it is difficult to ensure the authenticity of the data, the format of the data is not unified, and the data is lack of real-time. The main reason for these shortcomings is that domestic ERP software does not pay attention to the organization and management of enterprises, lack of investigation and research in this respect. Although the software has made improvements in the planning management function, there are still deficiencies in the organization and management function. For example, the separation of financial management function and financial accounting of some ERP software is the best embodiment. So the domestic financial management system is far less than the foreign financial management system.

With the development and application of computer technology, network technology and communication technology, enterprise informatization has become an important

I. Ahmad et al. (Eds.): STSIoT 2021, LNDECT 122, pp. 614–622, 2023.
https://doi.org/10.1007/978-981-19-3632-6_72

guarantee for brand to achieve sustainable development and improve market competitiveness. It covers five main processes: project startup, planning, implementation, monitoring and final management, including project management, plan management, schedule management, cost management, quality management, procurement management, risk management and other functions with specific business logic.

To sum up, OA system and project management system are the only way for enterprise information construction. By means of integration, the "two main management systems" are integrated to form a comprehensive operation control platform.

First, the remote company used its advanced "V6 application design platform" to build "two main management systems", namely, basic business management system and operation analysis and evaluation system. The basic business management system focuses on the main business and gradually integrates all the business activities of the company, so that the enterprise can operate at low cost and efficiently. Through intelligent analysis of basic business information, the operation analysis and evaluation system forms valuable management information and decision-making information analysis, so that managers can quickly obtain various types of information for operation analysis and provide the basis based on business decisions. Integrate business and establish a business system of enterprise management. Bring all the company's business operations into the financial control system, form a comprehensive operation and management pattern with financial management as the core, and take budget management as the means to realize the coordinated operation of business processing, business monitoring, financial accounting and financial supervision, reflect the operation status in real time and provide effective business decision support. Based on the integrated enterprise management concept, the remote company has established a unique system construction mode.

The reason why we can innovate the system construction mode and "Consulting + system design" quickly realizes the construction of integrated operation management system because there is a professional "V6 application design platform" in the distance, so that business oriented design can be carried out. The shared implementation scheme is "main departments first, step by step", and the industry promotion plan is a business process reform solution of "optimization first, then consolidation". Establish personalized and controllable cost management; Personalized controllable cost management. This paper consists of the following parts. The first part introduces the relevant background and significance of this paper, the second part is the related work of this paper, and the third part is data analysis. The fourth part is example analysis. The fifth part is conclusion.

2 Related Work

To study consumer perceived value of online financial products, Ye et al. study the objects based on grounded theory [1]. In order to decrease the delay of financial product information update among multi institutions, manage multi-dimensional and diversified financial product information, enhance the properties traceback ability, Chen et al. propose a blockchain-based financial product management platform [2]. According to the natural ecosystem structure model, Wang et al. summarize the composition of the supply chain financial ecosystem, designed the financial eco-mode of the supply chain, cleverly integrated the financial tree theory, and vividly linked the relationship among the environment, big data platform and enterprises [3]. Haberly et al. apply it to examine the impact of the digital platform model on asset management [4]. Ref [5] uses AHP to construct the AHP model of the P2P network lending platform. Based on ERP supply chain management in cloud computing environment, Wang et al. explore the main value of the implementation of China's financial sharing platform with the case of specific enterprise groups [6]. Setiawan et al. use qualitative methods by conducting literature studies, observation, and semi-structured interviews [7]. Therefore, the objective of Ref [8] is to establish a financial leasing value model of debt cession with an optimal economic pattern and an analysis of the risk assessment to improve the management of the asset value docking quality of both parties. Chen et al. analyze the teaching environment and platform of financial management, constructs the remote teaching platform based on B/S mode, designs the financial management platform module from four aspects: financial voucher management, financial account management, financial statement management and user management, and analyzes the specific content of the remote teaching of financial management course in detail from three aspects: preparation before class, classroom analysis and discussion, summary and evaluation and writing report [9]. Other influential work includes Ref [10].

On this basis, the management consulting report is prepared from the aspects of business system, analysis system, process system and system interface, which comprehensively explains the management needs of the enterprise. System design: design an "information" operation management system.

2.1 Common Language Runtime

In the 1990s, Microsoft began to develop a product called .Net strategy. This policy covers all Microsoft product lines. All Apple product lines covered by this strategy. In 2000, the .Net framework was named the next generation windows service "NGWS" The first beta version of the .Net framework was released at the end of 2000, while the first version of 1 .net 1.0 was released on February 13, 2002. The following are the subsequent versions and their features: .NetFRAME 2.0 (November 2005): includes general collections, iterators, and nullable types. .NetFRAME 3.0 (November 2006): includes WPF, WCF, and WWF. .NetFRAME 3.5: November 2007): includes Ajax, LINQ, and asp.netmvc Netframework April 2010: April 2010: including MEF, DLR task parallel library, razor view engine and new c8.0 new functions, so as to realize .Net standard 2.1. Make windows desktop support Windows Forms and WPF.

Net is cross platform, it can be used not only to develop desktop applications, but also to develop web based applications. At present, it is suitable for the development of windows .Net framework, such as windows .Net framework. This system uses net and ADO net framework .Net development platform from top to bottom structure is shown in Fig. 1.

The common language runtime is at the bottom of the structure, abbreviated as CLR, which is the foundation of net framework. In general, the common language runtime is regarded as the manager agent of program execution code. It provides the most basic management for program running, such as memory management, compilation, etc. In addition, it can ensure that strict type safety detection is implemented and the reliability and correctness of the code are ensured.

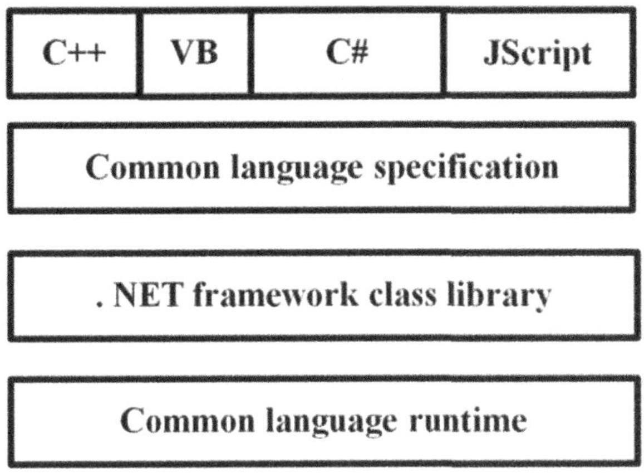

Fig. 1. Structure of net development platform

2.2 B/S Mode and Three Tier Architecture

The three-tier architecture is generally composed of three independent units: presentation layer, business logic layer and data access layer. As the interface of user interaction, presentation layer is mainly responsible for the dialogue between user and application system. For example, the user's input and output data displayed to the user are displayed in the presentation layer. In order to facilitate the user's operation, the graphical interface is used for programming. For different users, only the relevant attribute values of the display control can be changed, and the structure of other layers will not be affected. The structure of graphical interface has flexible and changeable design style, which can design different interface according to different aesthetic views of programmers.

The application of the three-tier architecture reduces the redundant and complex transaction processing functions of the client, while more business logic processing is implemented on the web server side, which greatly reduces the requirements for the client software, so that programmers can focus on updating the functions on the server instead of debugging back and forth between various clients for the maintenance of application programs. The three layers are independent and closely related to each other, and can effectively customer service the shortcomings of the two-tier framework (Fig. 2).

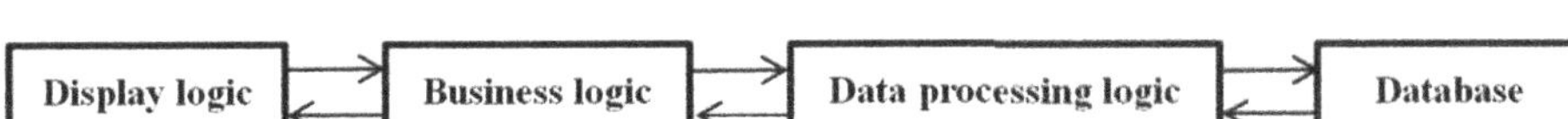

Fig. 2. Three layer system structure

The distance calculation in the algorithm generally uses European distance or Manhattan distance, which is calculated by formula (1) and (2).

$$d(x, y)\sqrt{\sum\nolimits_{k=1}^{n}(x_k - y_k)^2} \tag{1}$$

$$d(x, y)\sqrt{\sum\nolimits_{k=1}^{n} |x_k - y_k|} \quad (2)$$

In the above three operations, they all implement .Net standard library. They are also the specification of the .Net API. Therefore, code created during an operation can also be executed by other runtimes. And all operations use tools and infrastructure to compile and run code. The languages involved include c# and Visual Basic. The compiler includes Roslyn, garbage collection, msbuild or "CLR" and other construction tools.

$$I(S) = I(p) = -\sum\nolimits_{i=1}^{n} p_i log_2 p_i \quad (3)$$

$$p = (S_{i1}/S_i, S_{i2}/S_i, \ldots, S_{im}/S_i) \quad (4)$$

Below, we will focus on the three main operations of .Net: the following are the main operations in this article: the following three main running times: net.

Net template is a software development framework for building and running applications on windows. At present, it is partly open source Net template consists of common language runtime (CLR), net Linux database and application load "WPF, windowsforms and ASP .Net". CLR is a part of the common infrastructure. It can not only run code, but also execute processes (such as JT + just in time), garbage collection "C", etc.

The code managed by the CLR is called managed code. These codes are compiled into the common intermediate language "common entry for communication, lis" and stored in an assembly with the extension. When the application runs, the CLR will execute assembly and use the Jin compiler to convert its machine code into code that can operate on a specific computing architecture. The B/S structure frame is shown in Fig. 3 below.

3 Data Analysis

3.1 Technical Feasibility Analysis

From the perspective of technology, computer has been widely used, especially in the enterprise management system. In addition, the use of computers is becoming more and more simple, so it is completely feasible. At the same time, the system is network oriented. At present, most computers are connected to the Internet. In addition, with the improvement of domestic width, the network speed has also been greatly increased. And network security, the level of related equipment is also improved, network security issues are guaranteed, from the perspective of network technology, the system is feasible. Finally, Microsoft Visual Studio 2008 is used as the development tool. It has advanced technology, strong operability, relatively easy development process, relatively high quality software, and in the field of programming based on Microsoft platform development software occupies a large proportion, therefore, from the development tool technology, it is feasible.

Net template library contains interfaces of various function types, interfaces and data types (such as string, file system support, etc.). It allows users to create different types

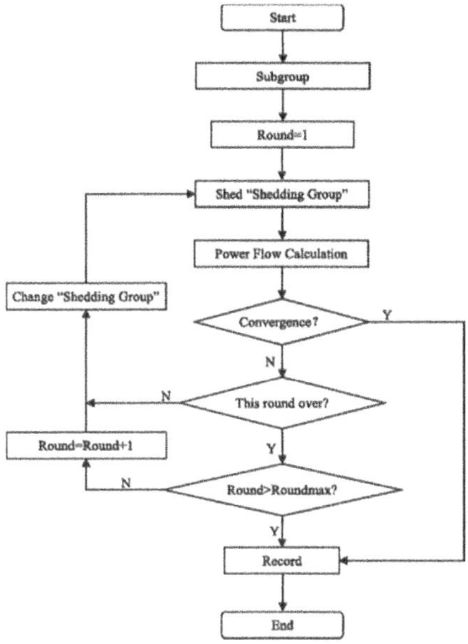

Fig. 3. B/S structure frame

of applications, such as console applications, windowsForm, windows communication foundation, windows Workflow Foundation, windows ventures foundation WPF, asp .net applications (such as various forms and web APIs) and azure applications. For example, webjobs and cloud services. However, most of them are limited to the windows platform and use the windows API.

As an operation of the .Net ecosystem, .NETCORE was released and open source in 2016. It is neither a new version of the .Net template nor a substitute .Net template. On the contrary, it is built independently and designed for cross platform application development. .NETCORE consists of a runnable CLR and library apphome hutnet.exe. It has the common language runtime "coreclr" and .Netcoreclasslibrary o Coreclr uses JT excel and garbage collection to run code. It supports C. It supports c.ts. as a subset of the .Net Linux database, the .Net coreclasslibrary includes and provides different functions. It supports different types of application loads, including ASP .Net core (for example, MVC and API), various console applications and uwp. Among them, uwp provides a common system, API and application model for all devices running on Windows 10. Starting from the .NetCORE 3.0 SDK, it can support various windowsforms applications (including windowsformsdesigner).

3.2 System Function Requirement Analysis

According to the demand of financial management of limited company, the system can be divided into nine modules, which are voucher management module, report management module, period end management module, transaction management module, account book management module, cashier management module, financial analysis module, salary management module and fixed assets management module. These modules are independent of each other and have no overlapping functions. In terms of system design, the main feature of the system is that it integrates a large amount of accounting information, and processes these information, such as recording income funds, fixed assets, analyzing financial indicators, etc., so as to reduce the workload of financial accounting personnel, improve work efficiency, and help enterprise managers straighten out the company's capital status, so as to provide the necessary basis for enterprise decision-making, As far as possible to avoid the subjectivity of managers in decision-making, and ultimately achieve the purpose of maximizing the profits of enterprises. Scientific and reasonable classification of fixed assets is the premise of its management and accounting. Because there are many kinds of fixed assets and their values vary greatly, if we don't divide them systematically, there will be no way to implement such operations as summary query. At present, there are different standards for asset classification, which must be determined according to the management requirements of the enterprise itself. The system divides fixed assets according to their functional attributes, such as fixed assets used for production and operation, unused fixed assets, fixed assets leased in and leased out, etc.

4 Example Analysis

As the digital transformation enters a more in-depth stage, the demand for charge control management of enterprises surges. It is not as good as the Huatian power collaborative OA system to be introduced. For example, for the more practical bank enterprise direct contact function for users, the number of open payment channels of Pan micro OA is less than the surplus of Huatian power collaborative OA system. As an "intelligent process" expert, the biggest advantage of Huatian power collaborative OA system is that it can realize the real-time interaction between process approval and business data. From the perspective of the charging control management platform of Zhiyuan OA system, there are still many deficiencies.

Huatian power cooperates with OA system to provide comprehensive charge control management support for the financial supervisor. The one-stop big data service platform meets various data analysis and application needs, so that the boss can understand the company's financial situation and make accurate decisions at any time; The whole process of the intelligent reimbursement platform is online and closed-loop, with standardized management. The control process system of Huatian power collaborative OA system with risk management as the core can effectively help the financial supervisor solve this problem. Huatian power collaborative OA system realizes the online closed loop of the whole process through electronization, and the data storage is convenient for verification, helping enterprises build a standardized management system. Figure 4 shows the financial management system.

Huatian power cooperates with OA system to establish a process driven and efficient charging control system, which realizes the whole process management from budget declaration, approval, use and adjustment to expense application, reimbursement, payment and voucher docking. It supports more than ten different types of intelligent programs, which can meet various complex process management needs, improve and optimize

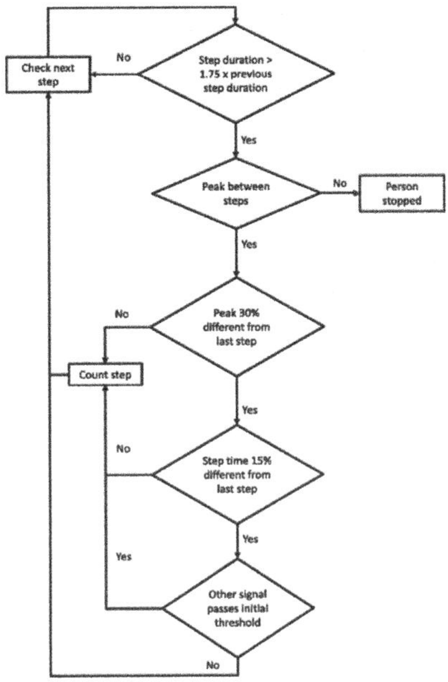

Fig. 4. Financial management system

enterprise business processes and improve the efficiency of financial approval. Huatian power cooperates with the document center in the OA system to help the financial supervisor classify and separate all vouchers, bills and documents by intelligent means for subsequent investigation.

Huatian power collaborative OA system supports invoice verification and electronic invoice, so that employees can bid farewell to the trouble of invoices, automatically generate docking vouchers, and greatly provide the work efficiency of financial personnel. Huatian power collaborative OA system uses intelligent and whole process electronic management methods to pre budget, post control or analysis, which can help the financial supervisor improve work efficiency and create more value for the enterprise. Huatian power collaborative OA system is a very complete intelligent collaborative office platform. Huatian power collaborative OA system integrates the integration of integrated systems to support the integration of management, business and finance, realize the whole process closed-loop management, integrate electronic images, data trace and classified storage. Efficient process approval: according to the company's business, the customized business approval process and daily office procedures can be flexibly adjusted to improve efficiency.

5 Conclusion

This paper analyzes the internal and external environment of the enterprise, determines the overall requirements of the financial management system, and establishes the development principles of the financial system. The financial system is divided into several independent sub modules, such as salary management module, financial analysis module, etc., and further subdivides the sub modules to determine the non functional requirements of their detailed functional requirements analysis platform, The former determines the security policy that the system should adopt, and the latter affects the choice of system development tools. According to the functional requirements of the system, the conceptual model and tables of the database are designed and planned. Using the selected development tools to achieve the functions of each independent sub module, and test to ensure that the system can run correctly and has a certain degree of fault tolerance. Based on the analysis of the structure and function of each sub module, the system is coded and tested with C programming language. At present, the salary management module, fixed assets module, account book management module and voucher management module have been applied in the financial management of the limited company, and they are in good condition.

References

1. Ye, Q.-W., Kang, W.-Y., Luo, Y.-M., Ma, X.: An empirical study on the consumer perceived value of online financial products based on grounded theory. TIST **12**(2), 1–22 (2018)
2. Chen, B., Tan, Z., Fang, W.: Blockchain-based implementation for financial product management. In: 2018 28th International Telecommunication Networks and Applications Conference (ITNAC), pp 555–610 (2018)
3. Wang, B., Lan, W., Fan, C.: Supply chain financial ecology model design and risk. Manage. Res. **12**(2), 1–22 (2018)
4. Haberly, D., MacDonald-Korth, D., Urban, M.A., Wójcik, D.: Asset Management as a digital platform industry: a global financial network perspective. Geoforum **14**(10), 2497–2529 (2019)
5. Cao, X.: Risk management and control countermeasures of P2P network lending platform under internet. Financ. Environ. **67**, 14–23 (2019)
6. Wang, L.M., Song, J.: Research on the construction of financial sharing service platform based on ERP supply chain management in the cloud computing environment. In: Proceedings of the 2019 International Conference on Economic Management and Cultural Industry (ICEMCI 2019), pp. 6521–6526 (2019)
7. Setiawan, A., et al.: The role of IT in transforming a platform business model: a case study in PT XYZ. In: 2020 International Conference on Informatics, Multimedia, Cyber and Information System (ICIMCIS), pp. 2214–2219 (2020)
8. Qu, J., Liu, H., Zhu, H., Gao, H.: An asset value evaluation for docking finance lease problems in the peer-to-peer platform. J. Bus. Econ. Manage. **29**(1), 61–70 (2020)
9. Chen, X., Zhang, S.-Y.: Financial management course distance teaching method based on B/S model. In: Fu, W., Liu, S., Dai, J. (eds.) eLEOT 2021. Lecture Notes of the Institute for Computer Sciences, Social Informatics and Telecommunications Engineering, vol. 389, pp. 68–83. Springer, Cham (2021). https://doi.org/10.1007/978-3-030-84383-0_7
10. Shen, H.: A review of P2P. Financ. Res. **105**, 87–95 (2018)

Design of IOT Sensor for Horizontal Well Development in Thin and Poor Reservoir Based on 5G Communication

Xiaolu Huang(✉)

Geological Brigade in No.7 Oil Production Plant of Daqing Oilfield Company Ltd., Daqing 163458, China
uangxiaolu@petrochina.com.cn

Abstract. Horizontal well increases the relative contact area of underground reservoir. Horizontal well has the advantages of large controlled reserves, large oil drainage area and high production. Horizontal well development technology is suitable for the whole process of oilfield development, and is an important technology to improve oil well production, oilfield recovery and development benefits.

Keywords: Low production reason · Existing problems · Stable production · Horizontal well development

1 Introduction

In January 2012, 77 wells were opened in an oil production plant, with a daily liquid production of 698 m^3, a daily oil production of 360 t, an average single well productivity of 36, a basic comprehensive water cut of 436%, an average dynamic liquid level of 1210 m, corresponding to 11 wells, and a daily water injection of 1213 m^3, The average daily injection of a single well is 96 m^3 1.1 points, and the production composition and productivity classification of the reservoir, The daily oil production of 41 relatively high-yield horizontal wells is 153T. According to the data classification analysis of productivity records, the production of 50% of wells is less than 3, in which the share of high-yield wells is small. The number of 14 horizontal wells is more than 8t, and the data is 143% of the total number of wells. The number of horizontal wells between 3 and 6 is 36, and the total number of wells is 367%. The number of wells less than 3 is 45, According to the analysis of water cut data classification data, 38 wells with water content less than 21%, accounting for about 45.1% of the total number of wells, 18 wells with water content more than 80%, accounting for about 235% of the total number of wells.

In order to further explore the technical means of enhancing oil recovery of tight conglomerate reservoir in Mahu oil area, Baikouquan oil production plant of Xinjiang Oilfield introduced group pressure on the basis of the obvious results of natural selection and repeated fracturing. The direct underground energy storage of 12 wells in ma18

I. Ahmad et al. (Eds.): STSIoT 2021, LNDECT 122, pp. 623–631, 2023.
https://doi.org/10.1007/978-981-19-3632-6_73

vertical well area and Aihu 1 fault block vertical well area in Mahu oil area of the plant is broken, and the output of single well is increased through "energy increasing" underground oil production.

During the implementation, the planning researchers of Xinjiang Oilfield Exploration and Development Research Institute made full use of logging and seismic materials to carry out spatial exhibition of target oil layer, stress distribution of target layer and overlying formation, and fine describe the target point of design horizontal well. It has important guiding significance for the development and efficient production of ultra deep conglomerate tight oil, and promotes a new stage of the development of mahulu tight oil. It has important guiding significance for expanding the scale of reservoir evaluation and development deployment. This paper consists of the following parts. The first part introduces the relevant background and significance of this paper, the second part is the related work of this paper, and the third part is data analysis. The fourth part is example analysis. The fifth part is conclusion.

2 Related Work

Wu et al. provide a reference for enhanced oil recovery for thin heavy oil reservoirs after steam injection [1]. Jaoua et al. present a concept for the optimization of oil production from an oil rim reservoir by numerical simulation [2]. Close collaboration between the operator and the directional drilling/logging-while-drilling (LWD)/geosteering service provider was a key component of developing a fit-for-purpose solution [3]. An integrated well-pattern optimization frame in adaption to the geological setting of Ordos basin is systematically proposed by Ref [4]. Taking the thin and poor reservoir sand body at the edge of a development zone as an example, Zhou introduced the horizontal well geosteering technology in detail, five techniques have been developed, including marker layer selection, seismic software aided guidance, target identification, lwd data analysis and rate of penetration control [5]. Carpenter summarized the design processes, selection criteria, challenges, and lessons learned during design and execution phases [6]. Ref [7] invested hot-water flooding after steam injec-tion to improve oil recovery in thin heavy-oil reservoir, according to the investion, the team explored a new reservoir for heavy-oil, so the method is better than the other methods. However, for the low permeability tight sandstone gas reservoirs, the paper share the development of gas field, according to the experiment, we know the method is effective [8]. Ref [9] solve the liquid loading in gas development, by normal work, the remote-controlled automated foam injection is working, so if the digital controller is effective. The new production capacity of the plant reached a record high, with a new production capacity of more than 1 million tons. However, due to the influence of the characteristics of low porosity and ultra-low permeability of Mahu tight lava reservoir, its production wells showed the characteristics of declining production rate and low predicted harvest rate. How to further improve reservoir recovery and realize energy increase, pressure maintenance and long-term stable production in Mahu oil area has become the focus of plant researchers.

2.1 Technical Difficulties in Drilling Horizontal Wells in Thin and Poor Reservoirs

(1) The error of reservoir depth is large, and it is difficult to accurately hit the target
The fundamental reason for the underdevelopment of thin and poor reservoirs is that the vertical depth of reservoirs often changes greatly. In some places, when the operators complete the drilling work, the data are not particularly detailed, which brings many unnecessary troubles to the efficient development of horizontal drilling construction, resulting in the inability to reasonably analyze the oil reservoir, which adds many unnecessary difficulties to the smooth development and implementation of follow-up work.

(2) The reservoir is thin and the trajectory control is difficult
The effective thickness of thin and poor reservoirs is less than 1 m. In some extreme cases, the reservoir thickness is even less than 0.5 m, and it is also very common that the reservoir is suddenly undeveloped. Facing this situation, when drilling a horizontal well for wellbore trajectory control, the operator should pay attention to the necessary control of its vertical depth range. Only in this way can the effect be more remarkable. Through calculation, the actual vertical depth range of well trajectory should be strictly controlled within 0.5 m, which meets the construction requirements of horizontal wells in thin and poor reservoirs.

(3) There is a blind area between the measuring instrument and the bottom hole, and the trajectory control is difficult
At present, when using the existing LWD system and bottom hole bit, there is often a certain distance blind area. In the face of this situation, we can only rely on the experience of relevant operators to estimate the distance of this blind area. This makes it very difficult to control the well trajectory. For example, when entering the reservoir, good results can be obtained only by accurately entering the target, but this work is very difficult and is not conducive to the actual operation.

2.2 Calculation Method of Dynamic Programming

In this paper, a dynamic programming algorithm for τ UE and false information diffusion ability (DPA) is proposed for the night broadcast model of true and false information, This method includes two steps: finding the shortest path and calculating the forwarding probability. In the experimental part, the calculation method based on dynamic programming proposed in this paper is compared with the evaluation method based on Monte Carlo in many different topological networks, The method proposed in this paper can quickly get the approximate results with Monte Carlo simulation method, taking into account the time efficiency and accuracy, and provides a new feasible method for analyzing the propagation influence of true and false information in different networks and the evaluation of information filtering ability.

Index of information filtering ability ty.TTA Message transmission ability (FTA) is defined as formula (1) respectively:

$$F_T = \frac{N_T{}'}{N} F_F = \frac{N_F}{N} \tag{1}$$

Information filtering ability (FA) is defined as formula (2):

$$F = F_T - F_F \tag{2}$$

Generally speaking, the propagation probability of a node should be the sum of the probabilities of all its neighboring back in edges forwarding information to it. This paper adopts a single access propagation model, that is, when a node is first accessed by an information, it decides whether to forward or not. Even if the node is accessed again later, it will not change its forwarding state, Only when there is no other incoming side forwarding the information to the node before, the new incoming side can forward the information to the node. The simple probability sum can not restore the propagation process under the model. Taking the network shown in Fig. 1 as an example, the probability step to which the source information is sent and propagated is calculated, and its alse represents the information type.

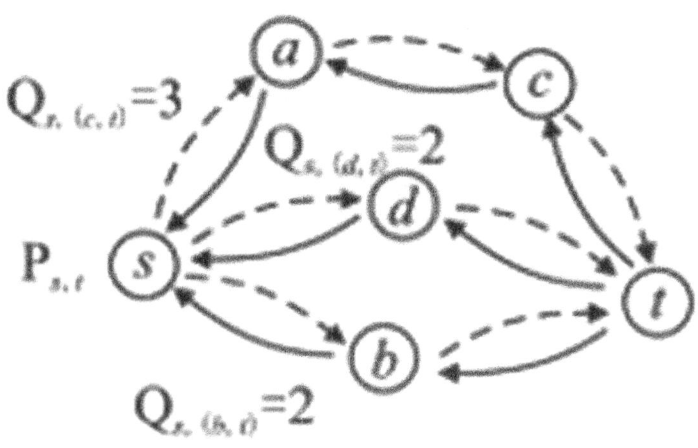

Fig. 1. Example of calculating forwarding probability

3 Data Analysis

In November 2020, Hanzhong development and Reform Commission mentioned the exploration and development of natural gas and shale gas in Zhenba County, southern Shaanxi in response to the proposal of CPPCC members. The proposal on further paying attention to and accelerating the exploration and development of natural gas and shale gas in Zhenba County, southern Shaanxi, put forward at the fourth session of the fifth CPPCC Hanzhong Municipal Committee, has been approved. The reply is as follows, Analysis formula of water well thin layer data:

$$\begin{cases} E(t)\dot{x}_k(t) = f(t, x_k(t)) + B(t)u_k(t) \\ y_k(t) = C(t)x_k(t) \end{cases} \tag{3}$$

$$\begin{cases} E(t)\dot{x}_d(t) = f(t, x_d(t)) + B(t)u_d(t) \\ y_d(t) = C(t)x_d(t) \end{cases} \tag{4}$$

Zhenba area belongs to the Daba Mountains and the geological structure belt in front of the mountains. The structural belt is rich in natural gas reserves and is one of the main fields of mountain exploration in China. It is a natural gas exploration and development block registered by Sinopec in Sichuan Basin, with an area of about 6980 km^2, including 1351.7 km^2 in Zhenba. According to the theoretical data, the natural gas reserves are 3.061 in the upper group $\times$ 1011 m^3 + 306.1 billion m^3, lower group 1.3235 $\times$ 1014 m^2 3 + 132.35 trillion US dollars. The total gas content of shale gas is 1.47–5.44 m^3t, with an average of 3.71 mm^3t. According to the national oil and gas geological evaluation data and the preliminary exploration results of Sinopec, Zhenba County has large reserves of natural gas and shale gas resources. It has high exploration and development prospects and is expected to become a new base of Shaanxi energy industry.

In May 2015, the oil and gas center of the Ministry of land and resources deployed the shale gas investigation well "i.e. Township 1 well" in Yongle Town, Zhenba County. The exploration task was completed at the end of November. The drilling depth is 1771 m, major discoveries have been made, and high-quality shale layer of 80m has been encountered. The page rock gas shows well, and the shale gas resources in towns and Ba County have been preliminarily explored. On January 21, 2016, the China Geological Survey Bureau of the Ministry of land and resources released the 2015 China Geological Survey Report on CCTV. It is expected that Zhenba will become the next important base for shale gas exploration and development in China. The learning law is as follows:

$$u_{k+1}(t) = u_k(t) + \Gamma_{l1}\dot{e}_k(t) + \Gamma_{l2}\dot{e}_{k+1}(t) + \Gamma_{p1}\Delta\dot{e}_k(t) + \Gamma_{p2}\Delta\dot{e}_{k+1}(t) \tag{5}$$

With the strong promotion of provincial and municipal governments at all levels, after full contact and demonstration, Shaanxi coalfield geology Group Co., Ltd. "Shaanxi Energy Group" signed a joint investigation and development agreement with Sinopec exploration branch in October 2018. According to the Zhenba shale gas resources cooperation and development agreement, Shaanxi Energy Group plans to invest more than 86 million yuan to carry out two-dimensional seismic survey in Zhenba County and deploy a parameter well of shale gas. In order to ensure the smooth entry of exploration equipment, Shaanxi Provincial Development and Reform Commission issued RMB 10 million for shale gas exploration road project in September 2019. A 15.9 km long shale gas exploration road will be built from Yongle Town to Dazhu village to facilitate the mobilization of large exploration equipment. On June 23 this year, shale gas parameter wells were officially drilled in Dazhu village, Yongle Town, Zhenba County.

"Shaanzhen Ye 1 well" is the first shale gas parameter well in Zhenba area, marking that the exploration and development of shale gas resources in Zhenba area has entered a new stage. According to the overall arrangement of the project, the drilling and evaluation of vertical well section is planned to be completed on September 31, and the drilling and evaluation of horizontal well will be completed on December 31, with fracturing test.

3.1 Main Difficulties in Development

Identification of interlayer by field logging guidance.

1. Judgment of drilling time
   Whether it is physical interlayer or lithologic interlayer, the drilling time is generally large. If the average drilling time of the whole reservoir is small, you can refer to micro drilling or instantaneous drilling, judge the change of lithology through the comparison of drilling time, stop drilling for cyclic observation, and take rock cuttings to confirm whether the interlayer is drilled. Figure 2 shows the operation flow of well layer data.

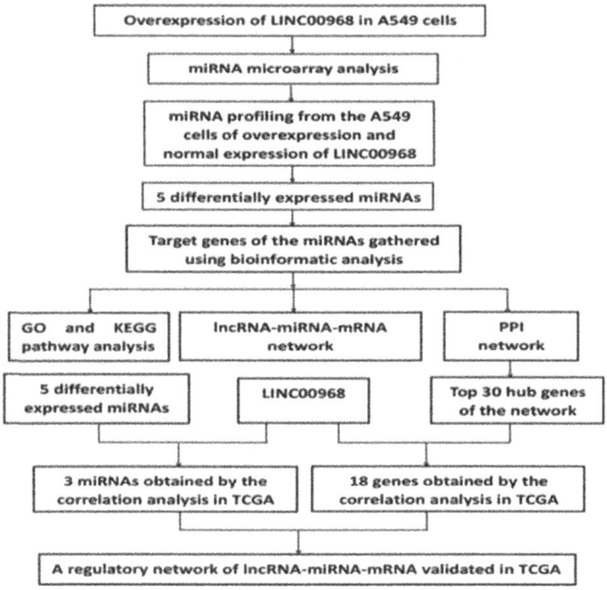

Fig. 2. Operation flow of well layer data

2. Physical properties and oiliness
   As interlayer, its physical properties and oil content are poor. This can be judged from the change of whole meridian content. 3. Lwo curve while drilling.
   The change of lithology can be judged from LWD curve. Generally speaking, the natural gamma of argillaceous intercalation is high, the bilateral curves show low resistance and the deep and shallow survey line curves are close to coincidence, and the resistance value of calcareous sandstone intercalation generally shows high resistance, and the deep and shallow survey line curves are close to coincidence.
3. Study on quantitative fluorescence IC index
   On site, we can judge the permeability of rock cuttings by secondary analysis of lithology to obtain IC index to judge the permeability of rock, so as to judge whether it is a dense interlayer.

Adjustment strategy for interlayer trajectory.

1. Sublayer correlation
   In order to provide the penetration rate, small layer comparison shall be carried out for the target horizon of adjacent wells, the horizontal variation shall be analyzed, and the actual drilling profile of construction wells shall be predicted. In the process of tracking while drilling, in combination with the feasibility analysis of drilling technology, interlayer shall be avoided or quickly passed through interlayer as far as possible to ensure the penetration rate of horizontal well reservoir.
2. Comparative logging
   Analyze and compare the logging data, compare the lithology and electrical properties with adjacent wells, identify the interlayer that may be drilled, and adjust the well trajectory as soon as possible.
3. Altitude difference
   In stratigraphic correlation, the difference of bushing altitude should be fully considered, which is particularly important for the avoidance of interlayer and the adjustment of trajectory.
4. Profile correction
   It is mainly to determine the difference between the predicted formation lithology and the actual drilling lithology through the projection of the actual drilling trajectory on the predicted profile, determine the actual depth and thickness data of each horizon, repeatedly modify and correct, predict the change of reservoir profile, then predict the next drilling profile, determine the horizontal and vertical change characteristics of reservoir, and improve the coincidence rate between the predicted profile and the actual profile, At the same time, the effective penetration rate of horizontal well reservoir is improved.

   1) After water breakthrough, the period of analyzing the direction of water inflow in horizontal wells is relatively long
   2) The matching technology of horizontal well is not perfect.
   3) The development of supporting facilities is still at the primary level of exploration
   4) The decline value of horizontal wells is relatively large. In 2014, the drawdown value of horizontal wells in an oilfield was 47.8%. In contrast, the decline value of directional wells in Chang 1 + 4 reservoir in d33 area is 291% and that of horizontal wells is 471%.

4 Example Analysis

In the process of shale gas horizontal well drilling, it is necessary to continuously improve the overall work efficiency of drilling construction, solve a series of technical difficulties of drilling fluid, adopt the best technology to select the most appropriate drilling fluid system, achieve the expected optimization goal, improve the overall quality of work, and meet the shale development effect and production demand. Based on the shale gas itself has a certain compactness, in the process of horizontal drilling, a series of problems such as the technical difficulties of drilling fluid need to be solved in order to ensure the safety

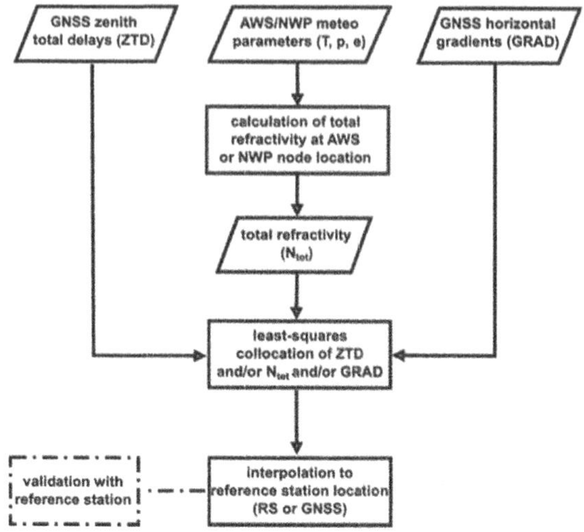

Fig. 3. Solve the technical difficulties of drilling fluid

of drilling construction. As shown in Fig. 3, solve the technical difficulties of drilling fluid.

Before using drilling fluid technology for horizontal wells, it is necessary to analyze what problems exist in shale gas development to be solved, which can be predicted in advance to find the direction of construction in the follow-up technical construction. The data investigation on improving the overall quality of construction shows that the borehole in shale gas development is complex, one of the reasons is that the wellbore stability is relatively poor, Shale gas horizontal well drilling fluid technology also faces a series of problems. For the technical analysis of shale gas horizontal well drilling fluid, it will interact with the underlying fractures and the weak underground layer in the process of use, resulting in very obvious changes in the pressure and strength of shale gas itself. Due to the interaction between shale and drilling fluid, both the wellbore and the stability of shale gas will change, In order to improve the quality of shale gas horizontal well drilling fluid technology, it is necessary to further analyze the causes of the problems. The main reason is that there is a certain dimension gap between the wellbore and the soil layer. Under the effect of the gap pressure, the liquid will have a certain interaction force with shale gas, The result is that the swelling stress and hydration stress in the middle of the underground clay layer will change to a certain extent, so that there will be a lot of liquid flowing in between the formations. The emergence of liquid will lead to the clay minerals contained in shale gas react directly with water, showing the overall expansion phenomenon. The direct result is that the tension will gradually expand, For the wellbore, its own stability has changed.

5 Conclusion

To sum up, the analysis of the technical difficulties and selection principles of horizontal well drilling fluid can improve the overall quality of shale gas horizontal well drilling fluid construction technology, and meet the relevant characteristics and requirements of current shale gas exploration in China. In the study of shale gas horizontal well drilling fluid technology, we need to learn from the past work experience, summarize, according to the actual situation of the whole construction process, constantly optimize the shale gas horizontal well drilling fluid construction technology, meet the whole construction requirements of shale gas horizontal well drilling fluid in the construction, and improve the use efficiency of shale gas horizontal well drilling fluid. Drilling high-quality shale gas horizontal wells can achieve the efficiency of shale gas exploration and development, and promote China to further achieve ideal scientific research results in shale gas research and development.

References

1. Wu, X., Li, J., Zhao, Z., Jia, J., Du, P., Hou, K.: Key technology and application of thin-bedded tight sandstone gas reservoir in horizontal well development. Springer Ser. Geomech. Geoeng. **63**(11), 139–144 (2018)
2. Jaoua, M., Rafiee, M.: Optimization of oil production in an oil rim reservoir using numerical simulation with focus on IOR/EOR application. Day 3 Thu, 19 September 2019 (2019)
3. Baslaib, M., et al.: A step change in field development drilling; a case study, onshore Abu Dhabi. J. Geogr. **3**(1), 98–105 (2020)
4. Zhang, T., et al.: An integrated well-pattern optimization strategy to unlock continental tight gas reservoir in China. Energy **20**, 123–138 (2020)
5. Zhou, Y.: Technical analysis of improving sand body penetration rate in horizontal well. J. Environ. Sci. 260–272 (2021)
6. Carpenter, C.: First multistage fracturing of a horizontal well drilled in a tight carbonate reservoir in UAE. J. Petrol. Technol. **64**(5), 402 (2021)
7. Wu, Z., Liu, H.: Investigation of hot-water flooding after steam injection to improve oil recovery in thin heavy-oil reservoir. J. Petrol. Explor. Prod. Technol. **53**(2), 1487–1509 (2018)
8. Wang, X., Qiao, X., Mi, N., Wang, R.: Technologies for the benefit development of low-permeability tight sandstone gas reservoirs in the Yan'an gas field, Ordos Basin. Nat. Gas Ind. B **7**(4), 353–383 (2019)
9. Bo, J.W., et al.: Remote-controlled automated foam injection: a digital solution to liquid loading in a China unconventional gas development. Day 1 Tue, 12 October 2021 (2021)

Application of Edge Computing in Data Dissemination Innovation

Ying Liu[1](✉), Reem Atassi[1], and Surbhi Mashat[2]

[1] XianYang Normal University, Xianyang 712000, China
MAOgen2007_a@163.cnm
[2] King Abdulaziz University, Jeddah, Saudi Arabia

Abstract. In the big data environment, the innovation of music communication path should not only pay attention to the powerful function of new media, but also can not ignore the important role of traditional media such as radio, television, newspapers and periodicals. The two should be combined to achieve coordinated development, so as to better promote the development of music industry.

Keywords: Big data · Music communication · Innovation path

1 Introduction

At present, the concept of cloud storage "cloud computing" mobile terminal gradually set off a new upsurge, the whole society is in an information age with data network as the core, all kinds of new media have emerged one after another, the fusion and convergence of massive information promote the development of various fields. Big data is characterized by huge information resources and wide distribution. It integrates numerous information to form a huge multi angle and all-round database, which makes human society enter a new era - big data era. It can collect information to the maximum extent and mine deep-seated relationships within things through massive original information, For the ability of data collection and collation is particularly importantp [1]. From the perspective of music communication, compared with the simple way of "oral teaching" of original music communication, or the early traditional media such as records, tapes and radio and television images, today's new media has a wider communication path and scope, and more flexible communication methods, which brings new opportunities and challenges to the communication and development of music art.

On September 6, the inaugural meeting of the world's first international research center for big data for sustainable development and the 2017 International Forum on big data for sustainable development opened in Beijing, announcing its formal establishment. This is a substantive measure to support the United Nations in playing a central role in international affairs, and provides a new platform for global scientists to participate in and promote the sustainable development goals "sdgs". According to Guo Huadong, director of the International Research Center for big data for sustainable development and academician of the Chinese Academy of Sciences, on the basis of the special earth big data science project of the Chinese Academy of Sciences, the newly established

I. Ahmad et al. (Eds.): STSIoT 2021, LNDECT 122, pp. 632–640, 2023.
https://doi.org/10.1007/978-981-19-3632-6_74

center will meet the main needs of sdgs. It is understood that the "2025 agenda for sustainable development" adopted 193 member states at the 70th United Nations General Assembly, including 17 documents with "sdgs" and 169 specific goals as the core content. Among them, 17 sustainable development goals aim to comprehensively solve the development problems in the three dimensions of society, economy and environment, so that mankind can move towards the road of sustainable development [2]. Therefore, the International Research Center for big data for sustainable development will establish a global sustainable development goal monitoring and evaluation system to provide data sharing, technical support and decision-making support for relevant United Nations agencies and Member States.

2 Related Work

In this paper, we aim to application of edge computing in data dissem-ination innovation.

Ref [1] present the social marketing literature as a supplement to the Prevention Synthesis and Translation System (PSTS), the system responsible for dissemination. Using data on firms operating in France, Cette et al. highlight that, at the technological frontier, productivity has accelerated, especially over the recent period, which contradicts the hypothesis of a decline in innovation [2]. Ref [3] review of some relevant literature and research results. The objective of Ref [4] is to analyse whether vocational training centres are indeed relevant agents in the articulation, knowledge exchange and dissemination of a local innovation system. The analytical stages of the mental process of diffusion do not determine adoption of innovation [5]. Sanders et. al present an exploration of public discourse surrounding the use of artificial intelligence (AI) in agriculture, specifically related to precision agriculture techniques [6]. The contribution of Ref [7] is to review the operation of the online makerspace upon disseminating innovation and ideas. The main essence of innovations in the social and ecological sphere of the organization of cities of the Western region of Ukraine is considered [8]. Other influential work includes Refs [9, 10].

This paper consists of the following parts. The first part introduces the relevant background and significance of this paper, the second part is the related work of this paper, and the third part is data analysis. The fourth part is example analysis. The fifth part is conclusion.

Compared with traditional media, media with Internet as the core have the following changes in music communication. Firstly, the difference in communication mode is different from the linear one-way communication mode of traditional music communication, which is mainly two-way and interactive communication. Secondly, with the increase of communication content, traditional media will be limited by many factors when they spread music. In the big data environment, the Internet media will be limited by many factors, Massive information sources make the content of communication increase significantly; finally, with the expansion of the scope of communication, traditional media can only carry out promotion and communication in its origin, while under the background of new media, music communication has got rid of the constraints of time and space. In the big data environment, the emergence of new media has brought unprecedented opportunities and challenges to music communication [3]. Considering that the complex data is the initial received frequency domain signal data, which can not be directly detected and processed, it can be input into the detection network after preprocessing. The so-called preprocessing refers to the imaginary part and real part of the series data. It is worth noting that the complex data should be extracted before the series connection, and after the preprocessing is completed, the transmission data can be obtained. The input is often connected with the hidden layer [4]. According to the above, the hidden layer of signal detection network is composed of neurons. Therefore, we can accurately express the output data of multi-layer DNN network with mathematical expression as follows:

$$\widehat{S} = f(y, \theta) = f_s^L(F_r(W_1 y + b_1)) \tag{1}$$

$$m = -\frac{n \cdot ln\, p}{(ln\, 2)^2} \tag{2}$$

2.1 Understand the Audience Needs, Spread a Wide Range

Mobile phones, computers and other mobile terminal devices are more and more widely used, which provides a broader space for people to communicate with each other. In particular, mobile devices based on smart phones have become an indispensable part of people's lives. Their users cover people of different ages, and the communication activities between people have become more simple and convenient. To a certain extent, it also makes the dissemination and sharing of music more efficient, and gradually become the main way of music dissemination. All kinds of music software carried on the mobile client can provide convenience for people to obtain different types of music and meet their basic needs for music. More importantly, these music software can record the listening frequency and type selection of music in people's daily life, collect massive data and apply big data technology, We can further analyze people's personality, hobbies and habits, fully understand their psychological needs, so as to recommend personalized music solutions to different users according to different needs [5]. Through a variety of interactive software platforms, such as wechat, QQ, microblog, etc., information can be shared on the network, so that music communication can break through the limitation of time and space, and realize large-scale interactive communication in virtual space. An influential music work can reach tens of millions of listeners in just a few minutes. By forwarding and sharing in cyberspace, people can make the propagation speed grow exponentially and realize large-scale global dissemination.

Since it was successfully held for the first time in July 2018, the national music education conference has shared advanced international experience and discussed the road of innovative development of Chinese style music education with the attention of the booming emerging social music education industry. The 2017 National Music Education Conference and international music life exhibition were grandly opened in Tianjin auditorium. And national music education cooperation platform. During this period, 100 special music education activities were also held, including famous forums, master workshops, on-site interactive teaching, brand musical instruments display, on-site production of hand-made musical instruments, and wonderful activities such as symphonic concert and "Wanye Cup" music education paper award ceremony jointly contributed by Beijing, Tianjin and Hebei artists. "The Tianjin municipal Party committee and government attach great importance to national music education. It is listed as a popular project, and has organized experts, scholars and teachers of music education in the city to actively organize meetings, gather with their peers, exchange and learn from each other, show Tianjin's style and bloom Tianjin's charm", Zhang Chonghe said [6].

The content of each branch venue includes special theme activities of different parts, such as basic music education, college music education, children's music education and out of school music education, "children's Palace", social and artistic accompaniment institutions, piano, folk music, chorus, integration of aesthetic education, thesis publicity and so on. Display a variety of high-end brand musical instruments, intelligent music teaching equipment, music training courses and comprehensive solutions for musical instruments entering the campus, providing most companies with a one-stop service platform for contact, cooperation, learning and exchange with educational channels. At the same time, members of the "Tianjin" charity training course of the vocational training center of China Musical Instrument Association opened during the music education conference will also be invited to participate in the opening ceremony of the 10th China future international children's Art Festival. Tianjin Municipal Education Commission selected excellent grass-roots music teachers and teaching researchers in Tianjin to carry out exhibition and exchange of music education topics and organize observation at the meeting; The first charity training course of "Tianjin" vocational training center of China musical instrument society, jointly sponsored by China Musical Instrument Association and the cooperation and exchange office of Tianjin Municipal People's government, will be held during the conference.

In three days, it is expected that there will be more than 100 keynote speeches, round table forums, master classes and music education workshops; It focuses on many advanced music education concepts, teaching methods, teaching materials, achievements, musical instruments, teaching materials and teaching equipment. It aims to further implement the spirit of the two documents, use music to build an art platform to realize the popularization of aesthetic education, promote the development of music education, expand the music population, guide the consumption of musical instruments, help an all-round well-off society, and jointly sing the theme song of the times "music makes life better".

$$\lim_{k\to\infty} \|\Delta u_k(t)\|_\lambda \leq \frac{1}{1-\breve{\rho}} m_5 d \tag{3}$$

$$\begin{cases} E(t)\dot{x}_d(t) = f(t, x_d(t)) + B(t)u_d(t) \\ y_d(t) = C(t)x_d(t) \end{cases} \tag{4}$$

2.2 The Communication Content is Rich and the Transmission Efficiency is Improved

With the improvement of living standards, people pay more and more attention to the spiritual needs. As an important form of spiritual civilization, music art is more and more popular. Under the big data environment, we can give full play to the advantages of network information. We can show the endless music resources and various kinds of music beauty in various network platforms. Different styles of music art can also be covered. The Internet can integrate multiple elements, such as images, text and video, and so on [7]. The establishment of relevant databases, through data analysis, the classification of different music resources, the scientific and effective dissemination of music art, the audience through a simple search can obtain the desired information, more targeted and efficient to provide convenient services for people [8]. At the same time, music delivery saves a lot of time and resources. Music resources can be used repeatedly, greatly improving the efficiency of communication.

Since it was successfully held for the first time in July 2018, the national music education conference has shared advanced international experience and discussed the road of innovative development of Chinese style music education with the attention of the booming emerging social music education industry. The 2017 National Music Education Conference and international music life exhibition were grandly opened in Tianjin auditorium. And national music education cooperation platform. During this period, 100 special music education activities were also held, including famous forums, master workshops, on-site interactive teaching, brand musical instruments display, on-site production of handmade musical instruments, and wonderful activities such as symphonic concerts jointly dedicated by Beijing, Tianjin and Hebei artists and the awarding ceremony of "ten thousand Leaf Cup" music education papers.

The content of each branch venue includes special theme activities of different parts, such as basic music education, college music education, children's music education and out of school music education, "children's Palace", social and artistic accompaniment institutions, piano, folk music, chorus, integration of aesthetic education, thesis publicity and so on [9]. Display a variety of high-end brand musical instruments, intelligent music teaching equipment, music training courses and comprehensive solutions for musical instruments entering the campus, providing most companies with a one-stop service platform for contact, cooperation, learning and exchange with educational channels.

3 Data Analysis

3.1 Change the Traditional Communication Concept and Establish the Consciousness of Innovation and Development

In the big data environment, music communication presents many new characteristics, realizing the transformation of communication mode from single to diversified. In the traditional sense, the whole process of music production, transmission and audience is unidirectional, and the transmission process presents a simple, top-down linear process. In the environment of big data, the mode of music transmission has undergone profound changes. Whether the creator, manager, communicator or audience of music can be a part of the process of music transmission. The whole process is closely linked and intertwined with each other. The communication and exchange between the three subjects is instant and equal, which enriches the transmission path of music. Therefore, in order to realize the innovation and development of music communication, we should first establish a new communication concept, no longer limited to the traditional way of communication, from all aspects of the communication process, in-depth thinking, into the new means of communication. In the complex and changeable network environment, we should actively create a healthy and harmonious development environment for music communication, pay attention to the pursuit of music creation quality, from the perspective of the audience, strive to create music works close to life and full of positive energy, make full use of the interactive development platform of network resources, change the traditional communication concept, and establish the awareness of innovation and development. Establish a sense of innovation and development, as shown in Fig. 1.

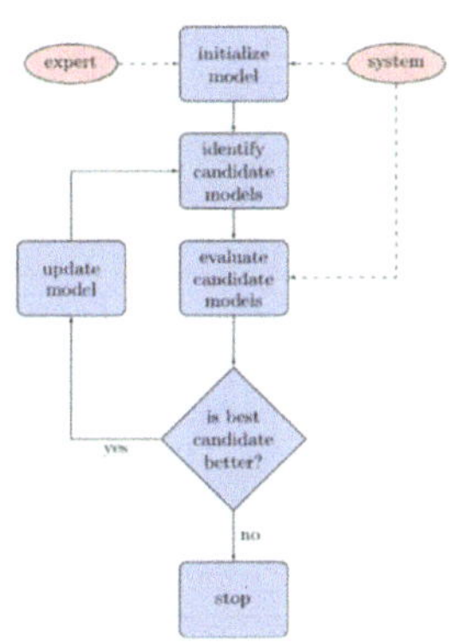

Fig. 1. Establish a sense of innovation and development

3.2 Pay Attention to Communication and Interaction, Strengthen the Supervision of Network Environment

In the process of music communication, we should make full use of the interactive characteristics of the network platform, expand the path of music communication, and make full use of various forms of social networking software to promote music. Because social networking software has the advantages of high transmission efficiency and strong interaction, it can speed up the spread of music. Some minority music art is shared by different people through the release of social networking software, It can also promote the transmission of the music resources, so that it can be welcomed by more people. We should pay attention to the in-depth development of music software in mobile terminal devices, and provide personalized music lists for different music audiences, which can not only meet the needs of different groups of people, but also cultivate audience loyalty and satisfaction, and realize the innovative communication of music. At the same time, we should pay attention to the interaction between the two links in the broadcasting process, find out the problems in time, and understand the feedback of the audience. For the virtual cyberspace, we also need to strengthen the supervision, maintain a healthy and harmonious network order, and give warnings and related penalties for plagiarism [10]. At the same time, we should also pay attention to purifying the network environment, cleaning up unhealthy and negative music works, and providing people with positive music resources. The network environment supervision is shown in Fig. 2.

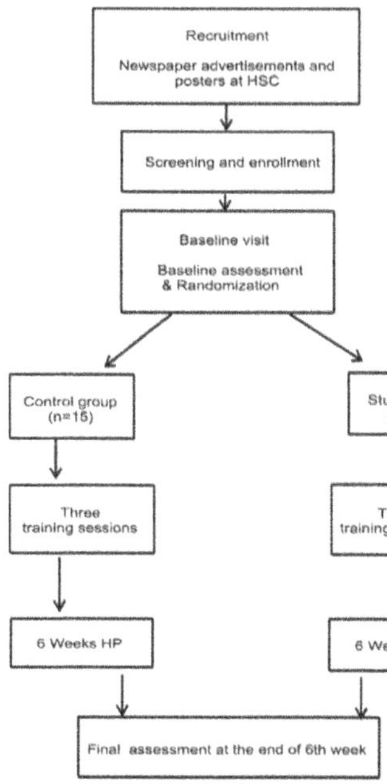

Fig. 2. Network environment supervision

4 Example Analysis

The convenient and fast network platform expands the scope of music communication, improves the speed of communication, brings unprecedented opportunities to music communication, and expands the space of music development. Everyone can participate in it, whether it is star singers, professional creators or "grassroots" people in the virtual and huge cyberspace can carry out different role transformation, the network gives people equal development opportunities and platform. Although the emergence of new media has brought a lot of convenience for music communication, we can not ignore the traditional way of music communication. We should combine the traditional media with the new media. Radio programs should break the shackles of traditional concepts, grasp the needs of the audience in time, and no longer be limited to the traditional types of music programs. In music radio stations, the music elements that the audience likes should be increased, the interaction with the audience should be increased, and the music development situation and content should be innovated. In recent years.

Hong Kong stocks were decoded, and the original trump column of the Hong Kong Financial News Agency gathered financial celebrities.

This advantage lies in that the company cuts into the copyright operation from the b-end, staggers the C-end market of pop music, and gets involved in the universality of music quality education to avoid confrontation with the pop music platform. Therefore, cook music's differentiated business model is the reason for its survival in the environment of industry giants, and it also makes it a leader in the field of industry segmentation. On the other hand, although cook music's profit margin is very high, "reaching 36.54% in 2018", the problem of over reliance on upstream music producers and music platform customer base can not be ignored. The revenue structure of foreign music streaming media giant and spotify, the third largest shareholder of Tencent music, is relatively single, but in recent years they have been trying to increase advertising service revenue and attract members to listen to music in advertising; At the same time, it is trying to deploy broadcasting companies and pan music entertainment businesses such as k-song. Cook music has also been trying to achieve diversified development in recent years, and one of its main strategic priorities is to enter the music education business.

According to the data in the prospectus, in addition to Tencent music, Netease cloud music and other streaming media platforms, in 2016, Kux music studied in the music education market and began to independently develop smart music. Starting from providing high-quality classical music education to kindergarten children, American stock online education benchmarking and upgrading. According to the public information on American listed companies in the market, tal and who to learn from the two companies are highly valued by the market at the same time. The extremely high multiple of the valuation of leading online education enterprises, combined with the company's respective professional characteristics, shows that American stock investors attach great importance to the user scale and stickiness of online education platform, the quality of independent production content of the platform and the diversification of business scenarios.

With such a huge market scale and growth of the education market, online music education still has a long way to go. As the largest revenue source of KUKE group, "copyright" has established a very high barrier in this business and won 95% of the world's classical music copyright; From the perspective of customer attributes, b-end

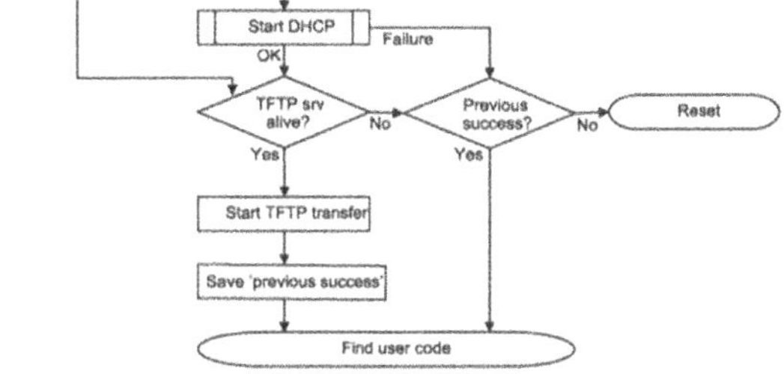

Fig. 3. Example analysis

customers such as universities, libraries and music platforms have strong stickiness. The company's core value and growth space are reflected in the field of intelligent education services. The company's products and models are relatively mature, the curriculum is modular, automated, large-scale, and the curriculum pricing is low. Music activities will become a new business highlight to help the company gradually open its brand awareness from the niche market of classical music, attract more potential customers and continuously expand its market share. Example analysis is shown in Fig. 3.

5 Conclusions

In the big data environment, the innovation of music communication path should not only pay attention to the powerful function of new media, but also can not ignore the important role of TV, newspapers and other traditional media. Instead, the two should be combined to achieve coordinated development and better promote the development of music industry.

References

1. McAlindon, K.: Selling innovations like soap: the interactive systems framework and social marketing. Am. J. Commun. Psychol. **60**, 242–256 (2017)
2. Cette, G., Corde, S., Lecat, R.: Firm-level productivity dispersion and convergence. Econ. Lett. **166**, 76–78 (2018). (IF: 3)
3. Indraningsih, K.S.: Strategi diseminasi inovasi pertanian dalam mendukung pembangunan pertanian (2018). (IF: 3)
4. Porto Gómez, I., Zabala-Iturriagagoitia, J.M., Larrakoetxea, U.A.: Old wine in old bottles: the neglected role of vocational training centres in innovation. Vocat. Learn. **11**, 205–221 (2018). (IF: 3)
5. Morrato, E.H., Lennox, L., Schuster, A.: 2564 designing for dissemination: characteristics of clinical and translational science award (CTSA) hubs as adopters of clinical and translational science innovation. J. Clin. Transl. Sci. **2**, 8–9 (2018)
6. Petry, J.F., Sebastião, S.A., Martins, E.G., de Azevedo Barros, P.B.: Innovation and the diffusion of technology in agriculture in floodplains in the state of Amazonas (2019). (IF: 3)
7. Sanders, C.E., Mayfield-Smith, K.A., Lamm, A.J.: Exploring Twitter discourse around the use of artificial intelligence to advance agricultural sustainability. Sustainability **13**, 12033 (2021). (IF: 3)
8. Kamaruzaman, A.F., Ismail, A.R., Daud, K.A.M., Jusoh, N.: Innovation dissemination in online makerspace learning. In: Proceedings of 8th International Conference on Advanced Materials Engineering & Technology (ICAMET 2020) (2021)
9. Asgari, E.: Emergence and evolution of an industry: exploration of distinct roles of demand pull and supply push. In: Academy of Management Proceedings (2021)
10. Basarab, V., Novosad, D., Pavlova, O., Shabala, O.: Implementation of social and ecological innovations of cities of the western Ukraine. Galic'kij Ekon. Visnik 112–119 (2021)

The Study Plan Design of Geography Research Based on Reinforcement Learning

Yanxin Bi(✉)

Yunnan Land and Resources Vocational College, Kunming 650000, Yunnan, China
yanxin0001985@163.com

Abstract. The design of excellent geography research and study plan is conducive to the efficient development of research and study activities, the improvement of students' comprehensive quality and the professional growth of geography teachers. Taking a research and study activity as an example, this paper discusses the design strategy of research and study plan preparation and research theme, goal, content, form and process.

Keywords: Deep study · Geography study · Study plan design

1 Introduction

From the perspective of education, Yanwu and Ziban society schools and nature explore human history and natural knowledge, connect society, schools and families, and jointly train Qiao people. Li K D ba a people's travel promotes the further integration of education and cultural tourism.

Sichuan has made full use of the rich Mu ran and Ba Wen Xuan o, u Mu right "Bai Na into the primary and secondary school education and teaching plan. It has established a number of "natural ecological beauty and colorful cultural charm" The scientific research, low-1 art, culture, land, curriculum and route, around the four Wang interests, carry out the nine core S-line suitable for school collective learning and family parents and children, such as shangshaosheng live Bashu culture, Three Kingdoms culture, non genetic inheritance, Poetry Road, etc., and build a multi-level tourism product system of scenery, customs and style.

From the perspective of travel, the study tour promotes the integration of J Gongzi Gongzhuang's practice, and more cultural tourism products emerge in the practice of studying travel agencies. Therefore, the study and research travel agency has become an important starting point for the transformation and upgrading of tourism industry.

The content covers the "camp" of the R & D base, scientific research tutor training, route product development, Yungao Pingkou version 'xiuha center takes advantage of 5g + dynamic geological structure changes and long-standing humanistic, geographical and cultural charm. Moruo drama town takes Moruo culture as the core, parent-child leisure as the theme and green leather train as the carrier to create a new type of research and study integrating catering, accommodation, travel, tourism, shopping and learning Form.

I. Ahmad et al. (Eds.): STSIoT 2021, LNDECT 122, pp. 641–649, 2023.
https://doi.org/10.1007/978-981-19-3632-6_75

Through the selection basis, the graduate schools in the whole province will be guided to develop towards the standard. Dunhuang research tourism is becoming a research and development of various forms of the Silk Road for domestic and foreign tourists, college teachers, students and business groups to perceive the Silk Road civilization and inherit the new Shidan landform and other cultural and natural resources of Chinese culture. As shown in Fig. 1 below, the geography learning framework.

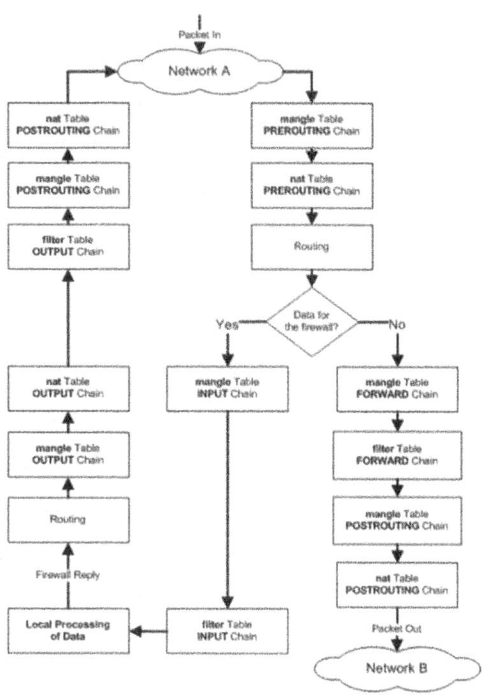

Fig. 1. Geography learning framework

With the continuous development of geography research travel, research activities must be changed from random to standard, from coarse to fine, from shallow to deep. The design of research and study plan is the core link of geography research and study curriculum design. It is the "blueprint" for the school to organize students to carry out research and study travel and carry out geographical exploration activities. To carry out in-depth research and study, we must design scientific and standardized research and study plan. This paper consists of the following parts. The first part introduces the relevant background and significance of this paper, the second part is the related work of this paper, and the third part is data analysis. The fourth part is example analysis. The fifth part is concluding remarks.

2 Related Work

L Against the background of a review of international research using a spatial perspective on accommodation this paper analyses the geography of tourist bednights in South Africa [1]. It makes a case for slowing down modes of abstraction and proposes some contribution of Ref [2] on sustainability transitions in contact zones. Knight discussed issues of decolonization and transformation of Geography curricula at different universities in South Africa, and whether issues such as decolonization and transformation are being addressed in these curricula [3]. Ref [4] discuss the significance of some of these new topic areas for South African Geography, and the application of such research to address twenty-first century local to global issues. Facing the UN's goal of eradicating poverty by 2030, poverty geography research in the new period should focus on the complexity, spatial heterogeneity and mechanism of poverty, and designs anti-poverty paths and models suitable for different countries [5]. Lee et al. tried to analyze the recent research trend and topics of geography by analyzing the abstracts submitted to the AAG Conference from 2018 to 2020, using the LDA-based Topic Modeling [6]. Suciani et. al aimed to determine the character of students in the development of character-based learning media [7]. Zaedun aimed to describe the effect of the Snowball Throwing learning model on the learning interest of students at SMA Negeri 2 Labuapi in Geography [8]. In recent years, Nanchong has adhered to the combination of going out and introduction, strengthened the interaction and exchange with the surrounding cities "states", and created a series of high-quality research tourism routes with diverse themes and flexible forms, such as the cultural research tour of the Three Kingdoms. Nanchong has developed a number of research and learning routes or bases with outstanding educational effects and Nanchong characteristics, so as to promote the integration of urban literature and tourism. In November 2016, the Ministry of education and other 11 departments issued the "opinions on promoting primary and secondary school students' research and study travel" (hereinafter referred to as the "opinions") clearly put forward that "it is necessary to carefully design the research and study travel activity curriculum, so as to achieve high intention, clear purpose, vivid activities and effective learning, and avoid the phenomenon of only travel but not study or only study but not travel.

2.1 The Meaning of Reinforcement Learning Theory

Reinforcement theory, also known as behavior modification theory, is a new behaviorism theory proposed by Skinner, an American psychologist and behavioral scientist, based on the in-depth study of the characteristics of conscious behavior. He is an extreme behaviorist in the academic point of view of psychology. His goal is to predict and control people's behavior through people's internal psychological process and state's response to the external environment. He believes that human behavior has the characteristics of conscious conditioned reflex, that is, it can act on the environment and promote its change, and the change of the environment (behavior result) in turn affects the behavior. Therefore, when a certain behavior is consciously affirmed and strengthened, it can promote the repetition of this behavior; Negative reinforcement of a certain behavior can correct or prevent the repetition of this behavior. Therefore, people can use this positive reinforcement or negative reinforcement method to affect the consequences of behavior, so as to modify their behavior. According to this principle, different reinforcement methods and means can achieve the purpose of effectively stimulating positive behavior. The reinforcement theory advocated by Skinner is a theory about understanding and modifying human behavior based on the reinforcement principle of learning. The so-called reinforcement, from its most basic form, refers to the positive or negative consequences (reward or punishment) of an act. It will determine whether this act will repeat in the future at least to a certain extent.

2.2 It is Conducive to the Efficient Development of Research and Learning Activities

Compared with classroom teaching, geography research activities are characterized by changeable location, long time, complex process, obvious openness, generation, irregularity, uncontrollability and discontinuity. If we don't plan the theme, goal, content and method of research and study activities in advance, research and study activities will become mere formality and the effect will be difficult to guarantee. Geography research and study plan is the "teaching plan" of off campus research and study activities. A reasonably designed research and study plan can make up for the deficiencies of school curriculum and classroom teaching, ensure the standardized, efficient and smooth development of research and study activities, and is conducive to the completion of various research and study objectives. Nonlinear generalized iterative learning system:

$$\begin{cases} E(t)\dot{x}_k(t) = f(t, x_k(t)) + B(t)u_k(t) + d_k(t) \\ y_k(t) = C(t)x_k(t) \end{cases} \tag{1}$$

$$\begin{cases} E(t)\dot{x}_d(t) = f(t, x_d(t)) + B(t)u_d(t) + d_d(t) \\ y_d(t) = C(t)x_d(t) \end{cases} \tag{2}$$

$$\sup_{0 \le t \le T} \|d_k(t)\| \le \varphi \tag{3}$$

For industry practitioners eager to quickly turn their ideas into AI products, pytorch does not really solve their pain points. This shows that during the "14th five year plan" period, scientific and technological innovation will be promoted to a higher strategic position to change the current technology situation highly dependent on the United States and become an important force to promote future economic development. This change has injected impetus into China's innovation power. Especially in the past few years, due to the development of trade protectionism and anti globalization, many fields have been affected by foreign industrial chains, and the self-improvement of science and technology is indeed imminent. This paper explores the importance of the deep learning framework called "operating system" in the Al era. As shown in Fig. 2, the importance of deep learning framework.

2.3 It is Conducive to the Improvement of Students' Geographical Core Literacy

Compared with classroom teaching, the autonomy and practicality of geography research and learning activities are more prominent. Students are the main body of research

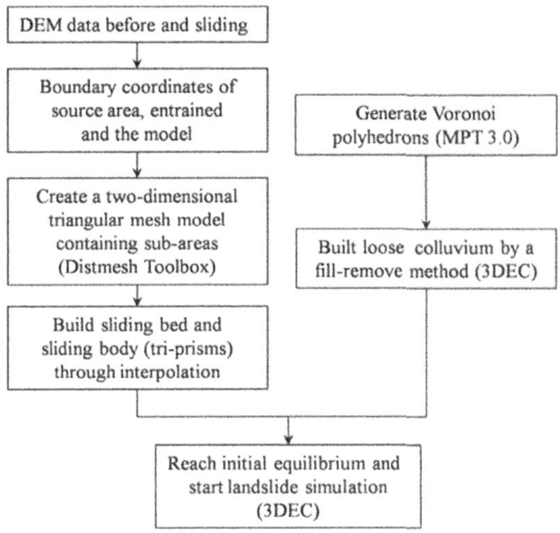

Fig. 2. Importance of deep learning framework

and learning, and also the participants of research and learning plan design. Students participate in the design of the program, choose the theme of the activity from daily life, society and contact with nature, and put forward reasonable suggestions on the goal, content, mode and steps of the activity. The designed program is more in line with the actual needs of students' growth, and is more conducive to students' active participation and willing to explore. In the design and development of research and learning activities, students can also find personal value, improve their awareness and ability of observation, analysis and problem solving, and cultivate their sense of responsibility, innovative spirit and practical ability, which is conducive to the implementation of core literacy. In the final of the first "China Cup" national research and study travel competition, the team of the middle school affiliated to Luoyang Institute of technology from topic selection to research and study activity design were all completed by the students themselves. What the students gained was not only the first prize of the group, but also the improvement of their comprehensive quality.

As one of the important choices of autonomous and controllable core technologies, the key value of deep learning in AI system is that it is not only the technical barrier required for the further development of artificial intelligence, but also the core content of the transformation and competition of technology giants. For our real world, its technical value is mainly reflected in the transplantation and replication of expert experience, so that the machine can further use low-level repetitive labor channels to improve the overall efficiency. Chinese companies that will be deeply integrated into the ecology of foreign in-depth research framework must face l the stagnation of R & D. to switch to a new in-depth learning framework, we have to experience a run in of overlapping new ecology. Each time the system runs iteratively, the initial value of the system shall meet the following requirements:

$$x_k(0) = x_d(0), k = 1, 2, 3, \cdots \tag{4}$$

The learning law is as follows:

$$u_{k+1}(t) = u_k(t) + L(t)(\dot{e}_{k+1}(t) + e_{k+1}(t)) \quad (5)$$

2.4 It is Beneficial to the Professional Development of Geography Teachers

The geography teacher is not only the tutor of the activity, but also the main designer of the study plan. Teachers should design excellent research and study plans. They should not only study the relevant national documents, consult the literature, study the curriculum standards and understand the situation of students, but also go to the research and study base in advance to find out the resources of the base. This process is actually the process of curriculum development, subject research and teacher growth. In recent years, many of the teachers who have participated in the activities of "seven color eyes Heluo research and Study Workshop" and the design of research and study plan have grown into excellent geography research tutors and subject leaders, and have been invited to exchange experience in academic activities at home and abroad, and even in the whole country. As shown in the figure below, the three participants exchanged experiences in academic activities (Fig. 3).

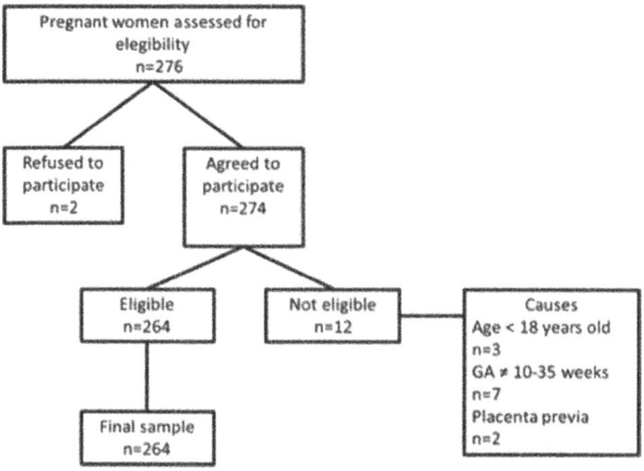

Fig. 3. Exchange of experience in academic activities

Today, as China's first independent and controllable industrial level in-depth learning platform, it is open-source and fully functional. In the report on the market share of in-depth learning framework platform in the second half of 2020 released by authoritative data research institution IDC, Baidu, Google and Facebook ranked among the top three in the comprehensive share of in-depth learning platform market in China. Among them, Baidu's overall market share ranks second, almost the same as the first Google, breaking the limitations of domestic al developers' over reliance on foreign open source deep learning frameworks. At this stage, the rapid development of artificial intelligence technology has promoted the global technological revolution and industrial change. Baidu PaddlePaddle and Baidu Kunlun, which are independently developed, are not only the representatives of China's self strengthening, but also become the important infrastructure of AI's new infrastructure and the foundation of China's intelligent economy, leading the trend of industrial intelligence in China.

3 Data Analysis

The research plan of geography can be classified from different angles. According to the scale involved, it can be divided into the overall plan of the school, the grade (class) plan, the school year (semester) plan, a specific activity plan, etc.; according to the activity mode, it can be divided into the field investigation type, the social investigation type, the theme exploration type, etc.; according to the design subject, it can be divided into the plan designed by the school, the research group, individual teachers, students, etc.; According to the use of the main body can be divided into teachers, students, travel agencies and other programs. This paper discusses the scheme design of research and learning activities in a research and learning point. According to the design intention and use time, it can be divided into preview plan and activity design.

Activity design is a task plan designed to avoid students' blind and disordered activities when formal activities are carried out, which is mainly used by students. Its main contents include research and learning objectives, time and place, research and learning process, task list (design focus) effect detection, etc.

The input of each layer of the density NTE network is the union of the output of all previous layers, and the feature map learned by this layer will also be directly transmitted to the listening layer as the input. The problem of vanishing gradient is alleviated, feature propagation is strengthened, feature reuse is encouraged, and the number of parameters is greatly reduced.

$$x_i = H_i([x_0, x_1, \cdots, X_{i-1}]) \tag{6}$$

Where, H means that the nonlinear transformation function includes batch normalization (BN), activation function reu and convolution operation. With this design, the number of parameters in the density NTE network layer is greatly reduced, and the transfer between features is enhanced. When the size of the feature map changes, the join operation in Eq. (3) is not feasible. To solve this problem, a transition layer is added between different dense blocks to perform convolution and merging operations. The transition module also includes batch standardization, activation function convolution layer and pool layer. The spatial and temporal features of vehicle behavior can be extracted from video through dual stream convolution neural network, and then the two features are fused by concat algorithm.

4 Example Analysis

Excellent in-depth research and learning plan has the characteristics of distinct theme, complete structure, reasonable process, deep integration of resources and subject knowledge, matching of learning stages, rich content, diverse forms, efficient activities, scientific evaluation and so on. The design process and content of geography research and learning plan include: preparation before design, determination of research and learning theme, setting research and learning objectives, design of research and learning activities (including activity content, form and process, etc.), design of research and learning evaluation scheme, etc. (see Fig. 4 below).

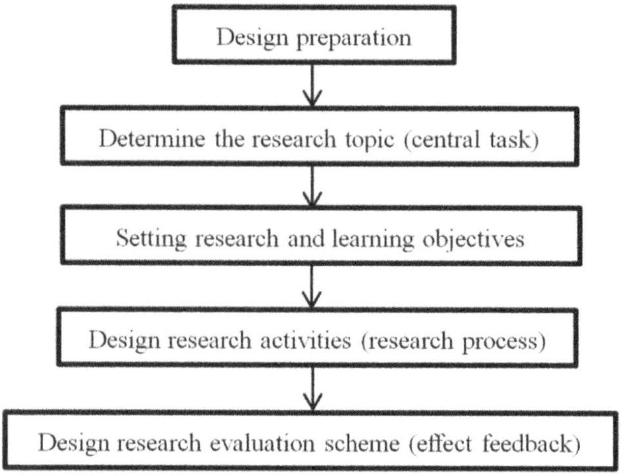

Fig. 4. Design process and content of learning plan

(1) Connecting with the geography curriculum standard and core literacy, the geography curriculum standard and core literacy are not only the "Outline" of classroom teaching, but also the "foundation" of in-depth study outside school. The geography curriculum standard for senior high school (2017 Edition) puts forward the basic concepts of "cultivating students' essential core literacy of geography discipline, innovating the learning methods of cultivating core literacy of geography discipline", and puts forward the curriculum goal of "implementing the fundamental task of moral education from the perspective of geography education through the cultivation of core literacy of geography discipline". It is necessary to carry out this basic idea and curriculum goal, and promote the integration of research and study travel, curriculum standard and school curriculum.
(2) Docking base is an important carrier of research and learning activities. Deep research and learning must promote the deep integration of subject knowledge, students' life experience and research and learning base resources. The design of learning plan must fully understand the situation of base resources. For example, before the research of Longmen Grottoes, the research tutors visited Longmen scenic spot many times, identified eight main research sites, such as yuwangtai, Qianxi temple, Moya Sanfo temple and Fengxian Temple, designed theme research activities based on practical problems (the actual situation of research site resources), and asked students to consult the relevant materials of Longmen Grottoes in the preview plan, At the same time, students are encouraged to find and put forward innovative and in-depth research and learning problems from the base resources, comprehensively

use existing knowledge and experience for analysis, carry out subject research, and put forward solutions.

5 Concluding Remarks

In order to save time and improve efficiency, the selection, connection, combination and activity arrangement of research and study sites should be scientific and reasonable, and the space-time sequence of research and study sites should be reasonably arranged according to the research and study objectives, research and study site location and other elements. We can put the activities that can stimulate the curiosity in the front, the relatively boring activities in the middle, and leave the "answer" of the problem to the last, so as to stimulate students' curiosity and improve the effect of research activities. In the research and study activities of Xiaolangdi water control project, first lead the students to study each part of the water control project, and then visit the museum, which is helpful for the students to rise from perceptual knowledge to rational knowledge, and improve the overall research effect. The design of learning plan also includes the design of research effect evaluation. After the research and learning practice, according to the actual situation and the new cognitive experience of teachers and students, timely improve and adjust the design of research and learning activities, which is conducive to the development of future activities.

Acknowledgements. Scientific Research Fund project of Yunnan Education Department in 2020: "Study on the Evaluation and Application of SERVQUAL Model on Higher Vocational Tourism Management Teaching – Taking Yunnan Land and Resources Vocational College as an example", 2020J0967.

References

1. Rogerson, J.M.: The geography of tourist bednights in South Africa. GeoJ. Tour. Geosites **289**, 112420 (2018)
2. Schwanen, T.: Thinking complex interconnections: transition, nexus and geography. Trans. Inst. Br. Geogr. **39**(2), 957–972 (2018)
3. Knight, J.: Decolonizing and transforming the geography undergraduate curriculum in South Africa. South Afr. Geogr. J. **100**, 271–290 (2018)
4. Knight, J.: New perspectives on the discipline of geography in South Africa. World Regional Geogr. Book Ser. **14**(7), 1–11 (2018)
5. Zhou, Y., Liu, Y.: The geography of poverty: review and research prospects. J. Rural Stud. **38**(10), 12608–12618 (2019)
6. Lee, Y., Lee, Y.-H., Seong, J., Stanescu, A., Ji, S.-H., Hwang, C.: An analysis of the latest trends and topics in geography research using topic modeling. Int. J. Eng. **25**(3), 618–627 (2020)
7. Suciani, A., Effendi, D.I., Soedirman, Z.: Geography learning media based on character education in the digital revolution era. J. Geogr. Inf. **54**(2), 233–242 (2021)
8. Zaedun, Z.: Pengaruh model pembelajaran snowball throwing Terhadap Minat Belajar Geografi Siswa SMA Negeri 2 Labuapi. J. Teknol. Pendidikan: J. Penelitian Dan Pengembangan Pembelajaran **18**(4), 1018–1027 (2021)

Design of Improved Genetic and BP Hybrid Algorithm and Neural Network Economic Early Warning System

Qing Li(✉)

Huali College of Guangdong University of Technology, Guangzhou 511325, China
liqing100862020@163.com

Abstract. This paper analyzes some key problems in the design of neural network economic alarm system, and puts forward an improved Liao Chuan algorithm. On this basis, it introduces a hybrid algorithm neural network economic prediction system based on the improved genetic algorithm and BP algorithm. In order to optimize BP neural network comprehensively and make it have better generalization performance, a genetic algorithm is improved and designed. The comparison test shows that the improved genetic algorithm reduces the memory consumption, ensures the diversity of population, and improves the running speed and convergence effect of the algorithm.

Keywords: Early warning · BP algorithm · Genetic algorithm · Neural network

1 Introduction

Artificial neural network is a mathematical model, which has some characteristics of human neural network and the ability of self-learning. The training model can be used to solve the same type of problems. The wrong signal output sample and network outlet will be adjusted according to the weight and threshold before connecting the path channel, while the error signal will describe the method of using the general gradient descent method to calculate the reverse connection to adjust the restoration.

BP neural network has the ability of nonlinear mapping, which essentially realizes the mapping function from input to output. Mathematical theory proves that the three-layer neural network can approach any nonlinear continuous function with any accuracy. The convergence speed of BP neural network algorithm is slow: because BP neural network algorithm is essentially a gradient descent method, its optimization objective function is very complex. Because BK algorithm is used to optimize the weighting coefficients of multilayer feedforward neural networks, BT algorithm usually refers to multilayer feedforward neural networks without topology feedback. In the process of learning and propagation, the input signal is input from the input layer, transmitted layer by layer and finally output.

The application of artificial neural networks (ANN) in the early warning (EW) system is a kind of expansion and breakthrough to the traditional EW both in thought and

I. Ahmad et al. (Eds.): STSIoT 2021, LNDECT 122, pp. 650–659, 2023.
https://doi.org/10.1007/978-981-19-3632-6_76

technology. It solves the problem that the traditional EW model is difficult to deal with highly nonlinear model, and focuses on quantitative indicators, It is difficult to deal with the warning limits of qualitative indicators. The way to determine the warning limits does not have the characteristics of time-varying, self-adaptive, self-learning ability, indirect acquisition of EW information and knowledge, and low efficiency. Thus, it lays the foundation for the early warning to be realized. Some scholars at home and abroad have also made some theoretical and practical discussions on ANN early warning system, However, due to the professional reasons, they have put into solving several major problems in the design of ANN economic EW system, such as the determination method of ANN economic early warning system network topology structure, the control of convergence speed, the selection of learning factor? And the method of escaping from local minima, etc. Therefore, it is still far from practical application. Based on the experience of Jiaxing Economic EW system, this paper proposes a design method of neural network economic early warning system based on genetic algorithm and BP hybrid algorithm. In other words, the improved genetic algorithm is used to optimize the first three layers of BP network economic electronic system. This paper consists of the following parts. The first part introduces the relevant background and significance of this paper, the second part is the related work of this paper, and the third part is data analysis. The fourth part is example analysis. The fifth part is conclusion.

2 Related Work

The subject of Ref [1] is to provide bibliometric analysis of early warning system, their development and usage in different environment especially in economic and finance sector. Fathani et al. describe the achievements and the current activities of the IPL-158 Project "Development of Community-based Landslide Early Warning System" [2]. Fan et al. introduce the successful landslide early warning system [3]. The subject of Ref [4] is to diminish the deficiency in the strategic cost management and prediction of economic crises. Based on the above background, the aim of Ref [5] is to design an economic early warning system based on improved genetic and BP hybrid algorithm and neural network. Bazhenova et al. deal with the early warning system that allows monitoring the external sustainability of an economy due to external economic shocks [6]. The contribution of Ref [7] is to provide a complex embedded system analysis of early warning systems and their development and use in a variety of environments, especially in the economic and financial fields. The development of an Internet of Things (IoT) technology based Intelligent Tracking, Early-warning & Management (ITEM) tool is elaborated, which is used to track the route of the tanker and temperature of the milk [8]. This work is concerned with the development of an early warning system that can act as a predictive tool for public health preparedness and response [9].

In the first mock exam, the economic EW is a pattern classification process: from the mapping relationship between the mega index to the alert index and the alert degree, the economic EW is a function approximation process. From the voice alarm and the alarm accuracy processing mode from the mega index to the universal index to the alarm degree, the economic EW is also an optimization process. Pattern recognition, function approximation and optimization are the application fields of ANN. Therefore, Ann is very suitable for economic EW.

Genetic algorithm is a global probability search algorithm based on natural selection and genetic mutation. Like the derivative based analytical method and other heuristic search methods (such as mountain climbing, simulated annealing, etc.), genetic algorithm is a formal iterative method. It starts from the selected initial solution and improves the current solution through continuous iteration until the optimal solution or satisfactory solution is finally found. In evolutionary computation, the evolutionary mechanism of simulated organisms is adopted in the iterative calculation process. Starting from a group of solutions (population), the population with better performance index is generated by adopting the method similar to natural selection and sexual reproduction and inheriting the original excellent genes.

2.1 Brief Introduction of Forward Three Layer BP Network Economic EW System

Artificial neural network is the most widely used three-layer BP network composed of input layer, hidden layer and output layer. The hidden layer is located between the read layer and the write layer used for internal display in the read mode.That is to say, it extracts the features which are different from other types of input patterns in a class of input patterns, and transmits the extracted features to the output layer. The output layer makes the final judgment on the categories of input patterns, so the hidden layer is regarded as the feature extraction layer. The process of feature extraction in hidden layer is actually the process of "self-organizing" the connection weight between input layer and hidden layer. In the process of network training, the connection weight between the layers plays the role of "transfer characteristics". The process of gradual evolution from the initial mode to the random representation is the process of the organization's self representation.

The exponential EW method is the basis of other traditional EW methods. This is also the most commonly used electronic warfare method. Obviously, EW system is very suitable for selecting three-layer forward BP network.

2.2 Determination of Network Topology of Forward Three Layer BP Network Economic EW System

The number of nodes in the input layer depends on the number of warning indicators. The number of output layer nodes is determined by the number of alarm stages (for example, the alarm stage is divided into 5 stages, and the number of light, medium, light and no alarm output layer nodes is 5).

On the basis of a large number of experiments and the application examples of three-layer BP network at home and abroad, we get the empirical formula of the upper limit of the hidden node.

$$l_h \leq \sqrt{m(n+3)} + 1 \tag{1}$$

(M is the number of input layers and N is the number of output layers).

Therefore, the lower limit of hidden nodes can be defined as the number of types of warning indexes.

New research shows that artificial intelligence network based on human brain can effectively perform cognitive tasks. The research team examined MRI data through a large open scientific database, reconstructed the brain connection mode and applied it to the artificial neural network "NN". The research team hopes to combine brain connectivity histology with related ATL structures, understand how brain structures support specific cognitive functions, and introduce new design principles into artificial networks.

Ducks can swim soon after hatching. Human babies will naturally be attracted by human faces. Even if the brain evolves to the point where it dares to face the world with little experience, many researchers hope that AI also has this natural ability.

New research has found that artificial neural networks can evolve to the extent that they can perform tasks without learning.

Such networks usually learn many tasks by adjusting the "weight" or strength of connections between neurons, such as playing games or recognizing images. The network is mutated by adding neurons, increasing the connection or changing the sensitivity of neurons to the sum of inputs. After gradually improving the weight of the standard network architecture, you can skillfully complete these three simulation tasks: driving the car, making the biped robot walk and controlling the two wheeled cart to balance the strut.

For example, he has amazing insight and observation ability in external affairs, is good at thinking and research, and dares to explore. Psychology and computer are his strengths. At the same time, he has made great achievements in the field of artificial intelligence neural network, which is also known as the "father of neural network".

In 1960, Michael mett and Papert, the "father of artificial intelligence", published an article pointing out that the composition of the concept of neural network is too simple and can only be used to solve current problems.

Based on the research results of medicine and psychology, he decided to use computer science methods to simulate the brain and continue to explore the direction of neural network. He is committed to using artificial neural networks to simulate human brain storage and thinking. Although the academic circles thought the problem was unrealistic at that time, Geoffrey chuon and his colleagues still made some achievements in such a harsh environment in the 1980s when computer computing was slow and data could only be processed in small batches. A year later, they improved the model and made challenging suggestions to limit the Boltzmann machine. In the following years; Geoffrey Hinon and his team NCAP gradually realized some ideas of early neural network research due to the great improvement of computer computing and data processing ability.

2.3 A Method to Determine Whether Hidden Nodes are Redundant

For the network with $m \times l \times n$ topology, after N training samples, the output matrix Z of hidden layer is as follows:

$$Z = \begin{pmatrix} Z_{11} & Z_{21} & \cdots & Z_{41} \\ Z_{12} & Z_{22} & \cdots & Z_{42} \\ & & \vdots & \\ Z_{1N} & Z_{2N} & \ldots & Z_{3N} \end{pmatrix} \tag{2}$$

When new China was founded, in many newly liberated cities, the influence of hyperinflation during the Kuomintang rule continued. From July 27 to August 15, the central finance committee held a financial conference in Shanghai and decided to take measures such as issuing bonds to strengthen the strength of the state-owned economy in the market struggle. A large number of illegal speculators took the opportunity to rise a storm, and many big price fluctuations continued to appear in the market, leading to social unrest. Establish and consolidate a new political power and maintain economic and social stability? People don't trust the RMB. They rush to buy silver, gold and foreign currencies as soon as there is a disturbance, which hinders the RMB from occupying the market and stimulates the price rise. Does the people's government led by the Communist Party have the ability to stabilize the economic situation, stop hyperinflation and restore production, and make the new regime stand firm economically and politically?

The outline points out that since the people's revolutionary war is winning nationwide, in order to restore and develop the national economy as quickly and systematically as possible, and make use of the current needs to supply the people's revolution. For the purpose of war and improving life, we should establish central financial and economic institutions with working ability, so that the financial and economic system of local governments and the committees of people's governments at all levels have established some financial and economic departments, and many financial and economic institutions have been established under the leadership of the central and higher financial and economic organs.

While preparing for the establishment of the central finance committee, the Party Central Committee also began to consider and deal with the national financial and economic problems, strive to stabilize prices and establish RMB as the only currency. In the face of the acts of speculators and ourselves, and against the rampant actions of the new regime to maintain economic stability, the East China Bureau of the CPC Central Committee decisively decided to take base salary measures and begin to seal up the concentration point of the financial speculative market. Finally, the instructions of the CPC Central Committee on cracking down on banks and the occupation of seats by RMB were issued. On June 8, the instructions of the East China Bureau and the Shanghai municipal Party committee put forward a series of economic measures, including explicitly requiring railway transportation and municipal utilities to accept the opening of RMB and the exchange between each release zone.

On behalf of the central government, Chen yundai drafted the telephone newspaper of the East China Finance Committee and decided to first attack the power of speculation in Shanghai with severe means. Seeing that the government did not sell materials, the speculative forces firmly believed that the Communist Party did not have enough material strength to compete with them. At the same time, the people's Government pressed for taxes, tightened the root causes of banks, frozen loans and marched on speculators. Speculators suffered from the enemy and went bankrupt. Only more than ten days later, the people's government won a fruitful victory, and the capital of illegal speculation came from here.

3 Data Analysis

3.1 Introduction of Genetic Algorithm

The main characteristics of genetic algorithms (GAs) are simple data processing, strong robustness and giant parallel computing. Its essence is a cyclic process composed of four operators: reproduction, exchange, mutation and selection. GAs does not need any gradient information and does not need any calculus calculation in the process of searching for the global optimum. Only through the above four operators, the global optimal or suboptimal solution can be found in the solution space with great probability, thus effectively reducing the probability of falling into the local minimum. From the ability to find the global optimal solution, gas is far superior to simulated annealing algorithm, and is an ideal tool to improve the forward three-layer BP network. The introduction of genetic algorithm is shown in Fig. 1.

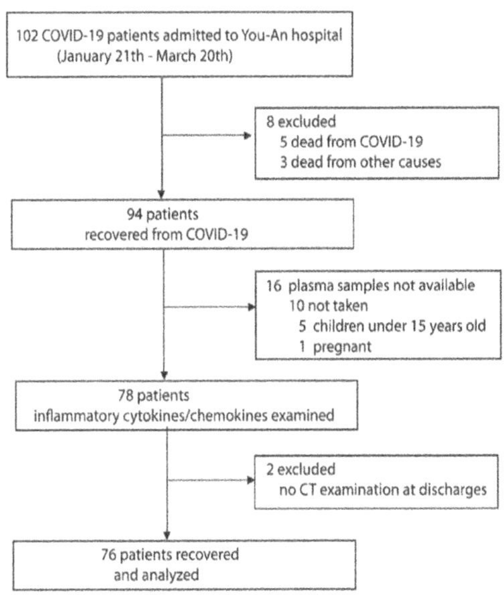

Fig. 1. Introduction of genetic algorithm

The degree of individual population can be defined by the generalized Hamming distance between individuals. For the individual coding in this paper, the generalized Hamming distance between individuals is defined as follows:

$$H_X = (X_i, X_j) = \frac{1}{(2n+1)} \sum\nolimits_{x=1}^{2n+1} H_G(G_i^x, G_j^x) \tag{3}$$

3.2 Basic Idea and Concrete Method of Accelerating GAs Convergence Rate

Suppose an individual is composed of two chromosomes. After several cycles of reproduction, exchange, variation and selection, the value of one chromosome becomes the ideal value, while the value of the other chromosome deviates from the ideal value. According to the individual fitness of conventional GAs, this individual should be eliminated from the sea. If we don't take the individual as the research object, but take the chromosome of the individual as the research object, we can carry out the reproduction, selection, exchange and laughing operation of the individual chromosome one by one in order to find and save the chromosome that can improve the individual fitness, so as to avoid the phenomenon of abandoning the single chromosome which improves the individual fitness, The convergence rate of GAs can be greatly accelerated.

The network error of BP neural network model is in the form of mean square error, and the complexity of network structure is expressed by the number of nodes that can be connected between layers. In order to optimize the weight and structure of the network at the same time, the number of fitness networks is designed as follows:

$$f(i,t) = \frac{\lambda}{1 + E_1(i,t) + \beta E_2(i,t)} + \frac{(1-\lambda)}{T(i,t)} \tag{4}$$

GAs regards a weight a of a forward three-layer BP network as a chromosome, and regards the set of all weights in the network as an individual: a large number of individuals are generated in the initialization stage, and some individuals with higher fitness are selected as parents according to the fitness of each individual, According to the probability, select two individuals for exchange operation to generate two offspring, or select an individual for mutation operation to generate one offspring. When the adaptability of all individuals in an individual is almost the same, it is transferred to the first three layers of BP network. At the end of the current three-layer BP network, the mean square error of the network can not meet the accuracy requirements. At this time, it may be reduced to the local minimum. Therefore, it is transferred to gas and a new whole is generated based on the volume at this time. Repeat the above steps until the balance root error of the network meets the accuracy requirements.

4 Example Analysis

On August 18, Sichuan disaster reduction center, emergency command center of Sichuan emergency management department, Wenchuan County People's government and Chengdu Institute of high tech disaster reduction jointly announced that the multi disaster early warning demonstration project in Wenchuan has been started.

According to the Wenchuan earthquake early warning network built in 2012, relying on the implementation of multi disaster early warning demonstration project, Wenchuan will establish nearly 30 data sources covering "sky and underground", covering earthquakes and other diseases. The system is interconnected with the disaster early warning network of surrounding cities and counties, and transmits warning information to the people, emergency departments and towns through comprehensive research and judgment in the affected areas of Wenchuan province.

$$u_{k+1}(t) = u_k(t) + L(t)(\dot{e}_{k+1}(t) + e_{k+1}(t)) \tag{5}$$

$$\sup_{0 \leq t \leq T} \|I - L(t)C(t)P^{-1}(t)B(t)\| \leq \rho_1 < 1 \tag{6}$$

It is reported that the Institute and Sichuan disaster reduction center jointly take the lead in technical coordination. At the same time, the early warning system will also synchronize disaster information to the city and county emergency management command platform, realize the connection between disaster early warning and emergency management, and provide emergency command. On the basis of the earthquake early warning network built in 2010, the multi disaster warning project was launched in September this year, covering 51 multi disaster early warning and monitoring points. The launch of the project will certainly promote the disaster early warning work throughout the province, better provide strong support for basic public services in Sichuan, further improve the technical service capacity of disaster prevention and reduction, improve the level of disaster prevention and reduction, promote the upgrading of disaster prevention and reduction industry, and help Sichuan's economic and social development and safe development. Through cooperation with Chengdu High Tech Institute of disaster reduction in technology and application, a multi disaster, all region and all channel natural disaster early warning service system has been established in Wenchuan County, which has comprehensively improved the modernization level of earthquake prevention and control capacity. Including 63 destructive earthquakes and 180 cases of landslides, debris flows, mountain torrents, wildfires and other Wenchuan multi disaster early warning demonstration projects, marking the first time that "Sichuan intelligent manufacturing" has fully applied the multi disaster early warning results within the county.

Xian Sheng, deputy director of Sichuan disaster reduction center, said that the launch of the Wenchuan multi disaster early warning demonstration county project will take the lead in promoting disaster relief and early warning in Sichuan into a basic public service. "Multi disaster early warning is a natural extension of earthquake warning, from a single natural disaster to a response to multiple disasters. Before that, we carried out some experiments, tests and preliminary services of catastrophic operation technology in Chengdu. The project has coordinated with more than 15 departments in Wenchuan County, realizing a better linkage between disaster early warning information and government departments, so as to better serve the people Wenchuan city serves". Wang Yi, director of Chengdu multi disaster early warning engineering technology research center and director of Chengdu Hi Tech Institute of disaster reduction, said that the multi-party linkage of the project will build a disaster early warning system. Example analysis is shown in Fig. 2.

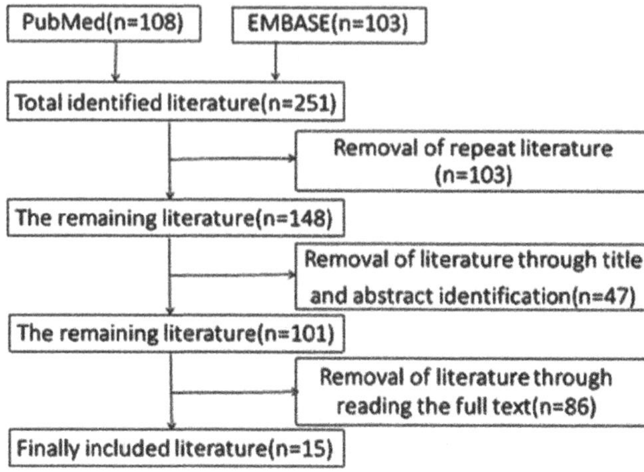

Fig. 2. Example analysis

5 Conclusion

The improved back application which reduces the population size to the minimum can greatly reduce the memory consumption of genetic algorithm and improve the running speed. As a special evolutionary operator, BP algorithm can adjust the network weight and change the network structure according to the knowledge of the leading city. The introduction of generalized Hamming distance between individuals can ensure the diversity of the population; BP operator, mutation operator and crossover operator compete together to adapt to adjust the operation probability and promote the convergence of the algorithm. Compared with BP algorithm and hybrid genetic algorithm, this algorithm can optimize the network structure, make it have better generalization performance, effectively ensure the diversity of population, improve the running speed and convergence effect of the algorithm, which is stable and reliable; moreover, compared with BP algorithm and hybrid genetic algorithm, this algorithm has smaller internal storage and faster running and convergence speed than BP algorithm and hybrid genetic algorithm.

References

1. Klopotan, I., Zoroja, J., Meško, M.: Early warning system in business, finance, and economics. Int. J. Eng. Bus. Manage. **63**(3), 581–592 (2018)
2. Fathani, T.F., Karnawati, D.: TXT-tool 2.062-1.1: a landslide monitoring and early warning system. J. Syst. Manage. **32**(17), 13233–13244 (2018)
3. Fan, X., et al.: Successive landsliding and damming of The Jinsha River in Eastern Tibet, China: prime investigation, early warning, and emergency response. Landslides **179**(6), 764–774 (2019)
4. Ayvaz, E., Kaplan, K., Kuncan, M.: An integrated LSTM neural networks approach to sustainable balanced scorecard-based early warning system. IEEE Access **173**, 150–162 (2020)
5. Yin, X., Li, J., Huang, S.: The improved genetic and BP hybrid algorithm and neural network economic early warning system. Neural Comput. Appl. **169**(9), 1133–1139 (2021)

6. Bazhenova, O., Chornodid, I., Yarmolenko, Y., Golubev, O.: Early warning system of external sustainability of an economy: case of Ukraine. Financ. Credit Act. Probl. Theory Pract. **27**(9), 2610–2626 (2021)
7. Lin, J.: Design of enterprise financial early warning model based on complex embedded system. J. Syst. Sci. **10**(6), 363–377 (2021)
8. Avhale, G.V., Shaligram, A.D., Gharpure, D.C.: IoT based intelligent tracking, early-warning & management (ITEM) tool for efficient milk transportation in dairy industry. J. Syst. Design Dyn. Dev. **8**(3–4), 279–292 (2021)
9. Pergantas, P., et al.: Towards a semi-automatic early warning system for vector-borne diseases. Int. J. Environ. Res. Public Health **30**(4), 377–399 (2021)

Analysis of an Intelligent Optimization Algorithm for Automatic Generation of Computer Software Test Data

Liping Li and Xiaoyan Zhang(✉)

Yunnan College of Business Management, Kunming 650300, Yunnan, China
z316819595@163.com

Abstract. Software testing can guarantee the quality of software products, but it also takes up nearly half of the cost and resources of the entire software development cycle. The traditional test data acquisition requires manual design, but as the scale and complexity of software increases, manual design of test data can no longer meet the requirements of testing, therefore, automatic test data generation has become a hot spot and focus of many scholars' research. In this paper, we will study and analyse the automatic generation of computer software test data based on intelligent optimisation algorithms.

Keywords: Software testing · Automatic data generation · Intelligent optimisation algorithm

1 Introduction

Software testing is a method and means of evaluating software products by designing test data and using it to find defects or errors in the software in order to guarantee the quality of the product. The importance of software testing is evident in the fact that it runs through almost the entire software development process, and the design of test data is the most central and important part of testing, which determines the quality and efficiency of testing. The traditional way of designing test cases and test data is to design them manually, relying mainly on the experience of testers, but However, as the size of software increases, the traditional manual data design mode of generating test data is inefficient, laborious, prone to omissions, long testing cycles and high testing costs, so more and more researchers are beginning to study the automatic generation of test data.

In this paper, the research of automatic test data generation is carried out by optimising some shortcomings of the algorithm in the application of the problem, improving the performance of the algorithm so as to achieve the goal of fast and effective test data generation, reducing the time cost and resources consumed in software testing due to the design of test data, and improving the efficiency of test data generation while reducing unnecessary costs in testing. This paper is not only of theoretical significance, but also of practical application.

I. Ahmad et al. (Eds.): STSIoT 2021, LNDECT 122, pp. 660–669, 2023.
https://doi.org/10.1007/978-981-19-3632-6_77

Achieve usability and ease of use; The core production management software has leading advantages, and the material base, knowledge base, model base, process base, algorithm database and other basic elements are constantly improved; Cultivate new industrial software platform; Strengthen the enabling of industrial data and improve the intelligent level of industrial software. As shown in Fig. 1, the intelligent level of industrial software.

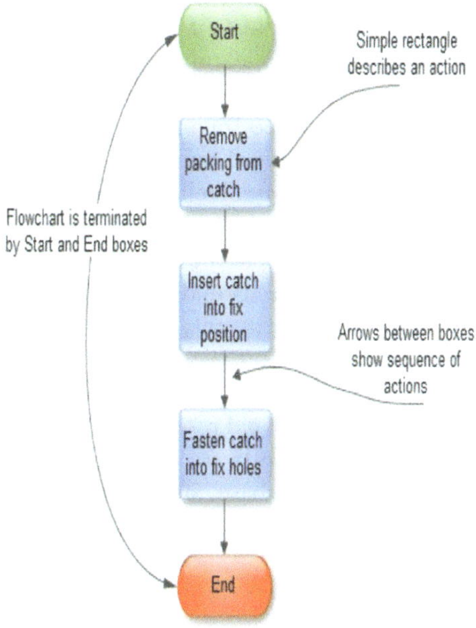

Fig. 1. Intelligent level of industrial software

Focus on the development of core industry software. For the process industry, support the research and development of planning and scheduling, production scheduling and real-time optimization software, overcome the deep integration of scheduling and planning and the development of rolling optimization technology, and promote the application of petrochemical, chemical and non-ferrous metals. Applications in light industry and other fields accelerate the breakthrough of core technologies such as industrial software and hardware, intelligent algorithm and industrial mechanism model, focusing on the whole process of optimization design, production, operation and maintenance, quality improvement, intelligent manufacturing, intelligent detection, real-time scheduling, optimization decision-making, predictive protection and other industry scenarios.

Focus on industrial Internet, intelligent manufacturing and production services. This paper consists of the following parts. The first part introduces the relevant background and significance of this paper, the second part is the related work of this paper, and the third part is automatic generation of single-path test data. The fourth part is automatic generation of multi-path software test data. The fifth part is conclusion.

2 Related Work

Koparan aimed to examine the effect of dynamic data analysis software-supported learning environments on secondary school students' achievement and attitude [1]. Su et al. take the touch screen characteristic test as an example,through the multi-dimensional automatic motion platform, the test pen is driven to move on the touch screen according to the set trajectory [2]. Transformer routine tests have been analyzed by using the generated firefly algorithm [3]. The subject of Ref [4] was to improve students understanding of introductory oceanography with the aid of a computer program. Rahmawati et al. aim to find out if computer self-efficacy, learning motivation, and accounting knowledge affect the computer anxiety of accounting students in using accounting software [5]. To cooperate with the research on the fragmentation pre-conditioning technology of large hard rock in natural caving mining, Jingjie et al. perform a large flow hydraulic fracturing test in Tongkuangyu Copper Mine to investigate the relationship between the occurrence and expansion of fractures in ore bodies and the pressure and flow of water injection [6]. The research results show that the system constructed has certain practical effects [7]. Huang et al. introduce the secondary development process of simcenter Test [8]. Lab software. CBCT images of twenty patients with UCCLP were included Ref [9]. Other influential work includes Ref [10]. Relying on universities and scientific research institutes, build industrial software adaptation, testing, verification and pilot test platforms for production, University and research. Vigorously promote the promotion of new technologies for industrial software integration; Promote the autonomy of industrial software of industrial Internet base, support micro service architecture such as aggregation tools, algorithms and models, promote industrial software components and services, and improve the comprehensive integration, test and verification, quality control, life cycle management and service capabilities of industrial software.

2.1 Overview of Intelligent Optimisation Algorithms

The so-called intelligent optimisation algorithm is a kind of intelligent search calculation method developed according to some principles of the real phenomena in nature, and the intelligent optimisation algorithm can be regarded as a kind of reference and simulation of the laws of nature. At present, in addition to traditional algorithms such as genetic algorithms and particle swarm algorithms, there are also algorithms such as firefly algorithms and firework explosion algorithms. The genetic algorithm is an algorithm invented by simulating biological genetics and biological evolution in nature, which is characterised by its adaptive structure and global optimisation effect. The particle swarm algorithm is based on the interaction between particles and is able to find optimal regions in a complex search space. In the process of computer software testing, these two more basic intelligent optimisation algorithms have been used in large numbers.

$$\begin{cases} E(t)\dot{x}_k(t) = f(t, x_k(t)) + B(t)u_k(t) \\ y_k(t) = C(t)x_k(t) \end{cases} \tag{1}$$

$$u_{k+1}(t) = u_k(t) + \Gamma_{l1}\dot{e}_k(t) + \Gamma_{l2}\dot{e}_{k+1}(t) + \Gamma_{p1}\Delta\dot{e}_k(t) + \Gamma_{p2}\Delta\dot{e}_{k+1}(t) \tag{2}$$

2.2 Applications of Genetic Algorithms

Genetic algorithms have a global probability search function. In using this algorithm, the problem is transformed by replacing the problem of generating software test data with a functional optimisation problem, and then designing the fitness function, while the population of software test data is coded to facilitate the application of genetic operations. The populations of software test data are evolved over many generations to obtain the corresponding test data results. In the automatic generation of software test data, the first step in the application of genetic algorithms is to encode the data entered through the software program in order to facilitate the formation of different individuals.

$$\dot{x}_{k+1}(t) = f(t, x_d(t)) - f(t, x_{k+1}(t)) \tag{3}$$

$$\begin{aligned} \Delta x_{k+1}(t) &= \int_0^t Q^{-1}(f(t, x_d(\tau)) - Q^{-1}f(t, x_{k+1}(\tau)))d\tau \\ &+ \int_0^t Q^{-1}B\Delta u_k(\tau)d\tau - \int_0^t Q^{-1}Z\dot{x}_d(\tau)d\tau + \int_0^t Q^{-1}F\dot{x}_{k-1}(\tau)d\tau \end{aligned} \tag{4}$$

The second step is to generate the initial population. The initial population is composed of N individuals, each of which is made up of N initial strings of structural data, the production of which is random. The initial population is an important starting point for iterative updating of the genetic algorithm.

The third step is to select the population. The selection operation is a simulation by the genetic algorithm of the survival of the fittest principle. Selection begins by selecting a certain number of individuals in the initial population that meet the criteria of being well adapted. These selected individuals will become the new parents and from this a new generation of individuals will be generated.

The fourth step is to carry out crossover. The main purpose of the crossover operation is to facilitate the exchange of data and information. Without crossover, the new generation of individuals would not be available to the parents selected by the algorithm. Almost every individual of the new generation has some of the characteristics inherited from its parents.

The fifth step is the mutation operation. When the mutation operation is performed, the elite individuals are kept out of the mutation operation. The first step is to select individuals at random in the population and to change them in a random way.

The sixth step is to calculate the value of the fitness function. If the result of the calculation satisfies the termination condition of the algorithm, the operation of the algorithm is terminated, and if the result of the calculation does not meet the requirements of the termination condition, the genetic algorithm is repeated until the result of the data satisfying the termination condition is found.

2.3 Firefly Algorithm

The proposed firefly algorithm is based on the luminous properties of fireflies, and summarises the relevant laws of their luminous activity in order to find the optimal solution.

The first step is to determine the population size of the fireflies, and then to initialise the population to determine the initial position of the individuals.

The second step is to calculate the absolute brightness of individual fireflies by using the objective function based on their position. As shown in Fig. 2 below, initialize the population to determine the initial position of the individual.

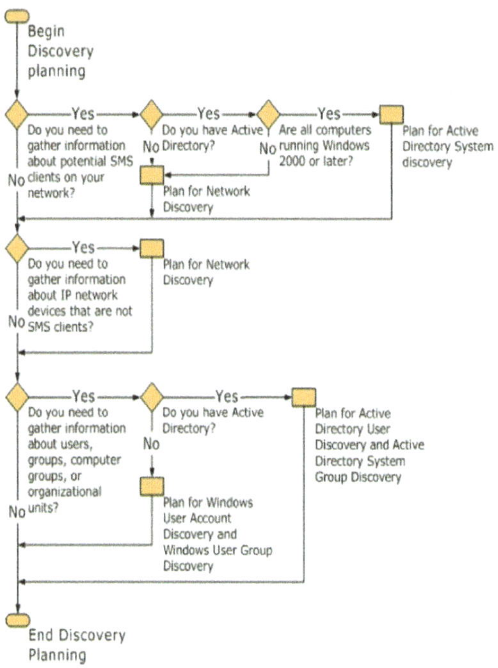

Fig. 2. Initialize the population to determine the initial position of the individual

The third step is to calculate the relative brightness and attraction of each of the two individuals based on the formula for the relative brightness of the individuals and the formula for the attraction between the individuals.

In the fourth step, according to the individual position change formula, the update of individual positions in the population can be calculated. The brightest individual firefly in the population, however, moves irregularly and its position change can be calculated using a separate formula.

In the fifth step, the luminous intensity of each new individual firefly is recalculated for each individual firefly in the population at the end of the update.

Step 6, determine if the algorithm meets the conditions for the end of the algorithm, if so, end the algorithm and produce the final result, if not, repeat step 3 until the algorithm is finished.

3 Automatic Generation of Single-Path Test Data

There are many classical optimization algorithms, such as genetic, ant colony, particle swarm, simulated annealing, etc., which have been used in single-path applications, so there is less and less room for development and it is difficult to further improve the efficiency of data generation by improving classical algorithms such as genetic and particle swarm algorithms. In solving the single-path testing problem, this paper chooses to apply the Firefly algorithm (FA) to it and optimise it in order to find new ideas and solutions for the study of single-path testing.

3.1 FA Algorithm Mathematical Model

For the establishment of the FA algorithm mathematical model, the first thing to understand is the concept of absolute luminance and relative luminance, the absolute luminance of fireflies, is for a firefly individual *i*, its initial luminance value is called absolute luminance, can be recorded as I_i. And for the definition of relative luminance, and the absolute luminance I_i is different, it refers to for two individuals *i* and *j*, firefly *i* in firefly *j* position of the luminous intensity, can be be expressed as Iij.

At position Xi(xi1,xi2,...,xid), the expression for the absolute luminance

$$I_i = f(X_i) \tag{5}$$

The brightness between fireflies is not fixed. It will be gradually weakened with the increase of the interval between them and the absorption of some substances in the air. Therefore, the relative brightness formula of firefly individual *i* to firefly individual *j* can be expressed as

$$I_{ij}(r_{ij}) = I_i e^{-\gamma_{ij}^2} \tag{6}$$

3.2 Basic Flow of FA Algorithm

Step l: Assume that the whole population of fireflies is *N*, and initialize the positions of individuals in the whole population;

Step 2: According to the position of firefly X_i, we can know the value of the objective function $f(X_i)$, so as to calculate the absolute brightness $I_i(X_i)$ of firefly *i* according to the formula (1);

Step 3: Calculate the relative luminance I_i; and attractiveness of fireflies *i* and *j*, respectively;

Step 4: Update the positions of the individuals in the population, and for the brightest fireflies, move them randomly in an irregular manner;

Step 5: After the iterative update of the individuals in the population, the luminous intensity of each individual firefly is recalculated;

Step 6: Determine whether the algorithm meets the conditions of the end, if not, then return to Step3, if so, the algorithm will end and the results will be output.

FA algorithm basic flow chart is shown in Fig. 3.

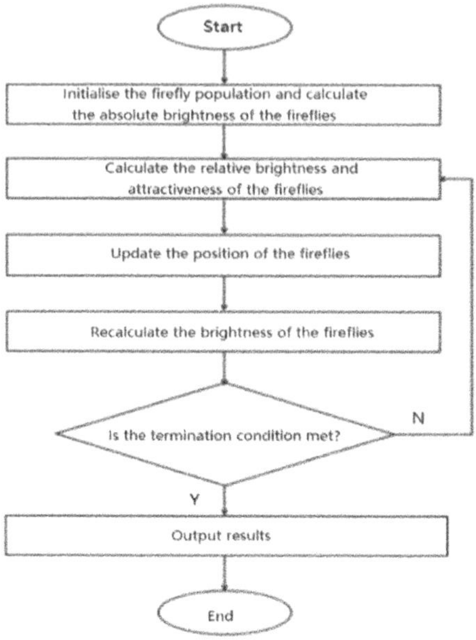

Fig. 3. Basic flow chart of the FA algorithm

3.3 Construction of the Fitness Function

The bridge between algorithms and real application problems is the fitness function, which can be used to assess the quality of the data generated. A well-designed fitness function can better guide the algorithm towards the optimal solution region, covering the target in fewer iterations and less time path and find the most optimal solution. Therefore, it is important to construct a suitable fitness function, which is related to the efficiency of data generation.

In this chapter, the fitness function is constructed using the branching function superposition method devised by Korel, which converts branching predicates into branching functions, and then uses these functions, which are formed by superposition of branching functions, as the final objective function to be optimised.

This is the concept of a branch predicate, which indicates when a branch can be overwritten. A branch predicate can be expressed in the form of "E1 op E2", where E_1 and E_2 are mathematical expressions and op is a relational operator containing $<$, $\leq$, $=$, $>$, $\geq$, $\neq$. When the predicate does not contain logical operations, then a branching predicate in the form of "E1 op E2" can be converted to "f rel0", where f is the mentioned "branching function" and f is used to quantitatively evaluate the degree to which the input value satisfies the target function, when the branching predicate is true, the function value is negative. When the branch predicate is true, the function value is negative, when $f = 0$; conversely, when the branch predicate is false, the function value is positive, $f > 0$.

The relationship between branch predicates and branch functions is shown in Table 1.

Table 1. Relationship between branch predicates and branch functions

Branch predicates	Branching functions
E1 > E2	E2-E1
E1 ≥ E2	E2-E1
E1 < E2	E1-E2
E1 ≤ E2	E1-E2
E1 = E2	|E1-E2|
E1 ≠ E2	|E1-E2|

4 Automatic Generation of Multi-path Software Test Data

The automatic generation of computer software test data for multiple paths involves two techniques. The first is path similarity. In computing, path similarity refers to the degree to which the path of software test data matches the target path. In order to determine path similarity more accurately, a more appropriate measure of path similarity is needed. At present, there are three main factors that affect path similarity: the number of identical nodes, the number of consecutive identical nodes, and the weight of different nodes. These three factors must be fully considered in designing the path similarity measure. The second is the design of the multi-path fitness function. In multi-path testing, the design of the fitness function is usually based on the idea of mean value. The first step is to calculate the value of the similarity of all paths according to a professional calculation method, and then take the average value. This mean value can be used as the fitness value of the data for the target set of paths. By comparing the matching of different paths with the mean value, it is more intuitive to see the strengths and weaknesses between the test data.

The path selector selects the least number of paths in the program control flow diagram for use by the test data generator. The path selection must meet certain coverage principles. Here are some common principles:

(1) Statement coverage: any statement can be covered by the selected path.
(2) Branch coverage: all conditional branches in the procedure must be covered. For example, in an IF statement, the condition predicates that are true or false must be covered by the selected path.
(3) Conditional coverage: when each clause in the condition of a branch statement is true or false (single condition coverage) and all combinations of the true values of each clause must be covered by the selected path (multi condition coverage).
(4) Path coverage: traverse all paths in the program control flow graph.

Among the above four principles, multi condition coverage and path coverage are difficult to achieve, because with the growth of program scale, the combination of clauses in conditions and the number of paths in program control flow graph will generally increase exponentially. Therefore, statement coverage and branch coverage are widely used as the basic measures of software testing.

Static test data generation is not based on the input data of the program, but adopts the method of program symbol execution and expression digestion and transformation. Dynamic test data generation is a method of executing the program by using the actual input data of the program. In the early research of automatic generation of program test data, the symbol based method is basically used. Because the problem of program test data generation is NP hard, the symbol based method occupies a lot of computer resources and has certain restrictions on the program. The advantage is that there is no need to check the truth value of the branch predicate.

The method based on the actual operation of the program is to take the value of the input variable to actually execute the program, and determine whether the value of the selected input variable can traverse the path selected by the path selector by observing the data flow in the program. Using different search algorithms, we can find the value of the input variable traversing a path, but it often takes a lot of time. It combines the symbol based method and the actual operation method based on program to generate test data, which saves the workload.

5 Conclusion

Software testing is an important part of the computer software development industry and a major need for the industry. The use of intelligent optimization algorithms to automatically generate computer software test data can be studied to obtain important results to improve the efficiency of automatic software testing, which is of great value to ensure the quality of software testing, improve the efficiency of software development, and help the development of China's software development industry.

References

1. Koparan, T.: Examination of the dynamic software-supported learning environment in data analysis. Int. J. Math. Educ. Sci. Technol. **83**, 80–98 (2018)
2. Su, M., Zhu, N., Huang, L., Xu, H.: Research on multi-degree-of-freedom and high-precision touch screen characteristic test instrument. J. Test Syst. **16**(8), 1204 (2019)
3. Zile, M.: Routine test analysis in power transformers by using firefly algorithm and computer program. IEEE Access **26**, 401–417 (2019)
4. Firdaus, M.L., Parlindungan, D., Elvia, R., Swistoro, E., Sundaryono, A., Rahmidar, L.: Teaching oceanography using ocean data view software. In: Proceedings of the International Conference on Educational Sciences and Teacher Profession (ICETEP 2018), pp. 3177–3184 (2019)
5. Rahmawati, A., Abidin, F.I.N.: The influence of computer self-efficacy, learning motivation, and knowledge of accounting on accounting students computer anxiety in using accounting software. Acad. Open **93**, 395–422 (2021)
6. Jian, J., Peng, H., Ma, X., Sun, Y.: Software design of a data acquisition system for the hydraulic fracturing experiment of rock at Tongkuangyu copper mine. J. Earth Environ. Sci. **282**, 232–247 (2021)
7. Narengerile, L., Di, L.: Framework and performance analysis of college English testing system based on data mining technology. J. Intell. Fuzzy Syst. **26**, 401–417 (2021)

8. Huang, J., Chen, B., Tan, M., Liu, M., Jia, C.: Exploration and research of noise automatic processing algorithm based on multi-scale convolution neural network. J. Neurnalnet Appl. **209**, 16–36 (2021)
9. Phienwej, K., Chaiworawitkul, M., Jotikasthira, D., Khwanngern, K., Sriwilas, P.: Comparison of preoperative measurement methods of alveolar cleft volume using cone beam computed tomography between computer simulation and water displacement methods. Cleft Palate-Craniofac. J. **34**, 2677–2684 (2021)
10. Zhang, H., Lin, F., Zhang, X., Wen, X.: Design of torque motor characteristic test system. J. Exp. Manag. **1786**, 108322–108326 (2021)

Application of Multisensor Data Acquisition in Reservoir Heterogeneity

Min Li(✉)

Daqing Oilfield Limited Company, No. 4 Oil Production Company, Daqing 163511, China
935748819@qq.com

Abstract. The triadic compound flooding test of three types of oil layers in the North East block of Lamadian oilfield is a key test project of tertiary oil production in Daqing Oilfield. The test area is planned to be put into operation in 2011. Through the analysis of logging data of 32 newly drilled oil wells in the experimental area, the heterogeneity of the layers, interlayer and plane and the influence on the remaining oil are studied. According to the water flooded condition of the reservoir in the test area, the types and distribution of remaining oil are analyzed, The results show that the reservoir conditions of high II 1–18 in polymer flooding test area of three types of reservoirs are compared. Combined with the design and experience of polymer flooding test parameters, it can provide important guidance for the design of three-dimensional test parameters of three types of oil layers.

Keywords: Keywords three types of reservoir · ASP flooding · Heterogeneity · Remaining oil

1 Sedimentary Characteristics of Test Target Layer

Gaoii reservoir belongs to delta front facies deposition. During deposition, the water environment is relatively deep, mainly composed of stable sheet sand with low permeability. There are many calcareous layers. The color of mudstone is mostly gray green to gray black. The plane of sand body is stable and the difference of heterogeneity is small. The sand bodies of high delta front can be divided into two types of high delta units, i.e., high delta units, high delta units and high delta units.

On the afternoon of September 2, a special video conference on the analysis of the city's economic operation was held [1]. All district and county departments are required to strengthen the research and analysis of main economic indicators, determine the prominent links between prominent contradictions and weak links, strengthen operation scheduling, promote the implementation of various objectives and tasks, and improve the quality and efficiency of economic development.

On the afternoon of July 16, a special dispatching promotion meeting was held to analyze the city's industrial economic operation and accelerate the investment in industrial "technological reform" to cultivate new scale industrial enterprises. Further deploy the leading task of "six comparisons" industry and information technology system

I. Ahmad et al. (Eds.): STSIoT 2021, LNDECT 122, pp. 670–678, 2023.
https://doi.org/10.1007/978-981-19-3632-6_78

and the current urban key industrial economic work, and spare no effort to promote the return of urban industrial economy to a reasonable scope as soon as possible.

The meeting stressed the need to unify thinking, enhance confidence, focus on the municipal Party committee and municipal government, highlight the deployment requirements of industry to break through industry, further take effective measures and strive to overcome difficulties [2]. Ensure that the annual industrial economic growth target is achieved. He analyzed the current economic situation, studied and deployed the next work, and stressed the need to accurately grasp the economic development trend, firmly grasp key areas and key links, and take effective measures to accelerate high-quality development. The following Fig. 1 shows the economic data study of reservoir heterogeneity.

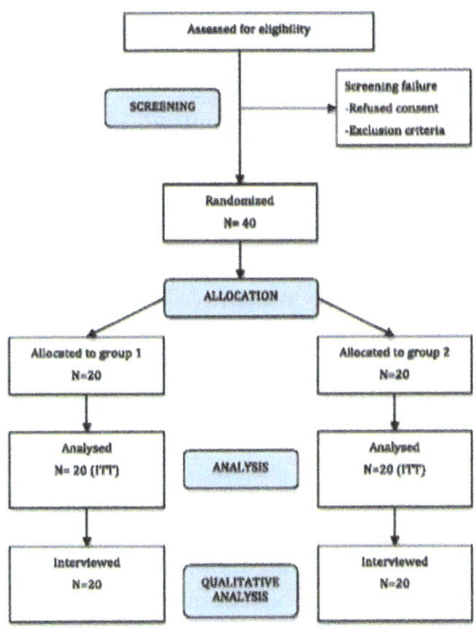

Fig. 1. Study on economic data of reservoir heterogeneity

2 Related Work

In this paper, we aim to application of multisensor data acquisition in reservoir heterogeneity.

Ref [1] reveal that pore type is a crucial cause for strong reservoir heterogeneity of carbonates. Thus, Ref [2] study the effect of wettability heterogeneity and reservoir temperature on the vertical CO2 plume migration, and capillary and dissolution trapping capacities. Accordingly Ref [3] show that the main part of the Mishrif Reservoir is affected by diagenetic processes related to subaerial exposures, resulting in zones with higher storage capacity and fluid flow rates. Petrographic, cathodoluminescence and ultra-violet light fluorescence microscopy analyses of the Arab-D reservoir in an oil field of Saudi Arabia, Ref [4] have been combined with isotope geochemistry (δ13C, δ18O and 87Sr/86Sr) to decipher the sequence of diagenetic events affecting the reservoir before and during oil emplacements, and the impact of depositional mineralogies and their modifications during the diagenesis on the reservoir quality and heterogeneity [5]. Ref [6] investigate a nonroutine methodology to predict the external and internal distribution of PFUs. These advances provide a platform for introducing a practical approach for introducing the Risk of Commercial Failure (RCF) due to reservoir heterogeneity in hydraulic fracturing projects. Ref [7] define such a parameter and the methodology to calculate it in a time-efficient manner. Ref [8] aim to use electrical image logs in the carbonate Asmari Formation reservoir in Zagros Basin, SW Iran, in order to evaluate natural fractures, porosity system, permeability profile and heterogeneity index and accordingly compare the results with core and well data. Despite its successful application in conventional reservoirs, significant errors arise when extending the concept to unconventional reservoirs. Ref [9] aim to clearly demonstrate such errors when using the traditional square–root–of–time model for DOI calculations in unconventional reservoirs, and to develop new models to improve the DOI calculations. Ref [10] present integrated static reservoir modeling and basin modeling to better characterise the reservoir rock; hydrocarbon-bearing sandstones of the Cenomanian Bahariya Formation.

This paper consists of the following parts. The first part introduces the related background and significance of this paper, the second part is the related work of this paper, and the third part is data analysis. The fourth part is example analysis. The fifth part is conclusion.

2.1 Overview of Data Fusion Theory

Multi sensor data fusion is a new subject developed from 1970s. It has been widely used in C3I system and various weapon platforms. With the rapid development, modern warfare has developed into a five-dimensional structure of land, sea, air, space and electromagnetism. In order to obtain the best combat effect, in modern CI combat system, relying on a single sensor to provide information can no longer meet the demand. It is necessary to use multi-sensor to provide observation data or information, real-time target detection, optimization and comprehensive processing to obtain state estimation The data itself robustness, high measurement dimension, good target space resolution, strong fault tolerance and system reliability Since data fusion was put forward, it has attracted great attention of developed countries and listed it as an important topic in the field of military high-tech research and development [3]. The organize and coordinate systematic research on this key national defense technology, so that in the mid-1980s, data fusion technology first made great progress in the military field.

The observation data of Ruqian sensors obtained in a certain spatio-temporal order are processed under a certain criterion by using computer technology Automatic analysis and synthesis, so as to produce new meaningful data, which can not be obtained by any single sensor.

Today, data fusion technology has been widely used in robotics, traffic management and military fields [4]. At present, the application effect in military field is the most remarkable.

2.2 Reservoir Parameters of Each Well

The plane imbalance is mainly to study the plane distribution characteristics of sand, the flexibility caused by the differences in geometric shape, thickness, continuity and permeability of sand, and the formation and distribution of oil wells.

$$\begin{aligned} & f(t, x_d(t)) + B(t)u_d(t) - f(t, x_k(t)) - B(t)u_k(t) - d_k(t) \\ & = f(t, x_d(t)) - f(t, x_{k+1}(t)) + B(t)\Delta u_{k+1}(t) - d_{k+1}(t) \end{aligned} \tag{1}$$

$$\begin{aligned} & \Delta\dot{x}_{k+1}(t) = P^{-1}(t)(f(t, x_d(t)) - f(t, x_{k+1}(t)) + B(t)\Delta u_k(t) \\ & -\big(B(t)L(t)\dot{C}(t) + B(t)L(t)C(t)\big)\Delta x_{k+1}(t) - d_{k+1}(t)) \end{aligned} \tag{2}$$

Using the reservoir parameters of each well obtained by logging interpretation of 32 newly drilled wells, the thickness contour map of sandstone reservoir sand body unit show that the sandstone thickness of high II 1–18 unit in the test area is larger in the west, smaller in the East and more uniform in the middle; from the perspective of permeability distribution, the distribution is extremely uneven, generally low in the middle and East. From the micro structure distribution map, it is high in the West and low in the East, and there are small undulating structures in the middle. Under the 106 m five point well pattern, the sand body control degree of the effective thickness well points developed in gaoП 1–18 reservoir in the test area is counted according to the sedimentary unit. The results show that the first class connectivity effective thickness of non channel sand is 97.5%, which is 19.8% and 1.0% test area. Among them, the first class connectivity rate of four-way well pattern is 374%, that of sand body is 63.7%; that of three-way well pattern is 68.6%, that of sand body is 90.1%; that of two-way well pattern is 82.6%, and that of sand body is 97%. Through the analysis of various factors of plane heterogeneity, it is judged that the reservoir in the East is poorly developed, low permeability and prone to residual oil [5].

3 Data Analysis

Interlayer heterogeneity refers to the vertical difference between research units. In the test area, there is a good interlayer condition between Gao 1-18 reservoir. The average thickness of the interlayer is about 1.9 m. There is no adhesive layer between. The average thickness of the upper and lower interlayer is 1.8 m and 1.6 m, respectively.

There are 18 units in the target layer in the test area. According to the oil layer development of each unit, 6 units of Gao II 2, 3, 4 + 5, 12, 13 and 16 are well developed, with large area of channel sand body and remaining oil, which should be the main target of tapping potential. The 4 units of Gao II 1, 9, 182 and 183 are mainly off the surface, with poor reservoir development and less remaining oil. The interlayer heterogeneity in the test area is shown in Fig. 2 below.

Intraformational heterogeneity refers to the vertical change of reservoir properties within a single sand body. It is a key factor that controls and affects the swept volume of injectants and the formation and distribution of remaining oil in a single sand layer [6].

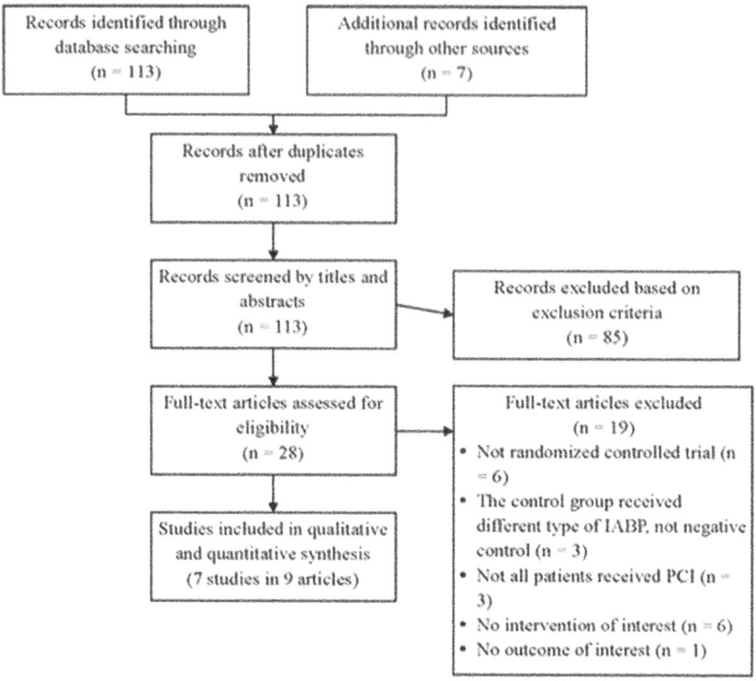

Fig. 2. Interlayer heterogeneity in test area

3.1 Permeability Heterogeneity Model

The high II 1–18 unit is a complex positive rhythm pattern. The permeability at the bottom of the unit is larger than that in the upward direction. However, there are several rhythms of high to low due to the inflow and recession of lake water. Therefore, from the perspective of the whole Gao II 1–18 unit, the rhythm is more complex, and the remaining oil in each unit is different, but on the whole, the remaining oil is less in the lower part and more in the upper part. The complex positive rhythm pattern is shown in Fig. 3 below.

3.2 Improved Neural Network Algorithm and its Application in Reservoir Identification

Multi feedback BP neural network is composed of input layer, output layer and their invisible layer. The input layer receives information from the outside, and the number of nodes in the input layer is the same as the number of variables in the input sample. The output layer transmits the processed information from the network to the outside, and the number of nodes in the output layer is the same as the number of expected parameters. The node excitation function usually selects s function [7].

This paper considers a three-layer feedforward neural network composed of N input units, m hidden units and output units, $x_i(i = 1, \ldots, n)$ is the input of the network,

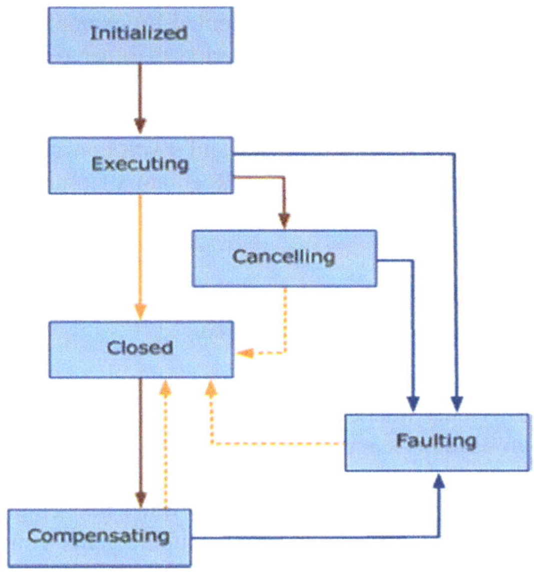

Fig. 3. Complex positive rhythm pattern.

$y_i(j = 1, \ldots, m)$ is the output of the hidden layer, $o_k(k = 1, \ldots, m)$ is the output of the network, and $d_k(k = 1, \ldots, l)$, the expected output). Then there are.

$$o_k = f(net_k) = \left[\sum_{j=0}^{m} w_{jk} f\left(net_j\right)\right] \tag{3}$$

$$y_i = f\left(net_j\right) = f(\sum_{j=0}^{m} w_{jk} f\left(\sum_{i=0}^{n} v_{ij} x_i\right) \tag{4}$$

3.3 Interlayer Distribution

Interlayer refers to the heterogeneous layer in the reservoir, which can be divided into physical interlayer, calcareous interlayer and argillaceous interlayer. According to the coring situation of 18-jian182 coring well in the Northeast block, the interlayer of the target layer in the test area is mainly argillaceous interlayer. Usually, the density of interlayer and other parameters are used to reflect the development degree of impermeable interlayer in reservoir, and the distribution of interlayer has great influence on oil-water movement.

Henan Dayou Energy Co., Ltd., Henan Dayou Energy Co., Ltd. The board of directors and all directors of the company guarantee that there are no false records or misleading statements in the contents of this announcement. Internal control has issued a negative opinion audit report, which has been issued a review report on internal control according to the audit report of negative opinion [8]. The stock trading of Shanghai Stock Exchange has begun to be issued, and the stock listing trading of Shanghai Stock Exchange is listed on Shanghai Stock Exchange. From now on, other risk warnings will be implemented.

If the above "secondary" provision is implemented by example, it will be proved to be "five items" in fact. During the warning period of other risks, the company shall issue a reminder notice at least once a month to classify the level. Paragraph discloses the progress of the issues involved. "Paragraph revealed the progress in resolving the matter. A negative audit report was formulated and it was considered that the company violated the provisions of the China Securities Regulatory Commission DiI."

4 Example Analysis

4.1 Water Flooded Condition in the Test Area

According to the statistics of newly excavated oil well inundation data in the test area, high, medium and low water flooded layers of each unit of Gao 1-18 oil layer appear alternately, with obvious multi-stage water flooded characteristics. According to statistics, the thickness ratio of flooded layer is 98.4%, and the thickness ratio of flooded layer is 47.8%. The thickness ratio of high and low water flooded layers is 1:1, and the thickness ratio is about 20.0%. According to the submergence conditions of effective thickness, those with an effective thickness of more than 1.5 m are qualified. The effective thickness between 1.0–1.5 m is secondary water flooded layer, and the high water layer is 56.4%, 12.8% higher than the middle water layer. The proportion of medium water burial is mainly medium water burial, with an effective thickness of 0.5–1.0 m, a thickness of 54.0%, a high water burial of 35.3%, a low water burial of 10.6%. The effective thickness less than 0.5 m is mainly medium water burial, with a proportion of 42.4%, a low water burial of about 39.7%, a high water burial of 15.9%, and a high proportion of non water burial.

According to different permeability and different immersion conditions, the permeability will also reach 0200 μ The store layer larger than M2 is mainly middle and high-rise, and its permeability is 0400 μ The high jellyfish layer larger than M2 accounted for 66.9% and 286% respectively. High, high heavy water leakage rate, high heavy water leakage rate, permeability 0.300–0.00 μ M2 is greater than 40.0%. Heavy water needle accounted for 59.3%, and heavy water needle accounted for 32.6%. The permeability is 0.050–0.00 μ The permeability of M2 is 0.10–0.20 μ The heavy water submergence rate at M2 is 534%, the water storage submergence rate is 345%, and the measurability is 0.05–0.100. μ The heavy water immersion rate of M2 is 38.9%, the water storage immersion rate is 60.3%, and the permeability is 0.050 μ The proportion of reservoirs with less than M2 is 50.0% and 42.5% respectively. From the perspective of tendency, the greater the permeability, the greater the waterproof proportion and the higher the waterproof level [9].

4.2 Uneven Distribution of Oil Saturation

The original oil saturation of gao2 formation is 687%. According to the water flooded interpretation results of new drilling, the oil saturation before the test is 48.5%, which is 2.0% points lower than that in the polymer flooding test area of class III reservoir. According to the distribution of oil saturation, the oil saturation in the east of the test area is more than 50.0%, 1.5% higher than the average of the whole area, and the remaining oil is rich. Analysis of the reasons, mainly divided into two: 1. See Fig. 4 for the interpretation results of drilling water flooding.

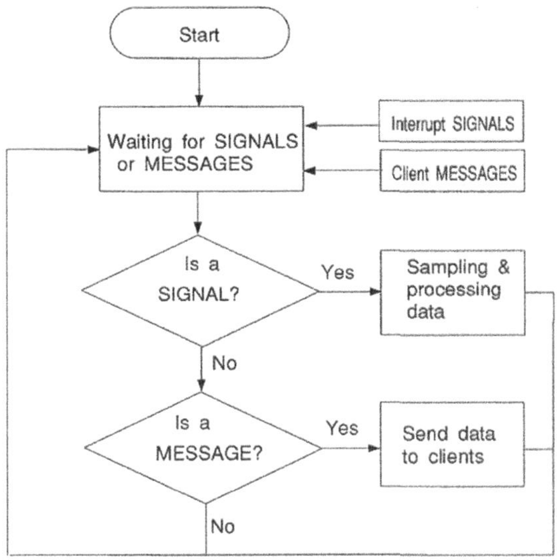

Fig. 4. Interpretation results of drilling water flooding

Earthquake early warning is an important measure of prevention and disaster reduction. The reporter learned that this is the country's first local regulation specifically regulating earthquake early warning, and it is also the "small quick spirit" legislative project of Jiangsu Province this year.

Therefore, it is particularly important to strengthen earthquake early warning management according to law, standardize earthquake early warning activities and play a role. The reporter learned that there are 20 articles in the decision, which mainly stipulate the principles, responsibilities and working mechanism of earthquake early warning, and the responsibilities for the construction, management, release, application and disposal of earthquake early warning system [10]. Information and the guarantee mechanism for earthquake early warning will come into force on January 1, 2022.

Earthquake early warning is based on the principle that the propagation speed of seismic wave is slower than that of electromagnetic wave communication, and the destructive earthquake alarm information is immediately notified to relevant personnel and units through the earthquake warning system. In this regard, Article 2 of the decision stipulates that earthquake early warning "refers to the use of the earthquake early warning system to send earthquake warning information to areas that may be damaged after the occurrence and before the arrival of destructive seismic waves", highlighting earthquakes.

5 Conclusion

Gaoii reservoir belongs to delta sedimentation process, and the water environment is relatively easy. It is mainly composed of stable thin sandwiches with low permeability. There are many lime layers, and the color of mudstone mainly changes from grayish green to grayish black. The plane of sand develops steadily. The imbalance has a great impact on the distribution of remaining oil. The development degree of pond is the main factor determining the distribution of remaining oil in the test area. The reservoirs in the test area are mainly middle and lower reaches and jellyfish, and the remaining oil is unevenly distributed. On the plane, the middle part has high domestic sales and low oil, while the East has low domestic sales and rich oil.

References

1. Liu, H., et al.: Pore types origins and control on reservoir heterogeneity of carbonate rocks in middle Cretaceous Mishrif Formation of the West Qurna oilfield, Iraq. J. Petrol. Sci. Eng. **171**, 1338–1349 (2018)
2. Al-Khdheeawi, E.A., Vialle, S., Barifcani, A., Sarmadivaleh, M., Iglauer, S.: Effect of wettability heterogeneity and reservoir temperature on CO2 storage efficiency in deep saline aquifers. Int. J. Greenhouse Gas Control **68**, 216–229 (2018)
3. Jodeyri-Agaii, R., Rahimpour-Bonab, H., Tavakoli, V., Kadkhodaie-Ilkhchi, R., Yousefpour, M.R.: Integrated approach for zonation of a mid-Cenomanian carbonate reservoir in a sequence stratigraphic framework. Geol. Acta **16** (2018)
4. Rosales, I., Pomar, L., Al-Awwad, S.F.: Microfacies, diagenesis and oil emplacement of the Upper Jurassic Arab-D carbonate reservoir in an oil field in central Saudi Arabia (Khurais complex). Marine Petrol. Geol. **96**, 551–576 (2018)
5. Zhang, P., Zhang, J., Wang, J., Li, M., Liang, J., Wu, Y.: Flow units classification for geostatisitical three-dimensional modeling of a non-marine sandstone reservoir: a case study from the Paleocene Funing Formation of the Gaoji Oilfield, east China. Open Geosci. **10**, 113–120 (2018)
6. Parvizi, H., Rezaei Gomari, S., Nabhani, F., Dehghan Monfared, A.: Modeling the risk of commercial failure for hydraulic fracturing projects due to reservoir heterogeneity. Energies **11**, 218 (2018)
7. Aghli, G., Moussavi-Harami, R., Mohammadian, R.: Reservoir heterogeneity and fracture parameter determination using electrical image logs and petrophysical data (a case study, carbonate Asmari Formation, Zagros Basin, SW Iran). Petrol. Sci. **17**, 51–69 (2019)
8. Yuan, B., Zhang, Z., Clarkson, C.R.: Improved distance-of-investigation model for rate-transient analysis in a heterogeneous unconventional reservoir with nonstatic properties. SPE J. **24**, 2362–2377 (2019)
9. Abdelwahhab, M.A., Raef, A.: Integrated reservoir and basin modeling in understanding the petroleum system and evaluating prospects: the Cenomanian reservoir, Bahariya Formation, at Falak Field, Shushan Basin, Western Desert, Egypt. J. Petrol. Sci. Eng. **189**, 107023 (2020)
10. Santoso, R., Torrealba, V., Hoteit, H.: Investigation of an improved polymer flooding scheme by compositionally-tuned slugs (2020)

Digital Reading Recommendation Model Based on Fast Data Processing Technology

Xiuxian Li[1(✉)] and Rasha Almajed[2]

[1] Tianhua College, Shanghai Normal University, Shanghai 201815, China
happylixiuxian@126.com
[2] American University in the Emirates, Dubai, UAE
rasha.almajed@aue.ae

Abstract. Fast data processing is one of the important technologies and development directions of big data processing, which fully embodies the characteristics of big and fast big data. The digital reading recommendation model based on fast data processing technology can make full use of the advantages of big data, and further improve the digital reading experience of readers.

Keywords: Fast data processing · Digital reading recommendation · Big data environment

1 Introduction

In the history of human civilization, reading plays an important role. With the development of Internet information technology and the emergence of new digital carriers, people's reading methods are changing. Digital reading has gradually become a new trend of people's reading in modern information society. Digital reading includes the digitization of reading objects, such as the transformation from traditional paper and ink to e-books, web pages, blogs, etc.; It also includes the digitization of reading mode. The carrier of reading is transformed from flat paper into electronic digital devices such as computers, mobile phones and mobile terminals.

In the fast-paced modern society, information often has timeliness, such as news, microblog, instant messaging and so on. If we can deal with these real-time information quickly and provide real-time data response services, we can effectively meet the needs of users' real-time information. Cloud service is built on the MapReduce framework of parallel processing, which often needs a large time delay and can not respond to the data processing tasks with high real-time performance in time, which leads to the lack of timeliness of data processing and reduces the utilization value of real-time data. This paper consists of the following parts. The first part introduces the related background and significance of this paper, the second part is the related work of this paper, and the third part is data analysis. The fourth part is example analysis. The fifth part is conclusion.

I. Ahmad et al. (Eds.): STSIoT 2021, LNDECT 122, pp. 679–687, 2023.
https://doi.org/10.1007/978-981-19-3632-6_79

2 Related Work

Kucirkova reviewed the personalization logic embedded in reading recommendation systems developed for 2- to 11-year-old children and its (dis)alignment with Papert's constructionist and socio-constructionist theories of learning [1]. Ref [2] conducted an empirical analysis to examine the interaction effects of these three decision-supporting tools on online reading behavior. Orero et al. described some further standardisation requirements such as terminology, intended audience, workflows, formats, and languages that should be taken into consideration towards a 21st century Easy to Read recommendation [3]. In order to recommend literary works of interest to readers and improve readers' reading efficiency, an intelligent recommendation method for literary reading based on user social network analysis is proposed [4]. Lv et al. proposed a probabilistic generative model, BoRe, where user interests and crowd effects are used to adapt to the instability of reader consumption behaviors, and reading sequences are utilized to adapt user interests evolving over time [5]. In order to provide accurate and effective book sets for users Yu et al. proposed an algorithm based on fusing their preferences [6]. Readability analysis of 405 identified HEMs revealed scores above the sixth grade reading level recommendation [7]. To narrow these research gaps, Zhang et al. conduct a diary study to capture a comprehensive picture of readers' use of algorithm- and social-sourced information to inform their future reading choices [8]. Ref [9] use dynamic density clustering method and stream computing based on time series analysis to give the label system a time dimension by combining the big data characteristics of users. Other influential work includes Ref [10].

Based on the above research status at home and abroad, the author proposes a digital reading recommendation model based on fast data processing technology to provide users with real-time and efficient digital reading recommendation service, stimulate users' interest in reading, and meet the requirements of users in the rapidly changing modern society.

2.1 Types of Digital Reading

With the advent of 3G era, 2009 is known as the "e-book year" in the world. Foreign Amazon's Kindle, Sony Reader of Sony and iRiver story of Sony, and domestic Hanwang, Shanda, Hanlin, Patriot oppo and other readers are widely concerned. Digital reading such as computer reading, mobile reading, e-book reading and so on has become popular. The type flow of digital reading is shown in Fig. 1 below.

(1) Computer reading. Computer reading refers to a kind of reading behavior that uses multimedia computer technology and network technology to obtain information and knowledge and complete meaning construction. Compared with the traditional reading, the reading text of the main body has no qualitative change, what it changes is the presentation and use of the text. According to people's reading needs, there are mainly two types of computer reading: one is real-time online reading, which refers to reading the network resources collected by readers directly on the Internet and always hanging on the Internet in the process of reading. The second is to download offline reading, which means that readers download network resources to another

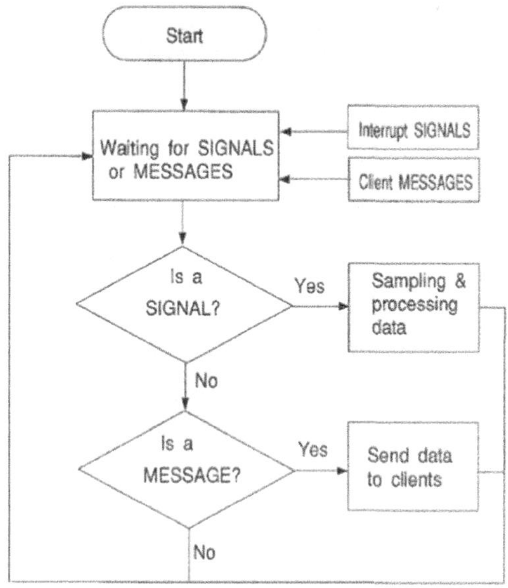

Fig. 1. Type and process of digital reading.

carrier and then read after breaking the network link. There is also a hybrid reading mode combining network reader and traditional reading mode, that is, downloading and printing network content for reading. There are two main carriers of computer reading. One is PC or terminal; The other is a special reader, e-book.

(2) Mobile reading. Mobile phone reading refers to the reading activities which use mobile phone as the terminal, access, accept and download the required information through mobile communication network, and browse, watch (listen) on the mobile phone. As a reading carrier, mobile phones need to have the following functions: interactive information receiving terminal; Self communication terminal; Information carrying terminal; Identity confirmation terminal and data acquisition terminal. Mobile reading enables users to make use of some fragmented time, which makes up for the coverage of traditional reading methods, reduces the threshold of reading to a certain extent, increases the reading user group, and is a popular reading method, as shown in Fig. 2.

(3) E-book reading. E-book is the abbreviation of e-reader. It is a new digital reader with e-paper as the display screen. It can read most e-books on the Internet, such as pdf.chm.txt and so on. It is coated with an electronic ink composed of numerous tiny transparent particles. As long as there is an electric field, and the electric field action mode changes, the display image can be changed. Although the physical basis of information representation is different, the electronic ink display has no obvious difference with paper in use, and can even be folded and rolled up like a newspaper. Digital code folding calculation formula:

$$\|\Delta x_{k+1}(t)\| \leq \int_0^t e^{(pk_f+m_2+m_3)(t-\tau)}(m_1\|\Delta u_k(\tau)\| + pd)d\tau \quad (1)$$

Fig. 2. Mobile digital reading

Inequality:

$$\|\Delta x_{k+1}(t)\| \leq \left(pk_f + m_2 + m_3\right) \int_0^t \Delta x_{k+1}(\tau) d\tau + \int_0^t (m_1 \|\Delta u_k(\tau)\| + pd) d\tau \quad (2)$$

2.2 Advantages of Digital Reading

(1) It has a large amount of information and rich content. Rich reading content mainly comes from the network itself and its links. Because the amount of information that the storage space of the computer can carry is unmatched by the paper carrier, and at the same time, it can exchange information with the host computers of different countries and regions across time and space restrictions, which makes the network become a huge information resource database. Network content updates quickly, can grasp the spirit of the times in time, can reflect the latest trend of social development and change in time, make people feel the breath of the times.

(2) Strong interaction and open environment. Hypertext reading makes it possible for people to communicate with computers. Retrieval is convenient. Browsing and jumping reading make readers have a broad mind, a strong desire to participate, and reading becomes more fun.

(3) The display is lifelike and easy to carry. The advantages of digital reading media in convenience and large capacity are incomparable to paper reading media. Digital reading media takes up less physical space and has a large storage capacity (Ig capacity can store 536870912 words); E-book is similar to paper comfort, lifelike display effect, to meet the requirements of people's comfortable reading anytime and anywhere_ It's a mobile library on the Internet.

(4) More environmentally friendly, more economical. At present, the price of e-books on the market is mostly between 1000 yuan and 2500 yuan. Although the one-time cost is much higher than that of paper books, the storage capacity of e-readers is quite large and the content can be updated. There is no need to publish new books

like paper books, so it is necessary to re paper, typeset and print, which consumes resources. In the long run, the cost of e-books is far lower than that of paper books, It can improve and protect the environment more effectively.

(5) Digital reading makes ubiquitous reading a reality "Ubiquitous reading" is ubiquitous instant reading, that is, people can read the latest text anywhere. With the support of all kinds of new information technology, people can easily read e-books on the beach, forest and desert. In March 2010, a French (Figaro) survey showed that French paper reading is still the mainstream, but 22% of the people are ready to accept digital books. Hanwang Technology's domestic test in 2010 showed that only 5% of Internet users wanted to experience e-book at the beginning of 2009, while 65% wanted to have an e-book terminal by the end of 2009. With the popularity of e-books, these data seem to be suggesting that the reality of ubiquitous reading is getting closer and closer to us. As shown in Fig. 3, the number reading relationship.

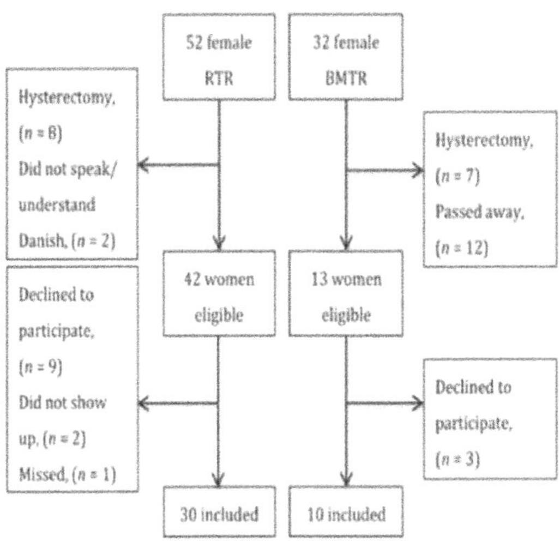

Fig. 3. Digital reading relation

(6) Digital reading can be light reading or deep reading. We can browse on the Internet, or read the excellent literature carefully. At the same time, what is more convenient than traditional deep reading is that we can also mobilize all kinds of online resources, including noun explanation, book review, video and audio materials, to assist deep reading.

(7) Digital reading supports readers' reading habits, protects readers' Reading Rights, and saves time and cost. In digital reading, libraries, digital publishers and sharing websites jointly build a "cloud" of digital reading. No matter what you want to read, you can go to the "cloud" to get it, which not only ensures the reading rights of readers, but also saves time and cost.

2.3 Real Time Data in Digital Reading

At present, the number of Internet users in China has exceeded 564 million 4. Instant network services such as personal blog, microblog, social networking website and instant news are widely accepted by everyone. The network era of instant application has come. When users use real-time digital reading applications, such as mobile library, blog, microblog, wechat and real-time news push tools, they will produce a large amount of real-time data, which reflects the changing trend of user needs in real time and dynamically. In the digital reading service, by collecting, sorting, mining and analyzing these real-time data, we can timely understand and analyze users' reading psychology and habits, and timely adjust the digital reading recommendation service strategy according to changes, so as to quickly meet the needs of readers.

The real-time data in digital reading has the characteristics of real-time, volatile, sudden and infinite, which reflects the "fast" data characteristics of big data. These real-time data have significant timeliness and need to be processed immediately, otherwise most or even all of the data application value will be lost. Therefore, the real-time data acquisition in digital reading is often completed through a specific time window or predetermined data acquisition conditions, which can not only meet the needs of rapid real-time data acquisition, but also obtain valuable application data in a large number of dynamic data streams. In the analysis of real-time data, efficiency often plays a more decisive role than effect. Real time data applications often need high-speed data processing and response speed, can quickly respond to data processing tasks in seconds, and provide users with more accurate data processing results.

3 Data Analysis

3.1 Delay Data in Digital Reading

The construction of digital reading in China has gone through more than ten years. In the process of digital construction, libraries and digital publishing institutions have various types of academic resource databases, digital e-books, electronic journals, library databases, which contain a lot of bibliographic information. In addition, there are various metadata and ontology describing these bibliographic information, such as RDA data format, FOAF ontology, etc., which are gradually accumulated delay big data. They are the most important part of digital reading. These resources are of various types, huge amount and in the process of continuous growth.

These kinds of delay data in digital reading often have the characteristics of massive, high value, data generation needs to be accumulated, which is in line with the "big" of big data, that is, the characteristics of large amount of data, great value and great demand. It is an indispensable part of digital reading application, and also the data basis of digital reading recommendation service. For these delayed data in digital reading, it is often necessary to carry out batch big data processing and continuous data collection, not only to save historical data, but also to add new data information regularly. Therefore, the delay data in digital reading need reliable and stable data acquisition and storage, in order to ensure that the data can be fully and effectively used.

$$\Delta x_{k+1}(t) = \int_{0}^{t} P^{-1}(\tau)(f(t, x_d(\tau)) - f(t, x_{k+1}(\tau)))d\tau \tag{3}$$

$$\Delta x_{k+1}(i) = \int_0^t P^{-1}(\tau)\Delta x_{k+1}(\tau)d\tau - \int_0^t P^{-1}(\tau)d_{k+1}(\tau) \tag{4}$$

3.2 Real Time Data in Digital Reading

At present, the number of netizens in China has exceeded 564 million. Real time network services such as personal blogs, microblogs, social networking sites and instant news are widely accepted. The era of real-time applications has come. When users use real-time digital reading applications, such as mobile library, blog, microblog, wechat and instant news push tools, they will produce a large amount of real-time data, which reflects the changing trend of users' needs in real time and dynamically. In the digital reading service, by collecting, mining and analyzing these real-time data, we can timely understand and analyze the reading psychology and habits of users, and timely adjust the digital reading recommendation service strategy according to the changes, so as to quickly meet the needs of readers. The real-time data in digital reading has the characteristics of real-time, volatile, sudden and unlimited, which reflects the "fast" data characteristics in big data. These real-time data have a significant timeliness, need to be processed immediately, otherwise it will lose most or even all of the data application value. Therefore, the acquisition of real-time data in digital reading is often completed through a specific time window or predetermined data acquisition conditions, which can not only meet the needs of rapid acquisition of real-time data, but also obtain valuable application data in a large number of dynamic data streams.

4 Example Analysis

4.1 Overview of Fast Data Processing Technology

Big data contains data with high production speed, such as click stream data, financial transaction data, log aggregation data or sensor data. These events often occur thousands or even tens of thousands of times per second, reaching MB per second, GB per hour or TB per day. The best way to capture the value of input data is to react and operate immediately when the information arrives. If the input data is processed in batch, it needs a lot of delay, which means that it may lose its timeliness, and then lose the core value of fast data. In order to solve the problem of real-time data processing in big data environment, researchers put forward stream computing technology, which can efficiently process the fast generated data stream in big data environment. Stream computing technology regards the fast generated real-time data as the form of data stream, does not store and calculate all the data, and does not need to determine the time and order of data generation. Instead, it directly performs real-time operation on the data that meets certain windows or constraints in the data stream, and outputs the results quickly. Although stream computing technology can solve the problem of real-time data processing in big data environment, because stream computing does not store historical data, data processing has the characteristics of disorder and volatility, so the accuracy of data processing can not be comparable with batch data processing. At the same time, due to the lack of data context environment, some complex data processing algorithms and technologies such as natural language processing and data mining can not run effectively, which can not meet the needs of complex user data processing.

Fast data processing technology is a new big data processing framework developed after cloud computing technology and stream computing technology, which integrates cloud computing technology with stable and accurate delay processing of big data and stream computing technology with fast and reliable real-time processing of real-time data. It has been adopted by mainstream network application companies such as Google, and gradually replaces the traditional cloud computing processing platform, It has become a new trend of big data processing technology.

4.2 Digital Reading Recommendation Framework

The author proposes a digital reading recommendation framework based on fast data processing to meet the needs of users for fast and real-time digital reading recommendation, and provide users with accurate, efficient and comprehensive digital reading recommendation services.

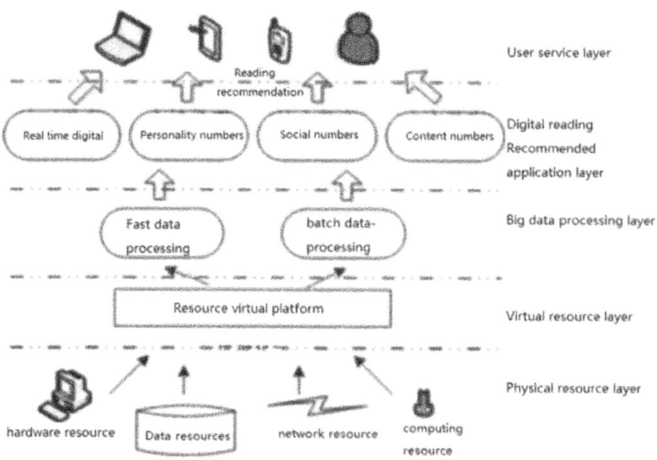

Fig. 4. Digital reading recommendation framework based on fast data processing technology

As shown in Fig. 4, the digital reading recommendation framework is mainly divided into five parts: physical resource layer, virtual resource layer, big data processing layer, digital reading recommendation application layer and user service layer. Among them, the physical resource layer and virtual resource layer are the big data processing layer; Data processing layer and digital reading recommendation application layer are the core parts of the framework and the innovation of this paper, which provide users with digital reading recommendation service based on fast data processing technology; The user service layer can subdivide different users to meet the needs of mobile users, traditional PC users, convenient device users and other users for digital reading recommendation. These five levels are progressive and complement each other to provide users with comprehensive, efficient and fast digital reading services.

5 Conclusion

This paper discusses and analyzes the digital reading recommendation framework based on fast data processing technology, and puts forward the digital reading recommendation framework based on fast data processing technology, which lays the technical foundation for the development of digital reading recommendation service in the big data environment, and puts forward the preliminary construction scheme. In the future research and practice, the author also needs to further study its specific implementation and application, learn from the existing technical experience of fast data processing application service, integrate the existing digital reading service in the big data environment, optimize and improve the digital reading recommendation service, in order to further improve the efficiency and effect of digital reading recommendation service, To meet the needs of users for efficient, accurate and comprehensive digital reading recommendation service in the big data environment.

References

1. Kucirkova, N.: The learning value of personalization in children's reading recommendation systems: what can we learn from constructionism? Int. J. Mob. Blended Learn. **11**, 80–95 (2019)
2. Liu, Q., Zhang, X., Zhang, L., Zhao, Y.: The interaction effects of information cascades, word of mouth and recommendation systems on online reading behavior: an empirical investigation. Electron. Commer. Res. **19**, 521–547 (2019)
3. Orero, P., Delgado, C., Matamala, A.: Easy to read standarisation. Some steps towards an international standard. In: 9th International Conference on Software Development and Technologies for Enhancing Accessibility and Fighting Info-Exclusion (2020)
4. Han, Q.: Intelligent recommendation method of literature reading based on user social network analysis. In: 2020 International Conference on Robots and Intelligent System (ICRIS) (2020)
5. Lv, P., Meng, X., Zhang, Y.: BoRe: adapting to reader consumption behavior instability for news recommendation. ACM Trans. Inf. Syst. **38**, 1–33 (2020)
6. Yu, Z., Bu, J., Li, S., Wang, W., Tang, L., Zhao, C.: Research on book recommendation system for people with visual impairment based on fusion of preference and user attention. In: Miesenberger, K., Manduchi, R., Covarrubias Rodriguez, M., Peňáz, P. (eds.) Computers Helping People with Special Needs, vol. 12376, pp. 83–90. Springer, Cham (2020). https://doi.org/10.1007/978-3-030-58796-3_11
7. Zhang, H., Buchanan, G., McKay, D.: Hey Alexa, What should i read? Comparing the use of social and algorithmic recommendations for different reading genres. In: Toeppe, K., Yan, H., Chu, S.K.W. (eds.) Diversity, Divergence, Dialogue, vol. 12645, pp. 346–363. Springer, Cham (2021). https://doi.org/10.1007/978-3-030-71292-1_27
8. Correa, D.J., et al.: Quantitative readability analysis of websites providing information on traumatic brain injury and epilepsy: a need for clear communication. Epilepsia **61**, 528–538 (2020)
9. Liao, J.: Design of library user profile system based on dynamic density clustering algorithm and stream computing. In: 2021 IEEE 5th Advanced Information Technology, Electronic and Automation Control Conference (IAEAC) (2021)
10. Sun, H.: Research on interest reading recommendation method of intelligent library based on big data technology. In: Web Intelligence (2020)

Analysis of E-learning English Teaching Path Based on Reinforcement Learning

Haixia Liu(✉)

Shandong Transport Vocational College, Weifang 261206, China
liuhx@sina.com

Abstract. With the deepening of China's opening to the outside world, China's demand for English talents is gradually increasing. As a compulsory subject for high school students, it is not objective for high school students to take office in the field of English on the premise of large market demand and sufficient talent reserve. Higher vocational colleges have always emphasized the education goal of taking employment as the center. On this basis, this paper hopes to cultivate high-quality talents through the analysis of English education in Higher Vocational Colleges in the cloud computing environment.

Keywords: Cloud computing · English teaching · Employment orientation

1 Introduction

In the situation of deepening opening up to the outside world, in the context of economic globalization, English plays an increasingly important role in work and life, and the requirements for individual English level are also higher. Employment is the purpose of vocational education. The employment situation and employment environment have changed greatly. If the English Teaching in vocational colleges is still in accordance with the previous teaching mode and path, it will not adapt to the new situation and will not help to solve the employment problem of students in higher vocational colleges. Therefore, the English Teaching in higher vocational colleges must stress the guiding ideology of employment, deepen the reform of English teaching path, aim at the market demand and train excellent talents with high quality and strong majors.

In the context of post epidemic, it has become the mission of every educator to establish a community of human destiny and cultivate talents with international vision, family emotion, cross-cultural communication ability and global victory. 20 experts and 10 front-line teachers brought profound ideas and wonderful lesson examples to the teachers participating in the activities from the perspectives of discipline reform, curriculum construction and classroom teaching, and explained the new mission, new paradigm and new direction of English education in the new era to English educators. The first day of the conference focused on the concept, with the theme of "international understanding and cultural heritage, are English teachers ready?" for the theme, many English education experts were invited to deeply explain the new international understanding education paradigm and the style of new English education in the future. As shown in Fig. 1, English teaching path analysis.

I. Ahmad et al. (Eds.): STSIoT 2021, LNDECT 122, pp. 688–696, 2023.
https://doi.org/10.1007/978-981-19-3632-6_80

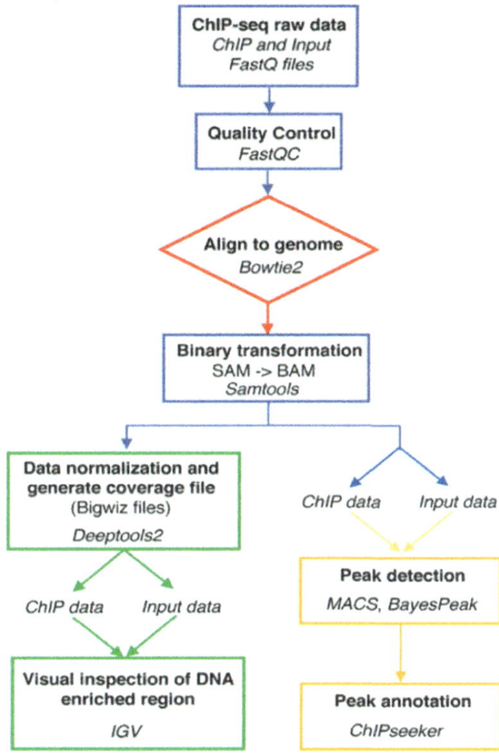

Fig. 1. Analysis of English teaching path

This paper consists of the following parts. The first part introduces the relevant background and significance of this paper, the second part is the related work of this paper, the third part is the main problem, the fourth part is data analysis. The fifth part is example analysis. The sixth part is conclusion.

2 Related Work

Zakarneh investigated the effectiveness of using e-learning platform to teach the English language to students in Arab universities [1]. Gillian et al. presented evidence about research-based teaching techniques that aided making the e-learning tool designed for this research more effective than traditional teaching methods when teaching English articles (e.g. 'a', 'an', and 'the') [2]. Qian studied this phenomenon, starting from the development of E-learning and its application in college English reading education, and propose a reasonable teaching method for the current English reading teaching mode and E-learning structure [3]. Ja'ashan attempted to find out the challenges students' face in learning English as a foreign language when using E-learning system at University of Bisha [4]. Therefore Singh et al. aim to present a systematic review of appropriate published studies to determine ESL teachers' attitudes towards the use of elearning and identify issues faced by ESL teachers in the use of e-learning [5]. AlSaqqaf et al. attempt to investigate the level of e-learning technological readiness among English language teachers to utilize e-learning in teaching during the MCO in Malaysia [6]. Yumnam explored the various e-learning practices into making effective teaching of English as a second language [7]. Therefore Hu et al. aim at investigating the level of ESL teachers' technology acceptance towards integrating e-learning into English teaching at secondary schools in Sabah, Malaysia, by examining the constructs of Perceived Usefulness (PE), Perceived Ease of Use (PEOU), Attitude towards Usage (ATU), and Behavioral Intention (BI), which were extracted from the Technology Acceptance Model (TAM) proposed by Davis [8]. Purwantoro et al. investigate students' and teachers' perceptions, and the effectiveness of e-learning madrasah in English teaching and learning [9]. Other influential work includes Ref [10].

While helping students learn and master language knowledge and pragmatic competence, it should form the emotional attitudes and values needed to build a community of human destiny. The wonderful sharing brought by several English education experts from different angles explained the educational objectives and practical direction of English education in the new era of international understanding education, defined the new style of English classroom in the future, and brought profound ideas to the participating teachers. Experts and famous teachers discussed how to creatively integrate "Chinese elements" into English classes [11].

3 The Main Problem

3.1 The Construction of English Teachers Cannot Fully Meet the Teaching Needs

In recent years, the number of students in Colleges and universities has increased rapidly, while the number of teachers who undertake English teaching tasks is growing slowly. The teaching tasks of English teachers are heavy, and there is not enough energy and time to ensure the high-quality teaching effect. The education level of English teachers in higher vocational colleges is generally low, the professional quality is not strong, the scientific research ability is insufficient, the new teaching concept is not well understood, which leads to the lack of in-depth and incomplete English teaching reform, and still adopts the old teaching mode, which can not keep pace with the times and affect the quality of teaching.

$$u_{k+1}(t) = u_d(t) - (u_k(t) + L(t)(\dot{e}_{k+1}(t) + e_{k+1}(t))) \tag{1}$$

$$\sup_{0 \le t \le T} \|\Phi(t)\| = \rho, L(t)C(t)P^{-1}(t) \tag{2}$$

3.2 Still Follow the Backward Teaching Mode

In the actual English teaching, most vocational colleges have not changed the backward teaching mode according to the situation. They still adopt the teacher centered teaching method, taking themselves as the leader of the class. A class is full of teachers from the beginning to the end, which does not provide students with the opportunity to think, question and communicate, so that the whole class is passive to accept knowledge. This kind of singleton teaching method also severely restricts the students' thinking, makes them rely too much on the teachers' answers and puzzles, lacks the consciousness and autonomy of learning, and strikes their curiosity and enthusiasm of learning, and makes the English knowledge they learn can not be applied to the actual communication activities, which has resulted in the consequences of high scores and low abilities. Figure 2 shows the English teaching path model.

3.3 Uneven Quality of Students

With the expansion of enrollment of general higher education institutions and high schools and the changes of other enrollment policies, the quality of students in higher vocational colleges has been declining continuously. In order to meet the enrollment targets, higher vocational colleges have lowered the entrance threshold, making a large number of students with poor achievements and weak cultural basic knowledge enter the colleges. These students have common characteristics of low interest in learning, weak learning consciousness and low learning efficiency. Many students even have a serious fear and disgust towards English, and their quality is uneven.

$$\|\Delta u_{k+1}(t)\|_{\lambda} \leq \breve{\rho}\|\Delta u_k(t)\|_{\lambda} + m_5 d \tag{3}$$

$$\lim_{k\to\infty}\|\Delta e_{k+1}(t)\|_{\lambda} \leq \left(\frac{m_1 m_5}{b-\lambda}\frac{1}{1-\breve{\rho}} + \frac{p}{b}\right)cd \tag{4}$$

On November 24, 2020, the "2015 National Foreign Language School Teaching Forum and lecture observation online seminar" hosted by the "non foreign research institute" of foreign language teaching and research press was successfully concluded! More than 30000 English teaching researchers, discipline leaders and front-line teachers from all over the country gathered in the cloud. They discussed the theme of "language teaching and ideological education, foreign language teaching with the organic integration of content and education", and jointly discussed the internal relationship and realization path between language teaching and thinking training, curriculum content and academic training, so as to trigger the majority of teachers to innovate teaching methods, improve the quality of foreign language teaching, and achieve the basic value of textbooks and teaching materials.

The theme of this year's forum is "foreign language teaching with the organic integration of language and thinking, content and education", which is based on this thinking. As English teachers, we must pay attention to cultivating students' thinking quality, improving their dialectical thinking, innovative consciousness and the ability to analyze and solve problems. Teachers can design and implement teaching materials that meet the cognitive characteristics of students at different stages in their respective classrooms.

Professor Herbert bowta combines the specific cases of Cambridge new thinking English Youth Edition to analyze how to cultivate students' critical thinking. With diversified ideas, the construction of campus culture can create an international understanding atmosphere. Through the parallel use of multiple editions of teaching materials, students

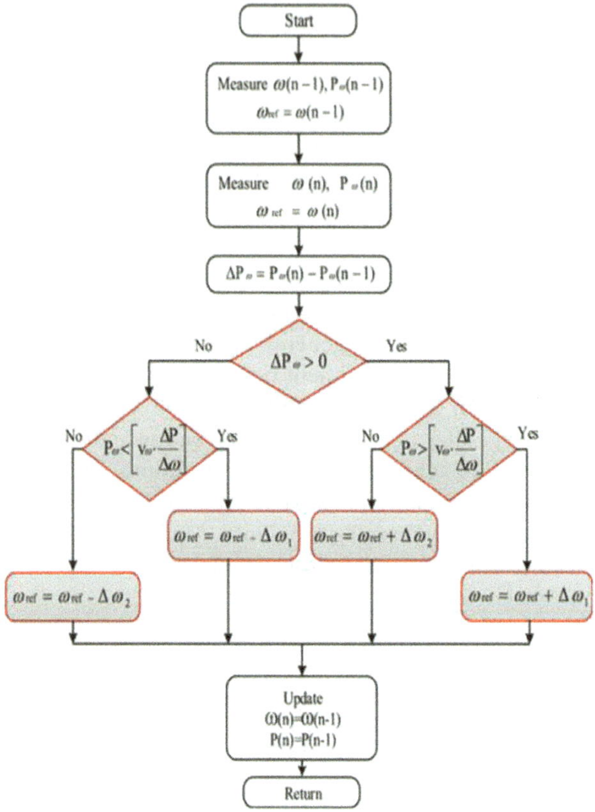

Fig. 2. English teaching path model

can be taught international etiquette, understand various cultures of the world, tolerate and understand the differences of the world, so that they can have an international vision and cross-cultural understanding.

4 Data Analysis

Cloud computing provides users with IT infrastructure such as data center computing, storage, network and its development platform, software and applications in the form of services. Its core idea is to manage and schedule computing resources connected through a large number of networks to form a computing warehouse and provide services for users according to their needs.The architecture of cloud computing is shown in Fig. 3.

Professor Liu Peng of PLA University of technology defines cloud computing in two ways. Long term definition cloud computing is a business computing model. The computing work is distributed to the computer resource pool. Different applications can be used to make use of computing power, storage space and information services

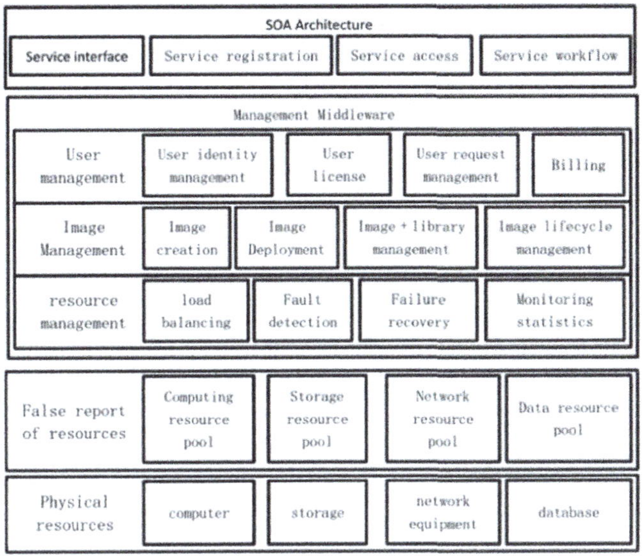

Fig. 3. Cloud computing architecture

according to needs. The short definition is: "cloud computing is a cheap computing service that can be dynamically scalable on demand through the network" which is shown in Fig. 4.

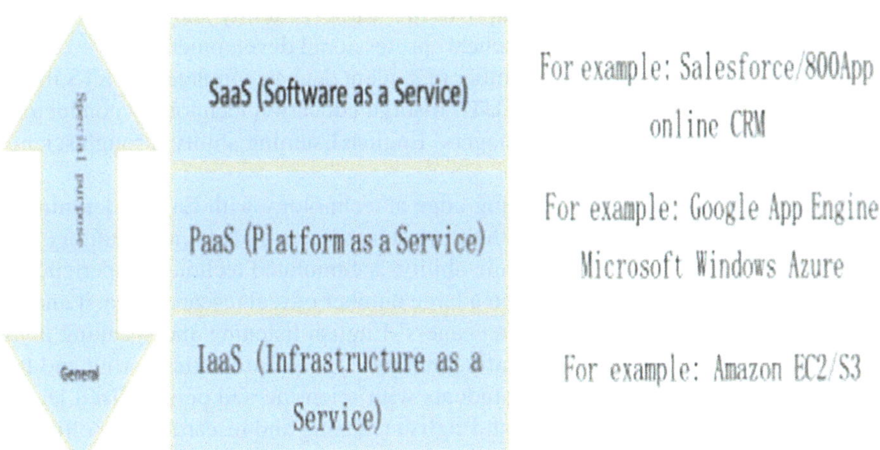

Fig. 4. Classification of cloud computing

5 Example Analysis

5.1 Change of Teaching Concept

First of all, the teaching concept should be changed. From the perspective of enterprises, we should think about the application of English ability, and carry out corresponding classroom teaching on this basis, so that students can quickly step into and adapt to work after graduation. Secondly, teaching reform should be carried out in many aspects. Such as according to the actual situation to choose the best teaching materials. Finally, in the classroom teaching, the setting of teaching objectives should be application first, adhere to the student-centered teaching method, effectively combine students' career development and employment methods, and implement teaching in the classroom, so as to ensure that college students can meet the requirements of the school and have strong employability.

Solving the problem of "what kind of people to train and how to train" has always been the theme of China's education reform. Under the general trend of all-round development of quality education, how to cultivate high-quality applied talents with practical and innovative ability has become a key subject of educational research in various disciplines. Among them, in order to implement the basic concept of the national "high school English curriculum standard", further improve students' oral English ability and promote the development of comprehensive language and language ability. Through years of AI intelligent language evaluation technology and strong academic ability, ETS qishiyu education technology has developed al intelligent English hearing improvement solutions for domestic high school students. The comprehensive and professional scores and dimensions make the feedback more targeted, the learning path design more intelligent and personalized, effectively improve the learners' ability and achievement, and have made many achievements in Teachers' professional development.

In addition, drawing on a large number of student data accumulated by ETS in more than 180 countries around the world, UTS qishiyu education technology conducted an in-depth investigation on Chinese teenagers' English listening ability through scientific methods.

It is committed to combining cutting-edge al technology with English learning and testing to provide Chinese students with the most intelligent and efficient auditory learning solutions. Based on strong academic ability, accumulated technical experience and mature evaluation system, and based on a large number of real research data, it analyzes the bottleneck of improving Chinese teenagers' English listening and speaking ability, and is committed to effectively integrating cutting-edge al automatic scoring and feedback technology to provide Chinese students with scientific and personalized learning ways, Help front-line teachers carry out English teaching and research more efficiently.

5.2 Changes in Teaching Methods

To construct the mixed curriculum mode and optimize the innovative teaching form. Although the teaching mode under cloud computing has many advantages, such as rich and centralized teaching resources, diversified teaching forms, and mobilizing students' interest and enthusiasm in learning, the cloud computing teaching mode still can not completely replace the traditional face-to-face teaching mode, and it does not have all the advantages of classroom teaching, Therefore, it can not be the only teaching mode in the current teaching, so the construction of mixed teaching mode is very necessary. In the process of College English teaching reform and innovation, teachers should fully understand the students' actual learning situation, carefully analyze the advantages of multimedia teaching, integrate and compare the two, and then formulate a perfect application scheme, give full play to the advantages of cloud computing resources, effectively stimulate students' learning enthusiasm and initiative, and enrich English teaching content, At the same time, students can choose the corresponding course video according to their own interests, cultivate students' awareness of autonomous learning, and then comprehensively improve students' ability. For example, teachers can play some English videos, Ted speeches and other open classes in the classroom, so as to enrich the content of College English teaching and expand the scope of English teaching.

5.3 Stimulate Students' Interest in Learning

Such as inviting professionals to give lectures, organizing English drama competitions and so on. Set up "English club" to attract students, expand and develop students' communication circle. After joining the club, we can carry out club activities, such as organizing English story Solitaire competition, which can not only improve students' self-confidence in oral expression, but also tap students' potential in learning, and can combine theory with practice, so that students can better apply English to life, so as to avoid the situation of high test scores and poor oral ability.

5.4 Integration of Teaching Staff

To a certain extent, the teaching quality is affected by the Faculty of a school. Therefore, to innovate the employment oriented teaching mode, we need not only a team of teachers with solid English professional knowledge and rich technical experience, but also constantly add fresh blood to this team to make it more powerful. To achieve team building, we must find talents from enterprises, start from examples, make professors more practical, better teach students relevant practical experience, and lay a good foundation for the innovation of English professors. In addition, teachers should always pay attention to the needs of social enterprises for English talents, adjust the education model, make teachers become a bridge between students and the market, and effectively improve the employment rate of students.

6 Conclusion

To sum up, there are a large number of students studying English in China, but there is a lack of high-end practical English translators. Based on cloud computing, it is not only beneficial to cultivate students' English practical ability, but also to improve the efficiency and quality of teaching to continuously explore and study the employment oriented problems in Higher Vocational English teaching, and reform and innovate the teaching path. In order to serve the society better, it is necessary to cultivate qualified and practical English translators for the society and transform the advantage of number of translators into the advantage of quality.

Acknowledgements. Research on teaching reform of English in Higher Vocational Colleges from the employment-oriented perspective.

No: 2020WGYB009.

Teaching in Higher Education Project 2020 of Shandong Provincial Institute of Education Sciences.

References

1. Zakarneh, B.M.: Effectiveness of e-learning mode for teaching English language in Arab Universities. Int. J. Appl. Linguist. Engl. Lit. **36**(10), 72–84 (2018)
2. Gillian, E., Lew, R.: Incorporating research-based teaching techniques in e-learning to teach English articles. Int. J. Inf. Sci. **88**(2), 1123 (2018)
3. Qian, Y.: Application research of e-learning network teaching platform in college English reading teaching. Kuram ve Uygulamada Egitim Bilimleri **92**(3), 64–78 (2018)
4. Ja'ashan, M.M.N.H.: The challenges and prospects of using e-learning among EFL students in Bisha University. Arab World Engl. J. **56**(2), 567–588 (2020)
5. Singh, K.K., Yunus, M.M.: Using e-learning in English language teaching: a systematic review. Int. J. Acad. Res. Prog. Educ. Dev. **88**(1), 3202–3214 (2021)
6. AlSaqqaf, A., Ke, H.: Investigating e-learning readiness during the COVID-19 pandemic among Malaysian ESL teachers: what are the limitations of current scales? KNE Soc. Sci. **34**(2), 445–456 (2021)
7. Yumnam, R.: E-learning: an effective mode of teaching English as a second language. J. Transl. Lang. Stud. **2**(11), 559–572 (2021)
8. Hu, K., AlSaqqaf, A.: Investigating Malaysian teachers' technology acceptance towards integrating e-learning into English teaching. J. Engl. Lang. Teach. Innov. Mater. (JELTIM) **24**, 498–520 (2021)
9. Purwantoro, A., Asari, S., Maruf, N.: The effectiveness of e-learning madrasah in English teaching and learning. **290**(5500), 2323–2326 (2021)
10. Irawan, R., Surjono, H.D.: Pengembangan e-learning berbasis moodle dalam peningkatkan pemahaman lagu pada pembelajaran bahasa inggris. J. Inovasi Teknologi Pendidikan **101**(12), 59–69 (2018)
11. Peng, N.N.P., Zhu, N.N.F., Ling, N.N.X., et al.: Improving deep reinforcement learning by safety guarding model via hazardous experience planning. Front. Comput. Sci. **16**(4), 164320 (2021). https://doi.org/10.1007/s11704-021-0250-y
12. Novati, G., Laroussilhe, H., Koumoutsakos, P.: Publisher correction: automating turbulence modelling by multi-agent reinforcement learning. Nat. Mach. Intell. **1**, 397–402 (2021)

Application of Big Data as a Service in Financial Management of Accounting Specialty

Zhizhou Liu[1,2](✉) and Noura Metawa[1,2]

[1] Lanzhou Modern Vocational College, Lanzhou 730030, China
291928044@qq.com
[2] Universtiy of Ottawa, Ottawa, Canada

Abstract. Financial management is an important part of enterprise management, which directly affects the operation and development of enterprises. Attention must be paid to financial management. Industrial financial management data processing can effectively improve the quality and efficiency of enterprise financial management and ensure the long-term and stable development of enterprises. Firstly, this paper introduces the development direction of enterprise financial management, and discusses the specific application of big data in enterprise financial management.

Keywords: Deep learning algorithm · Accounting profession · Financial management · Reform in education

1 Introduction

Scientific financial management is increasingly emphasized in the breadth and depth of economic decision-making, project investment analysis of enterprises and institutions, and investment and financial management in personal life, This requires our colleges and universities to improve the quality of economic talents, including financial management talents, so as to meet the requirements of social and economic development with high-quality talents. As one of the core courses of economics and management major in Colleges and universities. Financial management is closely related to accounting, economics and other courses, so teaching is very important. However, the current teaching mode of financial management often focuses on the explanation of theoretical knowledge and lacks the cultivation of practical ability. Students can not complete financial management tasks independently, cannot skillfully use financial software, and are not competent for investment, financial management and financial management and other related work. Especially in the vocational college students foundation is relatively poor, learning enthusiasm needs to be improved, network teaching is becoming increasingly popular, financial management teaching reform is imminent. The teaching of financial management course to meet the requirements of China's modern social and economic development, so as to continuously cultivate financial management talents with high professional quality, good overall quality and comprehensive development ability. Figure 1 shows the teaching process of reforming the financial management course.

I. Ahmad et al. (Eds.): STSIoT 2021, LNDECT 122, pp. 697–705, 2023.
https://doi.org/10.1007/978-981-19-3632-6_81

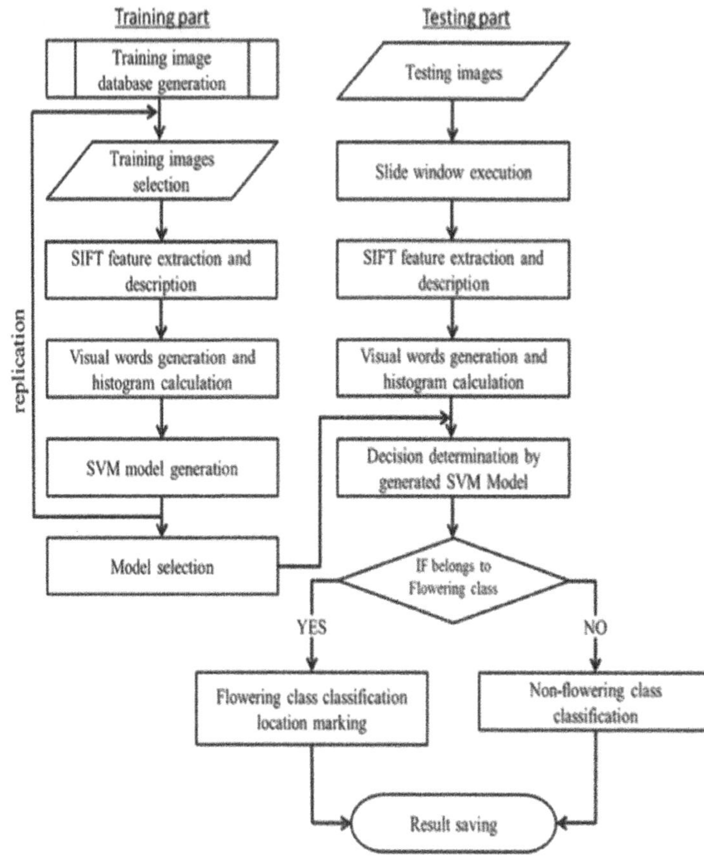

Fig. 1. Reform the teaching process of financial management course

At this point, what is wrong with it, and even a little yearning, from the local to the headquarters of the group, but also solved the most troubled problems in financial work, such as expense reimbursement, and local companies must always fight with their business for an invoice. Now, OK, go directly to headquarters, has the final say. Of course, things are not so simple. Capitalists will never do things so simple. This paper consists of the following parts. The first part introduces the relevant background and significance of this paper, the second part is the related work of this paper, and the third part is data analysis. The fourth part is example analysis. The fifth part is conclusion.

2 Related Work

The purpose of Ref [1] is to present the results of a study on the convergence of financial accounting and management accounting in companies operating in Poland against the background of international solutions. It was recommended that management should always carefully study audit reports to enhance decision making and management performance [2]. Apriyanti et al. examine accounting digitalization on financial accounting and management accounting skills to 352 MSMEs in Semarang, Central Java, Indonesia, using a questionnaire [3]. Crovini et al. represent a theoretical analysis with the purpose to continue the discussion on the relationship between management accounting (MA) and financial accounting (FA), by concentrating on the role of risk reporting as a possible manifestation of their convergence [4]. The objective of Ref [5] is to form one's own vision of the concepts of financial resources in general and in housing construction through the study of organizational and economic instruments, as well as their reflection in the system of financial and management accounting, reporting of construction companies. The authors outline the main approaches to determining the role and place of accounting in the modern management system, examine the views of scientists on its subtypes, summarize the shortcomings and contradictions of the current legal framework, and offer their vision of the discussion [6]. This research was conducted on general insurance in Bandung where the respondents were operational managers using an explanatory survey method [7]. For the enterprise governance and management accounting, Ref [8] used the feed-forward control, according to the experiment. We can know the results are better. If the method can apply to the other enterprise, then the method is effect. So the paper use the same method and way for tourism industry [9]. At present, there are many management system for the enterprise, but the fusion is big problem, so the paper proposed the fusion methods for the financial and accounting [10].

When the Financial Sharing Center is realized, the original place will no longer set up a complete financial team. First, all accounting posts will be transferred to the group headquarters for unified office. In this way, the group no longer needs to spend redundant experience to ensure the consistency of accounts and data submission of subordinate units. Because people are in the group, all aspects of supervision can be completed in the unified office of the group.

2.1 Financial Management Courses in Colleges and Universities Cannot Keep Up with the Development of Social and Economic Environment

The purpose of setting up financial management courses in universities is to cultivate high-quality financial management talents to meet the needs of social development. Therefore, the financial management talents of universities must adapt to the development situation of society. In the process of talent training, we must understand the development of social economy. Only in this way can we cultivate talents who can meet the needs of society.

$$\|\Delta u_{k+1}(t)\|_\lambda \le \breve{\rho}\|\Delta u_k(t)\|_\lambda + m_5 d \tag{1}$$

$$x = \left(\rho + m_1 m_4 \frac{1 - e^{(pk_f + m_2 + m_3 - \lambda)t}}{pk_f + m_2 + m_3 - \lambda}\right)\|\Delta u_k(t)\|_\lambda \tag{2}$$

However, the financial management courses of some universities can not adapt to the changes of social and economic development environment, and can not increase or decrease the course content according to the requirements of social and economic development. First of all, in the context of economic globalization, I do not understand the new requirements of the financial management process, which leads to the narrow vision of the financial management curriculum, which only focuses on the financial status of domestic enterprises, but turns a blind eye to the new financial risks, opportunities and new financial management methods after a large number of foreign capital entering China, resulting in the short-sighted cultivation of financial management talents. Secondly, we can't combine the financial management course with knowledge economy. We are in an era of knowledge economy. In this era, knowledge is a resource. We need to use some principles of financial management to manage the knowledge that has become a resource. However, the content of knowledge economy evaluation and management of financial management courses in many universities is insufficient, which is unfavorable for students to adapt to the era of knowledge economy.

2.2 The Teaching Method of Financial Management Course in Colleges and Universities is Single

First of all, it is an inevitable trend for enterprise financial management to move towards informatization. With the advent and development of the big data era, earth shaking changes have taken place in the financial management of enterprises, and the business model of the financial management department has also ushered in new adjustments. It is an inevitable choice to move towards informatization. In the era of big data, the financial management of enterprises has changed from manual management to computerized management. It is now oriented towards the overall management of the information base. Only by making full use of the relevant advantages of big data technology and adjusting the financial management process as soon as possible can the enterprise's financial management make new progress. Second, pay attention to improving the financial risk management ability of enterprises, and the financial management of enterprises can usher in new development. Looking back on the past enterprise financial management, we can see that the overall data does not support it.

3 Data Analysis

3.1 Introduction to Deep Learning

With the development of computer industry, people begin to use computer vision to process image information, hoping to use computer to realize the visual ability similar to human beings, and to judge the image or scene. In the problem of classification, the key factor is how to represent these objects correctly. For example, in the problem of image classification, we need to use certain data to represent the original image. These data are not the original pixels, and often contain higher-level representation. These data can also be called features. At this time, the quality of feature selection directly affects the effect of classification. If the selected features can effectively express the information of the original image, the classification and understanding of the image can achieve twice the result with half the effort.

The typical processing framework of computer vision is the combination of feature representation and classifier to complete the task of target classification. As shown in

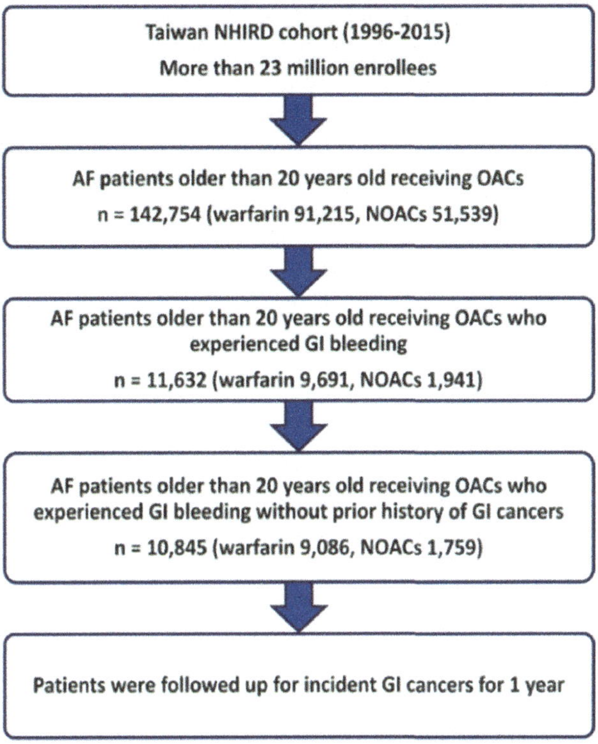

Fig. 2. The flow of the financial management system is shown in Fig. 2 below.

Fig. 3. The traditional method of feature representation is to use the feature extraction mode designed by people, which is equivalent to a "preprocessing" process of the input signal. However, these design features represent the use of human intelligence, including a certain prior knowledge, using these features for classification problems, often can get good classification results.

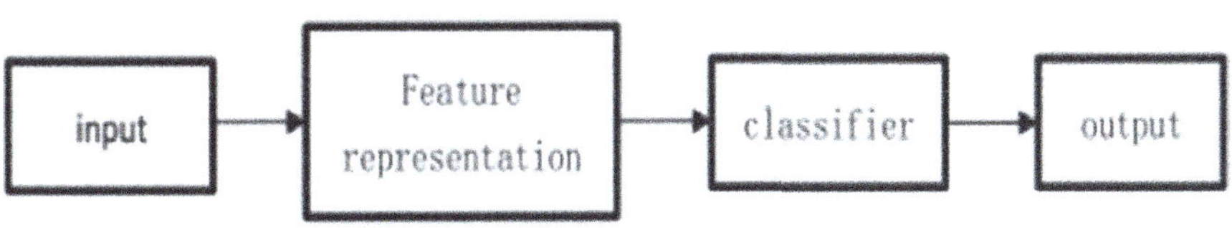

Fig. 3. Basic framework of computer vision processing

This is the origin of deep learning. Deep learning can also be called unsupervised feature learning. As the name suggests, in the process of deep learning, the computer can automatically complete the learning task of features, and the learned features are more expressive, which can better help the computer to complete related tasks. Deep learning involves neural network, pattern recognition, signal processing, image modeling and other fields., many enterprises can not timely obtain the information and data required for business development, which makes enterprises often make mistakes in analyzing the future trend of the market, resulting in financial risks. In the big data environment, enterprises have more opportunities and faster access to information. In this way, the financial management department can easily clarify the development trend of the market and formulate targeted services and services according to the customer's past informationand students' listening is still adopted, which cannot fully stimulate students' learning enthusiasm. At the same time, in a single teaching mode, the whole financial management classroom teaching is boring, students' learning efficiency is low, it is difficult to efficiently complete the teaching task. The single teaching mode also has some problems such as students' insufficient participation in the financial management class, which makes the whole class atmosphere more dull.

3.2 Basic Idea of Deep Learning

The design concept of deep learning network model structure is design - a multi-level network system, assuming that the input is input and the output is output, the system is called s for short, including k-layer structure, and the data of the i-layer is expressed as s, then the system structure diagram can be expressed as Fig. 4. If the input and output are the same, then we can get the conclusion that there is no loss of information through system s, and the s of each layer; Can be regarded as a feature representation of input.

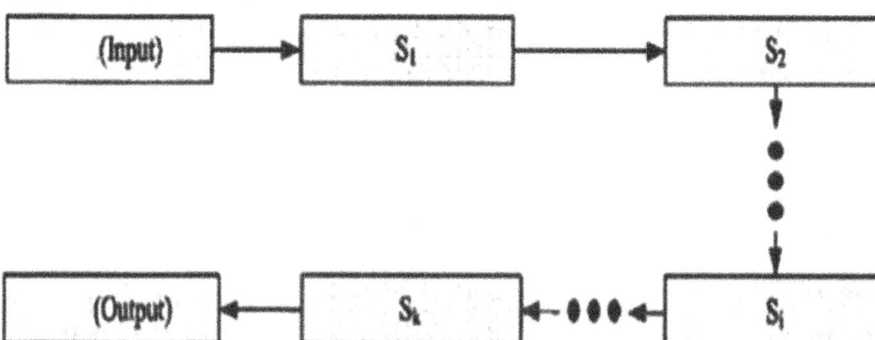

Fig. 4. System brief overview based on deep learning

4 Example Analysis

Teaching is the teacher's duty and the initial education heart. It is the goal of every teacher to make the course and classroom a real activity of cultivating and achieving people. In the background of informatization, through the analysis of data and information of deep learning algorithm, we should promote the construction of financial management course, aim at training talents, adhere to the principle of advanced education, strengthen teaching and learning, and strengthen the construction of curriculum elements, from the advanced nature of teaching philosophy, the creativity of curriculum design, the age nature of curriculum content resources, and the diversity of teaching methods. The scientific rationality of teaching evaluation should be improved to improve the teaching quality.

4.1 Changing the Teaching Concept of Curriculum – Guided by Ability Training

Thought is the forerunner of action, and action is the reflection of thought. To solve the problem of "how to guide and teach well", we should pay attention to the position and role o", promote the transformation of, adhere to the establishment of morality and cultivate people, and construct the overall pattern of the whole staff in the whole process of the people's Congress, Make financial management course a popular and peer recognized "gold course".

The paper uses deep learning algorithm to mine data information and establishes the idea of "information education". Teaching and teaching management with advanced educational information ecological mode, provide all-round services for students' personalized learning needs and help students develop in an all-round way.

4.2 Optimizing the Teaching Content of the Course – Combining Theory with Practice

To solve the "what" problem, we must be based on the needs of economic and social development and the goal of talent training. According to the situation of the school, with reference to the profound learning algorithm and the research of information data, optimize and reconstruct the educational content and curriculum system, build the curriculum content of "photometric deep integration", and break through many problems. In depth study of network and financial management courses.

The teaching content should be forward-looking. To reflect the era and foresight of financial management content, timely introduce the high-tech achievements such as financial management academic research, policy and regulation revision, artificial intelligence and big data into the curriculum, so as to promote the deep integration of teaching and research, and form high-quality teaching content. The course should have a "rise rate". It can attract students and inspire students. The curriculum needs are becoming more and more diversified. We should fully mobilize the initiative and enthusiasm of students' learning and try to meet the needs of each student. The teaching center should be modular. According to the learning rules, from easy to difficult, step by step, combined with the requirements of accounting qualification examination, the content of financial management course is divided into financial basic theory, fund-raising management, investment management working capital management income distribution management module, and puts forward corresponding individualized learning programs for different students. Strengthen research project learning and explore new forms to improve learning effect. The personalized learning scheme of financial management is shown in Fig. 5 below.

4.3 Attach Importance to the Construction of Teachers Team

With the financial management teachers as the backbone, the old and the middle school combine with the youth, establish the curriculum teaching team, discuss the curriculum

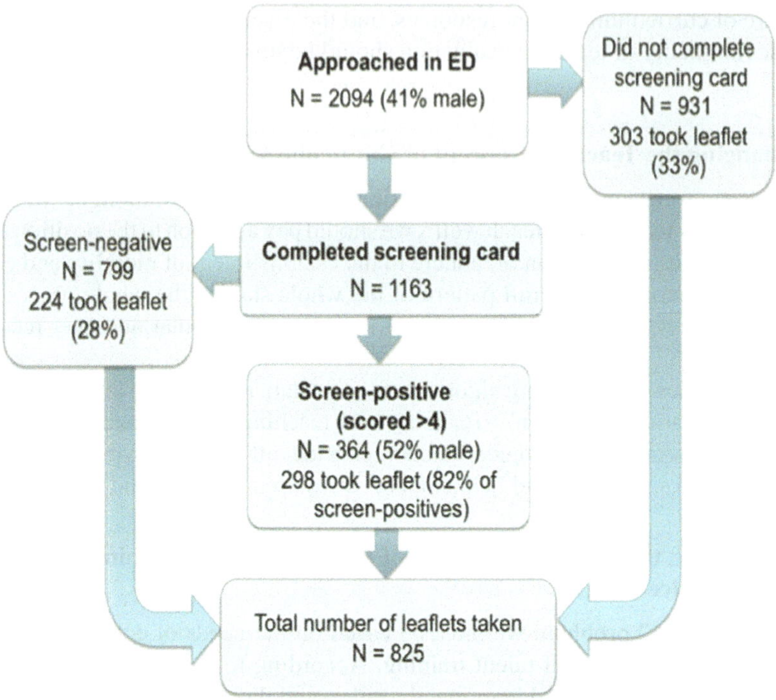

Fig. 5. Personalized learning program of financial management

design, organize the collective preparation regularly, carry out teaching research, and play a good role of "guiding".

Strengthen the training of teaching ability. In addition to obtaining the qualification certificate, we must take the teaching assistant, trial lecture and assessment to teach the course of financial management. Establish lifelong learning concept, continuously "charge" themselves, carry out regular teacher professional training, establish a "long-term continuous line" training system for teachers' teaching ability, improve relevant assessment system, and promote teachers' teaching ability to improve continuously.

5 Conclusion

In a word, the teaching of financial management course should keep pace with the times, guided by modern education and teaching concepts, guided by new ideas and new practices of in-depth learning and research, and with the help of network teaching platform, implement hybrid teaching, cultivation of students' practice and innovation ability, and strengthen the practical teaching of the course. In the limited teaching time, it not only imparts financial theoretical knowledge to students, but also improves the quality of teaching, What is more important is to teach students to master the methods of financial management, cultivate correct values, make them have "learning ability" and career development potential, and constantly improve the quality and level of financial management teaching.

References

1. Kabalski, P., Zarzycka, E.: The convergence of financial and management accounting in Poland. Financ. Sci. Nauki O Finans. **6**(4), 1–19 (2018)
2. Ugoani, J.: Accounting function as management performance tool in organizations. ERN: Other Organ. Mark. Policies Process. (Topic) **42**(4), 767–799 (2020)
3. Apriyanti, H.W., Yuvitasari, E.: The role of digital utilization in accounting to enhance MSMEs' performance during COVID-19 pandemic: case study in Semarang, Central Java, Indonesia. In: Barolli, L., Yim, K., Enokido, T. (eds.) Complex, Intelligent and Software Intensive System, vol. 278, pp. 2619–2625. Springer, Cham (2021). https://doi.org/10.1007/978-3-030-79725-6_49
4. Crovini, C., Ossola, G.: Is risk reporting a possible link between financial and management accounting in private firms? Financ. Rep. **5**(4), 1–19 (2021)
5. Gumenna-Deriy, M., Ivasechko, U.: Financial resources in housing construction: accounting and reporting aspect. World Financ. **22**(1), 5–53 (2021)
6. Holovai, N., Sysoieva, I.: Place of accounting in the management of the enterprise. J. Enterp. Manag. **32**(7), 2141–2164 (2021)
7. Nuraliati, A., Sianturi, T.S.S.: Analysis of the effect of business strategy on the quality of management accounting information systems. J. Inf. Manag. **32**(7), 2487–2506 (2021)
8. Nishimura, A.: Enterprise governance and management accounting from the viewpoint of feed-forward control. Manag. Uncertainty Account. **33**(4), 1257–1264 (2018)
9. Sariannidis, N., Garefalakis, A., Ballas, P., Grigoriou, E.: Eco-efficiency, sustainable development and environmental accounting in the tourism industry during a crisis. Corp. Board: Role Duties Compos. **42**(21), 7386–7398 (2018)
10. Yue, W.: The fusion of enterprise financial and management accounting in the new situation. J. Fuzzy Syst. **553**, 110–128 (2021)

Mobile English Learning Platform Based on Collaborative Filtering Algorithm

Boyun Qi(✉)

Henan University of Animal Husbandry and Economy, Zhengzhou 450000, Henan, China
qiboyun@126.com

Abstract. With a large number of digital resources as the carrier, mobile learning breaks through the shortage of resources and the limitation of time and space under the current learning mode. However, its rich resources also bring information overload, which greatly affects the learning efficiency. The mobile English learning platform based on collaborative filtering algorithm not only makes full use of the advantages of mobile learning, but also recommends learning resources according to the learning needs of different learners to meet the learning needs of different learners, saving learners' time and effort to a certain extent all. It has certain practical significance.

Keywords: Flow teaching algorithm · Information education · Recommendation system

1 Introduction

Internet shows an explosive growth. A large amount of useless and redundant information seriously interferes with and hinders netizens from obtaining and analyzing correct and valuable information efficiently and quickly. How to efficiently and quickly obtain useful personalized information in the vast ocean of information has become an urgent demand of the majority of Internet users. Traditional information systems, such as search engines, can search the information required by users according to the input keywords, so as to retrieve some high-quality and specific information. For example, academic journals, papers or commodities can be retrieved by using search engines. Therefore, the traditional information system can partially solve the problem of "information overload". However, for the same users, the results returned by search engines are the same, so the traditional information system is lack of personalization. In order to solve this problem, information recommendation system came into being. The core idea of information recommendation system is: firstly, collect and analyze various information characteristics of users through the information system, then use various machine learning methods to learn users' personalized interests and behavior patterns, and finally recommend personalized goods or services for the user according to the user's interests and behavior patterns obtained from learning analysis. Information recommendation system. Figure 1 shows the process of personalized recommendation algorithm.

I. Ahmad et al. (Eds.): STSIoT 2021, LNDECT 122, pp. 706–715, 2023.
https://doi.org/10.1007/978-981-19-3632-6_82

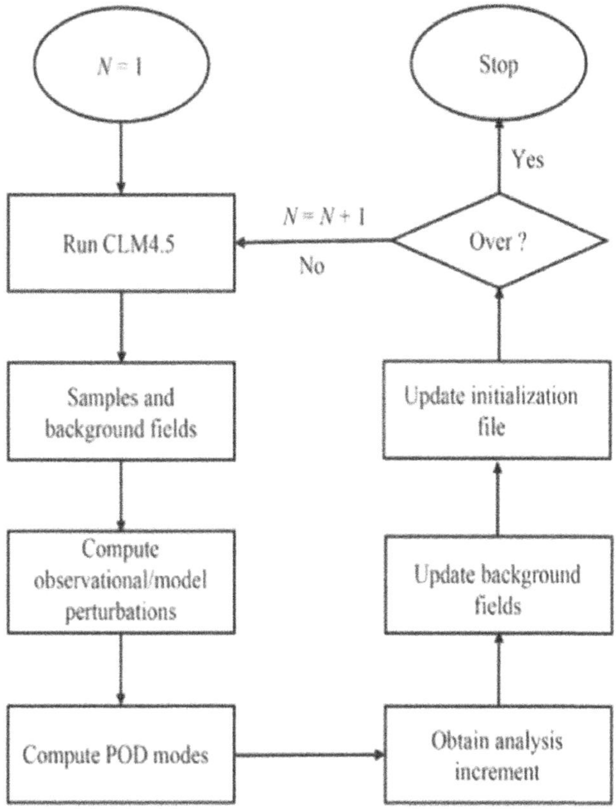

Fig. 1. Personalized recommendation algorithm process

In order to meet the needs of students at all stages of learning and continuously improve their English level, I will focus on English to meet the needs of learning intelligent content. Autonomous Learning Platform Based on collaborative filtering recommendation algorithm, build voice and video resource database and semantic database, integrate fragmented time, and give full play to students' learning initiative from listening, speaking, reading and translation and other dimensions. This paper consists of the following parts. The first part introduces the relevant background and significance of this paper, the second part is the related work of this paper, and the third part is data analysis. The fourth part is example analysis. The fifth part is conclusion.

2 Related Work

Firstly, the background of mobile learning is introduced and a Conversation-Activity-Distribution theoretical model, CAD, is proposed with activity theory, situation cognition theory and distributed cognition theory as the cognitive basis [1]. Ref [2] aim to identify whether this platform could significantly improve the proficiency of English as a foreign language (EFL), yield learner satisfaction, and reduce learners' cognitive loads in EFL classes. The subject of Ref [3] is to investigate the mobile tools (based on Web 2.0 platform) for language activities in English lessons for non-linguistics students as well for linguistics specialized group of the two universities. Therefore Ref [4] aim to explore the incorporation of collaborative learning in an English course using WhatsApp and to identify the students' perception toward that tool in an English course. Li discussed the flipped classroom English translation teaching model based on the fusion algorithm of network communication and artificial intelligence [5]. Based on this module, the platform software is designed according to the three steps of database design, platform encryption technology, and learning recommendation algorithm to create the English mobile learning platform [6]. Therefore, from the perspective of English vocabulary deep learning investigate the main mobile vocabulary learning in the functional architecture and its advantages and disadvantages in promoting English vocabulary, and designs and develops a mobile platform oriented English vocabulary deep learning system [7]. The overall framework of the English mobile learning platform is designed [8]. Ref [9] present the English language teaching applications created based on the interface. The background management system adopts the latest SSM framework of Java EE, which effectively solves the problem of server-side development [10]. The classic recommendation process for mobile English platfor design is generally divided into two steps: first, obtaining user information, in this paper, it refers to obtaining the user's (student's) score information on some.

2.1 Overview of Collaborative Filtering Algorithm

English materials, for example, the user's score on some learning materials according to their preferences is an integer from 1 to 5 [11], of which 5 points means like and 1 point means don't like; second, the user's score on some learning materials is an integer from 1 to 5, The similarity of ratings among users is analyzed and the preference of target users for a certain data is predicted. Figure 2 shows the general process of harmonious influence.

As you can see in Fig. 1, the basic recommendation strategies of collaborative filtering include prediction and top-N recommendation. In the predictive recommendation strategy, the recommender system learns the user's preference according to the scored information, and predicts the non scored items; In this strategy, we don't need to know the user's rating of each item, but only need to generate the recommendation list of the items most related to the user's preference, That is to say, it is equivalent to ranking items by learning users' preferences; In this strategy, classification accuracy and sorting accuracy are generally used to evaluate the effectiveness of the algorithm.

$$m_1 = \int_0^t e^{(pk_f+m_2+m_3-\lambda)(t-\tau)} e^{-\lambda t} \|\Delta u_k(\tau)\| d\tau \tag{1}$$

$$\lim_{k\to\infty} \|\Delta u_k(t)\|_\lambda \le \frac{1}{1-\breve{\rho}} m_5 d \tag{2}$$

There are many kinds of collaborative filtering algorithms. Breese et al. Divided collaborative filtering algorithms into storage based algorithms and model-based algorithms. As shown in Fig. 3, the memory algorithm and the model-based algorithm.

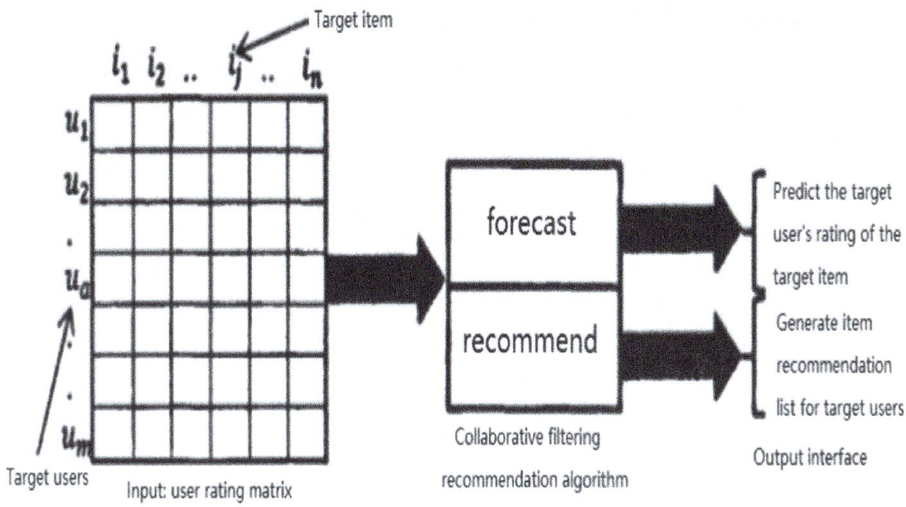

Fig. 2. General process of collaborative filtering recommendation

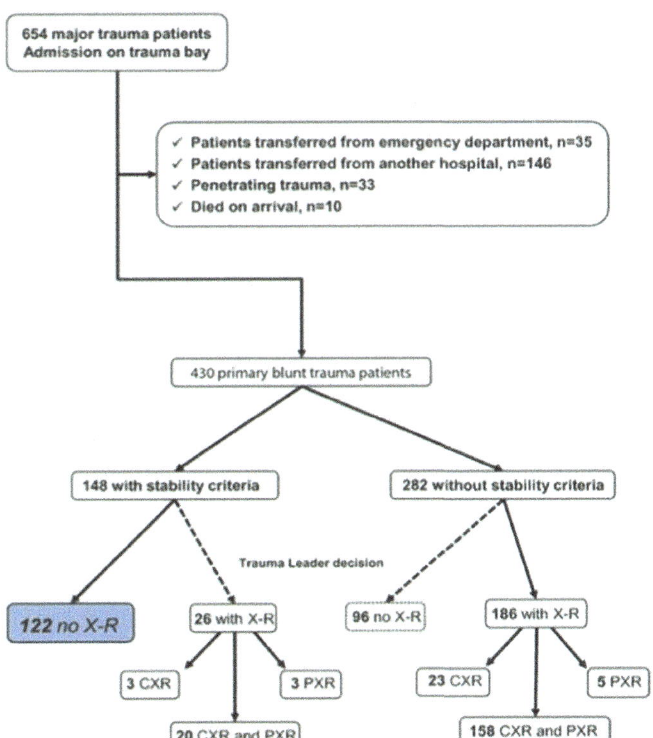

Fig. 3. Memory algorithm and model-based algorithm

2.2 Storage Based Collaborative Filtering Algorithm

The storage based algorithm uses a database of the whole user to complete the prediction. These systems use statistical techniques to find the nearest neighbor set of active users, who have the same history as the active users (they have similar evaluation of different items with the active users). Once the nearest neighbors of active users are found, these systems will use different algorithms to synthesize the evaluation of these nearest neighbors, and thus provide prediction for active users or recommend n highest evaluation items. This algorithm has the characteristics of simple calculation method and high accuracy. At present, most of the collaborative filtering algorithms used in practice belong to this type. However, any prediction must be based on all records of the database, which undoubtedly greatly increases the amount of calculation. In today's situation where the scale of users is often up to millions, this problem is particularly serious.

2.3 Model Based Collaborative Filtering Algorithm

Similarly, collaborative filtering algorithms are mainly divided into two categories: one is to process explicit preference data, such as explicit scoring data and the other is to process implicit feedback data =, such as whether to click on the web page diagram of explicit scoring (left) and implicit feedback (right) data). In the real application environment, users' implicit feedback data is more extensive and easier to obtain, such as whether users have seen a movie, whether users have heard a song and so on. This kind of data does not require users to provide clear scores, so it is easier to obtain. Because the negative cases in implicit feedback data are uncertain and only positive cases can be clearly distinguished, the collaborative filtering problem based. The core task of collaborative filtering based on implicit feedback is to use various machine learning methods to learn and analyze these implicit feedback data, and learn the user's behavior and interest patterns, so as to sort the recommended object set according to the user's preference according to the user's preference. Considering that in the information recommendation system represented by e-commerce transaction system, most of the data processed by the recommendation algorithm is implicit data; At the same time, in academia, implicit feedback has become a recommendation system.

3 Data Analysis

Principle: put forward suggestions according to similar users or projects. Now the mainstream is around the two parts of the most typical advertising. That's user based advertising. Project based advertising strategy. Interview question: both methods have a common core. Since calculation is involved, there must be various formulas. Otherwise, we can't say algorithm at all. Therefore, there are relevant technical details below, which may be understood by PM with strong technology and algorithm logic. But it doesn't matter if there is no foundation in this aspect. I will tell the principle in vernacular and as much language as I can understand. (Supplement 1: let's talk about a very realistic thing. There are still some lost technical terms in this article. It's normal for PM not to understand or understand, but if Baidu and even ask for search are unwilling, how can we learn and grow? You know, there are still a lot of PM competing with you in terms of salary increase and offers from large companies). As shown in Fig. 4, the learning platform adopts the collaborative filtering algorithm of memory.

Although collaborative filtering recommendation algorithm is widely used, it is rarely used in information-based teaching system. Although many students do a lot of exercises every day, they have no targeted exercises. Therefore, in the information-based teaching system, collaborative filtering recommendation algorithm can be used to provide personalized exercise recommendations for students, which can significantly improve students' grades.

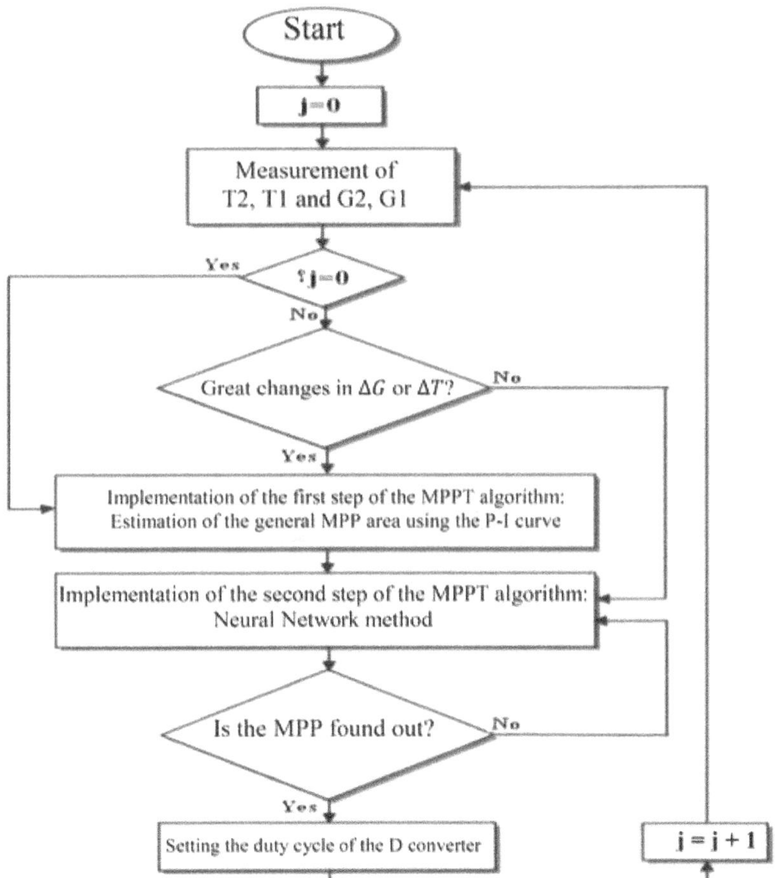

Fig. 4. The learning platform adopts memory collaborative filtering algorithm

Using the idea and principle of collaborative filtering algorithm, this paper applies it to the information teaching system, mainly focusing on exercise recommendation. The teaching system records students' problem making records, the algorithm predicts the exercises that students have not done, and recommends the exercises with the prediction result of "doing wrong" to this student. Figure 5 shows the algorithm prediction process.

4 Example Analysis

Collaborative filtering, also known as collaborative filtering in English, is called CF for short. Note that this does not refer to game CF. For example, you want to go out to a movie with your girlfriend, but you don't know what to watch. At this time, you will think, why don't I ask my friends who have similar interests to me what's worth watching?

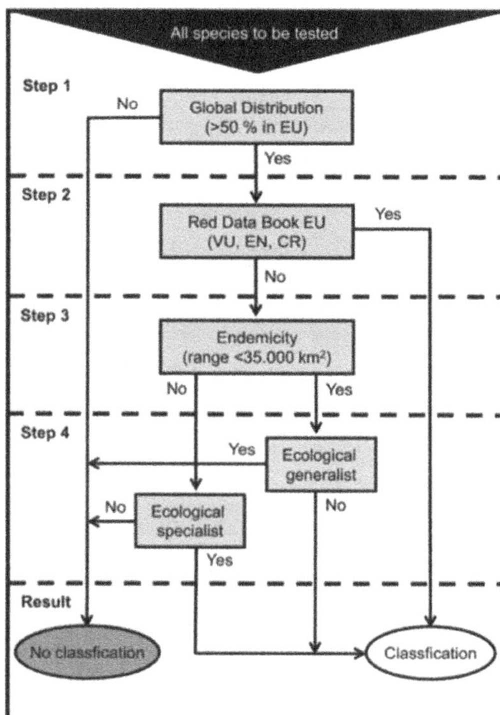

Fig. 5. Algorithm prediction process

Algorithm advantages:

Collaborative recommendation is the most widely used recommendation algorithm, which can filter out the construction of concept tags that are difficult to quantify;

Only user behavior is used for recommendation, which greatly improves the speed and accuracy;

Users' potential interests and preferences can be well found.

Algorithm disadvantages

For new users or new items, the recommended quality will be poor, which is often called the cold start problem.

Calculation of similarity

In the calculation of similarity, SIM is used to replace English similarity, and sim (a, b) is used to represent the similarity between a and B. The following are three classical algorithms about similarity. 1. Cosine similarity: cosine theorem similarity measure cosine distance.

Traditional collaborative filtering collaborative filtering algorithm first obtains the user set most similar to the target user by calculating the score similarity between users, then uses these nearest neighbors to predict the non scored items of the target user, and then recommends a group of items of most interest to the target user.

However, due to the increasing number of users and commodities, the problems of data sparsity, cold start and scalability in traditional collaborative filtering algorithms are becoming more and more serious. Therefore, to solve these problems, many researchers have proposed many new methods to improve the recommendation efficiency. Based on the sparsity of scoring data, a dynamic similarity calculation method is proposed. The algorithm is studied and optimized from the aspects of project similarity measurement method and project nearest neighbor selection, which effectively alleviates the problem of data sparsity. The non fixed k-nearest neighbor algorithm and conditional probability are combined, and the idea of step-by-step filling matrix is used to effectively improve the recommendation quality. The value of similarity is dynamically adjusted according to the common scoring times, and the weight affecting the recommendation result is adaptively adjusted according to the scoring support, so as to obtain a better recommendation effect.

In the mobile English learning platform designed in this paper, personalized recommendation module is the core module. The difficulty of this module lies in which way and which rules are used to recommend interested materials to learners. In order to be able to more accurately recommend English video materials to learners in line with personal preferences.

Step 1: let $L = \{l_1, l_2, \ldots, l_i, \ldots, l_N\}$ be the learner set; $M = \{m_1, m_2, \ldots, m_\mathrm{i}, \ldots, m_\mathrm{n},\}$ is the video data set; The interest function $g_{l,m}$ can be used to predict the user l_i rating of video m_i.

Step 2: which is recorded as *sim* (x, y)

$$sim(x, y) = \frac{\sum_{m \in m_{xy}} (g_{x,m} - \bar{g}_x)(g_{y,m} - \bar{g}_y)}{\sqrt{\sum_{m \in m_{xy}} (g_{x,m} - \bar{g}_x)^2 \sum_{m \in m} (g_{y,m} - \bar{g}_y)^2}} \tag{3}$$

Step 3: select the learner learner set, record it as L, and predict that is $g_{x,m}$

$$g_{x,m} = \bar{g} + \frac{\sum_{a=1}^{k} (g_{a,i} - \bar{g}_a) sim(x, a)}{\sum_{a=1}^{k} sim(x, a)} \tag{4}$$

5 Conclusion

This filtering algorithm on the English learning platform. From traditional media to modern intelligent media, production and distribution of content more personalized and intelligent. The user platform based on artificial intelligence technology has an impact on media channels, and so does the intelligent English learning platform. The dissemination of English learning content has also changed, resulting in corresponding changes in English learning resources, English learning channels and English learning methods. How long does the English network teaching platform course take and how effective is it? For young children, 35–40 min will take too long to give feedback. Children's attention is not focused. Different children are different. Some children are chatting in full swing. Suddenly, the time is up, and the meaning is still incomplete. However, for young children, they can't sit still and move around for too long, so they should know their children's character and have a choice.

Who is the right home for so many online foreign teachers? Some parents only choose the well-known and popular English online teaching platform, and think that the hot teachers have good comprehensive strength. Some parents don't choose the hot ones, and think that a large scale means a large number of enrollment, and teachers will recruit a large number, so it is difficult to guarantee the teachers and the service can't keep up. I suggest that in the early stage of selection, choose more English online teaching platforms to finish the audition, and then list the advantages and disadvantages for comprehensive judgment.

Do you want to investigate Chinese teachers and customer service? The experience of people from the past tells you that Chinese education and customer service are very important. Chinese education has good business ability. It will introduce you very clearly. It will help you select appropriate courses according to the child's level and situation. After signing up for classes, if you want to refund, adjust the level, etc., all these need the help of Chinese education teachers. Another point is not to be very sticky customer service, Otherwise, calling you every day will make you unbearable.

References

1. Zhang, X., Bi, J.: Design of a college English mobile learning system based on CAD model. IJET **763**(1), 50 (2018)
2. Tong, L.: Research on English cloud classroom blended teaching mode based on mobile app. In: 2021 International Conference on Intelligent Transportation, Big Data & Smart City (ICITBS), pp. 5880–5888 (2021)
3. Nguyen, T.L.Y., Ton, Q.C.: Integration mobile technology into English lessons: a case study at VNU-ULIS and VNU-UED. **139**(3), 1242–1253 (2018)
4. Syah, M.N.S.: Incorporating collaborative learning in an English course using Whatsapp. **17**(1), 168–192 (2020)
5. Li, L.: English translation teaching model of flipped classroom based on the fusion algorithm of network communication and artificial intelligence. Wirel. Communi. Mob. Comput. **11**(PB), 81–90 (2021)
6. Liu, X., Liu, H.: Design of English mobile learning platform based on GSM-R wireless network communication system. Int. J. Antennas Propagat. **143**, 33–40 (2021)
7. Shen, N.: A deep learning approach of English vocabulary for mobile platform. Mechatron. Autom. **143**(1), 9–22 (2021)
8. Ling, Z.: An English mobile learning platform embedded in GSM-R wireless network communication. Secur. Commun. Netw. **143**(5), 123 (2021)

9. Cho, Y., Kim, J.: Production of mobile english language teaching application based on text interface using deep learning. Electronics **182**(2), 543–558 (2021)
10. Yu, Z., Zhu, Y., Yang, Z., Chen, W.: Student satisfaction, learning outcomes, and cognitive loads with a mobile learning platform. Comput. Assist. Lang. Learn. **559**, A74 (2018)
11. van Vuuren, J.J., Tang, L., Al-Bahadly, I., Arif, K.M.: A benchmarking platform for learning-based grasp synthesis methodologies. J. Intell. Rob. Syst. **102**(3), 1–16 (2021). https://doi.org/10.1007/s10846-021-01410-5

Two Layer Model and Algorithm of Traffic Network Design Based on Multi-sensor Fusion Technology

Jia Tan(✉)

College of Transportation, Chongqing Vocational College of Transportation, Chongqing, China
tanjia12180117@163.com

Abstract. This paper studies the optimization model and algorithm of the hybrid transportation network design problem considering sustainable development. The bi level programming model is used to describe the problem, and the link level decision variables are used to discretize the problem, and the algorithm is solved by the model. The example shows that the traffic congestion of the optimized transportation network is significantly alleviated, Moreover, the reduction of vehicle emissions in the road network is also very obvious. All these prove that the bi level programming model and algorithm proposed in this paper is an effective method to study the traffic network design problem in the environment of sustainable development.

Keywords: Traffic engineering · Sustainable development · Traffic network design · Bilevel programming model

1 Introduction

Transportation network planning and design is an important part of land planning and comprehensive transportation network planning, and transportation network design has always been a hot issue. Sustainable development refers to a development mode that not only meets the needs of contemporary people, but also does not harm the ability of future generations to meet their needs. However, the traditional development mode of transportation system is obviously unsustainable, This is mainly reflected in its resource consumption, environmental and ecological protection, which is not in harmony with the sustainability of social and economic development. Therefore, it is particularly important to study the sustainable development of transportation system, especially in the system planning stage, how to optimize the transportation network system under the limitation of limited resources is one of the key problems to be solved urgently.

Since morlok first proposed the quantitative traffic network design problem in 1973, a lot of relevant theories have been studied. According to whether the optimized variables are continuous or not, the problem can be divided into three types: continuous type, discrete type and mixed type. In the continuous problem, it is relatively easy to design algorithm because the decision variables of link capacity are continuous variables. Gao

I. Ahmad et al. (Eds.): STSIoT 2021, LNDECT 122, pp. 716–725, 2023.
https://doi.org/10.1007/978-981-19-3632-6_83

Ziyou, Chiou Jeff, Zhang Guoqiang, Yang Jin and Xu all studied the model and algorithm of continuous problem Liu Canqi, Gao Z.Y. and others discussed the algorithm for solving discrete problems. In mixed problems, there are both discrete variables and continuous variables, so it is difficult to solve. Nie Wei and Sun Yang analyzed the algorithm for solving mixed problems. The algorithm for solving the hybrid problem is shown in Fig. 1 below.

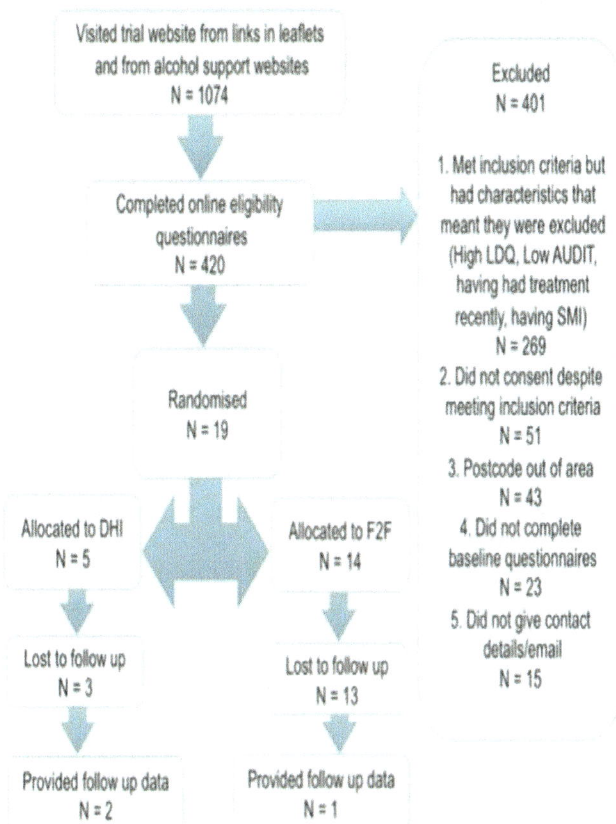

Fig. 1. Algorithm for solving mixed problems

In this paper, the problem of mixed traffic network optimization design considering sustainable development is studied, and the corresponding bi level programming model of network design is established. In the model, the factors of sustainable development such as vehicle exhaust emission, land occupation scale and road section load are considered. After the mixed problem is discretized by using road section grade decision variables, the algorithm of the model is designed. This paper consists of the following parts. The first part introduces the relevant background and significance of this paper, the second part is the related work of this paper, and the third part is data analysis. The fourth part is example analysis. The fifth part is conclusion.

2 Related Work

Ramakrishnan et al. propose several Recurrent Neural Network (RNN) architectures (the standard RNN, Long Short Term Memory (LSTM) networks, and Gated Recurrent Units (GRU)) to solve the network traffic prediction problem [1]. To address this challenge Cui et al. propose a novel deep learning framework, Traffic Graph Convolutional Long Short-Term Memory Neural Network (TGC-LSTM), to learn the interactions between roadways in the traffic network and forecast the network-wide traffic state [2]. Ding et al. discuss its applications in urban traffic network studies in several directions [3]. Based on the graph Markov process, Cui et al. propose a new neural network architecture for spatial-temporal data forecasting, i.e. the graph Markov network (GMN). Simulation results obtained using a traffic simulation model of the network Chania [4], Greece, an urban traffic network containing many varieties of jun [5]. Lim et al. generate packet-based datasets through the own network traffic pre-processing [6]. To capture the complex spatial-temporal dependencies in network-wide traffic data, Cui et al. propose a graph wavelet gated recurrent (GWGR) neural network [7]. Huo et al. propose an AI-based Lightweight Adaptive Measurement Method (ALAMM) for SDN to reduce the traffic measurement overheads and improve the measurement accuracy [8].

Based on the concepts and achievements of urban planning and traffic planning, using the basic theories and principles of traffic engineering, aiming at traffic safety, smoothness, efficiency, convenience and harmony with the environment, the "resources" of the traffic system (including time, space resources and investment level) as constraints, optimize the design of existing and future traffic systems and facilities, seek the best scheme to improve traffic, and scientifically determine the time and space elements and traffic conditions of the traffic system.

In fact, the problem of traffic network design is to solve the contradictions among people, vehicles, roads and environment. Therefore, these four aspects can be considered in the study of traffic network design.

2.1 Optimization Objective

1) The total travel time of the system is the minimum.

   For the transportation network system, the first thing is to ensure the minimum travel time of the network.

$$min\, Z_1 = \sum_{a\in A} x_a \cdot t(x_a, y_a) \tag{1}$$

2) Minimum construction cost.

   When decision-makers make decisions, they all hope to solve the problems most effectively with the least cost

$$min\, Z_2 = \sum_{a\in A} g_a \tag{2}$$

2.2 Constraints

1) Service level constraints.

In the future, the service level of traffic network is in a certain range, and if the service level is too small, it will waste resources; Too much service level will cause traffic congestion.

$$Ra_{aamaxmin} \tag{3}$$

2) System atmospheric environmental capacity constraints.

Vehicle exhaust pollution is one of the main sources of urban air pollution. From the perspective of sustainable development, it is necessary to restrict the air pollution caused by urban road system. The proportion of carbon monoxide in vehicle exhaust reaches 80%, so this study takes the CO emission capacity limit of road system as a constraint.

The limit model of regional road traffic air pollution is as follows:

$$Q\bar{7}\sqrt{S}_{max} \tag{4}$$

3) The constraints of road section acoustic environment capacity.

Environmental noise seriously affects people's study, work and life. Road traffic noise is the main source of urban environmental noise, so it needs to be restricted.

The prediction model of mixed traffic noise at 10 m away from the lane is as follows:

$$L_a = -40.7 + 10\,lg\,x_a + 33\,lg\left(v_a + \frac{500}{v_a} + 40\right) + 10\left(1 + \frac{5P_a}{v_a}\right) + 0.2G \tag{5}$$

According to the national outdoor standard of environmental noise, combined with the situation that the daytime traffic volume accounts for the vast majority of the daily traffic volume, this study selects the daytime standard of 55 dB (residential areas, cultural and educational areas and other areas sensitive to noise), namely:

$$0 \le L_a \le 55 \tag{6}$$

2.3 Mathematical Model

The above analysis can be used to get the two-layer model of traffic network design based on sustainable development.

1) upper model:

$$min\,Z = \sum\nolimits_{a \in A} x_a \cdot t(x_a, y_a) + a\sum\nolimits_{a \in A} g_a \tag{7}$$

2) Lower model:

$$min\sum\nolimits_{a \in A}\int_0^{xa} t(\theta, y_a)d\theta \tag{8}$$

3 Data Analysis

The model constructed in this study is a nonlinear programming problem with multiple day scales and constraints, and the genetic algorithm has a good effect on solving this kind of problem. In order to avoid the slow convergence and local convergence of traditional genetic algorithm, the crossover and mutation operators of genetic algorithm are improved.

3.1 Improved Genetic Algorithm

1) Coding.
   Real number coding is adopted, and the coding length is the number of roads to be expanded.
2) Fitness function.
   The fitness function adopts the upper objective function, i.e.

$$min\, Z = \sum\nolimits_{a\in A} x_a \cdot t(x_a, y_a) + a\sum\nolimits_{a\in A} g_a \quad (9)$$

3) Constraint processing.
   For the decision variables that do not meet the constraint conditions, a penalty factor is added on the basis of the fitness function,
4) Operator design.

   ① Selection operator: in order to speed up the convergence speed, the selection strategy combining the best individual reservation and competition selection is adopted.
   ② Crossover operator: use multivariate arithmetic crossover. The parent population is randomly divided into m/2 pairs, which are crossed as follows.
   Fork:

$$\begin{cases} y_i^1 = \gamma(x_i^1 - x_i^2) + ax_i^1 + \beta x_i^2 \\ y_i^2 = \gamma(x_i^1 - x_i^2) + ax_i^2 + \beta x_i^1 \end{cases} \quad (10)$$

3.2 Algorithm Steps

The algorithm flow chart is shown in Fig. 2.

Step 1: set the corresponding parameters of genetic algorithm, and generate the initial population pop randomly under the constraint condition, namely, the increment of road capacity y_a.
Step 2: bring the generated section capacity increment y_a into the lower level model, calculate the lower level traffic distribution model by F-W method, and get the section flow x_a.
Step 3: the corresponding fitness function of the section flow x_a and the generated section capacity increment y_a is taken. For a group of x_a and y_a which satisfy the constraints, the fitness function value is obtained directly; For a group of x_a and y_a which do not meet the constraints, a penalty factor is added on the basis of the fitness function.

Step 4: genetic operation. According to the above operators, the selection, crossover and mutation operations are carried out.
Step 5: after a round of genetic operation, Gen = Gen + 1. If gen > t, go to step 6; Otherwise, go to step 2.
Step 6: the algorithm ends and outputs the optimal individual and corresponding fitness value.

4 Example Analysis

Design capacity: refers to the maximum number of vehicles (standard vehicles on mixed traffic roads) that a component of traffic facilities in a design can pass through in one hour under the selected design service level on a uniform section or a cross section where a lane of the component is representative of the above conditions under the predicted road, traffic, control and environmental conditions.

Service level: the service level or service quality that road users may get from road conditions, traffic conditions, road environment, etc. Service traffic volume: the traffic volume required by different service levels. (the capacity should be related to the service level. The service level is high and the service traffic volume is small).

Service level division indicators: 1) driving speed and running time; 2) Degree of freedom (patency) when the vehicle is running; 3) The degree of traffic obstruction or interference, as well as driving delay and parking times per kilometer; 4) Safety of driving (accident rate and economic loss, etc.); 5) Driving comfort and passenger satisfaction; 6) Maximum density, the maximum density of vehicles per lane per kilometer; 7) Economy (driving cost).

Traffic conflict: refers to the traffic phenomenon that two or more traffic travelers are close to each other to a certain extent in a certain time and space. If their operation state is not changed at this time, a collision risk may occur. (diversion, confluence and intersection: traffic flows enter the intersection from two different directions and then leave the intersection in different directions. When vehicles cross each other, the place where they may collide).

Constituent elements of transportation system: people/objects, means of transportation, transportation facilities, traffic environment, traffic rules and information. Traffic elements: Mobile subject, traffic mode and traffic access. The four elements of a traffic mode: traffic power, means of transportation, traffic access and operation management.

Applicability of various transportation modes: railway has the advantages of long-distance transportation, high speed, low cost, but it is not flexible enough for short-distance transportation, poor response to short-term changes in transportation plan, huge initial investment and long construction cycle. Road: small single vehicle traffic volume and strong flexibility. Door to door transportation can be realized, but the cost is high. Water transportation: natural water transportation resources can be used only by regulation. It is the cheapest transportation mode with high carrying capacity, large transportation volume and low energy consumption, but it is greatly restricted by natural factors. Pipeline: large transportation volume, small land occupation, short construction period of pipeline transportation, low cost, safe and reliable, but poor flexibility and single cargo. Aviation: high speed, high mobility, comfort and safety, short construction period, but high transportation cost, limited by weather and low punctuality rate.

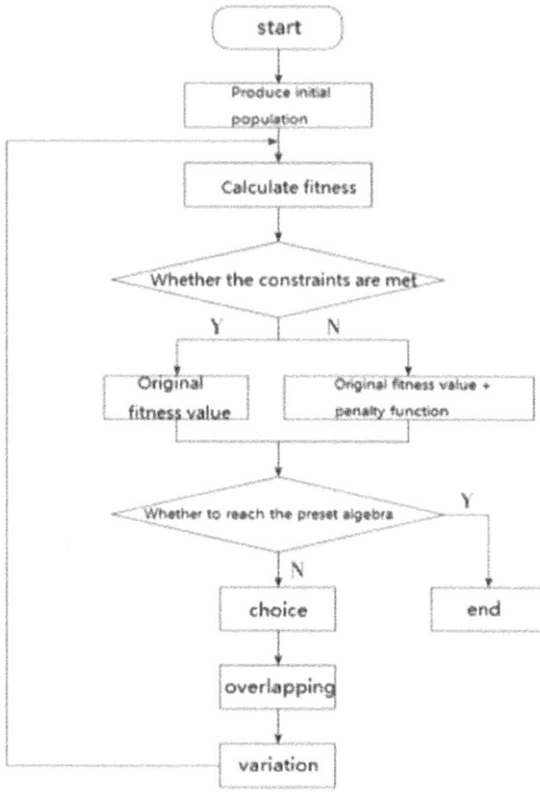

Fig. 2. Algorithm flow chart

In this study, the classic network proposed by Tianze Xu is used to test the above model algorithm. As shown in Fig. 3, there are 6 nodes in the network diagram, 18 road sections to be expanded, with OD flow of Q16 = 15 and q61 = 20.

In this study, the genetic algorithm is programmed by MATLAB, the population size is 80, the individual length is 18, the crossover probability is 0.8, the mutation probability is 0.01, and the maximum genetic algebra is 500. When a is 1, we can get the evolution curve as shown in Fig. 4.

Figure 3 shows the evolution curve of the improved genetic algorithm. Figure 4 shows the evolution curve of solving the model by using the improved genetic algorithm. Compared with Fig. 3 and Fig. 5, the improved genetic algorithm has better convergence performance.

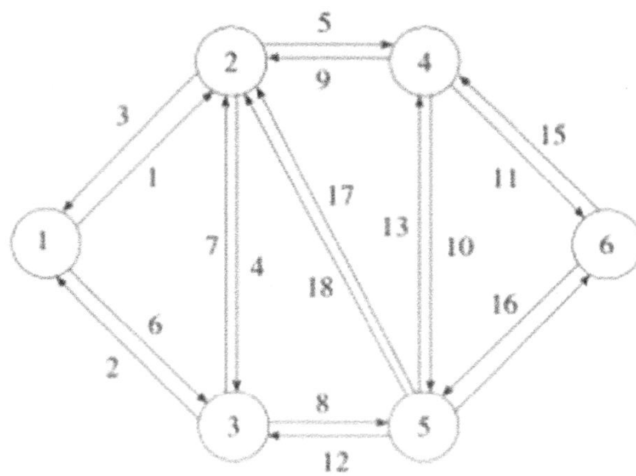

Fig. 3. Node network diagram

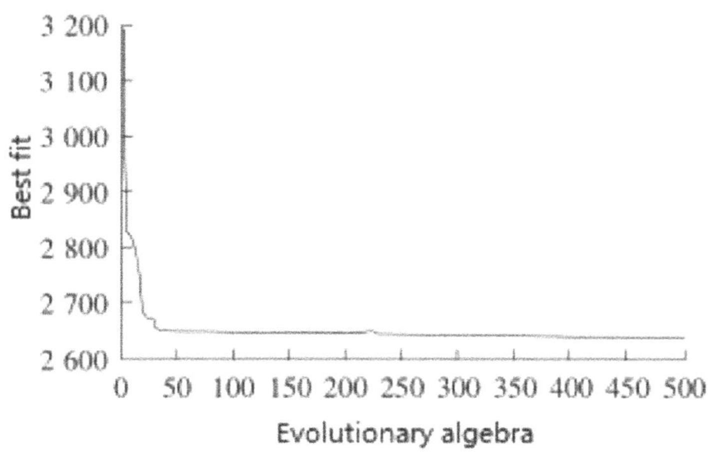

Fig. 4. Evolution curve 1

In transportation planning and management, transportation network design is an indispensable and important link. This is because the transportation network design has an important impact on the planning and management of other stages of the transportation system. Its rationality and feasibility are the preconditions to ensure the normal operation of the whole transportation system. More importantly, there is a complex relationship between road network and other urban subsystems, which promotes and restricts each other. Traditional traffic planning theories and network design methods focus on the degree star of traveler mobility, such as moving speed and travel time. However, the improvement of mobility will lead to the increase of travel demand and the occurrence of traffic owners. In this situation, the method aiming at minimizing the system travel time is not suitable for the traffic network planning process based on accessibility, nor

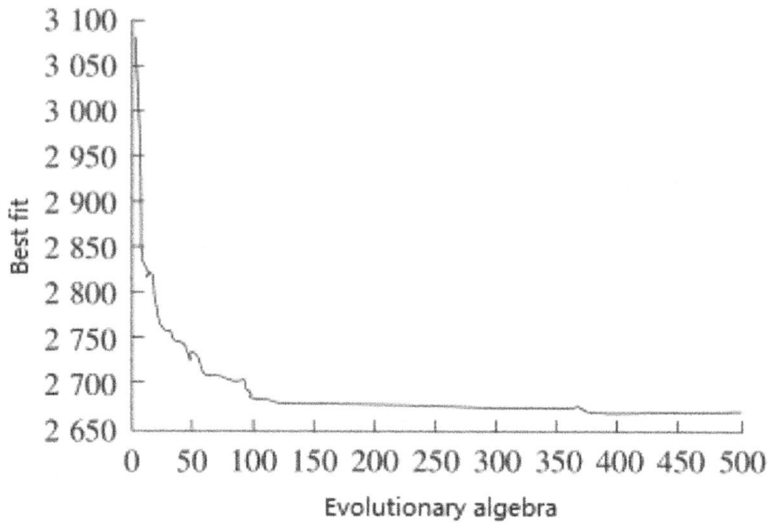

Fig. 5. Evolution curve 2

can social fairness be considered. As we all know, the main goal of transportation network design, organization and management is to improve the service level and travel accessibility of transportation network. Therefore, the method of maximizing network accessibility expands the strategic scope of traffic planning and network design to a certain extent. Based on the idea of system optimization, taking the traffic wind network design method as the main line and accessibility as the comprehensive evaluation index of traffic network, this paper aims at the design of discrete traffic network under determined demand, discrete traffic network under random demand The discrete traffic network design considering the temporal and spatial attributes of activities is studied theoretically, and the corresponding research results are applied to urban road traffic management and rail transit organization optimization.

5 Conclusion

Traffic design undertakes traffic planning, guides the construction and management of traffic facilities, and feeds back to each other. Traffic design should first understand the relevant basic conditions and constraints such as traffic planning. For the reconstructed traffic system, it should also investigate the current situation and use of traffic infrastructure, so as to determine the objectives of traffic design and further determine the basic needs and methods of traffic design.

References

1. Ramakrishnan, N., Soni, T.: Network traffic prediction using recurrent neural networks. In: 2018 17th IEEE International Conference on Machine Learning and Applications (ICMLA), pp. 1863–1873 (2018)

2. Cui, Z., Henrickson, K., Ke, R., Pu, Z., Wang, Y.: Traffic graph convolutional recurrent neural network: a deep learning framework for network-scale traffic learning and forecasting. Arxiv Application, vol. 177, pp. 114952–114963 (2018)
3. Ding, R., et al.: Application of complex networks theory in urban traffic network researches. Netw. Spat. Econ. **18**(1), 307–309 (2019)
4. Cui, Z., Lin, L., Pu, Z., Wang, Y.: Graph Markov network for traffic forecasting with missing data. Arxiv (2019)
5. Baldi, S., Michailidis, I., Ntampasi, V., Kosmatopoulos, E.B., Papamichail, I., Papageorgiou, M.: A simulation-based traffic signal control for congested urban traffic networks. Transp. Sci. **44**(1), 113–125 (2019)
6. Lim, H.-K., Kim, J.-B., Heo, J.-S., Kim, K., Hong, Y.-G., Han, Y.-H.: Packet-based network traffic classification using deep learning. J. Neurnalnet Appl. **39**(1), 87–101 (2019)
7. Cui, Z., Ke, R., Pu, Z., Ma, X., Wang, Y.: Learning traffic as a graph: a gated graph wavelet recurrent neural network for network-scale traffic prediction. Transp. Res. Part C-Emerg. Technol. **12**(1), 79–90 (2020)
8. Huo, L., Jiang, D., Qi, S., Song, H., Miao, L.: An AI-based adaptive cognitive modeling and measurement method of network traffic for EIS. Mob. Netw. Appl. **28**(2), 190–200 (2021)

Application Risk Analysis of Artificial Intelligence in Public Management Based on Cloud Computing

Jiaxin Tang[1](✉), Yiran Wang[1], Lei Ning[1], Yong Luo[1], and Deepmala Karki[2]

[1] Hainan University, Haikou 570228, Hainan, China
Tangjiaxin777@163.com
[2] LBEF Campus (In Academic Collaboration With APU Malaysia), Kathmandu, Nepal

Abstract. The current wave of artificial intelligence is mainly the performance of the prosperity of deep learning algorithm based on big data. The process of artificial intelligence acting on public management practice roughly needs to go through three processes: intelligent infrastructure construction → intelligent algorithm design → intelligent application landing, and correspondingly forms three levels: infrastructure layer → algorithm layer → application layer. This paper analyzes the corresponding data security risks of the three levels, technical accuracy, algorithm bias and algorithm supervision risks, as well as the risk of intelligent transformation of public sector, and discusses the general logic of risk transmission from the bottom to the surface, which provides a holistic perspective for analyzing the risk of artificial intelligence application in public management.

Keywords: Data security risk · Technical accuracy risk · Algorithm bias risk · Algorithm supervision risk · Risk of intelligent transformation

1 Introduction

New technologies such as artificial intelligence are influencing public management practice at an unprecedented speed and scale. However, the value of disruptive technology is often accompanied by its hidden risks" "Colingridge dilemma" shows that when people create a new technology and put it into application, they usually do not have the ability to control the risk of the technology. In the early stage of the application of artificial intelligence technology in public management practice, managers may pay more attention to the dividend of new technology, and often neglect to estimate the many risks caused by it. On the contrary, when artificial intelligence technology and public management practice are deeply integrated, the hidden risks will gradually appear and arouse the vigilance of managers, it is likely to have been rooted in the national governance system and become an important part of the whole governance structure. Figure 1 below shows an important part of the governance structure.

This paper analyzes the risks of the application of artificial intelligence in public management from two aspects: first, clarify the connotation of artificial intelligence, dig

I. Ahmad et al. (Eds.): STSIoT 2021, LNDECT 122, pp. 726–735, 2023.
https://doi.org/10.1007/978-981-19-3632-6_84

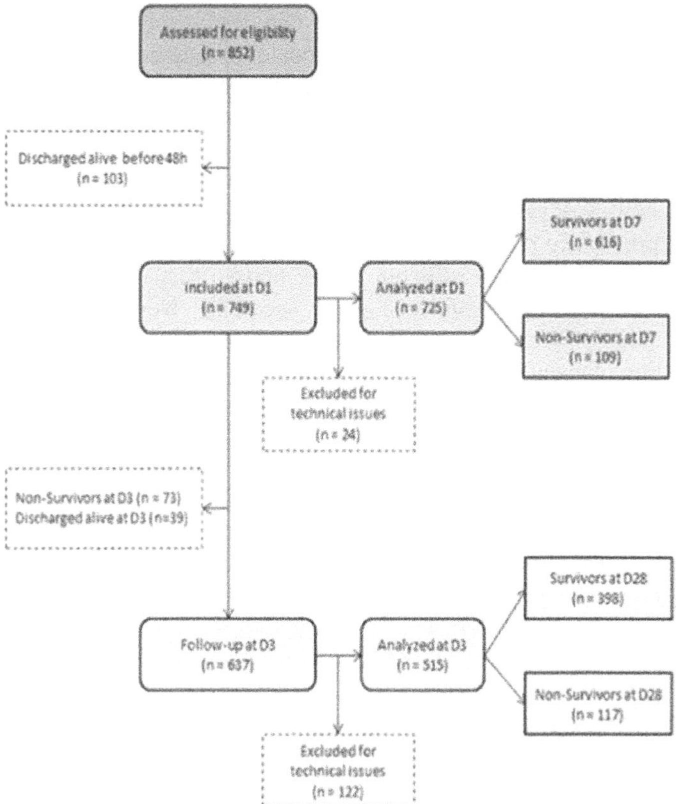

Fig. 1. An important part of the governance structure

deep into the process and mechanism of its role in public management practice, and analyze the possible risks in different stages; The second is to start with the risk events that have erupted in the field of public management and caused by artificial intelligence, summarize the general logic of risk generation, and then confirm the previous risk analysis. This paper consists of the following parts. The first part introduces the relevant background and significance of this paper, the second part is the related work of this paper, and the third part is data analysis. The fourth part is example analysis. The fifth part is conclusion.

2 Related Work

Laihonen et. al. show how knowledge management can support public management.,A case study on the application of an action research process was conducted to study how the City of Tampere in Finland aimed to overcome challenges in utilizing performance information by applying the ideas of knowledge management [1]. Laihonen et. al. suggest that a holistic knowledge management strategy promotes the use of performance information by providing a systematic management framework for gathering and utilizing the information [1]. Four factors appear critical for strategic knowledge management in local government. examine the impact of management upon employee outcomes (perceptions of discretionary power, well-being, engagement, and affective commitment), comparing public and private sector nurses in Australia, the United Kingdom, and Italy [2]. González presented an overview of the book, as well as its relevance and contribution to the discipline of public management [3]. Yotawut reviewed a new three-volume collection of previously published articles on how public or private organizations are operationalizing the principles of public value: firstly, 'Measuring the public value of e-government: A case study from Sri Lanka'; secondly, 'New Public Management to public value: Paradigmatic change and managerial implications', and lastly, 'Developing an understanding of result-based management through public value theory' [4]. The aim of Ref [5] is to present the role of individual concepts of public management in Polish and Dutch municipal (public) real estate management. Borgonovi et. al. examine the contingent decision-making arguments stimulating output instead of outcome measurement in public management [6]. Rajala et. al. examine the contingent decision-making arguments stimulating output instead of outcome measurement in public management and according to the experiment way to explain the outcome, by the cure, we can get the effect is very good [7]. Michael Barzelay tackles the challenge of making public management into a true professional discipline by Barzelay, and according to his design, we have new professional discipline, it will conduct the public management in the future [8].

In the reference [9], the author analyzed the public service logic, and got the the public service organizations can get the corresponding the value, so in this paper, the mathematic model is constructed. When we bulided the model, we need evalutate the effect and quality, so the author thinked the effect and quality is working according to his model for evalutate the quality [10]. At this time, it often costs a lot of social and economic costs to repair. In view of this, when artificial intelligence technology is initially applied to public management practice, it is of great urgency and practical value to comprehensively study and judge the possible risks.

3 Data Analysis

Generally speaking, artificial intelligence is a technology used to simulate, extend and expand human intelligence. However, it is difficult to have a clear definition of "intelligence", which leads to the generalization of the concept of artificial intelligence. The same term can be understood from different dimensions such as future vision and current reality. The current mainstream view is that in the foreseeable future, there will be no "artificial general intelligence" with emotional perception, spontaneous imagination and autonomous purposes. Moreover, while human reason promotes the development of artificial intelligence, it also constantly optimizes its adaptability and expansibility. Therefore, from the perspective of technological progress and human quiescence, it proposes that "strong artificial intelligence surpasses human beings in an all-round way", and analyzes the human-computer ethical risks and human survival risks, which may face "conditions are not tenable, Conclusion can be arbitrary "logic challenge.

$$\|\Delta u_{k+1}(t)\| \leq \rho\|\Delta u_k(t)\| + m_4\|\Delta x_{k+1}(t)\| + m_5 d \tag{1}$$

$$\lim_{k\to\infty} \|\Delta u_k(t)\|_\lambda \leq \frac{1}{1-\breve{\rho}} m_5 d \tag{2}$$

$$\lim_{k\to\infty} \|\Delta e_{k+1}(t)\|_\lambda \leq \left(\frac{m_1 m_5}{b-\lambda}\frac{1}{1-\breve{\rho}} + \frac{p}{b}\right) cd \tag{3}$$

Therefore, the author intends to base on the actual connotation of the current artificial intelligence, combined with specific examples to analyze its potential risks in the application of public management. At present, the wave of artificial intelligence is mainly based on the performance of the prosperity of deep learning algorithm based on big data. Deep learning algorithm can show close to human characteristics in some target behaviors through pattern recognition and rule mining of massive data. In fact, this behaviorist intelligence belongs to weak artificial intelligence, and its ability in some domain specific areas surpasses that of human beings, but it has not yet formed independent thinking and autonomous emotion, so it still needs to be subordinated to human beings as an aid.

According to the difference of technology direction, the specific scenarios of artificial intelligence application in the field of public management can be divided into three categories: one is to use computer programs based on computer vision and natural language processing technology to replace public managers to complete personnel screening, business consulting and other processes, The marginal cost of each task will tend to zero with the passage of time, which will greatly reduce the administrative cost and improve the administrative efficiency, such as chat robot in government call center and government service hotline; The second is big data mining technology, which can help improve the quality of public decision-making and realize the customized supply of public services by discovering imperceptible rules and patterns from the complex and multidimensional public big data. The third is intelligent planning and decision-making system, which can help public managers choose the optimal decision-making scheme and avoid risks by simulating the effective scene of public decision-making and predicting the possible results, As shown in Fig. 2.

From the process and mechanism of artificial intelligence acting on public management practice, the above three types of application realization need to go through the following processes: intelligent infrastructure construction → intelligent algorithm design → intelligent application landing, and three levels of function are formed correspondingly: infrastructure layer → algorithm layer → application layer public manager directly contacted, usually the most obvious application layer, Including intelligent hardware and software. However, from the path of risk generation, infrastructure layer and algorithm layer are at the bottom, which is more likely to cause risk conduction effect: if there is a problem at the bottom, the surface application supported by this will almost inevitably appear abnormal or even collapse. Therefore, in the application of public management, the risk generated by artificial intelligence is conducted from the underlying infrastructure and algorithms to the surface application step by step. Risk analysis should follow this transmission path, combined with the outbreak of risk events, and analyze it layer by layer. The propagation path of risk analysis is shown in Fig. 3 below.

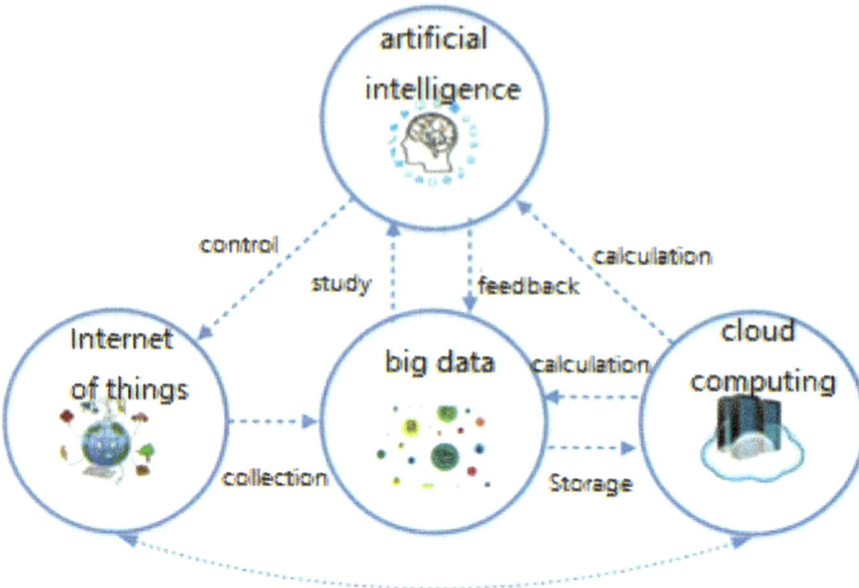

Fig. 2. Intelligent planning system

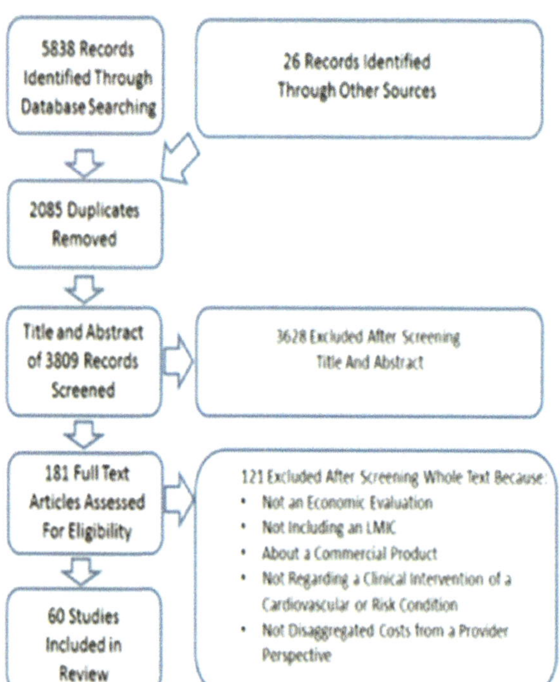

Fig. 3. Risk analysis propagation path

In public management applications, the infrastructure based on artificial intelligence refers to the computing platform of public big data and the storage, scheduling and operation of public big data. Public big data is the basis for the effectiveness of artificial intelligence technology. The more large-scale, high-quality and multimodal public big data, the more complex relationship between public affairs can be objectively reflected, and the more accurate intelligent algorithm results can be trained, and more accurate and reliable policy suggestions are provided. The public management application strategy is shown in Fig. 4 below.

Therefore, public management behavior and many important information attached to it are all transformed into online quantitative data in advance, and this process is also called datalocation. For example, to build a smart government affairs platform, transfer the government service that needs to be handled face-to-face to online website or mobile phone application, and the public service behavior is transformed into electronic track by data; Novel coronavirus pneumonia is a new technology for the public identification of public health. The establishment of face recognition gate machine requires the collection of facial images of public officers.

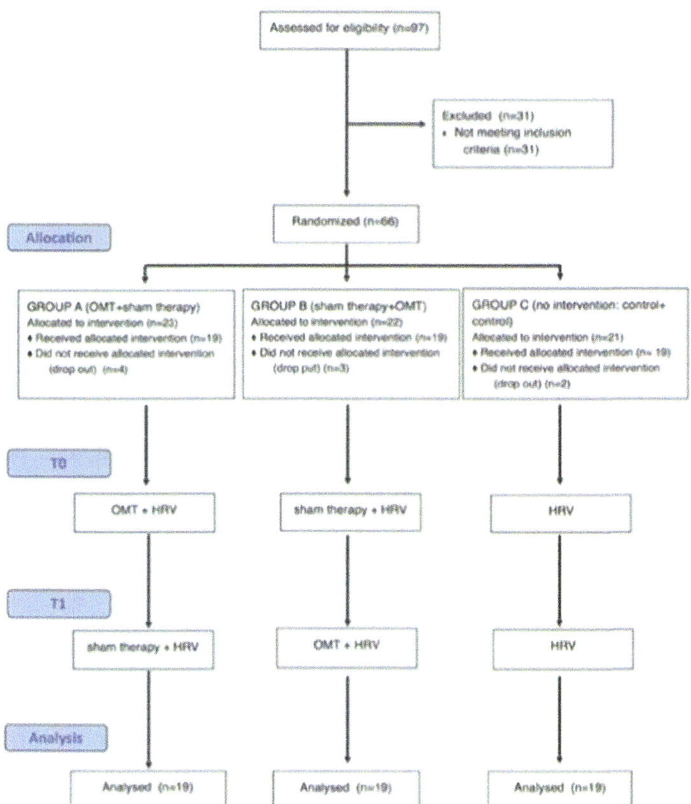

Fig. 4. Public management application strategy

The personal information of public officials is converted into binary data. Under the special situation of new crown pneumonia, the implementation of online office, online meeting and online consultation in public sector has speeded up the process of data mining from the decision point to the server side.

$$\lim_{k \to \infty} \|\Delta e_{k+1}(t)\| = 0 \tag{4}$$

$$\lim_{k \to \infty} \sup_{0 \le t \le T} \|\Delta e_{k+1}(t)\| \le \lim_{k \to \infty} e^{-\lambda T} \|\Delta e_{k+1}(t)\|_{\lambda} \le e^{-\lambda T} \left(\frac{m_1 m_5}{b - \lambda} \frac{1}{1 - \breve{\rho}} + \frac{p}{b} \right) cd \tag{5}$$

However, this new data form also tests the data security work. The platform security vulnerability may be exploited by network hackers or digital fraud groups, which may cause public big data leakage or hijacking and tampering by unknown third parties, and data security risk will break out. This part comes from the fact that AI technology itself is mastered by hackers, which makes security work more difficult; But what needs to be paid more attention to is the barrel effect of data security, that is, the vulnerable link of public big data platform (such as individual ports without protection due to lack of adequate budget) is often attacked, which leads to the security risks of leakage, stagnation and out of control of the whole government platform. Although these are not unique in the application of artificial intelligence in public management, the particularity of public domain and the sensitivity of public big data make the data security risk easily upgrade to public security risk, resulting in incalculable losses.

4 Example Analysis

Application layer refers to intelligent application and solution based on infrastructure layer and algorithm layer to realize specific requirements of public management. There are two generating paths for application layer risk:

One is generated by the risk conduction of infrastructure layer and algorithm layer. Taking the risk of privacy leakage as an example, privacy leakage can be caused by hacker attacks or digital fraud on the privacy big data in the infrastructure layer, or by the manipulation of interest groups in the algorithm layer, which intercepts privacy through "binding overlord clause", while the regulator has not established a mature process to regulate it. As shown in Fig. 5 below, the intelligent applications and solutions of infrastructure layer and algorithm layer are shown.

Second, the penetration of intelligent applications by external environment has produced adaptive risk, which is mainly manifested in the risk of intelligent transformation of public sector. The adaptability of the public sector where the public managers are located to the application of artificial intelligence is relatively behind that of the private sector. Compared with the private sector, which can quickly complete the intelligent transformation through survival of the fittest and structural adjustment, the risks of intelligent transformation faced by the public sector are much more complex: first, if the organization leaders lack full awareness of the underlying risks of artificial intelligence and are too optimistic about intelligent applications, they may promote some unnecessary artificial intelligence projects, In the absence of technical evaluation, artificial intelligence is used in areas that it is not good at, resulting in a waste of public resources. Then, the intelligent administrative mode will challenge the previous work mode, and the organization members may lack sufficient information, resources and technical ability to adapt to this change. With the spread of AI application in public sector, if the technical adaptability of organization members is not upgraded in time, it may not only hinder public management practice to benefit from technological progress, but also lead to structural redundancy; Finally, for those organization members whose positions are replaced by intelligent technology, they will face the risk of job content adjustment and loss of discretion.

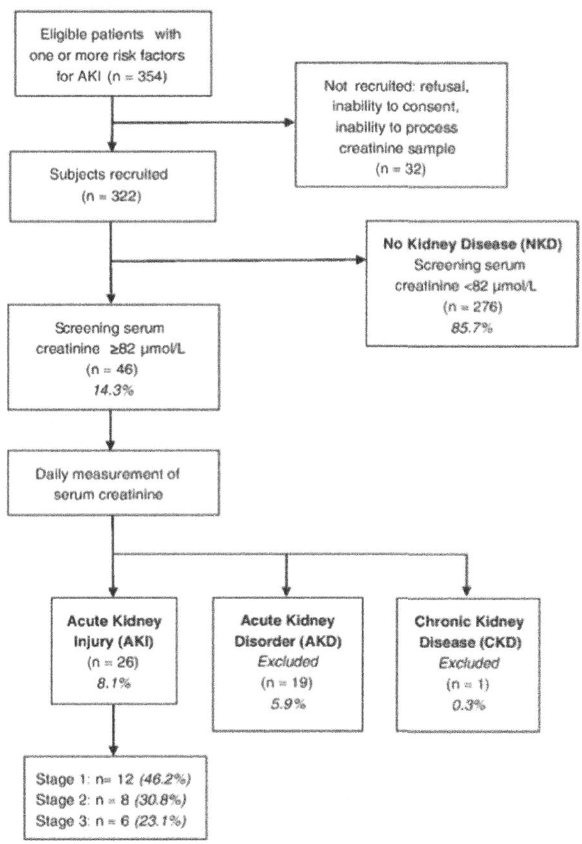

Fig. 5. Intelligent applications and solutions in infrastructure layer and algorithm layer

It is estimated that the application of artificial intelligence in the public sector will replace 30% of the working time of human labor in five to seven years. For example, part of the administrative consulting work can be replaced by government chat robots, while the original consulting staff will be transferred to other jobs. In addition, the public sector may be affected by its own level, personnel composition, acceptance of external innovation, and support from superior leaders, showing different levels of intelligent transformation risk. The risk of intelligent transformation will ultimately affect the internal relations of the organization, the identity and cohesion of the organization members. Therefore, we should conduct a comprehensive investigation on the readability of the public sector before the implementation of artificial intelligence technology.

5 Conclusion

The current wave of artificial intelligence is mainly the prosperity of deep learning algorithms based on big data. The process of artificial intelligence acting on public management practice roughly needs to go through three processes: intelligent infrastructure construction → intelligent algorithm design → intelligent application landing, and three action levels are formed accordingly: infrastructure layer → algorithm layer → application layer. This paper analyzes the corresponding data security risk, technical accuracy, algorithm bias and algorithm supervision risk of the three levels, as well as the intelligent transformation risk of the public sector, and discusses the general logic of risk transmission from the bottom to the surface, which provides an overall perspective for analyzing the risk of the application of artificial intelligence in public management.

Under the background of information technology, great changes have taken place in public resource management. In order to solve the problems of low efficiency and waste of resources in public resource management, the introduction of cloud computing and application theory is a way to improve the efficiency of public resource management and further improve the level of scientific research. This paper aims to put forward measures to strengthen the information construction of public management through cloud computing, so as to provide a valuable reference for the development of colleges and universities.

References

1. Laihonen, H., Mäntylä, S.: Strategic knowledge management and evolving local government. J. Knowl. Manag. **76**(3), 1053–1068 (2018)
2. Brunetto, Y., et al.: Comparing the impact of management on public and private sector nurses in the UK, Italy, and Australia. Public Manag. Rev. **22**(1/2), 379–392 (2018)
3. Campos González, S.A.: Public management in transition: the orchestration of potentiality. Revista de El Colegio de San Luis **12**(6), 172–179 (2018)
4. Yotawut, M.: Examining progress in research on public value. Kasetsart J. Soc. Sci. **65**(5), 655–665 (2018)
5. Marona, B., van den Beemt-Tjeerdsma, A.: Impact of public management approaches on municipal real estate management in Poland and The Netherlands. Sustainability **36**(1), 61–74 (2018)
6. Borgonovi, E., Pessina, E.A., Bianchi, C.: Outcome-based performance management in the public sector. J. Syst. Sci. **23**(10), 1107–1117 (2018)
7. Rajala, T., Laihonen, H., Vakkuri, J.: Shifting from output to outcome measurement in public administration-arguments revisited. J. Public Manag. **56**(12), 23–34 (2018)

8. Barzelay, M.: Public management as a design-oriented professional discipline. J. Inf. Manag. **34**(7), 556–568 (2019)
9. Osborne, S.P.: From public service-dominant logic to public service logic: are public service organizations capable of co-production and value co-creation? Public Manag. Rev. **12**(7), 331–345 (2018)
10. Lapuente, V., Van de Walle, S.: The effects of new public management on the quality of public services. Governance **56**(11), 2719–2729 (2020)

Design and Application of a Public Management System Based on Edge Cloud Computing

Quanzhou Tao[1](✉), Chaoyi Lv[1], Lei Ning[1], Zhongwen Chen[1], Yongan Cheng[1], Kailun Li[1], and H. Alsharif[2]

[1] Hainan University, Meilan District, 58 Renmin Avenue, Haikou 570228, Hainan, China
taoquanzhouacca@163.com
[2] South Valley University, Qena 83523, Egypt

Abstract. With the rapid development of IT industry, network sharing has been unable to meet the current resource management. Cloud computing has become an important tool for the development of various software systems. Using cloud computing to manage resources can effectively solve the current massive resource management problems. This paper analyzes the cloud computing technology, and then designs the resource management system on the basis of the technology, divides the whole system into three layers, and finally gives the core code of some modules.

Keywords: Cloud computing · Resource management · System

1 Introduction

With the rapid development of network, the business volume and data volume increase at explosive speed, which leads to the increase of the cost of data storage center. In this era of high computing speed and data explosion, for most enterprises, the performance of their computer equipment may never meet the needs, because the expansion speed of internal information of enterprises is far faster than imagined. The simple way is to purchase more and more advanced equipment to ensure the computing speed and storage capacity [1]. However, for some enterprises with less it budget, it is not a complete strategy to constantly replace the equipment. Moreover, the more and more additional costs caused by these more and more devices are also a problem. And even if there is enough money to buy more devices, with the increase of the number of devices, the differences between various storage architectures increase, and the fusion becomes poor, so it is difficult to manage and fully use storage resources in the network. Cloud computing is a computing mode, which is mainly used to solve the problem of sharing storage resources and data between servers and personal computers, so that the storage resources in the network can be fully utilized and managed easily. Figure 1 below shows how to use and manage storage resources in the network.

Cloud computing emerges as the times require.

I. Ahmad et al. (Eds.): STSIoT 2021, LNDECT 122, pp. 736–745, 2023.
https://doi.org/10.1007/978-981-19-3632-6_85

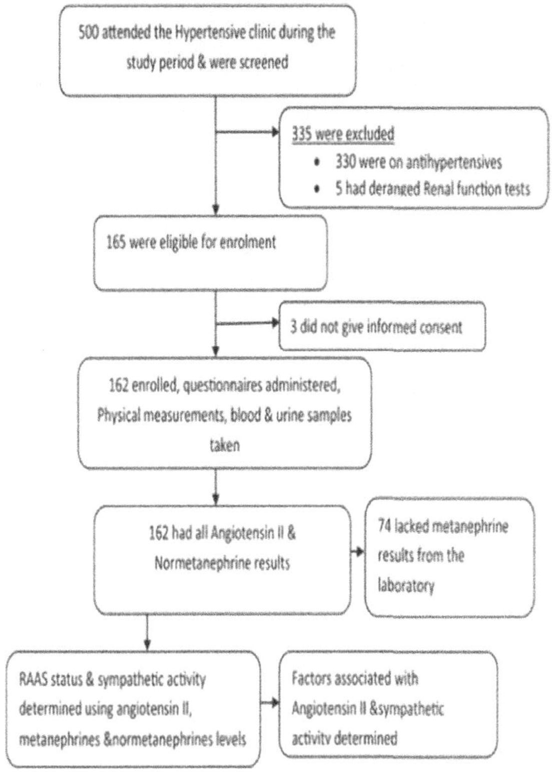

Fig. 1. Leverage and manage storage resources in the network

The development of network makes people from all over the world close to each other, and people's activities cannot be carried out without the network. With the continuous expansion of network software and resources, the network is full of massive resources. With the continuous development of computer hardware, the computing ability of computer is greatly improved, but with the increase of resources, the computing capacity defects of a single computer are enlarged. Cloud computing is based on grid computing, and it dynamically expands the virtualization resources by increasing the correlation services [2].

2 Related Work

In this paper, we aim to design and application of a public management system based on edge cloud computing.

Therefore, performance audit is not performed. To this end, it is foreseen that perceptions of personnel be measured and survey be applied to a provincial unit of a public organization [1]. The purpose of Ref [2] is to analyse the effectiveness of reforms based on new public management concept in the United States and identify key elements for implementation in Ukraine. The contribution of Ref [3] is to describe the New Public

Management model, the implementation of which helps public institutions to respond to challenges posed by external and internal stakeholders. Ref [4] discuss challenges and innovation that public management encounters in the big data era and analysis the problems of public management under the influence of the big data, using the development characteristics of the big data era to increase the service ability and management ability of public management, to sum up, the reasonable innovative channels and to build a public management system that regards the big data as the core and all aspects can take coordinating cooperation. Analyzing local government legislation, the discretion of central government in local governance, and the changes in the status of local government in public governance, Ref [5] present the evolution of the local government system in Turkey during the Justice and Development Party government. Ref [6] argue for the thesis, the systematic interaction between state and society in the sphere of Islamic education is extremely useful. Other influential work includes Refs [7–10].

This paper consists of the following parts. The first part introduces the relevant background and significance of this paper, the second part is the related work of this paper, and the third part is data analysis. The fourth part is example analysis. The fifth part is conclusion.

2.1 Cloud Computing Overview

As a key technology in the era of interconnection of all things, edge computing has a wide range of application scenarios. Firstly, this paper analyzes the problems faced in the promotion of edge computing platform; Then, starting from the architecture, this paper analyzes the typical edge computing platforms, lists the requirements of edge computing application scenarios, and participates in the classification model of edge computing platforms [3]. The development of Internet of things applications has brought an exponential increase in terminal devices.

As shown in Fig. 2 below, "cloud" is the center of data storage and application services.

According to the Cisco network index, the number of network device connections will reach 50 billion by 2022, Among them, the proportion of IOT terminals will reach [4]. Due to the resource limitations of IOT terminals, remote cloud computing resources need to be used for services. If all data of the terminal is transmitted to the cloud Center for unified processing and then returned to the terminal, it is bound to bring great pressure to the network link and data center. It is also very easy to lead to cloud center overload and denial of service (DOS), affecting the end user experience.

$$\begin{aligned}
&\|\Delta x_{k+1}(t)\|e^{-\lambda t} \le e^{-\lambda t}\int_0^t e^{(pk_f+m_2+m_3)(t-\tau)}(m_1\|\Delta u_k(\tau)\| + pd)d\tau \\
&\le m_1\int_0^t e^{(pk_f+m_2+m_3-\lambda)(t-\tau)}e^{-\lambda t}\|\Delta u_k(\tau)\|d\tau \\
&+ pd\int_0^t e^{(pk_f+m_2+m_3)(t-\tau)}d\tau
\end{aligned} \tag{1}$$

$$\|\Delta x_{k+1}(t) = \|e^{-\lambda t} \le e^{-\lambda t}\int_0^t e^{(pk_f+m_2+m_3)(t-\tau)} \tag{2}$$

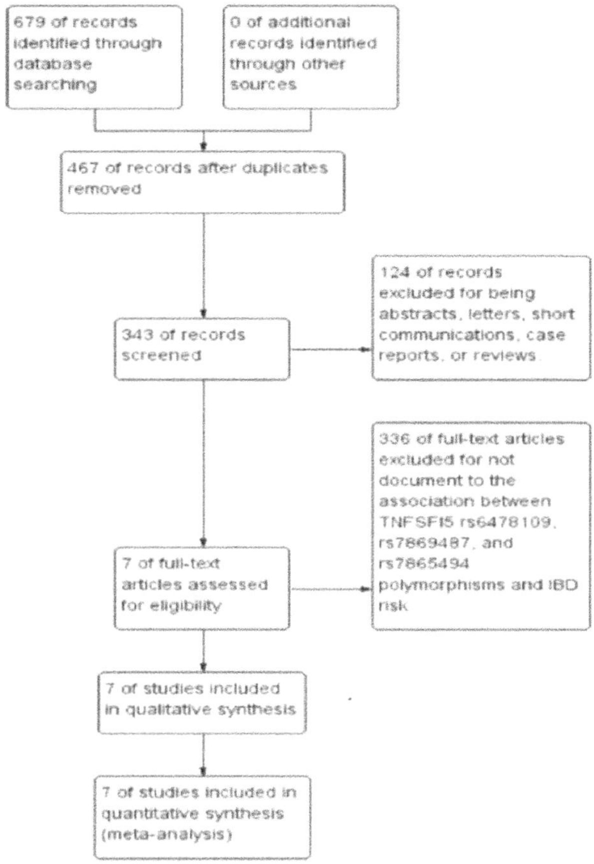

Fig. 2. "Cloud" as the center of data storage and application services

The edge calculation is studied η Some edge computing application cases are analyzed. Several current edge computing platforms are studied, and an improved architecture of mobile integration is proposed to enhance the adaptability of edge computing. However, the diversity of application scenarios leads to the lack of unified standards and a wide variety of edge computing, which limits the popularization and application of edge computing. Based on this, this paper analyzes and proposes the edge application parameters and the classification module of edge calculation and take "cloud" as the center of data storage and application services [5].

2.2 Cloud Computing Features

To become cloud computing, we must have the following five characteristics:

The network layer represents the connection mode supported by mobile edge computing. It consists of mobile cellular network, local network and extranet related hardware. The mobile edge host layer includes mobile edge hosts and mobile edge layer management devices [6]. Mobile Edge hosts can be further divided into Mme applications and virtual infrastructure. The system layer is the most important layer of MEC. It consists of the me host in the operator network and the me components required to manage me applications. It is responsible for controlling MEC task allocation and system 12 fog calculation. Fog computing is a research achievement of Cisco, "Fog computing is a highly virtualized technology that can provide computing, storage and network services between terminal devices and traditional cloud computing data centers, but it is usually located at the edge of the network. Fog computing enables communication between different communication protocol layers and different communication protocol devices. Therefore, fog Computing supports various types of infrastructure, including but not limited to However, the rich equipment support makes the scheduling, operation, maintenance and deployment of fog computing platform more difficult.

User transparency. User transparency is an indispensable feature of cloud computing. User transparency greatly facilitates the use of users [7].

3 Data Analysis

ET in 2009, satyanarayanan m, a professor at Carnegie Mellon University, proposed eight resource rich micro data centers near edge devices with high-speed connection to the Internet (d) provide computing services to directly connected LAN users. Et saves user data directly in the cloud [8]. Even if the user moves to his cloudlet LAN, he can quickly rebuild application services for the user through cloud backup data. Cloudlet provides services to users through a high one hop network to minimize latency computing tasks and related properties, etc. The index file can be updated, and the query of all drawings can be provided in multiple ways; For example, according to the serial number, classification number, region, map name and scale of topographic map, we can carry out fuzzy query of a certain type of map, accurate query of a certain map, and comprehensive retrieval. This retrieval speed is several times faster than manual query, and it is accurate and reliable.

3.1 System Architecture Design

Under the guidance of the current goal of cloud computing resource management system, on the basis of several mature system architectures, the hierarchical architecture mode is used to design the system. The system design is shown in Fig. 3.

3.2 User Interaction Interface Layer

Edgex foundry is an open source project hosted by the Linux foundation ⑦], which aims to build a general and open platform for the development of edge computing, and has won del! With the support of more than 50 enterprises such as Baidu and ntel, edgex foundry is compatible with a variety of operating systems, supports a variety of hardware architectures, and supports communication between devices with different protocols [9]. As shown in Fig. 4, edgex foundry effectively improves the efficiency of application and service development based on micro service architecture design. Its micro service is

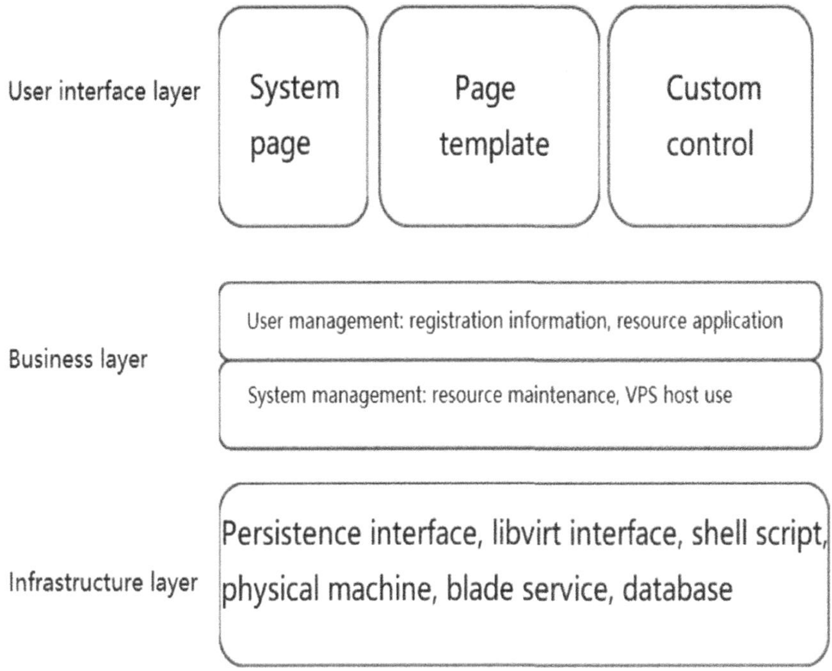

Fig. 3. The architecture model of cloud computing public management system

divided into four service layers and two basic system services. The four service layers of dgex foundry include core service layer, support service layer, export service layer and device service layer; System services include security services and management services. Based on the tailorability of microservices, the edgefoundry service can be run on low-performance devices. Therefore, edgex foundry supports heterogeneous devices such as embedded PCs, hubs, gateways, routers and local servers. Edgex foundry supports container deployment and effectively improves the operation efficiency of the platform.

3.3 Business Layer

Business layer is the core of the whole system [10]. For resource management, resources mainly refer to computing services, virtual machines and user information. After entering the system, there are two roles: general user and operator. In this layer, although there are relatively many operable functions, the business of the whole business layer can be divided into four modules, namely system management, resource application, user management and VPS host usage. As shown in Fig. 4, the core process of the whole system.

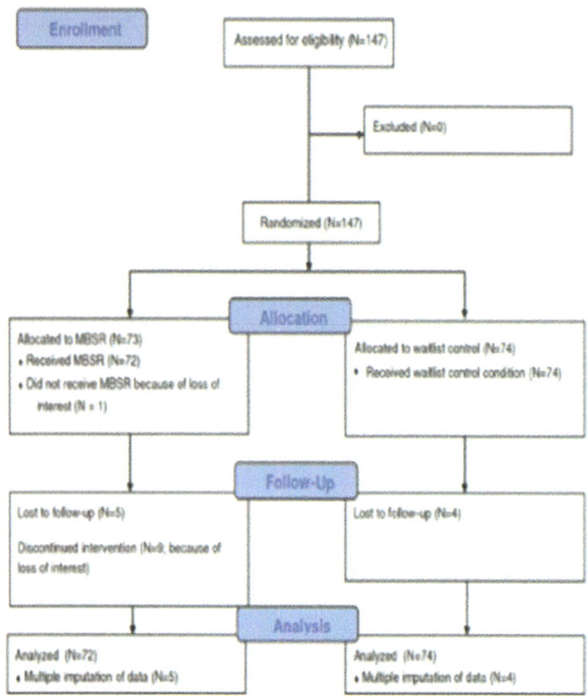

Fig. 4. Core process of the system

4 Example Analysis

Project planning. Information management system is an indispensable part of public management system [11]. It is a VF database development application program. It mainly includes user login, information input and modification, information query, score input, score query, system user management and so on. Therefore, the system has strong practicability.

System overview. After logging in, the system enters the main interface. The buttons on the left of the main interface are used for various queries, while the buttons on the right of the main interface are used for relevant management directly. It has the functions of adding, deleting and modifying information. Fully realize the main information management and query functions of the system [12].

Functional analysis. The following describes the functions of each module according to the demand analysis. According to the demand analysis, the system should have three modules and sixteen functions, namely management system module and information query module. Respectively deal with the addition, deletion and modification of student information and the query function of student information.

Login module (1) user login (2) user registration.

Management module (1) department management (2) class management (3) teacher management (4) information management (5) course management (6) score management.

The functions of each part of the system are described as follows:

User module: handle user login and visitor registration. Management module: the management module realizes the following 6 functions.

According to the design and analysis of the system, the realization of business layer module is the key point of the whole system.

4.1 Resource Application Module

To realize public management under the new normal, we must change the traditional management mode, change ideas and break through difficulties. The diversification of public management subjects is the characteristic and development trend of public management under the new normal. Compared with the past, the public management subject under the new normal highlights the position of the public in public management and increases the proportion of public governance [13]. The development of public management under the new normal should focus on strengthening the innovation of modern public management technical means, build a mechanism for the implementation of top-level design, and practice the socialist core values in the field of public management [14].

Resource application module is mainly used by the administrator of the system to allocate the resources applied by users after entering the system, and maximize the utilization of resources as much as possible. The specific functions of the module include virtual machine application, submission of calculation tasks, review progress results and approval of user application. Some of the core codes are as follows:

```
<?php
include_ once('./common.php');
//Whether to close the site
checkcloseO;
//Login required
checkloginO;
//Spatial information
Sspace = getspace($_ SGLOBAL'supe_ uid"]);
if(empty($_ SCONFIG[my _status'])) {
showmessage('no_ privilege. my_ status');
}
```

4.2 VPS Host Module

This module is used to determine which computing tasks can be performed after the user has applied to the VPS host [15]. The specific operation mainly includes the switch, restart and connection time reminder of VPS host; In addition, audit the computing resources. In order to ensure that a user will not occupy the resources for a long time, set the latest completion time, and view the progress of its implementation in real time. The specific core code is as follows:

```
Smy_ appld = $appid;
$my_ suffx = base64_ _decode (urldecode (S_ _GET['my_ suf:x"));
$my_ prefix = getsiteurlO;
if (!$my_ sufix) {
header (Location: userapp.php?id='. Smy_ appld.
'&my_ sufix=' .urlencode(base64_ encode(/));
exit;
}
if (preg. _match(/^V/", $my_ suffx)) {
Surl = http://apps.manyou.com/. $my_ appld.
Smy_ _suffix;
} else {
if($my_ suffix) {
$url = http://apps.manyou.com/*. $my_ appld./.
Smy_ _suffix;
}else{
$url = http://apps manyou.com/.$my_ appId;
}
```

5 Conclusion

This paper analyzes the resource management system of cloud computing. Firstly, it describes the basic concepts and characteristics of cloud computing, and analyzes its three service levels. Secondly, it designs the architecture of the system by analyzing the objectives of the system. Finally, it gives part of the source code of the core module of the system.

Computing delay and service capability: computing delay and service capability directly depend on the computing capability and resource scheduling strategy of edge computing nodes. In this regard, mobile edge computing and cloudlet use resource rich dedicated hosts as service nodes, which has significant advantages; The fog computing node is mainly composed of traditional equipment, so the computing performance is poor. Edgex foundry is compatible with different performance devices through the lightweight features of container and go language, and provides computing services on demand. Deployment and operation and maintenance: in the management and supervision of the edge computing platform, each platform follows the hierarchical management method. Usually, the supervision node acts as an agent, which is responsible for communicating with the underlying node and collecting the resources and status that can be provided by the underlying node. However, different edge computing platforms have different proxy settings and layered definitions. Due to the heterogeneity and diversity of nodes in fog computing platform, multiple scheduling layers need to be designed.

References

1. Hatikler, M.A., Çalıyurt, K.T.: Sustainability in personnel performance auditing: case from Turkey. In: Çalıyurt, K., Said, R. (eds.) Sustainability and Social Responsibility of Accountability Reporting Systems. Accounting, Finance, Sustainability, Governance & Fraud: Theory and Application, pp. 281–315. Springer, Singapore (2018). https://doi.org/10.1007/978-981-10-3212-7_17
2. Khomytskyi, V.: New public management: US experience for Ukraine. Bull. Taras Shevchenko Natl. Univ. Kyiv. Public Adm. 11, 46–52 (2019)
3. Taran, Ye.I.: Transforming The public management system in Ukraine (2020)
4. Stroińska, E.: New public management as a tool for changes in public administration. J. Intercult. Manage. **12**, 1–28 (2020)
5. Yi, S.: The discussion on the challenges and innovative development of public management in the era of big data (2020)
6. Tan, E.: Quo vadis? The local government in turkey after public management reforms. Int. Rev. Adm. Sci. **86**, 115–133 (2020)
7. Kang, I.-K., Yang, T., Kim, D.-M., Yeo, G.-T.: A study on priority determination of policies on the extension of public management system in coastwise passenger transport services. J. Digit. Converg. **18**, 1–12 (2020)
8. Saburova, H.B.: Main directions of developing the public management system in Ukraine (2021)
9. Shopsha, N., Yarovoi, T.: Mechanism of public management of the system of development of intellectual resources of the economy of Ukraine. Public Adm. Law Rev. (2021)
10. Yakovleva, A.V.: State-public system of management in islamic educational organization. Islam Mod. World **16**, 183–194 (2021)
11. Ijaz, M., Li, G., Lin, L., et al.: Integration and applications of fog computing and cloud computing based on the internet of things for provision of healthcare services at home (2021)

Research on Implementation Technology of High Performance Distance Education Management Platform Based on K-Means Algorithm

Weirong Wang(✉)

Xi'an Siyuan University, Xi'an 710061, Shaanxi, China
623653137@qq.com

Abstract. Distance education management platform is the core of network education, which undertakes most of the functions of teaching resources, organization and implementation of educational activities. Since 2010, more and more students have been served by the distance education management platform, which eventually causes the distance education management platform to face the situation that at a certain moment, many users access the system service together, and the system service is very easy to collapse. In recent years, in this case of multi-user and high concurrency, a variety of emerging technologies emerge in endlessly. Based on K-means algorithm technology, this paper aims to achieve a high-performance distance education platform.

Keywords: K-means algorithm · High performance · Distance learning

1 Introduction

With the development of science and technology, the rate of knowledge renewal is getting faster and faster. Traditional school education can not match the current knowledge update speed completely. How to do well in the education after university has been put in front of all educators. After many years of research and discussion, distance education came into being. It can pass on the latest knowledge almost equally to anyone who wants to learn. After the first decade of the 21st century, various new online learning has gradually entered people's vision. MOOCS, open classes, etc. are also an updated way of education. After more than ten years of development, by 2015, 68 universities have carried out distance education. We will open up our school's high-quality teaching resources for the whole society. All people have to admit that distance education based on Internet technology has become a new learning mode which can complement the traditional school education.

The distance education management platform was developed in 2015, and it has the functions of teaching resources, organization and implementation of teaching activities. But with the advent of the new wave of Internet applications with high concurrency of multiple users. The distance education institutions face the situation that many users access the system services together at a certain time, which eventually leads to the system

I. Ahmad et al. (Eds.): STSIoT 2021, LNDECT 122, pp. 746–754, 2023.
https://doi.org/10.1007/978-981-19-3632-6_86

service collapse easily. On the other hand, in order to solve the problem of concurrent access by multi-user in a short time, various emerging solutions are also emerging. Figure 1 below shows the distance education management platform.

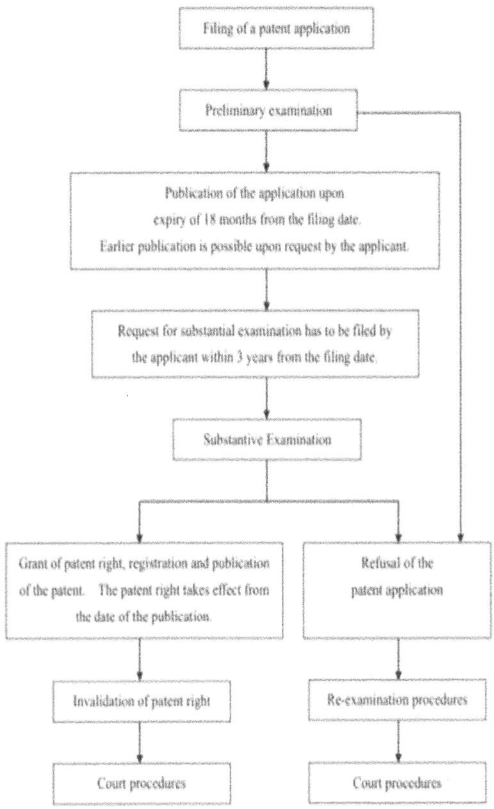

Fig. 1. Distance education management platform

Vertical stratification, horizontal partition, distributed, cache, cluster, asynchronous and other means of the system are all the best strategies to deal with high concurrency problems. Using high concurrency deployment strategy, the distance education management system can still provide services to the outside stably under multi-user. At the same time, because the system has been divided several times, the coupling degree of each module has been reduced. Once there is new business demand, it can also add new modules conveniently without stopping the existing business. From the above analysis, it is of great significance to study the high-performance distance education management platform for the development of distance education in Xi'an University of electronic technology and the smooth use of distance education resources for students in China. This paper consists of the following parts. The first part introduces the relevant background and significance of this paper, the second part is the related work of this paper, and the third part is data analysis. The fourth part is example analysis. The fifth part is conclusion.

2 Related Work

Analysis showed no significant differences in TTR between the three time periods ($p = 0.520$), the three groups ($p = 0.460$) or the groups over time ($p = 0.263$) [1]. Therefore, in the context of big data, Wang et al. proposed targeted the corresponding measures have served as a reference and reference for improving the management level of higher education [2]. The research results show that this model improves the information level of the higher education management model and promotes the wide application of big data in the higher education management model [3]. Using the correlation between MVC components of higher education management platform and the support of data mining technology, this paper constructs the MVC model 2 framework of higher education management platform which is a powerful framework used for developing large-scale projects with ease [4]. Zhou summarized the main problems existing in the process of education and teaching, and puts forward the construction method of university network ideology management platform based on big data [5]. The aim of Ref. [6] is to study the design of college English teaching platform based on artificial intelligence. Clustering is actually a classification process. One class cluster is the aggregation in the test space [7]. The distance between any two nodes in the same class cluster is smaller than that between any two points in different class clusters; Class clusters can realize multidimensional connected regions including high-density point sets, which are separated from other regions according to the regions including low-density point sets and other regions [8]. Different from traditional classification and clustering methods, the traditional classification is classified into conventional classification according to the specified attribute characteristics [9]. Other influential work includes Ref. [10].

For continuous unknown data clusters, the calculation difference based on the corresponding centroid or feature is clustering, and the premise of clustering operation is unknown.

Suppose $X = \{X_1, X_2 \ldots X_n\}$ is a collection of n objects $X, = \{X_{j,1}, X_{j,2} \ldots X_{j,m}\}$. Is an object with a m-dimensional variable. K-means algorithm gathers the object set X into K clusters in the process of clustering, which makes the objective function (i.e. the sum of error square criterion function) P minimum, and P is the sum of the distance between all points in each cluster and the cluster center.

$$P(W, Z) = \sum_{l=1}^{k} \sum_{j=1}^{n} \sum_{i=1}^{m} w_{l,j} d(z_{l,j}, x_{j,i}) \tag{1}$$

It is found that if the difference between clusters is obvious and the data distribution is dense, the sum of squares criterion function is more effective; However, if the shape and size of each cluster are very different, in order to minimize the sum of square error p value, it is possible to segment large clusters. In addition, when the sum of squared error criterion function is used to measure the clustering effect, the best clustering result corresponds to the extremum of the objective function. There are many local minima in the objective function, and every step of the algorithm is along the direction of the objective function decreasing. If the initial point falls near a local minimum, the algorithm will converge at the local minimum.

The algorithm flow chart is shown in Fig. 2.

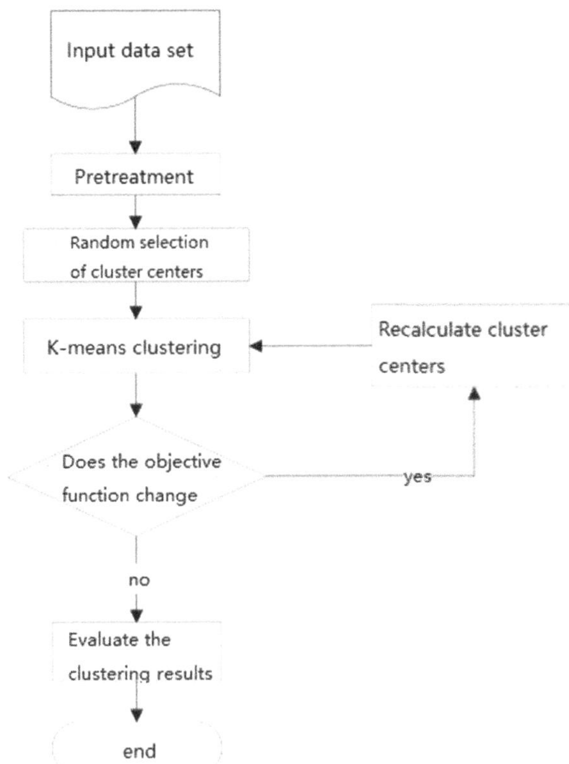

Fig. 2. Algorithm flow chart

When learning artificial intelligence, if it is found that there is no suitable learning method, then a simple and clear learning route is very important. Examples of emotion analysis, collaborative filtering, labeling and prediction are listed; For example, it cites emotion analysis, coordination, filtering, labeling and prediction.

The deep learning roadmap consists of four parts: thesis, neural network, network architecture and tools used. This book assumes that the reader has a background in computer science, is familiar with programming programs, and understands computational performance, complexity problems, graphics knowledge and entry-level calculus.

The second part is practice, which introduces the popular and easy methods of learning python programming, so as to gradually use the language to build neural networks to recognize human handwritten letters, especially to make them work like networks developed by experts. This paper introduces the basic theory of artificial intelligence, and also introduces hot topics such as machine learning, neural network and natural language processing, so as to help readers understand all aspects of the field of artificial intelligence.

Combining the basic theory and algorithm implementation, this book gradually introduces the common algorithms in the field of artificial intelligence, comprehensively and systematically introduces the use of Python to realize the use of artificial intelligence algorithms, and realizes in-depth learning through pytorch framework. Provide courseware, source code, teaching outline and teaching materials.

This paper systematically introduces the basic theory of automatic driving technology, and relies on Baidu Apollo automatic driving platform to provide services.

The main contents include: automatic driving vehicle hardware platform, automatic driving vehicle software platform, automatic driving vehicle development platform, automatic driving vehicle software calculation framework and automatic driving development platform, etc. this book can be used as a teaching material for college students in vehicle engineering and transportation engineering, and can also be used as an engineer in auto driving related industries.

Explanation - written in English with no coding experience. Oliver Theobald introduced the core algorithm and gave a clear explanation, and added an intuitive example to make it easy to learn at home.

With the help of this book, you will: explore the prospect of machine learning, especially neural networks. Use scikit learn end-to-end tracking example machine learning project to explore several training models, including support vector machines, decision trees and random forests, as well as integration methods. Tlaxflow library constructs and trains neural networks, and deeply studies the structure of neural network system, including convolution network, which is submitted to the network and deep reinforcement learning technology for training and expanding deep neural networks.

3 Data Analysis

Directory tree is the specific display of user permissions in the application. The directory tree of each user group is consistent. That is, all students share a set of directories, all exam administrators use a set of directories, and all accounting administrators use a set of directories. After users enter the system, they will get the directory tree first. The sequence diagram of getting the directory tree is shown in Fig. 3 below.

Distance education management platform is a high performance system based on K-means algorithm. It uses the popular open source development framework jeecg (J2EE code generation) in the industry. This framework is the integration of spring MVC, spring, hibernate and jdbctemplate. From the level of application itself, the whole application can be divided into three logical layers: front-end presentation layer, data logic layer and database persistence layer. In the presentation layer, the framework used is spring MVC, which has the advantage of separating the processing and display of system data. In the presentation layer, the specific implementation of the system is the repackaging of easyUI. After the user clicks a button in the foreground, the system background will receive the request and submit it to a controller. The controller belongs to the business layer, and the business layer of the whole system is controlled by spring. The major business modules of the system are divided into packages and sub packages, such as

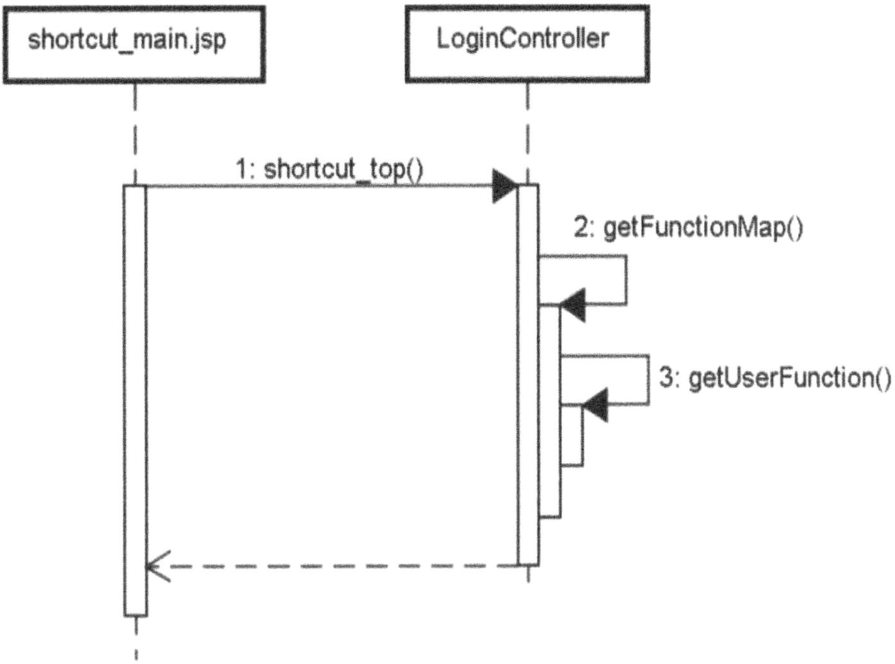

Fig. 3. Get the sequence diagram of the directory tree

student status module, enrollment module, examination module, accounting module, etc. the specific function implementation is reflected in one controller. Every time a request comes, spring will re apply for a new controller instance for the new request. The framework of persistence layer is hibernate and JDBC template. If you access regular data entities, you can use hibernate. The advantage is that it is convenient and fast. For some user-defined SQL operations, you can use the native SQL implementation of JDBC template to control every detail of the implementation.

After careful study of the code, we can find that getuserfunction mainly deals with the problem of obtaining the user's permission and mapping the permission into a directory tree. Further research shows that each user has executed all the processes shown in Fig. 2 when logging in. That is: first get the user's role, and then get the corresponding directory tree according to the user's role, and then the foreground displays these directories according to the corresponding JS. Here, especially getuserfunction, there is a lot of optimization space. As shown in Fig. 4 below, permissions are mapped to the directory tree.

The significance of optimizing the directory tree function is: this function point is the basis of any user and any operation in the system, and it is bound with the user's login. So since all users need to use this function, the optimization of this part can affect the overall effect of the system. If we save 3 s for each user in the generation of the directory tree, then if there are 25000 students, the system can save them 75000 s, about 20 h.

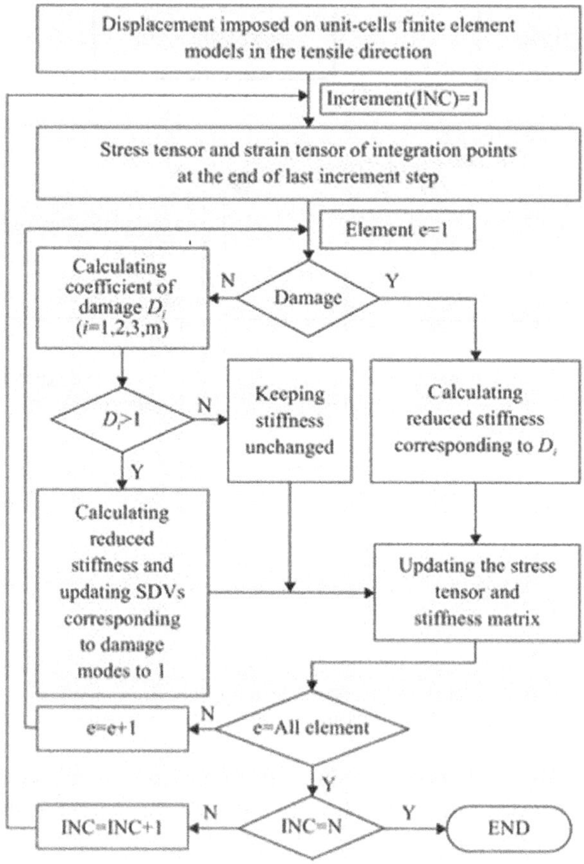

Fig. 4. Permissions are mapped to the directory tree

4 Example Analysis

Now J2EE system, in order to deal with a large number of concurrent requests, usually uses multiple application servers to form a cluster to provide services. In this case, there must be a problem of how requests are evenly allocated to each servlet container without bottleneck.

In the specific implementation of load strategy, there are both hardware implementation and software implementation. The advantage of hardware load lies in its high efficiency and stability, while the disadvantage lies in its high price; For software load, commonly used are nginx, LVS, Apache and so on.

The schematic diagram is shown in Fig. 5.

Configure multiple servers in DNS server www.mysite.com Records, such as www.mysite.com in a 114.100.80.1, www.mysite.com in a 114.100.80.2, www.mysite.com in a 114.100.80.3. The browser sends the domain name resolution request to the DNS server, and the DNS server will return www.mysite.com. The user's browser uses this IP address to access the network.

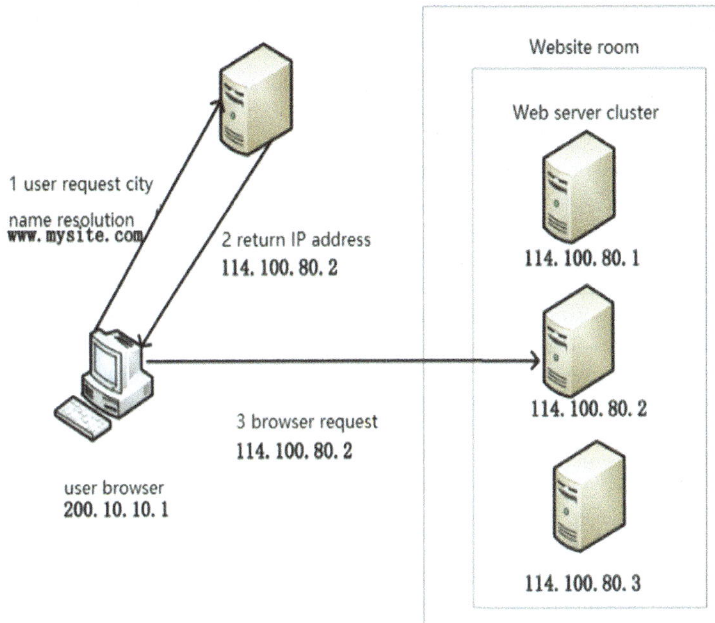

Fig. 5. Get the sequence diagram of the directory tree

5 Conclusion

Distance education management platform undertakes most of the functions of educational resources, organization and implementation of teaching activities. With the continuous development of network technology, a new wave of Internet applications with multi-user and high concurrency has come. Based on the k-means algorithm of high-performance distance education management platform implementation technology research, to achieve a high-performance distance education management platform. The classification criteria of online learning are vague and different. In the process of classification, the learning situation is not fully considered. In addition, because students' own needs and motivations have certain differences, resulting in different learning behaviors among learners, we use the results of behavior cluster analysis to analyze the similarity of learning attributes, realize the division of different categories, and carry out targeted analysis by using the behavior characteristics of different groups.

In the process of longitudinal design and analysis model, the online learning behavior data analysis model is used, and the data information is obtained through the platform through learning tasks, and preprocessing is carried out. In addition, online learning behavior information data have different types, and there are also new problems such as incomplete data, redundancy and noise in the collection process, which requires preprocessing before data analysis. In addition, the method of online learning is more important. Only by using reasonable learning methods can we ensure the correctness of the analysis results, so as to improve the teaching structure and optimize.

References

1. Talboom-Kamp, E.P., et al.: Effect of a combined education and eHealth programme on the control of oral anticoagulation patients (PORTALS study): a parallel cohort design in Dutch primary care. BMJ Open **7**(9), e017909 (2017)
2. Wang, H.: Research on the innovation of higher education management model. J. Big Data **313**(5786), 504–507 (2020)
3. Liang, J.: Research on the application of big data in the informatization of higher education management mode. In: 2020 International Conference on Intelligent Transportation, Big Data & Smart City (ICITBS), pp. 2201–2205 (2020)
4. Hu, J., Li, H.: Composition and optimization of higher education management system based on data mining technology. Sci. Program. **22**(2), 341–362 (2021)
5. Zhou, G.: Construction of university network ideology management platform based on big data education platform. In: 2021 4th International Conference on Information Systems and Computer Aided Education, pp. 4521–4525 (2021)
6. Huang, X.: The design of college english teaching platform based on artificial intelligence. J. Online Educ. **27**(5), 2408–2419 (2021)
7. Fengchao, C., Xuewen, P.: Research on the innovation mechanism model of education management under the background of artificial intelligence technology. In: 2020 5th International Conference on Smart Grid and Electrical Automation (ICSGEA), pp. 6621–6625 (2020)
8. Wei, Y.: Reflection on college education and teaching management under new media. Rev. Educ. Theory **3**(4), 9–13 (2020)
9. Li, Q.: The use of artificial intelligence combined with cloud computing in the design of education information management platform. Int. J. Emerg. Technol. Learn. (IJET) **30**(1), 197–215 (2021)
10. Lin, Y.: Construction of university network ideology management platform based on information technology. In: 2021 4th International Conference on Information Systems and Computer Aided Education, pp. 598–604 (2021)

Design of Online Auxiliary System for Action Teaching Based on Reinforcement

Xueqin Wu(✉)

Wuhan Institute of Physical Education, Martial Arts School, Wuhan 430079, Hubei, China
wuxueqin202101@163.com

Abstract. Through the combination of artificial intelligence technology and distributed technology, this paper puts forward a design scheme of Taijiquan action teaching online assistant system based on deep learning, which can be applied to Taijiquan teaching, broadcast gymnastics, fitness action and other fields, and provide intelligent action coach for users in the environment of no guidance. The platform adopts the front-end and back-end separation development, and uses the distributed and cluster technology to solve the problems of high concurrency and availability. Through tensorflow.js framework, the platform can identify, analyze and guide actions in real time on the front end, which can avoid privacy leakage and server overload in the front and back end data interaction. Through the scoring and comparison algorithm, users can constantly correct their actions in the process of using, and achieve the positive feedback effect of action learning.

Keywords: Action identification · Action teaching · Deep learning · Taiji boxing

1 Introduction

With the promotion of Internet plus, the combination of Internet technology and other industries is an inevitable trend in modern society. In this context, building an intelligent fitness service platform combined with Internet technology has important strategic significance for promoting the equalization of public sports services and building a modern, scientific and intelligent national fitness service system.

Due to the sharp rise in the number of students and the increasing complexity of teaching courses, the task of teaching management in Colleges and universities is becoming more and more important. Once the teaching managers make mistakes, they often waste teaching resources and even lead to serious teaching accidents. The construction of digital campus makes various network applications have a platform and rely on. Considering that now all colleges and universities have basically established a complete campus network, we can rely on the existing campus network to design a web-based online teaching management system, which can reduce the work pressure of teaching management departments and prevent mistakes in teaching management.

The existing fitness platform products have reliable video teaching courses. Users can learn professional sports knowledge, but lack of feedback in the learning process, users can not understand the gap between their own actions and standard actions. The

I. Ahmad et al. (Eds.): STSIoT 2021, LNDECT 122, pp. 755–764, 2023.
https://doi.org/10.1007/978-981-19-3632-6_87

new platform combined with AR/VR technology has high cost, great technical difficulty and high hardware requirements, which is difficult to achieve, and it is also difficult to achieve the effect of equalization of public sports services. In order to promote the innovation of fitness service platform, we need lower but more effective technical means to build it. This paper consists of the following parts. The first part introduces the relevant background and significance of this paper, the second part is the related work of this paper, and the third part is data analysis. The fourth part is example analysis. The fifth part is conclusion.

2 Related Work

Reference [1] proposed a mobile learning model from the perspective of distributed cognition. In comparison with common teaching auxiliary system, Jing designed an online English course teaching system [2]. The objective of Ref. [3] were 81 English lecturers of some colleges or universities in Indonesia. Mishra et al. seek to address the required essentialities of online teaching-learning in education amid the COVID-19 pandemic and how can existing resources of educational institutions effectively transform formal education into online education with the help of virtual classes and other pivotal online tools in this continually shifting educational landscape [4]. Sun et al. developed an online English teaching system in comparison with the common teaching auxiliary system [5]. And Miao proposed an English hybrid intelligent teaching assistant model based on mobile information system development, which integrates massive English teaching resources [6]. In view of the poor communication quality and flexibility of current streaming media technology, Wang built a selective streaming media online teaching architecture based on animation media service platform [7]. The system adopts three-tier architecture, pays attention to good scalability, adopts modular design method, divides it into four functional modules: teaching resources, online Q & [8]. Experimental data show that the stability of the design method is better than the other two traditional methods [9]. The platform adopts SSM framework based on Spring architecture as the cornerstone and is oriented to ordinary users and teachers and students Ref. [10].

The existing fitness platform product is the existing fitness platform (non product) concept, which mainly includes a new intelligent fitness platform based on physical perception technology, augmented reality and virtual reality.

2.1 Rationality of the Teaching Method of Taijiquan Movement Decomposition

1) The complicated Taijiquan style is divided into several movements to teach. Beginners are easier to understand, imitate and remember, so they can learn and practice well in a short time.
2) Through such practice, not only can the complete posture (fixed form) of each type be grasped perfectly, but also the transition movement can make the route and the arrived position of trunk and limbs correct according to the requirements.
3) During the practice of decomposition teaching, each action can be stopped in that position, and checked and corrected by teachers or students at any time.

Some people, after learning Taijiquan, insist on practicing, but some boxing styles will still be "out of shape", which is not in accordance with the standard, which also needs to be reviewed and corrected by the method of action decomposition. So the practice of decomposition method is not only suitable for beginners, but also necessary for the improvement of practitioners.

When teaching with the method of action decomposition, I insist on making students grasp one posture correctly before learning the next action. I don't advocate that kind of method that ignores quality and learns to "draw a path" first and then correct it. At the same time, each class uses the time of the fake preparation to arrange some basic movements such as step type, footwork, body method and technique, which will help to improve the teaching quality.

Some people worry that teaching Taijiquan with decomposition action will be practiced as "doing gymnastics" and lose the style of Taijiquan "continuous". In fact, this concern is unnecessary. When teaching new movements, we should use the method of action decomposition. When teaching the formula after teaching, we should review the formula learned before repeatedly and continuously, and gradually increase the content of explanation so that students can understand and master the characteristics and requirements of Taijiquan (equivalent to the "detailed explanation of the movement essentials" of the twenty-four types). With the improvement of proficiency, the decomposition movement can be gradually connected to meet the requirements of Taijiquan, such as round life, coherence, coordination and even speed. For example, the action decomposition is like a crutch. When you can walk, you can throw it away.

2.2 The Development of Deep Learning

In other words, deep learning is machine learning based on artificial neural network, which is often called AI (Artificial Intelligence). Its core idea is to use fewer network parameters, deeper network depth and more complex network model to complete the corresponding tasks, and use a large number of samples to replace more parameters used in traditional machine learning, Reduce a large number of manual labels in the sample to complete its target task. In short, it is to replace the high-cost manual labels with high-quality training data by increasing the number of data. Although deep learning is derived from machine learning, it is actually a solution, but it is very different from traditional machine learning algorithms. The inspiration of deep learning comes from neurons in human brain, and its core is data processing through computing units such as neurons as basic units, The idea of combining a large number of neurons to form a neural network to solve more complex problems can be traced back to an MP model proposed in a paper published by psychologist McCulloch and mathematician Pitts in 1943. This is the earliest working principle that imitates the physiological structure of the human brain in medical research and is introduced into mathematical scientific calculation, It can be seen from the figure that the following formula is obtained:

$$y_j = \sum_{i}^{n} w_{ij} - \theta_j \tag{1}$$

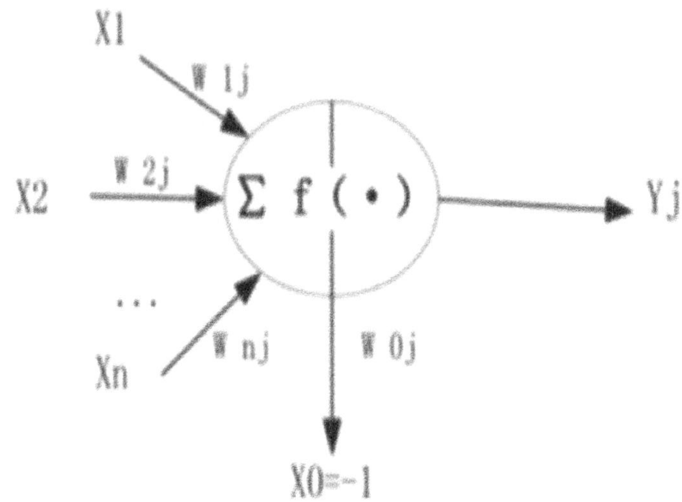

Fig. 1. MP network neuron

As shown in Fig. 1, the input data X is input into the neuron, calculated with the corresponding weight WJ, added with the set threshold, and finally summed to obtain the corresponding output value y. This is the simplest neural network model. The essence of deep learning is actually the process of constantly learning to calculate the weight and threshold of the neuron in the neural network, Only after training mature neurons to calculate can we get accurate output. After the introduction of the concept of neuron, its calculation process is a process that requires people to continuously participate in the operation. For different inputs, different weight parameters need to be manually adjusted to complete the corresponding output tasks. How to establish the relationship between operation unit neurons is an obstacle to the construction of neural network. Since the concept of bionic neuron is adopted, Therefore, we can learn the connection between neurons to find a solution. In 1949, Canadian psychologists proposed a rule Heb theory based on unsupervised learning. This theory is defined as follows. People can assume that continuous and repeated reflex activities will lead to a lasting improvement in the stability of neurons, When the axons of two neurons are very close and generate continuous stimulation, some growth processes or metabolic changes will occur in the two neurons or one or both of them, resulting in enhanced efficiency between the two cells. Therefore, MP network based on Heb theory lays the basic model of neural network.

3 Data Analysis

3.1 Distributed Architecture

Distributed architecture is a low cost method to solve high concurrency problems, and it has good scalability. The distributed architecture will process the service of business logic and the database of data storage are constructed in the way of nodes, reducing the load of a single node, and improving the overall performance of the system. It is a popular way to deploy services and databases by docker. Docker is container virtualization technology, similar to a lightweight virtual machine iw%. Docker can be used for the construction, migration and operation of distributed applications. It allows developers or operation and maintenance personnel to package applications and their dependencies into a standardized container to realize the application configuration and run everywhere. Docker virtualization is much less expensive than traditional virtual machines, and it is easy to increase or reduce the operation of node containers, so it can deploy services more effectively.

The use of artificial intelligence technology based on machine learning can effectively strengthen the feedback effect of Taijiquan movement teaching exercise fitness learning. Through the technique of action recognition and action guidance, the part of the function of artificial intelligence action coach can be realized instead of real person coach. For the problem of action recognition, the current mainstream real-time recognition analysis method is the attitude estimation algorithm based on convolutional neural network model. Attitude estimation I8 is a computer vision technology, which can detect the characters in images and videos, so as to determine the coordinate position of the key points in the image. Besides detecting the key point data of 2D image, attitude estimation algorithm can be used to detect 3D key data of characters from image video, and to recognize 3D actions and build 3D character model.

Based on the distributed architecture, this paper proposes a general action guidance technology based on the latest movement recognition technology, and proposes an online auxiliary system design scheme for Taijiquan movement teaching which can be realized, high concurrency and high availability.

3.2 Overall Architecture Design

The system adopts the design method of separation of front and rear ends. The front end adopts Ant Design Pro framework, which follows spa design style, and packages the functions of routing, menu, permissions, reverse proxy, asynchronous request, etc. The front-end system communicates with the back end through asynchronous request. According to the data results returned, the content of the page is updated in real time, and the user interaction experience is improved. The front end is started by NPM and deployed to a separate web server, decoupled from the back end. The front end uses webpack or nginx to agent, solve the cross domain request problem and interact with the back end. The back end adopts spring Webflux framework, which provides real restful httpapi to asynchronous process the requests from the front end. The spring session and spring security are used to manage user login information, security token, data cache, etc.

Mongodb is selected as the database system in the data storage part, and the spring mongodb real-tive framework is used to exchange data with the back end without blocking. All data processing in the architecture is asynchronous. The front end sends out asynchronous request, the back end responds and processes asynchronously. Finally, it performs asynchronous IO with the database. The whole process takes the form of non blocking. The specific overall structure is shown in Fig. 2.

In Fig. 1, the presentation layer is used as the front-end part of the system, and nginx is used as the HTTP server and deployed to the server cluster through NPM. The

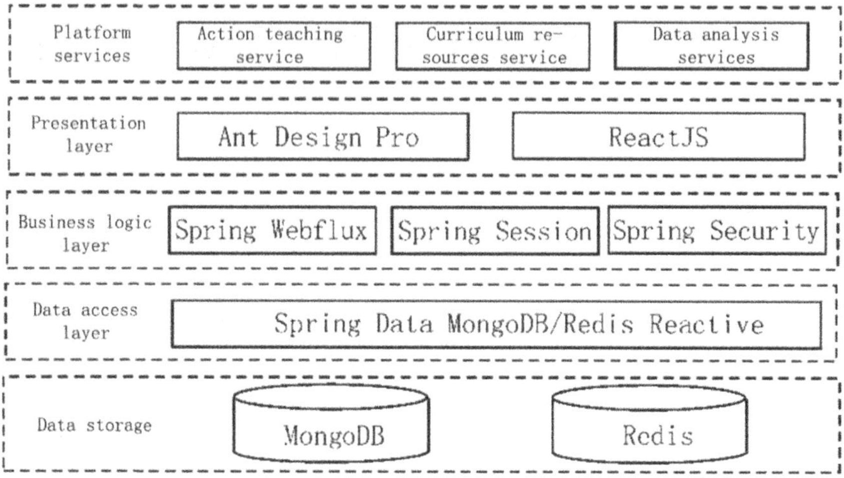

Fig. 2. Overall architecture design

back-end part of the system is composed of business logic layer and data access layer, which is deployed to the server cluster with the built-in netty of spring boot. Because the expected data reading load of the system is much larger than the expected data writing load, the data storage part of the system adopts the master-slave database cluster mode, which separates the database reading and writing and improves the data reading load capacity.

3.3 Action Teaching Algorithm

The action teaching service needs to realize the functions of action recognition, action evaluation, action guidance and so on. The action recognition function takes pictures, videos or live streams as input, performs bottom-up attitude estimation algorithm for input, obtains key point data, and then draws key points and action posture images in the source of transmission. If the input source is video stream, the input source is processed by frame, that is, the attitude estimation algorithm is performed once for each frame image.

After the action recognition, the similarity analysis algorithm is used to calculate the similarity cost of the two according to the key point data of the standard action and the user action, and the feedback results such as similarity evaluation score, standard action attitude projection image, user position offset direction and angle prompt information are calculated. The specific action teaching algorithm flow is as follows.

Firstly, the pose estimation of the characters in the standard action image is performed by posenet model, and the coordinate matrix of the key points of the standard action is obtained U_P and confidence matrix U_c. Where the coordinate vector of the K key is $U_{Pk} = (x_k, y_k)$, confidence degree is U_{ck}, K from 1 to 17.

$$U_{Pk} = U_{Pk} - U_{P0} \tag{2}$$

According to the standard action key point coordinate matrix U after linear transformation and normalization processing, the confidence matrix of the standard action key point U_P'', coordinate matrix of key points of user action V_P'', the cost of similarity is calculated as follows:

$$C = \frac{1}{\sum_{k=1}^{17} U_{ck}} \sum_{k=1}^{17} U_{ck} U_P'' - V_P'' \tag{3}$$

The standard action projection image is drawn on the user action image according to projection matrix w in real time by canvas, which provides reference for users to correct the action. Finally, according to the standard action projection coordinate matrix W and the UN normalized user action key point coordinate matrix V_P'', the vector difference and vector angle of the two parts are calculated to get the action offset prompt of each part.

Therefore, the final technical route of action teaching is shown in Fig. 3.

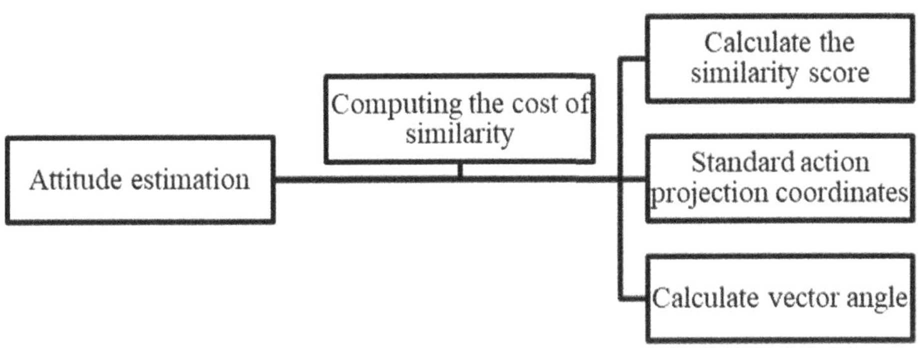

Fig. 3. Technical route of action teaching

4 Example Analysis

4.1 Quality Assurance in Each Stage of Software Project

According to the progress, software project development is divided into demand, design, implementation, testing and other stages. Quality assurance always runs through each stage, and corresponding measures must be taken according to the characteristics of each stage.

(1) System design and implementation stage

Implementation is the production process of code. This includes not only the generation of code, but also the generation of test cases. The detailed design is provided for the system design stage, the programmer starts coding and debugging the program, and the tester designs the test cases according to the product specifications. The designed use cases need to be approved by the project team members and reviewed by the project manager before entering the configuration library. At the same time, after debugging the program, the programmer submits it to the tester for program correctness detection.

At this stage, we adopt the RUP unified process idea: support iteration and increment, allow changing requirements, allow continuous integration, reduce risks as soon as possible, improve reusability, and generate stronger products, so as to ensure the quality of products.

(2) System test phase

Strict and standardized software testing is undoubtedly of great benefit to the quality assurance of software products. Software test engineers should be involved in the project as soon as possible, and put forward some problems from the perspective of testing for the requirements and design of the project. The Development Engineer shall communicate with the test engineer on the problems existing in the product in time. The test engineer shall carefully read the product requirements documents and product specifications, and formulate a detailed test plan.

Common software testing strategies in online teaching system include:

Software correctness test: it mainly tests whether the functions of the software are correctly realized. The test method is mainly to see whether there is a given output according to the given input according to the functional requirements, and whether the output is abnormal in case of non-standard input.
Software performance index test: the performance requirements of the project are different from those of general software projects. Performance testing often includes stress testing, aggression testing and so on.
Software usability test: when designing and implementing the software interface, we should try to separate it from the realization of functions. The reason for the separation is that ease of use is achieved through a friendly interface.
Practice has proved that only by continuously implementing quality management measures in each stage of the project can we find problems as soon as possible and ensure the success of the project.

4.2 Data Analysis Algorithm

Statistics of the past learning of Taijiquan teaching courses by users, and the statistical charts and graphical sports reports are generated by using eckarts. The statistical charts such as learning duration are shown in Fig. 4. The sports report gives the numerical results of the number of courses, the length of study, the scores obtained by each course, the overall score, the trend of increase and decrease of the scores, the prediction of future scores, the ability values of each part of the body and so on.

Users can fill in comments or score for courses they have learned. The scoring of users will affect the intelligent recommendation of the system. The learning situation of users and the scores of each course are converted into points, and the ranking of points is given.

In addition, the system generates statistical charts and graphical motion reports based on the previous course learning of all users. The total number of courses, the total length

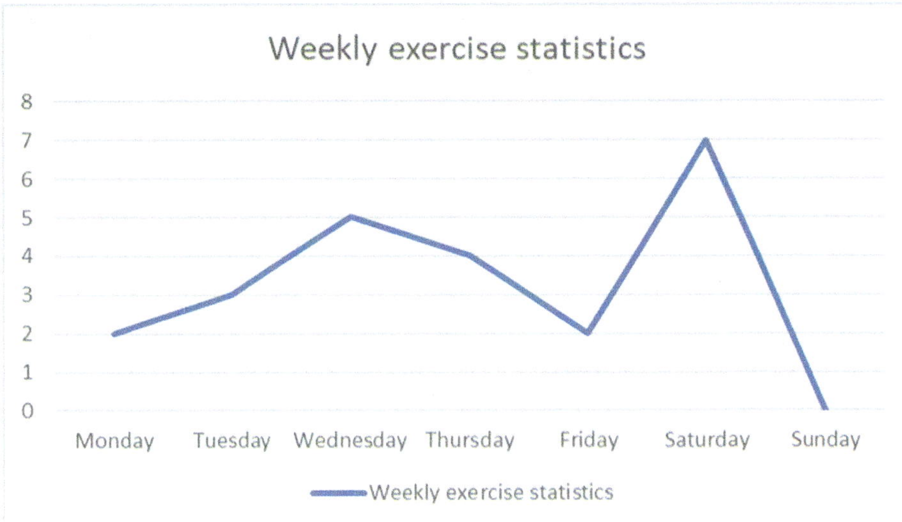

Fig. 4. Learning duration statistics

of study, the score of each course, the overall score, the average ability of each part of the body and other numerical results are given. It can analyze the user's course selection and learning, and get the user's characteristic portraits, such as the user's interest classification, the user's high click rate course category and other data.

Finally, the analysis of different user groups is realized, and the automatic division results of user groups are obtained by clustering algorithm, as well as the overall score increase and decrease trend and future score prediction obtained by statistics of all users.

5 Conclusion

Through the combination of artificial intelligence and distributed technology, this paper puts forward a design scheme of Taijiquan action teaching online assistant system based on deep learning. Its main purpose is to provide users with a large open online learning platform, including Taijiquan, martial arts, dance, gymnastics, sports and fitness. As an intelligent action teaching platform, intelligent technology is used to replace part of the functions of real action coach to a certain extent.

References

1. Bu, B.: Design and implementation of color teaching online platform in mobility environment. In: 2018 10th International Conference on Measuring Technology and Mechatronics Automation (ICMTMA), pp. 770–778 (2018)
2. Jing, L.: A teaching system of English online course based on artificial intelligence. In: 2019 11th International Conference on Measuring Technology and Mechatronics Automation (ICMTMA), pp. 112–116 (2019)

3. Fitria, T.N.: Teaching English through online learning system during COVID-19 pandemic. Int. J. Health **98**(11), 1227–1233 (2020)
4. Mishra, L., Gupta, T., Shree, A.: Online teaching-learning in higher education during lockdown period of COVID-19 pandemic. Int. J. Educ. Res. Open **14**, 1063–1070 (2020)
5. Sun, Z., Anbarasan, M., Praveen Kumar, D.: Design of online intelligent English teaching platform based on artificial intelligence techniques. Comput. Intell. **9**, 3840:1–3840:11 (2021)
6. Yanfei, M.: Online and offline mixed intelligent teaching assistant mode of English based on mobile information system. Mob. Inf. Syst. **51**(3), 248–257 (2021)
7. Wang, K.: Optimization of the online teaching system based on streaming media. Complexity **33**(33), 91–963 (2021)
8. Junwa, Y.: Development of physical education resources and design of college physical education teaching system from the perspective of data fusion. In: 2021 IEEE International Conference on Advances in Electrical Engineering and Computer Applications (AEECA), pp. 279–288 (2021)
9. Sun, H.: Research on remote control method of assisted instruction based on machine learning. Int. J. Contin. Eng. Educ. Life-Long Learn. **31**(1), 69–86 (2021)
10. Jiang, J., Xia, Y.: Optimization and simulation of literature aided reading system based on wireless sensor network. J. Sens. **49**(1), 1–7 (2021)

The Role and Mechanism of the Diabetes Control Based on the Association Rule Apriori Algorithm

Shuqi Yang(✉)

Xishuangbanna Vocational and Technical College, Jinghong, Yunnan, China
yangshuqi456@163.com

Abstract. This paper is based on the Apriori algorithm to obtain more efficient data mining methods, analysis of Eucommia Leaf prevention and treatment of diabetes and mechanism. The classic Apriori algorithm of association rule mining is used to analyze the medical data with the characteristics of privacy, polymorphism, incompleteness, timeliness and redundancy. Firstly, Apriori algorithm is used to find the frequent itemsets of data in the database, and then strong association rules are generated according to the frequent itemsets to find useful association relationships or patterns between itemsets in massive data. The final purpose is to analyze the application of association rules mining in clinical disease monitoring, evaluation of drug treatment effect and prevention of diabetes by Eucommia ulmoides leaves.

Keywords: Association rules mining · Apriori algorithm · Disease prevention · Data correlation

1 Introduction

Diabetes mellitus is a common endocrine and metabolic disease, which seriously affects patients' physical and mental health. Traditional Chinese medicine for diabetes has been a long-standing treatment. Eucommia ulmoides leaves have been used in the treatment of diabetes in our country. The experiments of relevant scholars have also confirmed that Eucommia ulmoides leaves have the effects of reducing blood fat and blood sugar. Eucommia ulmoides polysaccharides mainly improve the immune response ability and immune ability by improving the body's immune response ability. Related studies have shown that the aqueous extract of Eucommia ulmoides leaves can reduce the level of oxidative stress in diabetic rats, improve the ability of free radical scavenging in rats, inhibit the formation of lipid peroxide, and enhance the antioxidant capacity. Separation and purification is the basis of the study on the activity and structure of polysaccharides, and it is also an important step in the preparation of polysaccharides.

The generation, collection and storage capacity of scientific data and medical data has been greatly improved, which makes the amount of information in hospital database continue to expand. However, at present, the application of database in hospital is not

I. Ahmad et al. (Eds.): STSIoT 2021, LNDECT 122, pp. 765–774, 2023.
https://doi.org/10.1007/978-981-19-3632-6_88

sufficient, and the integration and analysis of data is lack. This topic is to introduce the new technology of data mining into the analysis of the data of prevention and treatment of diabetes, and find the internal correlation in the data, so as to provide scientific decision-making basis for the diagnosis and treatment of diseases in the field of medical and health care, and also find a breakthrough for the standardized management and scientific research of clinical.

Association rule mining is the most mature, active and important research field in data mining, and hospital information system is the main source of hospital data. By using computer technology, database resources and advanced data analysis methods, this paper explores the application of data mining technology in disease prediction, disease diagnosis and other data, so as to save resources and share resources, It has become an important means of scientific research in the field of medicine.

At the same time, as a highly practical, experimental and statistical subject, medicine makes full use of a large amount of information accumulated in clinical medicine and medical research, carries out data mining on disease data, and selects the best association rule algorithm, which has important practical value for improving the quality of medical treatment. This study is based on the association rules Apriori algorithm to explore the effect and mechanism of eucommia leaf in the prevention and treatment of diabetes. This paper consists of the following parts. The first part introduces the relevant background and significance of this paper, the second part is the related work of this paper, and the third part is data analysis. The fourth part is example analysis. The fifth part is conclusion.

2 Related Work

We use many method for prevetion disease, but the effect is very low, so the Ref [1] proposed phytoncides in the prevention and therapy of blackhead disease and their effect on the turkey immune system, the method solved some issues, but for part disease, the way was not working, so some doctor and scholar proposed if we contact the guidance effect, thus, we can prevent some disease by beta titanium alloy suface, its structure is organized by electron-beam technology [2], we know that some mechanism of AL(OH)3 is limited, so Lin S et al. used the thermal analysis way to do the prevention [3], it get the better effect. In the effect, we have also been used to alleviate the effect of quinoa, by the experiment results, the author think, the experiment is successful [4].

If we use the food mycotoxins to do the related experiment, we also get the better experiment results, but all setting are only experiment, so if we can set different experiment item and group, the effect is very clear [5]. By TLR4/NLRP3 signalling pathway in rats with nonalcoholic fatty liver disease, the Ref [6] also use thus techniqus to against inflammatory damage and know the effect of shenling baizhu powder is good [6]. Different methods have different effect, so Ref [7] used the different treatment of metabolic syndrome and forecast the effect successfully. For the auricular acupressure, Han R et al. think if the way they proposed can prevent and control the effect with children, then all problems will be solved [8]. Based above ways have some defects, so Ref [9] does thus a review for these ayurvedic plants. However some approach assumed that there is enough complete effect for these disease, but it only is for the experiment. Some medical institutions at home and abroad have also extracted and analyzed the medical record data of the hospitalized and discharged patients in the hospital information system, such as the medical expenses, disease examination and so on, obtained the demand rules of hospital beds, drugs, doctors and so on, and put forward the corresponding management strategies, which have achieved remarkable results in reducing the medical cost of patients and the operating cost of the hospital.

2.1 Diabetes Overview

Diabetes is one of the most serious endocrine disorders, mainly manifested as glucose, fat and protein metabolism disorders. According to the latest data from the International Diabetes Federation (IDF), the prevalence of diabetes was 9.3% (463 million people) in 2019, and 10.9% (700 million) in 2045. The urban sector (10.8%) exceeded the rural sector (7.2%). 1/2 of diabetic patients (50.1%) did not know that they had diabetes. China is already the "most devastated area" of diabetes in the world. According to Ningguang studies, the total number of diabetic patients in China has reached 92 million 400 thousand. The prevalence rate of diabetes in adults is as high as 11.6%, increasing by 3000 people per day, that is, every 10 people have diabetes. According to investigation and research, Type 2 Diabetes Mellitus (T2DM) has accounted for 95% of patients with diabetes mellitus (Mellitus). Type 2 diabetes is becoming a serious threat to global health. However, our understanding of the etiology and the best treatment of this disease is not completely clear. The current drug treatment can not well control the sustainable development of hyperglycemia, and even lead to adverse reactions such as hypoglycemia, which makes patients bear further economic burden. What is more terrible is the damage of large blood vessels, microvessels, heart, brain and kidney induced by hyperglycemia and hyperlipidemia, which brings great pressure to the regional health system and economy. Therefore, it is urgent to develop the best and effective treatment method, and to further study the prevention and treatment of diabetes.

Modern medicine has begun to study and focus on Eucommia ulmoides leaf, which is the dry fruit of morusalba L. in Moraceae. It is also called mulberry camp. It is a traditional dual-purpose resource for medicine and food in China. It was first published in the newly revised materia medica of the Tang Dynasty. It is now cultivated in central and Northern China from northeast to southwest provinces and regions, northwest to Xinjiang. This study proved that Eucommia ulmoides leaves in Xinjiang are rich in flavonoids, polysaccharides, alkaloids, vitamins and amino acids. They are medicinal mulberry germplasm resources in Xinjiang. Due to the unique natural environment of Xinjiang, the drought and barren Eucommia ulmoides leaves have become special and rare resources. It has the functions of tonifying blood, calming, reducing blood sugar and delaying aging.

In 1988, the Ministry of health listed mulberry as the first group of medicinal and food homologous plants, and was praised by the medical community as "the best health fruit in the 21st century". It is a special medicinal and edible plant used for the treatment of hypertension, hyperlipidemia and diabetes, and has been evaluated consistently by cardiovascular experts in China. The lipid-lowering, hypoglycemic and antioxidant activities of different polar extracts of Eucommia ulmoides leaves are gradually being widely recognized. Hassimotto and others found that flavonoids in medicinal mulberry have strong antioxidant capacity. Fu Daxu screened the antihypertensive activity of Eucommia ulmoides leaves in Xinjiang and found that its effective parts are anthocyanins and other flavonoids. It is considered that its effect is mainly related to the antioxidant effect of flavonoids. In the previous research, our research group successively adopted ultraviolet spectrophotometry, high performance liquid chromatography The contents of flavonoids, polysaccharides, alkaloids (DNJ) and amino acids in Eucommia ulmoides leaves were determined by high performance capillary electrophoresis and other spectral and chromatographic separation techniques, which were consistent with the results in the literature. In view of this, we believe that Eucommia ulmoides leaves have multiple components of anti diabetic activity. It is necessary to further study the effective ingredients and possible targets of Xinjiang Eucommia ulmoides leaves for preventing and treating diabetes, which will open up a new direction for the development of mulberry resources and provide a new breakthrough for finding new antidiabetic drugs.

2.2 Association Rules in Data Mining

Mining rules is an important research topic in the field of data mining, and it is also an important part of database knowledge discovery. With the continuous accumulation of data, many industry stakeholders are more and more interested in mining relevant rules in the database. Association rules and mining algorithms are the main contents of association rule mining.

Correlation analysis is an important method of data mining. If the values of two or more data items repeat with a high probability, they have a certain correlation, and relevant rules can be established for these data items. Data association is an important knowledge that can be found in database. Reflects the independence or relevance of one event relative to other events. If there is an as association between two or more attributes, the attribute value can be predicted based on other attribute values. For example, 90% of customers who buy bread buy milk. If you put these two products together and sell them in department stores, the sales will increase. In large databases, these related rules usually need to be filtered. Typically, support and trust thresholds are used to exclude unnecessary association rules. Supports the proportion of rules displayed in all cases. Confidence represents the proportion of cases represented by the rule when the preconditions are met. r. Apriori algorithm proposed by Agrawal et al. Is the most famous and important association rule discovery algorithm in the field of data mining. Therefore, the research direction of related rule discovery has shifted from single concept level related rule discovery to multi concept level related rule discovery. The effective algorithms of association rule mining and association rule mining in fuzzy association rules are further studied.The basic model of association rule mining is shown in Fig. 1.

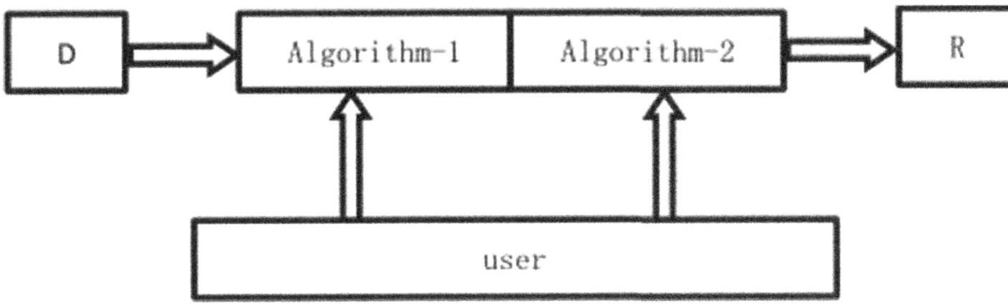

Fig. 1. Basic model of association rule mining

3 Data Analysis

3.1 Definition of Association Rules

Association rules collect specified data items and transaction databases (all transactions are data item sets), filter the frequency relationship of data item sets in the transaction database, and find valuable correlations between data item sets through a large amount of data. Association rules mainly reflect the relevance between things. For a record reflecting the same thing, if it has both characteristic attribute a and characteristic attribute B, then characteristic attribute a and B are related. A typical application example of association rule analysis is to analyze customers' shopping habits according to a large number of customers' shopping records (for example, when buying some items, they will also buy other items), so it is also called "shopping basket analysis". There are also related phenomena in the field of medicine. If a disease may induce a variety of complications at the same time, there is a correlation between the disease and its complications.

The confidence of association rules is the percentage of both y and X contained in transaction database d, which can also be regarded as conditional probability $P(Y/X)$, then the percentage is the confidence (c) of association rules $X \rightarrow Y$, which can be calculated as follows:

$$Confidence(X \rightarrow Y) = |T : X \cup Y \subseteq T, T \in D/T : X \subseteq T, T \in D| \tag{1}$$

The promotion degree life of rule (x = y) is the ratio of the probability of rule $X \rightarrow Y$ being established to the probability of feature attribute being independent of Y. it can be calculated as follows:

$$Life(X \rightarrow Y) = \frac{Confidence(X \rightarrow Y)}{Support(Y)} \tag{2}$$

It can be seen from the above formula that $Life(X \rightarrow Y) = Life(Y \rightarrow X)$.

3.2 Apriori Algorithm

Apriori algorithm is a fast mining algorithm proposed by R. Agrawal in 1994. It is also the most famous and influential frequent pattern mining algorithm. The most important property of frequent pattern mining algorithm is the property of Apriori algorithm. In practical research, in order to avoid the high computational cost of exponential search space, we use the nature of apriari algorithm to realize frequent pattern mining algorithm to prune the search space.

The basic idea of Apriori algorithm is to generate candidate itemsets of specific scale, and then scan the database and count them to determine whether these candidate itemsets are frequent itemsets. The specific implementation process is to first scan all transactions in the database, calculate the occurrence times of each item, generate 1-candidate set C, and then determine 1-frequent set l according to the preset minimum support, and then L × L performs connection operation to generate 2-candidate set C, scans all transactions in the database again, calculates the occurrence times of each element in C, and determines 2-frequent set l according to the preset minimum support. This process is repeated until the k-frequent set L is generated, and it is impossible to generate the (K + 1) itemset satisfying the minimum support. For the j-candidate set C (J = 3, *. K), if the (J − 1) subset of an element is not the (J − 1) frequent set, it will be deleted. Accordingly, we give its operation architecture, as shown in Fig. 2.

Apriori algorithm is simple and easy to implement. When the data set is small and the frequent patterns are short, most people choose Apriori algorithm. In the Apriori algorithm, the minimum support and minimum confidence are the key thresholds, which are also the constraints of the algorithm. However, only these two conditions are not enough when processing medical data. The algorithm generates candidates and tests the support of each candidate. When the number of candidates is large, interesting medical rules cannot be found quickly; In addition, Apriori algorithm needs to scan the data set the number of times equal to the length of the longest frequent pattern. If the data set is large and the frequent pattern is long, the execution efficiency is restricted. Therefore, restrictions should be used in the procedure if interesting medical rules are to be found quickly.

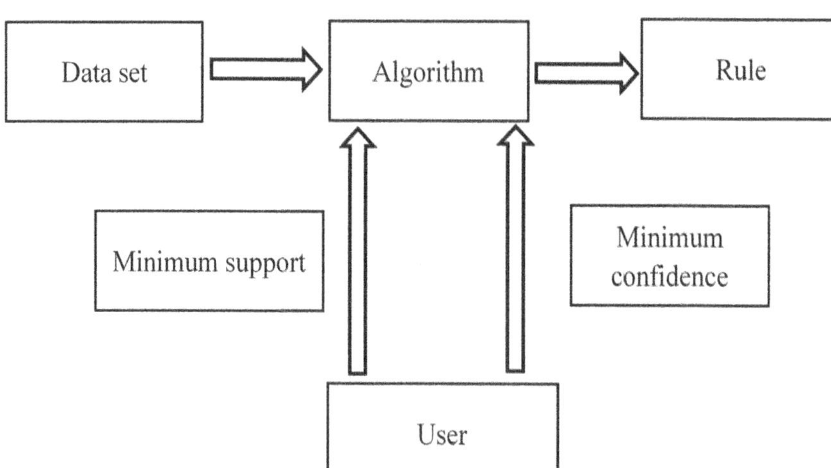

Fig. 2. Basic model of association rule mining

3.3 Application of Association Rules Mining in Prevention and Treatment of Diabetes Mellitus with Eucommia Ulmoides Leaves

Association rules mining was used to study the relationship between diabetes mellitus and Eucommia ulmoides leaf prevention. The results show that association rule mining can not only find the information correlation hidden in the data, but also quantify the strength of various correlations.

The pathogenesis of diabetes is complex, including oxidative stress, abnormal glucose metabolism and genetic factors. The interaction of different factors aggravates the development of the disease. The DPPH free radical scavenging rate of Eucommia leaves was 65.15%, and that of VC was 91.45%. Among them, the high-dose group of Eucommia ulmoides leaves polysaccharide had significant difference, as shown in Fig. 3. Zhong Shujuan et al. Showed that DPPH ICSO and abtsicso of Eucommia ulmoides leaves were 0.35 mg/L and 0.55 mg/L respectively. In vitro antioxidant activity test of different parts of Eucommia ulmoides leaves, the antioxidant activity of Eucommia ulmoides leaves was the strongest, and the content of total flavonoids in Eucommia ulmoides leaves and male flowers was higher. Eucommia ulmoides leaves contain a lot of protein, which can be used as a source of dietary protein, and it also has more carbohydrate and fat content, providing a certain capacity.

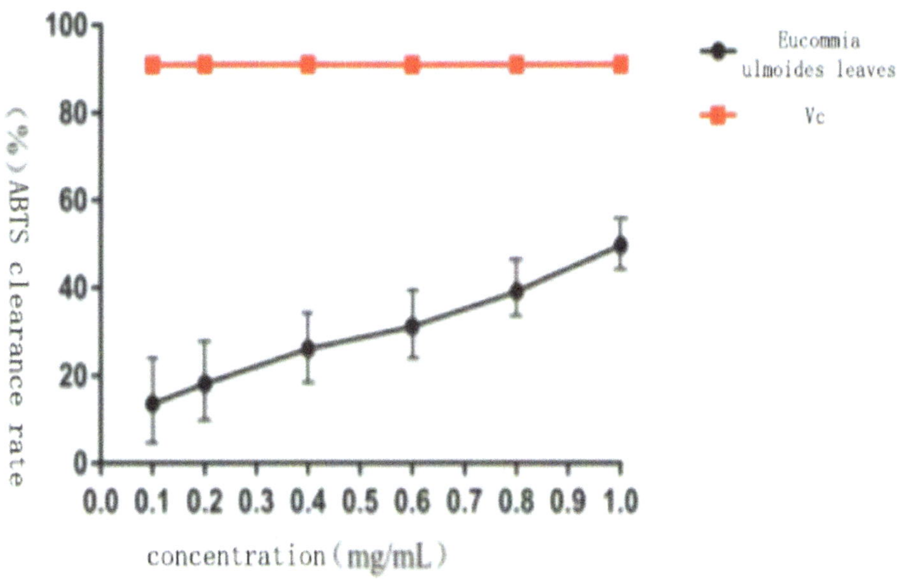

Fig. 3. Scavenging rate of Eucommia ulmoides leaves on ABTS free radical

Based on the Apriori algorithm of association rules, this paper discusses the effect of Eucommia ulmoides leaves on the prevention and treatment of diabetes. The main pharmacological effects of Eucommia ulmoides leaves polysaccharides are hypoglycemic, anti-aging, and hypotensive. Eucommia ulmoides leaves contain mineral elements, protein, vitamins, fat and other nutrients, which supplement the protein needed by the human body and can be used as a certain source of dietary protein. The content of crude protein and various amino acids in Eucommia ulmoides leaves are relatively high. It contains a variety of trace elements, vitamins and other nutrients It has the characteristics of great health function and high nutritional value. The results showed that the levels of FBG and BUN in irbesartan group, low-dose group, medium dose group and high-dose group were lower than those in model group. Among them, the levels of FBG and BUN in high-dose group were 9.21 mmol/L and 6.85 mmol/L, respectively. Xing Dongjie et al. Showed that the FBG level was 14.35 mmoll after high dose of flavone from Eucommia ulmoides leaves intervened in diabetes for 28 days. The water extract of Eucommia ulmoides leaves could play a hypoglycemic role by increasing glycolysis, and the flavonoids from Eucommia ulmoides leaves could stimulate the secretion of insulin from pancreas and play a hypoglycemic role. Qian zengkun and other studies showed that Alisma orientalis has the effect of reducing blood glucose, and its mechanism may be related to the regulation of IGF expression. The FBG of high-dose Alisma orientalis polysaccharide group was 19.92 mmol/L. Eucommia ulmoides leaves have lignans, polysaccharides, phenylpropanoids, cyclic ether mushrooms, flavonoids and other medicinal components, with antioxidant, immune regulation, hypoglycemic and a variety of health care effects.

4 Example Analysis

Insulin resistance at the cellular level is defined as the decrease in the number of insulin receptors on the cell surface or the defect after the receptor, resulting in a certain concentration of insulin can not make the cell reach the expected metabolic level. Liver and peripheral tissues, such as skeletal muscle and fat, are the main sites of insulin resistance. In this experiment, human hepatoma cell line HepG2 was selected, and the high affinity insulin receptor expressed on its surface met the standard required by typical insulin receptor. Some researchers have stimulated HepG2 cells with high concentration insulin (1 × 107 mol/L), there are functional defects of insulin receptor and post receptor, and the receptor self phosphorylation function and insulin stimulated glucose, lipid and egg self metabolism function are reduced. Therefore, HepG2 is an ideal cell model for studying insulin resistance 1-531. The results showed that when different concentrations of inducers induced insulin resistance in HepG2 cells, the palmitic acid induced model had the best effect and stability at 250 umol/L for 24 h. In the model induced by insulin and glucose, 1×10^{-7} mol/L and 55 mmol/L in 24 h were better. Therefore, based on the above situation, it is determined that the best scheme of insulin resistance HepG2 cell model is: 250 umol/L palmitic acid acts on cells for 24 h. At this time, the consumption of glucose by cells is reduced, that is, obstacles to glucose uptake and utilization occur, indicating that the establishment of insulin resistance model is successful. The established cell model was used to screen the multi components of Eucommia ulmoides leaves. It was found that the ethyl acetate layer (100, 200, 400 µG/ml), n-butanol layer (200, 400, 800 ug/ml), neochlorogenic acid (50, 100 µG/ml), isoquercetin (50, 100 µG/ml), Morin (50, 100 ug/ml), myricetin (50, 100 µG/ml) has good glucose utilization activity and no cytotoxicity, indicating that they are the main active compounds of Eucommia ulmoides leaves to improve insulin resistance.

The extract of Eucommia ulmoides leaves had a good effect on glucose metabolism and lipid metabolism of HepG2 insulin resistant cells. The indexes were determined by HK, PK, TC and TG kits. The activities of HK and PK could be increased by the intervention of drug administration at the extraction site ($P < 0.05$); The extracts from Eucommia ulmoides leaves could significantly reduce the accumulation of TG in cells ($P < 0.05$) and clear TC ($P < 0.05$). The results showed that the extract of Eucommia ulmoides leaves promoted the entry of glucose into liver cancer cells and accelerated the oxidative decomposition of glucose, so as to regulate glucose metabolism and improve insulin resistance. The extract of Eucommia ulmoides leaves may play a role in lipid clearance by reducing the accumulation of lipids in cells. Four small molecular monomer compounds in Eucommia ulmoides leaves increased HK activity and PK activity, and decreased TG content and TC content, but there was no significant difference. The improvement of glucose and lipid metabolism in Eucommia ulmoides leaves may be related to the synergistic effect of a variety of compounds and the activation of a certain pathway. When the three small molecule compounds were administered alone, there was no significant difference, which may be due to the dose or the need to cooperate with other monomers to activate PI3K Akt mTOR pathway. Traditional Chinese medicine materials have the advantages of multi-component and multi-target. Combined with network pharmacology methods, relevant pathways are predicted. WB experiment explains the activated gene protein pathway from the protein level. To study the gene and protein expression changes of key targets in PI3K Akt mTOR insulin signal transduction pathway can explain the causes of insulin resistance to a certain extent. If drug intervention can effectively regulate the key targets in this signal pathway, it can also improve the internal molecular mechanism of insulin resistance to a certain extent.

The marker proteins PI3K and Akt in this pathway are the objects we continue to observe. Studies have shown that PI3K/Akt mTOR pathway affects cell growth, proliferation, differentiation, movement, survival and metabolism through signal transduction. Insulin signaling has also been confirmed to be transmitted through this pathway. Insulin exerts its biological effects through receptors on target cells and post receptor signal transduction system. Insulin acts on cells, first binds to insulin receptor and transmits insulin signal from extracellular to intracellular. Insulin receptor binding consists of two a subunits and two β A heterotetramer composed of subunits. Once insulin binds to subunit A, it immediately causes β The tyrosine residues of subunits are self phosphorylated, and catalyze the disc acidification of multiple tyrosine residues of insulin receptor binding receptor substrate to transmit insulin signal down. This pathway mainly regulates glycogen synthesis. In conclusion, Eucommia ulmoides leaves can improve insulin resistance and regulate glucose and lipid metabolism through PI3K Akt mTOR pathway.

5 Conclusion

The role of association rule mining and classification analysis in mature data mining algorithms is studied. Firstly, the association analysis model is used to extract relevant rules and generate frequent itemsets. Then, highly relevant rules are generated from frequent project sets, and interesting relationships and models between project groups are found in a large amount of data. Based on the relevant rules of the Apriori algorithm, the role of the two intermediate leaves in the prevention and treatment of diabetes is analyzed, and the relevant rules are used to extract data for analysis, which has important reference value.

References

1. Tykaowski, B., Miaek, M., Kowalczyk, J., et al.: Phytoncides in the prevention and therapy of blackhead disease and their effect on the turkey immune system. Int. Metabol. **76**(11), 5701–5713 (2021)
2. Ferraris, S., Warkomicka, F., Barberi, J., et al.: Contact guidance effect and prevention of micro-fouling on a beta titanium alloy surface structured by electron-beam technology. Nanomaterials **11**(4), 21–34 (2021)
3. Lin, S., Liu, Z., Qian, J., et al.: Inertant effects and mechanism of Al(OH)3 powder on polyethylene dust explosions based on flame propagation behavior and thermal analysis. Fire Saf. J. Mater. **124**, 103392 (2021)
4. Chenwei, S., Wei, L., Yahui, L., et al.: Alleviating the effect of quinoa and the underlying mechanism on hepatic steatosis in high-fat diet-fed rats. Nutr. Metabol. **77**(4), 479–489 (2021)
5. Sharma, V., Patial, V.: Food mycotoxins: dietary interventions implicated in the prevention of mycotoxicosis. Int. J. Dietary **44**(3), 289–299 (2021)
6. Pan, M.X., Zheng, C.Y., Deng, Y.J., et al.: Hepatic protective effects of Shenling Baizhu powder, a herbal compound, against inflammatory damage via TLR4/NLRP3 signalling pathway in rats with nonalcoholic fatty liver disease. J. Diet **19**(5), 11 (2021)
7. Williams, D.A., Horsburgh, K.J., Schultz, D.M., et al.: Proudman resonance with tides, bathymetry and variable atmospheric forcings. Nat. Hazards. J. Int. Soc. Prevent. Mitigat. Nat. Hazards 106–119 (2021)
8. Han, R., Kong, X.H., Zhao, F., et al.: Auricular acupressure for myopia prevention and control in children and its effect on choroid and retina: a randomized controlled trial protocol. Trials **22**(1), 52–68 (2021)
9. Kumari, R., Verma, A., Sharma, D., et al.: Mechanism ethnomedicinal and phytochemical effect of ayurvedic plants in prevention and management of senile dementia of Alzheimer's disease: a review. Int. J. Med. **33**(11), 222–234 (2021)

Clustering and Evolution of International Sports Field Based on Multi-sensor Fusion Technology

Jun Yuan(✉)

JiangXi Police Institute, JiangXi, China
yinjiesen55@126.com

Abstract. Based on 549 literatures with the theme of "sports artificial intelligence" and other keywords in web of science database since 1995, this paper uses CiteSpace V software for visualization processing and analysis, and combs the country, discipline distribution, research hotspots and evolution trend of sports artificial intelligence research in recent 25 years by means of visual knowledge mapping, and discusses its research progress and development direction. 1) The research area of sports artificial intelligence is widely distributed, among which the United States, China and Germany are in the leading position. 2) Sports artificial intelligence research involves many disciplines, mainly using and learning from the research methods and theoretical perspectives of computer science, engineering, sports science and other disciplines. 3) The frequency and centrality of keywords confirm that machine learning is the main direction in the field of sports artificial intelligence, artificial neural network is the main algorithm, and data mining is the basis of practice and research. 4) Research hotspots include simple activity recognition and energy consumption research based on wearable accelerometer technology; action analysis and damage prevention and control research based on wearable sensor; computer vision scene classification research based on convolution neural network algorithm; analysis and prediction of physical fitness and technology and tactics based on computer vision; human posture recognition technology based on computer Deep learning.

Keywords: Artificial intelligence · Sports · Knowledge map · Machine learning · Computer vision · Neural network

1 Introduction

Artificial intelligence is based on computer science, physics, information theory, system science, philosophy and other disciplines to continuously improve human living standards and development level. The ability of human beings to use knowledge to discover, define and solve problems. Artificial intelligence is often used to solve problems that cannot be solved by computers. This is the peak of human understanding and ability transformation of the objective world in the era of information industry revolution. Through the analysis of the development of artificial intelligence technology in China in recent 40 years, some scholars believe that the main short-term goal of artificial intelligence is to use machines to simulate and perform some intelligent functions of the human

I. Ahmad et al. (Eds.): STSIoT 2021, LNDECT 122, pp. 775–783, 2023.
https://doi.org/10.1007/978-981-19-3632-6_89

brain. The long-term goal is to simulate human psychological activities and intelligent functions through machines. Yes, yes. With the development of science and technology, artificial intelligence is more and more used in the field of international sports training. This paper will use citespave software to sort out and analyze the relevant literature in wos database. Through the analysis of the research and application of artificial intelligence in the field of international sports, we can improve the relevant research structure, learn from each other and make up for the shortcomings of domestic research. Seize the opportunity of the times, promote the research of artificial intelligence in China's sports field, and make China realize the goal of becoming a sports power as soon as possible. This paper consists of the following parts. The first part introduces the relevant background and significance of this paper, the second part is the related work of this paper, and the third part is data analysis. The fourth part is clustering and evolution of sports artificial intelligence research. The fifth part is conclusion.

2 Related Work

Since it is difficult for traditional theoretical and practical paradigms to solve a series of new formats, mechanisms and logic brought about by new technologies, the development and operation have become the top priority for the development of media and a problem that Ref [1] need to think deeply at present. Based on this analyze the important role that artificial intelligence technology can play from the perspective of serving sports training activities and improving the effect of sports training [2]. Zhang et.al. [3] use the recognition algorithm based on the multilayer decision tree recognizer to identify the joint movement; the experiment shows that the method used accurately identified joint movement for football players in sports training. In the algorithm, the grey relational analysis method is introduced [4]. By expanding the standard action data, the standard database of score comparison is established, and the system architecture and the key acquisition module design based on 3D data are given [5]. Liang et.al. conduct a questionnaire analysis on the application of artificial intelligence technology in leisure sports courses [6]. Research and analyze the characteristics and functions of new intelligent information service tools, and explore the effects of artificial intelligence in optimizing university sports information services from the three aspects of intelligent evolution information service, intelligent push information, and intelligent retrieval information, and the connotation of intelligent environment Analyze characteristics and technical support to promote the optimization and upgrading of university sports information services, research on the transformation of evaluation methods from manual evaluation to intelligent evaluation, and from standardized evaluation to differential evaluation, and specifically analyze the connotation of intelligent evaluation and differential evaluation, Features and key technologies, analyze the general process of intelligent evaluation, and summarize the implementation suggestions for intelligent evaluation [7]. Wu et.al. [8] show that the node prediction model is established by using the method described. Other influential work includes [9, 10].

According above description, we know if we use the traditional mtheds to monitor the sport, it will have many issues, but when the sensors come, this issues will be easy. At present, high-leve sports need many sensors to monitor all the characteristics of athletes is the trend of development in the future.

2.1 Research on Artificial Intelligence Technology

As early as the 1950s, the research on artificial intelligence technology began. Alan Mathison Turing, the "father of artificial intelligence", believed that it would not be difficult to manufacture intelligent machines in the future human society. He pushed down the possibility of producing intelligent machines through the development of existing technologies, and proposed Turing standard as a method to verify whether machines are intelligent, This pioneered the research of artificial intelligence technology. In 1956, the term "artificial intelligence" was formally put forward for the first time in the summer artificial intelligence seminar in Dartmouth. This is the first seminar held for the study of artificial intelligence. This seminar has become the symbol of the birth of the concept of artificial intelligence. At this seminar, artificial intelligence is considered to be an intelligent machine, The key technology is to realize intelligent computer program [11].

The research on artificial intelligence technology in China has been gradually enriched in recent years. Huang Xu and Dong Zhiqiang believe that artificial intelligence technology is an intelligent machine or intelligent system, which simulates human's ability to perceive signals and make decisions through designed programs and algorithms, so as to assist or replace human beings to complete the work that only human beings could do in the past. Further analysis of artificial intelligence technology, Tang Yonghe and Zhang Xian believe that there are two important dimensions of cognition and practice in the use of artificial intelligence technology. In the cognitive dimension, artificial intelligence technology simulates the cognitive process of human beings from signal transmission to acceptance through the designed intelligent program, giving intelligent machines a certain cognitive ability; In the practical dimension, artificial intelligence technology can transfer the information obtained through cognition to the machine, so that the intelligent machine can complete the instructions given by human beings. According to the degree of intelligence realized by artificial intelligence technology, Zhang Xian's film divides the development stage of artificial intelligence into low artificial intelligence stage, high artificial intelligence stage and super artificial intelligence stage. Point out now. The development of artificial intelligence is still in the initial stage of "weak artificial intelligence". The cognitive and practical abilities of intelligent programs and intelligent machines were further improved.

2.2 On the Connotation of Internet Plus

Corporate leaders, scholars and politicians combine their professional backgrounds to explain "Internet plus", and analyze their connotation. Internet has defined the connotation of Internet plus mainly in the following situations:

(1) "Cross border integration" theory

Mr. Ma (Tencent Holdings)'s CEO has repeatedly mentioned and discussed Internet plus. In 2015, Internet plus became the engine of China's economic and social innovation and development. "Internet plus" is based on the Internet platform and adopts ICT and industry integration. This will promote industrial upgrading, create new products, new enterprises and new models, build a new ecology, and effectively promote China's economic and social development. Internet plus is the combination of Internet and other traditional industries. In recent years, with the increase of the number of Internet users in China, the Internet penetration rate has reached 50%. In particular, the vigorous development of mobile Internet has had a far-reaching impact on other Internet industries. Stronger and stronger. According to the founders, directors and CEO of American technology companies, Internet Plus combines Internet technology with internet thinking and the real economy to promote the transformation, value-added and efficiency of the real economy.

Deputy director of the national information center of Ying Jia County said that "Internet plus" refers to the dissemination and application of the next generation of information technology, such as the Internet, cloud, Internet of things, big data, and so on. It is a process of deep integration. In the field of economic and social life, it will have a huge, far-reaching and far-reaching impact on human economy and society. Ma Hua lanterns, Robin Li, Lei Jun and other enterprises are closely related to the Internet. Based on Internet technology, Internet managers emphasize the value of the Internet at the technical level and the integration and combination of the chemical reaction between the Internet and other industries. This is a traditional industry. It has a broad and far-reaching impact on the real economy and economic society.

(2) "Technology upgrading" theory

Professor Huang of Peking University said that the Internet includes not only manufacturing, but also e-commerce. Industrial Internet, network finance and Internet plus innovation are two upgraded versions of technology integration. Not only industrialization, but also the Internet, as an important feature of information development, is closely related to industry, commerce and finance. Internet plus Internet plus is Internet plus, Internet plus, Internet professor Fu Zhilong, Tsinghua University professor.

3 Data analysis

3.1 Data Sources and Research Methods

This paper studies the development and application of artificial intelligence in sports in a broad sense, and selects "Sports" as the explanation of "Sports". Artificial intelligence is widely used in the field of sports. The keyword "artificial intelligence movement" alone can not fully reflect the research status in this field. Through literature retrieval and expert consultation, the current research fields in the field of artificial intelligence include mechanical learning, deep learning, natural language processing, knowledge representation, machine reasoning and so on. Computer vision and robotics. Among them, mechanical learning, deep learning, natural language processing and computer vision have become research hotspots, and have penetrated into the field of sports.

$$P_{n-1}(k, L) = t_{n-1}(k, L) + qC_{n-1}(L) \tag{1}$$

$$P_{n-1}\left(k, m^{*}\right) = \min\{P_{n-1}(k, L)\} \tag{2}$$

According to the analysis of the city space V software, because the related literature first appeared in 1995, the time slicing is selected from 1995 to 2020, the time slice is

1 year, the node type is "account", the threshold item is "topn = 50", and the others are the default values to get the co-occurrence network map (Fig. 1). The size of circle radius and the thickness of node connecting lines are directly proportional to the number of papers and the degree of connection.

Fig. 1. Distribution of sports AI research countries (regions) from 1995 to 2020

3.2 Subject Distribution of Sports Artificial Intelligence Research

Artificial intelligence involves a wide range of disciplines, such as philosophy, cognitive science, mathematics, neuroscience, physiology, psychology, computer science, information theory, cybernetics, etc. When setting the analysis parameters of city space V, the time is divided into 1995–2020, sliced every year, the node type is "category", and the threshold item is "topn50". The results are shown in Fig. 2. The top five papers were Computer Science (219), engineering (15), Computer Science (Artificial Intelligence) (107), engineering (electronic and electrical) (104) and Sports Science (100). In terms of centrality, engineering (0.66) was significantly higher than other disciplines, followed by Computer Science (0.34), sociology (0.18), Sports Science (0.16), Economics (0.13) and Mathematics (0.13).

$$(E(t) - M_2C)\Delta\dot{x}_{k+1}(t) = f(t, x_d(t)) - f(t, x_{k+1}(t)) + B\Delta u_k(t) - \left(\Gamma_{p1}C + M_1C + M_2C\right)\dot{x}_d(t) + \Gamma_{p1}C\dot{x}_{k-1}(t) \quad (3)$$

According to this analysis, computer science and engineering firmly occupy the top two of the number and centrality of published articles, and sports science is also in the top five, which further, and based on engineering, develops artificial intelligence equipment and applies it to the field of sports science. Generally speaking, sports AI technology "comes" from computer science, "transforms" in engineering "and" goes "in sports science. In addition, it is worth mentioning that sociology, economics and mathematics, with only about 10 papers published in history, have attracted great attention in 2020. Among them, there was a sudden increase in the centrality of articles in Sociology in 2016. For example, alphago defeated Li Shishi, a professional nine stage chess player in 2016, and what role human beings will play in the field of sports in the future has aroused sociological thinking on the development of sports in the world. Mathematics and economics have attracted much attention in the field of sports artificial intelligence since 2017. Advanced mathematical algorithms provide the theoretical basis and guarantee for the development of sports artificial intelligence technology, while economics widens the capital channel for the future development of sports artificial intelligence.

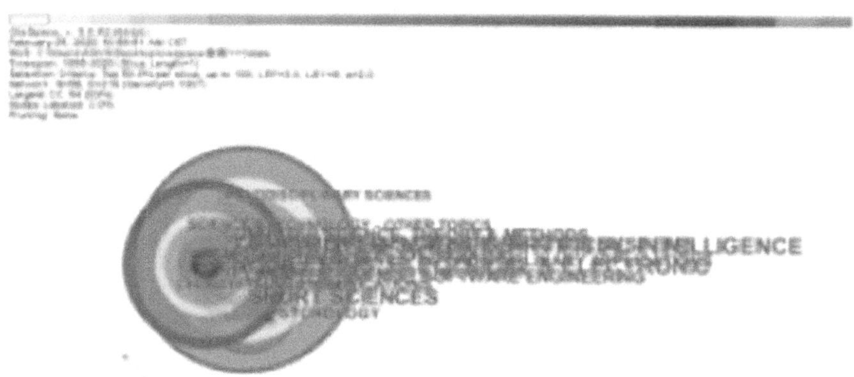

Fig. 2. Network Atlas of main disciplines of sports artificial intelligence research from 1995 to 2020

3.3 Key Words Analysis of Sports Artificial Intelligence

Based on the co-occurrence diagram of keyword research hotspots in Fig. 3, the time node surge diagram of six highly cited key words in recent years is generated by using cite-spvev software, as shown in Fig. 4. The six keywords with the highest burst intensity are tracking (2009–2013), physical activity (2011–2015), artificial neural network (2011–2013), valdiy (2011–2015) (human body), energy expenditure (2011–2013) and action recognition. In addition to gesture recognition, the other five words appeared intensively from 2011 to 2013. What is the relationship between these words? Analyzing the representative literature, such as staudenmayer, developing and testing two artificial neural networks and applying them to the uniaxial accelerometer to collect sports activity data, and analyzing low-intensity exercise, strenuous exercise and housework, it is considered that the application of neural network to the accelerometer in sports recorder has a wide application prospect. Based on the original data of genea (gravity estimator for normal daily activities), Zhang's team successfully developed a wrist accelerometer model suitable for walking, running, home or sedentary activities. Its performance is comparable to that of the traditional waist hip accelerometer, and finally used for physical activity evaluation.

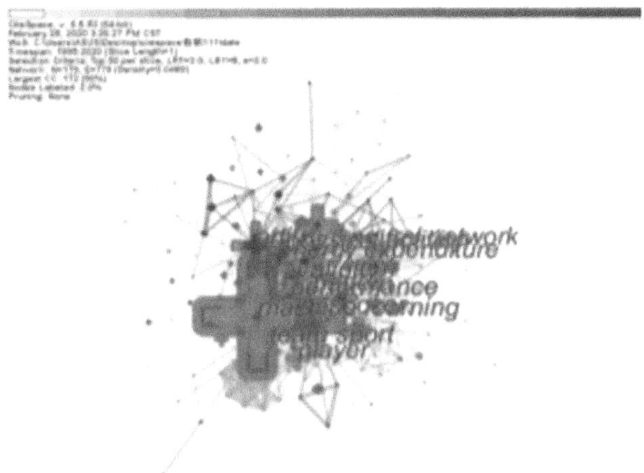

Fig. 3. Hot spots of sports artificial intelligence research keywords from 1995 to 2020

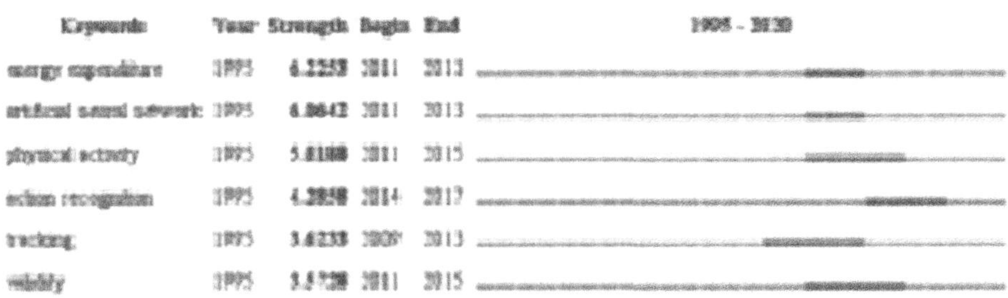

Fig. 4. Network Atlas of main disciplines of sports artificial intelligence research from 1995 to 2020

Through the above literature analysis, the emergence of these six keywords has a strong correlation. Since around 2011, based on the theory of artificial neural network technology and human energy consumption as the index, a series of research on efficient identification and tracking technology of physical activities have been carried out, and finally the terminal carrier of intelligent Tibet piercing equipment has been realized. Based on this, researchers have seen and learned, broadened the development direction of technology and theory, and driven the efficient combination and deepening development of sports and artificial intelligence.

4 Clustering and Evolution of Sports Artificial Intelligence Research

4.1 C1 Knowledge Group: Simple Activity Recognition and Energy Consumption Based on Accelerometer Technology

C1 knowledge group is the # 1 cluster "validity". The application of a technology, its reliability and effectiveness is the basic guarantee. In the early field of sports artificial intelligence, the simple and effective monitoring and evaluation of human motion state is a research hotspot. From the time line chart of highly cited literature, we can see that the relevant theoretical research focused on 2004–2012, and it was during this period that five prominent keywords appeared. Since then, the related research has gradually faded out of the field of vision, instead of the in-depth application of intelligent sensors in the analysis of technical characteristics and spatiotemporal parameters of action. In the two high school mind and highly cited papers of this knowledge group, Crouter et al. 19's paper "a novel method usngaccelerator data to predict energy expenditure" established a new binary regression model of artificial neural network, It is proved that the new algorithm is more accurate than the traditional actigraph accelerometer in predicting energy consumption. Staudenmayer et al. M's paper "Journal of Applied Physics" developed and successfully tested two kinds of artificial neural network models which can be used to collect physical activity data and applied to uniaxial accelerometer, and tested the behavior of 46 subjects. Finally, the recognition accuracy of human action through the amount of metabolism reached 88%. These two papers are innovative to establish the artificial neural network model, which makes the judgment and recognition of action type and energy consumption more efficient and practical.

4.2 C2 Knowledge Group: Computer Scene Classification Based on Convolutional Neural Network Algorithm

Related scholars also put forward many other artificial neural network implementation methods. For example, the depth image segmentation method based on Kohonen neural network can be applied to the selection of swimming far mobilization, handball tactical mode selection, swimming competition performance prediction and other competitive sports fields. Girshick applies the artificial neural network grammar model to human detection, which can recognize the part of the human body that is occluded or whose posture and appearance change. It can effectively solve the problem of human detection in pascal-voc data set. The variable detection model developed by Pedro can also realize the target location in chaotic images.

On the basis of the research of C2 knowledge group, the computer vision technology based on machine learning in the field of sports sports has begun to develop rapidly. This knowledge group began to use computer vision analysis technology to try to analyze various data in sports, especially team sports, and achieved good results. Osgnach et al., the author of the first central article, creatively used video matching analysis technology to evaluate the physical performance of professional football players. Based on the video analysis of 399 Serie A players running on the field, this paper evaluates the instant metabolism ability of top football players, so as to redefine the concept of "high intensity" according to the actual metabolic ability rather than just according to the speed. He also assumed a model combining video matching analysis of official matches with GPS of training, which can conduct a more in-depth study on the differences related to match position, ranking and fatigue degree in a single match or football season. This broadens the field of vision for the further study of wearable sensors in the future. In his subsequent research, he further improved the collection technology of complex terrain "acceleration" data.

5 Conclusion

This paper studies all the current databases and resource databases. From their various characteristics, we can find that if we combine the current college students' own characteristics to select the corresponding projects, the success rate will be higher. Moreover, the current e-commerce is a project integrating multiple advantages, which can not only improve the ability of College students to start their own business, And it can make the project land faster, which is also very good. It can apply their knowledge to practice.

References

1. Li, Y.: Media development in the background of artificial intelligence. In: 2019 International Joint Conference on Information, Media And Engineering (IJCIME), vol. 31(5), pp. 855–868 (2019)
2. Xianguo, S., Cong, W.: Research on the application of artificial intelligence tech-nology in physical training. In: 2021 2nd International Conference On Big Data And Informatization Education (ICBDIE), vol. 32(23), pp. 17209–17227 (2021)
3. Zhang, B., Lyu, M., Zhang, L., Yang, W.: Artificial intelligence-based joint movement estimation method for football players in sports training. Mob. Inf. Syst. **78**(3), 1007–1014 (2021)
4. Duan, X., Sun, C., Tian, D., et al.: Multi-sensor fusion detection method for vehicle target based on kalman filter and data association filter. Pattern Recognit **26**, 112–124 (2021)
5. McLean, S., Read, G.J.M., Thompson, J., Hancock, P.A., Salmon, P.M.: Who is in control? managerial artificial general intelligence (MAGI) for football. Soccer Soc. **35**(2), 485–503 (2021)
6. Liang, F., Yang,Y., Shan, Z., Kim, B.S.: Application of artificial in-telligence technology in leisure sports course. In: 2021 2nd International Conference On Computers, Information Processing and Advanced Education, vol. 323(11), pp. 533–536 (2021)
7. Li, X., Song, L., Wu, H., Wang, Y.: Optimization of ice and snow sports industry chain structure based on sensor network communication and artificial intelligence. Mob. Inf. Syst. **94**, 149–161 (2021)
8. Wu, G., Ji, H.: Short-term memory neural network-based cognitive computing in sports training complexity pattern recognition. Soft Comput. (2022). https://doi.org/10.1007/s00500-021-06568-6
9. Jiang, H., Tsai, S.B.: An empirical study on sports combination training action recognition based on smo algorithm optimization model and artificial intel-ligence. Math. Prob. Eng. **32**, 287–294 (2021)
10. Cong, C., Fu, D.: An AI based research on optimization of university sports information service. J. Intell. Fuzzy Syst. **94**, 122–134 (2021)
11. Wang, M., Krishna, C.V., Luhar, M., et al.: Model-based multi-sensor fusion for reconstructing wall-bounded turbulence. In: Proceeding of the Workshop on Language Technologies for Digital Humanities and Cultural Heritage, pp. 90–96, Hissar, Bulgaria (2021)

Design and Effect of Micro Nano Robot in Fracturing and Oil Displacement Technology of Three Types of Reservoirs

Lina Zou[1,2](✉) and Rasha Almajed[1,2]

[1] Daqing Oilfield Limited Company No.4 Oil Production Company, Daqing 163511, China
zouln@petrochina.com.cn

[2] American University in the Emirates, Dubai, UAE

Abstract. Aiming at the problems of many types of sand bodies, poor reservoir physical properties, low recovery degree, scattered residual oil, large adsorption capacity, and difficult chemical flooding in the development process of the third type reservoirs in Sazhong Development Zone, the "fracturing oil displacement" technology is proposed, which can form a high-speed channel by fracturing long fractures, and quickly send high-efficiency oil displacement agent to the remaining oil enrichment position through fractures, It can reduce the contact time and distance between the chemical agent and the formation, reduce the loss of chemical agent performance along the injection process, improve the utilization efficiency and enhance the oil displacement effect.

Keywords: Three kinds of reservoir · Fracturing and oil displacement · Alkali surface binary system

1 Introduction

With the growth of the world economy and the improvement of people's quality of life, the world's demand for oil is also growing. As a limited renewable resource, oil has become an important factor affecting the sustainable development of the world economy. At present, China's dependence on foreign crude oil has exceeded 55%, but the domestic old oil fields are in short supply due to the increase of exploitable burial capacity year by year. Except for the long-term diesel war, Xinjiang Oilfield and Qinghai Oilfield, other oilfields have entered the stage of production reduction year by year. The main development oilfields have entered the late stage of high or ultra-high function. In addition, the distribution of remaining oil is complex and it is difficult to develop and adjust.

According to the reservoir classification standard of Daqing Oilfield, the reservoirs in Xingbei development zone are divided into primary reservoirs and Tertiary reservoirs. The geological reserves of class I reservoirs account for 35.5% of the whole region. Some blocks have entered the subsequent water injection stage of polymer flooding, and strong alkali composite flooding has also been applied in industry. With the application of the

I. Ahmad et al. (Eds.): STSIoT 2021, LNDECT 122, pp. 784–793, 2023.
https://doi.org/10.1007/978-981-19-3632-6_90

tertiary oil recovery technology, the recoverable reserves of the main layer in Xingbei development zone are gradually decreasing, and the oilfield is facing the situation of insufficient recoverable reserves for continuous and stable production. However, 64.5% of the geological reserves in Xingbei development zone are located in class III oil layers, which are rich in reserves. However, the geological reserves are mainly distributed on the surface, accounting for 87.3%, with high mud content and poor physical properties. The third type of recovery is low, but the remaining oil is dispersed in each storage group. The tertiary recovery factor of the main reservoir in Shengli Oilfield is about 10%, and that of the main reservoir in Daqing Oilfield is more than 20%. It shows that the tertiary recovery of the main reservoir is good. The first part introduces the relevant background and significance of this paper, the second part is the related work of this paper, and the third part is design method of fracturing and oil displacement technology for three kinds of reservoirs. The fourth part is field effect of fracturing and oil displacement in class III reservoir. The fifth part is conclusion.

The third type of oil layer test area in Sazhong Development Zone is mainly composed of Pu I5 + 6_1−7 and Pu II formation, belonging to delta inner and outer front facies.

2 Related Work

Based on the theory and technology of combined stimulation, Su et al. presented an application of the combined stimulation technology to a low-deliverability CBM well in the Sunan Syncline, Anhui Province, China [1]. Wu et al. improved the fracturing effect of this kind of reservoir [2]. As the fracturing modeling and stimulation technology advances question: "Can use the fracturing modeling and reservoir simulation technologies to optimize well energy supplement and cluster spacing based upon Fracture Controlling Fracturing (FCF) technology, which is the latest concept for stimulation technology with successful applications in China's unconventional oil and gas development?" [3]. Due to the limited volume of reconstruction, the production decreases rapidly, which makes it difficult to achieve the purpose of economic and effective exploitation [4]. Theoretically, the mechanical model of the intersection of HF and NF is established, and some judgment criteria are put forward [5]. Therefore Wang et al. investigated the evolution of crack initiation and propagation in a hydraulic rock mass under various stress conditions [6]. In line with the complex geological characteristics of ultra-deep oil and gas reservoirs in China, seven technical development directions are proposed: (1) To establish systematic new techniques for basic research and evaluation experiments; (2) to strengthen geological research and improve the operational mechanism of integrating geological research and engineering operation; (3) to develop high-efficiency fracturing materials for ultra-deep reservoirs; (4) to research separated layer fracturing technology for ultra-deep and hugely thick reservoirs; (5) to explore fracture-control stimulation technology for ultra-deep horizontal [7]. for the multistage for flooding, there are many ways to handle the openhole, however, there is little dangerous for worker, in the openhole completion, the fracturing technology is used, so it will reduce the dangerous, so the author research the fracturing openhole technology for flooding, it will slow the pressure and get good reservoir for production [8]. Therefore, the research on EOR and optimized combined flooding technology of three flow reservoir is carried out to provide technical support for the production of three flow reservoir. After years of development, the total water content of the three reservoirs has reached 91.3%. However, the development of water cycle is difficult and the economic benefit is low. In the combined oil displacement test of three flow reservoir, the combined oil displacement of three flow reservoir can improve the oil recovery by 5%.

3 Strong Reservoir Heterogeneity

The research shows that the heterogeneity of Pu I5 + 6_1–7 layer is mainly manifested in the large permeability difference between single wells, mainly in the medium permeability layer; the reservoir of Pu II group is mainly in the low permeability layer between single wells. In general, the heterogeneity of the three types of reservoirs is more serious than that of the first and second types of reservoirs.

3.1 The Vertical and Upward Watered Out Thickness Proportion of Pu II Formation is Large, and the Remaining Oil Distribution is Scattered

The vertical immersion thickness ratio of the two groups was 89.5%, mainly concentrated at the middle nozzle. Reservoir immersion conditions vary with thickness. The effective oil layer thickness $\geq 0.5 \sim 1$ m is mainly medium jellyfish, and the effective oil layer thickness of 0.2–0.5 m is mainly oil layer. The average functional saturation of complex reservoir is 46.86%. There are two types of residual oil. The first type is incomplete residual oil, which is mainly distributed in layer 5 and 10 of Fuer 2. Secondly, the remaining oil with low water absorption is mainly distributed in fu24, fu6 and fu9 layers. The potential of the remaining oil in this reserve is difficult to develop.

3.2 Each Small Layer of the Third Type Reservoir has High Water Cut and High Production Degree

At the initial stage of production, the water cut of the third type reservoir test area is 91.8%. The test data in the second half of 2005 show that the average thickness of sandstone and the effective thickness of liquid are 82.9% and 87.9% respectively. In the range of 0.2–0.4 m, the number of non main sand producing layers accounts for 60.7%, the thickness of sandstone accounts for 83.8%, and the effective thickness accounts for 62.4%; in the range of 0.5–1.0 m, the number of small layer producing liquid layers accounts for 75.6%, the thickness of sandstone accounts for 89.8%, and the effective thickness accounts for 79.1%; and in the case of $\geq$1.0 m, the liquid producing condition of small layer is 100%. The test results show that each sub layer has high water content, only Pu III1 sub layer has water content lower than 90%.

$$||\Delta x_{k+1}(t)||e^{-\lambda t} \leq e^{-\lambda t} \int_0^t e^{(pk_f+m_2+m_3)(t-\tau)}(m_1||\Delta u_k(\tau)|| + pd)d\tau$$

$$\leq m_1 \int_0^t e^{(pk_f+m_2+m_3-\lambda)(t-\tau)}e^{-\lambda t}||\Delta u_k(\tau)||d\tau + pd \int_0^t e^{(pk_f+m_2+m_3)(t-\tau)}d \tag{1}$$

4 Design Method of Fracturing and Oil Displacement Technology for Three Kinds of Reservoirs

4.1 Well and Layer Selection Design Object and Method

4.1.1 Well and Layer Selection Criteria for Production Wells

The selection of injection wells (production wells) and injection layers for fracturing flooding in three types of reservoirs is based on the construction purpose, so as to solve the prominent contradiction of three types of reservoirs. The selection of production wells mainly includes the following three types.

(1) The wells and layers whose recovery degree is lower than that of the whole area, residual oil is highly scattered, and the effect of conventional fracturing measures is not good. The physical property development of most of these wells is lower than the average level of the whole area, and there are many types of sand bodies, serious interactive distribution and other problems. At the same time, in the late stage of chemical flooding, because the recovery degree is lower than the level of the whole area, the reserves are bound to be sealed after this round of chemical flooding, resulting in a certain degree of loss of recoverable reserves in chemical flooding stage.
(2) The wells with serious formation energy deficit are usually the ones with slow recovery rate of fluid volume and low formation energy after long-term shut in or drilling down. At this time, the water cut rises rapidly. The oil gas water three-phase flow is produced near the well due to the degassing of formation crude oil, so the fracturing oil displacement method can be used to quickly supplement the formation energy. The above two types of oil wells also need to have a certain number of connected directions. Because the later stage of fracturing oil displacement needs to further supplement the formation energy through water injection (or polymer flooding/ASP flooding), the oil and water wells need to have a certain connection relationship, at least two directions of connection.

$$\lim_{k\to\infty} ||\Delta x_{k+1}(t)||_\lambda \le m_1 \frac{1-e^{(b-\lambda)t}}{b-\lambda}\frac{1}{1-\breve{\rho}} m_5 d + pd\frac{1-e^{bt}}{b}$$
$$= \left(m_1\frac{1-e^{(b-\lambda)t}}{b-\lambda}\frac{1}{1-\breve{\rho}}m_5 + p\frac{1-e^{bt}}{b}\right)d \le \left(\frac{m_1 m_5}{b-\lambda}\frac{1}{1-\breve{\rho}} + \frac{p}{b}\right)d \quad (2)$$

(3) The isolated well point usually has no injection production relationship or the injection production relationship is very imperfect. For example, the production well at position (1) in Fig. 1 is only supplied by one injection well and the well spacing is far away. At this time, the remaining oil in the non mainstream direction is relatively rich. The oil displacement by oil well fracturing can supplement the formation energy, release the remaining oil in some isolated well points and increase the production degree. Fault edge well is also a kind of isolated well point, which usually displaces oil by formation pressure elasticity between fault and production well, but there is no follow-up energy supplement. As shown in Fig. 1, there is no effective supply between the production well at position ② and fault, and the fault edge is still rich in residual oil. Therefore, through

reverse fracturing, the formation energy can be supplemented, and the dual effects of huff and puff and oil displacement can be exerted.

4.1.2 Injection Well Selection and Well Layer Selection Criteria

In the selection of injection wells, the wells with low injection rate and high start-up pressure should be selected. Through fracturing and oil displacement, high efficiency oil displacement agent is injected into thin and poor oil layers at one time to improve formation pressure and quickly achieve start-up pressure. At the same time, in some areas of the north, when the injection volume of the well is less than 8 m^3/D, the well must be closed in winter to prevent the freezing of the surface system. Therefore, if fracturing and oil displacement are carried out before the well is closed in winter, the normal production of the oil well in the past half a year can be guaranteed, and the effect of advanced water injection can be achieved.

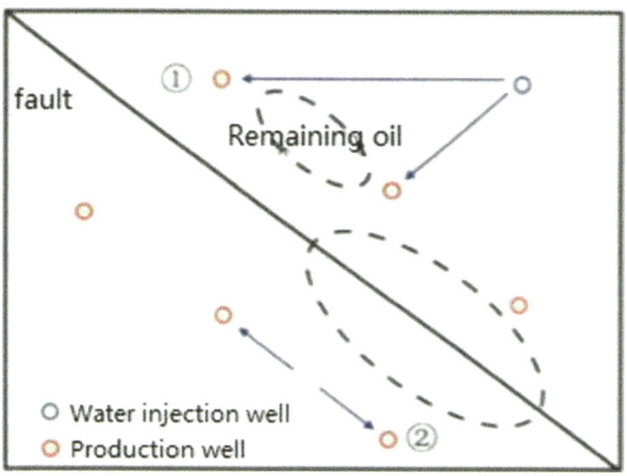

Fig. 1. Schematic diagram of well selection for fracturing oil displacement

4.2 Design of Injection Flooding Agent

4.2.1 Design Principles of Reagents

According to the above design idea of fracturing and oil displacement, the fracturing and oil displacement fluid should be filtrated up and down as far as possible to transport the fracturing fluid (oil displacement fluid) to the deep part of the reservoir, and the break-through problem caused by rapid fracture extension also needs to be reduced. Therefore, the injection agents should be selected according to the performance requirements of low viscosity system, high oil washing efficiency, good matching with three types of reservoirs, no wall building and enhanced filtration. At the same time, after the measure well is opened, some oil wells are supplemented with injection fluid, so the fracturing oil displacement fluid should be compatible with the subsequent injection fluid to avoid reservoir pollution caused by chemical incompatibility.

4.2.2 Drug Type Screening

The alkali surfactant binary system and surfactant binary system with mature technology in Sazhong Development Zone of Daqing Oilfield are optimized for evaluation. The time to achieve ultra-low interfacial tension and equilibrium interfacial tension are used to quantify the oil displacement performance index of the agent, the oil washing efficiency is used to evaluate the oil displacement performance of the agent, and the viscosity is evaluated at the same time. Finally, the injection compatibility of subsequent fluids is considered.

In the evaluation of oil washing efficiency, 5 g oil sand is weighed and put into 25 ml colorimetric tube, 2.5 ml crude oil is added to saturate for 48 h (at 45 °C), then the pressure drive fluid system is prepared respectively, and the colorimetric tube is added to 25 ml scale line, at the same time, the preparation water is also added to 25 ml scale line as blank, standing for 24 h, the upper oil washing volume is read, and the oil washing efficiency is calculated.

$$\lim_{k\to\infty}\ \sup_{0\le t\le T} \|\Delta e_{k+1}(t)\| \le \lim_{k\to\infty} e^{-\lambda T}\|\Delta e_{k+1}(t)\|_{\lambda} \le e^{-\lambda T}\left(\frac{m_1 m_5}{b-\lambda}\frac{1}{1-\breve{\rho}}+\frac{p}{b}\right)cd \tag{3}$$

All the indexes of the three fracturing fluid systems can meet the requirements of fracturing and oil displacement, and the indexes are relatively similar. Therefore, considering the adaptability of alkali surface binary and injection reservoir and injection system, the method of alkali plus surfactant is adopted in this test, in which the mass fraction of alkali is 1.2%, and the mass fraction of surfactant (heavy alkylbenzene sulfonate) is 0.3%. The viscosity of the binary system is 2.1–4.8 mpa-s, and the fluidity is strong, which can meet the performance requirements of fracturing oil displacement fluid with low viscosity and easy filtration. The strong alkali surfactant has good compatibility with crude oil, wide range of interfacial activity and strong oil washing ability, which can meet the demand of high oil displacement efficiency.

5 Field Effect of Fracturing and Oil Displacement in Class III Reservoir

5.1 Field Test of Pressure Drive

Taking two fracturing oil wells in Daqing Oilfield as an example, in the selection of production well A1, according to the subdivision of injection wells, the conditions of production wells and the requirements of stratification technology, the fracturing oil displacement layers are divided into six sections. Table 1 shows the design of fracturing oil displacement, in which the physical properties of the 1st to 3rd sections are slightly poor, the interlayer difference is improved and the formation energy is recovered by fracturing oil displacement, and the physical properties of the 4th to 6th sections are better, and the remaining oil in the unswept parts is exploited by fracturing oil displacement. According to the binary liquid property of alkali surface, according to the relation chart of injection system construction displacement and friction, combined with the calculation of horizontal fracture reservoir filtration rate, when the construction displacement is 4.5 m^3/min, the formation filtration is large, and the fracture can be effectively extended. In the actual construction process, it is strictly in accordance with the design, and the amount of pressure drive fluid is 555 m^3, which is close to the design amount (5949 m^3).

In the selection of injection well A2 horizon, according to the requirements of the current layering technology, the fractured oil displacement horizon is divided into five sections, and the first and second sections are drilled by fracturing to drive the remaining oil in the affected area. In the third to fifth sections, the interlayer difference is improved and the formation energy is recovered by fracturing. The actual stratification design is shown in Table 2. In the actual construction process, the consumption of pressure drive fluid is 6 297 m^3.

Table 1. The design of fracture-flooding of the oil Well A1

Fracturing interval number	Horizon	Effective thickness/m	Permeability/d	Actual injection volume/m^3
1	SII14	1.0	0.148	927
2	SII12	0.2	0.038	940
		0.3	0.061	
	SII11	0.8	0.125	
3	SIII2	0.4	0.113	829
	SIII3_1	0.4	0.072	
	SIII3_2	0.5	0.122	

Table 2. The design of fracture-flooding of the injection Well A2

Fracturing interval number	Horizon	Effective thickness/m	Permeability/d	Actual injection volume/m^3
1	SII10	1.9	0.522	1290
2	SIII3_1	0.9	0.126	1027
3	SIII5 + 6	0.7	0.112	1520

5.2 Oil Displacement Effect of Oil Well Fracturing

After fracturing and oil displacement in production well A1, the effect of increasing fluid and oil and reducing water cut is obvious, and the production curve is shown in Fig. 2.

Production well A1 has a daily fluid production of 15t, oil production of 0.5T and water cut of 96.6% before well pressure, and a maximum daily fluid production of 65t,

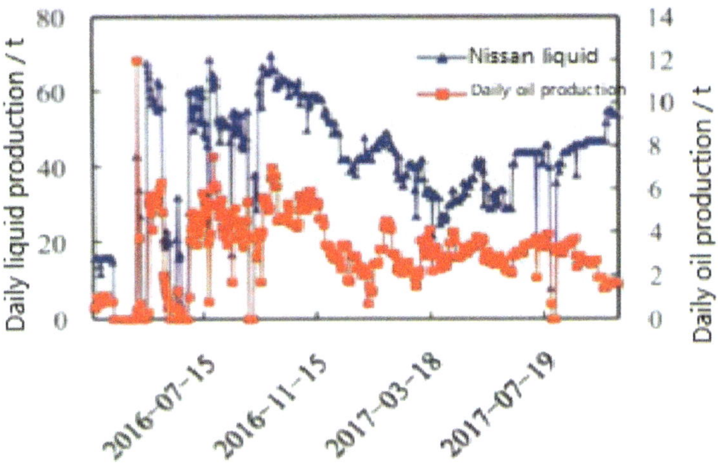

Fig. 2. Oil well production curve

oil production of 8.5t and water cut of 88.9% after well pressure. Up to now, the period of validity has reached 429d, the daily fluid production is 54t, the daily oil production is 1.6T, and the water cut is 96.0%. Due to the improvement effect of fracturing on the reservoir, the current fluid production is 3.6 times higher than that before fracturing, and the cumulative oil increase is 910t.

$$\begin{cases} E(t)\dot{x}_k(t) = f(t, x_k(t)) + B(t)u_k(t) + d_k(t) \\ y_k(t) = C(t)x_k(t) \end{cases} \tag{4}$$

Figure 3 shows the produced fluid curve. The polymer concentration in the produced fluid reaches a peak value of 499 mg/L at 126 days after fracturing, indicating that even if large-scale fracturing and oil displacement are carried out, rapid breakthrough of the reservoir is not caused.

The injection rate of conventional chemical flooding in the test area is about 50 m^3/d. according to the current fracturing oil displacement operation speed, it can reach 6480 m^3/d, and the agent transportation through fractures can reach about 100 times of the normal injection. At the same time, one-time injection of 0.05 PV plays a decisive role in the rapid recovery of formation pressure in the low permeability layer, and the purpose of further starting the reservoir is achieved by increasing the formation pressure. Because of the use of fracture direct transportation, the upper and lower filtration on the fracture surface can reduce the loss of agent performance caused by shear, adsorption and retention.

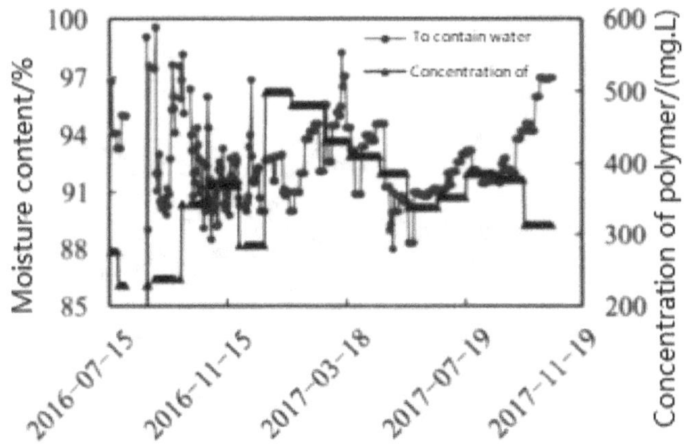

Fig. 3. The injection curve of water well

6 Conclusion

The three oil layers in the development zone of the company are mainly buried physical properties, poor reservoir physical properties, many main types and low recovery during development. The development characteristics of dispersed remaining oil are formed. At the same time, chemical oil displacement is difficult due to the large amount of reactants and adsorption. "Fracturing oil displacement" is to form a high-speed channel by fracturing long fractures, quickly send the high-efficiency oil displacement agent to the remaining oil enrichment position through fractures, use the oil displacement agent as fracturing fluid, and quickly fill the oil displacement agent into the pores while fracturing fractures, so as to reduce the contact time and distance between the chemical agent and the formation, In order to solve the problems of chemical agent performance loss along the way, low utilization efficiency and influence on oil displacement effect in the process of injection.The field test results show that the three oil products can realize autoclave and oil injection. The remaining oil potential determined by dispersion can be rapidly developed by using stimulation and water injection technology in three reservoirs in Central China IV development zone.

References

1. Su, X., Wang, Q., Lin, H., Song, J., Guo, H.: A combined stimulation technology for coalbed methane wells: part 2. application. Fuel **178**(1), 37–51 (2018)
2. Wu, Z.Y., Hu, Y.F., Jiang, T.X., Liu, J.K., Wu, C.F.: Research and application of horizontal well cross-layer fracturing technology in tight sandstone reservoir, vol. 5, p. 153 (2019). https://doi.org/10.1007/978-981-15-2485-1_41
3. Guo, Y., et al.: Optimization on well energy supplement and cluster spacing based upon fracture controlling fracturing technology. reservoir simulation, vol. 31, p. 23 (2019)
4. Zeng, L., Zheng, Y., Zou, L.: Fracturing technology of real time control guarantees highly efficient exploitation of shale Weiyuan Gas-Field, SW CHINA,vol. 43 p. 44–46 (2020)

5. Dong, K., Jiang, M., Li, J., Zhang, D.: Research progresses in formation mechanism of complex fracture network for unconventional reservoir. Arab. J. Geosci. **33**(1), 18 (2020). https://doi.org/10.1007/s12517-020-05724-w
6. Wang, H., Li, J., Zhao, F., Dong, J., Cui, Y., Gong, W.: Experimental study of volumetric fracturing properties for shale under different stress states. Geofluids **38**(12), 14609–14623 (2021)
7. Lei, Q., et al.: Progress and development directions of stimulation techniques for ultra-deep oil and gas reservoirs. Petrol. Explor. Dev. **6**(1), 68–77 (2021)
8. Bartko, K., et al.: Multistage fracturing openhole completion. J. Inf. Manag. **14**, 78 (2018)

Author Index

I. Ahmad et al. (Eds.): STSIoT 2021, LNDECT 122, pp. 795–798, 2023.
https://doi.org/10.1007/978-981-19-3632-6

The manufacturer's authorised representative in the EU is Springer Nature Customer Service Centre GmbH, Europaplatz 3, 69115 Heidelberg, Germany. If you have any concerns regarding our products, please contact ProductSafety@springernature.com

Printed and bound by CPI Group (UK) Ltd, Croydon, CR0 4YY
30/09/2026
02225251-0019